ENZYMES: Catalysis, Kinetics and Mechanisms

Narayan S. Punekar

ENZYMES: Catalysis, Kinetics and Mechanisms

Second Edition

 Springer

Narayan S. Punekar
Department of Biosciences and Bioengineering
Indian Institute of Technology Dharwad
Dharwad, Karnataka, India

ISBN 978-981-97-8178-2   ISBN 978-981-97-8179-9   (eBook)
https://doi.org/10.1007/978-981-97-8179-9

Life Sciences
Enzymology
Protein Science
Biotechnology
Applied Microbiology
Chemistry/Food Science, general

Editorial Contact: Aishwarya Thyagarajan, Swati Sharma

This Springer imprint is published by the registered company Springer Nature Singapore Pte Ltd.
The registered company address is: 152 Beach Road, #21-01/04 Gateway East, Singapore 189721,
Singapore

If disposing of this product, please recycle the paper.

To Sandhya,
Couldn't have done it without your warmth
and companionship

Preface to the First Edition

Any living being is a reflection of its enzyme arsenal. We are and do what our enzymes permit.—Christian de Duve

Enzymes are the lead actors in the drama of life. Without these molecular machines, the genetic information stored in DNA would be worthless. With increasing attention to the fashionable fields like molecular biology, genetic engineering, and biotechnology, the techniques to manipulate DNA have occupied center stage. Being popular, many concepts of molecular biology and genetic engineering are now introduced to undergraduates. Unfortunately, this has happened at the cost of other fundamental facets of biology, including enzymology. In the excitement to collate volumes of data for systems biology (and the various "omics" fashions), the beauty and vigor of careful analysis—one enzyme at a time—is neglected. It is an intellectual challenge to assay individual enzymes while avoiding complications due to others—an almost forgotten activity in modern biology. Many in the present generation assume that performing one standard assay will tell you everything about that enzyme. While biochemists spent lifetimes on a single native enzyme, the notion today is that one can characterize a mutant in the morning. Over the last three decades, devoted enzymologists have become a rare breed. Many biology teaching programs have expanded in the areas of molecular and cellular biology, while they manage with a makeshift enzymology instructor. New students who are attracted to the study of enzymes do exist, but they find themselves in a very bleak teaching environment. Not surprisingly, their numbers are dwindling. Reservoirs that are not replenished may soon run dry.

Mumbai, Maharashtra, India   N. S. Punekar

Preface

The devil may write chemical textbooks....because every few years the whole thing changes—J. Berzelius

This observation made by Berzelius is true for the domain of Enzymology as well. Much exciting research and progress is made in this field since the first edition (where the original literature was referred/collated up to the year 2016). Based on excellent feedback from many students and teachers, an upgrade and addition of few more topics was warranted. Also, the current research frontiers in Enzymology are expanded to bring the book up to date. This entailed addition of relevant references that are contemporary. Recent reviews on most topics are provided and the literature is cited mostly from easily available and open-access resources. An important feedback to the first edition was to introduce an index to this book; this is made available in the second edition. This edition also addresses certain editorial and printing errors that crept into the first edition.

The book chapters have been reorganized to improve the flow of the subject matter. The chapter on 'Enzyme Nomenclature and Classification' now forms the last section of Part I. The chapter on 'Exploiting Enzymes: Technology and Applications' is now moved to Part V of the book wherein a chapter on 'Enzyme Inhibitor Design: Drug Discovery' is freshly added. Also, another new chapter (Chap. 34 Free Radicals and Radical Enzymology) is added to Part IV. The two chapters on isotope applications in enzyme study are merged (Chap. 24, Part III). New material is included in many chapters after their revision. Extensive yet relevant literature (preferably open-source reviews) is provided at the end of each chapter. The uninitiated reader may benefit by going through Chaps. 8, 9, and 10 (covering chemical kinetics) and Chaps. 27, 28, and 29 (covering organic reaction mechanisms) before delving into the details. Chapter 23 arrives before a primer on acid-base chemistry in Part IV, hence it is suggested to read Chap. 28 before approaching the material in Chap. 23. A mechanistic synthesis on how enzymes act involves diverse experimental approaches. Few examples of these are found in Chaps. 26 and 35.

The first edition reached a very large number of interested readers as this book was made 'open resource' by Springer during COVID times. Excellent feedback received from the readers has helped improve the contents and correct some lacunae from the first presentation. Affiliation to IIT Dharwad was instrumental in providing the much-needed ambience while I worked on the second edition. This is gracefully acknowledged. Once again, I am driven towards better academic goals by the encouragement and time management enabled by my wife Sandhya. Unqualified support and love received from Sandhya and our daughter Jahnavi was invaluable when this work was being written up. Few figures in Chaps. 33, 34, and 36 were freshly added and help in preparing them by Mr. Yeshvanth S. (Department of Chemistry, IIT Dharwad) is much appreciated. The cryo-EM pictures were kind courtesy of Prof. P Bhaumik, IIT Bombay.

Dharwad, Karnataka, India   Narayan S. Punekar

Purpose of This Book

Genes for enzymes are routinely fished out, cloned, sequenced, mutated and expressed in a suitable host. Characterizing a new or mutant enzyme, however, requires a thorough mechanistic study—both chemical and kinetic. It is thus an exciting time to do enzymology. Hopefully, this book provides enough basic exposure to make this happen.

The ease with which sophisticated data are collected nowadays has dispirited the slow and burdensome approach of resolving and reconstituting a complex enzyme system. Microarrays that measure the transcription of many genes at a time disclose neither the abundance nor any attributes of the enzymes/proteins they encode. As F.G. Hopkins wrote in 1931 '..the biochemist's word may not be the last in describing life, but without his help, the last word will never be said'. This is true of enzymology as well. While the interest and expertise in teaching/learning enzymology has declined exponentially, working knowledge of enzymology remains indispensable. Enzymes have come to occupy vast areas of modern biology research and the biotechnology industry. Enzymes whether used as popular kits, mere research tools, or for their own sake require a minimal appreciation of their workings. A tome on enzymology that focuses and logically connects theory of enzyme action to actual experimentation is desirable. In contrast to the advances in the research laboratory, enzymology in the classrooms seems to have stagnated. One objective of this book is to bridge this gap and enable students to understand, design and execute enzyme experiments on their own.

Enzyme study can range from the simple to the most complicated. Approaches that can be performed in a modest laboratory setup and with no fancy equipment are needed. Conveying the excitement of enzymology within a modest budget and with few experiments is desirable. And hence, equipment intensive approaches—such as structural enzymology, sophisticated techniques like X-ray, NMR, ESR, fast reactions and isotope effects—have received a somewhat limited coverage. Readers interested in them will yet find sufficient background material here.

Audience and Their Background

Reasons for the cursory coverage of enzymology in most contemporary biology academic programs are twofold. Overemphasis and glamorization of molecular biology (later genetic engineering!) in the last few decades have captured a disproportionately large allocation of resources and time. Secondly, as a cumulative effect of this attitude, very few well-trained specialists in enzymology are available today. Therefore, study material that encourages students/researchers to understand, design and execute experiments involving enzymes on their own is needed. The contents of the present book are expected to serve this purpose.

Most biochemistry and molecular biology students are introduced to enzymes as commercial reagents and as faceless as buffers and salts. This has led to inadequate appreciation of enzymology and its practices. Standards for reporting enzymology data (STRENDA; available at http://www.strenda-db.org) are a recent effort to prescribe the best approaches to generate and report enzyme data. With an ever-increasing reliance on genomics and proteomics, enzymes are no longer isolated and/or assayed for activity. Often their role is inferred from sequence data alone. 'Molecular biology falters when it ignores the chemistry of the products of DNA blueprint—enzymes—the protein catalysts of the cellular machinery'. This philosophy was beautifully reiterated by Arthur Kornberg in his 'Ten Commandments of Enzymology' (J Bacteriol (2000) 182:3613–3618; TIBS (2003) 28:515–517). The present book is an attempt to sift through chemical sophistication and simplify it for an audience with a biology background. It will serve the curricular needs of senior undergraduates and postgraduates in biochemistry, biotechnology and most branches of modern biology.

Dealing with reaction rates, enzymology is a quantitative and analytical facet of biological understanding. Appreciation of rate equations and their meaning therefore becomes important. Minimal competence with algebra, logarithms, exponential relationships, equations to fit straight lines and simple curves is crucial. While one need not be scared of fearsome equations, the essence of the physical models they represent (or do not represent!) ought to be understood. 'Filling the blackboard with equations and reaction schemes is the last thing an instructor would want to do when addressing an audience who are predominantly biologists.'(Srinivasan B (2021) Words of Advice: teaching enzyme kinetics. FEBS J 288:2068–2083) To an extent, this book is my response to oust the fear of the quantitative in the students of biology. Because enzymes catalyse chemical reactions, chemical mechanisms are of great concern. They are best understood with adequate preparation in concepts like valency, movement of electrons and charges in molecules, acids and bases, etc. The study of mechanistic enzymology is meaningless without this background. We may recall from Emil Fischer's Faraday Lecture to the Chemical Society in 1907: '... the separation of chemistry from biology was necessary while experimental methods and theories were being developed. Now that our science is provided with a powerful armoury of analytical and synthetic weapons, chemistry can once again renew the alliance with biology, not only for the advantage of biology but also for the glory of chemistry'. Enzymology without chemistry (physical and organic) is a limited descriptor of surface (superficial!) phenomena. This requirement obviously puts some burden on students who have lost touch with chemistry for few years in the pursuit of 'Biology Only' programs.

Basic knowledge on amino acids, their reactivity and protein structure is a prerequisite to study enzymes. Protein (and hence enzyme) purification methods/ tools like various fractionation/separation techniques and chromatographies are not explicitly covered here. Also, essential techniques of protein structure determination do not find a dedicated treatment in this book. One may find such background material in the standard textbooks of biochemistry. Lastly, the reader is expected to be familiar with the concepts of concentrations, ionic strength, pH, etc., and exposure to biochemical calculations is essential.

Organization

This book endeavours to synthesize the two broad mechanistic facets of enzymology, namely, the chemical and the kinetic. It also attempts to bring out the synergy between enzyme structures and mechanisms. Written with self-study format in mind, the emphasis is on how to begin experiments with an enzyme and subsequently analyse the data collected. Individual concepts are treated as standalone short sections, and the book is largely modular in organization. The reader can focus on a concept (with real examples) with minimal cross-referencing to the rest of the book. Many attractive enzymes were consciously passed up in order to suit the 'Biology' audience. This error of omission painfully belongs to the author. A limited treatment on the applied aspects of enzymes (in Chaps. 36 and 37) is deliberate as one fully subscribes to Louis Pasteur's dictum: 'There are no applied sciences. . . . The study of the applications of science is easy to anyone who is master of the theory of it'. The book then would also have become unmanageably long.

Individual concepts (as chapters) are conveniently grouped into six broad parts. It all begins with an overview of enzyme catalysis (Part I) followed by a section (Part II) on kinetic practices and measurement of enzyme activity. Two major themes of mechanistic enzymology, namely, the kinetic (Part III) and the chemical (Part IV) occupy bulk attention. Besides two chapters on enzyme applications (in Part V), aspects of enzymology in vivo and frontier research themes form the last section (Part VI).

Useful Constants and Conversion Factors

Calorie (cal):
  (Heat required for raising the temperature of 1 g water from 14.5 °C to 15.5 °C)
  1 cal = 4.184 J
  1 kcal = 1000 cal = 4184 J
Joule (J):
  1 J = 0.239 cal = 1 kg $\times$ m^2 $\times$ s^2 = 2.624 $\times$ 10^{19} eV
Coulomb (C):
  1 C = 6.242 $\times$ 10^{18} electron charges
Avogadro's number (N):
  N = 6.022 $\times$ 10^{23} mol^{-1}
Faraday constant (F):
  F = 23.063 kcal $\times$ V^{-1} $\times$ mol^{-1} = N electron charges = 96,480 C $\times$ mol^{-1}
Boltzmann constant (k_B):
  k_B = 1.381 $\times$ 10^{-23} J $\times$ K^{-1} = 1.38 $\times$ 10^{-16} cm^2 $\times$ g $\times$ s^{-2} $\times$ K^{-1}
Plank's constant (h):
  h = 6.626 $\times$ 10^{-34} J $\times$ s = 6.626 $\times$ 10^{-27} cm^2 $\times$ g $\times$ s^{-1}
Gas constant (R):
  R = N k_B = 1.987 cal $\times$ mol^{-1} $\times$ K^{-1} = 8.315 J $\times$ mol^{-1} $\times$ K^{-1}
Absolute temperature (degree Kelvin, K):
  0 K = absolute zero = -273 °C; 25 °C = 298 K
RT at 25 °C:
  RT = 2.478 kJ $\times$ mol^{-1} = 0.592 kcal $\times$ mol^{-1}
Units for ΔG, ΔH, and ΔS:
  For ΔG and ΔH: cal $\times$ mol $\times$ ^{1} (or J $\times$ mol^{-1})
  For ΔS: cal $\times$ mol^{-1} $\times$ K^{-1} (or J $\times$ mol^{-1} $\times$ K^{-1})
Enzyme catalytic unit:
  1 U = 1 µmol $\times$ min^{-1} = 16.67 nkatal
  1 katal = 1 mol $\times$ s^{-1}
Curie (Ci):
  Quantity of a radioactive substance that decays at a rate of 2.22 $\times$ 10^{12} disintegrations per minute (dpm)

SI units for concentration ranges:

M, molar (10^0); mM, millimolar (10^{-3}); μM, micromolar (10^{-6}); nM, nanomolar (10^{-9}); pM, picomolar (10^{-12}); fM, femtomolar (10^{-15}); aM, attomolar (10^{-18}); zM, zeptomolar (10^{-21}); yM, yoctomolar (10^{-24})

Contents

Part VI Frontiers in Enzymology

About the Author

Narayan S. Punekar currently holds a Visiting Professor position and is also the Dean (Academic Programs), at Indian Institute of Technology (IIT) Dharwad, Karnataka, India. He superannuated in December 2021 from Professorship and as a Biology faculty at IIT Bombay, Mumbai, India.

The author obtained his Ph.D. from the Indian Institute of Science, Bangalore, India, in the year 1984, and subsequently worked as postdoctoral fellow at the Enzyme Institute, University of Wisconsin, Madison, USA, till 1987. After a short stint as Senior Research Scientist at Hoechst Research Centre, Mumbai, he joined IIT Bombay as Assistant Professor in 1988 and subsequently became a full Professor in 2001. He possesses more than 30 years of experience in teaching biochemistry and enzymology at IIT Bombay.

The major research interests of Dr. Punekar lie in biochemistry and molecular enzymology, microbial metabolism and regulation, understanding metabolism through biochemical and recombinant DNA techniques, and fungal molecular genetics and its applications to metabolic engineering. He has published over 70 papers in peer-reviewed journals and has 4 patents to his credit. He enjoys teaching Enzymology and Industrial Microbiology, for which he received the 'Excellence in Teaching Award' at IIT Bombay in the years 2000, 2012 and 2018 and also the 'Research Dissemination Award—2018'.

Dr. Punekar is associated with various societies and committees of the Government of India as an expert member. He also serves on the editorial board of some international journals.

Part I

Enzyme Catalysis: A Perspective

Enzymes: Their Place in Biology

One marvels at the intricate design of living systems, and we cannot but wonder how life originated on this planet. Whether the first biological structures emerged as the self-reproducing genetic templates (genetics-first origin of life) or the metabolic universality preceded the genome and eventually integrated it (metabolism-first origin of life) is still a matter of hot scientific debate. There is growing acceptance that the RNA world came first—as RNA molecules can perform both the functions of information storage and catalysis. Regardless of which view eventually gains acceptance, the emergence of catalytic phenomena is at the core of biology. The last century has seen an explosive growth in our understanding of biological systems. The progression has involved successive emphasis on taxonomy → physiology → biochemistry → molecular biology → genetic engineering and finally the large-scale study of genomes. The field of molecular biology became largely synonymous with the study of DNA—the genetic material. Molecular biology, however, had its beginnings in the understanding of bio-molecular structure and function. Appreciation of proteins, catalytic phenomena, and the function of enzymes had a large role to play in the progress of modern biology.

Enzymes and catalytic phenomena occupy a central position in biology. Life as we know is not possible without enzyme catalysts. Greater than 99% of reactions relevant to biological systems are catalyzed by protein catalysts. A few RNA-catalyzed reactions along with all the uncatalyzed steps of metabolism occupy the remaining one percent. While it may do to explain living beings as open systems that exchange matter and energy with their environment—thermodynamic feasibility alone is insufficient to be living! Kinetic barriers have to be overcome. Reactions with relatively fast un-catalyzed rates, like the removal of hydrogen peroxide or hydration of carbon dioxide, also need to be accelerated. Enzymes are thus a fundamental necessity for life to exist and progress. The key to knowledge of enzymes is the study of reaction velocities, not of equilibria. After all, living beings are systems away from equilibrium.

N. S. Punekar, *ENZYMES: Catalysis, Kinetics and Mechanisms*,
https://doi.org/10.1007/978-981-97-8179-9_1

Enzymology—the study of enzymes—has been an auto-catalytic intellectual activity. Apart from knowledge gained on their structure and function, the study of enzymes is a driving force in advancing our understanding of biological phenomena as diverse as intermediary metabolism and physiology, molecular biology and genetics, cellular signaling and regulation, and differentiation and development. The confidence in our experience with enzymes is so strong that they have found applications in a variety of industries including food, pharmaceuticals, textiles, and the environment.

We encounter enzymes in every facet of biology and are forced to admire their exquisite roles. Enzymes were excellent models and the earliest examples of understanding protein structure function. These include enzymes like hen egg-white lysozyme, bovine pancreatic ribonuclease A (RNase A), trypsin, and chymotrypsin. A few of these were encountered during the study of digestive processes. The selectivity of proteases was exploited, and they served as useful reagents to cleave and study protein structure. The field of molecular biology has benefited enormously from enzymatic tools to cut, ligate, and replicate information molecules like DNA and RNA. Metabolic and cellular regulation is unthinkable without involving enzymes and their response to various environmental cues. The complexity associated with life processes is largely due to their catalytic versatility, exquisite specificity, and ability to be modulated.

Current advances in X-ray crystallography, cryo-electron microscopy, NMR spectroscopy, mass spectrometry, and genetic engineering have made it possible to view an enzyme closely while in action. Reverse genetics and genomics have made enzymology more powerful. Enzymology begins with a defined function and its purification; after which homing on to the corresponding gene has become very easy. Picomoles of pure enzyme protein are enough to determine its partial peptide sequence and obtain a fingerprint. From here it is a well-beaten track of gene identification, cloning, over-expression, and manipulation.

Enzymes are superbly crafted catalysts of nature, and they are at the heart of every biological understanding. Life has literally preserved its past as chemistry. The book of life is written in the language of carbon chemistry and enzymes form a major bridge between chemistry and biology. Enzymology is the domain where chemistry significantly intersects biology and biology is at its quantitative best. From early history, the evergreen tree of enzymology was nurtured by chemical and biological thoughts. We will look at this rich history in the next section.

Suggested Reading

Cleland WW (1979) Enzymology-dead? Trends Biochem Sci 4:47–48
Khosla C (2015) Quo vadis, enzymology? Nat Chem Biol 11:438–441
Kornberg A (1987) The two cultures: chemistry and biology. Biochemistry 26:6888–6891
Kornberg A (1996) Chemistry—the lingua franca of the medical and biological sciences. Chem Biol 3:3–5
Zalatan JG, Herschlag D (2009) The far reaches of enzymology. Nat Chem Biol 5:516–520
Walsh CT (2019) Chemical biology: here to stay? Isr J Chem 59:7–17

Historically, the field of enzymology was born out of practical and theoretical considerations. A perusal of early enzyme literature indicates that the field has evolved from fundamental questions about their function, their nature, and their biological role. This chapter outlines the course of the historical development of enzymology and some of these landmarks are listed in Table 2.1.

2.1 Biocatalysis: The Beginnings

Past human industry like cheese making provided insights into some properties of enzymatic processes. The earliest recorded example of cheese making contains a reference to extracts of fig tree—a source of the proteolytic enzyme ficin. Only later did rennet (a source of another protease chymosin) become popular in cheese processing. Meat tenderizing is the other application that implicitly used enzymes over the years. Apart from the fig tree extract, the fruit and other parts of papaya (*Caryca papaya*; contains the now well-known proteolytic enzyme papain) have found early utility in meat tenderizing.

Indeed, the work on gastric digestion of meat—proteases in particular—by Rene Reaumur (1751) and Lazzaro Spallanzani (1780) laid a scientific foundation for the study of enzyme catalysis. Reaumur's experiments with the digestion of meat represent the first systematic record of the activity due to an enzyme. However, the term enzyme was yet to be coined then! Theodor Schwann used the word pepsin in 1836 for the proteolytic activity of the gastric mucosa. He also conducted careful quantitative experiments, to establish that acid was necessary but not sufficient for this reaction to take place. Among his many other contributions, Schwann also coined the term metabolism.

Parallel to the work on proteolytic enzymes, developments with fermentation and starch hydrolysis have equally contributed to the initial growth of enzymology.

N. S. Punekar, *ENZYMES: Catalysis, Kinetics and Mechanisms*, https://doi.org/10.1007/978-981-97-8179-9_2

Table 2.1 Landmarks in enzyme studies (Enzymology classics)

Author(s)	Year[a]	Contribution
R. Reaumur	1751	Gastric digestion in birds
L. Spallanzani	1780	Digestion of meat by gastric juice
A. Payen & J. Persoz	1833	Amylase (diastase) activity
J. Berzelius	1836	Catalysis as a concept
W. Kuhne	1867	"Enzyme" term defined
J. Takamine	1894	Patent on fungal diastase
E. Fischer	1894	Lock and key concept
G. Bertrand	1897	Co-ferment (coenzyme) conceived
P.E. Duclaux	1898	Enzyme names to end with the suffix "ase"
V. Henri	1903	Hyperbolic rate equation
S.P.L. Sorensen	1909	pH scale and buffers
L. Michaelis & M. Menten	1913	Equilibrium treatment for the ES complex
R.M. Willstatter	1922	Trager theory of enzyme action
G.E. Briggs & J.B.S. Haldane	1925	Steady-state treatment for ES complex
J.B. Sumner	1926	Urease: Purification and crystallization
H. Lineweaver & D. Burk	1934	Double reciprocal plot (1/v vs 1/S)
K. Stern	1935	First ES complex observed
M. Doudoroff	1947	Radioisotope use in enzyme mechanisms
A.G. Ogston	1948	Asymmetric interaction with substrate
L. Pauling	1948	Enzyme binds TS better than S
F. Westheimer	1951	Enzymatic hydride transfer (^{2}H, ^{3}H used)
D.E. Koshland Jr.	1958	Induced fit hypothesis
C.H.W. Hirs et al.	1960	First enzyme sequenced: RNase A
Enzyme Commission	1961	Enzyme classification and nomenclature
D.C. Phillips et al.	1962	First enzyme structure: Lysozyme
W.W. Cleland	1963	Systematization of enzyme kinetic study
J. Monod et al.	1965	Model for allosteric transitions
R.B. Merrifield	1969	Chemical synthesis of RNase A
S. Altman & T.R. Cech	1981	Catalysis by RNA molecules

[a]Year of discovery and/or publication

Gottlieb Kirchhoff discovered plant amylase (later identified as α-amylase) activity while characterizing the hydrolysis of starch to sugar. He demonstrated the acid-facilitated conversion of starch to sugar and clearly recognized that the formation of sugar from starch during germination of grain is akin to chemical hydrolysis (1815). The work of Kirchhoff on starch hydrolysis was extended by Anselme Payen and Jean Persoz (1833). They enriched (first attempts of enzyme purification!) the hydrolytic activity from malt gluten and termed it as diastase. The name diastase (Greek; *diastasis*—to make a breach) has significantly influenced the development of the field of enzymology since then (see below). Yet another source of starch hydrolyzing activity was identified in saliva by Erhard Leuchs (1931). Remarkably, this report also invoked the possible practical utility of this activity.

The two nonhydrolytic enzyme activities reported early include the peroxidase activity from horseradish and catalase. These two enzymes were recognized much ahead of the study of oxidative enzymes in the early twentieth century. The work on catalase reaction by Louis Thenard (1819) is the first quantitative study of an enzymatic reaction. He also anticipated that such activities may be found in other animal and vegetable secretions.

Enzymology finds its roots in some of the greatest names since the eighteenth century, both in chemistry and biology. Clearly, this subject is a true and sturdy bridge between contemporary chemistry and biology. Among the greats who contributed to its early development include Reaumur, Spallanzani, Thenard, Schwann, Berzelius, Liebig, Berthelot, Pasteur, Buchner, and Fischer. Many fundamental contributions were made to enzymology by chemists of fame like Berzelius, Liebig, and Berthelot. It is however important to note that historically, the idea of catalysis arose because of the study of enzymes and their action. Jons Jacob Berzelius was the first to define the term "catalyst" in 1836. In his view, a catalyst was a substance capable of wakening dormant energies, merely by its presence. He was also the first to recognize the similarity of catalysis in a chemical reaction and inside a living cell. However, Berzelius made no distinction between the catalytic phenomenon occurring in the animate and inanimate world. He also used the now famous words isomer, polymer, ammonium, protein, and globulin. Those were the times when a "vital force" was associated with living cells and biocatalysts were part of this explanation. Only much later did the concept take root that ordinary physical and chemical principles apply to enzyme catalysis.

Pierre Berthelot was the first to derive a second-order rate equation which influenced the publication by Guldberg and Waage on the law of mass action leading to chemical kinetics.

As a part of their study on catalytic phenomena, Wohler and Liebig discovered "emulsin" (a β-glucosidase) from almonds in 1837. Indeed, this enzyme was cleverly used by Fischer subsequently (almost fifty years later!) to define enzyme specificity.

2.2 "Enzyme": Conceptual Origin

Swedish chemist Berzelius (1779–1848) proposed the name catalysis (from the Greek: *kata*: wholly and *lyein*: to loosen) in 1836. When Berzelius first invoked the term "catalysis," he did not make any distinction between chemical catalysis and catalysis in (or by) biological systems. He used a generic term "contact substance" for a catalyst. The origin of the word "enzyme" dedicated to biological catalysts has a convoluted history. Much of this drama was played out during a vigorous debate on whether there is a special force (the "vital force") associated with reactions occurring in living systems. Ever since Payen and Persoz (1833) introduced this name for the starch hydrolytic activity, "diastase," it has often been used to generally mean a catalyst of biological origin. Victor Henri in his 1903 book on enzyme kinetics (an early classic on enzyme action) used diastase to mean an "enzyme." Many other French scientists including Pierre Duclaux and Gabriel Bertrand did use diastase to mean what we now call enzymes. The suffix "ase"—arising out of diastase—was subsequently recommended for all enzyme names (by Duclaux in 1898).

Ferment as a term was used to describe both living yeast as well as the action of its cellular contents. Berthelot's extraction of "ferment" (1860) from yeast cells marks the beginning of the action of enzymes outside of a living cell. This also dealt a blow to vitalistic thinking in biochemistry. The analogy between ferment-catalyzed and acid-catalyzed hydrolysis of starch was well recognized by the successive contributions of Kirchhoff, Payen and Persoz, and Berzelius. Schwann had used a similar analogy for pepsin. Willy Kuhne in 1867 extended this further to pancreatic digestion of proteins and called this activity trypsin in 1877. The essential meaning of "ferment" was consolidated by Kuhne; subsequently, the word **enzyme** (*in yeast*) was first used by him in 1877. And trypsin was the first candidate "ferment" to be called an enzyme.

The evolution of and acceptance of the word enzyme has taken its time. Both the descriptions—"diastase" (mostly in French scientific literature) and "ferment"—were used occasionally well into the early twentieth century.

The vitalistic theory was firmly laid to rest with Eduard Buchner's conclusive demonstration that a suitable extract from yeast cells could convert sucrose to alcohol. This was revolutionary in 1897 since fermentation was shown to occur "without living yeast" for the first time. The activity was ascribed to a single substance which was named "zymase" (and alcoholase by Emile Roux). It is now history that this activity in fact represents the entire glycolytic sequence of reactions. Out of controversy on the nature of alcoholic fermentation, the word "enzyme" was born. This word reminds us that yeast ("zyme") and its activities were resolved through the prisms of biology and chemistry to create the rich domain of enzymology.

2.3   Key Developments in Enzymology

Protein nature of enzymes: Early progress on enzymes was impeded because not much was known about the chemical nature of proteins. Also, much less was known about the chemical nature of enzymes. One approach to understanding them was to purify them for a detailed analysis. Kuhne and Chittenden extensively used the technique of protein fractionation by ammonium sulfate and also introduced the use of dialysis and dialysis tubing (1883). Powerful methods to purify enzymes were developed by Richard Willstatter—the first introduction of alumina Cγ gel was made. Peroxidase was taken to such a high level of purity that the preparation failed in the prevailing tests for protein. This unfortunately led him to wrongly conclude that enzymes are not proteins (1926). The seminal discovery by James Sumner in the same year, proving that urease is a protein therefore assumes great significance (Sumner 1933). This view was further confirmed by purification and crystallization of three more enzymes—pepsin, trypsin, and chymotrypsin—by Northrop and Kunitz (between 1930 and 1935). It must impress anyone to note that all this was accomplished by just two simple purification techniques—fractional precipitation of proteins by ammonium sulfate and pH changes.

Laccase is one of the early examples of the use of an enzyme that was not a hydrolase. Bertrand (1895) described it as an "oxidase" and suggested that this enzyme contained a divalent metal. His description of "co-ferment"—a nonprotein component of laccase—is the first descriptor of an enzyme cofactor.

More powerful yet gentler procedures of protein purification (and dialysis) hastened the progress of enzymology by providing many pure enzyme preparations. The end of the nineteenth century saw an increase in the number of reports on enzymes. By 1955, the number of enzymes reported was so large that their proper organization into categories became necessary. Under the auspices of the International Union of Biochemists (IUB), an International Commission on Enzymes was established to systematize the classification and naming of enzymes. As a result, the Enzyme Commission (EC) produced guidelines on enzyme nomenclature and brought out its recommendations in 1961.

Kinetic foundations: Because they are excellent catalysts, enzyme kinetic behavior could be studied regardless of meager knowledge of their composition. Even after their protein nature was established, it has taken a long time to relate the structural basis of enzyme kinetic behavior.

As early as 1898 the reversibility of an enzyme reaction was reported (Hill 1898). The enzymatic synthesis of a glucoside (maltose from glucose) by the yeast maltase established some key features: (a) an enzyme being a catalyst speeds up the reaction in both directions of a reversible reaction, (b) at least some steps in metabolism may go in either direction, and (c) enzymes may be involved in the cellular biosynthetic processes.

The reversibility of enzyme catalysis brought it within the ambit of thermodynamic analysis and physical chemistry. The thermodynamic constraints imposed upon catalyzed and uncatalyzed reactions were set forth by J van't Hoff. This subsequently led JBS Haldane to relate enzyme kinetic parameters with reaction thermodynamics and arrive at the famous Haldane relationship (Enzymes 1930).

Yeast invertase has the singular distinction as the working example for early work on enzyme reaction kinetics and thermodynamics. AJ Brown (1902) deduced the formation of the invertase-sucrose complex (the ES complex) from initial rate measurements. It was in 1903 that V. Henri for the first time derived the hyperbolic rate equation for a single-substrate enzymatic reaction (Cornish-Bowden et al., 2014). He provided the general process used to derive such rate equation—an exercise central to any enzyme kinetic study. It goes to Henri's credit that the validity of a rate equation is necessary, but not sufficient to prove the postulated kinetic mechanism to be recognized. In fact, the now famous Michaelis–Menten equation, based on the equilibrium treatment of the system, was published about ten years later in 1913. A more general form of the Henri–Michaelis–Menten equation to describe enzyme kinetics was derived by Briggs and Haldane via the steady-state approach in 1925. We continue to use this fundamental equation even today to describe the substrate saturation phenomenon of an enzyme reaction. A very popular linear form of this hyperbolic relation between initial velocity and substrate concentration is attributed to Lineweaver and Burk (1934).

A systematic study of enzyme reaction rates dictated that buffers be used to control hydrogen ion concentrations. This was indeed the impetus for the work published in 1909 by Sorenson on the pH scale and buffers. Subsequently, Leonor Michaelis and others emphasized the importance of pH on enzyme activity and routinely controlled it in all their studies.

The ES complex formation was a kinetic concept to begin with. The first direct observation of an enzyme substrate complex of catalase was made by KG Stern (1935); he monitored the catalase–HOOEt complex using spectroscopy.

Mechanistic studies: Emil Fischer was an unusual organic chemist of high caliber. He was responsible for establishing the rigor of synthetic and analytical skills of organic chemistry to biological problems. As early as 1894, he observed that substrates for invertin (now the invertase or sucrose hydrolase) are not substrates for emulsin (a β-glucosidase) and vice versa. Fischer opined that—"enzymes are fussy about the configuration of their object of attack." For example, the enzyme and the glucoside on which it acts must fit each other like a "lock and key" to be able to catalyze the chemical reaction. The future, as we know it, confirmed the genius of Fischer. This laid the foundation for describing fundamental properties of enzymes like specificity, stereo-selectivity, and the famous lock-and-key analogy for enzyme–substrate interactions (Eschenmoser 1995).

In an attempt to explain how enzymes work, the "Trager" or carrier theory was proposed by Willstatter (in 1922). According to him, enzymes contain smaller reactive groups that have affinity toward specific groups on the substrate—leading to enzyme specificity. Of course, these reactive groups were thought to be attached to an inert colloidal carrier to form the enzyme. Clearly, the fact that enzymes are proteins was not yet established then.

The hypothesis by AG Ogston (1948) attempted to explain how enzymes achieve chemical asymmetry through three-point contact with their substrates. This paved the way for further experiments in elucidating enzyme chemical mechanisms. Redox reactions involving pyridine nucleotides and the mechanism of hydride transfer followed shortly thereafter. Frank Westheimer and his colleagues, working with alcohol dehydrogenase and lactate dehydrogenase as examples, showed that the substrate hydrogen was transferred selectively to one side of the nicotinamide ring. This pioneering research in 1953 made use of deuterium- and tritium-labeled substrates to establish the stereo-specificity of these hydride transfers.

Work by Michael Duodoroff's group (1947) on bacterial disaccharide phosphorylases forms an early and brilliant example use of radioisotopes (^{32}P phosphate) in the study of enzyme mechanisms. Two similar reactions involving disaccharide phosphorolysis, namely sucrose phosphorylase and maltose phosphorylase were shown to follow completely different mechanisms. This led directly to the notion of single displacement versus double displacement reactions and subsequently the $S_N 1$ and $S_N 2$ reaction pathways.

The theory of kinetic criteria to distinguish enzyme mechanisms was elaborated by WW Cleland in three seminal papers (1963). At the least, this provided a common language to present enzyme kinetic data, for an otherwise confusing variety of notations found in enzyme kinetic literature. The impact of systematizing enzyme kinetics served two useful purposes—(1) it provided a common kinetic notation for presentation and (2) provided a summary of criteria on how to relate kinetic data with reaction mechanisms.

The rigidity of enzyme active site structure became untenable over time. Complementarity of enzyme active site to accommodate the transition state structure (rather than the substrate or the product) by Linus Pauling (1946) was prophetic; this clearly anticipated the need for protein motion, however subtle, at enzyme active sites. The idea of conformational flexibility of protein molecules as a prerequisite for enzyme activity superseded the earlier lock-and-key concept. This theme further matured into the concept of induced fit hypothesis as proposed by Koshland (1958). The conformational flexibility of a protein and ligand binding through induced fit later became key elements of allosteric transitions. The plasticity of protein structure for regulation thus became inescapable (the famous Monod–Wyman–Changeux model to explain cooperative interactions in oligomeric proteins).

Recognition that enzymes bring about enormous rate accelerations quickly led to a search for underlying principles of such catalysis. Attempts to demystify and explain enzyme catalysis in physicochemical terms were made. Different contributory factors were dissected out through model chemical reactions as well as enzymes. The work of TC Bruice, WP Jencks, ML Bender, DE Koshland Jr., and others are significant in this query. A combination of factors—intermolecular/conformational effects, general acid/base catalysis, nucleophilic/electrophilic catalysis, etc.—contributed to accomplishing remarkable rate accelerations observed with enzymes. It is now well recognized that a combination of many factors produces an enzyme. The search for novel catalytic tools evolved by nature continues unabated even today.

Structure and synthesis: The stamp of the chemist's contribution to the study of enzymes is obvious from the progression—isolation, structure elucidation followed by total synthesis. Insulin was the first protein whose complete chemical structure was determined. However, among enzymes, this credit goes to bovine pancreatic ribonuclease A (RNase A)—it was the first enzyme whose primary sequence was elucidated. But lysozyme (this was followed later by RNase A) is the first enzyme whose three-dimensional structure was made available through X-ray crystallography. In a typical organic chemist's approach, the total synthesis of a molecule completes the structure elucidation process. In this sense, RNase A was the first enzyme whose total synthesis was achieved (RB Merrifield) and it culminated in a catalytically active protein.

In summary, the history of enzymology is a rich source of factual and conceptual discoveries. Developments in this field were accelerated by chemists and biologists in equal measure. Once established, enzymology revolutionized both the parent disciplines—biology and chemistry. This is amply evident from the list of Nobel laureates (Table 2.2) and their work recognized by the two scientific communities that nurtured the study of enzymes.

Table 2.2 Nobel laureates who contributed to the growth of enzymology

Scientist(s)	Year[a]	Enzymology: topic of study
E. Fischer	1902-C	Stereochemistry and lock and key concept
S. Arrhenius	1903-C	Activation energy and catalysis
E. Buchner	1907-C	Cell-free extracts and fermentation
A. Harden & H. von Euler	1929-C	Coenzymes and fermentation
C. Eijkman & F.G. Hopkins	1929-M	Vitamins, nutrition, and coenzymes
O. Warburg	1931-M	Respiratory enzymes
A. Szent-Gyorgyi	1937-M	Fumarate catalysis of the TCA cycle
R. Kuhn	1938-C	Vitamins and coenzymes
A. Fleming	1945-M	Penicillin and lysozyme
J.B. Sumner, J.H. Northrop & M. Kunitz	1946-C	Purification and crystallization of enzymes
C. Cori & G. Cori	1947-M	Enzymes of glycogen metabolism
H.A. Krebs & F. Lipmann	1953-M	TCA cycle and Coenzyme A
L. Pauling	1954-C	Secondary structure: α helix; concept that enzyme binds the transition state
H. Theorell	1955-M	Oxidative enzyme mechanisms
A.R. Todd	1957-C	Nucleotides and nucleotide coenzymes
F. Sanger	1958-C	Insulin sequence through proteases
S. Ochoa & A. Kornberg	1959-M	Nucleic acid biosynthesis enzymes
M.F. Perutz & J.H. Kendrew	1962-C	Crystal structure of globular proteins
D. Crowfoot Hodgkin	1964-C	Structure of vitamin B12
K. Bloch & F. Lynen	1964-M	Cholesterol and fatty acid enzymes
F. Jacob, A. Lwoff & J. Monod	1965-M	Genetic control of enzyme synthesis and allostery
L.F. Leloir	1970-C	Sugar nucleotides and carbohydrate biosynthesis
E.W. Sutherland Jr.	1971-M	Enzyme and metabolic regulation by cAMP
C.B. Anfinsen, S. Moore & W.H. Stein	1972-C	Chemical structure: catalytic activity of RNase A
J.W. Cornforth	1975-C	Stereochemistry of enzyme reactions
W. Arber, D. Nathans & H.O. Smith	1978-M	Restriction endonucleases
F. Sanger	1980-C	DNA sequencing (ddNTP method) with enzymes
R.B. Merrifield	1984-C	Chemical synthesis of RNase A
J.W. Black, G.B. Elion & G.H. Hitchings	1988-M	Inhibitors (enzyme) as drugs
S. Altman & T.R. Cech	1989-C	Catalysis by RNA molecules
E.H. Fischer & E.G. Krebs	1992-M	Protein kinases and protein phosphorylation
K.B. Mullis	1993-C	Polymerase chain reaction
P.D. Boyer, J.E. Walker & J.C. Skou	1997-C	ATP synthase and Na,K-ATPase
A.H. Zewail	1999-C	Detection/ existence of transition state
I. Rose (and others)	2004-C	Ubiquitin-protein degradation and isotope exchanges in enzymology

(continued)

Table 2.2 (continued)

Scientist(s)	Year[a]	Enzymology: topic of study
A. Warshel (and others)	2013-C	Computational enzymology
F.H. Arnold	2018-C	Directed evolution of enzymes
E. Charpentier & J.A. Doudna	2020-C	CRISPR/Cas9 genetic scissors
D. Baker	2024-C	De novo enzyme design

[a]Prize awarded this year for: C, chemistry; M, physiology and medicine

References

Cornish-Bowden A, Mazat J-P, Nicolas S (2014) Victor Henri: 111 years of his equation. Biochimie 107:161–166

Eschenmoser A (1995) One hundred years lock-and-key principle. Angew Chem Int Ed Engl 33: 2363

Hill AC (1898) Reversible zymohydrolysis. J Chem Soc Trans 73:634–658

Sumner JB (1933) The chemical nature of enzymes. Science 78:335

Further Reading

Bugg TDH (2001) The development of mechanistic enzymology in the 20th century. Nat Prod Rep 18:465–493

Cornish-Bowden A (1999) The origins of enzymology. Biochemist 19:36–38

Cornish-Bowden A (2013) The origins of enzyme kinetics. FEBS Lett 587:2725–2730

Friedmann HC (1981) Enzymes: benchmark papers in biochemistry. Hutchinson Ross Publishing Company, Stroudsburg

Johnson KA (2013) A century of enzyme kinetic analysis, 1913–2013. FEBS Lett 587:2753–2766

Neidleman SL (1990) The archeology of enzymology. In: Abramowicz DA (ed) Biocatalysis. Van Nostrand Reinhold, New York, pp 1–24

All reactions relevant to biology belong to one of the two groups: those with a kinetic barrier and those with a thermodynamic barrier. The kinetic barrier can be overcome by employing a catalyst. Reactions with a thermodynamic barrier (uphill, endergonic reactions) require the provision of energy (such as ATP hydrolysis) in addition to a catalyst. Regardless of the nature of barriers faced—greater than 99% of all the reactions occurring in biological systems are catalyzed. With the minor exception of a few RNA catalysts (ribozymes) all the enzymes are built from a protein scaffold. These catalysts par excellence are at the very foundation of life. The three hallmark features of these biocatalysts are—*Catalysis, Specificity,* and *Regulation.*

3.1 Catalysis

It is possible that a reaction (say A→P) may proceed by more than one reaction path under given experimental conditions. One of these routes may exhibit a faster reaction rate. Many reactions are known where rates can be enhanced by the addition of a reagent. Such entities that increase the reaction rates but are not included in the stoichiometry of that reaction are called *catalyst*s. A catalyst provides a facile, additional route to the formation of the same products. Catalysis can occur with the catalyst and the reaction occurring in the same phase (homogeneous catalysis) or in different phases (heterogeneous catalysis). Heterogeneous catalysis on metal surfaces is common in many industrial processes. Most biological catalysis (by enzymes) is *homogeneous catalysis* in solution. This forms the main focus of this book. In some cases, like the conversion of chymotrypsinogen to chymotrypsin, the reaction product also acts as a catalyst. This is an example of *autocatalysis.* With membrane-bound enzymes we often encounter interfacial catalysis.

First and foremost, an enzyme is a catalyst and is responsible for the rate acceleration of a reaction that is inherently slow. Some of the chemical reactions are relatively fast even without a catalyst. For instance, reversible hydration of

N. S. Punekar, *ENZYMES: Catalysis, Kinetics and Mechanisms,*
https://doi.org/10.1007/978-931-97-8179-9_3

carbon dioxide, dismutation of superoxide anions, and elimination of hydrogen peroxide are all very rapid. These rates are however not fast enough, so nature has evolved enzymes (carbonic anhydrase, superoxide dismutase, and catalase, respectively) to further accelerate them. There should be—and indeed there is—nothing magical about how enzymes bring about fantastic rate accelerations. They do obey the basic physical and chemical principles—but are simply better at it. Like any other catalyst, an enzyme

(a) achieves rate acceleration by bringing down the activation energy barrier. It may convert a complex reaction into a number of simpler ones—with each step having its own smaller activation energy barrier. But the tricks enzymes use to do this are time-tested by evolution and are exquisite. We will describe these tools a little later (Chap. 4).

(b) does not change the equilibrium constant for a particular reaction but hastens the approach to equilibrium. Since the equilibrium position can be reached from both directions (forward or reverse), in principle, enzymes can accelerate the rates in either direction. At the least, micro-reversibility of catalytic steps (even if the overall reaction is largely unidirectional and practically irreversible!) is expected at the enzyme active sites.

Thermodynamics of catalysis: For any reaction to occur, it should be accompanied by a decrease in free energy. Accordingly, the equilibrium constant and the corresponding standard free energy of reaction are related by the equation $\Delta G° = -RT \ln K_{eq}$. A catalyst cannot displace the equilibrium of the reaction and therefore K_{eq} (and $\Delta G°$) remains unaffected. Enzymes increase the *rate* at which equilibrium is reached—by bringing down the activation energy barrier. Biological catalysis obeys the same general rules observed for nonenzymatic catalysis. The transition state theory of reaction rates has been meaningfully extended to enzyme catalysis. For a reaction going from reactant (S) to product (P), the highest point along the imaginary reaction coordinate is called the *transition state* (TS in Fig. 3.1). This transition state is of the highest free energy and is an extremely unstable ephemeral species, involving bond-breaking and bond-forming events. Thus, by definition, $\Delta G^{\neq}$ is the standard free energy of activation for this reaction—schematically represented in Fig. 3.1.

Heat released during enzyme catalysis: The ΔG should be negative for any reaction to proceed spontaneously. Without exception, this is true for any enzyme-catalyzed reaction as well. Then what about the heat generated/released (the ΔH component of ΔG) during the reaction at the enzyme active site? The temperature rises during enzyme catalysis of highly exothermic reactions. Several explanations have been put forward to account for the dissipation of heat from the enzyme active site to the bulk medium. Dissipation of heat generated by catalysis, through the protein, may propel the enzyme molecules (Riedel et al. 2015), and that "enzymes surf the heart wave" as observed through their enhanced molecular diffusion (see Chap. 40 for more details). This increased diffusion of enzymes is not due to collective heating (and the consequent temperature rise of the buffer).

Fig. 3.1 Free energy
diagram—schematic
comparing the catalyzed
versus uncatalyzed reactions.
The standard free energy of
reaction ($\Delta G°$) and the free
energy of activation for
uncatalyzed ($\Delta G^{\neq}_{uncat}$) and
catalyzed ($\Delta G^{\neq}_{cat}$) reactions
are shown

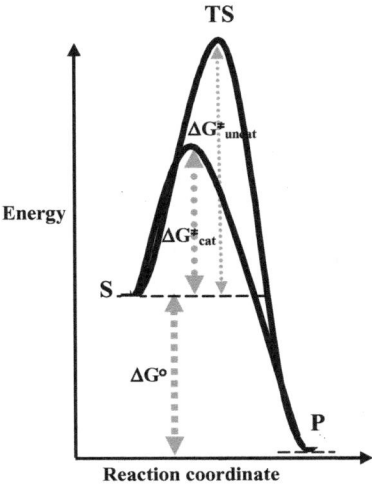

Relation between rate constant (k) and free energy of activation ($\Delta G^{\neq}$): The
frequency with which the TS decomposes (ν) is the same as the vibrational frequency of the bond that is breaking. From quantum theory calculations, $\nu = k_B T/h$.
Transition state theory predicts that the rate of chemical reaction is related to the
transition state concentration as

$$\text{reaction rate} = \frac{k_B T}{h}\,[\text{TS}]$$

where k_B is Boltzmann's constant (1.38×10^{-16} cm^2 g s^{-2} K^{-1}), h is Plank's
constant (6.626×10^{-27} cm^2 g s^{-1}), and T is the absolute temperature. At 300°K,
the frequency factor $k_B T/h = 6.3 \times 10^{12}$ s^{-1}. By the very definition of TS, its
concentration is not a measurable term. However, this can be indirectly substituted
by invoking a quasi-thermodynamic equilibrium between reactant and the TS, with a
hypothetical equilibrium constant ($K^{\neq}_{eq}$).

$$K^{\neq}_{eq} = \frac{[\text{TS}]}{[S]}\quad\text{and}\quad\therefore[\text{TS}] = K^{\neq}_{eq}[S]$$

On substituting the value of [TS] in the rate equation above we get

$$\text{reaction rate} = \frac{k_B T}{h}\,K^{\neq}_{eq}[S]$$

Comparing this equation with the equation for reaction rate, $-(d[S]/dt) = k\,[S]$,
we obtain

$$k = \frac{k_\mathrm{B}T}{h} K^{\neq}_\mathrm{eq}$$

where k is the rate constant.

By analogy to the relation between K_eq and ΔG°, we establish that

$$k = \frac{k_\mathrm{B}T}{h} e^{-\frac{\Delta G^{\neq}}{RT}} = \frac{k_\mathrm{B}T}{h} e^{-\frac{\Delta H^{\neq}}{RT}} e^{+\frac{\Delta S^{\neq}}{R}}$$

From this equation, it is obvious that as the free energy of activation increases (i.e., $\Delta G^{\neq}$ becomes larger), the rate constant (k) for the reaction will decrease in an exponential fashion. Also, the value of the rate constant is directly related to the temperature (T). We note that this equation is similar to the Arrhenius equation (from Collision Theory):

$$k = A e^{-\frac{E_a}{RT}}$$

where A is the Arrhenius constant (pre-exponential factor) and E_a is the activation energy. A comparison of the two treatments (transition state theory versus Arrhenius theory) provides the following equivalences:

$$E_a = \Delta H^{\neq} + \mathrm{RT} \quad \text{and} \quad A = \frac{k_\mathrm{B}T}{h} e^{+\frac{\Delta S^{\neq}}{R}}$$

Although the Arrhenius equation (and collision theory) is of great historical importance, most modern treatments of reaction rates follow the transition state theory. We can now appreciate how enzymes bring about rate accelerations by decreasing the activation energy barrier. The magnitude of reduction in the free energy of activation (see Fig. 3.1) can be translated quantitatively into the extent of rate enhancement. This is shown with an example in the box below.

We know that $k = \frac{k_\mathrm{B}T}{h} e^{-\frac{\Delta G^{\neq}}{RT}}$

Rearranging this equation and writing it for catalyzed as well as uncatalyzed reactions, we get

$$k_\mathrm{uncat} = \frac{k_\mathrm{B}T}{h} e^{-\frac{\Delta G^{\neq}_\mathrm{uncat}}{RT}} \quad \text{and} \quad k_\mathrm{cat} = \frac{k_\mathrm{B}T}{h} e^{-\frac{\Delta G^{\neq}_\mathrm{cat}}{RT}}$$

Taking ratios and simplifying

$$\frac{k_\mathrm{cat}}{k_\mathrm{uncat}} = 10^{\frac{\Delta G^{\neq}_\mathrm{uncat} - \Delta G^{\neq}_\mathrm{cat}}{2.303\,RT}}$$

(continued)

For example, consider a reaction where $\Delta G^{\neq}_{uncat} = 25.7$ kcal/mol and $\Delta G^{\neq}_{cat} = 11.0$ kcal/mol. Substituting ($R = 1.987$ cal/mol K) and simplifying,

$$k_{cat}/k_{uncat} = 10^{(25.7-11.0)/1.36} = 5 \times 10^{10}$$

A decrease in free energy of activation ($\Delta\Delta G^{\neq}$) of about 15 kcal/mol can result in 10^{10}-fold increase in rate accelerations. Catalase (with $\Delta G^{\neq}_{uncat}$ of 18.2 kcal/mol and $\Delta G^{\neq}_{cat}$ of 7.2 kcal/mol) accelerates its reaction rate by a factor of 10^8. This is how enzymes bring about their magic!

The question of how enzymes achieve decrease in free energy of activation ($\Delta G^{\neq}_{uncat} - \Delta G^{\neq}_{cat} = \Delta\Delta G^{\neq}$) for a given reaction is detailed in the next chapter (Chap. 4).

3.2   Specificity

Specificity, at the molecular level, is the hallmark of most biological interactions. Molecules like receptors and antibodies specifically interact with their cognate counterparts. But discrimination while performing catalysis is of paramount importance to biology—and is unique to enzymes! Specificity is a virtue when two similar reactions are to be kept separate, at times in the same compartment. Most biosynthetic reactions are catalyzed by $NADP^+$ requiring enzymes while catabolic reactions use NAD^+. For instance, glutamate dehydrogenases (GDHs) with two distinct specificities for pyridine nucleotide are known—the biosynthetic NADP-GDH (EC 1.4.1.4) and the catabolic NAD-GDH (EC 1.4.1.2).

Range of specificities: Enzyme specificity has been the subject of study from the beginning. Reiner (1959) pointed out that—"Some enzymes are more discriminating than others, but it seems fair to say that any enzyme can be fooled if one goes to enough trouble." The esterase activity (on p-nitrophenyl acetate) of chymotrypsin—an endopeptidase is well known. While enzymes can be quite discriminatory with respect to the substrates they act on, a range of specificity is observed with different enzyme examples. At one extreme they can be *absolutely specific* like glucose oxidase—acting on glucose but not on galactose or mannose. Glucokinase (acting on glucose alone) and succinate dehydrogenase (acting on succinate) are other examples of high substrate specificity enzymes. A comparison of glucokinase with that of hexokinase is instructive. Hexokinase exhibits *broad substrate specificity* and also acts on hexoses other than glucose; it is however less efficient in handling other sugars. An enzyme catalyst may be selective in acting on specific groups. Examples of *group-specific* enzymes include—alcohol dehydrogenase (acting on primary as well as secondary alcohols), esterases (hydrolyzing various carboxylic esters), and phosphatases (hydrolyzing many phosphate esters). Stereo-selectivity of enzyme action was recognized very early by Emil Fischer (Table 2.1). Many enzymes act on only one optical isomer—such as—glucose oxidase acts on β-D-glucose (and not on α-isomer) and β-galactosidase acts on β-galactosides alone. L-Amino acid oxidase, specific to L-amino acids, is yet another example of a *stereospecific* enzyme. Even identical groups on a prochiral center of a molecule can be discriminated by an enzyme. Aconitase (the second enzyme from the Krebs cycle) clearly distinguishes the two $-CH_2COO^-$ groups of the prochiral citrate—selectively attacking the *pro-R* carboxymethyl group. Similarly, yeast alcohol dehydrogenase distinguishes the two H atoms on the $-CH_2-$ group of ethanol (CH_3–CH_2–OH); only the *pro-R* hydrogen is transferred to NAD^+.

Enzyme specificity is not just limited to the substrates on which they act. Molecules that satisfy the specificity criteria for an enzyme—but do not possess the susceptible chemical bond(s)—may act as potential *inhibitors* of that enzyme. Structural variation among inhibitors can lead to different degrees of enzyme inhibition.

Limits of enzyme discrimination (and biological specificity!): As seen in the previous paragraph, enzyme specificity can be absolute or quantitative. It may be useful to assess the levels to which enzymes (and hence biological systems!) can resolve and discriminate molecules. Recognition at the molecular level is easy to appreciate. Gross structural differences like those between glucose and galactose, cAMP and cGMP, or NAD^+ and $NADP^+$ provide sufficient molecular handles to grip them differently by the enzymes.

With some enzymes, the substrate to be selectively bound is a very small entity indeed. They may be mere diatomic blobs of different atoms! Cytochrome oxidase (EC 1.9.3.1) binds oxygen (O=O) but cannot exclude cyanide (CN^-) or carbon monoxide (C=O); these are however potent inhibitors of respiration. Guanylate cyclase (EC 4.6.1.2) has to contend with accommodating nitric oxide (NO) or carbon monoxide (CO) at its regulatory site. Similarly, nitrogenase (EC 1.18.2.1) should receive and reduce N_2 but avoid the more reactive O_2. This is quite a challenge as the two gases predominate in the atmosphere and both are diatomic molecules of electronegative atoms. In fact, nitrogenase can also bind and reduce acetylene to ethylene—a reaction used to assay this enzyme. The inability of nitrogenase to keep oxygen away (from its reaction center) often results in its inactivation (Gallon 1981). Nature has made allowance for this loss by means other than nitrogenase specificity, however. Nitrogenase is (a) rapidly turned over and (b) protected by mechanisms to reduce the local concentrations of oxygen and to eliminate reactive oxygen species formed.

It is fascinating to note that enzymes can be made to discriminate even at the atomic level. RuBP carboxylase (EC 4.1.1.39) is the key enzyme responsible for the entry of CO_2 into the biosphere. Its acronym, rubisco, stands for ribulose-1,5-bisphosphate carboxylase-oxygenase because the enzyme can confuse O_2 for CO_2 as its substrate. Carbon dioxide is devoid of chemical hand- or footholds for the enzyme to grip it and help in its reaction with a fickle enediol intermediate. The difficulties associated with selectively binding a nearly featureless CO_2 molecule are obvious. Both CO_2 and O_2 are gaseous electrophiles and RuBP carboxylase struggles to discriminate between them. This is particularly relevant when we consider that the solution concentration of O_2 (250 μM) far exceeds that of CO_2 (10 μM). The oxygenase activity of this catalyst is thus an expected outcome. Nevertheless, the enzyme (from C3 plants) selectively favors CO_2 to O_2 by 3:1. As a tradeoff, however, this CO_2/O_2 specificity comes at the cost of relatively poor catalytic rates (Griffiths 2006).

The resolution/discrimination achieved in the above examples is the outcome of a single interaction event of the enzyme with its substrate. For higher stringencies, nature resorts to building multiple recognition events and sieves. Amino acyl tRNA synthetase is a good example of this. The amino acyl-AMP derivative is made first and then in a second event the tRNA is charged. In all, the amino acid being charged is discriminated twice by the same enzyme. Additional opportunities for specificity are offered by proof-reading processes, which remove incorrect products. All such proof-reading mechanisms cost energy—but are well worth the effort in preventing an error. The side chains of L-leucine and L-isoleucine are hydrophobic blobs of almost similar volumes. But leucine tRNA synthetase (EC 6.1.1.4) discriminates against isoleucine by about 1000:1. This is the basis of high fidelity of the translation machinery in ensuring that Ile and Leu are correctly incorporated into polypeptides. Additional check for specificity is afforded by proofreading processes, which remove incorrect entries. All such proofreading costs energy—but are well worth the effort in preventing errors. The evolutionary pressures drive the stringent molecular recognition and selectivity of these enzymes (Tawfik and Gruic-Sovulj 2020).

Cofactor tuning: An interesting aspect of specificity is the selective interaction of the polypeptide component with its cognate cofactor. Each apoenzyme provides a unique chemical environment to the cofactor thereby modulating its reactivity. For instance, the redox potential for the two-electron reduction of free FAD is about −200 mV. This value measured for flavoenzymes ranges from −450 mV to +150 mV. The redox potential of flavin is thus fine-tuned at the active site. Catalytic protein milieu achieves this by (a) placing suitable positive charge (increases redox potential) or negative charge (decreases redox potential) and (b) possibly forcing the FAD (tricyclic isoalloxazine ring system) to adopt a planar or nonplanar conformation. Other examples of fine-tuning chemical properties include cofactors like metal ions and heme. The reactivity of a common heme cofactor (iron protoporphyrin IX) is modulated and adjusted for specific biological functions by covalent attachment, axial ligation, hydrogen bonding, and distortion from planarity imposed by different protein environments. Cytochromes of various redox potentials (participating in enzymatic electron transport) are a manifestation of this purposeful fine-tuning by nature. Enzyme proteins control the active site metal ion reactivity by selective provision of the number and nature of coordinating ligands. Despite unfavorable concentrations of various metal ions in the cytoplasm, correct metallation of an apoenzyme is largely favored in vivo (Foster et al. 2014). The nature and geometry of metal ion binding residues provide significant selectivity toward the metal ion.

Enzyme promiscuity: Enzymes display a range of substrate specificities—they can be highly specific or broadly specific. By and large, enzymes are exquisitely specific for their substrates and the reactions they catalyze. However, examples of enzymes diverging from this statement are accumulating over time. Such enzymes are sometimes called promiscuous, and their promiscuity may manifest as (a) relaxed substrate specificity, (b) catalyzing distinctly different chemical transformations, or (c) distinct catalytic activity under unnatural conditions of low water activity (anhydrous media), extreme temperature, or pH. Relaxed substrate and reaction specificities can have an important role in divergent enzyme evolution. Nature may exploit primordial enzymes to evolve and hence catalyze novel reactions in metabolism; thus, enzyme evolvability may be related to promiscuity (Copley 2003; Gupta et al. 2011). Enzymes have been extensively used outside of their natural context in the industry. For instance, reactions run under conditions of low water activity favor ester synthesis instead of hydrolysis (see Chap. 37). The capacity to promiscuously catalyze reactions other than the ones they evolved for can be an advantage. Directed evolution of such enzymes (see Chap. 37) promises improvements in existing catalysts and provides synthetic pathways (green chemistry) that are currently not available (Hult and Berglund 2007).

3.3   Regulation

The third important feature of enzymes is their ability to act as regulatory points of metabolism. Marveling at the metabolic complexity, Jacques Monod observed that "From a glance at the drawing condensing what is now known of cellular metabolism we can tell that even if at each step each enzyme carried out its job perfectly, the sum of their activities could only be chaos were they not somehow interlocked so as to form a coherent system." A cell is not a bag of enzymes and obviously regulating their enormous catalytic potential is a necessity. We will discuss different strategies to control reaction rates by regulating enzyme activity in a later section (Part VI).

Each enzyme is a structural microcosm capable of catalysis, specificity, and regulation. Chemical reactions performed by an enzyme occur at a specific location on the enzyme protein called—the *active site* (or catalytic center). Enzyme active sites are small relative to their total molecular volume. The architecture of the active site is responsible for imparting specificity and catalytic potency to each enzyme. Active sites are usually clefts and crevices in the protein; they create a unique three-dimensional micro-environment by (a) aligning an array of amino acid side chains and cofactors and (b) excluding bulk solvent, i.e., water. Reactants (generally referred to as *substrates*) are assembled at the enzyme active site. Once the reaction is complete, the products leave the active site. For a reversible reaction, however, products become substrates and vice versa. Enzyme active sites may also accommodate inhibitors—molecules structurally resembling substrates or products.

Just as there may be a compromise between catalytic potency and specificity (seen with rubisco, above), the need to optimize regulation may also feature in enzyme design. Very little comes free in nature—regulation too has its costs. Apart from the active site, some enzymes may display an additional site—the *allosteric site* (*allos* in Greek means another). Such sites serve a regulatory function when bound by *ligands* (a general term used to describe small molecules like substrate, product,

Fig. 3.2 Schematic of an enzyme showing different ligand binding sites. Possible binding to the substrate (S), activator (A), and inhibitor (I) is shown. The susceptible bond in the substrate (S) is shown in gray

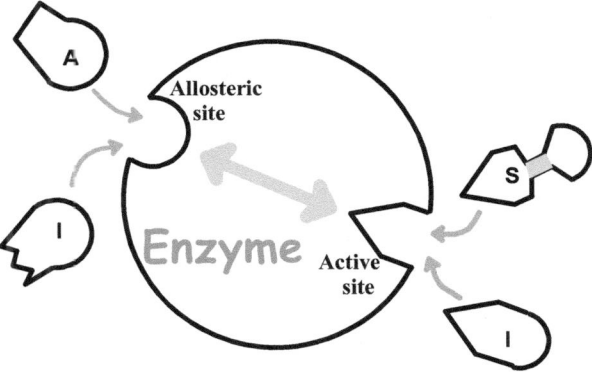

inhibitor, or activator). Allosteric sites and active sites could exist on the same subunit (example—phosphofructokinase) or on distinct subunits (example—aspartate transcarbamylase) of an oligomeric protein. Ligand binding to an allosteric site may influence the active site geometry and function. Communication between sites can be achieved by conformational coupling across the protein matrix. A schematic representing all these structural features of an enzyme is shown in Fig. 3.2.

We will end this chapter by recalling the performance of enzyme catalysts *par excellence*. Reduction of N_2 to ammonia serves as an exquisite example. The conditions used in Haber's chemical process for making ammonia, employed in the fertilizer industry, are drastic, to say the least! This direct reduction of N_2 by hydrogen gas to form ammonia requires a pressure of 300 atmospheres, a temperature of 500 °C, and an iron catalyst.

$$N_2 + H_2 \xrightarrow{\text{300atm, 500°C, Fe}} 2NH_3$$

$$N_2 + 8H^+ + 8e^- + 15ATP \xrightarrow{\text{Nitrogenase}} 2NH_3 + H_2 + 16ADP + 16P_i$$

This difficult reaction is performed at ambient conditions of temperature and physiological pH in the root nodules of leguminous plants by nitrogenase. A fantastic achievement and a catalytic fete indeed!

References

Copley SD (2003) Enzymes with extra talents: moonlighting functions and catalytic promiscuity. Curr Opin Chem Biol 7:265–272

Foster AW, Osman D, Robinson NJ (2014) Metal preferences and metalation. J Biol Chem 289: 28095–28103

Gallon JR (1981) The oxygen sensitivity of nitrogenase: a problem for biochemists and microorganisms. Trends Biochem Sci 6:19–23

Griffiths H (2006) Designs on rubisco. Nature 441:940–941

Gupta MN et al (2011) Isozymes, moonlighting proteins and promiscuous enzymes. Curr Sci 100: 1152–1162

Hult K, Berglund P (2007) Enzyme promiscuity: mechanism and applications. Trends Biotechnol 25:231–238
Reiner JM (1959) Behavior of enzyme systems. Burgess Publishing Co., Minneapolis
Tawfik DS, Gruic-Sovulj I (2020) How evolution shapes enzyme selectivity. FEBS J 287: 1284–1305

Further Reading

Bretz SL, Linenberger KJ (2012) Development of the 'enzyme-substrate interactions' concept inventory. BMBEd 40:229–233
Lienhard GE (1973) Enzymatic catalysis and transition-state theory. Science 180:149–154
Riedel C et al (2015) The heat released during catalytic turnover enhances the diffusion of an enzyme. Nature 517:227–230

Origins of Enzyme Catalytic Power

4

*Before ascribing (this) combination of techniques to any
undue wisdom of Mother Nature or her various male
consorts, we must recall that she has had 10^9 years to evolve
enzymes while man has about 10^2 years to comprehend and
duplicate them.*
Albert S Mildvan

Understanding enzyme function is exciting research because diverse and often
unpredictable solutions are developed to perform seemingly impossible tasks.
Enzymes are such powerful catalysts as they lower the height of the activation
energy barrier. How do enzymes bring about this decrease in $\Delta G^{\neq}$? Many excellent
attempts to dissect this into discrete enthalpic and entropic factors have been made.
Rate accelerations are favored when the enzyme specifically binds and assembles
substrates at the active site and then provides an optimal arrangement of catalytic
groups. This is clearly an entropic $(\Delta S^{\neq})$ contribution. Stabilization of the transition
state through enthalpic $(\Delta H^{\neq})$ factors like select hydrogen bonds, salt links, acid–
base groups, and covalent interactions is another feature. Various tricks that enzymes
employ in achieving catalytic prowess are shown in Fig. 4.1. These components are
best understood through case studies, and we will do this through representative
examples for each.

4.1 Proximity and Orientation Effects

Entropic contributions in accelerating the enzymatic reaction rates are substantial.
This is described by enzymologists variously as—approximation, coming together,
spatial relationship, pre-organization, propinquity effect, Circe effect (character from
Odyssey of Homer!), orbital steering, restricted motion, loss of degrees of
freedom, etc.

N. S. Punekar, *ENZYMES: Catalysis, Kinetics and Mechanisms*,
https://doi.org/10.1007/978-981-97-8179-9_4

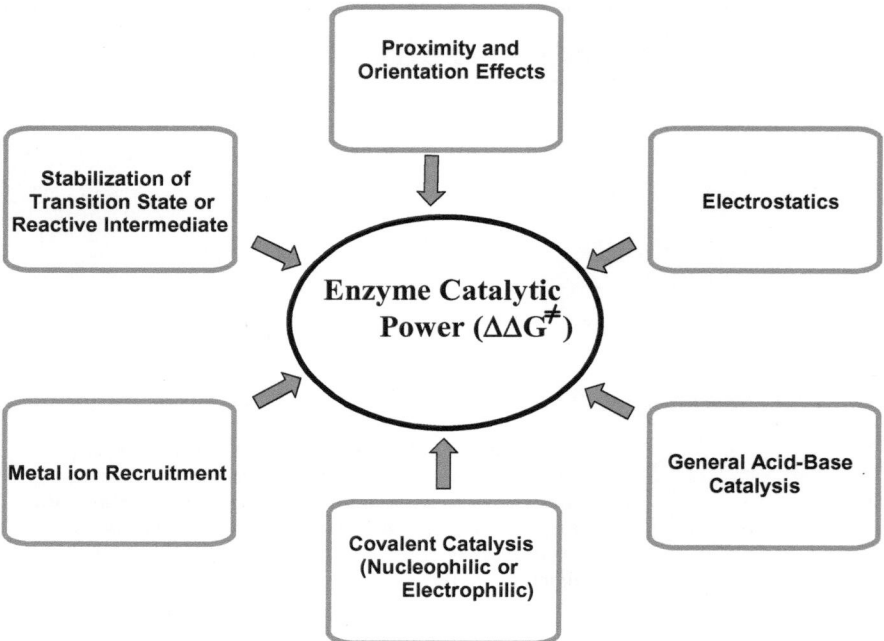

Fig. 4.1 Various features that enzymes use to achieve reaction rate acceleration

The reactants in the solution have substantial degrees of rotational and translational freedom. This means a given molecule is an ensemble of many conformational states. Only one (or few) of these conformations is capable of reaching the transition state and beyond. By selectively binding to the enzyme's active site such a reactant conformation is frozen out of many. Clearly on binding to the enzyme active site the substrate loses many degrees of freedom—becomes ordered. Recall that entropy is a measure of disorder or randomness. This entropy loss ($\Delta S^{\neq}$) is reflected in the lowering of $\Delta G^{\neq}$. This entropic cost is paid in part by the substrate binding energy—the entropic cost of a chemical event is thus shifted to the binding (physical) event! Apart from *freezing the reactant conformation* other consequences of substrate binding to the enzyme active site are important. These are discussed below.

Every enzyme active site is endowed with functional groups (amino acid side chains or from the cofactor or both) that are involved in catalysis. The three-dimensional scaffold of the enzyme protein ensures that these groups create a well-defined *pre-organized* active site. Binding to such a site brings these catalytic groups in close proximity and proper orientation with the susceptible substrate bonds. Any enzyme catalytic group acting on the substrate will now be an *intra-molecular* event. Intra-molecular reactions generally proceed much more rapidly than their inter-molecular counterparts. An excellent application of this concept is the Cleland reagent (Cleland 1964). Dithiothreitol (HS-X-SH) is a better reducing

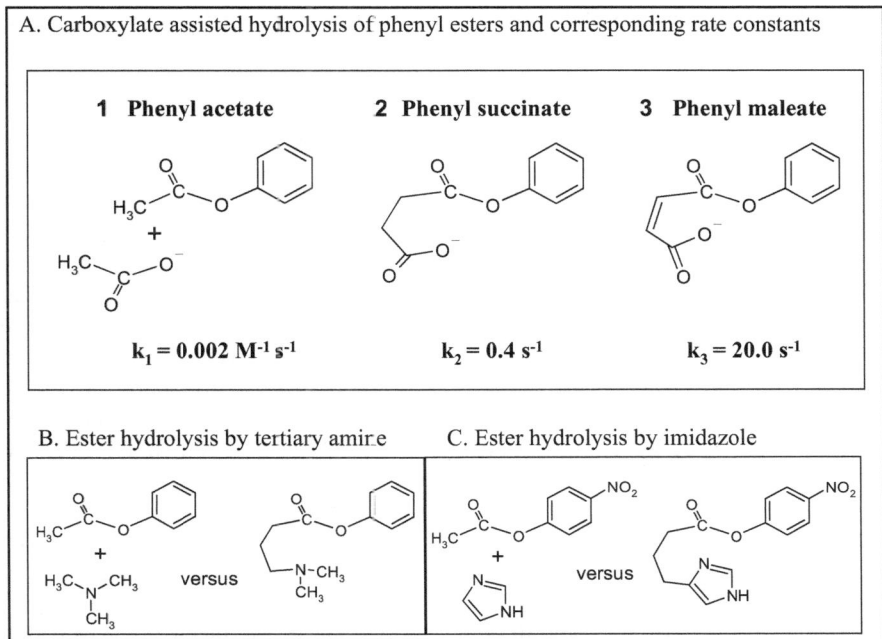

Fig. 4.2 Organic model reactions exemplifying the proximity and orientation effects that lead to rate accelerations. Hydrolysis of phenyl esters by carboxylate (**a**), tertiary amine (**b**), and imidazole (**c**) is shown

agent than the two equivalents of a monothiol (X-SH). Here, the oxidation is made intra-molecular and the formation of more number of product molecules ($n = 3$) confers entropic advantage (more degrees of freedom!) for the overall reaction.

R_1 - S - S - R_2 + X - SH + X - SH $\rightarrow$ R_1 - SH + R_2 - SH + X - S - S - X	
($n = 3$)	($n = 3$)
R_1 - S - S - R_2 + HS - X - SH $\rightarrow$ R_1 - SH + R_2 - SH + (X - S - S)	
($n = 2$)	($n = 3$)

Substrate tethering (proximity) and proper orientation of catalytic group(s) for the attack lead to a large kinetic advantage. It appears as though the effective local concentration of the catalytic group is raised enormously. Organic model reactions have helped to understand many of these features of enzyme catalysis. Examples illustrating these concepts are discussed below.

The first example in Fig. 4.2a is the typical hydrolysis of phenyl acetate by sodium acetate (CH_3COO^- ions). This is a bimolecular reaction where the ester and the acetate ion are independent entities. Note that accordingly the rate constant k_1 has units of $M^{-1}\ s^{-1}$. However, when the $-COO^-$ group is built into the same molecule (phenyl succinate) the reaction becomes intra-molecular—and the absolute value of rate constant (k_2 with units of s^{-1}) increases by a factor of 200! A direct comparison of the two rate constants is difficult because they represent two different orders of reaction (see Chap. 8 for details). By fixing the acetate concentration (say $[CH_3COO^-] = 1.0$ M) and assuming pseudofirst-order condition in the first case, this can be worked out. An effective $-COO^-$ concentration of 200 M may be calculated for phenyl succinate, while this would be an enormous number of 10,000 M for phenyl maleate! Achieving such high concentrations by acetate addition is practically impossible—concentration of glacial acetic acid itself is around 17 M!

We know from the above discussion that when a reaction is made intra-molecular, rate accelerations take place as if the apparent reactive group concentration felt at the site is enormously raised. Reactions in Examples 2 and 3 above are intra-molecular, then why is k_3 50 times larger than k_2? Because of two methylene groups (no double bond) phenyl succinate has many rotational degrees of freedom. The *cis* double bond in phenyl maleate restricts this motion and the $-COO^-$ is always favorably oriented for attack—the reaction is entropically more favorable. However, if the double bond is *trans* (as with phenyl fumarate) no rate acceleration occurs. Another elegant demonstration of rate acceleration by stereo-population control is the acid-catalyzed lactonization of hydroxy coumaric acid (Milstien and Cohen 1970). Appropriate alkyl substitution in both aromatic ring and side chain results in rate constant increase by a factor of 10^{11}; and in comparison, with the bimolecular esterification of phenol and acetic acid by almost 10^{16}.

Two other examples of such rate accelerations due to intra-molecular catalysis are ester hydrolysis by tertiary amine (Fig. 4.2b) and imidazole (Fig. 4.2c). The apparent concentrations felt, by making the reaction intra-molecular in these two cases, are 1300 M and 24 M, respectively.

Enzyme active site generally occupies a small region compared to the total protein volume. With respect to reactions involving multiple substrates, their proper assembly on to enzyme active site takes place by a series of bimolecular collisions. Ultimately, when the reaction occurs at a fully occupied active site—it represents an intra-molecular chemistry. This is an entropically favored situation as more degrees of freedom are lost! All the reactants (substrates) are bound very close to each other with their appropriate groups suitably aligned. Then the reactions take place through near-attack conformers (NACs) that clearly resemble the transition state. This proximity and orientation of reactants effectively increase their local concentration and profound rate accelerations are achieved (Jencks 1975; Page and Jencks 1971).

We note that the proximity and orientation of reactive groups (both from the enzyme and the substrate) contribute substantially to rate enhancements. But what tools are employed by nature to gain this advantage? The formation of a noncovalent enzyme–substrate complex is the first step in enzymatic catalysis. Substrates are bound to enzymes by multiple weak interactions—often mediated by van der Waals forces, hydrophobic interactions, and hydrogen bonds. Hydrogen bonds are prominent among these as they are directional (Fersht 1987). In fact, the strength of a hydrogen bond is a function of its length, orientation, linearity, and the micro-environment. When multiple hydrogen bonds occur, the effect can be cooperative and dramatic! Apart from their cooperative effect observed between two strands of DNA, five well-directed hydrogen bonds define the strength of the avidin-biotin complex. A special kind of hydrogen bond—the low barrier hydrogen bond (LBHB)—has been implicated in some enzymic catalysis (Cleland et al. 1998). LBHBs are short, very strong hydrogen bonds formed when the partner electronegative atoms sharing the hydrogen have comparable pK_a values. Evidence for LBHB formation exists for chymotrypsin catalysis and this hydrogen is shared between His57(NH) and Asp102(COO^-). This is proposed to help stabilize the tetrahedral intermediate. Similarly, an LBHB between the –OH group (of the substrate—lactate) and His195(N: of the enzyme) is observed in lactate dehydrogenase catalysis. Whatever the individual pK_a of such hydrogen bonding partners, the two pK_as should transiently match (and the H is equally shared between the partners!) during the course of each catalytic event for LBHB to form. The transient formation of this LBHB may provide a convenient intermediate step for the large pK_a change (from +15 for >CHOH of lactate to −5 for >C=O of pyruvate) during the catalytic trajectory of the lactate dehydrogenase reaction.

4.2   Contribution by Electrostatics

Enzyme active sites are clefts, crevices, or pockets formed by the overall three-dimensional structure of the protein. Substrate binding to the active site accompanies—(a) de-solvation of the substrate by replacing the water positions with the pre-organized polar framework and (b) exclusion of water from the pocket unless it is a reactant. Active sites generally exclude bulk water and thus create a unique local dielectric environment. The protein cage provides a structured micro-environment such that the enzyme-bound substrates bear little resemblance to the chaotic structure of substrates dissolved in water (Richard et al. 2014). This has profound consequences for functional group reactivity. The pK_a of a carboxylate may be elevated by a low local dielectric constant (Glu35 of lysozyme has a pK_a of 6.3!). Similarly, the pK_a of an amino group may vary by several units from its normal value because of the proximity of charged groups in the neighborhood (Lys-NH_2 of acetoacetate decarboxylase displays an unusually low pK_a of 5.9). Triose phosphate isomerase exploits one of its α-helix dipoles to modulate the pK_a of the active site His95. Although not exhaustive, the following examples further illustrate the contribution of electrostatics in enzyme catalysis (Warshel et al. 2006).

Decarboxylation of hydroxyethyl thiamine pyrophosphate adduct: The decarboxylation of pyruvate is facilitated by the formation of the initial hydroxyethyl thiamine pyrophosphate (HETPP) adduct between thiamine pyrophosphate and pyruvate. This HETPP adduct is charged but the charge is diffuse due to the delocalization of electrons (Fig. 4.3a). Since the active site of pyruvate decarboxylase is nonpolar (hydrophobic), this adduct tends to lose charges by expelling CO_2. This electrostatic feature is supported by studies with an analog of the HETPP adduct. The HETPP analog (Fig. 4.3a, where R1 = CH_3 and R2 = H) was prepared; its decarboxylation rate in ethanol is 10^5-fold faster than in water. This rate is even

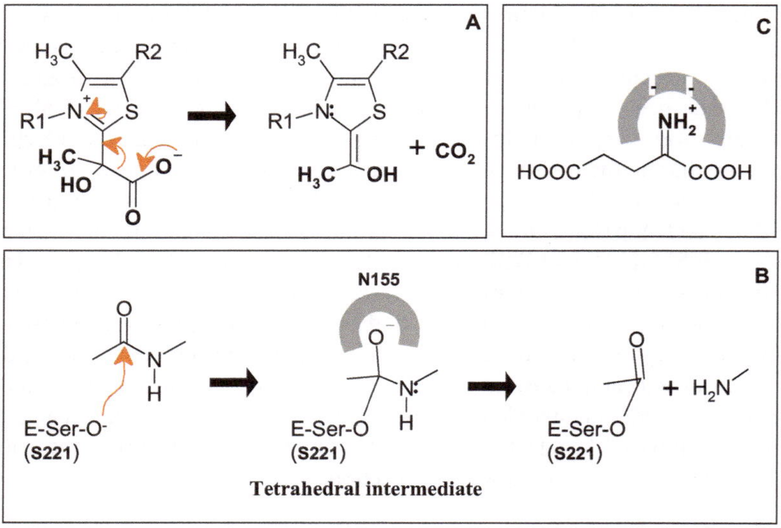

Fig. 4.3 Electrostatic effects at the enzyme active sites. (**a**) Hydrophobic active site of pyruvate decarboxylase facilitates expulsion of CO_2 from the HETPP adduct. (**b**) The oxyanion hole stabilizes the tetrahedral intermediate in subtilisin catalysis. (**c**) Negatively charged groups of glutamate dehydrogenase electrostatically stabilize the bound 2-iminoglutarate

faster in other polar aprotic solvents. The model decarboxylation reaction points to a striking catalytic effect of de-solvation and subsequent charge destabilization at the enzyme active site.

Oxyanion hole of subtilisin: During each catalytic cycle, subtilisin (and all other serine proteases like chymotrypsin and trypsin) goes through an attack of enzyme-Ser-OH onto the carbonyl carbon (sp² hybridized) of the scissile peptide bond. This leads to an initial "oxyanion" formation where the carbon is sp³ hybridized (Fig. 4.3b). Stabilization of the "oxyanion" in a special pocket (oxyanion hole) is a strategy for rate acceleration. Site-directed mutagenesis was used to evaluate the contribution of this oxyanion stabilization at the active site. Mutant enzyme forms showed that even after a triple replacement (where the catalytically essential residues—D32, H64, and S211—were replaced by Ala), the mutant subtilisin retained the ability to hydrolyze peptide bonds (at 10^3–10^4-fold above the uncatalyzed rates). This significant residual rate is attributed to the stabilization of the oxyanion intermediate. An Asn residue (N155) contributes significantly to this oxyanion hole of the subtilisin active site.

Selective enrichment of 2-iminoglutarate at glutamate dehydrogenase active site: There is strong evidence that the reductive amination of 2-oxoglutarate catalyzed by glutamate dehydrogenase proceeds through an enzyme-bound 2-iminoglutarate intermediate. However, the equilibrium (2-oxoglutarate + NH_3 ⇄ 2-iminoglutarate + H_2O) in solution is largely in favor of 2-oxoglutarate. It follows that the enzyme electrostatically stabilizes the bound

2-iminoglutarate complex, by utilizing negatively charged groups and hydrogen bonding basic groups at its active site (Fig. 4.3c). This results in substantially increased levels of 2-iminoglutarate ready for reduction. The same negatively charged enzyme groups while stabilizing the iminium ion ($>C=NH_2^+$) decrease the population of bound 2-oxoglutarate ($>C=O^{\delta-}$) due to charge repulsion. Better interaction with 2-methyleneglutarate ($>C=CH_2$, which is uncharged) at this pocket may account for its efficacy as a good inhibitor (Choudhury and Punekar 2007). Electrostatics permits glutamate dehydrogenase to discriminate between iminium and carbonyl groups and thus forms the chemical basis of ammonia recognition by the enzyme.

Guidance of charged substrates: Charge distribution about the active site to stabilize transition states and/or intermediates is termed electrostatic catalysis. Besides this, in several enzymes, the overall charge distribution of the protein matrix serves to guide polar substrates to their active sites. Superoxide dismutase (SOD) and acetyl cholinesterase are two well-studied examples of this "torch of charge guidance" effect (Getzoff et al. 1992; Silman and Sussman 2008).

The electrostatic contributions to SOD catalysis were analyzed through calculations of electrostatic potential and the electrostatic field vectors in the active site channel. The arrangement of global electrostatic charges in SOD promotes productive enzyme–substrate interaction through substrate guidance and charge complementarities. The extensive electrostatic field directs the negatively charged superoxide ($O_2^{\cdot-}$) substrate to the bottom of the active site channel (Fig. 4.4). Charge

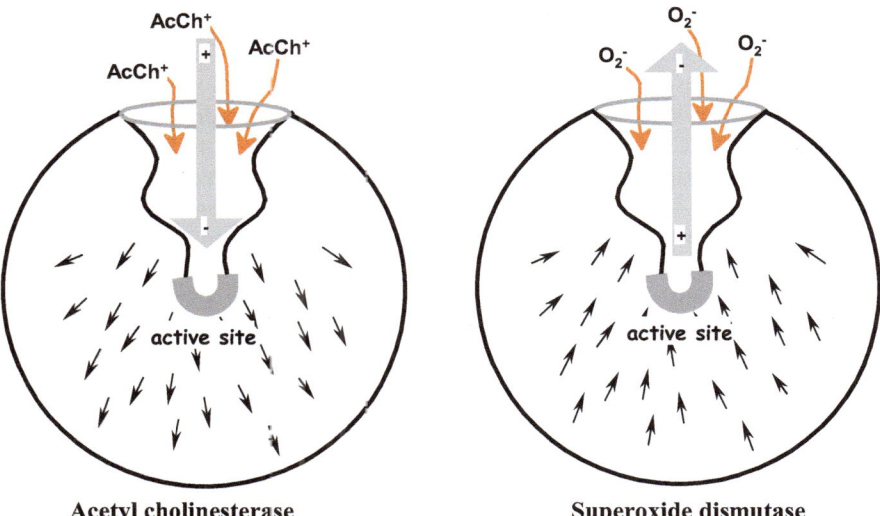

Acetyl cholinesterase   **Superoxide dismutase**

Fig. 4.4 Electrostatic torch of guidance effect in superoxide dismutase and acetyl cholinesterase. The gray block arrows indicate the overall enzyme electrostatic field vectors with respect to the active site present at the bottom of the channel/gorge. Small arrows represent various contributing dipoles to the overall field vector

guidance is also suggested from studies of the enzyme's dipole and overall electro-static potential. The maximal rate of SOD reaction is $2 \times 10^9 \, M^{-1} \, s^{-1}$, very close to the diffusion-controlled limit. This is remarkable since the active site channel forms only about 10% of the enzyme surface! Electrostatic forces possibly guide the charged substrate and enhance oriented diffusion. The electrostatic field vectors indicate that the attraction of $O_2^{\cdot-}$ is a long-range process, occurring even beyond the active site channel!

Similarly, acetyl cholinesterase also has a strong electrostatic dipole (but of opposite orientation to that of SOD!). This dipole is aligned with the deep gorge leading to its active site so that the positively charged acetylcholine is drawn to the active site by an electrostatic field. By comparison, a structurally related lipase (from *Geotrichum candidum*) has a poor and markedly different dipole orientation. Charge guidance and electrostatics may not operate in the case of lipase because the lipase substrate is neutral.

Electrostatic forces that are felt at the enzyme active site are the effects of the extended environment of that protein—including dipoles from the second shell and much beyond. Mutations at locations remote to the active site, yet significantly affect enzyme activity, implicate a role for electrostatics in catalysis.

4.3   Metal Ions in Catalysis

A third of all the known enzymes require metal ions for their function. Such metal ions may function as determinants of protein structure but more importantly, they could directly participate in the catalytic process. Metal ions may bind to the substrate and as a consequence enhance their interaction with the enzyme—by stabilizing and presenting one specific substrate conformation for catalysis. Aspects of metal chemistry and various roles played by metal ions in redox biochemistry are found in Part IV (Chaps. 30 and 31). In this section, however, the emphasis will be on their role in rate accelerations. Metal ions may be viewed as "super acids" because they can—(a) have charge density $>+1$, (b) be present at concentrations higher than those achieved by protons (H^+ ions) around neutral pH, and (c) coordinate with several groups to act as templates. They are recruited in enzyme-active sites as tools for electrostatic catalysis (see above). In terms of enzyme catalysis, metal ions contribute to reaction rate accelerations in the following ways.

Charge shielding: Several anionic (negatively charged) compounds act as substrates only when present in their divalent metal ion complexes. Enzymes acting on citrate, ATP, etc. are some examples of this kind. The requirement of Mg^{2+} for most ATP reactions (nucleotide triphosphate reactions in general!) led to the recog-nition that the Mg-ATP complex is the true substrate for such enzymes (also see Chap. 30). Apart from orienting the highly flexible oligomeric phosphates of ATP, the divalent metal ion partly neutralizes (and shields!) the negative charges on the polyphosphate chain. Charge shielding becomes particularly important when an anionic nucleophile has to attack the substrate during catalysis. Charge repulsion

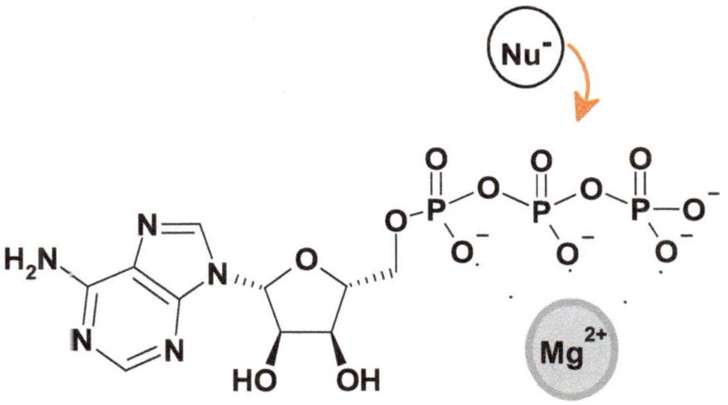

Fig. 4.5 The structure of ATP with Mg^{2+} shielding the polyphosphate negative charges. An anionic nucleophile can favorably approach such a polyphosphate

is expected to reduce the efficacy of negatively charged attacking groups, while neutral nucleophiles are not always feasible (especially due to pH constraints). For example, a kinase active site nucleophile (anionic) can easily approach the negatively charged polyphosphate of ATP (or any other NTP)—when it is complexed with Mg^{2+} ions (Fig. 4.5).

Charge stabilization: An important role for metal ions in catalysis is to serve as an electrophilic catalyst (Lewis acid) by stabilizing a negative charge on the reaction intermediate. This is better achieved by multivalent metal cations than mere protons. Three different enzyme chemistries exemplify this aspect of metal catalysis: (a) decarboxylation of oxaloacetate is catalyzed by divalent metal ions. For instance, Mn^{2+} chelated by oxaloacetate electrostatically stabilizes the developing enolate ion during reaction (Fig. 4.6a). Indeed, most oxaloacetate decarboxylating enzymes require a divalent metal ion for activity. Decarboxylation of oxalosuccinate by isocitrate dehydrogenase similarly requires Mn^{2+}. (b) Increased electron delocalization to stabilize the enolate intermediate also occurs during aldol cleavage. Class II aldolases (from fungi) usually contain Zn^{2+} to polarize the carbonyl oxygen of the substrate (Fig. 4.6b). (c) Zn^{2+} ion of carboxypeptidase A is coordinated to the carbonyl oxygen of the scissile peptide bond (Fig. 4.6c). This facilitates the polarization of the carbonyl group and stabilizes the negative charge on the oxygen during the reaction. The zinc ion in carboxypeptidase A also enhances the nucleophilicity of coordinately bound water for attack (see below).

Enhance nucleophilicity of water: Metal ions may generate a nucleophile by increasing the acidity of a nearby molecule. When this molecule happens to be water—a reactive hydroxide is generated. A water molecule bound to a transition metal, in its inner coordination sphere, is a better source of OH^- than bulk water. Metallo-hydrolases exploit this feature for catalysis and impart a better nucleophilic nature to water. Carbonic anhydrase provides a good example of this effect. When bound to the positively charged Zn^{2+} ion, the pK_a of the water molecule is reduced from 15.7 to 7.0. Thus, a substantial concentration of hydroxide ion (metal bound!) is generated at near neutral pH (Fig. 4.7a). Carbonic anhydrase utilizes the reactivity intrinsic to a zinc-bound hydroxide ion in its catalysis.

Arginase employs a bimetallic (Mn^{2+}) cluster to hydrolyze arginine to form ornithine and urea. It appears that two metal ions are better than one in generating the hydroxide. A water molecule is sandwiched between two Mn^{2+} ions (Fig. 4.7b)

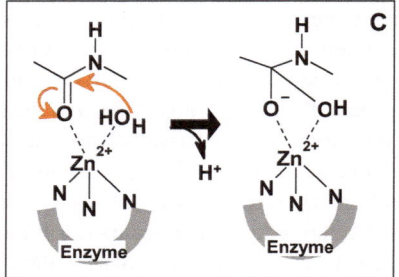

Fig. 4.6 Metal ions involved in charge stabilization during catalysis. (**a**) Oxaloacetate decarboxylase—Mn^{2+}; (**b**) Class II aldolase—Zn^{2+}; and (**c**) Carboxypeptidase A—Zn^{2+}

Fig. 4.7 Metal ions enhance the nucleophilicity of water during catalysis. (**a**) Carbonic anhydrase—Zn^{2+} and (**b**) arginase—Mn^{2+} bimetallic cluster

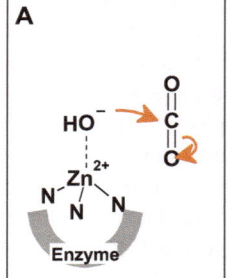 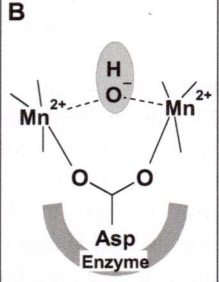

and is made sufficiently nucleophilic to attack the guanidinium group. A variation on this theme is the *H. pylori* enzyme with a bimetallic Co^{2+} cluster. In accordance with differences in the chemical reactivity of Mn^{2+} and Co^{2+}, this cobalt-arginase has a different pH optimum. Two other examples of bimetallic centers are jack-bean urease with two Ni^{2+} ions and the *E. coli* alkaline phosphatase with two Zn^{2+} ions.

4.4 General Acid–Base Catalysis

The majority of the enzyme-catalyzed reactions involve one or more proton transfers and hence general acid–base chemistry permeates most of the enzymology. Almost all these proton transfers are catalyzed. Regardless of what other tools an enzyme catalyst exploits, general acid–base catalysis is a common ploy. Rate accelerations effected by this mode of catalysis may involve the abstraction, donation, or movement of H^+ ions. These proton transfers—*the prototropic shifts*—may involve—(i) a single general base or (ii) multiple ionizable groups—that relay the proton from one atom to another. Among numerous examples of general acid–base catalysis include—glucose isomerization, enolization of pyruvate, intramolecular hydrolysis of substituted aspirins, lactam-lactim inter-conversion of purine-pyrimidine bases, pyridoxal phosphate chemistry with amino acids, etc. While almost every enzyme employs acid–base catalysis, the concept is best illustrated through a few case studies.

Mutarotation in glucose is facilitated by an acid or a base. The inter-conversion of α-D-glucose and β-D-glucose (via the linear form) is subject to the acid–base catalysis as shown in Fig. 4.8a. The ring closure to form the hemi-acetal may occur through the attack of oxygen on either face of the C1 carbonyl. Similarly, a *keto-enol tautomerization* is facilitated by the general base, general acid, or both. For instance, the developing negative charge on carbon (the carbanion-like transition state) can be stabilized by the protonation of the oxygen (Fig. 4.8b). In both these examples proton donation (acid catalysis) or proton abstraction (base catalysis) may occur independently. However, the two events could occur in concert. It is also possible that the same acid–base group is involved in the prototropic shift—a net proton transfer from one atom to the other. *1,3-prototropic shifts* (where a proton is moved from the first atom to the third) are quite commonly observed during enzyme catalysis. Enolization of pyruvate (as represented in Fig. 4.8b) is one such case where the proton shifts between the C3 of pyruvate and the carbonyl oxygen.

That biological catalysis has primarily/predominantly originated in an aqueous environment is consistent with this. According to the Bronsted definition of acids and bases, any species of a functional group that has a tendency to lose a proton is an acid. This definition eminently suits our understanding of the role of acid–base groups at the enzyme active site. Therefore, the pH dependence of enzyme activity is a reflection of its ionizable groups involved in binding and/or catalysis (see Chap. 23 for a detailed treatment). Ionizable amino acid side chains of the enzyme protein are typically involved in such catalysis. These could be through a concerted action of acid–base groups on the enzyme. Most hydrolytic enzymes rely on general

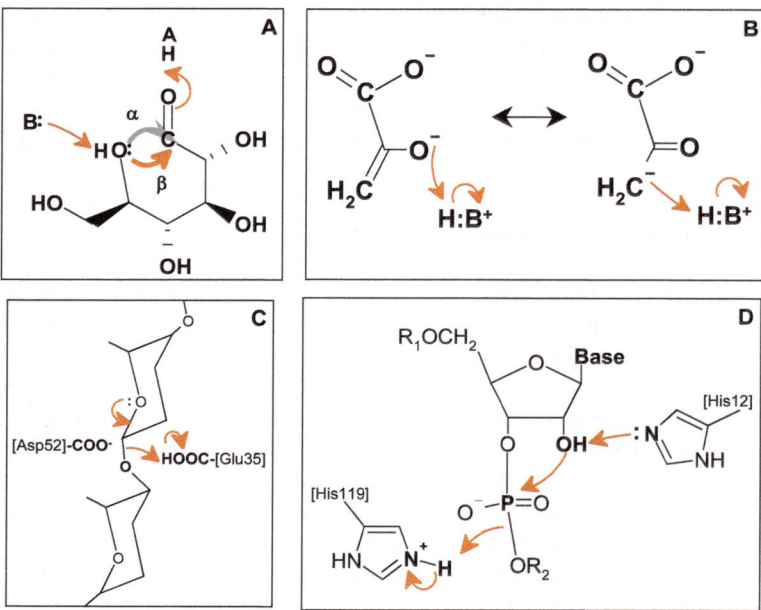

Fig. 4.8 General acid–base catalysis in rate accelerations. (**a**) Mutarotation of glucose is both acid and base catalyzed; (**b**) keto-enol tautomerization, as shown for pyruvate here, could involve a single acid–base group in the 1,3-prototropic shift; (**c**) schematic of the lysozyme chemistry involving the two active site carboxylates; and (**d**) RNase A catalysis is initiated by His12 abstracting the proton from 2′ OH. His119 acts as a general acid to donate a proton to the leaving group. The cyclic intermediate generated is subsequently attacked by water where the roles of two His residues are reversed

acid–base catalysis as a tool. Acid proteases (such as pepsin) and glycosidases (like lysozyme, cellobiohydrolase, and amylase) are some well-documented examples. In accordance with two carboxylate groups as catalytic residues, these enzymes exhibit acidic pH optima. The Glu35 (–COOH) and Asp52 (–COO⁻) form the catalytic residues of the lysozyme active site (Fig. 4.8c). Glu35 (acting as a general acid) facilitates bond cleavage by protonating the glycosidic oxygen. Ribonuclease A (RNase A) active site however displays two His residues—His12 as the general base and His119 as the general acid (Fig. 4.8d). Their roles are reversed during 2′,3′--cyclic intermediate hydrolysis, and the enzyme regains its original ionization state. RNA is alkali labile—hydrolyzed by general base catalysis. However, the rate accelerations achieved by the combined action of two His residues (at the RNase A active site) are many orders of magnitude higher.

The nature and chemical reactivity of important acid–base groups in enzyme catalysis are extensively covered in a later section (see Chap. 28).

4.5   Covalent Catalysis

In many reactions, at some stage during catalysis, an enzyme-substrate covalent intermediate may occur. This catalytic trick may even follow a different, more facile reaction path than the uncatalyzed one. An enzyme could break down a complex reaction into two or more simple cnes—each step with its own activation energy barrier. However, these new barriers are lower than that of the uncatalyzed reaction. This is the crux of covalent enzyme catalysis.

A covalent bond could be established between the enzyme and the substrate in one of the two ways—(a) an enzyme nucleophile may attack an electron-deficient center on the substrate or (b) an enzyme-bound electrophile could be attacked by the electron-rich center of the substrate. Side chains of many amino acid residues are known to participate in nucleophilic catalysis by forming covalent enzyme-substrate intermediates (Table 4.1). It has been possible to trap and directly demonstrate the existence of a few of these intermediates. The role of nucleophilic catalysis by Ser195 (–OH) of chymotrypsin is very well established (Carter and Wells 1988). Schiff base formation in fructose 1,6-bisphosphate aldolase is another classic example. More recently, the "Phillips mechanism" for lysozyme was revised based on the electrospray ionization mass spectrometry (ESI-MS) evidence in combination with an E35Q mutant of lysozyme (Kirby 2001). Active site residue Asp52 acts as a nucleophile and forms a covalent bond to the C1 carbon of the substrate glycoside.

Enzymes (being proteins) have a choice of many nucleophilic groups (side chains of amino acid residues) but have little to offer in terms of good electrophiles. This is one of the reasons why a number of small molecules (cofactors and prosthetic groups) are recruited by nature to complement an apoenzyme—resulting in a functional holoenzyme. These molecules act as temporary electron sinks during catalysis. Electrophilic recruitment may involve—(a) cofactor molecules like pyridoxal phosphate and thiamine pyrophosphate or (b) simple apparatus like Schiff

Table 4.1 Catalysis involving covalent enzyme-substrate intermediates

Amino acid [side chain]	Intermediate	Examples
Ser [–CH$_2$OH]	Acyl-enzyme Phospho-enzyme	Chymotrypsin, lipase, acetylcholinesterase Phosphoglucomutase
Thr [–CH(CH$_3$)OH]	Phospho-enzyme	Phosphotransferases
Tyr [–PhOH]	Phospho-enzyme	DNA integrase, topoisomerase
Cys [–SH]	Acyl-enzyme	Glyceraldehyde 3-phosphate dehydrogenase, acyltransferases, papain
Asp/Glu [–COOH]	Enzyme-ester Glycosyl-enzyme	Epoxide hydrolase, haloalkane dehalogenase Lysozyme
His [–imidazole-NH]	Phospho-enzyme	Glucose 6-phosphatase, succinyl-CoA synthetase
Lys [–NH$_2$]	Schiff's base	Acetoacetate decarboxylase, aldolase (type I), transaldolase

base, protein-bound pyruvate, or dehydroalanine. The nature and chemical reactivity of important nucleophiles and electrophiles are discussed in detail, in a later section (Chaps. 29 and 33).

4.6   Transition State Binding and Stabilization

The substrate is held in a unique active site environment by enzyme groups through proximity orientation and electrostatics. It may be argued on similar grounds, that the active site discriminates between the substrate and the transition state. Catalysis is thus a consequence of the preferential binding (and therefore stabilization) of the transition state (Lienhard 1973). Linus Pauling lucidly stated it first—"I think that enzymes are molecules that are complementary in structure to the activated complexes of the reactions that they catalyze,.." in his 1948 discourse (Pauling 1948).

Enzymes may be viewed as a device that preferentially binds/stabilizes the transition state (TS) rather than the ground state of the reactant (substrate). Although the TS is of the highest free energy and is extremely unstable, we can visualize the consequences of its preferential binding to the enzyme. The two binding equilibria and relevant kinetic and thermodynamic parameters are shown in Table 4.2.

If an enzyme binds (and/or stabilizes) its TS better than S, then one expects ΔG_{TS} to be more negative than ΔG_S. The two ΔG values are related to their corresponding $\Delta G^{\neq}$ values and the ratio of the rates of catalyzed versus the uncatalyzed reaction (k_{cat}/k_{uncat}) is related to K_S/K_{TS}. The more tightly an enzyme binds its reaction TS relative to the substrate (that is, the smaller the K_{TS} compared to K_S), the greater is the rate of acceleration. The magnitude of ΔG_{TS} − ΔG_S therefore significantly contributes to the overall decrease in activation energy ($-\Delta\Delta G^{\neq}$) during catalysis. Recalling (from Chap. 3) that $\frac{k_{cat}}{k_{uncat}} = 10^{\frac{\Delta\Delta G^{\#}}{2.303RT}}$, a rate enhancement factor of $\sim 10^6$ may be estimated for an enzyme that binds its TS complex with $8.0\,\text{kcal mol}^{-1}$ has greater affinity at 25 °C than its substrate. This is worth two hydrogen bonds that can form only in "E.TS" but not in the "ES" complex! LBHBs may be one such tool for the TS to make better contact with the enzyme. There are at least two consequences of the tighter binding of an enzyme to its TS. Devices that are designed to appreciably bind TS should catalyze the corresponding reaction! The concept of antibodies against TS mimics—called *abzymes* or *catalytic antibodies*—arose from the seminal idea of Linus Pauling. Secondly, *TS analogs* (stable molecules that resemble the TS) should be potent competitive inhibitors of the enzyme. Such TS analogs, as a corollary, provide insights into catalytic mechanisms (Mader and Bartlett 1997; Radzicka and Wolfenden 1995).

Table 4.2 Differential binding of the enzyme to substrate and transition state

Equilibrium	Dissociation constant	ΔG	Rate constant
E + S $\rightleftharpoons$ E.S	K_S	ΔG_S	k_{uncat}
E + TS $\rightleftharpoons$ E.TS	K_{TS}	ΔG_{TS}	k_{cat}

It is not necessary for a good substrate to have a high affinity for the enzyme as long as the corresponding TS form does so. The concept "underestimation of binding energy (in substrate binding) is utilized for catalytic rate acceleration" has also been widely recognized in the enzyme literature. These include—(a) destabilization of the ground state by the enzyme, (b) rack mechanism leading to strain and distortion in the substrate molecule, (c) induced fit versus nonproductive binding, and (d) sequestration of the TS (intermediate) at the active site. The following selected examples illustrate the concept of transition state binding and stabilization by the enzyme.

Strain and distortion: Ferrochelatase catalyzes the formation of heme by inserting Fe^{2+} into protoporphyrin IX. The iron entry requires that the planar porphyrin be bent. Indeed the enzyme binds the substrate in a distorted form to facilitate iron entry. Lysozyme is another example of this kind. The substrate glycosidic sugar (the N-acetyl muramic acid, ring D) is bound by the enzyme in a distorted/strained half-chair conformation (Fig. 4.9a). It is worth noting that the sugar residues at the other five sub-sites of lysozyme are in the normal, chair conformation. Introducing strain in the substrate glycosidic residue appears to be a feature of many glycohydrolases.

Preferential TS binding: According to Richard Wolfenden—enzymatic reaction transition state presents a "moving target." The TS develops and disappears very rapidly along the reaction coordinate. One could however build stable structures that resemble the TS— and enzyme may bind them in preference to either the substrate or the product of that reaction (Radzicka and Wolfenden 1995). Seeking TS mimics as powerful enzyme inhibitors is an active enterprise in drug discovery. Some of them like the α-glucosidase inhibitor acarbose (Fig. 4.9b) have matured into useful drugs. Pepstatin inhibits the aspartyl proteases (such as pepsin and rennin, in the nM range) because of the unusual amino acid statine in its structure. It is thought that the statine structure mimics the tetrahedral intermediate (TS) formed during catalysis (Fig. 4.9c). Pyrrole-2-carboxylate resembles planar TS of proline racemase and is a competitive inhibitor (Fig. 4.9d); it binds the enzyme with 160-fold better affinity than proline itself. Similarly, 2'-carboxy-D-arabinitol 1,5-bisphosphate is a powerful TS inhibitor of ribulose bisphosphate carboxylase (Rubisco) as it is the analog of carboxyketone intermediate (Fig. 4.9e). As a last example, we consider adenosine deaminase that catalyzes the irreversible deamination of adenosine to inosine. The enzyme is strongly inhibited by the analog of an unstable tetrahedral intermediate— formed with a change of hybridization from sp^2 to sp^3 at C-6 of adenosine. Nebularine 1,6-hydrate (Fig. 4.9f) binds the enzyme with a K_I of 3×10^{-13} M, whereas the K_M for adenosine is 3×10^{-5} M. In comparison, 1,6-dihydronebularin is a poor inhibitor with a K_I of 5.4×10^{-6} M. The lone –OH on the tetrahedral C-6 of the purine contributes approximately 10 kcal/mol (at 25 °C) in enzyme binding!

Although precise mechanisms are still debated, enzymes are thought to catalyze reactions by stabilizing transition states or destabilizing ground states. The two features are not contradictory to each other and can contribute to reducing the $\Delta G^{\neq}$s of reactions. The extent of their contribution may depend on the nature of the interactions of substrate(s) at the enzyme active site. For instance, OMP decarboxylase almost entirely exploits the TS stabilization by electrostatic interactions of

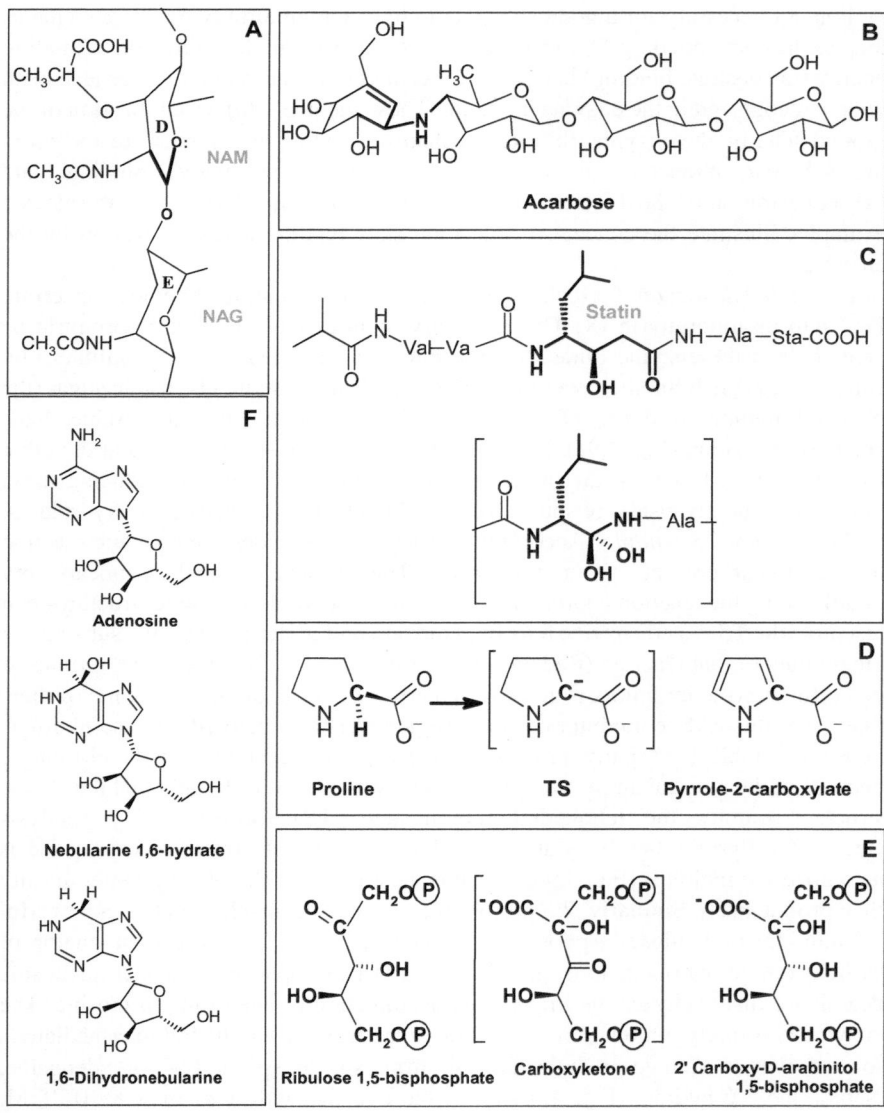

Fig. 4.9 Transition state binding and stabilization. (**a**) Half-chair conformation of the NAM sugar bound to lysozyme active site; (**b**) acarbose—α-glucosidase inhibitor; (**c**) structure of statin found in pepstatin and the corresponding enzyme bound tetrahedral intermediate; (**d**) the planar TS of proline racemase and pyrrole 2-carboxylate; and (**e**) the carboxyketone intermediate of Rubisco and its inhibitor analog 2′-carboxy-D-arabinitol 1,5-bisphosphate; and (**f**) nebularine 1,6-hydrate is a potent inhibitor of adenosine deaminase because of the –OH group on the tetrahedral C-6 of purine

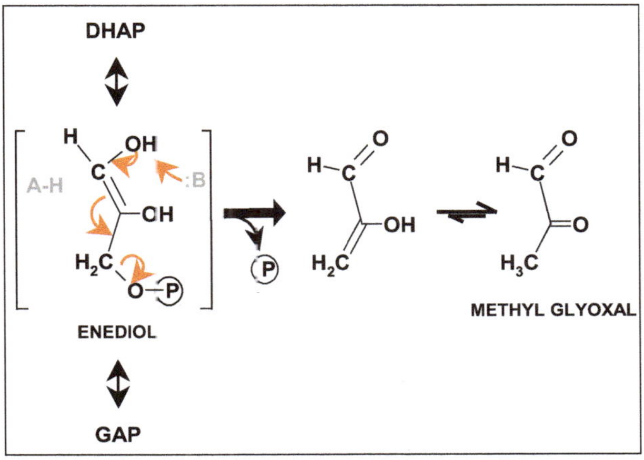

Fig. 4.10 Triose phosphate isomerase protects the *cis*-enediol intermediate. In the absence of this feature methyl glyoxal is the undesired product

the enzyme with charges spread over the substrate (Schneider et al. 2022). The electrostatic stress on the substrate carboxylate leaving group is negligible (no ground-state destabilization) as it is protonated and forms a favorable LBHB with a negatively charged residue.

Protection of TS (or intermediate): Intermediates in some chemical reaction paths are either reactive or unstable. Enzymes protect/stabilize such species by embedding them in their active sites. One such reactive intermediate occurs during the inter-conversion of dihydroxyacetone phosphate and glyceraldehyde 3-phosphate by triose phosphate isomerase (TIM). The active site of TIM (and many other TIM-barrel enzymes) is constructed by the folding of the eight $\beta\alpha$ front loops into a structured cage for the substrate. The *cis*-enediol formed in the triose phosphate isomerase reaction has a tendency to eliminate phosphate and form methyl glyoxal (Fig. 4.10). A short loop of polypeptide closes the active site and protects the reactive intermediate from bulk solvent—closing of this lid prevents methyl glyoxal forma-tion. A mutant enzyme without this loop is a poor catalyst and indeed produces significant quantities of methyl glyoxal.

4.7 Summing Up

Evolutionary selection has led to the development of enzymes that use a wide range of molecular mechanisms to facilitate reactions. The power of enzyme catalysis relies on lowering the height of the activation energy barrier through several tricks. Decreasing $\Delta G^{\neq}$ plays a dominant role in rate accelerations. A rigorous transition state theory approach also includes a transmission coefficient—a fraction of the TS that proceeds to products. While this is generally close to 1.0 in most simple reactions, it may vary significantly in the case of tunneling (see Chap. 25) and can lead to a sizable contribution to rate enhancements. Overall, the lowering of the activation energy barrier accounts for rate acceleration by a factor of 10^{11}, whereas the transmission coefficient contributes around a factor of 10^3 to the rate.

References

Carter P, Wells JA (1988) Dissecting the catalytic triad of a serine protease. Nature 332:564–568

Choudhury R, Punekar NS (2007) Competitive inhibition of glutamate dehydrogenase reaction. FEBS Lett 581:2733–2736

Cleland WW (1964) Dithiothreitol, a new protective reagent for SH groups. Biochemistry 3:480–482

Cleland WW, Frey P, Gerlt JA (1998) The low barrier hydrogen bond in enzymatic catalysis. J Biol Chem 273:25529–25532

Fersht AR (1987) The hydrogen bond in molecular recognition. Trends Biochem Sci 12:301–304

Getzoff ED et al (1992) Faster superoxide dismutase mutants designed by enhancing electrostatic guidance. Nature 358:347–351

Jencks WP (1975) Binding energy, specificity, and enzymic catalysis: the Circe effect. Adv Enzymol Relat Areas Mol Biol 43:219–410

Kirby AJ (2001) The lysozyme mechanism sorted—after 50 years. Nat Struct Biol 8:737–739

Lienhard GE (1973) Enzymatic catalysis and transition-state theory. Science 180:149–154

Mader MM, Bartlett PA (1997) Binding energy and catalysis: the implications for transition-state analogs and catalytic antibodies. Chem Rev 97:1281–1301

Milstien S, Cohen LA (1970) Rate acceleration by stereo-population control: models for enzyme action. Proc Natl Acad Sci 67:1143–1147

Page MI, Jencks WP (1971) Entropic contributions to rate accelerations in enzymic and intramolecular reactions and the chelate effect. Proc Nat Acad Sci U S A 68:1678–1683

Pauling L (1948) Nature of forces between large molecules of biological interest. Nature 161:707–709

Radzicka A, Wolfenden R (1995) Transition state and multisubstrate analog inhibitors. Meth Enzymol 249:284–312

Richard JP et al (2014) Enzyme architecture: on the importance of being in a protein cage. Curr Opin Chem Biol 21:1–10

Schneider T et al (2022) Ground-state destabilization by electrostatic repulsion is not a driving force in orotidine 5'-monophosphate decarboxylase catalysis. Nat Catal 5:332–341

Silman I, Sussman JL (2008) Acetylcholinesterase: how is structure related to function? Chem Biol Interact 175:3–10

Warshel A et al (2006) Electrostatic basis for enzyme catalysis. Chem Rev 106:3210–3235

Further Reading

Bruice TC (2002) Efficiency of enzymatic catalysis. Acc Chem Res 35:139–148

Chen D et al (2022) Key difference between transition state stabilization and ground state destabilization: increasing atomic charge densities before or during enzyme–substrate binding. Chem Sci 13:8193–8202

Fersht AR, Kirby AJ (1967) Structure and mechanism in intramolecular catalysis. The hydrolysis of substituted aspirins. J Am Chem Soc 89:4853–4857

Garcia-Viloca M et al (2004) How enzymes work: analysis by modern rate theory and computer simulations. Science 303:186–195

Graham JD et al (2014) Strong, low-barrier hydrogen bonds may be available to enzymes. Biochemistry 53:344–349

Gutteridge A, Thornton JM (2005) Understanding nature's catalytic toolkit. Trends Biochem Sci 30:622–629

Hansen DE, Raines RT (1990) Binding energy and enzymatic catalysis. J Chem Edu 67:483–489

Kahyaoglu A et al (1997) Low barrier hydrogen bond is absent in the catalytic triads in the ground state but is present in a transition-state complex in the prolyl oligopeptidase family of serine proteases. J Biol Chem 272:25547–25554

Kirby AJ, Hollfelder F (2008) Biochemistry: enzymes under the nanoscope. Nature 456:45–47

Knowles JR (1991) Enzyme catalysis: not different, just better. Nature 350:121–124

Kraut J (1988) How do enzymes work? Science 242:533–540

Scott LT (2014) Enzyme catalysis, basic principles in one easy lesson. Nat Chem 6:177–178

Richard JP, Zhai X, Malabanan MM (2014) Reflections on the catalytic power of a TIM-barrel. Bioorg Chem 57:206–212

Rindfleisch S et al (1995) On low-barrier hydrogen bonds and enzyme catalysis. Science 269:102–106

Which Enzyme Uses What Tricks?

5

Enzymes bring about reaction rate acceleration through several tricks. Of the tricks used to bring down activation energy—some are entropic, and others are enthalpic in nature. Various tools are recruited and relied on by each enzyme in different proportions. All these are within the realm of simple physical and chemical explanations—the combined effect, however, is quite dramatic! While there is no common formula, each enzyme uses a combination of these tricks to achieve the objective (Fig. 5.1). Indeed—*each enzyme is a biological experiment*—just the same way evolutionary biologist Ernst Mayr described the evolution of each species.

Lysozyme: Lysozymes are defined as 1,4-β-N-acetylmuramidases cleaving the glycosidic bond between the C-1 of N-acetylmuramic acid and the C-4 of N-acetylglucosamine in the bacterial peptidoglycan. They are ubiquitous enzymes found in birds, mammals, plants, and bacteriophages (Jolles and Jolles 1984). Hen egg white lysozyme is a well-dissected example of catalytic principles. A larger number of contacts to the substrate (peptidoglycan) provide good binding. Through these specific interactions, both Glu35 and Asp52 are brought in proximity to the susceptible glycosidic bond and properly oriented for catalysis. Being present at the same active site, the two –COOH groups behave quite differently—Glu35 is a general acid–base while Asp52 is a nucleophile. The pKa of Glu35-COOH is attenuated to 6.0 (from the expected pKa of 4.3). The notion of Asp52-COO$^-$ stabilizing the developing carbenium ion on the C1-anomeric carbon was recently revised to indicate its role as a nucleophile in covalent catalysis (see Sect. 4.5 in the previous chapter) (Kirby 2001; Vocadlo et al. 2001). Straddling the -(NAM-NAG)$_n$-polymer across the lysozyme active site cleft involves binding sub-sites for each sugar residue (Fig. 5.1). The binding of the fourth sugar residue is unfavorable and the weakest. It is thought that interactions at other sub-sites are exploited to accommodate the D sugar in a distorted, half-chair conformation—resembling the TS. Thus, except for metal ion catalysis, all other tricks are brought to bear on its substrate by lysozyme.

N. S. Punekar, *ENZYMES: Catalysis, Kinetics and Mechanisms*, https://doi.org/10.1007/978-981-97-8179-9_5

Fig. 5.1 Substrate binding at the lysozyme active site. The binding free energy for each sugar residue at each sub-site is given in kcal mol^{-1}. Lysozyme cleaves the substrate between sugar residues occupying sub-sites D and E

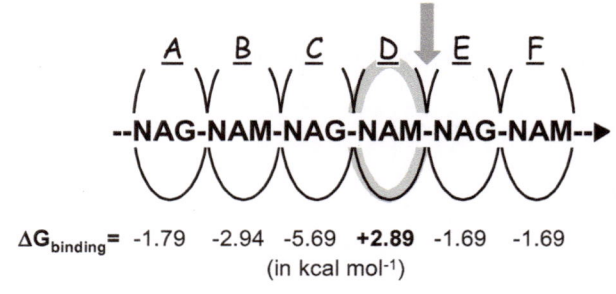

$\Delta G_{binding}$= -1.79   -2.94   -5.69   **+2.89**   -1.69   -1.69
(in kcal mol^{-1})

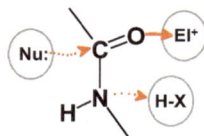

Fig. 5.2 A schematized protease active site. It typically displays a nucleophile (Nu:), an electrophile (El$^-$), and a general acid–base group for proton transfers

Proteases: Each catalytic solution in biology is unique to the reaction in question. But there can be different possible solutions to the same problem. Hydrolysis of a peptide bond is a case in point. Proteins and peptides constitute significant biomolecular components in all life forms. Consequently, peptide bond hydrolysis is also a universal requirement.

The problem associated with peptide bond hydrolysis is three-fold: (a) water being a poor nucleophile needs to be activated for attack, (b) the amine product resulting from peptide hydrolysis is a poor leaving group, and (c) amide (peptide) bonds are quite stable due to resonance (partial double bond character). In comparison, an ester is about 3000 times more reactive and a *p*-nitrophenyl ester is 300,000 times more so! Nature has invented suitable tools to overcome these three hurdles. Rate accelerations of up to 10^{10} times the uncatalyzed rates have been achieved. Peptide bond hydrolysis is an addition-elimination reaction that goes through a tetrahedral reaction intermediate. In fact, the formation of tetrahedral intermediate ensures that the peptide bond is weakened by eliminating resonance stabilization (no partial double bond!). The chemical apparatus at the protease active site provides suitable functional groups to interact with C, N, and O atoms of the peptide bond. These include—(i) a nucleophile like water (H-OH), Ser-OH, or Cys-SH to attack the carbonyl carbon, (ii) a general base to accept the proton from the nucleophilic – OH, (iii) an electrophilic group(s) to stabilize the oxyanion formed, and (iv) a general acid to protonate the amine—which is a poor leaving group. These features create the catalytic forces common to all proteases and are schematically shown in Fig. 5.2.

As expected, the protease active site extensively interacts with the substrate and freezes one of its conformations. Some proteases have distinct sub-sites to accommodate and bind substrate amino acid residues around the scissile peptide bond. The

Table 5.1 Catalytic tricks in four different protease groups

Feature	Serine Protease	Cysteine protease	Aspartyl protease	Metallo-protease
Proximity and orientation				
	Yes	Yes	Yes	Yes
Electrostatics				
	Oxyanion hole	Oxyanion hole	Asp-COOH to polarize carbonyl	Zn^{2+} to polarize carbonyl
Acid–base catalysis				
	Yes	Yes	Yes	Yes
Nucleophile used				
	DHS catalytic triad; Ser-OH	His-Cys-SH	H-OH activated by Asp-COO$^-$	H-OH activated by Zn^{2+} (or Glu-COO$^-$)
Covalent catalysis				
	Acyl enzyme	Acyl enzyme	No	No (Yes)
Transition state binding				
	Tetrahedral intermediate	Tetrahedral intermediate	Tetrahedral intermediate	Tetrahedral intermediate
Examples	Chymotrypsin, Subtilisin, Carboxypeptidase II	Papain, Caspase, Cathepsin C	Pepsin, Renin, HIV-1 protease	Carboxypeptidase A, Thermolysin, Leucine aminopeptidase

binding specificity may feature binding pockets for aromatic (chymotrypsin), positively charged (trypsin), small R (elastase) groups, or C-terminal (carboxypeptidase A) and N-terminal (leucine aminopeptidase) amino acid residues. Protease active sites typically include features that allow for—the activation of water or another nucleophile, the polarization of the peptide carbonyl group, and subsequent stabilization of a tetrahedral intermediate. Table 5.1 lists the different catalytic forces acting at the active sites of four major protease classes.

The problem of peptide bond hydrolysis has been solved ingeniously by nature. The result is four different major protease classes, namely, serine proteases, cysteine proteases, metalloproteases, and acid proteases. The nonisolable high-energy intermediate—the tetrahedral transition state—is generated and stabilized in different ways in these enzymes. The catalytic triad (Asp → His → Ser-OH) of serine proteases must be an effective apparatus to hydrolyze peptide bonds (Blay and Pei 2019; Hedstrom 2002). It has been independently selected, three different times throughout evolution. Chymotrypsin, subtilisin, and carboxypeptidase II exhibit very different protein architecture. But they all contain the active site catalytic triad—an excellent example of convergent evolution (the intrinsic chemical constraints to build a catalyst have led evolution to converge on equivalent solutions independently and repeatedly). Serine proteases like chymotrypsin, trypsin, and elastase, on the other hand, represent similar ancestry and are examples of divergent evolution (all three derived from a common ancestral homologous gene but diverged to perform different functions). More recently, enzymes with the known natural variation of the chymotrypsin-like triad in which a carboxylic acid is replaced by (1) a neutral hydrogen-bond acceptor (Wei et al. 1995) and (2) a chloride ion (Wan et al. 2019), are also reported.

References

Blay V, Pei D (2019) Serine proteases: how did chemists tease out their catalytic mechanism? ChemTexts 5:19

Hedstrom L (2002) Serine protease mechanism and specificity. Chem Rev 102:4501–4523

Jolles P, Jolles J (1984) What's new in lysozyme research? Molec Cell Biochem 63:165–189

Kirby AJ (2001) The lysozyme mechanism sorted- after 50 years. Nat Struct Biol 8:737–739

Vocadlo DJ et al (2001) Catalysis by hen egg-white lysozyme proceeds via a covalent intermediate. Nature 412:835–838

Wan Y, Liu C, Ma Q (2019) Structural analysis of a *Vibrio* phospholipase reveals an unusual Ser–His–chloride catalytic triad. J Biol Chem 294:11391–11401

Wei Y et al (1995) A novel variant of the catalytic triad in the *Streptomyces scabies* esterase. Struct Biol 2:218–223

Structure and Catalysis: Conformational Flexibility and Protein Motion

While enzymes can be studied from the viewpoint of (a) the active site environment, (b) the kinetic mechanism, and (c) the chemical mechanism, an enzyme is a unit and all these approaches tell valid but partial tales. A complete understanding of enzyme mechanisms requires a correlation of these and other information, like enzyme structure and its dynamics. Although chemical mechanisms have been elucidated for many enzymes, how they achieve a catalytically competent state has become approachable only recently through experiments and computation. The synergy between structure and plasticity results in the unique power of enzymes. Structural enzymology aims to address these catalytic motions in detail.

6.1 Structural Enzymology

This aspect of enzymology is concerned with the molecular architecture of enzymes, especially how they acquire their unique catalytically competent structures and how alterations in these structures affect their function. This subject is of great interest to enzymologists because it is only when proteins fold into specific three-dimensional shapes that they are able to perform catalytic functions. Amino acids are joined via peptide bonds; this bond has partial double bond characteristics and is almost always in the *trans* conformation. The *primary structure* (amino acids covalently joined in a particular order through peptide bonds in the polypeptide) dictates the higher order structures—as elegantly demonstrated with RNase A folding by C. Anfinsen (see Table 2.2) (Anfinsen 1973). The polypeptide sequence locally folds into *secondary structure*s like α-helices, β-sheets, or random coils. These in turn fold into the polypeptide *tertiary structure*. The complete tertiary structure of an enzyme may consist of a single domain, or a few domains juxtaposed in a suitable arrangement. Many proteins are oligomers consisting of subunits and this defines their *quaternary structure*. Hemoglobin is a α2β2 tetramer (Fig. 6.1), while aspartate

N. S. Punekar, *ENZYMES: Catalysis, Kinetics and Mechanisms*, https://doi.org/10.1007/978-981-97-8179-9_6

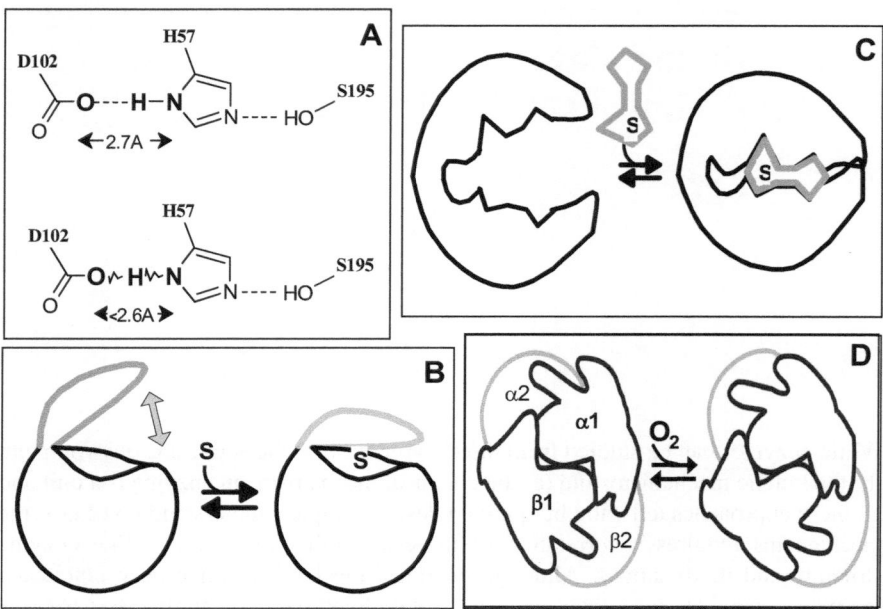

Fig. 6.1 Range of conformational changes relevant to enzyme function. (**a**) An LBHB between Asp102 and His57. (**b**) Flexible loop of triose phosphate isomerase to lock up the substrate. (**c**) Domain movement and substrate-induced fit. (**d**) The R (Oxy) and T (Deoxy) states of hemoglobin tetramer with two distinct positions of its α1β1 dimer relative to the α2β2 dimer

transcarbamylase has a $3R_2$-$2C_3$ architecture (with 12 subunits where R is a regulatory subunit and C is a catalytic subunit; see Chap. 38 for more details).

Structural details of an enzyme, the largely unchanging three-dimensional forms as also the conformational flexibility with associated mechanical motions, provide valuable clues in understanding the basis of catalysis. Toward this end, structure determination methods contribute immensely to the understanding of enzyme mechanisms. The X-ray crystallography has provided many insights into enzyme structure function and a few historically important ones are listed below.

- Lysozyme (from hen egg white) was the first enzyme whose crystal structure (X-ray diffraction) was solved by David Phillips in 1965 (Johnson and Petsko 1999). Co-crystallization with its substrate analog (GlcNAc)₃ provided considerable insight into the enzyme mechanism. This was the first example of structure providing clues to the mechanism of enzyme action. The so-called Phillips mechanism proposed the role of Glu-35 and Asp-52 in catalysis. This "Phillips mechanism" was further sorted and revised after almost 50 years using another structural technique (electrospray ionization mass spectrometry, ESI-MS) in combination with an E35Q mutant of lysozyme (Kirby 2001; Vocadlo et al. 2001). It was shown that active site residue Asp52 acts as a nucleophile and forms a covalent bond to the C1 carbon of the substrate glycoside.

- Another model enzyme whose three-dimensional structure was established (in 1967 by David Blow's group) was bovine pancreatic α-chymotrypsin (Matthews et al. 1967). With this structure, the catalytic triad (Asp-His-Ser) typical of so many serine proteases was discovered (Perona and Craik 1995). Whether the N-3 hydrogen on the imidazole of His-57 is actually transferred to Asp-102 during catalysis could not be ascertained by X-ray crystallography— because H atoms are too small to be resolved by this technique. Additional structural tools—like neutron diffraction studies with deuterated His-57 and ^{15}N NMR analysis—confirmed later that N-3 hydrogen remains attached to His-57. It is this H atom that participates in the low-barrier hydrogen bond during chymotrypsin catalysis.
- A super-secondary structure consisting of a parallel sheet formed by three extended parallel β-sheets connected by α-helices (the β-α-β fold) was first discovered in 1970 in Michael Rossmann's Laboratory. This motif forms the nucleotide-binding domain of NAD-dependent lactate dehydrogenase. Similar alternating motifs of β-α-β strand secondary structures (known as the Rossmann fold) are found in most enzyme proteins that bind nucleotide cofactors FAD, NAD^+, and $NADP^+$ (Laurino et al. 2016).
- The glycolytic enzyme triosephosphate isomerase (TIM) is the first example of an eight-fold β/α-barrel (reported by the Phillips group in 1975). The classical TIM-barrel fold $(\beta\alpha)_8$ consists of eight repeating $(\beta\alpha)$ units arranged so that eight parallel β-strands form a central protein core that is covered by α-helices on the outside. The TIM barrel is the most common protein fold in the Protein Data Bank and accounts for about 10% of all known three-dimensional structures (Berman and Gierasch 2021). The TIM barrel is the progenitor of a vast array of enzymatic activities through divergent evolution (Richard et al. 2014).

Over the years, many more enzyme structures have been solved and mechanisms better understood. Improvements in X-ray diffraction methodology and the ease of heterologous expression of almost any protein in *E. coli* (with or without an affinity tag for purification!) have accelerated the enzyme structure elucidation since the 1980s. Besides X-ray crystallography, other structural methods have also significantly contributed to our understanding of enzymes. Many proteins do not always crystallize easily, especially when parts of the structure are flexible, or the complex has structural heterogeneity. For such cases, methods like cryo-electron microscopy (cryo-EM) developed to obtain high-resolution 3-D images of proteins and structures of many enzymes like glutamine synthetase, β-galactosidase, isocitrate dehydrogenase, and glutamate dehydrogenase (see figure on back cover) are available now (Vonck and Mills 2017). It is beyond the scope to elaborate on all the tools available to study the structural aspects of enzymology. However, Table 6.1 lists the more commonly used methods along with comments on their strengths and constraints.

Many of the structural methods listed above provide snapshots of the enzyme protein. Catalysis being a dynamic kinetic process such information, while useful, is of limited value. Therefore, time-resolved spectroscopic tools are being developed with many of these methods to capture the dynamics of the enzyme catalysis.

Table 6.1 Methods for structural analysis of enzymes

Technique/method	Information/outcome	Constraints
High resolution		
X-ray diffraction	Complete 3-D structure at atomic resolution	Requires high-purity crystalline protein; snapshots with limited time resolution
Cryo-electron microscopy	3-D structure with detailed subunit arrangement; works with proteins in solution, no crystallization needed	Resolution of small protein structures needs further improvement
Nuclear magnetic resonance (NMR) spectroscopy	Complete 3-D structure at atomic resolution; structural dynamics, loop flexibility; side chain movements and ionization of residues; ligand binding	Limited to small-sized proteins; larger sample size; kinetics possible but at longer time scales
Global features		
UV-Visible spectroscopy	Gross structural changes and ligand binding; kinetics possible	Sample purity and composition (availability of intrinsic chromophores)
Fluorescence spectroscopy	Gross structural changes and ligand binding; kinetics possible; excellent sensitivity	Sample purity and composition; availability of fluorophores (intrinsic or extrinsic)
Circular dichroism (CD) spectroscopy	Overall secondary structure details and dynamics; environment change effects	Sample purity and size
Mass spectrometry	High-resolution molecular mass characterization and sequencing of proteins	Sample processing and purity
Dynamic light scattering	Size distribution profile; aggregation behavior; effective particle diameter	Presence of small impurities
Calorimetry	Protein folding (differential scanning calorimetry, DSC); thermodynamic parameters of interactions in solution (isothermal titration calorimetry, ITC)	
Hydrodynamic		
Analytical ultracentrifugation	Shape and molecular mass of the enzyme protein	
Gel filtration chromatography	Molecular weight of the native enzyme protein	
Gel electrophoresis	Molecular weight and gross quaternary structure (native, denatured, and cross-linked proteins)	
Chemical		
Chemical modification	Amino acid residues relevant for structure/function	Supporting evidences required
Site-directed mutagenesis	Amino acid residues important in structure, active site binding, etc.	Supporting evidences required

(continued)

Table 6.1 (continued)

Technique/method	Information/outcome	Constraints
Hydrogen-deuterium exchange	Solvent accessibility of various parts of the enzyme molecule; protein tertiary structure; folding pathways characterization	
Computational		
	Protein structure prediction, viz., AlphaFold	Tools improve with better database and availability of computational time
	Molecular dynamics simulation; molecular docking; animation of experimental structures (morphing)	Computational time intensive

6.2   Conformational Flexibility and Enzyme Catalysis

The native structure of a protein is simply the most thermodynamically stable conformation accessible to the folding polypeptide chain. As such polypeptide chains have many degrees of freedom, it is likely that the molecule has several other accessible conformations with almost as low $\Delta G°$ of formation as the native structure. Consequently, a population of rapidly equilibrating stable conformations for the same protein is possible. For instance, a protein could have two (or more!) conformations—the native structure and another conformation—in equilibrium with each other. Crystallographic evidence is available for two conformational states of hemoglobin, aspartate transcarbamylase, hexokinase, citrate synthase, and triose phosphate isomerase. Often the two states represent the unliganded conformation (most stable in the absence of a specific ligand) and liganded conformation (most stable with one or more specific ligands bound to it). In any case, if two conformations of a protein exist in a definite equilibrium then we could also define an equilibrium constant for the same. This was invoked as one of the mechanisms for allostery in regulating enzyme activity (the R and T states of the Monod–Wyman–Changeaux model; see Chap. 38).

An enzyme active site can accommodate either the substrate or the product of the reaction it catalyzes. Also, S and P are distinct chemical entities but are structurally related. For instance, glucose and glucose-6-phosphate (in hexokinase reaction) differ by a phosphate group—the rest of the sugar structure is by and large identical. Whatever their individual affinities, S and P have to interact with (and hence be complementary to) the enzyme pocket. Therefore, an enzyme cannot be a rigid structure ("lock for a key") but exhibits local conformational changes—in the vicinity of the active site at least! For the enzyme to participate in catalysis, protein motion (however small!) and conformational plasticity are a must. Most enzymes handle a substrate that is larger than an electron (except for cytochromes!). In all these cases the space-filling needs of reactants and products are obviously different in the enzyme active site. Since the enzyme must reach the TS, starting from either

Table 6.2 Conformational flexibility and enzyme catalysis

Conformational change	Effect	Example
Orbital steering and small structural changes	Large changes in kinetic property, co-operativity	Chymotrypsin; NADP-isocitrate dehydrogenase; keto-steroid isomerase; most enzymes
Flexible loops	Hold or protect the substrate (ligand)	Triose phosphate isomerase; adenylate kinase; HIV-protease
Domain movements	Induced fit, generation of active site	Hexokinase; citrate synthase; transglutaminase; cAMP-dependent protein kinase; calcium/calmodulin-dependent kinase II; adenylate kinase
Subunit communications	Allostery, co-operativity	Aspartate transcarbamylase; hemoglobin

reactant or product, some things have to move. Conformational flexibility and mechanical motion of the enzyme protein are thus a necessity. Enzymes may therefore be also viewed as *dynamic mechanical devices*. Almost any enzyme can, in principle, function as a molecular machine converting chemical energy to mechanical force (see Chap. 40 for details). It may become feasible to power motion with enzymes.

For many enzymes, snapshots of conformations that are sampled during catalysis have been obtained using ligands, substrates, and inhibitors. The protein motions promoted by such ligand binding are most interesting. Intrinsic motions along an enzymatic reaction trajectory could be monitored through X-ray crystallography, NMR, single-molecule FRET (fluorescence resonance energy transfer), and molecular dynamic simulations. A range of conformational changes are observed and are relevant to enzyme function (Table 6.2). These differ in the extent to which the changes transmit and extend from the active site.

Over three billion years of biochemical evolution, most enzymes have optimized rates such that the chemical reaction steps have become fast enough to be only partly rate limiting. Often conformational change is a significant rate determining step in enzyme catalysis (Cleland 1975). More likely the conformational changes that permit the product release are rate-limiting, viz., the release of pyridine nucleotide in a dehydrogenase catalysis. Homologous enzymes often exhibit different catalytic rates despite a fully conserved active site. The residues away from the active site also undergo distinct dynamics that are correlated with its catalytic activity and thus allow for different catalytic rates. Clearly, conserved dynamics drives the enzymatic activity (Torgeson et al. 2022). Secondly, intrinsic structural dynamics of an enzyme underlies catalysis. The pre-existence of collective dynamics in enzymes before catalysis is also seen as a common feature; enzymes have evolved under the combined pressure of structure and dynamics. Conformational heterogeneity is emerging as a defining characteristic of enzyme function. For example, dihydrofolate reductase exists in at least four ground-state conformers with different affinities for its different ligands. Depending on ligand binding the conversion between the conformers is accelerated by molecules that stabilize the transition state—facilitating lowering of the energy barrier and hence the product release.

Orbital steering and small structural changes: Local protein motion at the enzyme active site must occur. Consequences of small conformational changes are profound and are easily detected by the discriminatory power of enzymes. Orbital overlap produced by optimal orientation of reacting orbitals play a major quantitative role in the catalytic power of enzymes. In a large measure, the ability of enzyme to maximize orbital steering contributes to catalysis. Often such conformational changes are barely detectable by the best physical tools and structure elucidation methods! An interesting question was posed by Koshland— "How small a conformational change is big enough?" A fraction of an Å shift in an active site group of NADP-isocitrate dehydrogenase is "big enough" to be functional through large catalytic consequences (Koshland Jr 1998; Mesecar et al. 1997). Chymotrypsin active site has a catalytic triad of Asp-His-Ser. The normal hydrogen bond between His57-Asp102 goes through a low barrier hydrogen bond (LBHB) during the catalytic cycle (Fig. 6.1). A short contact distance is necessary for an LBHB to form. Thus, the two heteroatoms (N of His and O of Asp) are drawn close together. This is again an example of movement, of atoms/groups during catalysis, in sub-Angstrom scale. Keto-steroid isomerase binds to its substrate through hydrogen bonds that tighten up as the TS is approached. One of them is an LBHB. Tiny variations of the order of 10 picometer (about 0.1 Å) make a remarkable difference in efficiency of enzymatic catalysis (Kirby and Hollfelder 2008).

Flexible loops: Examples of enzymes in this group display movement of a relatively small loop upon ligand binding. This movement encloses the ligand in a cage-like structure and excludes it from the bulk solvent water. Often such loop regions are ill-defined in the X-ray or 2D NMR data indicating their conformational flexibility. The best example of this type of conformational motion may be found in triose phosphate isomerase (Table 6.1). Upon substrate binding, a short loop (residues 168–177) of this protein closes over the substrate to lock it in the active site—unstable intermediate formed during the enzymatic reaction is protected from decomposition by solvent water. This lid has to open for the product to leave the active site after each catalytic cycle. In fact, this conformational change (closing and opening motion of the loop) has to be much faster than the overall rate of catalysis (Rozovsky and McDermott 2001). Similar functionally significant motions occur in some proteins containing nuclear localization signals. Here, the signal sequence may occur in exposed state or be tucked in to prevent nuclear entry.

Conformational changes of a transmembrane helix may be coupled to a catalytic domain in the cytoplasm or outside. The transmembrane H-bond pattern can tilt the equilibrium favoring alternative conformations of a shared catalytic cytoplasmic domain, thereby sensing environmental cues like ambient temperature (Inda et al. 2020).

Domain movements: Member enzymes of this group show large-scale structural movements. Two large domains of the same polypeptide chain may move in relation to each other about a flexible hinge region. Subsequent substrate binding and rearrangement of various amino acid residues occur to produce a functional active site. This ***induced fit*** and ***productive binding*** is an important manifestation of protein flexibility (Koshland Jr 1958). Hexokinase active site is functionally assembled by closing two large domains, only upon binding glucose (Fig. 6.1). This ensures that ATP cleavage and phosphate transfer to glucose are strictly coupled—transfer to water cannot occur (nonproductive binding). Similar domain closure is observed when oxaloacetate binds to citrate synthase.

Adenylate kinase is yet another classic example of "induced fit" and is a representative member of NMP kinases. These enzymes contain a glycine-rich sequence (known as Walker A motif) forming the P-loop. The P-loop typically contains an amino acid sequence of the form Gly-X-X-X-X-Gly-Lys-(S/T) and interacts with the phosphoryl groups of the bound nucleotide. Interaction of the nucleotide substrate with adenylate kinase results in the movement of P-loop—this in turn closes the two domains to engulf the substrate. The P-loop NTPase domains (and Walker motifs) are encountered in several enzymes (Laurino et al. 2016) that undergo substantial conformational changes upon NTP binding and/or hydrolysis.

Subunit communications: It is possible, but not necessary, that the conformational change observed upon ligand binding may be restricted to in/around the enzyme active site. In multimeric enzyme proteins, conformational changes may be communicated across the subunits. These may lead to profound biochemical consequences such as cooperativity and regulation. Examples of subunit communication of this kind are observed in hemoglobin (an honorary enzyme!) and aspartate transcarbamylase. Apart from conformational changes within a subunit, an alteration in the spatial relationship among the subunits in an oligomer is also possible (Fig. 6.1).

Precise orientation of catalytic groups is required for enzyme action. Substrate binding causes an appreciable change in the three-dimensional relationship of amino acids of the protein—at least at/around the active site. In contrast to the rather rigid key-lock concept of Fischer, ***induced fit theory*** by Koshland proposes protein flexibility as an essential characteristic of enzymes (Koshland Jr 1958). Such a perspective for an enzyme explains many important phenomena like—the ability of enzymes to exclude omnipresent water, regulation outside the active site, and noncompetitive inhibition. Clearly, conformational changes are at the root of feedback inhibition, enzyme activation, co-operativity, etc. (refer to Chap. 38 for a detailed treatment on these topics). Small conformational changes having large effects explain the process of evolution of proteins and why enzymes are large.

6.3   Summing Up

It is a golden era of structural biology and understanding enzyme catalysis has gained much because of it. In the last two decades, structural biology has embraced an integrative approach employing a variety of tools such as X-ray crystallography, nuclear magnetic resonance spectroscopy, cryo-electron microscopy, and computer modeling. Protein structure prediction algorithms like AlphaFold (Jumper et al. 2021) have opened new avenues to address challenging problems. While

AlphaFold2 currently predicts static structures, enzymes display structural dynamics closely tied to their function. In many cases they—(a) contain intrinsically disordered regions that do not adopt a folded 3-D structure and (b) may be described as ensembles of closely related structures. Advances in instrumentation and software have made cryo-EM comparable to X-ray crystallography but without the need for crystallization. While most of the data gleaned through these techniques are very valuable, they provide snapshots of the structures. It may be impossible to solve all crystal structures that are important on the catalytic reaction path of an enzyme. Movies to represent enzyme action are created through animations using experimental structures by "morphing." These structures are pretty, yet that structure is a model and represents the average of many moving molecules. Molecular dynamics calculations are useful and to some extent capture enzyme dynamics. Representing dynamics of enzyme in action is hard stuff. As Richard Feynman famously said, "everything that living things do can be understood in terms of the jigglings and wigglings of atoms." This includes what and how enzymes do their bit.

References

Anfinsen CB (1973) Principles that govern the folding of protein chains. Science 181:223–230

Berman HM, Gierasch LM (2021) How the Protein Data Bank changed biology: An introduction to the JBC Reviews thematic series, part 1. J Biol Chem 296:100608

Cleland WW (1975) What limits the rate of an enzyme-catalyzed reaction? Acc Chem Res 8:145–151

Inda ME et al (2020) Driving the catalytic activity of a transmembrane thermosensor kinase. Cell Mol Life Sci 77:3905–3912

Johnson LN, Petsko GA (1999) David Phillips and the origin of structural enzymology. TIBS 24:287–289

Jumper J et al (2021) Highly accurate protein structure prediction with AlphaFold. Nature 596:583–589

Kirby AJ (2001) The lysozyme mechanism sorted– after 50 years. Nat Struct Biol 8:737–739

Kirby AJ, Hollfelder F (2008) Enzymes under the nanoscope. Nature 456:45–47

Koshland DE Jr (1958) Applications of a theory of enzyme specificity to protein synthesis. Proc Natl Acad Sci U S A 44:98–104

Koshland DE Jr (1998) Conformational changes: how small is big enough? Nat Med 4:1112–1114

Laurino P et al (2016) An ancient fingerprint indicates the common ancestry of Rossmann-Fold enzymes utilizing different ribose based cofactors. PLoS Biol 14:e1002396

Matthews BW et al (1967) Three-dimensional structure of tosyl-α-chymotrypsin. Nature 214:652–656

Mesecar AD, Stoddard BL, Koshland DE (1997) Orbital steering in the catalytic power of enzymes: small structural changes with large catalytic consequences. Science 277:202–206

Perona JJ, Craik CS (1995) Structural basis of substrate specificity in the serine proteases. Protein Sci 4:337–360

Richard JP, Zhai X, Malabanan MM (2014) Reflections on the catalytic power of a TIM-barrel. Bioorg Chem 57:206–212

Rozovsky S, McDermott AE (2001) The time scale of the catalytic loop motion in triosephosphate isomerase11Edited by P. E. Wright J Mol Biol 310(1):259–270. https://doi.org/10.1006/jmbi.2001.4672

Torgeson KR et al (2022) Conserved conformational dynamics determine enzyme activity. Sci Adv
    8:eabo5546
Vocadlo DJ et al (2001) Catalysis by hen egg-white lysozyme proceeds via a covalent intermediate.
    Nature 412:835–838
Vonck J, Mills DJ (2017) Advances in high-resolution cryo-EM of oligomeric enzymes. Curr Opin
    Struct Biol 46:48–54

Further Reading

Akdel M et al (2022) A structural biology community assessment of AlphaFold2 applications. Nat
    Struct Mol Biol 29:1056–1067
Carugo O, Djinović-Carugo K (2023) Structural biology: a golden era. PLoS Biol 21:e3002187
Eisenmesser EZ et al (2005) Intrinsic dynamics of an enzyme underlies catalysis. Nature 438:117–
    121
Erlandsen H, Abola EE, Stevens RC (2000) Combining structural genomics and enzymology:
    completing the picture in metabolic pathways and enzyme active sites. Curr Opin Struct Biol 10:
    719–730
Fraser JS, Murcko MA (2024) Structure is beauty, but not always truth. Cell 187:517–520
Galenkamp NS, Biesemans A, Maglia G (2020) Directional conformer exchange in dihydrofolate
    reductase revealed by single-molecule nanopore recordings. Nat Chem 12:481–488
Kempner ES (1993) Movable lobes and flexible loops in proteins: structural deformations that
    control biochemical activity. FEBS Lett 326:4–10
Miller C (2007) Pretty structures, but what about the data? Science 315:459
Nashine VC, Hammes-Schiffer S, Benkovic SJ (2010) Coupled motions in enzyme catalysis. Curr
    Opin Chem Biol 14:644–651
Ouzounis CA et al (2003) Classification schemes for protein structure and function. Nat Rev Genet
    4:508–519
Pinkas DM et al (2007) Transglutaminase 2 undergoes a large conformational change upon
    activation. PLoS Biol 5:e327
Zhao X et al (2018) Powering motion with enzymes. Acc Chem Res 51:2373–2381

On Enzyme Nomenclature and Classification

7

7.1   What Is in the Name?

The word **enzyme** (ενζυμη meaning "in yeast" in Greek), first used by Kuhne in 1877, is now well accepted to describe a biological catalyst. The majority of enzyme names today carry the suffix "-ase"—as recommended for all enzyme names by Duclaux in 1898. Proteolytic enzymes are a significant exception to this generally accepted norm. Some of them have retained the older tradition of usually ending with "-in"; for example, trypsin, chymotrypsin, papain, and subtilisin.

All enzymes are proteins but not all proteins are enzymes. Catalysis by RNA molecules (the so-called ribozyme) has expanded the realm of biological catalysts to beyond proteins. While few RNA catalysts have been recognized, the vast majority of enzymes that we come across in biology are proteins. Evolution has selected L-isomers of 20 amino acids to build proteins. This has put limits on the number and nature of available reactive chemical groups/functions that could be recruited for catalysis. Proteins are rich in nucleophilic groups, but electrophiles are poorly represented. In some instances, the protein component alone is inadequate to catalyze a given reaction. Nature therefore recruited many nonprotein components called *cofactor*s to generate a functional catalyst. In such enzymes, the inactive protein component without the cofactor is termed the *apoenzyme,* and the active enzyme, including the cofactor, the *holoenzyme*. The cofactors may be either metal ions (for example—Mn[II] in arginase, Ni[II] in urease, and Ca[II] in DNase I) or coenzymes (organic molecules like pyridine nucleotides—NAD^+ and $NADP^+$; flavin adenine dinucleotide—FAD; pyridoxal phosphate—PLP; thiamine pyrophosphate—TPP; biotin; cobamide; and heme). The binding of a cofactor to its cognate apoenzyme could exhibit a range of binding strengths. A very tightly bound cofactor—which is difficult to remove without damaging the enzyme—is also known as a *prosthetic group*. Quite often prosthetic groups are covalently bound to the apoenzyme—lipoamide of transacylase is an example (Cronan 2024).

N. S. Punekar, *ENZYMES: Catalysis, Kinetics and Mechanisms*, https://doi.org/10.1007/978-981-97-8179-9_7

$$\text{Holoenzyme} = \text{Apoenzyme} + \text{Cofactor}$$

Noncovalent interactions between a protein (such as an apoenzyme) and a cofactor may be weak or strong. This is amply obvious from the list of enzymes that require divalent metal ions for activity. For instance, enzymes with a tightly bound metal ion are termed *metallo-enzymes* (such as urease), while those with a weakly bound metal ion are grouped as *metal-activated enzymes* (such as Fe [II]-catechol dioxygenase). Enzymes with metal ion dissociation constants (K_D) of the order of 10^{-8} M or higher are generally grouped as metal-activated enzymes, while those with K_D values lower than 10^{-8} M are considered metallo-enzymes. This boundary is artificial and obviously there is a continuum of binding strengths observed in nature. It may be noted in passing that even noncovalent interactions can be very strong—essentially irreversible in a few cases like the avidin-biotin complex ($K_D = 10^{-15}$ M; $t_{1/2}$ of 2.5 years) or for that matter the two strands of a double-stranded DNA!

7.2 Enzyme Diversity and Need for Systematics

The middle of the twentieth century saw an exponential increase in research on enzymes. Sooner than later the number of new enzymes reported crossed manageable limits for an individual. As a consequence, in some cases, the same (or similar) enzyme activities were given different names. Names like catalase give very little indication of the nature of the reaction they catalyze. Systematic classification, cataloging, and nomenclature of enzymes therefore became a necessity. This was easier recognized than done. Enzymes could be grouped according to any of the following considerations.

(a) Occurrence and/or source of the enzyme: Laccase obtained from the Japanese lacquer tree, papain from papaya, and horse-radish peroxidase—are examples of three such plant enzymes. Similarly, several digestive enzymes are isolated from pancreatic juice such as trypsin, chymotrypsin, carboxypeptidase, and lipase. The common source of lysozyme is hen egg white (HEW).

(b) Nature of the substrate on which the enzyme acts: They could be classified into enzymes hydrolyzing (or acting on) proteins, carbohydrates, lipids, etc.

(c) Based on cofactor requirement: Many enzymes are simply proteinaceous in nature. But those depending on a cofactor could be listed into separate groups like—thiamine pyrophosphate (TPP) enzymes, pyridoxal phosphate (PLP) enzymes, and metalloenzymes.

(d) Common functional context: One could, in principle, group enzymes belonging to discrete pathways like glycolytic enzymes and enzymes of histidine biosynthesis. They may also be grouped as soluble, membrane-bound, or belonging to organelles such as mitochondria.

(e) Nature of the overall reaction catalyzed: An enzyme can be assigned to a group by considering the type of reaction it catalyzes. For instance, they may catalyze oxidation, hydrolysis, etc.

(f) The mechanism of reaction: The intimate mechanism of the reaction at the enzyme active site and the nature of intermediate complexes with the enzyme may be considered. For example, proteases may be classified depending on whether an enzyme-bound acyl-enzyme intermediate is formed or not.

It should be obvious from the above list that systematic, meaningful classification and cataloging of all the enzymes has not been easy. The problem is compounded by the enormous diversity of enzyme structures and activities. A typical RNA hydrolysis is achieved through a protein (RNase A), a protein-RNA complex (RNase P), or RNA alone (ribozyme). The peptide bond hydrolysis is possible with enzymes that are efficient in acidic pH or alkaline pH, require a divalent metal ion, contain a serine -OH or cysteine -SH, etc. (see Table 5.1). Enzymatic decarboxylation of histidine may recruit pyridoxal phosphate or in another form may just use a bound pyruvate moiety. Pyridoxal phosphate bound to glycogen phosphorylase serves more of a structural role rather than function as a cofactor. Alkyl-dihydroxyacetone phosphate synthase (an enzyme involved in the biosynthesis of ether phospholipids) uses FAD for a nonredox reaction. Clearly, no single criterion listed above would be satisfactory. To address these issues, an international commission was set up in 1955 which presented its first report in 1961 (Table 2.1).

7.3   Enzyme Commission—Recommendations

Considering the diversity of enzyme sources, reactions, and mechanisms, it became apparent that a formal system of nomenclature and classification was required. "Enzyme Commission" was appointed by the International Union of Biochemistry to address this issue. Its first report, published in 1964, forms the basis of the present system of classification. This system of classification is being updated periodically with updates made in 1972, 1978, 1984, and 1992 (Enzyme Nomenclature 1992). There are also many electronic supplements such as Supplement 14 of 2008. Most recent information and guidelines on enzyme nomenclature may be found at the official website of the International Union of Biochemistry and Molecular Biology (IUBMB)—http://www.chem.qmul.ac.uk/iubmb/enzyme/. The EC classification is universally accepted with a unique name and EC number for each enzyme. By this means, names for every enzyme (listed with a trivial name) could be rationalized and also given an EC catalog number (McDonald et al., 2009).

The nature of the overall reaction catalyzed by the enzyme—expressed by the formal equation—forms the basis of EC classification. Clearly, the intimate mechanism of the reaction and the formation of intermediate complexes with the enzyme, if any, are not considered. The Enzyme Commission defined six general categories of reactions thereby assigning an enzyme to one of the six classes. The number of unique enzymes, represented in each of these six classes (BRENDA database release

Fig. 7.1 Distribution of
enzymes into six different
classes according to EC
classification. Data from the
BRENDA database (January
2009 release) are shown
schematically

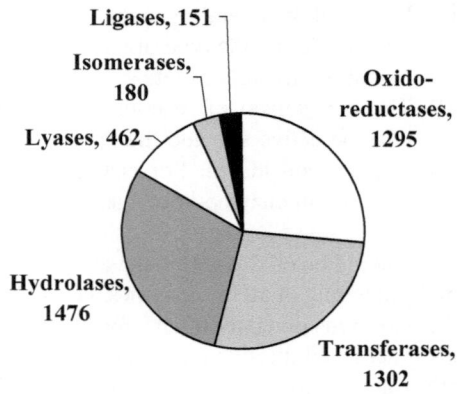

January 2009), is shown in Fig. 7.1. According to the recent count (March 2024, in http://www.enzyme-database.org/stats.php), there are a total of 6757 enzyme entries (class 1, 1975; class 2, 2031; class 3, 1339; class 4, 758; class 5, 312; class 6, 245), including 97 from the newly created seventh class of Translocases (see below).

Expectedly, enzymes for the redox and hydrolytic reactions are the most represented group. These six classes are Oxido-reductases, Transferases, Hydrolases, Lyases, Isomerases, and Ligases. Each of these six classes is further divided into several sub-classes and sub-sub-classes, according to the nature of the reaction catalyzed. In this classification each enzyme is given a code number consisting of a four-number system—the first number indicates the main class, and the second and third show the sub-class and sub-sub-class, respectively, thus defining the type of reaction. The fourth number is the actual number of that enzyme within its sub-sub-class. For example, alcohol dehydrogenase is given the code "EC 1.1.1.1." The first number indicates that it belongs to the oxidoreductase class (EC 1. x.x.x). Within this class, enzymes acting on the CH–OH group of donors bear the same sub-class number (EC 1.1.x.x). Within this sub-class, enzymes that use NAD^+ or $NADP^+$ as electron acceptors are given the number EC 1.1.1.x. Since alcohol dehydrogenase is the first enzyme in this category, it gets its fourth number (EC 1.1.1.1). Similarly, all carboxylesterases have the same first three digits in their EC code (EC 3.1.1.x). The fourth digit, however, distinguishes them by the actual carboxylic ester they hydrolyze.

A systematic name is assigned to each enzyme by the Commission, in addition to an accepted trivial name. The systematic name includes the name or names of substrates followed by a reaction name that ends in "-ase.". Because such systematic names can at times be too long and unwieldy, the Commission has also made recommendations for the use of trivial names. However, for the group (EC 3.3.3. x) of common proteases like pepsin, trypsin, and papain, it has not yet been possible to find acceptable systematic names. Enzyme Commission nomenclatures for enzymes representative of each class and those enzymes commonly referred to in this book are given in Table 7.1.

L

Table 7.1 Enzyme commission nomenclatures for representative/common enzymes

EC No.	Systematic name	Trivial name	Reaction
1. OXIDOREDUCTASES: Loss or gain of electrons by substrates			
$AH_2 + B \rightleftarrows A + BH_2$			
1.1.1.1	Alcohol:NAD$^+$ oxidoreductase	Alcohol dehydrogenase	Alcohol + NAD$^+$ $\rightleftarrows$ Aldehyde + NADH + H$^+$
1.1.1.27	L-Lactate:NAD$^+$ oxidoreductase	Lactate dehydrogenase	L-Lactate + NAD$^+$ $\rightleftarrows$ Pyruvate + NADH + H$^+$
1.2.1.2	Formate:NAD$^+$ oxidoreductase	Formate dehydrogenase	Formate+NAD$^+$ $\rightleftarrows$ CO$_2$ + NADH + H$^+$
1.4.1.4	L-Glutamate:NADP$^+$ oxidoreductase (deaminating)	NADP-Glutamate dehydrogenase	L-Glutamate + H$_2$O + NADP$^+$ $\rightleftarrows$ 2-Oxoglutarate + NH$_3$ + NADPH + H$^+$
1.11.1.6	Hydrogen peroxide:hydrogen peroxide oxidoreductase	Catalase	H$_2$O$_2$ + H$_2$O$_2$ $\rightleftarrows$ O$_2$ + 2H$_2$O
2. TRANSFERASES: Transfer of a reactive group from a donor substrate to an acceptor substrate			
$AX + B \rightleftarrows BX + A$			
2.1.2.1	L-Serine:tetrahydrofolate 5,10-hydroxymethyl-transferase	Serine hydroxy-methyltransferase	L-Serine + tetrahydrofolate $\rightleftarrows$ Glycine + 5,10-Methylenetetrahydrofolate
2.1.3.1	Methylmalonyl-CoA: pyruvate carboxytransferase	Methylmalonyl-CoA carboxytransferase	Methylmalonyl-CoA + pyruvate $\rightleftarrows$ Propionyl CoA + oxaloacetate
2.4.1.8	Maltose:orthophosphate glucosyltransferase	Maltose phosphorylase	Maltose + orthophosphate $\rightleftarrows$ β-D-Glucose 1-phosphate + D-Glucose
2.6.1.1	L-Aspartate:2-oxoglutarate aminotransferase	Aspartate aminotransferase	L-Aspartate + 2-Oxoglutarate $\rightleftarrows$ Oxloacetate + L-Glutamate
2.7. 1.1	ATP:D-hexose 6-phosphotransferase	Hexokinase	ATP + D-Hexose $\rightleftarrows$ ADP + D-Hexose 6-phosphate
2.7. 7.16	Ribonucleate pyrimidine-nucleotido-2'-transferase (cyclizing)	Ribonuclease	Transfers 3'-phosphate of a pyrimidine nucleotide residue of polynucleotide from 5' position of the adjoining nucleotide to the 2' position of the pyrimidine nucleotide itself, forms a cyclic nucleotide

(continued)

Table 7.1 (continued)

EC No.	Systematic name	Trivial name	Reaction
3. HYDROLASES: Introduction of elements of water into a substrate			
A-B + H$_2$O ⇄ AH + BOH			
3.1.1.1	Carboxylic ester hydrolase	Carboxylesterase	Carboxylic ester + H$_2$O ⇄ Alcohol + carboxylate
3.1.3.1	Orthophosphoric monoester phosphohydrolase	Alkaline phosphatase	Orthophosphoric monoester H$_2$O ⇄ Alcohol + orthophosphate
3.2.1.1	α-1,4-Glucan 4-glucanohydrolase	α-Amylase	Hydrolyzes α-1,4-Glucan links in starch
3.2.1.17	Mucopeptide N-acetylmuramylhydrolase	Lysozyme	Hydrolyzes β-1,4-Glucan links in peptidoglycan
3.4.4.4	Not possible yet	Trypsin	Hydrolyzes peptides, amides and esters of aromatic L-amino acids
3.5.3.1	L-Arginine amidinohydrolase	Arginase	L-Arginine + H$_2$O ⇄ L-Ornithine + Urea
3.7.1.1	Oxaloacetate acetylhydrolase	Oxaloacetase	Oxaloacetate + H$_2$O ⇄ Oxalate + Acetate
4. LYASES: Elimination of a group with double bond formation or addition of group to double bond			
A-BX ⇄ A=B + XH			
4.1.1.1	Pyruvate carboxy-lyase	Pyruvate decarboxylase	Pyruvate ⇄ Acetaldehyde + CO$_2$
4.2.1.1	Carbonate hydro-lyase	Carbonic anhydrase	H$_2$CO$_3$ ⇄ CO$_2$ + H$_2$O
4.3.1.1	L-Aspartate ammonia-lyase	Aspartate ammonia-lyase	L-Aspartate$^+$ ⇄ Fumarate + NH$_3$
4.3.2.1	L-Argininosuccinate arginine-lyase	Argininosuccinate lyase	L-Argininosuccinate ⇄ Fumarate + L-Arginine
4.99.1.1	Protoheme ferro-lyase	Ferrochelatase	Protoporphyrin + Fe^{2+} ⇄ Protoheme + 2H$^+$
5. ISOMERASES: Intramolecular rearrangements			
A ⇄ B			
5.1.1.4	Proline racemase	Proline racemase	L-Proline ⇄ D-Proline
5.1.3.2	UDPglucose 4-epimerase	UDPglucose epimerase	UDPglucose ⇄ UDPgalactose
5.3.1.1	D-Glyceraldehyde-3-phosphate ketol-isomerase	Triosephosphate isomerase	D-Glyceraldehyde 3-phosphate ⇄ Dihydroxyacetone phosphate
5.3.1.5	D-Xylose ketol-isomerase	Xylose isomerase	D-Xylose ⇄ D-Xylulose
5.4.99.2	Methylmalonyl-CoA CoA-carbonylmutase	Methylmalonyl-CoA mutase	Methylmalonyl-CoA ⇄ Succinyl-CoA

6. LIGASES (SYNTHETASES): Joining of molecules by covalent bond formation

$A + B + NTP \rightleftarrows AB + NDP + P$ (or $NMP + PP$)

6.1.1.1	L-Tyrosine:tRNA ligase (AMP)	Tyrosyl-tRNA synthetase	ATP + L-Tyrosine + tRNA $\rightleftarrows$ AMP + pyrophosphate + L-Tyrosyl-tRNA
6.3.1.1	L-Aspartate:ammonia ligase (AMP)	Asparagine synthetase	ATP + L-Aspartate + NH_3 $\rightleftarrows$ AMP + pyrophosphate + L-Asparagine
6.3.1.2	L-Glutamate:ammonia ligase (ADP)	Glutamine synthetase	ATP + L-Glutamate + NH_3 $\rightleftarrows$ ADP + orthophosphate + L-Glutamine
6.3.4.5	L-Citrulline:L-aspartate ligase (AMP)	Argininosuccinate synthetase	ATP + L-Citrulline + L-Aspartate $\rightleftarrows$ AMP + pyrophosphate + L-Argininosuccinate
6.4.1.2	Acetyl-CoA:carbon dioxide ligase (ADP)	Acetyl-CoA carboxylase	ATP + Pyruvate + CO_2 + H_2O $\rightleftarrows$ ADP + orthophosphate + Malonyl-CoA

Enzyme examples selected in this table are those often referred to in this book

Table 7.2 Databases with enzyme EC numbers

Database	Website
BRENDA (The Comprehensive Enzyme Information System)	http://www.brenda-enzymes.info/
ExPASy (Enzyme nomenclature database)	http://www.expasy.ch/enzyme/
KEGG (Kyoto Encyclopedia of Genes and Genomes)	http://www.genome.ad.jp/kegg/kegg2.html
EcoCyc (genome/ biochemical machinery of *E. coli*)	http://EcoCyc.org
MetaCyc (Pathway/Genome Databases)	http://metacyc.org/ and http://biocyc.org/
SYSTERS (Protein Family Database)	http://systers.molgen.mpg.de/
InterPro (database of protein families, domains and functional sites)	http://www.ebi.ac.uk/interpro/
Protein Mutant Database	http://pmd.ddbj.nig.ac.jp/
BioCarta (Pathways of Life)	http://www.biocarta.com/
ExplorEnz (The Enzyme Database)	https://www.enzyme-database.org/
IUBMB (Enzyme Nomenclature)	http://www.chem.qmul.ac.uk/iubmb/enzyme/

The universally accepted EC classification and enzyme codes are finding place (and direct utility) in many databases describing enzymes, genes, genomes, and metabolic pathways. Enzymes identified by EC numbers are valuable for relating the information to other databases. Some of these databases are listed in Table 7.2.

ExplorEnz is the primary source for all enzymes classified by the IUBMB. It was developed to facilitate the curation of the enzyme nomenclature data, to maintain correct data formatting, and to facilitate its use by other databases. As and when new information/interpretations become available, the database is updated by adding new enzymes, creating new classes/subclasses, etc. A new subclass on "isomerases altering macromolecular conformation" (numbered EC 5.6) has recently been added to the isomerase class. This subclass includes two sub-sub-classes. EC 5.6.1 contains enzymes altering polypeptide conformation or assembly, viz., chaperonin ATPase (EC 5.6.1.7). EC 5.6.2 houses enzymes altering nucleic acid conformation, viz., DNA topoisomerase (EC 5.6.2.1). In a major revision, a seventh class of translocases (EC 7, see below) was added in 2018.

7.4   Translocases—EC 7 Class

We recall that enzymes have evolved to bind tightly to their reaction transition states. Through this, the enzyme lowers the activation energy barrier and works by "flattening" the energetic landscape between the substrate and product. Similarly, membrane transporters can be thought of as systems that flatten the energetic landscape for the solute to cross the membrane. However, transporters do not bind preferentially to a transition state of the substrate because the substrate remains chemically unchanged during the transport process. Instead, one can visualize that the substrate binds preferentially to a high-energy conformation of the transporter (analogous to the transition state in an enzyme-catalyzed reaction). It is apparent that transporters are "catalysts" but cannot be accommodated in the six EC classes. Translocases are thus catalysts that facilitate the movement of ions or molecules across membranes or their separation within membranes. Several of these involve the hydrolysis of ATP and were previously classified as ATPases (EC 3.6.3.x), although the hydrolytic reaction is not their primary function (Shilton 2015). These enzymes have now been classified under a new EC class of translocases (EC 7). The reactions they catalyze are shown as transfers from "side 1" to "side 2" of a phospholipid bilayer membrane (Tipton 2018). The six subclasses designate the types of ions or molecules translocated:

EC 7.1—the translocation of hydrons (hydron being the general name for H^+)
EC 7.2—the translocation of inorganic cations and their chelates
EC 7.3—the translocation of inorganic anions
EC 7.4—the translocation of amino acids and peptides
EC 7.5—the translocation of carbohydrates and their derivatives
EC 7.6—the translocation of other compounds

The sub-sub-classes of EC 7 concern the reaction that provided the driving force for the translocation, where these are relevant. For instance, they may be coupled to oxidoreductase reactions (EC 7.x.1), the hydrolysis of nucleoside triphosphate (EC 7.x.2), and the hydrolysis of a diphosphate (EC 7.x.3) or a decarboxylation reaction (EC 7.x.4). Pores that change conformation between open and closed states in response to phosphorylation or some other catalyzed reactions are classified under EC 5.6 (Macromolecular conformational isomerases).

7.5   Some Concerns

The EC system of classification and nomenclature arrived at a broad consensus with a clear emphasis on the total reaction in question. The systematic names in each class may be based on a written reaction, even if only the reverse reaction is experimentally demonstrated. This has created some situations that are less than perfect. While all enzyme-catalyzed reactions are reversible in principle (at least micro-reversibility at the active site!), the reaction-based classification for the forward direction would not be the same as that for the reverse direction. This was recognized very early by JBS Haldane and according to him—calculation from thermodynamic data shows that catalase may act in the direction of H_2O_2 synthesis only under an O_2 pressure of many billions of atmospheres! It would, therefore, be perverse, if logical, to describe it as water oxidase; or for that matter the peroxidase as water dehydrogenase. To address such issues, the Enzyme Commission has recommended that the more important direction of the overall reaction, from a biochemical viewpoint, be used. It may be noted from Table 7.1 that reactions involving interconversion of NADH and NAD^+ are all written in the direction where NAD^+ is reduced by the other substrate. Also, when an overall reaction involves two types of reactions, then the second function is indicated in brackets. For instance, an oxidoreductase (decarboxylating) means the enzyme catalyzes an oxidation–reduction reaction in which one of the substrates is being decarboxylated.

Apart from issues related to the direction of the overall reaction considered, a major concern is that the reaction mechanism is given less importance or completely ignored! Functionally distinct reactions are catalyzed by the following four enzymes:

Fumarate + H_2O ⇄ Malate	(Fumarate hydratase; EC 4.2.1.2)
Fumarate + NH_3 ⇄ Aspartate	(Aspartate ammonia-lyase; EC 4.3.1.1)
Fumarate + Arginine ⇄ Argininosuccinate	(Argininosuccinate lyase; EC 4.3.2.1)
Fumarate + AMP ⇄ Adenylosuccinate	(Adenylosuccinate lyase; EC 4.3.2.2)

All four enzymes are part of an evolutionary super-family of proteins; mechanistically they add elements of H-X to fumarate, using the Michael reaction. This is a typical case where EC classification matches both in terms of the overall reaction as well as the reaction mechanism. While UDP-glucose epimerase—based on the overall reaction—rightfully belongs to "Isomerases" (Class 5), its intimate reaction mechanism tells a different story. The enzyme has tightly bound NAD^+ and the reaction at C4 of the hexose involves redox chemistry. The reaction mechanism of glucosamine 6-phosphate isomerase is similar to that of ketose-aldose isomerase:

$$D\text{-Glucosamine 6-phosphate} + H_2O \rightleftarrows D\text{-Fructose 6-phosphate} + NH_3$$

It was therefore formerly grouped under Class 5 (with the code EC 5.3.1.10) as an isomerase. Subsequently, this reaction was recognized as a C–N bond hydrolysis and reassigned as glucosamine 6-phosphate deaminase (EC 3.5.99.6).

Multiple enzyme forms and isozymes: At times we find that the same enzyme activity in an organism (or different organisms) is displayed by different protein forms. Enzymes that catalyze the same overall reaction, but follow different mechanistic paths, are not uncommon. Some examples of different protein forms are listed in Table 7.3.

When these multiple molecular forms are coded by different but related genes (having the different primary structures—amino acid sequence!) they are termed isozymes. Different isozyme forms of an enzyme are easily distinguished through their characteristic electrophoretic mobilities. The muscle and the heart forms of lactate dehydrogenase are the best-studied examples of isozymes. All isozymes are examples of multiple forms of an enzyme, but all multiple forms need not be isozymes.

Table 7.3 Enzyme examples with different mechanistic forms

Enzyme	EC number	Mechanistic difference
Ribonucleotide reductase	EC 1.17.4.1	An iron protein
	EC 1.17.4.2	Requires a cobamide coenzyme
Methionine synthase	EC 2.1.1.13	Contains cobalamin
	EC 2.1.1.14	Does not contain cobalamin
Proteases	EC 3.4.4.x	Serine, cysteine, carboxylate, or metal ion at the active site
Histidine decarboxylase	EC 4.1.1.22	Pyridoxal-phosphate (mammalian)
		Pyruvoyl prosthetic group (bacterial)
Fructose bisphosphate aldolase	EC 4.1.2.13	Class I—forms a protonated imine
		Class II—zinc polarized (microbial)
Dehydroquinase	EC 4.2.1.10	Type I—forms a protonated imine
		Type II—forms an enolate intermediate

Multiple molecular forms of the same enzyme may also arise due to other reasons. They may occur as—(a) interconvertible forms through covalent modifications (examples: phosphorylation of glycogen phosphorylase and adenylylation of *E. coli* glutamine synthetase), (b) proteolytic variants (chymotrypsinogen and chymotrypsin), (c) different oligomeric states of the same monomer (examples: bovine liver glutamate dehydrogenase and avian liver acetyl CoA carboxylase), and (d) distinct conformational states of the same enzyme protein (for example, R and T states of aspartate carbamoyltransferase). The occurrence of multiple enzyme forms is often associated with their role in metabolic regulation.

All multiple enzyme forms share the same EC number and a formal name. However, these names need to be suitably prefixed or suffixed to indicate the modification, organ source, or organelle source. Much effort goes into characterizing any new enzyme. Great care must be taken to define what it does and not repeat the name or indicate another reaction. In the final outcome, one need not get bogged down by issues of semantics, for the excitement of enzymology beckons the uninitiated and the specialists alike.

References

Cronan JE (2024) Lipoic acid attachment to proteins: stimulating new developments. Microbiol Mol Biol Rev 88:e00005-24

Enzyme Nomenclature 1992 [Academic Press, San Diego, California, ISBN 0-12-227164-5 (hard back), 0-12-227165-3 (paperback)] with Supplement 1 (1993), Supplement 2 (1994), Supplement 3 (1995), Supplement 4 (1997) and Supplement 5 (in Eur J Biochem, 223:1–5 (1994); Eur J Biochem, 232:1–6 (1995); Eur J Biochem, 237:1–5 (1996); Eur J Biochem, 250:1–6 (1997) and Eur J Biochem, 264:610–650 (1999); respectively), IUBMB, Academic Press, Orlando

McDonald AG, Boyce S, Tipton KF (2009) ExplorEnz: the primary source of the IUBMB enzyme list. Nucleic Acids Res 37:D593–D597

Shilton BH (2015) Active transporters as enzymes: an energetic framework applied to major facilitator superfamily and ABC importer systems. Biochem J 467:193–199

Tipton K (2018) Translocases (EC 7): a new EC class. Enzyme Nomenclature News (August 2018)

Further Reading

Chang A et al (2021) BRENDA, the ELIXIR core data resource in 2021: new developments and updates. Nucleic Acids Res 49:D498–D508

Ghodke VM, Punekar NS (2022) Environmental role of aromatic carboxylesterases. Environ Microbiol 24:2657–2668

Karp PD et al (2004) The *E. coli* EcoCyc database: no longer just a metabolic pathway database. ASM News 70:25–30

Karp PD et al (2023) The EcoCyc database. EcoSal Plus 11. https://doi.org/10.1128/ecosalplus.esp-0002-2023

Part II

Enzyme Kinetic Practice and Measurements

Chemical Kinetics: Fundamentals

8

Any given step of metabolism is in essence a chemical reaction. The vast majority of reactions in living systems are catalyzed. The complex web of metabolism is made possible through enzymes. Whether enzyme catalyzed or not, a chemical reaction is best understood through fundamental tenets of chemical kinetics. Therefore, an overview of this field is provided here.

For a given reaction A→Products, where exactly this system is located at a given point in time is determined by thermodynamic and kinetic considerations (Fig. 3.1). Thermodynamics does not inform us about the rate at which a chemical change will occur or how this rate will vary with conditions. The magnitude of ΔG (free energy of reaction) tells us *how far* the reaction can go while $\Delta G^{\neq}$ (free energy of activation) gives an indication of *how fast* this reaction can go! Chemical kinetics deals with the rates of chemical reactions and the mechanism by which they occur.

8.1 Measurement of Reaction Rates

The speed with which a chemical reaction takes place is called the reaction "*rate.*" For a hypothetical reaction A→Products, we may define the change occurring in the reactant or product per unit time. It may therefore be expressed as the change in their concentration during the time interval Δt as

$$\text{Rate of reaction} = \frac{-\Delta[A]}{\Delta t} = \frac{\Delta[P]}{\Delta t}$$

where $-\Delta[A]$ is the decrease in the concentration of A and $\Delta[P]$ is the increase in the concentration of P. By convention, square brackets are used to express the molar concentration (mol/L) of the reactant or product. Also, since the concentration of the reactant decreases with time, a negative sign is used to denote this. The reaction rate may be best visualized schematically (Fig. 8.1). If $[A]_1$ and $[A]_2$ are the

N. S. Punekar, *ENZYMES: Catalysis, Kinetics and Mechanisms*, https://doi.org/10.1007/978-981-97-8179-9_8

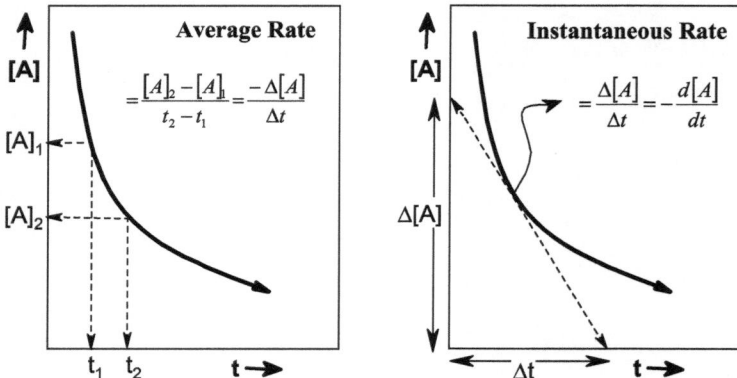

Fig. 8.1 Depiction of average rate and instantaneous rate for a reaction

concentrations of the reactant at time t_1 and t_2, then the ***average rate*** may be expressed as

$$\text{Average reaction rate} = \frac{[\text{A}]_2 - [\text{A}]_1}{t_2 - t_1} = \frac{-\Delta[\text{A}]}{\Delta t}$$

This is similar to expressing mechanical speed. As the reaction proceeds, the concentration of reactant(s) keeps decreasing. Because of this the rate of reaction may not be constant in the time interval between t_1 and t_2. Therefore, a better representation of reaction rate may be ***instantaneous rate***—the rate of change of reactant (or product) concentration at a given time.

We obtain the instantaneous rate by making the time interval (Δt) as small as possible. Mathematically (according to differential calculus) this is represented by

$$\text{Instantaneous reaction rate} = -\frac{d[\text{A}]}{dt} = \frac{d[\text{P}]}{dt}$$

Here, d[A] or d[P] means infinitesimally small changes in the concentration of A or P, in infinitesimally small intervals of time, dt. Graphically, this is nothing but the slope of a tangent drawn to the "[A] versus t" curve at that time point (Fig. 8.1).

Reaction rates can be measured by plotting a graph between the concentration of the reactant (or product) as a function of time. In practice, the changing concentration is recorded by corresponding changes in a measurable property such as—volume, pressure, pH, UV/visible absorbance, optical rotation, and refractive index. For instance, the time course of sucrose hydrolysis by invertase is monitored through changes in optical rotation.

8.2 Factors That Influence Chemical Reaction Rates

Several factors influence the rate of a reaction. Significant among them include:

1. Concentration of the reactant
2. Temperature of the system
3. Presence of a catalyst
4. Available surface area, radiation, etc.

The larger the available surface area of reactant(s), the faster the reaction rate. As the particle size decreases, the total surface area increases. This permits more reactant molecules to participate in the reaction. Most biological reactions occur in solution and therefore their reaction rate is directly proportional to the concentration of the reactant. In certain reactions, the interaction of reactant species with photons (radiation of suitable energy; $E = h \times \nu$) leads to rate enhancement. As enzymes are exquisite catalysts, catalytic phenomena form the central theme in enzymology and recur throughout this book! How chemical reaction rate is affected by reactant concentration and temperature forms an important basis for understanding enzyme catalysis. These two aspects will therefore be dealt with in some detail here.

8.3 Reaction Progress and Its Concentration Dependence

Reactant(s) get converted to product(s) in a chemical reaction. As a result, the concentration of the reactant decreases while that of the product increases with time. Concentration changes that occur as a function of reaction progress are shown in Fig. 8.2. Two representative examples, namely, the isomerization of glucose and hydrolysis of sucrose are considered.

Inspection of such data gives important information—about the stoichiometry and the endpoint of the reaction. At any time point after $t = 0$, the molar concentration of fructose formed equals (and corresponds) to the molar concentration of glucose disappeared. This permits us to establish the **reaction stoichiometry** and hence the balanced equation (glucose $\rightleftarrows$ fructose) for the reaction. On the other hand, at $t = \infty$ this reaction reaches equilibrium, and the ratio of [product]/[reactant] gives us the **equilibrium constant**. For the glucose isomerase reaction, the K_{eq} is close to one (actually, [glucose] to [fructose] ratio at equilibrium is 51:49). No matter how much time passes, one cannot fully convert glucose to fructose unless of course the product (fructose here) is removed from the system. A similar analysis of sucrose hydrolysis is also revealing. For every molecule of sucrose hydrolyzed, one molecule each of glucose and fructose is formed. A balanced equation would therefore be—sucrose $\rightarrow$ glucose + fructose. At the endpoint of this reaction, however, there is no sucrose left—the equilibrium is far to the right. The reaction is **unidirectional**, and sucrose is completely converted to products! Why water has not figured in the equation (as a reactant) will become obvious from the discussion below.

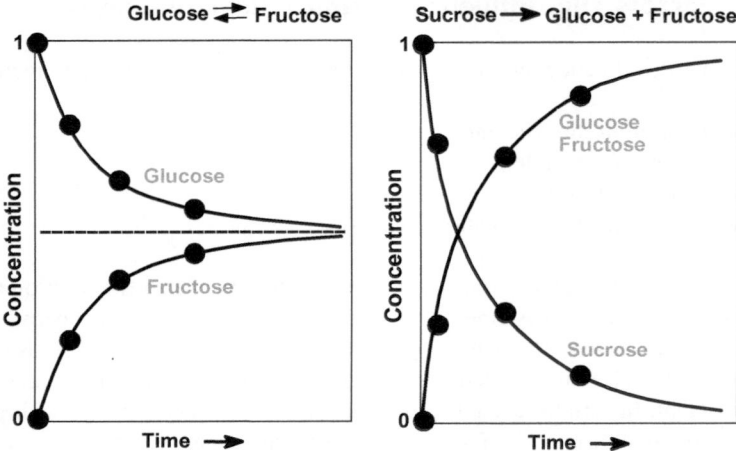

Fig. 8.2 Concentration changes as a function of reaction time. Isomerization of glucose to fructose (left) and hydrolysis of sucrose (right) are shown. The decrease in the concentration of water in sucrose hydrolysis is negligible compared to its bulk concentration (55.5 M)

Typically, at time zero, only the reactant is present while the product concentration is zero. The concentrations of various species change rapidly in the beginning, but this slows down at later time points. It thus follows that the reaction rate is the highest in the beginning and decreases with time. The reaction rate is directly proportional to the concentration of the reactant(s). As early as 1867, Guldberg and Waage proposed a quantitative relationship between the molar concentration of the reactant(s) and the reaction rate (Voit et al. 2015). According to this *Law of Mass Action*—"the rate of chemical reaction is directly proportional to the product of the molar concentration of the reactants." Obviously, all other factors (including temperature) must be kept constant for this proportionality to hold.

Consider the balanced reaction for the hydrolysis of sucrose.

$$\text{Sucrose} + H_2O \rightarrow \text{Glucose} + \text{Fructose}$$

We can write the rate expression for this reaction according to the law of mass action as

$$\text{Reaction rate} \propto [\text{Sucrose}]\,[H_2O]$$

$$\text{Reaction rate} = k\,[\text{Sucrose}]\,[H_2O]$$

The constant of proportionality "k" is called the reaction *rate constant*. This rate constant is nothing but the rate of reaction when the concentration of each of the reactants is unity (i.e., Rate = $k \times 1 \times 1$). The magnitude of "k" depends on the type (nature) of reaction and temperature; it is however independent of the reactant concentration. The law of mass action gives the rate expression on the basis of an overall balanced equation for the reaction. It therefore provides us with the *molecularity* of that reaction—which cannot be zero. It can only be whole numbers like 1, 2, 3..., etc. On the other hand, the *rate equation* (or the *rate law*) gives the experimentally observed dependence of rate on the concentration of reactants. This concentration dependence may be expressed in terms of the *order* of the reaction. A reaction order (denoted as "*n*") is the sum of powers to which the concentration terms are raised in the rate equation. The order of a reaction is thus an experimentally determined parameter and is reflected in the rate equation. Reaction order can be zero (as in catalyzed reactions) or a positive number including the fractional values (as in mixed-order reactions). We will come across such examples when dealing with enzyme-catalyzed reactions (Chap. 14) and the Henri–Michaelis–Menten rate equation.

The order of a reaction is an important descriptor in defining the mechanism of that reaction. There are a number of methods to determine the order of a reaction (and hence to define the rate equation). These are briefly described below.

The graphical method is useful when the reaction involves a single reactant. For such reactions, we can write a general rate equation of the type, **rate** $= k$ **[A]**^{n}. If we obtain a straight line by plotting a graph of rate versus $[A]^n$ then the order of the reaction is "*n*." Graphical analysis of data for zero-order ($n = 0$), first-order ($n = 1$), and second-order ($n = 2$) reactions is schematically summarized in Table 8.1. A variation of this is to determine the reaction order based on the half-life period. One determines the time interval by which the concentration of the reactant is reduced to half of its initial value. Subsequently, such data are analyzed graphically or by the best fit to the relevant equation (Table 8.1).

The initial rate method is suitable for reactions involving more than one reactant. Here, the initial rate (rate at the beginning of the reaction—instantaneous rate extrapolated to time zero) is measured. At a time, only one reactant concentration is varied (while keeping all others constant) and the order with respect to that particular reactant is calculated. The same procedure is repeated with all other reactants. The individual orders so obtained are summed up to obtain the reaction order. As we will see later (Part III), most enzyme kinetics experiments are designed this way.

Ostwald's isolation method exploits the fact that the concentration of the limiting reactant maximally influences the reaction rate. Here, except one reactant all others are taken in excess—so that the order so determined will be for that reactant alone. This process is repeated for each reactant and the order is determined. The overall order of the reaction will then be the sum of individual orders.

The rate expression (according to the law of mass action) and rate equation for a given reaction may or may not be the same. For example, sucrose hydrolysis is a *bimolecular* reaction. However, water is in large excess (55.5 M in the aqueous phase) and hence its concentration change does not figure in the rate equation. Experimentally, this hydrolysis follows a *first-order* rate with respect to [sucrose]. We call such a reaction *pseudo-first order* reaction.

Table 8.1 Reactions of different orders: a kinetic summary

Feature	Zero order	First order	Second order
Reaction type	A→Products	A→Products	A + A → Products
Rate expression	$Rate = k[A]^0$	$Rate = k[A]^1$	$Rate = k[A]^2$
Direct plot			
Integrated rate equation	$[A] = -kt + [A]_0$	$\log[A] = \frac{-kt}{2.303} + \log[A]_0$	$\frac{1}{[A]} = kt + \frac{1}{[A]_0}$
Linear plot			
Half-life expression	$t_{1/2} = \frac{[A]_0}{2k}$	$t_{1/2} = \frac{0.693}{k}$	$t_{1/2} = \frac{1}{k[A]_0}$

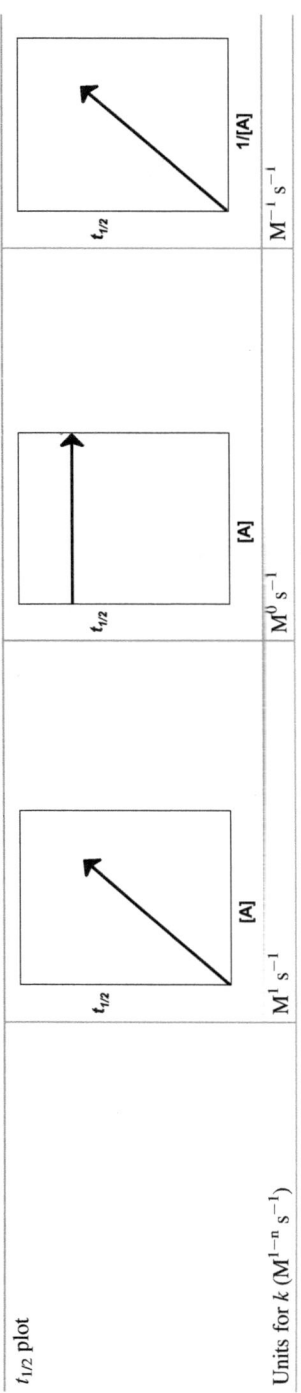

$$\text{Reaction rate} = k \times [\text{H}_2\text{O}] \times [\text{Sucrose}]$$
$$= k' \times [\text{Sucrose}]$$

where

k = second order rate constant and $k' = k \times [\text{H}_2\text{O}]$ = pseudo-first-order rate constant.

It is obvious that in most hydrolytic reactions in biology, water is a reactant (and counts in the molecularity of that reaction). But being present in such a large excess, it hardly contributes to the reaction order. It is a different matter, however, when the same hydrolytic reaction is conducted in a nonaqueous solvent where water is limiting!

8.4   Temperature Dependence of Reaction Rates

We have seen earlier that the reactant concentration is a significant factor in determining the reaction rates. Accordingly, the reactant concentration does feature in the rate equation. Then how does temperature influence the reaction rates? Rate constant (k) by definition is independent of the reactant concentration. However, it is strongly dependent on temperature. Elevated temperatures reduce the kinetic barriers of the reaction (Wolfenden and Snider 2001). In general, the higher the temperature, the greater the rate of reaction. The reaction rate is approximately doubled for every 10-degree rise in temperature. We may thus define the ***temperature coefficient (Q_{10})*** of a reaction as

$$Q_{10} = \frac{k_{(T+10)\,°C}}{k_{T\,°C}}$$

For a reaction with a Q_{10} of 2, an increase in temperature from 25 °C to 100°°C results in a 70-fold increase in the rate. More recent data show that "$Q_{10} = 2$" applies to only a few reactions that are rapid and easy to study at ordinary temperatures. However, slow reactions are vastly more sensitive to temperature (Table 8.2). The more sluggish a reaction is, the more temperature sensitive it tends to be. Heats of

Table 8.2   Temperature coefficients (Q_{10}) for some reactions

Q_{10}	Chemical reaction	Enzyme example
2	Hydration of CO_2	Carbonic anhydrase
4	Peptide bond hydrolysis	Chymotrypsin
6	O-Glycoside hydrolysis	α-Glucosidase
13	Orotidine 5-phosphate decarboxylation	OMP decarboxylase
16	P–O cleavage of phosphate monoester dianions	Inositol phosphatase

activation ($\Delta H^{\neq}$) play a much more pronounced role in determining the relative rates of these uncatalyzed reactions than do their entropies of activation (contribution by $T\Delta S^{\neq}$). A catalyst, if it employs lowering of enthalpy of activation as a catalytic ploy, then the rate enhancement that it produced would have increased automatically as the temperature is lowered. Thus, a primitive nonenzymatic catalyst might have achieved its effect entirely by lowering the enthalpy of activation. It appears that the enthalpic basis of catalysis for the great majority of enzymes reflects their origin in more primitive catalysts; a reduction of $\Delta H^{\neq}$ rather than an increase in $T\Delta S^{\neq}$ offered them a selective advantage in a gradually cooling environment (Wolfenden 2014).

An early explanation for the effect of temperature on reaction rates came from **Collision Theory** for reactions. This approach had its origin in the kinetic theory of gases and its theoretical basis may be summarized as follows.

Reactant molecules must collide with each other (come together!) for the reaction to take place. However, not every collision that occurs at a given temperature is fruitful. Only a fraction of these are **effective collisions**—and actually result in products. Two important barriers to effective collisions are—an energy barrier (largely enthalpic in nature) and an orientation barrier (mostly entropic in nature). Accordingly, the colliding molecules must possess a minimum amount of energy (called **threshold energy**) for them to react. Furthermore, these colliding molecules must come together in a proper orientation to achieve effective collisions. A rise in temperature increases the number of effective collisions and hence leads to an increased rate of reaction.

All reactant molecules do not possess the same energy. Instead, there is a distribution of energy among reacting molecules. This Maxwell's distribution of energies (Fig. 8.3) explains the effect of the rise in temperature on the reaction rate.

The energy distribution curve shifts to the right at higher temperatures ($T_1 \rightarrow T_2$). Because of this, the number of molecules possessing threshold energy is increased

Fig. 8.3 Distribution of molecular energies at two different temperatures. The number of molecules possessing threshold energy increases with temperature (shaded region)

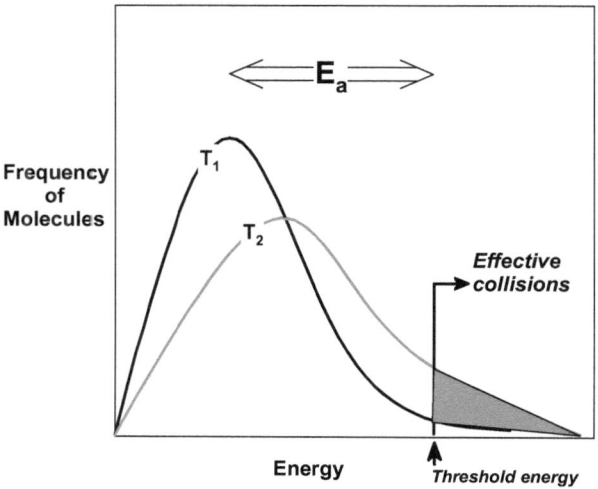

Fig. 8.4 Plot of ln k versus
$1/T$ to obtain the value of
activation energy. The gas
constant $R = 8.31$ J mol^{-1} K^{-1} (or 1.98 cal mol^{-1} K^{-1})

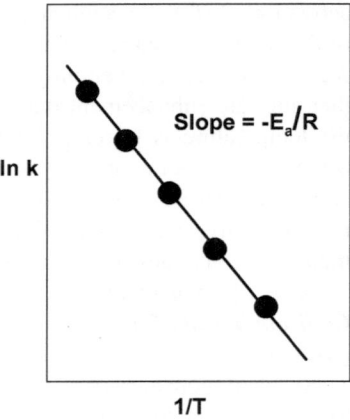

(shaded region in Fig. 8.3). The greater the number of such molecules, the greater the number of effective collisions and the higher the reaction rate. The excess energy, over and above the average energy of the reactants, must be supplied to reach threshold energy. This is called the ***activation energy***. It is denoted as E_a and represents $E_{threshold}$-$E_{average}$. A purely empirical relation for the temperature dependence of reaction rates was deduced by Svante Arrhenius. He showed that activation energy is the parameter that relates temperature dependence of the rate constant. According to the Arrhenius equation,

$$k = Ae^{-\frac{E_a}{RT}} \text{ and } \ln k = \ln A - \frac{E_a}{RT}$$

where A is the Arrhenius constant (pre-exponential factor) and E_a is the activation energy. This equation tells us that (a) the rate constant k increases exponentially with the increase in temperature and (b) a smaller value of E_a corresponds to an increase in k (and therefore an increased reaction rate). Lastly, for a reaction with $E_a = 0$ or when $T = \infty$, the rate constant equals the pre-exponential factor "A."

We can calculate the activation energy for a given reaction in a couple of ways. A plot of ln k against $1/T$ gives a straight line (Fig. 8.4). In practice, the value of activation energy is obtained from the slope ($E_a = -R \times$ slope) of this line. Extrapolation to $1/T = 0$ (i.e., $T = \infty$) is not feasible as most reactions are studied in a limited range of temperatures. Therefore, the value of "A" must be calculated by plugging in the value of E_a in the equation.

Alternatively, the values of E_a and "A" can be calculated by solving the simultaneous equations:

$$\ln k_1 = \ln A - \frac{E_a}{RT_1} \text{ and } \ln k_2 = \ln A - \frac{E_a}{RT_2}$$

And hence,

$$\ln \left(\frac{k_2}{k_1}\right) = \frac{E_a}{R} \left(\frac{1}{T_1} - \frac{1}{T_2}\right)$$

This can be evaluated by substituting the values of two rate constants (k_1 and k_2) at temperatures T_1 and T_2 respectively. For a typical reaction with an activation energy of 12 kcal/mol (~50 kJ/mol), a 10-degree rise in T (i.e., $T_2 - T_1$) will result in an approximate doubling of the rate ($Q_{10} = k_2/k_1 \approx 2$).

We have come across the Arrhenius equation once before (Chap. 3) while dealing with barriers to catalysis. Collision theory and the Arrhenius equation are useful in the understanding of simple gas phase reactions. For reactions involving more complex molecules (and in solution), Transition State theory provides a better conceptual framework. According to this theory, the reactant molecules must come together to form an *activated complex*—the unstable transition state (Fig. 3.1). Subsequently, a relationship between the rate constant (k) and free energy of activation ($\Delta G^{\neq}$) is established.

$$k = \frac{k_B T}{h} e^{-\frac{\Delta G^{\neq}}{RT}} = \frac{k_B T}{h} e^{-\frac{\Delta H^{\neq}}{RT}} e^{+\frac{\Delta S^{\neq}}{R}}$$

In this sense, $\Delta G^{\neq}$ is a composite term that reflects the overall ease of forming the activated complex. It conveys the same information as the rate constant to which it is directly related by the transition state theory. We will refrain from further repetition as this has already been dealt with before (in Chap. 3). The contribution to $\Delta G^{\neq}$ by $\Delta H^{\neq}$ and $\Delta S^{\neq}$ may however be evaluated through a variant of this equation, the Eyring equation.

$$\ln \frac{kh}{k_B T} = \frac{\Delta S^{\neq}}{R} - \frac{\Delta H^{\neq}}{RT}$$

A plot of $\ln \frac{kh}{k_B T} \to 1/T$ is linear, and one obtains $\Delta H^{\neq}$ (from slope) and $\Delta S^{\neq}$ (from intercept) by this analysis. The lower the value of $\Delta H^{\neq}$, the faster the reaction—the easier the bond-breaking/forming step. The more negative the $\Delta S^{\neq}$, the slower the reaction rate; entropy of activation plays a less prominent role in unimolecular reactions. In bimolecular reactions, however, $\Delta S^{\neq}$ is always negative and unfavorable. This has to be compensated by a favorable and negative $\Delta H^{\neq}$.

8.5   Purpose of Kinetic Studies: Reaction Mechanism

Elucidation of the mechanism of a chemical reaction is the most important purpose of kinetic analysis. A reaction mechanism may be understood at many different levels.

1. Determination of reaction stoichiometry from the number of moles of each reactant consumed to give the number of moles of each of the final products. For simple reactions, this is usually followed by writing a balanced equation and assigning molecularity.
2. Experimental confirmation of the predicted kinetic order provides strong support to the proposed chemical mechanism. Since several hypothetical mechanisms could lead to the same rate equation (and reaction order) such kinetic evidence may not be unambiguous. Nevertheless, kinetic data permit us to exclude many plausible alternatives and help narrow down the choice of mechanism.
3. Complex reaction mechanisms may include a sequence of two or more consecutive steps. These individual steps—*elementary processes*—are almost always either uni-molecular or bimolecular. Often in a complex reaction, the rate of the overall reaction is determined by the slowest step in the sequence. Such a step is called the *rate-determining step*. If a rate constant and kinetic order can be ascribed to a complex reaction, then the composition of the most unstable activated complex (TS of the rate-limiting step) can be deduced from the overall rate equation. However, such a rate equation does not tell us how many intermediates or steps are involved before or after this rate-determining step. It is also not possible to determine the number of solvent molecules involved in the activated complex in solution (solvent large excess!).

It is obvious from the above discussion that there are severe limitations to kinetics as a method to study reaction mechanisms. While kinetic evidence does not provide clinching evidence for a mechanism, it does provide excellent support. More importantly, it is frequently possible to exclude many alternative schemes by the kinetic approach. The kinetic study has to be supplemented and buttressed with various techniques to confirm the presence of intermediates and steps in the overall reaction mechanism.

8.5.1 Insights into Transition State Structure

The transition state is defined as the state corresponding to the highest energy along the reaction coordinate (Fig. 3.1). Any small displacement leads to its conversion into a more stable entity—the population at that point is zero. The transition state cannot be captured or directly observed. Chemical kinetics does provide an empirical approach to the characterization of this transition state structure. An early qualitative guide was the *Hammond–Leffler Postulate* derived from transition state theory—the structure of the transition state more closely resembles either the product or the reactant, depending on which is higher in enthalpy (Fig. 8.5). According to this structure-correlation principle, TS will be more product-like for an endothermic reaction (and occurs late on the reaction coordinate). Conversely, for an exothermic reaction, it will be reactant-like and occurs early.

On a quantitative (but semi-empirical) level, correlations between rate constants (kinetics) and equilibrium constants (thermodynamics) for a given reaction type

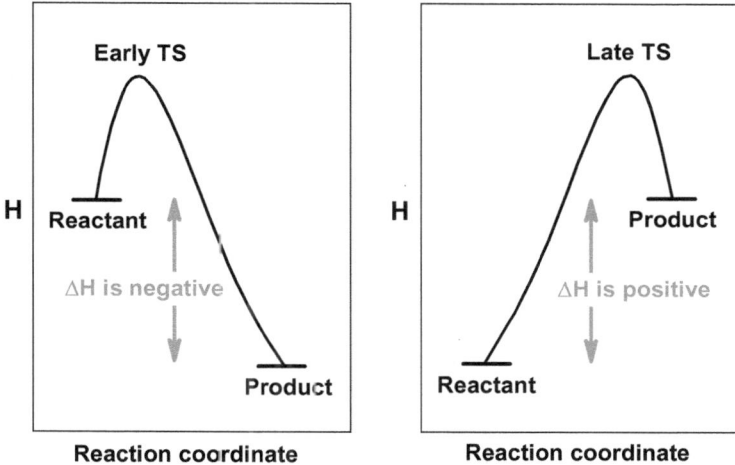

Fig. 8.5 Schematic energy diagram illustrating the Hammond–Leffler postulate. Profiles shown for an exothermic (left) and endothermic (right) reaction with early and late transition states, respectively

have helped develop this approach further. Two important *Linear Free Energy Relationship*s of this type are the *Bronsted relationship* and *Hammett equation*. The transition state theory similarly relates $\Delta G^{\neq}$ with the rate constant for that reaction. In a typical physical organic chemistry approach—reaction rates and corresponding equilibrium constants are determined for a series of substituted reactants. These structural variations are chosen so that they do not lead to gross chemical mechanism changes. Such data provide a measure of the location of the transition state along the reaction coordinate. Locating transition states by computational chemistry techniques has also gained ground recently.

Finally, the determination of substitution effects for enzyme-catalyzed reactions is only possible with enzymes of broad substrate specificity. For others, however, a systematic study of isotope effects would provide insights into the transition state. This is dealt with in some detail in Chap. 25 (Part III).

References

Voit EO et al (2015) 150 years of the mass action law. PLoS Comput Biol 11:e1004012

Wolfenden R (2014) Massive thermal acceleration of the emergence of primordial chemistry, the incidence of spontaneous mutation, and the evolution of enzymes. J Biol Chem 289:30198–30204

Wolfenden R, Snider MJ (2001) The depth of chemical time and the power of enzymes as catalysts. Acc Chem Res 34:938–945

Further Reading

Dill KA, Bromberg S (2003) Molecular driving forces: statistical thermodynamics in chemistry and
    biology. Garland Science, London/New York
Lienhard GE (1973) Enzymatic catalysis and transition-state theory. Science 180:149–154
Maskill H (1986) The physical basis of organic chemistry. Oxford Science Publications, Oxford
    University Press, Oxford/New York
Wolfenden R (2011) Benchmark reaction rates, the stability of biological molecules in water, and
    the evolution of catalytic power in enzymes. Annu Rev Biochem 80:645–667

Concepts of Equilibrium and Steady State

The key to understanding the catalytic action of an enzyme is the study of reaction velocities and not equilibria. Nevertheless, equilibrium and steady-state are two important states of any dynamic system. Both have much relevance to the understanding of enzyme mechanisms and hence metabolism. This chapter will elaborate on these concepts. There are many analogies/models to describe these states which include ponds and rivers. We will look at two simple setups to understand what is meant by equilibrium and steady state, before going into details.

Imagine two beakers connected via a stopcock. Suppose the stopcock is kept closed and water is filled into one of the beakers. What would happen if the stopcock were opened now? Water will move into the second beaker until the levels in the two beakers are the same (Fig. 9.1). Once the two water levels become equal there is no net flow of water, and the system as a whole becomes stable (and attains *equilibrium*). While water molecules continue to diffuse from one beaker into the other— the water level in the two beakers remains the same. This is an example of *dynamic equilibrium*. What happens if the stopcock is now closed? The water level on the two sides remains the same but there is no free exchange of water molecules across the two beakers. This stable state is an example of static equilibrium. We can distinguish between the static and the dynamic equilibrium by a simple test. A dye introduced in any one beaker will diffuse into the other over time only in the case of dynamic equilibrium. This two-beaker setup is an excellent analogy to "glucose $\rightleftharpoons$ fructose" isomerization. An equilibrium mixture of glucose and fructose defines a static equilibrium, as no inter-conversion occurs due to the prevailing activation energy barrier. The addition of glucose isomerase (enzyme catalyzing this inter-conversion, equivalent to opening the stopcock and opening a path to mix the two compartments!) makes it a dynamic equilibrium.

Let us now consider another situation. Suppose we have a beaker fitted with an inlet and an outlet for water as shown (Fig. 9.1). We start filling the beaker by letting water in (through the inlet) at a constant rate. Initially, water drains out (through the outlet) more slowly than it enters because of lower water level (and lower hydrostatic

N. S. Punekar, *ENZYMES: Catalysis, Kinetics and Mechanisms*, https://doi.org/10.1007/978-981-97-8179-9_9

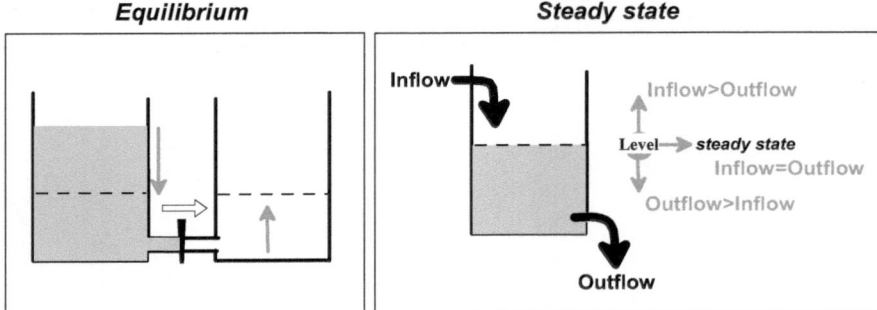

Fig. 9.1 Equilibrium and steady state. When the stopcock is opened water flows into the empty beaker until the two levels become equal—equilibrium is attained (left panel). Water level is maintained so long as the inflow equals the outflow—steady state (right panel)

pressure). However, this will cause the water level to rise in the beaker, generating more pressure and consequently water will drain more quickly. When the inflow of water becomes equal to the outflow—the water level in the beaker is maintained— and reaches a *steady state*. There is a constant flow of matter (and/or energy) through the system in a steady state.

A thermodynamic equilibrium indicates total randomness (heat death) while living beings represent systems at a steady state that are maintained away from equilibrium. Because biological systems are open systems, exchanging energy and matter with their surroundings, they are best represented by a steady-state model rather than an equilibrium one. Although an equilibrium assumption is simplistic, it is often invoked to approximate many living processes.

9.1   Chemical Reaction Equilibrium

All chemical reactions are reversible in principle. Consider the following equilibrium:

$$A \underset{k_{-1}}{\overset{k_1}{\rightleftarrows}} P$$

where k_1 and k_{-1} are the rate constants for the forward and reverse reactions, respectively. The position of this equilibrium is defined by the equation

$$K_{eq} = \frac{[P]_{eq}}{[A]_{eq}}$$

where K_{eq} is the *equilibrium constant,* while $[A]_{eq}$ and $[P]_{eq}$ represent the concentration of A and P at equilibrium, respectively. At any given instant, the overall rate of change in [A] will be the sum of its rate of disappearance and the rate of formation. This can be written as

$$\frac{d[A]}{dt} = -k_1[A] + k_{-1}[P]$$

The reverse reaction rate becomes negligible when either [P] = 0 or k_{-1} is much smaller than k_1. Under these conditions essentially a unidirectional reaction (A → P) is defined with the rate equation $-d[A]/dt = k_1[A]$. We have already come across a detailed kinetic treatment for such reactions (Chap. 8).

For a reversible reaction at equilibrium, the forward and the backward reactions cannot take different paths. This follows from the principle of detailed balance and *microscopic reversibility* of such phenomena. The forward and the reverse rates must be identical for a reaction at equilibrium. Accordingly,

$$k_1[A]_{eq} = k_{-1}[P]_{eq}$$

(We note that, at equilibrium, the two rates are equal and not the rate constants!) Rearranging and by definition of K_{eq} (as above) we obtain:

$$K_{eq} = \frac{[P]_{eq}}{[A]_{eq}} = \frac{k_1}{k_{-1}}$$

This equation links the equilibrium constant (K_{eq}, a thermodynamic parameter) with the corresponding rate constants (kinetic parameters) for a given reaction. The "Haldane relationship" is one such equation that relates enzyme kinetic constants with the corresponding reaction equilibrium constant (Chap. 14).

9.1.1 ΔG and Equilibrium

For any reaction to occur, it should be accompanied by a decrease in free energy. The change in free energy (ΔG) for a reaction at equilibrium is zero. During the course of a reaction, the composition of the reaction mixture changes with time (Fig. 8.2) and ΔG decreases. The actual ΔG is thus related to the composition of the reaction mixture and the standard free energy (ΔG°) of the reaction by the following expression:

$$\Delta G = \Delta G° + RT \ln \Gamma$$

where Γ, the *mass action ratio*, is the ratio of product concentration to substrate concentration ($\Gamma = [P]/[A]$). At equilibrium, $\Gamma = K_{eq}$ and ΔG = 0. Therefore,

Table 9.1 Variation of K_{eq} with $\Delta G°$

K_{eq} ([P]$_{eq}$/[A]$_{eq}$)	Percent [A] at equilibrium	$\Delta G°$ (at 25 °C and pH 7.0) kcal/mol	kJ/mol
10^{-5}	99.99	+6.82	+28.5
10^{-3}	99.90	+4.09	+17.1
10^{-1}	90.91	+1.36	+05.7
10^{0}	50.00	00.00	00.0
10^{1}	09.09	−1.36	−05.7
10^{3}	00.09	−4.09	−17.1
10^{5}	0.001	−6.82	−28.5

Table 9.2 $\Delta G°$ and its relation to the position of [P] $\rightleftarrows$ [A] equilibrium

$\Delta G°$	K_{eq}	At equilibrium	Reaction
Positive	<1.0	[P]$_{eq}$ < [A]$_{eq}$	Reactant(s) favored; endergonic
Negative	>1.0	[P]$_{eq}$ > [A]$_{eq}$	Product(s) favored; exergonic
Zero	=1 0	[P]$_{eq}$ = [A]$_{eq}$	At equilibrium; no net change

$$\Delta G° = -RT \ln K_{eq} \quad \text{and} \quad K_{eq} = e^{-\frac{\Delta G°}{RT}}$$

This is an important relationship—we can determine the value of $\Delta G°$ for a reaction if its K_{eq} is known (or vice versa). A small difference in $\Delta G°$ makes a big difference in K_{eq}—this is because of the log term in the equation (Table 9.1). Thus, for a given reaction with a favorable $\Delta G°$ of −4.0 kcal/mol, there will be 1000 times more molecules of P than A at equilibrium. Finally, $\Delta G°$ informs us about the position of equilibrium (Table 9.2).

In thermodynamic terms, ΔG of a given reaction is a **state function**. It is independent of the path (or molecular mechanism) of the reaction. For instance, ΔG for oxidation of glucose to CO_2 and water is the same regardless of whether it occurs by combustion in a bomb calorimeter or through cellular metabolism. Hence, ΔG provides no information about the rate of a reaction. A negative ΔG simply indicates that the reaction can occur spontaneously.

Lastly, all the rate constants (for example, k_1 for forward and k_{-1} for reverse) contributing to the equilibrium (and the reaction mechanism) vary independently with temperature. It follows that K_{eq} for a reaction need not be the same at different temperatures; for example, it is 1.00 at 55 °C and 1.17 at 80 °C for the glucose isomerase reaction.

9.1.2 ΔG and $\Delta G°$

It is important to recall that ΔG for a given reaction depends on the concentration of reactants and products. It can be numerically larger, smaller, or the same as $\Delta G°$. We will illustrate this with two examples (Fig. 9.2).

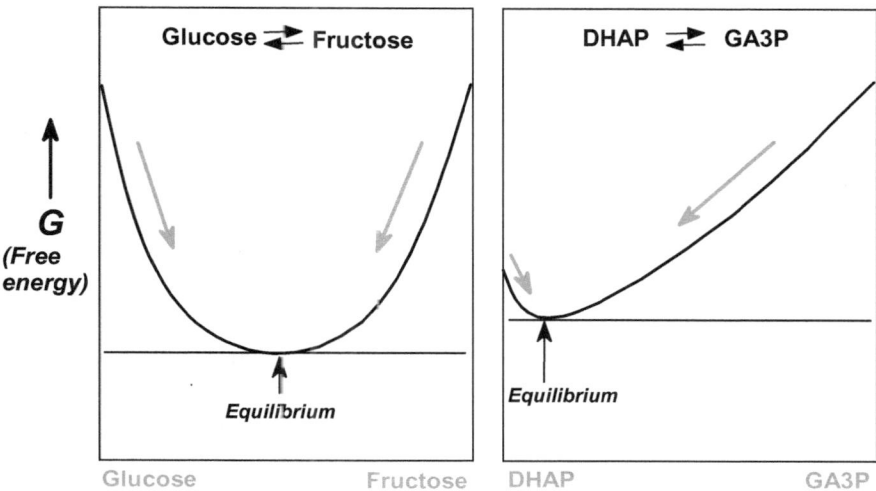

Fig. 9.2 Free energy and equilibrium constant of the reaction. The equilibrium composition of reaction mixtures corresponds to the lowest point on the curve for each reaction. Gray arrows indicate the direction of the spontaneous reaction

1. Isomerization of glucose to fructose (Fig. 9.2) has a K_{eq} of close to one, i.e., $[\text{Glucose}]_{eq} = [\text{Fructose}]_{eq}$. The standard free energy change for this reaction can be calculated by substitution ($\Delta G° = -RT \ln K_{eq} = 0$). Therefore, at equilibrium no net reaction takes place. However, with 1.0 M glucose and 0.1 M fructose as initial concentrations and at 25 °C, we obtain.

$$\Delta G = \Delta G° + RT \ln K_{eq} = 0 + RT \ln \frac{[\text{Fructose}]}{[\text{Glucose}]}$$

$$= 0 + RT \ln (0.1)$$
$$= 2.303 \times 1.987 \times 10^{-3} \times 298 \times (-1)$$
$$= -1363.67 \times 10^{-3}$$
$$= -1.364 \text{ kcal/mol}$$

   Since the ΔG is negative, under these conditions the reaction glucose→fructose is exergonic and can occur spontaneously. For initial concentrations of 0.1 M glucose and 1.0 M fructose, however, the ΔG will be +1.364 kcal/mol. Therefore, glucose→fructose is now endergonic, while fructose→glucose is exergonic and becomes spontaneous.

2. Isomerization of dihydroxyacetone phosphate (DHAP) to glyceraldehyde 3-phosphate (GA3P) occurs in glycolysis and has a K_{eq} of 0.0475. Corresponding $\Delta G°$ for this reaction (at 25 °C) is +1.80 kcal/mol and hence DHAP will not spontaneously convert to GA3P. However, when the initial concentration of DHAP is 200 μM and the initial concentration of G3P is 3 μM, ΔG becomes − 0.69 kcal/mol. Thus, under these concentrations (see Fig. 9.2) DHAP→GA3P becomes exergonic and can occur spontaneously.

The two examples drive home the message—*the criterion of spontaneity for a reaction is ΔG and not $\Delta G°$.* Continuous depletion of GA3P maintains the ΔG negative for the DHAP⇌GA3P reaction and feeds DHAP into glycolysis. Nature has exploited this principle to couple reactions of metabolic pathways; reactions are made spontaneous by adjusting the concentration of reactants and products. The direction of an equilibrium reaction is decided by suitably adjusting the mass action ratio (Γ).

9.2   Binding Equilibrium

Yet another type of equilibrium relevant to biological phenomena (including enzyme catalysis) is the binding equilibrium. To illustrate this let us consider the reversible interaction between an enzyme (E) and a small molecular ligand (L). Binding of L to E proceeds until an equilibrium is reached.

$$\text{E} + \text{L} \underset{k_{\text{off}}}{\overset{k_{\text{on}}}{\rightleftharpoons}} \text{EL}$$

As discussed before (for reaction equilibrium), at equilibrium the rates of formation of EL and dissociation of EL are equal.

$$\text{Association rate} = k_{\text{on}}[\text{E}][\text{L}] \text{ and}$$
$$\text{Dissociation rate} = k_{\text{off}}[\text{EL}]$$

At equilibrium,

$$\text{Association rate} = \text{Dissociation rate}$$
$$k_{\text{on}}[\text{E}][\text{L}] = k_{\text{off}}[\text{EL}]$$
$$\frac{k_{\text{on}}}{k_{\text{off}}} = \frac{[\text{EL}]}{[\text{E}][\text{L}]} = K_{\text{eq}}$$

This binding equilibrium is maintained by a balance between the two opposing reactions. The ratio of the rate constants for the association (k_{on}) and the dissociation (k_{off}) reactions is equal to the *equilibrium constant* (K_{eq}). Molecules of E and L must collide effectively with each other in order to produce EL. Hence, this bimolecular association rate is proportional to the product of [E] and [L].

Traditionally, the equilibrium constant is so defined that the concentration of product(s) appears in the numerator and the concentration of the reactant(s) appears in the denominator. In this sense, K_{eq} for the "E + L → EL" reaction is an *association constant* (K_{A}), also known as *affinity constant*. It has the units of M^{-1}. The larger the value of K_{A}, stronger is the binding between E and L. We may also define the K_{eq} for the "EL → E + L reaction"—accordingly called the *dissociation constant* (K_{D}). The K_{D} is the reciprocal of K_{A} and has the units of M. Obviously, the smaller the value of K_D, the stronger is the binding between E and L. This is illustrated with an example in the box below.

Suppose a fungal cell contains an enzyme with $[E] = 10^{-9}$ M and its ligand with $[L] = 10^{-6}$ M. Suppose the K_D for their binding be 10^{-7} M. Then,

$$\frac{[EL]}{[E][L]} = K_D$$

The ratio of unbound to bound [E] will be [E]/[EL] and therefore,

$$\frac{[E]}{[EL]} = \frac{K_D}{[L]} = \frac{10^{-7}\ M}{10^{-6}\ M} = \frac{1}{10}$$

Inside the cell, we thus expect one molecule of E to be free for every ten molecules of E in the bound (EL) form. What if for some reason (mutations or regulation!) the K_D changes to 10^{-4} M? We see that [E]/[EL] will then be 100. Only one molecule of E in a hundred is present as EL.

In most enzyme literature, the equilibrium constant is presented as the dissociation constant (K_D). Unless otherwise required, we will follow this convention for K_D throughout this book.

As discussed before (for reaction equilibrium), the binding equilibrium is also related to its ΔG. The free energy term is composed of entropic (ΔS) and enthalpic (ΔH) components—their individual contributions depend on the nature of interactions involved in the binding. Isothermal titration calorimetry (ITC) can be used to measure the heat effect (ΔH) of binding. A binding titration curve can be generated by varying the concentration of one of the interacting partners versus heat released (Thomas et al. 2012). This plot can be used to assess the dissociation constant (K_D) as well as the stoichiometry of binding. From K_D, the ΔG can be calculated (note that $\Delta G° = -RT \ln K_{eq}$) and from ΔG and ΔH, the entropic (ΔS) contribution toward binding can be calculated. This is an important tool in dissecting the roles of entropic and enthalpic factors in the binding of inhibitors, transition state analogs, etc., to the enzyme.

9.3 Complex Reactions Involving Intermediates

If the reaction A→P goes through several steps, then it is described as a *complex reaction*. Complex reactions may include one or more reversible steps and intermediates. Therefore, the kinetic investigation of such reactions is an analytical problem to determine the nature and number of constituent steps. Complex reactions are often described by complicated rate equations—that are not amenable to direct

Fig. 9.3 Concentration versus time curves for a reaction involving two first-order consecutive steps

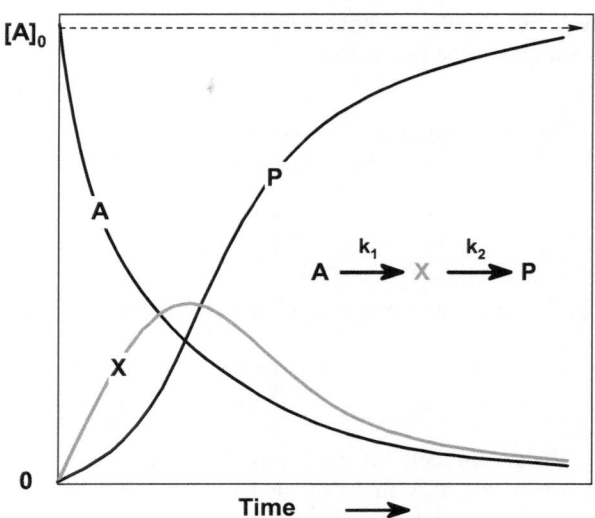

analysis. It is useful to introduce certain assumptions in order to simplify them. We will explore some of these approximations through examples.

Consider a reaction involving two first-order consecutive steps.

$$A \xrightarrow{k_1} X \xrightarrow{k_2} P$$

An intermediate X is produced in the first step and is consumed in the next. In other words, A yields P not directly but through X. Representative concentration versus time curves for A, X, and P are shown in Fig. 9.3.

As the reaction proceeds, we note that

– [A] decreases to zero at completion
– [P] increases from its initial value of zero and
– [X] builds up first, reaches a maximum, and then decreases to zero.

The position of the maximum in [X]—the extent of accumulation of X—depends on the relative magnitudes of k_1 and k_2. If $k_2 >> k_1$, then a significant accumulation of intermediate X will not occur. However, if $k_1 >> k_2$ then X is a relatively long-lived intermediate and an appreciable concentration of it may develop during the time course. This brings us to the concept of slow and fast steps of a complex reaction. We recall that rates and rate constants of an elementary step are not the same. Actual rates for the two consecutive reactions are as follows:

$$\text{For } A \rightarrow X, \text{rate} = k_1[A] \text{ and}$$

Fig. 9.4 Reaction profile for a sequential two-step overall process. When $k_2 >> k_1$ (the black curve), the first step is rate-limiting. But if $k_1 >> k_2$ (the gray curve) then the second step is rate limiting

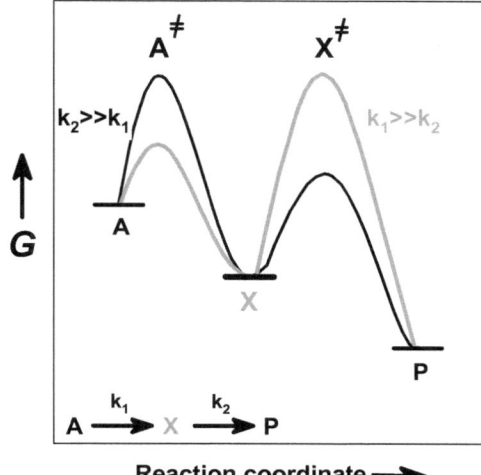

Reaction coordinate ⟶

$$\text{For } X \rightarrow P, \text{rate} = k_2[P]$$

The actual rates of the two steps therefore depend both on the rate constants as well as the concentration of the reactant species. When compared, the actual rate of the first step may be slower than the second or vice versa. If the first step (A→X) is slower than the second (X→P), then the barrier for the first step must be higher than that for the second (Fig. 9.4; black curve) (Sudi 1991). The overall reaction A→P can only be as fast as the slowest step in the sequence. Hence, the first slow step acts as a "bottleneck" and becomes the *rate-determining* (or *rate-limiting*) step of the reaction (see box below for a simple analogy). Consequently, the overall rate equation in such cases gets simplified to the rate equation for the slowest step itself. In general, ***the slowest step of a complex reaction mechanism will control the overall reaction rate***.

There are many interesting ways of illustrating the concept of the *rate limiting step* in a sequence of operations. These include—passing the baton during a relay race, assembling cycles in a factory, sequential steps in a laundry, liquid flow in pipes of different diameters, etc. We will use dishwashing as an example (Last 1985). Consider two significant operations in washing dirty dishes—scrubbing and rinsing.

Dirty Dishes → Scrubbed Dishes → Clean Dishes

In this sequence of operations, if rinsing (second step) is faster than scrubbing, then—(a) there will be very few scrubbed dishes at any given time and

(continued)

(b) overall dishwashing activity cannot be faster than that of scrubbing. For every ten dishes scrubbed in 10 min, we cannot have more than ten dishes cleaned—even if rinsing is done much faster. On the other hand, if scrubbing (first step) is faster than rinsing, then—(a) a pile of scrubbed dishes accumulates over time and (b) the overall dishwashing process cannot be faster than rinsing. For every ten dishes rinsed in 10 min, we cannot clean more than ten dishes—even if scrubbing is done much faster.

An interesting demonstration of the rate-determining step (kinetic control) is the co-translational folding of a polypeptide during protein synthesis in vivo versus in vitro. A chimeric yellow/blue fluorescence protein (YFP/BFP) gene construct was used to produce the corresponding protein. The N- and C-terminal half-domains of this chimera compete to interact with the central half-domain—the outcome of this competition determines the fluorescence properties of the resulting folded structure. Upon refolding, the denatured form of this polypeptide yields equimolar amounts of the N- and C-terminal folded structures (thermodynamic control). In contrast, when produced co-translationally in *E. coli*, the N-terminal folded structure (YFP) is formed in twice the amount compared to the C-terminal folded structure (BFP). Slowing down the translation of the C terminal half-domain (using rare synonymous codons) further increased the selection of the N terminal folded structure (YFP; increasing the amount of time available for the N-terminal and central half-domains to interact during translation). This is an example of translation rate-encoded protein folding under kinetic control (Sander et al. 2014).

9.4   How to Draw Reaction-Free Energy Diagrams

The profile shown in Fig. 9.4 provides a good visual description of the reaction in terms of the free energy (G) associated with—(a) each reactant, intermediate, and the product and (b) the various transition state species, along the reaction coordinate. However, drawing free-energy diagrams is quite challenging (Aledo et al. 2003). For enzymes, the chemistry may be understood, but a quantitative description of the free-energy profile is more difficult. For a given reaction, these steps may be followed to draw such diagrams.

1. Measure/estimate as many rate constants in the reaction coordinate as possible/required.
2. Place them sequentially on the individual steps of the reaction coordinate.
3. Relate rate constants (measured k values) to the respective $\Delta G^{\neq}$ of each step. Accordingly, draw a relative free energy profile.
4. Decide the rate-determining step(s) —such as the fastest and the slowest step in the reaction coordinate.

The entire process (and these steps) could be repeated by perturbing the reaction conditions to probe the system. If physical steps are rate-limiting, then studying the chemical steps along the reaction coordinate is a challenge. The rate-limiting step for alcohol dehydrogenase reaction is not the chemistry (the hydride transfer) but the release of the product NADH (Cleland and Northrop 1999). The mechanistic studies will now have to unmask such steps to see the chemistry. This could be done through approaches like the choice of different substrate/product concentrations, use of alternate substrates (viz., *p*-nitrophenyl acetate with chymotrypsin; see Chap. 35 especially Fig. 35.1), and studying various enzyme forms (viz., site-directed mutants of aminoacyl tRNA synthetases). Among the enzymes, triosephosphate isomerase is the best studied and the first example for its free energy diagram—with many rate constants of its individual steps estimated (Knowles and Albery 1977). A more recent view of catalysis considers the reaction coordinate as multidimensional, with a second coordinate representing the protein motion associated with catalysis.

9.5 Summing Up

As reactions become more complicated their exact kinetic solution becomes increasingly difficult. Some modifications to the reaction setup can simplify the situation to an extent. For instance, by taking a large excess of one reactant the working reaction order may be reduced (such as pseudo-first-order reactions). Introducing certain assumptions can also make the problem manageable. Two generally useful tools are the application of the *equilibrium assumption* and the *steady-state approximation*. They are helpful in deriving rate equations for complex reactions that involve simultaneous changes in three (or more) concentrations and two (or more) rate constants. In the case of the reaction A→X→P, deducing the rate equation is hampered by the fact that [X] is a continuous variable and is often difficult to measure directly. This is overcome either by assuming a rapid equilibrium between A and X (with a slow X→P step) or a steady state for [X] (after an initial induction period, the rate of formation and rate of disappearance of X are just balanced). When either of these assumptions is valid, we can express [X] in terms of initial [A]. The overall rate equation can then be deduced in terms of the initial concentration of A, the two rate constants (k_1 and k_2), and the single independent variable, time.

Equilibrium and steady state are general concepts broadly applicable at various scales of biological systems. They help us appreciate/analyze multi-step rate processes at the level of ecosystems, population growth, metabolic pathways, and enzyme forms along the reaction mechanism. We will revisit the two assumptions (equilibrium assumption and the steady-state approximation) and their utility in deriving the rate equation describing an enzyme-catalyzed reaction (Chap. 14, Henri-Michaelis-Menten equation). The trick in employing these assumptions is in appreciating their limitations and conditions under which they may not be used!

References

Aledo JC, Lobo C, del Valle AE (2003) Energy diagrams for enzyme-catalyzed reactions—concepts and misconceptions. BAMBED 31:234–236

Cleland WW, Northrop DB (1999) Energetics of substrate binding, catalysis and product release. Method Enzymol 308:3–27

Knowles JR, Albery WJ (1977) Perfection in enzyme catalysis: the energetics of triosephosphate isomerase. Acc Chem Res 10:105–111

Last AM (1985) Doing the dishes: an analogy for use in teaching reaction kinetics. J Chem Educ 62:1015–1016

Sander IM, Chaney JL, Clark PL (2014) Expanding Anfinsen's principle: contributions of synonymous codon selection to rational protein design. J Am Chem Soc 136:858–861

Sudi J (1991) How to draw kinetic barrier diagrams for enzyme-catalysed reactions. Biochem J 276:265–268

Thomas K et al (2012) Femtomolar inhibitors bind to 5′-methylthioadenosine nucleosidases with favorable enthalpy and entropy. Biochemistry 51:7541–7550

Further Reading

Klinman JP, Kohen A (2013) Hydrogen tunneling links protein dynamics to enzyme catalysis. Annu Rev Biochem 82:471–496

Patwardhan A, Marsh ENG (2007) Changes in the free energy profile of glutamate mutase imparted by the mutation of an active site arginine residue to lysine. Arch Biochem Biophys 461:194–199

Raines RT, Hansen DE (1988) An intuitive approach to steady state kinetics. J Chem Educ 65:757–759

ES Complex and Pre-Steady-State Kinetics 10

Enzymes are well-defined chemical structures at the molecular level. They cannot perform their "catalytic fete" from a distance—enzymes must come in intimate contact with their substrates (reactants). The interaction of an enzyme (E) and its cognate substrate (S) begins the moment the two come together through diffusion. Specific interactions and binding result in the formation of the enzyme–substrate (ES) complex. All the transition state(s) and intermediate(s) are represented in this simplified version of the ES complex.

$$E + S \rightarrow \text{'ES'} \rightarrow E + P$$

Changes in the concentrations of reaction participants with time, for an enzyme-catalyzed reaction, are shown in Fig. 10.1. When E and S are brought together, there is an initial buildup of the ES complex. The ES complex breaks down to form the product (P) and regenerates E. Being a catalyst, normally there will be fewer enzyme molecules in the reaction compared to those of the substrate. This sets up a distribution of enzyme molecules between E and ES forms. The amount of P formed increases with time. Eventually, the reaction equilibrium is attained. Although the enzyme continues to convert S→P and back, there will be no net change in the concentration of S or P, at equilibrium.

The system is more complicated as there may be many more intermediate forms present within the "ES" complex. The two-step scheme of enzyme catalysis in Fig. 10.1 is a grossly simplified picture. It is reminiscent of consecutive reactions (A → X → P) discussed in Chap. 9. We then noted that an intermediate cannot accumulate after the slowest step has been accomplished. However, when a step proceeds more slowly than the preceding ones, intermediate(s) do accumulate. The amount and buildup of the ES complex will thus depend on the relative rates of the two individual steps (E + S → ES versus ES → E + P) and the ratio of $[S]_{total}/[E]_{total}$. Two distinct phases of an enzyme-catalyzed reaction are seen with time. Immediately upon mixing E and S, we see a buildup of the ES complex (Grey box in

N. S. Punekar, *ENZYMES: Catalysis, Kinetics and Mechanisms*,
https://doi.org/10.1007/978-981-97-8179-9_10

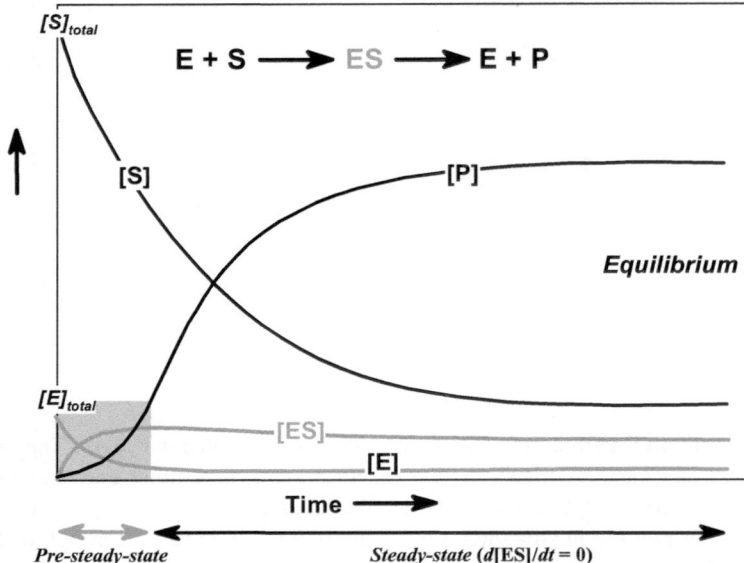

Fig. 10.1 Enzyme-catalyzed reaction: changes in the concentration of various participants as a function of the reaction time

Fig. 10.1). Thereafter, a steady state will be established where [ES] remains essentially constant with time—[ES] is maintained as long as its inflow equals the outflow (see Fig. 9.1, right panel). The steady-state region occupies an increasing fraction of the total reaction time as the $[S]_{total}/[E]_{total}$ ratio increases. Indeed, the most valuable information on enzymes has accrued through rigorous kinetic analysis in this steady-state region. The short time frame before the steady state is reached—the ***pre-steady-state***—is obviously of great importance in observing ES-complex(s), transient species, and intermediate(s).

Analysis of steady-state kinetics and the pre-steady-state kinetics complements each other in the complete understanding of an enzyme reaction mechanism.

10.1 ES Complex, Intermediates, and Transient Species

An ES complex was invoked in describing the enzyme-catalyzed reaction above. What is the compelling experimental evidence for this ES complex formation? Among many clues, the earliest of them was the ***saturation effect***. At a constant [E], the reaction rate increases with increasing [S] until it reaches a limiting, maximum value (Fig. 10.2). All the enzyme molecules occur as an ES complex at very high [S]. As the product is formed from the ES complex alone, the rate of formation of P cannot be further increased by increasing [S]. In this sense, although indirect, the saturation effect provides strong evidence for the existence of the ES

Fig. 10.2 Substrate saturation effect is a distinguishing characteristic of enzyme catalysts. Enzymatic reaction reaches a maximal velocity while the un-catalyzed rate (in gray) continues to increase with increasing reactant (substrate) concentration

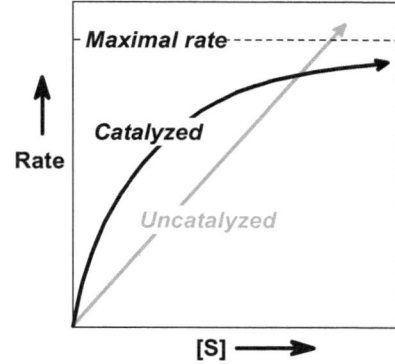

complex. Expectedly, such a saturation effect is not observed with un-catalyzed reactions (Fig. 10.2).

Apart from the characteristic saturation effect (discussed above), other lines of evidence have established the existence of the ES complex. At times, a direct *spectroscopic observation of complexes* and *intermediates* may be possible. Spectral properties of many enzymes (and substrates) change upon binding. Catalase and peroxidase (and their heme prosthetic groups) were the earliest examples studied; a catalase-H_2O_2 complex was first observed by spectroscopy. Yet another example is where the enzyme-bound pyridoxal phosphate (PLP) acted as a reporter. Tryptophan synthase upon L-serine (substrate) binding shows a marked increase in its PLP fluorescence. The E-serine complex was thus inferred from fluorescence spectroscopy. An "E-NADH-acetaldehyde" complex of alcohol dehydrogenase was deduced by monitoring the NADH absorbance. The presence of ES complex (and intermediates) may similarly be inferred through other forms of spectroscopy such as nuclear magnetic resonance (NMR), electron paramagnetic resonance (EPR), and Mössbauer. It is thus left to the imagination of the researcher to exploit one of the many forms of spectroscopy and the corresponding spectral properties of the substrate, enzyme, or both.

High-resolution *X-ray images of the enzyme active site* bound to substrates, substrate analogs, or inhibitors are now available in the public domain database (http://www.wwpdb.org/) (Berman and Gierasch 2021). Such structures provide a snapshot of what an ES complex may look like. The formation of 2-iminoglutarate as bound intermediates during the glutamate dehydrogenase catalyzed reaction was established through X-ray data by trapping them (Prem Prakash et al. 2018). Enzyme catalytic turnover (average time taken by an enzyme molecule to convert a single substrate molecule to its corresponding product) normally occurs in a fraction of a second—while the collection of X-ray diffraction data usually takes several hours. This limits the value of X-ray crystallography to a certain extent. Newer technical developments (such as time-resolved X-ray crystallography), however, permit the study of ES intermediates within seconds.

It is possible to *trap or stabilize the enzyme-bound reaction intermediate* in some cases. The Schiff base complex formed between muscle fructose bisphosphate aldolase and dihydroxyacetone phosphate (its keto-substrate) was trapped by reducing the bound imine with sodium borohydride. This classic experiment also implicated Lys as the active site residue. Dehydroquinase (3-dehydroquinate hydro-lyase) provides yet another example of trapping the ES complex as a Schiff base between the enzyme and its substrate. The acyl-enzyme intermediate of chymotrypsin was stabilized at acidic pH. Its rate of breakdown is so slow that the acyl-enzyme can be isolated and crystallized. According to X-ray data, the acyl group resides on Ser-195 of chymotrypsin. Information regarding ES complexes may also be obtained by deliberately slowing down the enzyme catalytic rates. Hydrolysis of 4-nitrophenyl acetate (a poor substrate for chymotrypsin) provides a case study. Because the formation of acyl-enzyme is fast (as compared to its breakdown!), the reaction exhibits a **burst phase** of 4-nitrophenyl acetate release followed by the slower steady state. It may be possible to **force one or more intermediate to accumulate**. Tricks used for this purpose include—(a) rapidly changing the reaction pH after the enzyme and substrate are mixed, (b) lowering the reaction temperature (cryo-enzymology), and (c) creating suitable mutant enzyme forms through site-directed mutagenesis.

10.2   Kinetic Competence of an Intermediate

In the course of the catalytic cycle, the ES complex may go through one or more intermediates. Occasionally such species may be stable enough to be isolated and characterized. Whether an enzyme reaction intermediate is stable or short-lived, it must be **kinetically competent**. The rates of formation and decay of a true reaction intermediate must be consistent with the overall reaction rate. The slowest step controls the overall rate of the reaction. Therefore, a true reaction intermediate cannot form or disappear at a rate slower than the overall reaction rate. If a postulated intermediate is so stable that it disappears more slowly than the overall rate of reaction then that intermediate is not kinetically competent. Such intermediates can at best be artifacts and are not part of the actual reaction mechanism.

10.3   Pre-Steady-State Kinetics

A great deal of kinetic insight was obtained by studying enzymes under steady-state conditions. Here, we work with conditions that permit the manual addition of enzymes to start the reaction and usually observe its progress in a spectrophotometer. This is normally carried out in a timescale of several minutes. The pre-steady-state region however is of great importance in observing ES-complex(s), transient species, and intermediate(s). Although exaggerated in Fig. 10.1 (gray box), it normally lasts for a very short time—often much less than a second. Timescales of various events associated with enzyme catalysis in general are shown in Fig. 10.3. Any event/ phenomenon is best studied by a detection technique whose response time (and the dead time of the instrument) is much shorter than the timescale of that event/

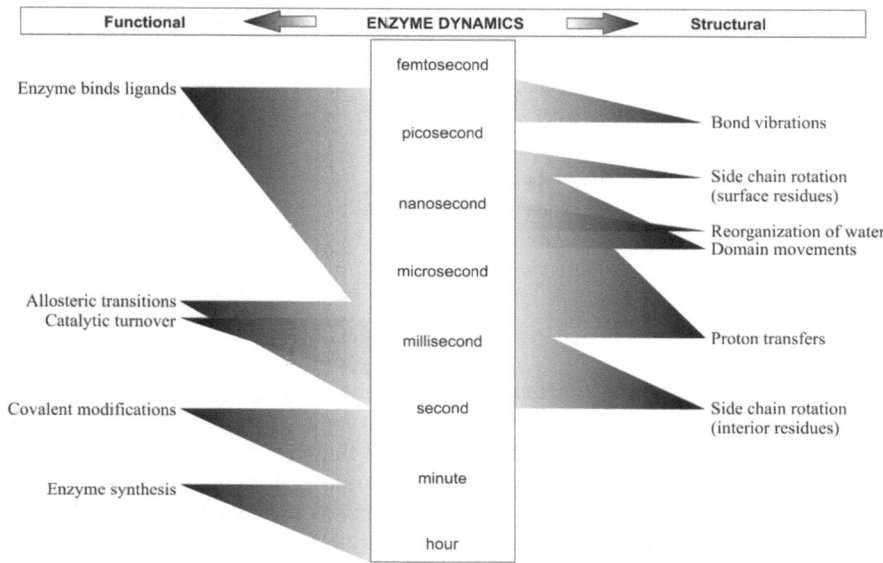

Fig. 10.3 Various events associated with enzyme catalysis. The timescales of these processes are indicated as approximate ranges

phenomenon itself. For instance, if a spectral tool (like NMR) takes several minutes to record a kinetic process, then we will miss all those events that may have occurred within a few seconds. We cannot clock a 100 m sprint using a stopwatch marked with a least count of 5 min (the Olympics record is under 10 s!). In order to observe processes that occur in the timescale of seconds, faster methods of observation are needed.

Interconversion of the reaction intermediates reflects the essential chemical steps of catalysis. Most often the steady-state kinetics is insensitive to these steps. *Fast reaction kinetics* is, however, well suited for this purpose. Ephemeral reaction intermediates that occur in the pre-steady-state stage are best monitored by suitable rapid techniques. For this reason, pre-steady-state kinetics is often synonymous with fast reaction kinetics (Fisher 2005). Special techniques are needed to examine processes that occur within seconds.

(a) **Detection methods**: In order to monitor fast events, suitable detection methods are required. Large absorbance changes, at a convenient wavelength, are often used for this purpose. Similarly, fluorescence or pH (via a pH indicator) may also be exploited. Rapid response time of detection is a prerequisite in fast reaction studies.

(b) **Techniques to reduce experimental dead time**: Technical advancements desirable to avoid manual steps are automated mixing and observation. Even with automation, there is an obvious time lapse between the first mixing of reactants (enzyme and substrate) and the arrival of the mixture in the observation chamber. This **dead time** is of the order of a millisecond.

Table 10.1 Fast reaction
kinetic techniques: a
comparison

Technique	Dead time	Time scale of operation
Flow methods		
Continuous flow	1 ms	Several seconds
Stopped flow	1 ms	Several seconds
Quenched flow	1 ms	Several seconds
Relaxation methods		
Temperature jump	1 μs	Up to few seconds
Pressure jump	1 μs	Up to few seconds
Flash photolysis	1 ps	Few μ seconds
Steady state kinetics	15–30 s	Several minutes/hour

In a stopped-flow apparatus, the enzyme and the substrate(s) are held in separate syringes. The two components are quickly brought together and mixed in an observation chamber. Change in absorbance or fluorescence is recorded and the kinetic trace is analyzed to obtain appropriate rate constants. This ***stopped-flow*** approach yields a complete time course of events as the reaction is continuously monitored. At times continuous monitoring may be difficult. In such cases, the reaction may be quickly quenched after mixing and the relevant components subsequently measured (Barman et al. 2006). Rapid cooling or denaturing agents may be used to quench the reaction. The ***quenched flow*** technique is a discontinuous method because each run yields only one point on the time course—larger amounts of enzyme are required for such experiments.

Mixing, stopping, and quenching are all steps that require finite time. A dead time of at least 0.5 ms cannot be escaped. Consequently, processes that are completed within 0.5 ms are beyond the observation limits of flow methods. We notice that some enzymes complete their single catalytic cycle within one millisecond (Fig. 10.3). Flash photolysis and relaxation methods however provide us access to sub-millisecond time scale (Table 10.1). In the ***flash photolysis*** approach, the active reactant form is generated in situ using a high-energy pulse of radiation (like a high-intensity laser beam). Suitable photosensitive precursors need to be prepared for this purpose. Caged ATP (2-nitrophenylethyl ester of ATP γ-phosphate group) is a well-documented photosensitive precursor of ATP. Another approach to overcoming the limitations of long dead times (related to mixing related issues) is through ***relaxation methods***. Here, the reaction mixture at equilibrium is subjected to a perturbation that alters its equilibrium constant. One then observes how the system relaxes to reach a new equilibrium. Temperature is the most common perturbation tool used. A relation between K_{eq} and temperature is given by the van't Hoff equation:

$$\frac{\mathrm{d} \ln K}{\mathrm{d}T} = \frac{\Delta H^\circ}{RT^2}$$

Therefore, the position of equilibrium will change due to ***T-jump*** provided the ΔH° of the reaction is not equal to zero. It is possible to produce (through electric discharge) an increase in the temperature of the reaction mixture by 10 °C in 1 μs. Other less common relaxation methods employ pH jump, pressure jump, etc. Relaxation kinetics is an elegant approach and rigorous mathematical treatment of such data is feasible. However, experimental observation of all the relevant steps may become difficult—as they might overlap. Two exponential terms (representing first-order decays!) that differ by a factor of seven or less cannot be distinguished. Unfortunately, enzyme-catalyzed reactions could contain many such steps. Assumptions of equilibrium or steady-state (Chap. 9) are often used to simplify and analyze relaxation kinetic data—a treatment that may not always be valid.

Rapid mixing techniques are useful both in the *determination of rate constants* and the *detection of transient species* during enzymatic turnover. An early application of the flow method was to study the binding of hemoglobin to oxygen. NADH binding to lactate dehydrogenase was studied by following NADH fluorescence enhancement. A $t_{1/2}$ of 2 ms was obtained for this event. This slowest step of the overall reaction was subsequently correlated to a loop movement that closes the lactate dehydrogenase active site. On the other hand, the interaction of NADH with malate dehydrogenase was titrated by a T-jump study. The magnitude of this association constant was closer to the diffusion rate—NADH binding was not rate limiting. A complete description of all the transient intermediates occurring in the dihydrofolate reductase (reduction of dihydrofolate to tetrahydrofolate by NADPH) reaction was made through stopped-flow experiments (Nashine et al. 2010).

10.4 Summing Up

The pre-steady-state kinetic analysis provides valuable information on ES-complex (s), transient species, and intermediate(s) of an enzymatic process (Fig. 10.1, gray box) (Barman et al. 2006; Fisher 2005). This, however, comes at a cost— requirements of specialized apparatus and lots of pure enzyme protein (Table 10.2). These are usually beyond the reach of most researchers. Fast reaction kinetics (and pre-steady-state experiments) are seldom used until after a basic understanding of the reaction mechanism has been obtained through steady-state kinetics and critical tests can be designed to elucidate the mechanism further. For

Table 10.2 Comparison of pre-steady-state kinetics with steady-state kinetics

Pre-steady-state kinetics	Steady-state kinetics
• Follows events leading up to the first enzyme turnover	• Reports on many enzyme turnovers
• These events are fast and kinetics tricky to interpret—Many kinetic events may be superimposed (overlaid) on each other	• Reasonable time scales; can be manipulated by suitably selecting conditions like [E], [S], and T; results simpler to interpret
• Expensive to perform and report on the events indirectly	• Cheap and directly monitors the reaction progress as it occurs
• Enzyme is viewed as a reactant and its concentrations are comparable to [S]	• Enzyme in catalytic amounts and its concentration negligible compared to [S]

these reasons, only an overview of fast reaction kinetics is given here. The practice of steady-state kinetics on the contrary is much simpler and is feasible in an average biochemistry laboratory. Therefore, steady-state kinetic tools are elaborately covered in this book. All one needs is catalytic amounts of the enzyme of interest and robust methods to assay it.

References

Barman TE et al (2006) The identification of chemical intermediates in enzyme catalysis by the rapid quench-flow technique. Cell Mol Life Sci 63:2571–2583

Berman HM, Gierasch LM (2021) How the protein data bank changed biology: an introduction to the JBC reviews thematic series, part 1. J Biol Chem 296:100608

Fisher HF (2005) Transient-state kinetic approach to mechanisms of enzymatic catalysis. Acc Chem Res 38:157–166

Nashine VC, Hammes-Schiffer S, Benkovic SJ (2010) Coupled motions in enzyme catalysis. Curr Opin Chem Biol 14:644–651

Prakash P, Punekar NS, Bhaumik P (2018) Structural basis for the catalytic mechanism and α-ketoglutarate cooperativity of glutamate dehydrogenase. J Biol Chem 293:6241–6258

Principles of Enzyme Assays

Enzymology is a quantitative and exact science. It is important to understand how enzyme activity is measured and presented. A robust and reliable measure of the progress of an enzyme-catalyzed reaction is the first and foremost requirement. Like with any other chemical reaction, the progress of an enzyme-catalyzed reaction can be monitored either by the product formed ($d[P]/dt$) or by the substrate consumed ($-d[A]/dt$). The two rates are of course related by the reaction stoichiometry. It is desirable and often safe to follow the formation of a product—a substance is better estimated when it is formed in a background where very little (or none) of it exists. On the other hand, measuring a decrease in the concentration of a reactant as it disappears—a small change in a large background—becomes daunting. In practice, a small decrease in substrate is relatively more difficult to observe than to follow a buildup of product from nothing. This is particularly relevant when we wish to record the initial rate (rate during the very early time after the reaction is initiated, abbreviated as "v"), which is given by $-d[A]/dt$ when $[P] \approx 0$. This is the rate at the beginning of the reaction or the instantaneous rate extrapolated to time zero.

11.1 Detection and Estimation Methods

Reliable methods of detection and estimation, of product formation or substrate depletion, are at the heart of a successful enzyme assay. Designing convenient and reliable assays is the first important step in studying a new enzyme activity. This, in turn, is limited by the creativity of the investigator alone. In principle, any signal that differentiates the substrate(s) or product(s) from other reaction components can form the basis for an enzyme assay. One usually looks for some physicochemical properties of the substrate or product as a handle. Spectral properties (unique to the substrate or product) are most often exploited for this purpose. The great majority of enzyme assays are based on absorption measurements. Detection methods available to follow the course of an enzymatic reaction are listed in Table 11.1. For many

N. S. Punekar, *ENZYMES: Catalysis, Kinetics and Mechanisms*, https://doi.org/10.1007/978-981-97-8179-9_11

Table 11.1 Detection methods used in enzyme assays

Technique	Detection of	Enzyme assay for
• Optical measurements		
UV spectroscopy	NADH, A_{340nm}	Alcohol dehydrogenase; Lactate dehydrogenase; Malate dehydrogenase
Visible spectroscopy	p-Nitrophenol, A_{405nm}	Alkaline phosphatase
Polarimetry	Optical rotation, $[\alpha]$	Invertase
Turbidimetry (Nephelometry)	Attenuation of incident light (intensity of scattered light)	Lysozyme
Fluorimetry	Fluorescein; $\downarrow$ at 470 nm and $\uparrow$ at 510 nm	Cholinesterase; Acylase; Chymotrypsin
Luminometry	Luciferin; $\uparrow$ at 562 nm	Luciferase
• Electrochemical measurements		
pH Meter/pH-Stat	$[H^+]$ change	Lipase; Cholinesterase; Urease; Glucose oxidase
	Carbon dioxide	Carbonic anhydrase
Potentiometry	Fe^{2+}/Fe^{3+}	Oxidase reactions (Cytochromes)
Amperometry	O_2	Oxygenases; Glucose oxidase
• Manometric measurements		
Warburg manometer	O_2 consumed, CO_2 released	Respiratory enzymes; Decarboxylases
• Radiotracer measurements		
Scintillation counter	β-emission	Dehydrogenases (^3H); Glutamate decarboxylase (^{14}C); Protein synthesis (^{35}S); Kinases, Enzymes of nucleic acid metabolism (^{32}P)

enzyme assays, these detection methods cannot be directly applied if the method does not discriminate between the substrate, product, or any of the reaction components. Then, their prior separation becomes necessary. One of the many separation techniques like chromatography or electrophoresis may be combined with one of the detection techniques. We shall refrain from describing separation techniques any further. However, a detailed treatment on this topic may be found in any text covering analytical biochemistry.

Two very commonly used tools are absorption spectroscopy and fluorescence spectroscopy. Absorption of a molecule at a particular wavelength can be related to its concentration by the Beer–Lambert law:

$$I = I_0 e^{-\varepsilon cl} \quad \text{and} \quad A = \varepsilon cl$$

where A is the absorbance ($-\log I/I_0$) of the sample at a fixed wavelength (in nm), c is the sample concentration (in molar units), and l is the path length of the light passing through the sample (usually 1 cm). The intrinsic property of a molecule ε is a constant known as the extension coefficient or molar absorption coefficient. This has

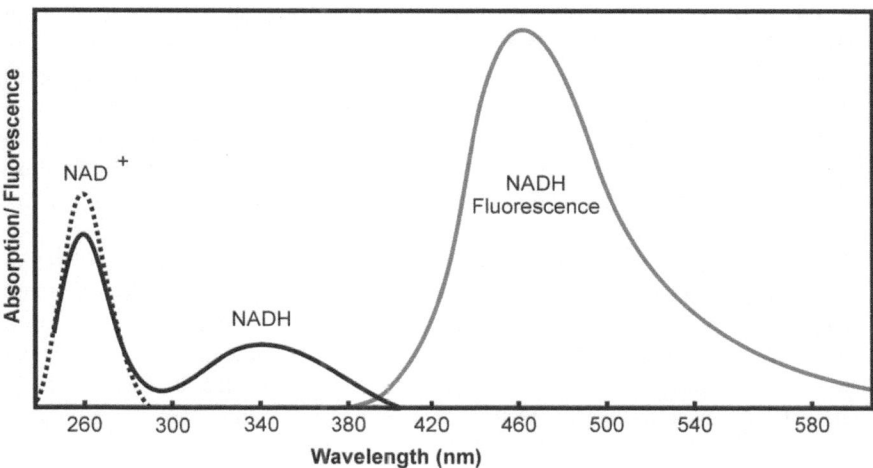

Fig. 11.1 Ultraviolet absorption spectra of reduced and oxidized NAD. Corresponding spectra for NADPH and NADP⁺, respectively, are almost identical

typically the units of M^{-1} cm^{-1}. Larger the value of ε greater is the sensitivity of that detection method. Knowing the value of ε for a given molecule and the path length (commonly the UV–visible spectrophotometer cuvettes have a path length of 1 cm), absorbance values can be directly related to concentration. For instance, many enzymes of biochemical importance involve inter-conversion of oxidized (NAD⁺) and reduced (NADH) forms of NAD. We can devise a spectrophotometric assay based on the large absorbance difference at 340 nm ($\varepsilon = 6220$ M^{-1} cm^{-1}) between them (Fig. 11.1). Suppose the absorption at 340 nm decreases (oxidation of NADH) by 0.0622 in a cuvette of 1 cm path length, in 1 min. The reaction velocity may then be expressed and calculated as shown below:

$$v = \frac{-d[S]}{dt} = \frac{-\Delta A}{\varepsilon l} \times \frac{1}{\Delta t} = \frac{0.0622}{6220 \times 1} \times \frac{1}{1} = 10^{-5} \text{ M min}^{-1}$$

The enzyme velocity is expressed here as molar per minute. By accounting for the total volume of the reaction mixture, this can also be given in units of moles per minute.

11.1.1 Suitability of a Detection Method

Each one of the methods listed in Table 11.1 has its own merits and demerits. Let us analyze them in some detail.

1. A common drawback associated with absorption measurements is a deviation from the Beer–Lambert law. The linear relationship between the absorption of the sample and its concentration holds only over a finite range of absorbance values. This has to be firmly ensured for the accuracy of the analysis. It is generally more difficult to measure a small absorption change for a sample with high initial (background) absorbance. Sample turbidity is another problem. Turbid samples show light scattering and vitiate the measurements—they need to be filtered beforehand.

2. Fluorometric assays are relatively more sensitive (by a factor of about 100) than the absorption-based methods. While absorption is measured at a single wavelength fluorescence inherently exploits two distinct wavelengths—an excitation wavelength (normally at the absorption maximum of a molecule) and an emission wavelength. Many fluorophores have large Stokes shifts—the fluorescence emission maximum is farther away (toward longer wavelengths) from the excitation maximum. For this reason, there will be fewer components that interfere with a unique fluorescence signal. Since not all molecules fluoresce, it may be necessary to attach a fluorescent tag to the substrate to develop a fluorescence-based enzyme assay. While most limitations of absorption spectroscopy also apply to fluorescence measurements there are additional caveats. Sources of interference in accurate fluorescence detection include—particulate matter, photodecomposition, polarity of the environment, temperature, and various quenching effects. Lastly, while the quantum yield of the fluorophore is an intrinsic constant for that molecule the signal obtained from a fluorimeter is relative and cannot be directly compared.

3. Any reaction leading to pH changes can be followed using a pH meter or a pH stat. Precautions are in order since enzyme reactions are strongly pH-dependent.

4. The oldest but a cumbersome manometric analysis is often substituted by other tools. For instance, decarboxylations are best measured by monitoring the release of $^{14}CO_2$ from a suitably labeled substrate.

5. The use of radioisotopes in enzyme assays invariably makes it a discontinuous method (see below). Since both the substrate and the product are radioactive an appropriate technique (chromatography or electrophoresis) to separate them becomes mandatory. The most common radioisotopes that find application in enzyme assays are β-emitters (^3H, ^{14}C, ^{32}P, and ^{35}S). Radiotracer analysis can be very sensitive but requires that suitably labeled substrates are available. Also, handling radioactivity requires experimental rigor and much care. The radioactivity is measured in a scintillation counter; the readout (in counts per min, cpm) can be converted into disintegrations per min (dpm) and this is related to concentration through specific radioactivity (such as μCi/μmol). Since scintillation counting measures light emission problems and precautions associated with fluorimetry also apply to this technique.

6. Whatever the detection method, before trusting the output provided by any instrument/machine, it is important for an experimenter to be certain that those numbers are properly calculated and are reliable.

11.1.2 Direct or Indirect Detection

In many enzyme assays, the detection techniques described in Table 11.1 can be directly applied. For example, lactate dehydrogenase catalyzes the oxidation of NADH while stoichiometrically reducing pyruvate to lactate. Concomitantly, NADH absorption (at 340 nm) decreases as a function of reaction time. Such assays are called *direct assays* because NADH (substrate) disappearance is directly measured. We may not always be lucky to devise such a direct enzyme assay. Neither the substrate nor the product of an enzymatic reaction may provide a distinct signal, convenient for measurement. It may, however, be possible to chemically convert the product into a convenient signal. Such a detection strategy is known as an *indirect assay*. Monitoring arginase and glutamate dehydrogenase activities provides two such examples. Urea, a product of arginase reaction, is converted to a yellow-colored complex (and measured at 478 nm) by reacting with dimethyl glyoxime reagent. Second, the electrons generated through glutamate oxidation (captured as NADH and transferred via phenazine methosulfate) are made to stoichiometrically reduce 2,6-dichlorophenolindophenol (DCPIP). A decrease in blue color with time is thus a good, indirect measure of glutamate dehydrogenase activity (DCPIP absorbs at 600 nm, but its reduced product does not!).

The above examples of indirect assays generate a detectable signal from the product by coupling to a nonenzymatic, chemical reaction. However, it may be possible to couple a second (sometimes even a third!) enzyme to the reaction to be observed. In such *coupled-enzyme assays* the second enzymatic reaction is chosen for its convenience of measurement. Monitoring hexokinase reaction provides a succinct example of how coupled assays are designed. These approaches are shown in Fig. 11.2. Neither glucose-6-phosphate nor Mg-ADP provides any direct means of detecting them in the background of the hexokinase assay. However, glucose-6-phosphate is a substrate for glucose-6-phosphate dehydrogenase (G6P dehydrogenase); $NADP^+$ is reduced to NADPH in the presence of this coupling enzyme and indirectly glucose-6-phosphate can be monitored at 340 nm. At times, it may be desirable to follow the other product of the hexokinase reaction. Mg-ADP can be detected by coupling two enzymes—pyruvate kinase and lactate dehydrogenase—and oxidation of NADH to NAD^+ (Fig. 11.2).

Because they involve multiple enzymes in the same reaction mixture, coupled enzyme assays are tricky to perform and require utmost care when high kinetic rigor is necessary. The following conditions have to be met for a successfully coupled enzyme assay: (a) it is necessary to identify conditions (such as a common pH) that are compatible with all the enzymes involved, (b) the reaction to be monitored should be the sole rate-limiting step and not the subsequent steps used for coupling, and (c) effects of inhibitors and other assay conditions to be tested on the enzyme of interest (like hexokinase above) should not interfere with the coupling reactions. For these reasons, rigorous kinetic analysis with coupled-enzyme reactions is difficult. However, coupled assays are convenient and simple to routinely follow enzyme activity when high accuracy is not needed—say during stages of purification. A final aspect of using coupled-enzyme assays is the quality and cost of coupling enzyme

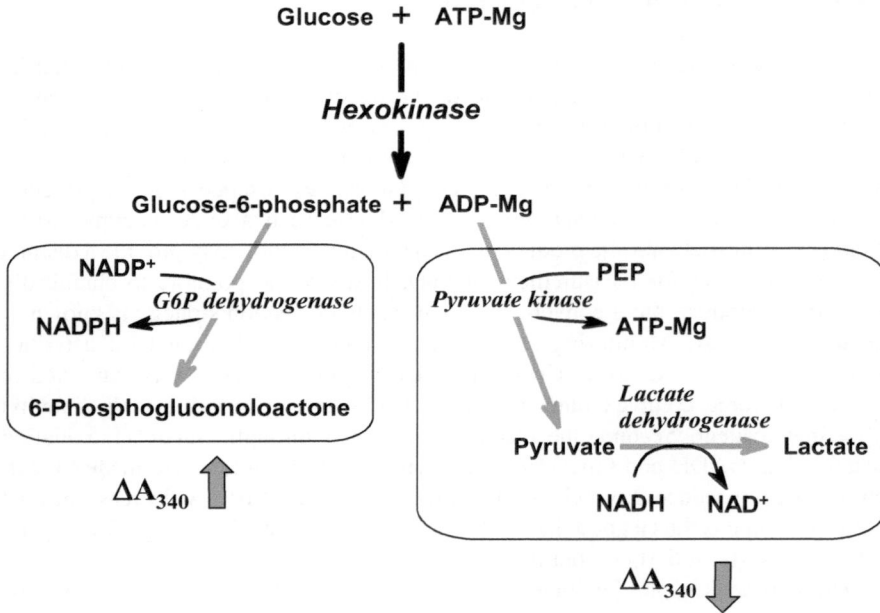

Fig. 11.2 Coupled enzyme assays to monitor hexokinase reaction. (1) Glucose-6-phosphate is detected indirectly as an increase in A_{340nm} due to NADPH formed. (2) Mg-ADP is converted to Mg-ATP while phosphoenolpyruvate (PEP) forms pyruvate. In the second coupled step, pyruvate is detected indirectly as a decrease in A_{340nm} due to NADH oxidization

(s) used. While it may be required in high amounts (this adds to the cost) for an efficient assay, its purity cannot be compromised. Simply stated, an enzyme chosen for coupling cannot have certain impurities that catalyze unwanted reactions. For instance, the estimation of glucose by glucose oxidase-peroxidase pair involves H_2O_2 as the stoichiometric intermediate. The presence of catalase as an impurity severely compromises this assay.

Nowadays, developing a method of detection from scratch is rarely required. Assays for almost every known enzyme may be found in the dedicated series *Methods in Enzymology* (Academic Press Inc.). Simplicity of operation is an important criterion for choosing a method of detection. After that, techniques that permit continuous monitoring of an enzyme reaction are desirable (see below).

11.1.3 High Throughput Enzyme Assays and Screens

A robust, simple, and sensitive enzyme assay forms the basis for designing high throughput enzyme screens (HTS). The conventional methods for evaluating enzymes—one assay at a time—are time-consuming and are considered a bottleneck in developing screens for the industry. High-throughput enzyme screens are very handy in—(a) the analysis of metabolites/inhibitors/activators, (b) screening for enzymes and enzyme-secreting cells, (c) kinetic characterization of enzymes, and (d) enzyme inhibitor and drug screening. Many enzyme assays are nowadays performed in HTS platforms to screen a large library of candidate compounds, the digitization of a single assay into millions of parallel reactions (Shao et al. 2023). Microtiter plates are the popular choice, but microfluidic arrays and droplet microfluidics are emerging as promising alternatives. Several HTS methods have been developed for fast evaluation and identification of hydrolases, dehydrogenases, and transaminases (Mathew et al. 2013). For this, it is often desirable to have enzymes with versatile features such as broad substrate specificity, high enantioselectivity, high turnover number, and the nonrequirement for regenerating cofactor. HTS approaches have also been developed to screen for specific enzymes using large pools of partially purified proteins or the proteome (Kuznetsova et al. 2005). Microarray platforms for activity profiling with a chemical strategy of active-site-directed probes are reported for proteases, oxidoreductases, and phosphatases.

Substrates (as analytes) may be assayed using suitable enzymes in two ways—the equilibrium method or the kinetic method. The equilibrium method has advantages in terms of changes in assay temperature and the presence of activators. Whereas the kinetic method scores better on the time required and the presence of products. We may compare these factors with another aspect of assay design—continuous assay versus discontinuous assay (see below). A major concern with HTS is to ensure that the number of false positives (more than the actual number of candidates selected for further study!) and false negatives (missing some good candidates!) arising out of the screen are minimized/eliminated. The substrate specificity of the chosen enzyme is a crucial parameter in the success of HTS. The primary assays could be to screen for relaxed substrate specificity followed by secondary screens with natural substrates. A grand attempt to forge a link between metabolome and genome was made by synthesizing some 2000 quenched fluorescent dye-metabolite compounds. They were used to create an array to obtain a global overview of the metabolic networks—but ignoring that different enzymes display varying degrees of substrate specificity led to its failure.

11.2 Enzyme Reaction Time Course

Armed with a convenient detection method, time courses for an enzymatic reaction are easy to conduct. Progress curves for the product formation may be generated in two distinct modes (Fig. 11.3). Reaction progress could be monitored continuously with a suitable signal and an automatic recorder. For example, the formation (or disappearance) of NADH is easily followed in a recording spectrophotometer at 340 nm. Such *continuous assays* are very desirable as they provide the safest mode of determining initial reaction rates. We may not always be lucky to establish a continuous assay for the enzyme of interest. Then, the reaction/assay must be quenched (stopped) at preselected time intervals to allow for subsequent product measurement. This second strategy of generating a progress curve is called

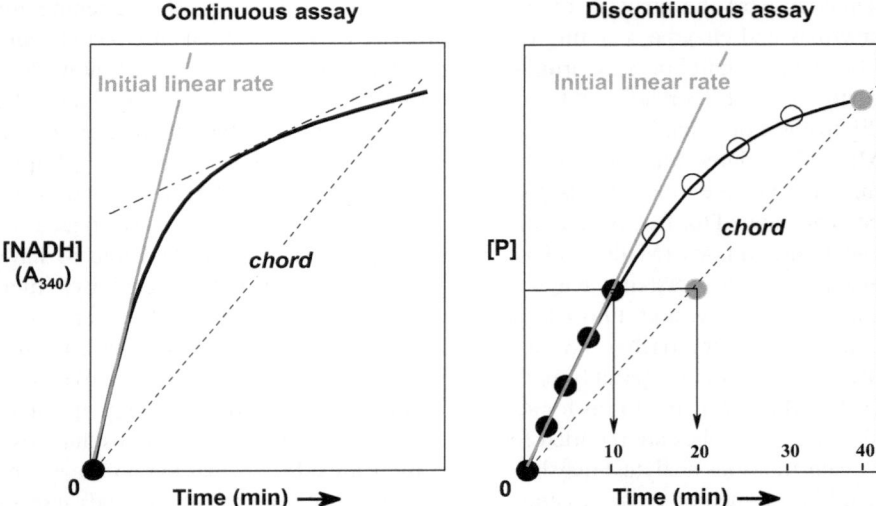

Fig. 11.3 Progress curve for an enzyme-catalyzed reaction. Estimation of the initial rate from a continuous assay is possible by aligning a tangent to the early phase of the progress curve. Tangents at any other point on the curve or a chord always underestimate the true initial velocity. In discontinuous assays, any data points beyond the linear portion of the plot (such as data beyond 10 min; open circles in the right panel) are unsuitable for rate measurements. Velocity estimates from a single point assay after 40 min (data point shown in gray, as defined by the chord) are just half of the true initial velocity

discontinuous assay (or *end-point assay*). An enzyme assay based on radioactivity invariably makes it a discontinuous assay. One has to separate the product from the remaining substrate, as both of them will contain the label.

Whenever used, the method of stopping (quenching) the reaction should be such that it completely stops the reaction. And it must not interfere with the subsequent product determination step.

A reliable progress curve is particularly relevant when we wish to record the initial rate, i.e., $-d[A]/dt$ when $[P] \approx 0$. This is the rate at the beginning of the reaction or the instantaneous rate extrapolated to time zero. It is possible to evaluate reaction velocity from the slope of a plot of signal versus time (Fig. 11.3). Obtaining a perfectly linear initial velocity for an enzyme-catalyzed reaction is a challenge. Progress curves are often *nonlinear*. This is because the reaction rate changes— usually decreases—due to the consumption of substrate(s), accumulation of product (s), loss of enzyme activity with time, etc. Uninterrupted monitoring (a continuous assay) gives a clear picture of the extent of nonlinearity. One attempts to find the initial rate (and not the average rate!) from such progress curves (Fig. 11.3). If there is nonlinearity then precise extrapolation to zero time is the only way out. This is done by drawing a tangent (and not a chord!) as close to the origin as possible. Manually, the tangent for a curve is best drawn using a glass rod. A straight line at right angles to the curve can be drawn (by aligning the glass rod such that when seen through it the curve appears continuous and without the two breaks) and then the required tangent is obtained as a line perpendicular to the first one. However, most recording spectrophotometers are equipped with programs to analyze the progress curves and provide the best estimates of initial velocity.

While continuous assays are desirable, discontinuous assays can be resorted to with due precautions. Ensuring the linearity of a reaction progress curve is much more demanding with discontinuous assays. The reason is that product measurements are made at preselected time points and these data are interpolated/ extrapolated to generate the entire progress curve. The problem is severe when single end-point assays are used. It is, therefore, necessary to exercise great care and attention—to ensure that the assay is indeed linear for the entire period. For example, proper initial velocity will not be estimated for any data point beyond 10 min in Fig. 11.3 (right panel)! If attempted, almost invariably this will be an underestimation of true initial velocity (i.e., $-d[A]/dt$ when $[P] \approx 0$). It is, therefore, mandatory to establish the extent of linearity before extracting rate information from such data.

11.2.1 Reasons for Nonlinearity

The time course of an enzyme-catalyzed reaction is initially linear, but the rate (slope of the curve!) starts to decline at later time points (Fig. 11.3). Since true initial velocity is best obtained from linear time courses it is important to understand the reasons for departure from linearity. Many of these were succinctly listed by Haldane in his early classics (Haldane 1965(1930)). If the velocity falls off during the reaction, then one or more of the following may be occurring:

1. As the substrate is continuously consumed the actual [S] falls with time. The period of linearity would, therefore, be expected to be longer at higher initial [S] values. Nonlinearity is more pronounced at low [S] and with high S→P conversion. In order to obtain linearity at lower [S] however, highly sensitive assays are required so that smaller values of [P] can be detected.
2. Increasing [P] with time leads to an increase in backward reaction (P→S) rates. The net forward rate will continue to fall until the forward and backward rates become equal, and the equilibrium is reached.
3. Product, by virtue of being derived from the cognate substrate, often interacts and reversibly binds to the enzyme to form inactive species. For instance, when NADH is a substrate for an enzyme, the corresponding product NAD^+ retains significant affinity to bind and inhibit the enzyme.
4. The assay pH may change during the course of the reaction. For instance, pH will decrease when an esterase continuously liberates an acidic product from a neutral substrate. Unless the assay system is adequately buffered this change of pH results in nonlinearity of the reaction time course.
5. Irreversible loss of one of the assay components may occur during the assay. It could be an unstable enzyme or a less stable substrate. By isolating the effect of

one component at a time, the probable cause of nonlinearity can be identified and excluded.

Increasing the straightness of a progress curve may be attempted by addressing the above issues. Fortunately, in many cases, the reaction's initial velocity remains constant for a relatively long period and hence can be measured accurately. It is desirable to observe the progress curve for a longer time period to discern the curvature properly. This helps better estimate the initial linear rate.

Despite taking precautions to overcome the different reasons for nonlinearity listed above, some enzymes show either a burst or a lag in the product formation rate before the linear phase is attained. When they are not artifacts of the assay, such a *burst phase* or a *lag phase* can give valuable information about the enzyme reaction mechanism. An excellent example of burst kinetics is the hydrolysis of *p*-nitrophenyl acetate by chymotrypsin. When *p*-nitrophenol formation is monitored there is an initial rapid formation (burst phase) followed by a slower linear (steady-state) rate. The slower linear rate is governed by the rate of hydrolysis of the acetyl-enzyme. The burst kinetics here provide useful kinetic evidence for the occurrence of an acyl-enzyme reaction intermediate during catalysis.

11.3 Precautions and Practical Considerations

Reliable kinetic data are a direct outcome of clean experimental design and good kinetic practices. To obtain meaningful results without artifacts and/or interferences—certain practical considerations are important. Although not meant to be exhaustive, some of these are listed below with suitable examples.

11.3.1 Purity of Assay Components

The purity of substrates, buffers, and other assay components plays a crucial role in the final outcome of an enzyme assay. At the least, unaccounted impurities result in overestimating the concentration of that component. For example, samples of NAD^+ purchased from suppliers often contain extraneous matter like buffer salts, moisture, etc. Obviously, the substance is not 100% NAD^+. A sample of NAD^+ from a vendor may be 95% NAD^+ the rest made up of alcohol, water, and phosphate buffer (carried over from the method of its preparation). While making stock solutions of such components one should exercise precaution. NAD^+ stocks may be routinely calibrated in a spectrophotometer directly at 259 nm and by reducing the same to NADH ($\varepsilon = 6220$ M^{-1} cm^{-1} at 340 nm). Obtaining accurate concentrations may require the use of primary standards—just the same way we make primary standards for acid-base titrations.

At times the impurity may be a potential inhibitor or activator. Vanadate (VO_4^{3-}) present in ATP samples was identified as an inhibitor of ATPase activity. It is thus desirable to use vanadate-free ATP for unbiased ATPase assays.

Mg-ATP is the true substrate of most kinases. Many of them are discriminatory and respond to submicromolar concentrations of Mn^{2+} in the presence of millimolar concentrations of Mg^{2+} ions. Magnesium salts quite often contain low levels of Mn^{2+}—this trace impurity may significantly interfere with Mg^{2+} studies.

11.3.2 Stability of Assay Components

Instability of substrate, product, or the enzyme itself can vitiate the outcome of an assay. If the substrate is unstable and is destroyed during the course of the assay, then the effective [S] would be significantly different from what is actually added. Accurate representation of the actual concentration is difficult with inherently labile molecules like superoxide anion (O_2^-) and carbamyl phosphate. Ferrous ions are inserted into porphyrins by ferrochelatase; Fe^{2+} species is susceptible to oxidation under assay conditions while Fe^{3+} is not a substrate. Measuring oxaloacetate reduction rates can be affected by its instability and loss due to decarboxylation. Similarly, if the product is labile, we may underestimate the true reaction rate. Many molecules like inorganic pyrophosphate (PPi) and H_2O_2 are unstable in the conditions of the assay—their actual concentration may be underestimated unless this loss is accounted for.

11.3.3 Nature of the True Substrate

The true substrate for a number of enzymes is a complex of the substrate and a divalent metal ion. The most widely studied example is Mg-ATP, the true substrate of most enzymes that are ATP-dependent. Similarly, other chelating substrates like citrate and isocitrate can exist in free or complex form—the two forms may be differently accepted by an enzyme. Citrate binds Mg^{2+} much more tightly than isocitrate. The presence of Mg^{2+} ions alters the apparent aconitase equilibrium since only the uncomplexed forms serve as substrates.

A compound in solution may exist in more than one form and only one of these is an effective substrate for the enzyme. The substrate aldehyde, of glyceraldehyde-3-phosphate dehydrogenase, also occurs as a hydrate. In fact, most of it in solution is in the form of the hydrate—the remaining 3% free aldehyde is the true substrate. Glutamate γ-semialdehyde (GSA)—an intermediate in the biosynthesis and catabolism of glutamate, proline, and ornithine—is another interesting example of this kind. Glutamate γ-semialdehyde exists in unfavorable equilibrium (Fig. 11.4) with its intramolecular cyclization product pyrroline-5-carboxylate (P5C). The aldehyde form itself suffers another equilibrium between the free and hydrated state. So, for an enzyme interacting with glutamate γ-semialdehyde, only a small fraction (0.05%) of

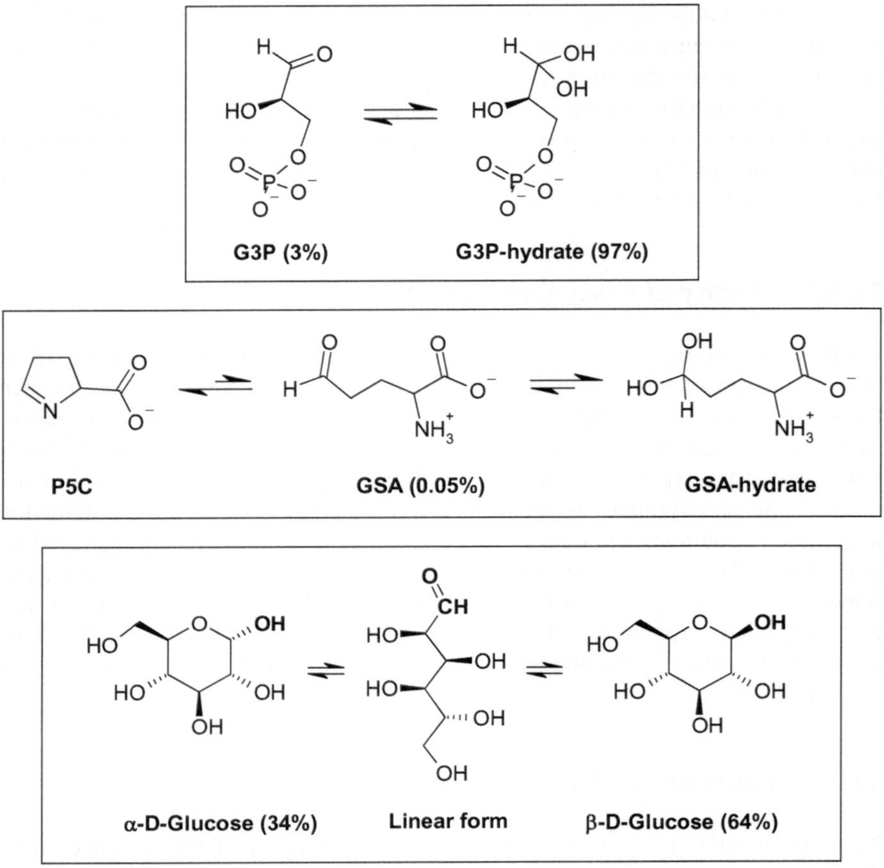

Fig. 11.4 Compounds that show inter-converting forms in solution. The true substrate for glyceraldehyde-3-phosphate dehydrogenase (G3P), glutamate γ-semialdehyde dehydrogenase (GSA), and glucose oxidase (β-anomer of D-glucose) form only a fraction of the total concentration present

the total is available in solution (Beame and Wolfenden 1995). Oxidation of glucose by glucose oxidase provides yet another illustration of substrate interconversions. In solution, three forms of D-glucose exist in equilibrium—linear chain (traces), the α-anomer (36%), and the β-anomer (64%). Only the β-anomer is acted upon by glucose oxidase.

Many enzymes act on substrates that are optically active. Racemic mixtures cannot be treated as the true substrate—only one stereoisomer (either D(R)- or L(S)-substrate) may be acted upon by the enzyme. It cannot be just assumed that effective substrate concentration is half of the total; in some cases, the wrong stereoisomer may inhibit the enzyme and complicate the kinetic data. Heterogeneity and nonspecificity of substrates should be carefully considered. Many protein kinases are routinely assayed by their capacity to phosphorylate casein; in almost all such cases the true physiological substrate is unknown.

So long as the nature of the true substrate (or inhibitor) for an enzyme is known the actual concentration can be and should be evaluated for accurate representation of data.

11.3.4 Contribution by Nonenzymatic Rates

Several substrates are inherently unstable and hence disappear with time. This contributes to the enzymatic rate when the reaction rate is measured as the disappearance of the substrate. Substrates like NAD(P)H, tetrahydrofolate, O_2^-, H_2O_2, thiol (RSH), p-nitrophenylacetate, and p-nitrophenylphosphate are unstable in solution. They get converted to the same end product as that formed by the corresponding enzymatic reaction. The reaction between different components of an assay mixture can also contribute to blank rates. For example, the nonenzymatic reaction of thiols with hydrogen peroxide is significant even in the absence of the peroxidase. Such nonenzymatic rates must be accounted for and suitably subtracted from the measured values to obtain true enzymatic rates. For instance, significant hydration of carbon dioxide occurs ($CO_2 + H_2O \rightleftarrows HCO_3^- + H^+$) in water; only after subtracting this rate from the rates obtained in the presence of carbonic anhydrase, actual enzyme-catalyzed rates are obtained.

11.3.5 Careful Examination of Interferences

Apart from their direct action on the enzyme, substrate, or product, reagents of an indirect assay can interfere with the measurements. Enzyme assay strategies based on indirect (and discontinuous) methods necessitate many controls. For instance, glucose assay using glucose oxidase—peroxidase coupled enzyme system is sensitive to many reducing/oxidizing compounds. Redox-active compounds, other than glucose, may interfere in a peroxidase reaction. Thiourea drastically reduces the color yield of Chinard's ninhydrin method (of ornithine estimation) thereby appearing as if it is an arginase inhibitor (Sudarshana et al. 2001). Therefore, a careful examination of all possible interferences in the chosen assay method becomes important.

11.3.6 Control of Assay pH, Temperature, and Ionic Strength

The activities of most enzymes are sensitive to changes in the assay parameters like pH, temperature, and ionic strength. Unless these are strictly maintained, the results of such enzyme assays are useless. We will have more to say later (Chap. 12) on how to control these parameters and design good enzyme experiments.

11.3.7 Nature of Enzyme Preparation

Several aspects of the enzyme sample used in the assay influence the measured rates. Some component(s) may be inadvertently carried into the assay along with the enzyme—this may be an activator or inhibitor. One possible outcome of their interference is the nonlinear enzyme concentration curve. A few illustrative examples of how the nature of enzyme sample matters are listed below:

1. Ammonium sulfate used to precipitate the enzyme protein may affect the enzymatic rates. While ammonium is a substrate/product of some enzymes, the high ionic strength (μ) contributed by it may activate/inhibit the enzyme activity.
2. The enzyme may have been so prepared that a significant fraction of it is in the apoenzyme form. Such subsaturated enzymes show suboptimal activity in the assay. For instance, yeast pyruvate decarboxylase loses thiamine pyrophosphate during purification and the addition of this cofactor is required to reconstitute full enzyme activity. It is not uncommon to find an enzyme that binds loosely to other cofactors like pyridoxal phosphate, and divalent metal ions.
3. A bound activator/inhibitor may be associated with or have co-purified with the enzyme. Despite extensive studies on glycolysis and phosphofructokinase over the century, an important allosteric effector (fructose-2,6-bisphosphate) was discovered only recently (in the early 1980s). While the loss of activator by dilution may decrease the measured enzymatic rate, removal of an inhibitor from the enzyme preparation results in a perceptible increase in observed rates.
4. Activities of the contaminating enzyme(s) at times interfere in the assay of the enzyme of our interest. There may be other activities in the preparation that compete for the same substrate or product. For instance, a nonspecific ATPase in a kinase preparation contributes to excess ATP hydrolysis. Similarly, the assay of dehydrogenases in crude tissue preparations becomes difficult because of the presence of nonspecific NADH oxidase activity. Monitoring glucose through a glucose oxidase–peroxidase coupled system requires that the two enzymes used are devoid of catalase contamination. Otherwise, the H_2O_2 formed is destroyed by catalase. In some cases, the contaminating enzyme may simply exploit the assay conditions and add its own rate to the true rate. Pyruvate decarboxylase activity is determined by measuring the reduction of acetaldehyde (coupled to alcohol dehydrogenase and conversion of NADH to NAD^+). Since the reduction of pyruvate by NADH can also occur, this assay system does not distinguish pyruvate decarboxylase from lactate dehydrogenase. The observed activity, therefore, should be corrected for controls performed in an identical manner but omitting alcohol dehydrogenase.
5. The discovery of DNA replication (and of DNA polymerase) is an interesting historical case of the nature of enzyme preparation leading to discovery. While the [^{14}C]thymidine incorporation into an acid-insoluble form led Severo Ochoa to polynucleotide phosphorylase, a similar approach with an *E. coli* extract steered Arthur Kornberg to DNA polymerase I. A purified enzyme and his

Commandment V (—*Do not waste clean enzymes on dirty substrates* (Kornberg 2000)) eventually resulted in the discovery of RNA priming for DNA synthesis.

Finally, one should be aware that in some cases the same enzyme may exhibit additional catalytic activities. This can be experimentally confirmed, however (see Chap. 13). For instance, the oxygenase activity of RuBP carboxylase is not due to contamination; but the two reactions are catalyzed at the same active site.

11.3.8 Enzyme Stability

Proteolysis often leads to enzyme inactivation over time. Apart from inactivation due to contaminating endogenous proteases, an enzyme preparation may lose its activity for other reasons. For instance, the presence of heavy metal ions (Hg^{2+}, Ag^{+}, or redox-active metal ions like Cu^{2+} and Fe^{2+}) may inhibit/inactivate the enzyme. Activation of fructose-1,6-bisphosphatase by EDTA (chelating agent) could be ascribed to its ability to chelate inhibitory heavy metal ions. Enzyme instability during the time course of the assay interferes and complicates initial velocity measurements. Whatever be the reason, it is of interest to know whether enzyme activity is being lost during the assay. Such enzyme inactivation can be detected by a simple test described by Selwyn (see Chap. 12).

11.4 Summing Up

A robust and reliable assay method is fundamental to the measurement of enzyme activity. It is often necessary to use different assay methods for characterizing the behavior of a single enzyme. The choice of a method also must take into account key features like sensitivity, convenience, economy, and reliability. Students are often initiated into an enzyme study by providing them with a published assay procedure. In most cases, the literature is so presented as to highlight the strengths of a method. The *controls* are taken for granted and on rare occasions, explicit mention is made of interfering factors. Careful controls, therefore, are "not extra" but absolutely essential in the use of an existing assay procedure as well as those being newly established.

While arriving at a suitable method to quantify enzyme activity the following points are borne in mind. (a) It is more convenient to precisely measure a finite increase (in the product concentration) from zero than a small decrease (in the substrate concentration) from a large initial concentration. (b) Continuous assay methods score over stopped assays. They provide a continuous readout as the reaction proceeds, thereby enabling one to detect any deviations from linearity. For single time point assays, it is mandatory to establish reaction linearity with time and enzyme concentration. (c) A detection method that directly measures the changing reactant concentration is desirable. Alternative assay strategies based on indirect measurements necessitate many more controls.

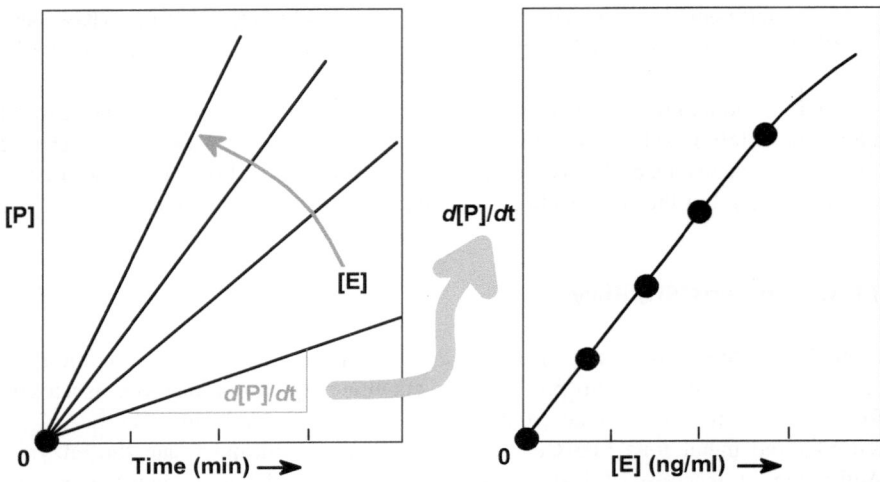

Fig. 11.5   Enzyme concentration curve. This may be constructed by plotting linear initial velocity (d[P]/d*t*) data at different [E] values

Developing an assay to make reliable initial velocity measurements is a prerequisite for the scientific and meaningful characterization of any enzyme activity. It is critical to collect good enzyme data in the first place because any degree of sophisticated analysis later will not transform bad data into good data. Keeping in view all aspects described in this chapter (and the next!), a robust assay should be chosen. The next most logical step is to generate a reliable time course for the enzyme-catalyzed reaction. From these data, at as early time points as possible, initial velocity (linear rates) may be obtained. Collecting initial velocity data at low [S] is always a challenge—as nonlinearity sets in early under these conditions. A plot of initial velocity versus [E] (the enzyme concentration curve) should then be constructed (Fig. 11.5). A linear relationship between initial velocity and [E] is a good control and a measure of the reliability of the assay. From all the iterated information, excellent data can be gathered for enzyme kinetic analysis.

Additional experimental measures, that constitute good kinetic practices, are described in the next chapter.

References

Beame SL, Wolfenden R (1995) Glutamate γ-semialdehyde as a natural transition state analogue inhibitor of *Escherichia coli* glucosamine-6-phosphate synthase. Biochemistry 34:11515–11520

Haldane JBS (1965) Enzymes. MIT Press Classics (originally publisher, Longmans, Green, 1930)

Kornberg A (2000) Ten commandments: lessons from the enzymology of DNA replication. J Bacteriol 182:3613–3618

Kuznetsova E et al (2005) Enzyme genomics: application of general enzymatic screens to discover new enzymes. FEMS Microbiol Rev 29:263–279

Mathew S et al (2013) High throughput screening methods for ω-transaminases. Biotechnol Bioprocess Eng 18:1–7

Shao F et al (2023) Emerging platforms for high-throughput enzymatic bioassays. Trends Biotechnol 41:120–133

Sudarshana S, Noor S, Punekar NS (2001) On the importance of controls in enzyme assays– an odd example. BAMBEd 29:76–78

Further Reading

Beloqui A et al (2009) Reactome array: forging a link between metabolome and genome. Science 326:252–257

Bisswanger H (2014) Enzyme assays. Perspect Sci 1:41–55

Goddard J-P, Reymond J-L (2004) Recent advances in enzyme assays. Trends Biotechnol 22:363–370

Halling PJ, Gupta MN (2014) Measurement and reporting of data in applied biocatalysis. Perspect Sci 1(1–6):98–109

Scopes RK (2002) Enzyme activity and assays. In: Encyclopedia of life sciences. Macmillan Publishers Ltd, Nature Publishing Group/www.els.net

Good Kinetic Practices

12

Generating reliable enzyme data requires a clean experimental design and good kinetic practices. Certain practical considerations are important in this quest. This chapter will describe many such aspects of experimentation.

12.1 How to Assemble Enzyme Assay Mixtures

12.1.1 Stock Solutions and Dilutions

Enzyme reaction rate, like any other chemical reaction rate, depends on the concentration of reactant(s), effector(s), pH, and ionic strength. Measurements, therefore, must begin with a precise definition of the various concentrations involved. Solutions for enzyme assays must be prepared accurately. Analytical precision may be achieved by good experimental practices like (a) differential weighing of chemicals in a calibrated balance, (b) volumetric transfers using precision pipetting aids, and (c) the use of reliable primary standards. A few standard solutions commonly employed in enzyme assays are listed in Table 12.1. Various buffers used to maintain reaction pH will be discussed a little later. Many assay components are required in very low concentrations; some of them may be hygroscopic. This necessitates the calibration of stock solutions before use. For instance, once prepared, the concentration of a stock NADH solution may be standardized by measuring its absorbance at 340 nm. From the knowledge of its molar extinction coefficient ($\varepsilon = 6220 \text{ M}^{-1} \text{ cm}^{-1}$), the actual concentration can be ascertained.

Good-quality water (double distilled or deionized) is always used to prepare solutions for enzyme assays. Most dilutions are also made in water. Few assay components are not readily soluble in water. They may have to be added to the assay as solutions in an organic solvent. In such cases, it is necessary to take suitable controls to check whether the solvent itself affects the enzyme activity or the assay method.

© The Author(s), under exclusive license to Springer Nature Singapore Pte Ltd. 2025
N. S. Punekar, *ENZYMES: Catalysis, Kinetics and Mechanisms*,
https://doi.org/10.1007/978-981-97-8179-9_12

Table 12.1 Standard solutions frequently used in enzyme assays

Component	Preparation/source[a]	Comments
NaOH	1.0 M; sodium hydroxide pellets (4.0 g) in 100 mL	Store in plastic (NOT glass!) bottle
Saline	0.9% NaCl (0.9 g in 100 mL)	Iso-osmotic with blood; commonly used as phosphate-buffered saline—With 20 mM Na, K phosphate, pH 7.4.
Ammonium sulfate	Saturated solution is 3.9 M (at 0 °C)	Used to precipitate proteins; highest ionic strength of 23.4 in water
Potassium chromate	Potassium chromate (20 mg) and KOH (1.6 g) to make a 500 mL solution	Spectroscopic standard with A_{375nm} of 0.991
Bovine serum albumin (BSA)	Crystalline BSA solution (1.0 mg/mL)	Protein standard with A_{280nm} of 0.66
Glycerol	5–50% solution (by volume)	Viscous, difficult to pipette; stabilizer, cryo-protectant
Ethylenediaminetetraacetic acid (EDTA)	100 mM stock; 373.2 mg of disodium salt in 10 mL	Chelating agent typically used at 0.1–1.0 mM
2-Mercaptoethanol	Pure liquid is 14.3 M	Thiol protectant; typically used at 1–5 mM
Dithiothreitol (DTT)	100 mM stock; 154.3 mg in 10 mL	Thiol protectant; typically used at 1–5 mM
Phenylmethanesulfonyl fluoride (PMSF)	10 mM; 1.74 mg/mL isopropanol	Stock store at −20 °C; typically used at 0.1 mM
NAD$^+$	1.0 mM; 6.63 mg in 10 mL	
NADH	1.0 mM; 7.09 mg in 10 mL	On 1:10 dilution should give A_{340nm} of 0.622
ATP	10 mM; 60.5 mg of the disodium salt in 10 mL	Kinase/synthetase substrate; used along with an excess of $MgCl_2$
Oxygen (O_2)	375 nM; solubility at 25 °C, 12 mg in 1000 mL	Substrate for oxidation and oxygenases; solubility depends on temperature and ionic strength
p-Nitrophenol	10 mM; 13.9 mg in 10 mL	Formed as a hydrolysis product of esterase and phosphatase; at pH 10.0 p-nitrophenol has $\varepsilon = 18{,}700$ at 405 nm
5,5'-Dithiobis-(2-nitrobenzoic acid) (DTNB)	10 mM; 39.6 mg DTNB in 10 mL. Prepare fresh, unstable in alkaline pH	Thiol estimation; 5-Thio-2-nitrobenzoic acid formed with $\varepsilon = 13{,}600$ at 412 nm

[a]All solutions are in water unless mentioned otherwise

It is generally desirable to prepare concentrated stock (from 5× to 200×) solutions. The idea is to provide sufficient space (volume) in the assay mixture to permit further additions. Concentrated stock solutions are also useful in minimizing significant changes (such as in terms of pH, temperature, and buffer content) to the reaction mixture upon their addition. Too large a dilution of buffer affects pH since ionization itself is concentration-dependent. Component additions made should not be more than 5–10% of the total reaction volume. Otherwise, special care is required to ensure that they are well mixed and equilibrated for pH, temperature, etc. It is usual to start the reaction by the addition of 10–20 μL of the enzyme per assay. Such volumes can be easily pipetted and do not essentially change the total reaction volume. A 10 μL addition to 1.0 mL reaction corresponds to a 1% increase in volume.

Often, it may be required to add different amounts of a component (such as while performing substrate or inhibitor saturation studies). This can be done in different ways once a stock solution is prepared. (1) Directly adding different required volumes by precision pipettes. This procedure should be discouraged because (a) different volume additions lead to volume changes, however small, and (b) pipettes come with volume ranges and are not uniformly accurate, particularly at the lower ranges. (2) Prepare a dilution series from the stock solution—such that a constant volume is added to the rest of the assay mixture. While making dilutions, it is desirable to independently pipette increasing amounts of stock solutions to each tube and make up the volumes. Serial dilution (stepwise from one dilution to the next!) should be avoided because a mistake in one tube is carried over to all the subsequent dilutions.

12.1.2 Use of Cocktails

An enzyme assay mixture may consist of several components like substrate(s), cofactor(s), metal ion, buffer, and protective agent (thiol compounds such as DTT or 2-mercaptoethanol). It makes practical sense to prepare a bulk mixture of many (or all) of these and take suitable aliquots for individual assays. Such *assay cocktails* should contain all components with the exception of one (quite often this is the enzyme) that is used to start the reaction. Assay methods employing cocktails are particularly useful—(a) to avoid pipetting mistakes and related scatter and (b) if a number of assays are to be performed under identical conditions. They are valuable when monitoring enzyme fractions from a chromatography column.

While the use of cocktails to assay enzymes can be convenient, certain precautions are in order. Adequate controls are required to ensure that (a) there are no instabilities or incompatibilities of various assay components and (b) preincubation with and/or the sequence of addition of some components have no significant contributions to the outcome of the assay.

However, mundane it may appear, accurate pipetting is a crucial determinant for good enzyme experiments. The largest contributions to measurement errors arise from pipetting mistakes.

12.1.3 Assay Dead-Time and Mixing

In a multicomponent reaction, it is essential to ensure that all the components are properly mixed at the start of the assay. Mixing in smaller volumes is not trivial. Too vigorous a shaking may lead to enzyme denaturation. When one of the components is more viscous (like a glycerol stock of an enzyme) mixing does require an effort. The contents can be satisfactorily mixed by covering the top with parafilm and inverting the tube (or the cuvette) a few times. Some enzymes (like *Bam*H I) are supplied as 50% glycerol stocks—repeated pipetting/ejecting of such samples into the small volume of an assay mixture facilitates their quick mixing. Otherwise, the dense enzyme solution settles quickly to the bottom of the tube. One other way to achieve gentle, but complete mixing is to add the initiating component to the side of the reaction vessel—as a droplet (of about 50 μL or less). At time zero, the reaction can now be initiated by simply repeated inversions of the closed vessel.

Regardless of how we mix the components to initiate the reaction, the process of mixing should be completed in a short period of time. With some practice, the fastest mixing time can be as short as 10 s. This is the ***dead-time*** of any manual assay—we cannot make any meaningful measurements before this! One can, however, adjust temperature and enzyme concentration to ensure that the reaction rate is slow enough to allow convenient (longer) time scales for measurement. One other trick is to use a substrate that is slow/poor. With *p*-nitrophenyl acetate as a substrate, the burst kinetics of chymotrypsin can be easily monitored—the acyl-enzyme accumulates significantly as the deacylation step becomes rate-limiting (acylation rate $= 37 \text{ s}^{-1}$ and deacylation rate $= 1.3 \times 10^{-4} \text{ s}^{-1}$).

In a spectrophotometric continuous assay, the reaction is started by mixing the solution, placing the cuvette in the holder, and then starting the detector by pressing a button. The time lapse between mixing and actually starting the measurement can be up to 20 s. For reaction rates occurring below the seconds scale, fast reaction kinetic tools (see Chap. 10) may be employed.

12.1.4 Order of Component Addition and Preincubations

Assays are routinely initiated by the addition of enzyme—as the last component—to the rest of the reaction mixture. However, this may not always be the best option. The reaction can also be initiated by adding (at time zero) other components like substrate(s) or cofactor(s). The actual choice of how to start the reaction will depend on one or many of the following factors:

(a) If the enzyme is unstable in the assay mixture, it should be the last component to be added.
(b) Due to compatibility issues, a component may not be suitable for inclusion in the assay cocktail. The assay format then would require that such a component be added last and made the initiating component.

(c) There may be significant blank rates—in one or more combinations of the incomplete reaction mixtures. Such combinations offer the best controls and should be used to measure blank rates.

(d) An enzyme may need enough time to establish a binding (and/or conformational) equilibrium with one or other substrate, cofactor, or inhibitor. This is best achieved by preincubating the enzyme with appropriate ligand(s) before starting the reaction with the missing component. Many enzymes bind their cofactors loosely. Therefore, a significant proportion of it upon purification is present as apoenzyme. The interaction of pyridoxal phosphate with serine hydroxymethyltransferase is one such example. Preincubation with the requisite cofactor is necessary to convert the apoenzyme fraction—into a fully active holoenzyme. The NADP-glutamate dehydrogenase from *A. niger* offers a different example. Upon preincubated with NADPH plus 2-ketoglutarate (and not individually!) and when the reaction started with ammonia, enhanced initial rates are observed. Clearly, ligand-dependent conversion to a more active form occurs during the preincubation step.

12.1.5   About Blanks and Controls

We have earlier mentioned that a common practice is to start the reaction with the addition of the enzyme. This, however, assumes that all controls and blanks have been taken into account. An assay system where all the components are present except the enzyme—whose volume is made up for with the buffer—allows the measure of nonenzymatic rates. For example, the CO_2 hydration rate in the absence of carbonic anhydrase in the assay. Any nonenzymatic rate has to be corrected for—actually subtracted from the rate recorded for the complete assay. Thus "enzyme minus" blank is an important and useful control. If blank rates occur in the absence of added enzyme, failure to subtract this blank rate results in an enzyme concentration curve that intersects the Y-axis—implying finite enzyme activity when no enzyme is present!

A "substrate minus" control should always be included. The substrate blank rate allows us to detect any time-dependent changes in the assay that are independent of the substrate conversion step (oxidation of NADH in the absence of pyruvate, in lactate dehydrogenase assay for instance). In principle, for multisubstrate reactions, "substrate minus" controls (and blanks) may be measured for each one of them. It is possible that a blank rate will only occur with certain components of an incomplete assay mixture. It is thus necessary to test for all such rates using different possible combinations.

A typical experimental design to meaningfully measure enzyme activity is shown in Table 12.2. This includes the two important controls mentioned above. While measuring the change in absorbance (ΔA) as a function of time, $\Delta A/t$ would represent the rate. Experimentally, we obtain three rates, namely, $(\Delta A/t)_{Test}$, $(\Delta A/t)_{-S}$, and $(\Delta A/t)_{-E}$. The true enzymatic rate would then be given by

Table 12.2 Design of a typical 1.0 mL enzyme assay with controls

Component	Addition in µL		
	Enzyme blank	Substrate blank	Test
Substrate (20×)	50	0	50
Buffer (10×)	100	100	100
Water	850	890	840
Enzyme	0	10	10
Rate measured	$(\Delta A/t)_{-E}$	$(\Delta A/t)_{-S}$	$(\Delta A/t)_{Test}$

$$\left(\frac{\Delta A}{t}\right)_{enzymatic} = \left(\frac{\Delta A}{t}\right)_{Test} - \left(\frac{\Delta A}{t}\right)_{-S} - \left(\frac{\Delta A}{t}\right)_{-E}$$

The two types of controls and blanks (substrate blank and enzyme blank) are essential so that their contributions can be corrected.

12.1.6 Enzyme Stability During Storage, Preincubation, and Assay

Enzymes are prone to inactivation like any other protein. They are optimally stable under specific conditions of temperature, pH, ionic strength, and the presence or absence of ligands. These parameters differ from enzyme to enzyme. Furthermore, conditions are not necessarily the same for optimal stability when compared to optimal enzyme activity. Hence, enzyme stocks should be maintained under conditions of their maximal stability. However, enzyme assays should be conducted under the conditions of optimal activity. Enzyme stocks are best stored at low temperatures but without repeated freezing and thawing. Their stability is greatly enhanced by additives like sucrose or glycerol (at 20–50% level). It is also useful to maintain suitable aliquots so that just enough enzyme is taken out to thaw before use.

At high enzyme concentrations, one may observe the "Zulu" effect—where inactivation of a significant proportion of the total enzyme may apparently go unnoticed; but below a critical concentration, there may be a sudden decrease. Immobilized enzymes may behave in this manner due to limits on substrate diffusion rates.

There are several reasons why an enzyme loses activity during the assay. The irreversible loss of enzyme may be due to specific adverse, conditions of assay like pH, temperature, ionic strength, and ligands. Apart from these, possible inactivation due to proteolysis (by contaminating protease activity, however, minor!) of the enzyme protein may occur. Enzymes, especially the intracellular kind, are found naturally at high concentrations. On dilution to much lower concentrations in an assay mixture, many of them lose activity rapidly. In dilute protein solutions, normally found in an assay with pure enzyme preparations, concentration effects come into play. Proteins bind avidly to glass or polystyrene (plastic) surfaces. Adsorption onto surfaces of containers, assay tubes, and pipette tips is a serious

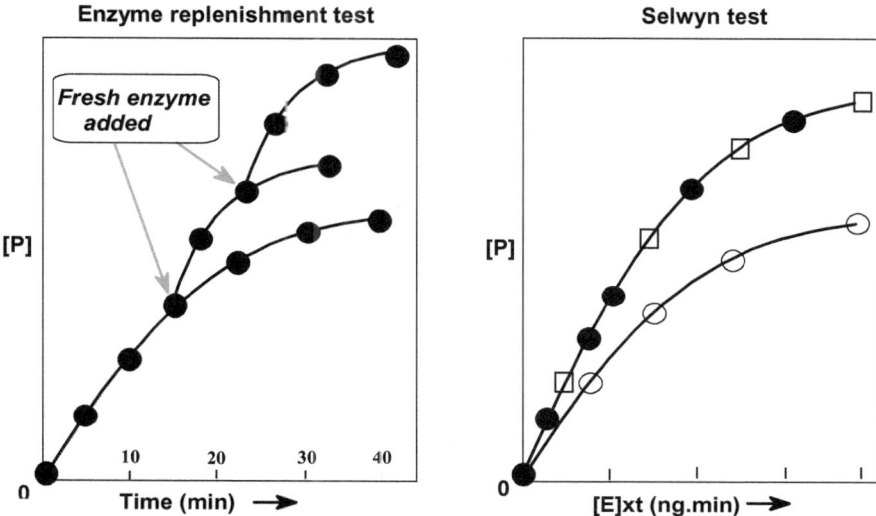

Fig. 12.1 Enzyme inactivation during assay. This may be tested by fresh addition of enzyme (left panel) and by Selwyn test (right panel). The curves for two different enzyme concentrations (● and □) superimpose when there is no inactivation. With inactivation, however, the curve for lower enzyme concentration (○) is distinct and falls off much faster

concern—particularly when the assay contains a dilute solution of the enzyme protein. Adsorbed enzyme may reflect as loss of enzyme activity—with an enzyme concentration curve intersecting the X-axis rather than passing through the origin. Therefore, it is prudent to use assay tubes and pipette tips made of low-protein binding material. Silicone-coated glassware may also be used. Another option is to add an inert carrier protein to pure enzyme samples. This ensures that potential protein binding surfaces are coated/saturated by the carrier protein. Noninterfering carrier proteins like bovine serum albumin (BSA) or gelatine (lacks aromatic amino acid residues!) are preferred as additives to enzyme stock solutions and/or assay mixtures. Many restriction enzymes are best used in a buffer fortified with BSA. For example, *Bam*H I is supplied as a 50% glycerol solution with 0.01% BSA.

Inactivation of the enzyme during assay could be a likely cause for nonlinearity. The addition of fresh enzyme aliquot, to an ongoing assay, should proportionately enhance the rate before it again falls off with time (Fig. 12.1; Punekar and Lardy 1987)). Enzyme inactivation makes the estimation of the initial rate difficult and inaccurate. The possibility of enzyme inactivation can be ruled out by a simple test described by Selwyn (1965). It is based on the fact that the value of [P] formed in an enzyme assay is determined solely by the product of time and enzyme concentration (that is: $[P] \propto [E] \times t$). A set of progress curves may be generated in which all the parameters are kept constant—except the amount of enzyme. These time courses when plotted with normalized [E] should be superimposable. The data points from all progress curves (at different [E]), when plotted as [P] versus $[E] \times t$ should fall on a single curve (**Selwyn plot**, right panel in Fig. 12.1). This implies—in order to form certain [P] in a standard assay, twice the amount of enzyme (i.e., 2 × [E]) should take half the time (i.e., $t/2$). If the enzyme is getting inactivated during the course of the assay, then the value of [E] itself decreases with time. Therefore, the data for different [E] should fall on different curves.

12.2   pH and Ionic Strength Considerations

Enzyme activity is profoundly affected by pH, buffer species used, ionic strength, and the dielectric constant of the solution. Enzyme-catalyzed reactions almost always involve ionizable groups on the enzyme and/or on the substrate. As proton transfers are crucial, maintaining a well-defined pH (H^+ concentration) for an enzyme assay is important. Since a range of pH values (between 0 and 14 in an aqueous environment) are in use, more than one kind of buffer ion may be required in an experiment. This is achieved by the judicious use of suitable buffers; some useful buffer components and their characteristics are listed in Table 12.3. Apart from the desired pH, and, hence, its pK_a, the choice of a buffer depends on many other factors. Some buffer components may have additional effects. For instance, phosphate may be an enzyme substrate/inhibitor or may chelate metal ions. Tris is known to inhibit some enzymes like succinic semialdehyde dehydrogenase. Amine buffers may form unwanted Schiff bases with substrate/product carbonyl groups. Counter ions (like SO_4^{2-}, Cl^-, K^+, Na^+, and NH_4^+) may influence the enzyme directly or through ionic strength effects.

Buffering of metal ion concentration is also a consideration in the design of enzyme assays. Besides substrates like ATP, many buffer species (for instance, phosphate and citrate) are significantly chelate divalent metal ions. This has to be

Table 12.3 Buffers frequently used in enzyme studies

Component (and its full name)	pK_a (at 25 °C)	$\Delta pK_a/\Delta$ °C
Acetate (with its Na or K salt)	4.76	+0.0002
MES[a]; 2-(N-morpholino)ethanesulfonic acid	6.15	−0.0110
Maleate (with its Na or K salt)	6.26 (pK_a2)	–
PIPES[a]; piperazine-N,N'-bis(2-ethanesulfonic acid)	6.80	−0.0085
Imidazole (with HCl)	6.95	−0.0200
Phosphate (with its Na or K salt)	7.21 (pK_a2)	−0.0028
HEPES[a]; N-(2-hydroxyethyl)piperazine-N'-(ethanesulfonic acid) (with NaOH)	7.55	−0.0140
Tris; tris(hydroxymethyl)aminomethane (with HCl)	8.06	−0.0310
Glycine (with NaOH)	9.78 (pK_a2)	−0.0250
Carbonate (with its Na or K salt)	10.33 (pK_a2)	−0.0090

[a]These pK_a values are reported at 20 °C

Table 12.4 Mixed buffer systems that span a broad pH range

pH range	Components of the mixed buffer		
5.2–8.5	MES (50 mM)	PIPES (50 mM)	HEPES (50 mM)
5.2–8.6	Maleate (50 mM)	Tris (50 mM)	–
6.0–10.0	Acetate (25 mM)	Tris (25 mM)	Ethanolamine (50 mM)
5.0–9.0	Acetate (25 mM)	MES (25 mM)	Tris (50 mM)

accounted for and suitable buffers that do not bind metal ions (like Mg^{2+}) should be used.

Effective buffering capacity is limited to ± 1.0 pH unit about the pK_a value of a given buffer species. The strength and choice of buffer used should take into account the required pH range and the nature of the reaction. Strong buffering is necessary to maintain pH with reactions that generate or consume H^+ ions (like urease and glucose oxidase). Buffers are usually prepared by dissolving the buffering component in a small volume, pH adjusted, and then making up to the desired volume—the pH of this solution is confirmed finally. Two considerations are very important in the use of stock buffer solutions: (a) ionization is affected upon dilution; it is, therefore, prudent to measure the final assay mixture pH and ensure that the required pH is reached, and (b) pK_a of a buffer species is temperature dependent; Tris is notorious for this (Table 12.3). It is thus necessary to measure/adjust the buffer pH at the same temperature at which it will be used. Finally, pK_as of ionizable species are subject to perturbation by ionic strength and dielectric constant changes. Extra caution is needed to measure/report pH values in aqueous-organic solvent mixtures (see Chap. 23).

Several buffer components were designed and synthesized (viz., HEPES) to minimize interference effects and provide a range of pH values (Good et al. 1966; Pielak 2021). Because of pK_a limitations, a single buffer system cannot be used over a wide range of pH values. Furthermore, buffer-specific effects (see above) may also exist. Effects on the enzyme due to switching of buffer species, if any, have to be eliminated. This is achieved by using either *single buffers with overlap* or *mixed buffer systems*. Two different buffers with overlapping pH ranges may be used to make measurements at the same pH. This informs us about the effects of buffer components other than those due to pH alone. Since different buffers contribute different ionic strengths, it is desirable to use mixed buffer systems (Table 12.4) that also provide for constant ionic strength.

Most enzymes display a bell-shaped pH-activity curve with maximal activity around neutral pH. However, there are enzymes with an extreme pH optimum in the acidic (such as pepsin) and alkaline (such as arginase) range as well. The decrease of activity on either side of the pH optimum may result from (a) instability of the enzyme and/or (b) changes in the kinetic parameters of the enzyme due to pH. It is important to know whether the effects of pH on enzyme activity are reversible or they result from irreversible changes leading to inactivation. Activity data in a pH range where the enzyme is rapidly losing activity are difficult to interpret—often meaningless. Enzyme pH stability can be evaluated by incubating it at different pH values (with and without substrate, effector, etc.) before readjusting to a pH where it is known to be stable. Subsequently, the activity remaining in these samples can be determined in a standard assay. Information about the stability of the enzyme over the pH range studied is necessary in designing correct kinetic studies (more on this in Chap. 23). Meaningful pH dependence of enzyme activity may be sought within a range of pH defined for stability.

Experimental determination of pH optimum (plots of pH versus activity) serves two purposes. It is of practical importance in enzyme assay optimization. Second, the ascending and descending limbs of such profiles give some idea about the range of pK_as and hence possible ionizable groups involved. Different enzymes have different pH optima—this pH optimum may be different from the physiological pH in which the enzyme functions (Bowman et al. 2020). If in vitro data are to be related to in vivo situations then it is relevant to assay the enzyme at physiological pH values.

12.3   Temperature Effects

The rate of a chemical reaction is directly affected by temperature. Usually, the rate doubles for every 10 °C rise in temperature (Chap. 8). While this is also true for enzymatic reaction rates, there is one major difference. Like any other protein, an enzyme undergoes thermal denaturation at higher temperatures. Beyond a particular temperature enzyme-catalyzed rate starts to decline—due to the inactivation of the catalyst. Hence, temperature optimum is a practically convenient expression with no absolute significance. It suggests something about the heat stability of the enzyme preparation—but is not a definite characteristic of the given enzyme sample, much less of the enzyme itself. At the temperature optimum, enzyme activity, and enzyme inactivation rates compete—leading to a maximum in the curve. *Temperature optimum* (T_{opt}) is a consequence of these two competing processes. The optimal temperature for the same enzyme may vary depending on the presence of stabilizers, pH, etc.

A temperature dependence curve (Fig. 12.2a) is more of a practical value in designing enzyme assays. Enzymes should be assayed at a temperature that is convenient, supports high activity, and does not significantly denature it. Most enzyme activities are measured at a standard temperature of 25 °C or 30 °C. However, in some cases, it may be desirable to use an appropriate physiological temperature. It is usually 37 °C for mammalian enzymes; and is upward of 72 °C for *Thermus aquaticus* DNA polymerase used in polymerase chain reaction (PCR), for example.

Assay temperatures are normally maintained by keeping the reaction mixture in a water bath. Ambient/room temperature should be better defined for the sake of reproducibility (Teixeira da Silva 2021). Temperature equilibration takes time and depends on the starting temperature of the mixture and also the nature of the reaction vessel used. Frequently, for reasons of stability, the enzyme (or some other labile component of the assay) is stored on ice. Adding such a cold component can lead to a drop in the effective temperature of the assay, and, therefore, a slower reaction rate.

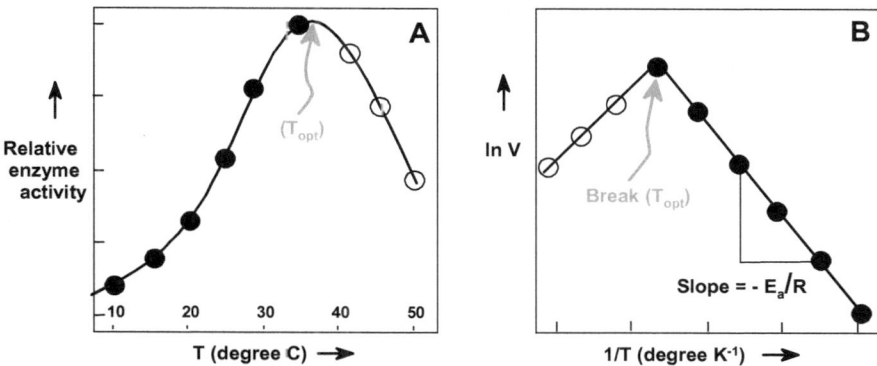

Fig. 12.2 Temperature dependence of enzyme activity. (**a**) Temperature optimum of an enzyme. Inactivation rate predominates in the descending limb (open symbols). (**b**) Arrhenius plot. The velocity data (V is maximal enzyme activity at a fixed enzyme concentration) are plotted against absolute temperature (°K) in $\ln V$ versus $1/T$ format. The gas constant $R = 8.31$ J mol^{-1} K^{-1} (or 1.98 cal mol^{-1} K^{-1})

Adequate time should be given in such cases for the assay mixture to reach the required temperature. Poor temperature equilibration can cause nonlinear reaction rates and spurious lags may be observed.

Temperature dependence of enzyme activity data can also be analyzed according to the Arrhenius equation (Chap. 8). In a temperature range over which inactivation is insignificant—a plot of $\ln V$ against $1/T$ gives a straight line (Fig. 12.2b). From its slope, we obtain the value of activation energy ($E_a = -R \times$ slope). A ***break in the Arrhenius plot*** is observed when data at higher temperatures, with loss of catalytic activity, are also included (open symbols, Fig. 12.2b). It should be possible to compare activation energies (E_a) for enzyme-catalyzed reactions with the corresponding uncatalyzed reaction. As expected, the catalyzed reactions have much lower E_a (also see Chap. 3).

The temperature stability of an enzyme can be characterized kinetically (rate of inactivation) or thermodynamically (inactivation treated as a reversible process with an equilibrium). For kinetic characterization of enzyme stability, enzyme solutions are incubated at different temperatures, and aliquots are removed at suitable time intervals. The enzyme activity in these samples is then measured (usually immediately), in a standard assay at its optimal temperature. A plot of relative activity versus temperature can be informative (Fig. 12.3). The temperature at the mid-point of inactivation (T_m) corresponds to that temperature at which half the enzyme has lost its activity. A high T_m implies a more thermo-stable enzyme form. Thermal stability could arise due to—(a) protection by a ligand or a stabilizing agent or (b) an inherently more stable enzyme (a mutant form).

Fig. 12.3 Decrease in the fraction of enzyme activity as a function of increase in temperature. The temperature at the mid-point of inactivation is shown as T_m. For an enzyme with higher thermal stability (curve with open symbols), the T_m is higher

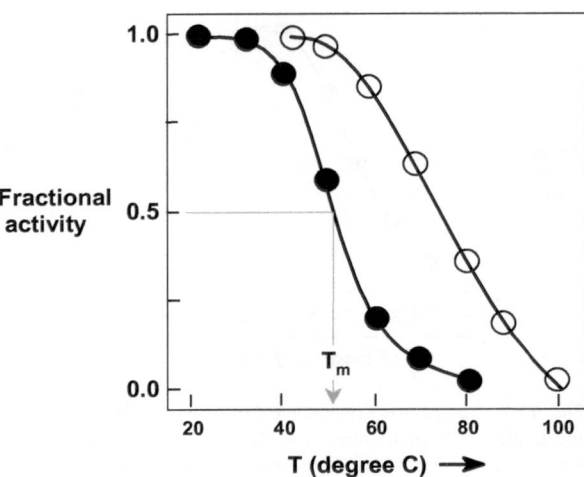

12.4   Summing Up

Many factors and a variety of artifacts influence the accuracy of enzyme activity measurements. Properly defined assay conditions of pH, temperature, and ionic strength go a long way in collecting reliable primary data on the enzyme of interest. Enzyme data without due attention to quantitative aspects are meaningless. The quality of these data begins with good kinetic practices and forms the foundation of further sophisticated analysis—with or without the use of computational support. Finally, the quality and completeness of enzyme data depend on reporting the essential metadata details (such as temperature, pH, ionic strength, concentrations of E and S, and the presence and concentrations of inhibitors/activators) and conditions under which the kinetic parameters were obtained. Standards for reporting enzymology data (STRENDA; available at http://www.strenda-db.org) are an effort to prescribe and follow the best approaches for reporting data in enzyme research (Tipton et al. 2014; Swainston et al. 2018).

References

Bowman L et al (2020) A simple and reliable method for determination of optimum pH in coupled enzyme assays. BioTechniques 68:200–203

Good NE et al (1966) Hydrogen ion buffers for biological research. Biochemistry 5:467–477

Pielak GJ (2021) Buffers, especially the good kind. Biochemistry 60:3436–3440

Punekar NS, Lardy HA (1987) Phosphoenolpyruvate carboxykinase ferroactivator: mechanism of action and identity with glutathione peroxidase. J Biol Chem 262:6714–6719

Selwyn MJ (1965) A simple test for enzyme inactivation of an enzyme during assay. Biochim Biophys Acta 105:193–195

Swainston N et al (2018) STRENDA DB: enabling the validation and sharing of enzyme kinetics data. FEBS J 285:2193–2204

Teixeira da Silva JA (2021) Room temperature in scientific protocols and experiments should be defined: a reproducibility issue. BioTechniques 70:307–308

Tipton KF et al (2014) Standards for reporting enzyme data: the STRENDA Consortium: what it aims to do and why it should be helpful. Perspect Sci 1:131–137

Further Reading

Cooper C (2010) Chapter 16. Laboratory data and SI units, in organic chemist's desk reference, 2nd edn. Boca Raton CRC Press

Visit http://www.chem.wisc.edu/areas/reich/pkatable/index.htm for comprehensive compilation of Bordwell pKa data

Quantification of Catalysis and Measures of Enzyme Purity

13

Exactness cannot be established in the arguments unless it is first introduced into the definitions.—Henri Poincare

Numerical precision is the very soul of science.—Sir D'Arcy Thompson

This is more so in enzymology, as it is a quantitative and exact branch of biology. It is necessary to understand how enzyme activity is measured, calculated, and presented.

13.1 Enzyme Units, Specific Activity, and Turnover Number

The molar concentration of an enzyme in a sample (at times even in the pure enzyme!) is often not known. However, with the help of a robust and reliable assay method reaction rates can be recorded for any enzyme sample. Assay conditions like concentration of the substrate, pH, and temperature should be well defined to obtain a reproducible initial velocity (v) (Halling and Gupta 2014). The relationship "$v\propto[E]$" holds only when true initial velocities are measured. Therefore, the first objective in any quantitative assay (and kinetic study) is to establish the two limits of linearity. These are the maximum [P] that can accumulate before the two responses, namely, $[P] \rightarrow t$ and $v \rightarrow [E]$, become nonlinear. Within these limits the measured initial velocity (v) can be used to express [E], the catalyst concentration.

Enzyme unit: To facilitate comparison of enzyme activities from various samples (and from values reported in the literature) an international unit is recommended.

The standard ***enzyme unit (U)*** is ***the amount that catalyzes the formation of one micromole of product per minute***, under defined assay conditions. This unit has the dimensions of $\mu mol \times min^{-1}$. The more the number of units in a sample means the more enzyme catalyst present in that sample. While one U of enzyme in a standard assay produces one μmol of product per min, two U of the same enzyme give

N. S. Punekar, *ENZYMES: Catalysis, Kinetics and Mechanisms*, https://doi.org/10.1007/978-981-97-8179-9_13

2.0 $\mu mol \times min^{-1}$ of the product—and so on. The enzyme concentration in a given sample is then expressed in terms of $U \times mL^{-1}$. A sample containing $2.0 \ U \times mL^{-1}$ is four times more concentrated enzyme than a sample with $0.5 \ U \times mL^{-1}$. We should note that, as defined, the catalysis unit by itself does not indicate anything about the purity of the enzyme sample.

The enzyme activity unit may also be expressed in terms of μmol substrate consumed per min. For any reaction with a defined substrate–product stoichiometry, this can be converted to the standard unit (U) described above. Some enzyme-catalyzed reactions may be relatively slow or fast. Accordingly, the unit may be redefined for convenience by suitably changing either the units for the product formed (from μmol to nmol, mmol, etc.) or the unit of time (from min^{-1} to h^{-1}, s^{-1}, etc.). If such changes are adopted for convenience, then they should be clearly documented. We find the most enzyme literature in clearly defined units. However, the International Union of Biochemists has recommended the use of katal—according to SI Units. *A katal corresponds to the amount of enzyme that produces one mole of product per second*. From the calculations (see box below), it is obvious that katal is a very large unit and hence is not in common use (Athavale 2020).

$$1 \ katal = 1 \ mol \times s^{-1} = 10^{6} \mu mol \times 60 \ min^{-1}$$
$$= 6 \times 10^{7} \mu mol \times min^{-1} = 6 \times 10^{7} U$$
$$Similarly, 1U = 16.67 \ nkatal$$

Specific activity: A way to express the amount and concentration of enzyme is through U and $U \times mL^{-1}$, respectively. These units reflect on the enzyme content of the given sample but do not tell us anything about the purity of the enzyme. The units of enzyme in a sample can be the same regardless of the quantity and diversity of other proteins present. We could, however, present the quantity (U) of enzyme present in a known amount of protein. Specific activity is thus defined as the *number of Units per mg of protein*. It is an index of the purity of the enzyme sample—the higher the proportion of enzyme protein in a given protein sample, the greater will be its specific activity. The purer the enzyme sample, the higher its specific activity. If this is extended logically to the stage of highest enzyme purity, then that sample must have every protein molecule representing only that enzyme. Beyond this point (of the limit of the highest $U \times mg^{-1}$ protein!) it is not possible to enhance the specific activity by any method of purification. Conversely, achieving the highest constant specific activity is considered a necessary criterion of enzyme purity.

Turnover number: The specific activity of an enzyme sample is expressed as $U \times mg^{-1}$ protein (note that $60 \ U \times mg^{-1}$ corresponds to 1 katal $\times kg^{-1}$). This is nothing but velocity per unit amount of catalyst protein—i.e., μmol product formed per min per mg protein. With a pure enzyme (possessing the highest limiting specific activity), the amount of enzyme protein (say in mg) can also be expressed as a number of moles of that enzyme (say in μmol). However, to do this we need to know one additional bit of information—the molecular mass of the enzyme. When this is available, we can present the specific activity (see box) of the pure enzyme.

Table 13.1 Range of enzyme turnover numbers

Enzyme (substrate)	Turnover number (s^{-1})
Catalase (for H_2O_2)	1.0×10^7
Carbonic anhydrase (for CO_2)	0.6×10^6
Ketosteroid isomerase	0.7×10^5
Urease	1.0×10^4
Triosephosphate isomerase	4.3×10^3
DNA polymerase I (*E. coli*)	6.0×10^2
Adenosine deaminase	3.7×10^2
Chorismate mutase	5.0×10^1

$$\text{Specific activity} = U \times mg^{-1}$$
$$= \mu mol \text{ product formed} \times min^{-1} \times mg^{-1}$$
$$= \mu mol \text{ product formed} \times min^{-1} \times \mu mol^{-1} \text{of enzyme}$$

This quantity—called the ***turnover number***—has the units of dimension "time^{-1}" (more commonly, s^{-1}). It indicates the number of times a single enzyme molecule converts the substrate into the product in one minute. In this definition, it is assumed that the substrate is saturating and that the enzyme has one active site per molecule. For enzymes with multiple active sites per molecule (such as lactate dehydrogenase, a tetramer with one active site per monomer) one could define a ***catalytic center activity*** by accounting for the number of active sites per subunit in the calculation. The turnover numbers of different enzymes (Table 13.1) could be compared under the best and optimal assay conditions (in terms of pH, temperature, saturating [S], etc.). For instance, an enzyme with a turnover number of 60 min^{-1} is ten times sluggish in comparison to another of 10 s^{-1}. The turnover number of catalase is among the highest known (1.0×10^7 s^{-1}). The reciprocal of the turnover number (sometimes given as k_{cat}) actually indicates the time required for a ***single catalytic cycle***—and for catalase, this is 100 ns! (Which is nothing but the reciprocal of 10^7 s^{-1}).

13.2   Enzyme Purification and Characterization

Most kinetic studies do not require a pure enzyme preparation provided there are no interfering activities. However, as we have noted from the turnover number calculations above, there are significant benefits to working with pure enzymes. The famous quote by Efraim Racker—*Don't waste clean thinking on dirty enzymes*—is at the core of molecular enzymology and good chemical practice. The availability of pure enzymes is very valuable in determining their molecular, mechanistic, and regulatory properties.

The objective of enzyme purification is to retain and enrich the enzyme protein of interest while eliminating most other proteins (and other biological macromolecules like DNA). This is typically achieved by a combination of protein separation techniques including fractional precipitation, ion exchange, size exclusion, and affinity chromatography. High throughput screening of various chromatography matrices, followed by scale-up to optimize the purification protocols, is also feasible (Kumar and Punekar 2021). An enzyme is best isolated from a source where it is abundant—for instance, lysozyme from egg white and chymotrypsin from the pancreas. With the advent of powerful molecular biology tools, most enzymes can now be produced in a suitable heterologous over-expression system, such as in *E. coli*. We also have the option of producing the protein/enzyme with or without a tag—and tags make purification a routine chore. Every tag is designed with a purification strategy in mind. A His_6-tagged enzyme is best purified on a metal affinity (Ni-NTA) column. It is a different matter, however, to ensure that enzymes with tags are—(a) active or not and (b) retain the original properties or are significantly altered. Clearly, for instance, the His_6 tag introduces the property to bind a divalent metal ion (such as Ni^{2+}, Co^{2+}, or other similar metal ion); this complicates further analysis whenever enzyme–metal ion interactions are to be studied. Proteolytic cleavage of the tag after purification is one solution—but adds an extra technical step in the process.

The theory and practice of protein (and hence enzyme) purification is a mature subject, and much literature has accumulated over the years. As the reader may access these tools through suitable books and references (e.g., Deutscher 1990; Burgess and Deutscher 2009), they are not covered here. Tools, such as fast protein (performance) liquid chromatography (FPLC), high-performance liquid chromatography (HPLC), and so on, have made protein purification much faster. With the use of multiple columns in tandem, for separation based on different principles (multidimensional chromatography) and extensive automation, protein purification became easy and reproducible. However, it must be borne in mind that whatever steps/protocols are used to purify them, it is highly desirable to consistently obtain a stable, concentrated enzyme preparation with well-defined cofactor content, etc.

Wherever possible, it is desirable to start with a sample that is intentionally enriched. Arginase is 10- to 15-fold induced in *Aspergillus niger* mycelia grown on L-arginine as the sole nitrogen source. Provided the induced enzyme form is not different, it would be prudent to start the purification with induced cells. Summary of a successful strategy for the purification of *A. niger* arginase (induced on L-arginine) is shown in Table 13.2 as an example. Excellent bookkeeping through a *purification table* is a must in monitoring the course of the purification process. Both the amount of protein (in mg) and activity (in U) are estimated in every step. With these experimentally measured data (given in Table 13.2, in black) all other parameters may be easily calculated. These derived parameters, after each step of purification, are shown in gray in the table.

Obtaining meaningful information about enzyme purification from *primary data* is at the heart of all calculations. An example of how the *derived parameters* (specific activity, fold purification, and yield) are obtained from experimental data (Table 13.2) is shown in the box below.

Table 13.2 Purification of arginase from *A. niger* mycelia[a]

Step	Total volume (mL)	Activity (U × mL^{-1})	Protein (mg × mL^{-1})	Specific activity (U × mg^{-1})	Yield (%)	Fold purified
Crude protein extract	72.5	19.4	2.47	**7.85**	**100.0**	**1.0**
(NH$_4$)$_2$SO$_4$ (30%–60%) fraction	50.0	28.8	3.15	**9.14**	**102.0**	**1.2**
DEAE-Sephacel	4.0	36.2	0.35	**103.43**	**10.3**	**13.2**
Hydroxylapatite	3.0	16.4	0.08	**205.00**	**3.5**	**26.1**

[a] Typically, this purification is initiated with 15 g of mycelia (wet weight); U of arginase is defined as the amount that produces one μmol of ornithine (product) in one min

The crude protein extract was obtained from 15 g of wet mycelial mat. Suppose 10 μL of this sample produced 0.194 μmol of ornithine (product) in one min in a standard assay. It thus contained 19.40 Units of arginase per ml and had a protein concentration of 2.47 mg × mL^{-1}.

$$19.40 \text{ U} \times \text{mL} - 1 \times 72.5 \text{ mL total volume} = 1406.5 \textbf{ total U}$$

and

$$2.47 \text{ mg} \times \text{mL} - 1 \times 72.5 \text{ mL total volume} = 179.1 \textbf{ mg total protein}$$

The specific activity of the crude extract will be

$$\frac{19.40 \text{ U} \times \text{mL}^{-1}}{2.47 \text{ mg} \times \text{mL}^{-1}} = \textbf{7.85 U} \times \textbf{mg}^{-1} \text{ protein}$$

(*The same number is obtained when we divide total U by total protein*)
After the final step, the specific activity of the purified arginase was

$$\frac{16.40 \text{ U} \times \text{mL}^{-}}{0.08 \text{ mg} \times \text{mL}^{-}} = \textbf{205.0 U} \times \textbf{mg}^{-1} \text{ protein}$$

The number of folds this enzyme got purified from the crude sample was

$$\frac{205.00 \text{ U} \times \text{mg}^{-1}}{7.85 \text{ U} \times \text{mg}^{-1}} = \textbf{26.1 fold}$$

The *final yield* of the pure enzyme was

(continued)

$$\frac{\text{Total U in pure fraction}}{\text{Total U in crude extract}} \times 100 = \frac{16.4\,\text{U} \times \text{mL}^{-1} \times 3.0\,\text{mL}}{19.4\,\text{U} \times \text{ml}^{-1} \times 72.5\,\text{mL}} \times 100 = \mathbf{3.5\%}$$

A couple of additional features can be gleaned from the purification data—provided the arginase sample after the final step is pure. If we know that all the protein is extracted from the cell mass, then the ***cellular abundance*** of arginase can be evaluated. A protein to be pure after 26.1-fold enrichment must form 3.83% of the total protein pool of *A. niger*. Second, knowing that arginase is a homohexamer (molecular mass of 219 kDa) we can calculate its turnover number. From Table 13.2, the specific activity of pure arginase is 205 U $\times$ mg^{-1} (or µmol$\times$min^{-1} $\times$ mg^{-1}). One milligram of arginase protein corresponds to 4.57×10^{-3} µmol of arginase (because 219 mg $\equiv$ 1.0 µmol). The turnover number for arginase is therefore

$$\frac{205.0\,\text{µmol} \times \text{min}^{-1} \times \text{mg}^{-1}}{4.57 \times 10^{-3}\,\text{µmol}} = 44858\,\text{min}^{-1} = 748\,\text{s}^{-1}$$

Since there are six active sites (per hexamer) the catalytic center activity should be one-sixth, i.e., 125 s^{-1}.

13.3   Interpreting a Purification Table: Criteria of Enzyme Purity

Reading a well-compiled purification table is very informative. The data provide a bird's eye view of the efficiency of each step and the progress of purification. Table 13.2 reports a small but significant increase in total activity (yield goes up from 100 to 102%). This may be because—(a) the $(NH_4)_2SO_4$ carried over from the fractionation process may be an activator of the enzyme or (b) some inhibitor(s) from the crude extract are removed by this step. The objective of each step is to take the enzyme to a higher level of purity (increase in specific activity). While this is achieved, the yield drops significantly in the DEAE-Sephacel step (leading to $\approx$90% loss). A good purification step should do both—recover most of the enzyme originally present and also enrich it. However, one may sacrifice yield for the sake of an excellent purification step. At times a good purification step is avoided because of poor yield. Overall, a well-established purification protocol should be robust, reproducible, and provide the enzyme of desired purity.

As the enzyme gets purified its specific activity will increase to a limiting value. No matter what additional steps of purification are used, this limiting specific activity cannot be bettered. A pure enzyme usually elutes from a chromatographic column as a symmetric peak with each fraction showing constant specific activity. This pure enzyme—of the highest specific activity—is a preparation, where all protein molecules present in the sample are of that enzyme alone. Native polyacrylamide gel electrophoresis is very popular among many criteria used to test the purity of an enzyme protein. Owing to its excellent resolution (based on charge and size) native

PAGE can resolve closely similar proteins—even isozymes. Often the presence of a single protein band (on native PAGE gels) is used to describe the electrophoretic homogeneity of the purified sample. An enzyme found electrophoretically pure, however, does not necessarily guarantee that all the molecules present are active. An *active homogeneous enzyme* means—only enzyme molecules are present and every protein molecule in the sample is enzymatically active. For example, in a sample containing 100 molecules of protein of which 50 are of the enzyme—then in principle—the specific activity of this sample can be doubled by eliminating those molecules which are not the enzyme. Similarly, even if a given enzyme sample is homogeneous, it can in principle contain both active and inactive forms of the same enzyme. There is thus scope to increase the specific activity of this sample by eliminating the inactive enzyme molecules from it. One way of achieving this is to use a *functional affinity separation* that selectively binds only the active enzyme species. Subsequently, bound molecules (of active enzyme) may be collected by suitable elution protocols. In many cases, it is possible to estimate the amount of active enzyme in the given sample by *active site titration*. Any agent that stoichiometrically reacts with the active site may be used to determine the number/concentration of active sites. The number of active molecules in an acetylcholinesterase preparation was determined using ^{32}P-labeled diisopropylfluorophosphate (DIFP). A highly reactive fluorogenic leaving group built into an unsaturated cyclitol ether (a slow substrate designed to form a covalent "glycosyl"—enzyme intermediate) labels the active site of α-amylases; the amount of fluorophore released can then be used to quantify active sites (Sweeney et al. 2020). The size of the burst observed, when chymotrypsin acts on *p*-nitrophenylacetate, can also be used as a measure of active enzyme in a homogeneous preparation of chymotrypsin.

A homogeneous, pure enzyme sample allows us to infer about the molecular details of the catalyst. Their oligomeric state can be deduced by a combination of techniques including denaturing PAGE (such as SDS-PAGE). Enzymes come in different designs. Lysozyme, RNase A, and chymotrypsin are straightforward examples of *single subunit-single active site* enzymes. Not every subunit of an oligomeric protein may contain one active site. HIV protease (and possibly proline racemase) is dimeric but contains a single active site. The active site of the homodimeric mammalian ornithine decarboxylase spans both subunits. The lone active site of *E. coli* RNA polymerase is generated from a holoenzyme made up of five ($\alpha_2\beta\beta'\sigma$) subunits. Mitochondrial ATP synthase consists of 22 subunits made from eight distinct polypeptide chains. It is thus possible to define an *enzyme equivalent weight*—it is the mass of a protein expressed in grams per mole of active sites. For lysozyme, it is the same as its molecular mass whereas it will be the mass of each subunit in the case of lactate dehydrogenase tetramer.

13.4   Unity of the Enzyme

It should be obvious from the above discussion that enzyme-specific activity, is an important measure of enzyme purity. The highest attainable specific activity coupled with electrophoretic homogeneity tells us that the enzyme sample is pure. There are limits to knowing how much of a protein contamination is present. A contaminant that is less than a fraction of a percent may not be noticed on native PAGE gels. Enzyme activity assays are inherently more sensitive and hence contaminating proteins may still be detected as additional interfering activities in the sample. It is equally possible that the "contaminating" activity may be an intrinsic property (side reaction!) of the enzyme itself (Table 13.3). There are documented examples where a single enzyme protein displays multiple activities—either at the same active site or on distinct sites (Kirschner and Bisswanger 1976). These also include multifunctional polypeptides (a single polypeptide harboring more than one distinct enzyme

Table 13.3 Enzymes exhibiting multiple activities

Enzyme example	Activities
One active site with many activities	
Glutamine synthetase	Glutamine synthesis γ-Glutamyl transfer
RuBP carboxylase	Carboxylase Oxygenase
Hexokinase	Phosphate transfer to sugar ATP hydrolysis (weak)
Sucrose phosphorylase	Sucrose phosphorylase Transglucosylase
Fructose-1,6-bisphosphate (FBP) aldolase/ phosphatase (from Archaea)	FBP aldolase FBP phosphatase
One enzyme with many active sites	
A. *Multifunctional polypeptides*	
Aspartokinase-homoserine dehydrogenase	Aspartokinase Homoserine dehydrogenase
DNA polymerase I (*E. coli*)	$5'$–$3'$ DNA polymerase $3'$–$5'$ exonuclease (proof-reading) $5'$–$3'$ exonuclease (repair)
Fatty acid synthase (type I; mammalian)	Seven different activities
6-Phosphofructo-2-kinase/fructose-2,6- bisphosphatase	Synthesis and degradation of F-2,6-BP
The *arom* complex	Five different activities (aromatic amino acid biosynthesis)
B. *Multienzyme complexes*	
Pyruvate dehydrogenase	Pyruvate decarboxylase Transacetylase Dihydrolipoamide dehydrogenase
Fatty acid synthase (type II; *E. coli*)	Seven different activities

activity) and multienzyme complexes (many polypeptides form oligomeric structures with more than one distinct enzyme activity).

Although not simple, it should be possible to demonstrate that the same polypeptide (or multienzyme complex) is responsible for the main reaction as well as the other reaction(s), if any. The criteria of unity of an enzyme may be confirmed by one or more of the following protocols:

(a) All the different activities exhibited by the same enzyme (with one or more active sites) co-purify during various chromatographic separations. The ratio of their specific activities remains constant through various steps of purification. Since a pure enzyme elutes as a symmetric peak of constant specific activity, other activities of the same protein also behave similarly. A contaminating activity can be resolved, in principle, by one or the other separation step.

(b) The two activities displayed by the same active site are not purely additive because one substrate becomes a competitive inhibitor of the other.

(c) A potent inhibitor affects two activities in parallel if they are due to the same active site. Inactivation (by heat, chemical agents, etc.) studies demonstrate a simultaneous loss of both activities. Similarly, both activities are affected in the same way by the presence/absence of cofactors, if any.

(d) Differential proteolysis sometimes provides clues to the relationship between various activities of an enzyme. If they are due to the same active site, their kinetics of inactivation will be superimposable. Activities residing in different domains of a protein (e.g., multifunctional polypeptide) may be separated by limited proteolysis. For instance, various activities of mammalian fatty acid synthase can be released as separate fragments. Also, the Klenow fragment (corresponding to 324–928 amino acid residues) from *E. coli* DNA polymerase I is obtained by the release of the first 323 residues. Accordingly, the Klenow fragment is missing the $5'–3'$ exonuclease (repair) function.

(e) Replacement of an essential amino acid residue from the enzyme active site, through a site-directed mutagenesis (SDM) approach, should in principle affect all the activities due to that active site. For instance, both the carboxylase and the oxygenase activities of RuBP carboxylase are knocked off simultaneously by the replacement of a single active site residue. With respect to enzymes having separate active sites, however, the situation will be different. The homoserine dehydrogenase activity of aspartokinase-homoserine dehydrogenase can remain unaffected by SDM at the aspartokinase site and vice versa. The SDM tool thus is a powerful approach to establishing the nature of multiple activities of the same protein—on the same active site or different active sites.

Moonlighting and promiscuous enzymes: The absolute enzyme specificity is more often an exception rather than a rule. Two major kinds of enzyme/protein multispecificities are promiscuity and moonlighting (Jeffery 2020). A single enzyme protein could have multiple activities—either at the same active site or on sites distinct from the active site. Such activities are closely related to each other—either as part of the same function or as related reaction steps in the metabolism (see

Table 13.4 Moonlighting activities of common metabolic enzymes

Enzyme example	Moonlighting activity
Structural components	
Lactate dehydrogenase, argininosuccinate lyase, enolase, GSH S-transferase	Lens crystallins
Transcriptional/translational regulation	
Aconitase	Iron-responsive-element binding protein (IRE-BP)
E. coli thioredoxin	Subunit of T7 DNA polymerase
Proline dehydrogenase (PutA)	Transcriptional repressor
Fructose 1,6-bisphosphatase 1 (FBP1)	Nuclear protein phosphatase
DNA repair and maintenance	
Glyceraldehyde 3-phosphate dehydrogenase, aconitase, fumarase	Cytosolic/nuclear component of the DNA damage response
Differentiation and maturation	
Phosphoglucose isomerase	Neuroleukin, autocrine motility factor, differentiation, and maturation mediator
Ribonuclease 5 (angiogenin)	Angiogenesis (new blood vessels)

Table 13.3). Many enzymes are known to "moonlight" and are found to serve additional functions that are not related to their catalytic aptitude. The protein structural features outside of the enzyme active site often participate in their moonlighting action. The moonlighting functions of enzymes/proteins were usually discovered by chance. Otherwise, in normal members of the intermediary metabolism, the "moonlighting" activities of enzymes may involve structural and/or regulatory role(s) within or outside the cell. Enzymes known to display moonlighting activities with well-defined functions include transcriptional/translational regulation, differentiation and maturation, DNA repair and maintenance, as growth factors and structural components. Aminoacyl-tRNA synthetases, which help translate the genetic code into a polypeptide, are known to also function in angiogenesis, fat metabolism, etc. A few well-known examples are listed in Table 13.4 and more may be found at the website http://moonlightingproteins.org/ and the reviews on this subject (Jeffery 2020; Sriram et al. 2005; Gupta and Uversky 2023).

Besides their moonlighting activities, it is being recognized that many enzymes are also catalytically promiscuous. An enzyme from the methionine salvage pathway is known to catalyze two distinct reactions depending on which metal ion is bound to the apoenzyme—the metal component dictating the reaction specificity (Dai et al. 1999). Due to their catalytic promiscuity, such enzymes are capable of catalyzing secondary (unrelated) reactions at an active site that is specialized to catalyze a primary reaction. The potential for catalytic promiscuity (see Chap. 3) can be an advantage in generating novel catalysts for industry. Both moonlighting by and catalytic promiscuity of enzymes are valuable playing fields for evolution to work (Copley 2003; Copley et al. 2023). They also help in our understanding of enzyme structure/function relationships and the directed evolution of new functions from existing protein scaffolds (see Chap. 37).

13.5   Summing Up

Quantitative tools to measure the amount of catalyst (enzyme) present in a given sample are a prerequisite to all enzymology. Well-defined units allow us to express the amount of enzyme in a sample and its relative purity. With good bookkeeping practices, it is possible to follow the course of enzyme purification. A qualitative and quantitative analysis of a pure enzyme gives us the first description of its molecular features. While it is desirable, a homogeneous preparation of an enzyme is not an absolute necessity for kinetic analysis. Ensuring that the contaminants are not interfering is sufficient. In fact, cell extracts are by their nature "dirty enzymes"; they contain other enzymes that act before and after the enzyme of our interest acts. The study of enzymes in intact cells and organisms is ultimately necessary to take the in vitro enzyme data into in vivo situations (see Chap. 39).

References

Athavale SV (2020) The katalytic speedometer. Nat Phys 16:704

Burgess R, Deutscher M (eds) (2009) Guide to protein purification. In: Methods enzymology, vol 463, 2nd ed. Academic, San Diego

Copley SD (2003) Enzymes with extra talents: moonlighting functions and catalytic promiscuity. Curr Opin Chem Biol 7:265–272

Copley SD, Newton MS, Widney KA (2023) How to recruit a promiscuous enzyme to serve a new function. Biochemistry 62:300–308

Dai Y et al (1999) One protein, two enzymes. J Biol Chem 274:1193–1195

Deutscher MP (ed) (1990) Guide to protein purification. In: Methods enzymology, vol 182. Academic, San Diego

Gupta MN, Uversky VN (2023) Moonlighting enzymes: when cellular context defines specificity. Cell Mol Life Sci 80:130

Halling PJ, Gupta MN (2014) Measurement and reporting of data in applied biocatalysis. Perspect Sci 1:98–109

Jeffery CJ (2020) Enzymes, pseudoenzymes, and moonlighting proteins: diversity of function in protein superfamilies. FEBS J 287:4141–4149

Kirschner K, Bisswanger H (1976) Multifunctional proteins. Annu Rev Biochem 45:143–166

Sriram G et al (2005) Single-gene disorders: what role could moonlighting enzymes play? Am J Hum Genet 76:911–924

Further Reading

Dance A (2020) Protein synthesis enzymes have evolved additional jobs. Scientist

Gerber SA, Kettenbach AN (2022) Metabolic phosphatase moonlights for proteins. Nat Cell Biol 24:1568–1570

Khersonsky O, Roodveldt C, Tawfik D (2006) Enzyme promiscuity: evolutionary and mechanistic aspects. Curr Opin Chem Biol 10:498–508

Rider MH et al (2004) 6-Phosphofructo-2-kinase/fructose-2,6-bisphosphatase: head-to-head with a bifunctional enzyme that controls glycolysis. Biochem J 381:561–579

Kumar S, Punekar NS (2021) High throughput screening of dye-ligands for chromatography. Methods Mol Biol 2178:35–47

Sweeney RP et al (2020) Development of an active site titration reagent for α-amylases. Chem Sci 12:1–5

Henri–Michaelis–Menten Equation

14

The only criterion of a model is usefulness, not its "truth"—Michael Dewar

Models are not descriptions of reality; they are descriptions of our assumptions about reality.—Gunawardena J (MBoC, 2012)

A rate equation (or the rate law) gives the experimentally observed dependence of rate on the concentration of reactants. Rate equations are at the heart of any kinetic study as they help us describe the system in a mathematical formalism. This is true for enzyme catalysis as well. Besides its esthetic beauty, the compact mathematical description of reaction kinetics serves the twin purposes of qualitative description of the system and quantitative evaluation of rate constants. An early attempt to capture the kinetics of enzyme catalysis was made by Victor Henri (in 1903, Chap. 2) and this was subsequently developed by Leonor Michaelis and Maud Menten (in 1913). The rate equation so described is a fundamental equation of enzyme kinetics and goes by the name Henri–Michaelis–Menten equation. It is more commonly referred to as the Michaelis–Menten equation. The derivation of the rate equation for a simple, single-substrate enzymatic reaction is especially instructive. In the process, it describes the general logic used to derive such rate equations—an exercise central to any enzyme kinetic study. This chapter will describe the development, significance, and salient features of the Michaelis–Menten equation.

14.1 Derivation of the Michaelis–Menten Equation

Initial clues to understand enzyme-catalyzed reaction rates came from the saturation effect. At a constant $[E_t]$, the reaction rate increases with increasing $[S]$ until it reaches a limiting, maximum value (Fig. 10.2). In contrast, the reaction rate increases linearly with increasing $[S]$ in un-catalyzed reactions. The ***saturation effect*** observed with enzyme-catalyzed reactions led to the postulation of an enzyme–substrate

N. S. Punekar, *ENZYMES: Catalysis, Kinetics and Mechanisms*, https://doi.org/10.1007/978-981-97-8179-9_14

$$E + S \underset{k_{-1}}{\overset{k_1}{\rightleftarrows}} ES \underset{k_{-2}}{\overset{k_2}{\rightleftarrows}} E + P$$

Fig. 14.1 Simplest enzyme reaction scheme for S → P conversion invoking the formation of a single ES complex. The forward rate constants for the first and the second steps are shown as k_1 and k_2, respectively. Reverse rate constants for the corresponding steps are shown as k_{-1} and k_{-2}

(ES) complex—formed when enzyme and substrate come together through diffusion. The product is formed and released from ES complex (this means, ES → E + P) and not directly from the substrate. We can then represent the simplest general scheme for one substrate–one product reaction as shown in Fig. 14.1.

According to this representation, the enzyme-catalyzed rate will be directly related to [ES] and indirectly related to [S]. Taking this into account we may write the rate of enzyme-catalyzed reaction (velocity "v") in terms of ES concentration as follows:

$$v = \frac{d[P]}{dt} = k_2 \times [ES]$$

Deriving the rate equation thus becomes an exercise in evaluating [ES]. In order to obtain a useful rate equation, it is necessary to obtain [ES] in terms of [S] at any given instance. But this is not a trivial matter! Because, as we have noted earlier, concentrations of S, P, E, and ES are all changing as a function of reaction time (Fig. 10.1, in Chap. 10). A conceptual breakthrough in simplifying this difficulty was made by setting up well-defined initial conditions and making certain clear assumptions. Three important experimental conditions are as follows:

1. All experimental conditions like pH, temperature, ionic strength, etc. remain constant throughout the course of the experiment.
2. Enzyme being a catalyst, its concentration is very much lower than the concentration of substrate. Typically, the concentration of the substrate is at least 1000 times higher than that of the enzyme. This permits us to approximate $[S_t] \approx [S]$, although $[S_t] = [S] + [ES]$.
3. Strictly, the initial rate (velocity "v") is recorded. This is the rate at the beginning of the reaction or the instantaneous rate extrapolated to time zero. This will be the unbiased rate when $[P] \approx 0$.

Obviously, additional assumptions are clearly needed to evaluate [ES] and then to obtain the rate equation.

The equilibrium assumption: Michaelis and Menten (and of course Victor Henri before them!) provided the conceptual breakthrough and derived the now famous rate equation for an enzyme-catalyzed reaction. They assumed that the formation of the ES complex from E and S is at equilibrium; accordingly, k_2 is considered to be much smaller in magnitude when compared to k_1 and k_{-1} (Fig. 14.1). With this equilibrium assumption, and the experimental conditions listed above, it was possible to evaluate [ES] as follows:

$$K_{eq} = \frac{[E][S]}{[ES]} = \frac{k_{-1}}{k_1} \text{ and hence } k_1[E][S] = k_{-1}[ES]$$

However, $[E_t] = [E] + [ES]$, and therefore substituting for [E] we get

$$k_1[E_t][S] - k_1[ES][S] = k_{-1}[ES]$$

On simplification,

$$[ES] = \frac{k_1[E_t][S]}{(k_{-1} + k_1[S])} = \frac{[E_t][S]}{\left(\frac{k_{-1}}{k_1} + [S]\right)}$$

Substituting this value of [ES] in the rate equation, $v = k_2 \times [ES]$, we get

$$v = \frac{k_2[E_t][S]}{\left(\frac{k_{-1}}{k_1} + [S]\right)}$$

By defining V_{max} (as $k_2 \times [E_t]$) and K_S (as k_{-1}/k_1) this takes the original form of the Michaelis–Menten equation:

$$v = \frac{V_{max}[S]}{K_S + [S]}$$

It is obvious that when all the enzyme is in the ES form (i.e., $[ES] = [E_t]$), the reaction velocity reaches a limiting maximum value (i.e., $v = V_{max}$). Also, the constant K_S (the Michaelis constant) is nothing but the equilibrium (dissociation) constant—in accordance with the equilibrium assumption. The rate constant k_{-2} (see Fig. 14.1) does not appear in the final form of the rate equation. The rate "$k_{-2} \times [P]$" equals zero as long as $[P] \approx 0$ (initial velocity conditions are met!).

The steady-state assumption: In the equilibrium assumption (described above), the binding step (E + S ⇄ ES) was set at equilibrium, by explicitly assuming that k_2 is quite small. It may not be always true that k_2 is negligible (when compared to k_1 and k_{-1} in Fig. 14.1). In such situations, the "equilibrium assumption" is invalid and cannot be used to evaluate [ES]. Briggs and Haldane (1925) overcame this limitation by suggesting a more general approach. They viewed the [ES] to be at steady-state (see Chap. 9 and Fig. 10.1). As the $[S_t]/[E_t]$ ratio increases, the steady-state region occupies an increasing fraction of the total reaction time. Accordingly, the concentration of ES remains unchanged—because its formation and disappearance rates are equal (i.e., $d[ES]/dt \approx 0$). With this assumption and from Fig. 14.1, we can set up a balanced equation and solve for [ES].

Rate of ES formation = Rate of ES disappearance

And, therefore,

$$k_1[E][S] = k_{-1}[ES] + k_2[ES]$$

However, $[E_t] = [E] + [ES]$, and therefore substituting for $[E]$ we get

$$k_1[E_t][S] - k_1[ES][S] = k_{-1}[ES] + k_2[ES]$$

On simplification,

$$[ES] = \frac{k_1[E_t][S]}{k_{-1} + k_2 + k_1[S]} = \frac{[E_t][S]}{\left(\frac{k_{-1}+k_2}{k_1}\right) + [S]}$$

Since velocity (v) is $=k_2 \times [ES]$ and substituting for $[ES]$

$$v = \frac{k_2[E_t][S]}{\left(\frac{k_{-1}+k_2}{k_1}\right) + [S]}$$

$$v = \frac{V_{max}[S]}{K_M + [S]} \text{ Michaelis–Menten equation}$$

As before, we define V_{max} (as $k_2 \times [E_t]$); however, K_M ($=(k_{-1} + k_2)/k_1$) is a lumped-up constant (commonly referred to as "Michaelis constant") arising from the three rate constants.

Equilibrium assumption is a limiting case of steady-state assumption: Regardless of whether one uses equilibrium assumption or steady-state assumption (or considers $d[ES]/dt$ as very small!)—we arrive at an equation that is isomorphic with the original Michaelis–Menten equation—which is an equation describing a rectangular hyperbola. The two representations of the enzyme rate equation differ in the nature of the Michaelis constant. According to the steady-state assumption, we see that,

$$\begin{aligned} K_M &= (k_{-1} + k_2)/k_1 \\ &= (k_{-1}/k_1) + (k_2/k_1) \\ &= K_S + (k_2/k_1) \end{aligned}$$

For an enzyme $K_M = K_S$, whenever the k_2 is very small compared to k_1 (i.e., $k_2/k_1 \approx 0$). Clearly, the equilibrium assumption is a limiting case of steady-state assumption.

Table 14.1 Order of an enzyme-catalyzed reaction changes with increasing [S]

[S]	Reaction order	Rate constant	Units
$[S] \ll K_M$	First	V_{max}/K_M	$M^0 \times s^{-1}$
[S] around K_M	Fractional	–	–
$[S] \gg K_M$	Zero	V_{max}	$M^1 \times s^{-1}$

14.2   Salient Features of the Michaelis–Menten Equation

The derivation of the Michaelis–Menten equation describes the general logic used to derive enzymatic rate equations. It establishes the rate law for an isolated enzyme-catalyzed reaction under clearly specified conditions. Despite the assumptions involved, this elegant equation has stood the test of time. We will analyze the salient features and limits of this equation.

Anatomy of the equation and its limits: The $v \rightarrow$ [S] plot form of the Michaelis–Menten equation is a rectangular hyperbola. This adequately describes the characteristic saturation effect (Fig. 10.2) observed with enzyme catalysis. Enzyme behavior in the two limiting cases of [S] may also be visualized.

1. When [S] is very small compared to K_M (i.e., $K_M + [S] \approx K_M$, in the denominator), the equation takes the form

$$v = (V_{max}/K_M) \times [S]^1$$

   For very small values of [S] therefore, the reaction is of first order with respect to "S." By analogy, V_{max}/K_M represents the first-order rate constant for this reaction (Table 14.1).

2. When [S] is very large compared to K_M (i.e., $K_M + [S] \approx [S]$, in the denominator), the equation takes the form
$$v = (V_{max}) \times [S]^0$$
   For very large values of [S] therefore, the reaction is of zero order with respect to "S." By analogy, V_{max} represents the zero-order rate constant for this reaction.

Obviously, the reaction order for an enzyme-catalyzed reaction changes when we move from sub-saturating to saturating levels of [S]. As we go from very low to very high [S], it follows the sequence: First order $\rightarrow$ Mixed (fractional) order $\rightarrow$ Zero order, as shown schematically in Fig. 14.2.

Enzyme kinetic constants and their units: Both V_{max} and K_M (and therefore V_{max}/K_M) are constants for a given enzyme. The units for K_M and V_{max} will be those in which [S] and v, respectively, are measured.

Fig. 14.2 The order of an
enzyme-catalyzed reaction
changes from first order to
zero order as [S] increases
from zero to infinity. A
schematic representation is
shown

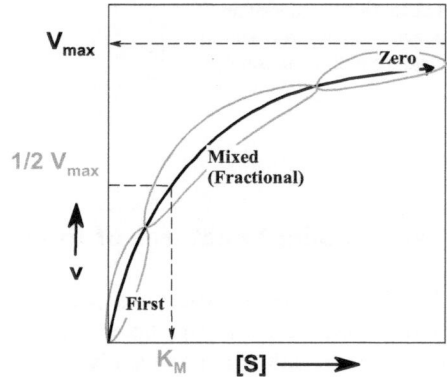

Since velocity "v" can reach a maximum of V_{max} (as $[S] \to \infty$, see Fig. 14.2), they
have similar units. Just like enzyme reaction velocity, V_{max} is expressed as μmol
product formed $\times$ min^{-1} $\times$ mg^{-1} protein or U $\times$ mg^{-1} protein. By definition,
$V_{max} = k_2 \times [E_t]$, and therefore V_{max} is a constant provided the amount of enzyme
catalyst ($[E_t]$) is fixed. We notice that the constant k_2 *is an intensive property* and is
intrinsic to a given enzyme. However, *$[E_t]$ is an extensive property* whose magnitude can be adjusted. The amount of enzyme protein can also be expressed as a
number of moles of that enzyme (Chap. 13). In this form, the constant k_2 (which
equals $V_{max}/[E_t]$) is more generally denoted as k_{cat}—the turnover number. It has the
units of time^{-1} as expected of a zero-order rate constant (Table 14.1).

The units for K_M and [S] are the same (i.e., concentrations such as mM and μM).
This is expected when we consider K_M purely as a dissociation constant (recall the
K_S of equilibrium assumption). Numerically, K_M is equal to [S] at which the reaction
rate is half its maximal value. This is obvious when we substitute velocity by $V_{max}/$
2 and simplify the Michaelis–Menten equation. By corollary, both K_M and [S] are
expressed in concentration units.

Relation to rectangular hyperbolic geometry: The Michaelis–Menten equation
describes an equation for rectangular hyperbola. Enzyme reactions that follow such a
rate law display $v \to [S]$ curves tracing a rectangular hyperbola (Fig. 14.3).

It is a geometric property of rectangular hyperbola (from mathematics) that—
despite the absolute magnitudes of X- and Y-axis scales, the ratio of X-axis
coordinates corresponding to any two Y-axis coordinates remains constant. Therefore, regardless of the absolute values of the kinetic constants (namely, V_{max} and
K_M), the ratio of [S] taken at any two fixed fractional saturation (v/V_{max}) remains the
same. For example, the ratio of [S] at 0.9 saturation (i.e., [S] when $v = 0.9\ V_{max}$) to
[S] at 0.1 saturation (i.e., [S] when $v = 0.1\ V_{max}$) is 81—a constant (as shown in the
box below). We can demonstrate this for many other substrate saturation ratios such
as $[S]_{0.8}/[S]_{0.2} = 16$ and $[S]_{0.9}/[S]_{0.2} = 36$.

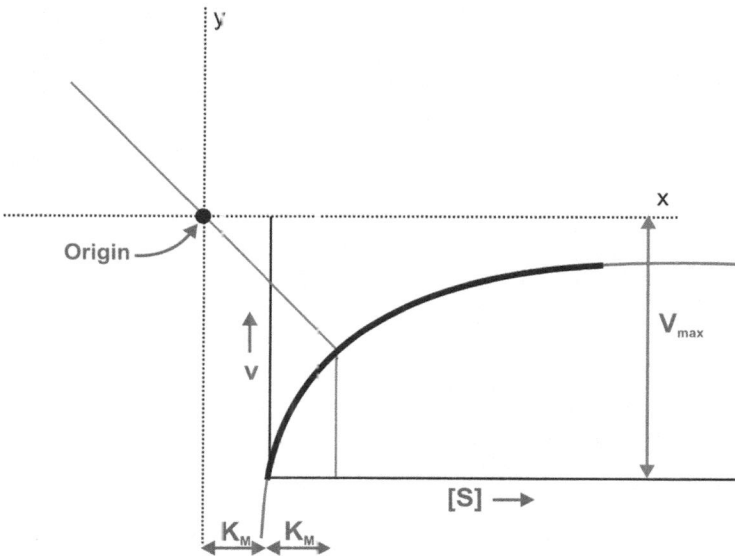

Fig. 14.3 The $v \rightarrow$ [S] curve described by the Michaelis–Menten equation is part of a rectangular hyperbola. The highlighted short arc (dark line) is the actual region where experimental observations are physically possible. The two limbs of this rectangular hyperbola form asymptotes of an infinite curve as shown. This makes direct estimation of both K_M and V_{max} values difficult

Calculating $[S]_{0.9}/[S]_{0.1}$

The Michaelis–Menten equation may be rearranged for [S] as follows:

$$[S] = \frac{v}{V_{max} - v} K_M$$

Substituting v by $0.9\,V_{max}$ and solving for $[S]_{0.9}$,

$$[S]_{0.9} = \frac{0.9 V_{max}}{V_{max} - 0.9 V_{max}} K_M = \frac{0.9}{0.1} K_M = 9 K_M$$

Similarly, substituting v by $0.1\,V_{max}$ and solving for $[S]_{0.1}$,

$$[S]_{0.1} = \frac{0.1 V_{max}}{V_{max} - 0.1 V_{max}} K_M = \frac{0.1}{0.9} K_M = \frac{1}{9} K_M$$

Taking ratios

$$[S]_{0.9}/[S]_{0.1} = 81$$

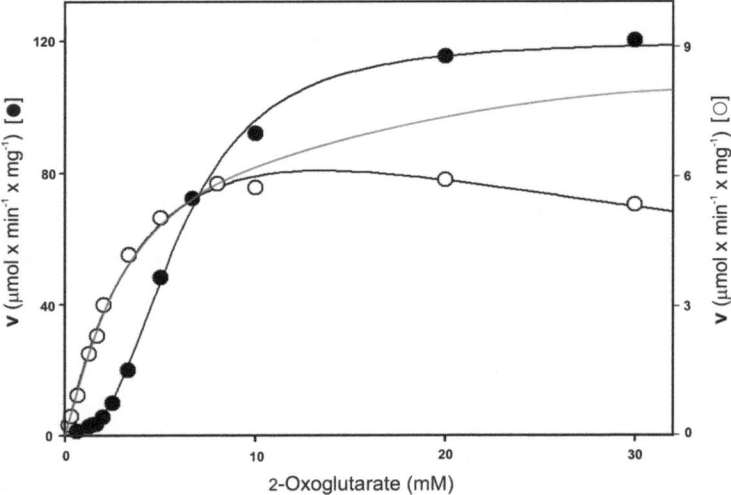

Fig. 14.4 Two examples of non-hyperbolic substrate saturation. NADP-Glutamate dehydrogenase from *Aspergillus niger* [●] shows sigmoid 2-oxoglutarate (substrate) saturation. The same enzyme from *Aspergillus terreus* [○] however exhibits substrate inhibition. A theoretical curve (in gray) describing the Michaelis–Menten kinetics is included for comparison

If $[S]_{0.9}/[S]_{0.1} = 81$, then that enzyme kinetic data fit the Michaelis–Menten equation. What if $[S]_{0.9}/[S]_{0.1}$ is not 81? Obviously, such an enzyme does not obey the Michaelis–Menten kinetics—and the data points do not trace a rectangular hyperbola. There are several possible causes of such behavior. Common examples of non-hyperbolic substrate saturation include—inhibition by high [S] and cooperative kinetics (Fig. 14.4) (Choudhury and Punekar 2009). Such departures from simple hyperbolic behavior (as predicted by the Michaelis–Menten equation) are also very informative. They are helpful in: (a) deducing the kinetic mechanism involved (Chap. 22; see Sect. 22.1) or (b) establishing the phenomenon of cooperativity and allosteric regulation (Chap. 38).

Cooperative kinetics: E and S interactions are considered cooperative when the binding of one molecule of substrate to the enzyme can either facilitate (*positive cooperativity*) or hinder (*negative cooperativity*) the binding of subsequent molecules of the same substrate. The binding of oxygen to hemoglobin (an honorary enzyme!) is an early example of positive cooperativity (Hill 1910). A sigmoid saturation curve (unlike the rectangular hyperbola traced by the Michaelis–Menten equation) is better described by the Hill equation:

$$v = \frac{V_{max}[S]^h}{K_{0.5}{}^h + [S]^h} \text{ Hill equation}$$

where [S] is the substrate concentration, v is the corresponding initial velocity, and h (also denoted as n_H) is the *Hill coefficient*. The constant $K_{0.5}$ is analogous to K_M of

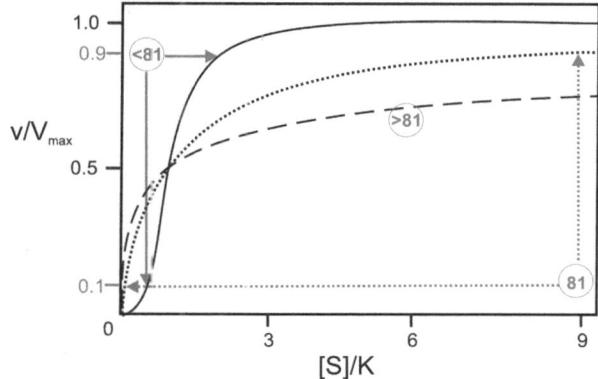

Fig. 14.5 The $v \rightarrow$ [S] curves for enzymes with different cooperativity indices. Schematic curves for (a) positive cooperativity (solid line; $R < 81$), (b) negative cooperativity (dashed line; $R > 81$), and (c) Michaelis–Menten kinetics (dotted line; $R = 81$) are shown. The $[S]_{0.9}$ value for negative cooperative enzyme lies far to the right (outside the X-axis scale) and is not shown. The two axes are plotted as dimensionless quantities with fractional velocity (Y-axis) and substrate concentration relative to K (X-axis). Here, K denotes K_M for the Michaelis–Menten kinetics and $K_{0.5}$ for the two cases of cooperativity

the Michaelis–Menten equation. It defines the value of [S] at which $v = V_{max}/2$, but it is not a Michaelis constant. The Hill coefficient is widely used as a measure of cooperativity. If the enzyme exhibits no cooperativity, then $h = 1$ and the above equation reduces to a simple Michaelis–Menten eq. A Hill coefficient greater than unity ($h > 1$) implies positive cooperativity, whereas negative cooperativity is indicated if it is less than unity ($h < 1$). A schematic of $v \rightarrow$ [S] curves for cooperative enzymes is depicted in Fig. 14.5.

For cooperative enzymes, we can show numerically that $[S]_{0.9}/[S]_{0.1}$ does not equal to 81. The $[S]_{0.9}/[S]_{0.1}$ ratio (also known as the *cooperativity index, R*) itself can be used as an alternative measure of cooperativity. The *R* values for positively cooperative enzymes are less than 81, while for negative cooperativity it is greater than 81. The two indices of cooperativity, namely *h* and *R*, are related to each other as shown in the box below.

_h_ and _R_: The Two Cooperativity Indices
The Hill equation may be rearranged as follows:

$$\frac{v}{V_{max}} = \frac{[S]^h}{K_{0.5}{}^h + [S]^h}$$

For $v/V_{max} = 0.9$, we obtain $[S]_{0.9} = 9^{1/h \times K_{0.5}}$

(continued)

Similarly, for $v/V_{max} = 0.1$, we get $[S]_{0.1} = \frac{1}{9^{\frac{1}{h}} \times K_{0.5}}$

Taking ratios,

$$[S]_{0.9}/[S]_{0.1} = R = 81^{1/h}$$

We can relate the two indices by substituting different values for h in this equation. For example, when a hyperbolic (Michaelian) kinetics is described $h = 1$ gives $R = 81$.

Type of $v \to [S]$ curve	h (or n_H)	R
Michaelis–Menten kinetics	1.0	81
Positive cooperativity ($h > 1$)	2.0	9
	4.0	3
Negative cooperativity ($h < 1$)	0.5	6560

Many cooperative enzymes also exhibit allostery and vice versa. Such enzymes have important roles in metabolic regulation. For instance, positive cooperativity enables them to respond with exceptional sensitivity to changes in metabolite concentration. We will revert to this topic in a later section (Chap. 38).

14.3 Significance of K_M, V_{max}, and k_{cat}/K_M

In the Michaelis–Menten world of enzyme catalysis, K_M and V_{max} are the fundamental constants. So long as the equilibrium assumption is valid (i.e., $K_M \approx K_S$), K_M may be viewed as an *apparent dissociation constant* for ES. Otherwise, it is not. It is observed that Michaelis constants recorded for most enzymes are in the range of their corresponding physiological substrate concentrations. Each metabolite (substrate) concentration in vivo is optimized by the evolutionary process for the efficient functioning of cellular metabolism. This in turn drives enzyme evolution to achieve a Michaelis constant of the same magnitude. The K_M therefore provides an approximation of [S] in vivo. In general, biosynthetic enzymes have much lower K_M values than their corresponding catabolic counterparts (more on this may be found in Chap. 38). As a thumb rule, enzymes best operate as catalysts with [S] around or above their Michaelis constants—otherwise, their catalytic potential is underutilized. Also, the substrate concentration region around the K_M is where the system exhibits fractional order!

The other constant V_{max} is the maximal velocity at saturating concentrations of substrate. In the classical derivation of the Michaelis–Menten equation, it was equated to $k_2 \times [E_t]$. Recall that k_2 was the rate constant for the "formation and release" of the product from the ES complex (Fig. 14.1). This simplistic mechanism involved a single ES complex. For more general cases, with many more intermediates and steps, k_2 may be replaced by **k_{cat}** (and therefore, $V_{max} = k_{cat} \times [E_t]$). Regardless of the number of steps/constants it describes, k_{cat} has the units of a first-order rate constant (i.e., $time^{-1}$). This is also known as ***turnover number***—and it defines the number of turnovers (catalytic cycles) the enzyme can undergo in unit time when the enzyme is fully saturated with substrate (also see Chap. 13). Whereas V_{max} depends on the enzyme concentration ($[E_t]$), k_{cat} does not. Therefore, k_{cat} is a fundamental property of the enzyme.

In light of the steady-state concept, a clean separation of binding (K_M) and rate of catalysis (V_{max}) may not be possible. The kinetic parameters K_M and k_{cat} (and therefore V_{max}) are algebraic aggregates of microscopic rate constants associated with many reaction steps. Recall that only in a simple mechanism (Fig. 14.1), k_2 equals k_{cat}. Otherwise, k_{cat} may also contain many more rate constants within it. Michaelis constant is a complex of at least three (and may be more if it is viewed as $(k_{-1} + k_{cat})/k_1$) rate constants and is conceptually difficult to grasp. Cleland therefore suggested that V_{max}/K_M (first-order rate constant) and V_{max} (zero-order rate constant) are the two ***fundamental kinetic constants*** for an enzyme-catalyzed reaction (Table 14.1). They represent apparent rate constants at very low and very high [S], respectively. According to this view, K_M is merely a derived constant obtained from the ratio of two rate constants (viz., $V_{max}/(V_{max}/K_M)$).

14.3.1 Kinetic Perfection and the Diffusion Limit

Enzymes are exquisite catalysts of nature. Nevertheless, can the catalytic efficiency of an enzyme be improved further (Albery and Knowles 1976)? What then is a "perfect" enzyme? Which fundamental kinetic constant(s) provide this information? We shall now attempt to answer these interesting questions. Imagine the events during enzyme catalysis as shown in Fig. 14.6.

In this depiction, the chemical steps in a single catalytic cycle—including all the bond-breaking and forming events and conformational changes—are sandwiched between two physical steps. The physical events of substrate colliding with the enzyme molecule and product dissociating from the enzyme surface are diffusion controlled. As a catalyst, an enzyme can do very little to overcome this diffusion-imposed limit. Regardless of how well an enzyme accelerates the chemical steps, ***the***

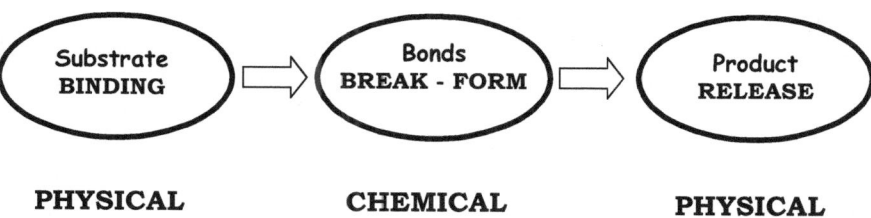

Fig. 14.6 Simplified view of events during a single cycle of enzyme catalysis. The chemical bond-breaking/forming steps are sandwiched between two diffusion-controlled physical events

Table 14.2 Enzyme catalytic efficiency (k_{cat}/K_M) compared to uncatalyzed reaction rate constant (k_{uncat})

Enzyme	Catalytic efficiency[a] (k_{cat}/K_M) $M^{-1} \times s^{-1}$	Uncatalyzed rate (k_{uncat}) s^{-1}	Catalytic proficiency (k_{cat}/K_M)/k_{uncat} M^{-1}
OMP decarboxylase	5.6×10^7	2.8×10^{-16}	2.0×10^{23}
Ketosteroid isomerase	3.0×10^8	1.7×10^{-7}	1.8×10^{15}
Triose phosphate isomerase	2.4×10^8	4.3×10^{-6}	5.6×10^{13}
Carbonic anhydrase	8.3×10^8	1.3×10^{-1}	9.2×10^8
Fumarase	1.6×10^8	–	–
Acetylcholinesterase	1.4×10^8	–	–
β-Lactamase	1.0×10^8	–	–

[a] These values are compared with corresponding bimolecular diffusion on-rate constant in water of about 10^8 and 10^9 $M^{-1} \times s^{-1}$, in water

upper bound for catalytic rate acceleration is the prevailing diffusion-limited on-rate

What kinetic parameter(s) of the enzyme should then be compared with diffusion-limited *on-rate* to evaluate its efficiency? At least under low [S] conditions, the Michaelis–Menten equation reduces to $v = (k_{cat}/K_M) \times [E] \times [S]$. The reaction is now effectively a bimolecular collision between free E and free S. The ratio k_{cat}/K_M is therefore a good measure of the enzyme's **kinetic perfection**. It has the same units as the bimolecular diffusion *on-rate* constant, namely, $M^{-1} \times s^{-1}$:

$$\frac{k_{cat}}{K_M} = \frac{k_{cat}}{\frac{k_{-1}+k_{cat}}{k_1}} = \frac{k_{cat}}{k_{-1} + k_{cat}} \times k_1 \quad \left(\approx \frac{s^{-1}}{M} = M^{-1} \times s^{-1} \right)$$

Consider a situation where every collision between the enzyme and a substrate molecule is productive: that is, k_{cat} (product formation rate) is much larger than k_{-1} (dissociation of ES back to E and S). Under these conditions, k_{cat}/K_M approximates to k_1 in the above equation. Thus, the upper limit to k_{cat}/K_M is set by k_1—the rate of formation of ES complex! Since k_1 is the bimolecular association rate constant between E and S, this rate cannot be faster than the rate of diffusion-controlled encounter of E and S. Notice that the units for k_1 ($M^{-1} \times s^{-1}$) are the same as those for the diffusion *on-rate* constant. Obviously, the upper bound for k_1 (and hence for k_{cat}/K_M, by the above argument) is diffusion-limited *on-rate* constant (Eisenthal et al. 2007). In this sense, k_{cat}/K_M **becomes diffusion controlled in a perfect enzyme**. For many enzymes, as noted by Cleland in 1975, the k_{cat}/K_M values approach the diffusion limit (between 10^8 and 10^9 $M^{-1} \times s^{-1}$). Triose phosphate isomerase was one of the early candidates to be characterized and termed "perfect" by this criterion. The k_{cat}/K_M ratios of superoxide dismutase and acetylcholinesterase are also between 10^8 and 10^9 $M^{-1} \times s^{-1}$. Any further gain in catalytic rate requires that the time for diffusion should decrease. The catalytic efficiencies (as k_{cat}/K_M) for a few representative enzymes are shown in Table 14.2. As can be seen, most of them have nearly attained kinetic perfection.

The concept of a kinetically perfect enzyme begs the question—*How to check whether an enzyme is diffusion limited?* We can probe this by varying the rate of diffusion through viscosity adjustments. The viscosity of aqueous solutions can be controlled by the addition of solutes like sucrose or glycerol. For instance, 30% sucrose increases the viscosity of water by a factor of 3—the diffusion rate also is decreased by the same factor. A decrease in the k_{cat}/K_M of an enzyme with increasing medium viscosity is indicative of a diffusion-limited enzyme. If the magnitude of k_{cat}/K_M is smaller, then the bimolecular collisions occur faster, and diffusion is not rate limiting. For such enzymes, k_{cat}/K_M is independent of viscosity changes. When such a study was conducted on carbonic anhydrase, indeed the hydration–dehydration rate of CO_2 decreased with an increase in viscosity; the esterase function displayed by this enzyme was however unaffected. Clearly, the chemical step (and not the diffusion) is rate limiting for the non-physiological esterase function of carbonic anhydrase (Hasinoff 1984). Experimentally, sucrose or glycerol may be used to raise the medium viscosity. Such viscosogens raise the micro-viscosity of water and influence diffusion rates. Whereas polymers like Ficoll® or polyethylene glycol (PEG) increase macro-viscosity (also termed bulk viscosity) but have no effect on micro-viscosity; they cannot be used to test if an enzyme is diffusion limited because they do not slow down diffusion. A word of caution with viscosity-dependence experiments: Viscosogen used and/or viscosity change may have more complex effects on the enzyme itself. Such effects (other than on k_{cat}/K_M) require more careful interpretation.

Proton diffusion on-rates in water are in the order of 10^{11} $M^{-1} \times s^{-1}$. Diffusion on-rates for bimolecular collisions in an aqueous environment are in the range of 10^{10} $M^{-1} \times s^{-1}$. When one of them is larger in size (say an enzyme), this is further diminished to 10^8–10^9 $M^{-1} \times s^{-1}$. Additional effects of local micro-viscosity and macromolecular crowding in vivo may reduce this rate further (see Chap. 39 for details). What then may be the meaning of observed k_{cat}/K_M values that are higher than the physiologically relevant diffusion-limited on-rate constant? Such abnormal high k_{cat}/K_M values could indicate one of the following:

(a) There may be something wrong with our interpretation or the estimation of second-order diffusion on-rate constant.
(b) Special enzyme features like charge guidance may be operating. For instance, strong electrostatic field gradients near the active site may cause an increase in the rate of association of superoxide dismutase with its anionic substrate (see Chap. 4, Sect. 4.2).
(c) The enzyme may be operating as a component of a multienzyme aggregate. Enzyme catalyzing a step in a multienzyme sequence may directly transfer its product to the next enzyme. Bulk is minimized and local concentrations of metabolites are enhanced by channeling. Such channeling may lead to apparently high k_{cat}/K_M values than for those enzymes acting on freely diffusible substrates. Multi-enzyme complexes and multifunctional polypeptides may lead

to physiologically relevant regulation (Chap. 38) by confining substrates and products in a limited volume (Chap. 39).

If an enzyme reaction is diffusion rate limited, then its kinetic efficiency cannot be further improved. We see that individually both k_{cat} and K_M can take a range of values, but their ratio (k_{cat}/K_M) can only reach the upper limit of diffusion rate. A tighter binding to the substrate (lower K_M; smaller denominator) has to be offset by a smaller turnover number (lower k_{cat}; smaller numerator). Conversely, large k_{cat} values are associated with poor substrate binding by that enzyme. Recall that enzymes are devices that utilize part of the substrate binding energy to accelerate reaction rates (Chap. 4). An example of this K_M-k_{cat} *compensation*, within the bounds of diffusion-limited rates, is shown in the box below.

The Tradeoff Between k_{cat} and K_M

Two small molecules diffuse toward each other in an aqueous environment with a bimolecular rate constant of 10^{10} $M^{-1} \times s^{-1}$. When one of the partners is large (an enzyme protein!), this rate constant is lowered to 10^8–10^9 $M^{-1} \times s^{-1}$. Now examine the kinetic parameters for triose phosphate isomerase:

$$K_M \text{for glyceraldehyde-3-phosphate} = 10^{-5}M \text{ and } k_{cat} = 10^3 s^{-1}$$

Accordingly,

$$k_{cat}/K_M = 10^8 M^{-1} \times s^{-1}$$

Triose phosphate isomerase therefore is almost at the "plateau of perfection."

Now consider I-*Ppo*I endonuclease. This intron-encoded endonuclease has a K_M of 4 nM for its substrate (42-bp duplex DNA). Then what should be the k_{cat} of this enzyme? Suppose if its k_{cat} value is the same as that for triose phosphate isomerase (i.e., 10^3 s^{-1}), then we obtain

$$k_{cat}/K_M = \left(10^3 s^{-1}\right)/\left(4 \times 10^{-9}M\right) = 2.5 \times 10^{11} M^{-1} \times s^{-1}$$

Such a value is beyond the limit set by diffusion-limited on-rate and hence is not feasible. A realistic k_{cat} for this enzyme at best should therefore be 4 s^{-1}.

Clearly, a smaller value of K_M in the denominator (of k_{cat}/K_M) constrains the upper limit that k_{cat} can approach. Both the enzymes will have reached *kinetic perfection* because their k_{cat}/K_M values are close to the diffusion limit. It is just that triose phosphate isomerase has evolved for a higher k_{cat} (10^3 s^{-1}) but with a reasonable K_M (10 μM) while I-*Ppo*I endonuclease is set to a lower k_{cat} but with tighter interaction with the substrate (a lower K_M of 4 nM). In fact, this k_{cat}-K_M compensation forms the basis for the functional existence of isozymes (discussed later in this chapter).

14.3.2 Kinetics of Interfacial Enzymes and the Sabatier Principle

Interfacial enzyme reactions are ubiquitous both in vivo and in process industries. Hydrolysis of water-insoluble polymers like cellulose is an example, but the kinetic analysis of such systems remains less understood. The conventional Michaelis–Menten theory, which requires a large excess of substrate, may not be appropriate in these cases. A quasi-steady-state assumption requires that the substrate is in excess with respect to the enzyme, and its validity is difficult to assess in heterogeneous systems. With interfacial substrates, the molar concentration of accessible surface sites cannot be defined unambiguously. The interfacial enzyme catalysis may reach its steady state in the opposite experimental limit, where the enzyme concentration far exceeds the molar concentration of accessible substrate surface sites. An "inverse Michaelis–Menten equation," with the roles of enzyme and substrate interchanged, may prove useful in this situation (Kari et al. 2017). The corresponding kinetic expression may look like this:

$$v = \frac{V_{max}[E_0]}{K_M + [E_0]}$$

In this case, V_{max} is the $^{inv}V_{max}$ and $[E_0]$ is the total enzyme concentration. One can also obtain the density of enzyme attack sites on the substrate surface as probed by that enzyme. This density of attack sites is analogous to the molar substrate concentration for interfacial reactions. Different enzymes may access the same substrate differently, reflecting on how an enzyme discriminates between local differences in substrate surface structure. The inverse Michaelis–Menten analysis is yet an approximation because of the possible substrate site heterogeneity (both qualitative and quantitative) and processive nature of some surface-active enzymes.

The trade-off between binding strength and catalytic rate is well-known within inorganic, heterogeneous catalysis—the so-called *Sabatier principle*. Accordingly, catalysis is most effective when the interaction between a catalyst and reactants is intermediate in strength. Tight binding (lower K_M) implies slow dissociation of ES complex, while weak binding (higher K_M) is associated with low [ES]. Only an intermediate binding strength balances off the two effects and hence leads to a faster turnover. This can be illustrated with the volcano plots—a plot of catalytic efficacy (turnover number) on the ordinate and the binding strength (K_M) on the abscissa axes. The Sabatier principle has found applications in interfacial enzyme reactions for industry (Baath et al. 2022).

14.3.3 Other Interpretations of k_{cat}/K_M

Specificity, besides excellent catalysis, is the hallmark of an enzyme catalyst (Chap. 3). How then should we define the specificity of an enzyme? Without question, the ratio k_{cat}/K_M provides insight into this aspect of an enzyme. This ratio has been referred to as the **specificity constant**—because it describes the

Table 14.3 Specificity constants for competing fumarase substrates

Substrate	K_M (µM)	k_{cat} (s^{-1})	k_{cat}/K_M (µM$^{-1} \times$ s^{-1}) (Specificity constant)
Fumarate	5	800	160
Fluorofumarate	27	2700	100
Chlorofumarate	110	20	0.18
Bromofumarate	110	2.8	0.025

relative velocities of two substrates competing for a single enzyme (Koshland Jr 2002). This competition between S_1 and S_2 is given by

$$\frac{v_1}{v_2} = \left(\frac{k_{cat1}}{K_{M1}} [S_1] \right) / \left(\frac{k_{cat2}}{K_{M2}} [S_2] \right)$$

The specificity constant, therefore, is the parameter that determines the ratio of rates when competing substrates are vying for the same enzyme. Thus, k_{cat}/K_M expresses the ability of an enzyme to discriminate in favor of any one substrate over the others. It can rank structurally similar substrates with respect to the catalytic power of an enzyme. This is illustrated by the kinetic data (Table 14.3) for fumarase and its substrates (Teipel et al. 1968).

Since its k_{cat} (zero-order rate constant) is higher (compared to other fumarates) at saturating concentrations, fluorofumarate appears to be a better substrate. At lower concentrations (k_{cat}/K_M; first order rate constant condition), fumarate is better. However, when the two are present together, k_{cat}/K_M best expresses the ability of fumarase to discriminate in favor of fumarate. In a similar analysis, the specificity constants for peptide bond hydrolysis by chymotrypsin are as follows: R group of Gly < Val < Tyr.

The k_{cat}/K_M goes up as the enzyme shows higher affinity for the substrate and/or higher catalytic rates. In general, the higher the k_{cat}/K_M, better the enzymatic performance. This is true for one enzyme acting on many substrates (for instance, fumarase data in Table 14.3) but is also valid when many mutant forms of the same enzyme, acting on a single substrate, are compared. Therefore, Koshland preferred the term *performance constant* to describe k_{cat}/K_M.

The catalytic power of an enzyme was related to the corresponding uncatalyzed reaction rate by Radzicka and Wolfenden (1995). The second-order rate constant for enzyme action on its substrate (i.e., k_{cat}/K_M) may be compared to the corresponding uncatalyzed rate. *Catalytic proficiency* defined this way measures an enzyme's ability to lower the activation energy barrier ($\Delta G^{\neq}$) for that reaction (Chap. 3). The spontaneous, uncatalyzed rate constants (k_{uncat} values) vary over 15 orders of magnitude (Table 14.2). This is also a reflection of the remarkable sensitivity of slow chemical reactions to temperature (Table 8.2; Wolfenden 2014). The k_{cat}/K_M values for the corresponding enzyme reactions however fall within a narrow range of 3 orders. And none cross the ceiling of $\approx 10^9$ M$^{-1} \times$ s^{-1}—the diffusion rate limit!

There is one other descriptor for k_{cat}/K_M. According to Northrop, k_{cat}/K_M (or V_{max}/K_M) actually provides a measure of the rate of substrate capture—by the free enzyme—into a productive complex (Northrop 1998). Thus, $k_{capture}$ *is the rate of substrate capture* into the ES complex (Fig. 14.1). Given this, k_{cat} (or V_{max}) may be viewed as a measure of the rate of release of the product ($k_{release}$), from the captured complex. The ratio $k_{release}/k_{capture}$ now defines K_M as a derived parameter. The perception of k_{cat}/K_M as $k_{capture}$ (and its physiological relevance!) is beautifully demonstrated by antibiotic resistance in bacteria (Radika and Northrop 1984). When present, aminoglycoside acetyltransferase confers antibiotic resistance to the host bacterium. This enzyme inactivates aminoglycoside antibiotics by acetylating them. A correlation between minimal inhibitory concentration (MIC) and the kinetic characteristics of this enzyme (i.e., V_{max}, K_M, and V_{max}/K_M) was attempted. No relationship exists between MIC values of nine different aminoglycosides with their corresponding V_{max} or K_M values. But a very tight correlation ($R^2 = 0.995$) is seen between their MIC values and V_{max}/K_M for the enzyme. Executing a successful capture (meaning a large $k_{capture}$) is vitally important for antibiotic resistance and hence bacterial survival. Once captured, the fate of the antibiotic is sealed—it does not matter how long it actually takes for the enzyme to form and release the product. Rates of product release ($k_{release}$) are not very impressive for these enzymes.

14.4 Haldane Relationship: Equilibrium Constant Meets Kinetic Constants

The classic form of the Michaelis–Menten equation (earlier in this chapter) was derived for a reaction proceeding in a single direction (S → P). For reasons practical and otherwise, the rate of P → S was set to zero. Many reactions of metabolism, however, are reversible, and significant amounts of both substrate and product are present at any given time. Therefore, it is important to include the reverse reaction as well. Although a bit more complicated, it is possible to derive the rate equation for a reversible reaction. Figure 14.1 describes the simplest case where k_{-2} is finite and [P] ≠ 0. For a single ES complex example of the reversible enzyme reaction (S ⇌ P; Fig. 14.1), we can derive the following Michaelis–Menten equation (for the actual derivation see Chap. 15):

$$v = \frac{\frac{V_{maxf}[S]}{K_{MS}} - \frac{V_{maxr}[P]}{K_{MP}}}{1 + \frac{[S]}{K_{MS}} + \frac{[P]}{K_{MP}}}$$

This equation is a more general form of the Michaelis–Menten equation where both forward and reverse reaction rates are considered. It collapses to the simple form (for S → P reaction) by putting [P] = 0 in this equation. The equation is symmetric with respect to S → P and P → S. If we put [S] = 0, then the equation collapses to the classic form of the Michaelis–Menten equation for the reverse direction. However, when both S and P are present, and when the system has reached equilibrium (see Chap. 9), the net velocity is zero. Putting $v = (v_f - v_r) = 0$ in the general rate equation above and simplifying,

$$v = \frac{\frac{V_{maxf}[S]}{K_{MS}} - \frac{V_{maxr}[P]}{K_{MP}}}{1 + \frac{[S]}{K_{MS}} + \frac{[P]}{K_{MP}}} = 0$$

Therefore, $\frac{V_{maxf}[S]}{K_{MS}} - \frac{V_{maxr}[P]}{K_{MP}} = 0$
and

$$\frac{V_{maxf}[S]}{K_{MS}} = \frac{V_{maxr}[P]}{K_{MP}}$$

Substituting the corresponding equilibrium concentrations and rearranging, we get

$$\frac{V_{maxf} \times K_{MP}}{V_{maxr} \times K_{MS}} = \frac{k_{catf} \times K_{MP}}{k_{catr} \times K_{MS}} = \frac{[P]_{eq}}{[S]_{eq}} = K_{eq} \; \text{Haldane relationship}$$

This equation relates enzyme kinetic constants to the overall equilibrium constant of that reaction. This relationship was first shown by JBS Haldane hence the name (Mellors 1971).

The Haldane relationship for multi-substrate and/or multiproduct can be similarly evaluated, but the equation would be more complex and accordingly have more constants built into it. For a *ter-bi* sequential mechanism (A + B + C $\rightleftarrows$ P + Q), the equation would be

$$K_{eq} = ([A][B][C]/[P][Q]) = (V_{maxr} \times K_A \times K_B \times K_C)/(V_{maxf} \times K_P \times K_Q)$$

Kinetic characterization of the catalyzed reaction in both directions would yield the required data. At times, it may not be feasible to determine all the constants due to practical reasons (viz., the limited range of substrate concentrations accessible, etc.). This was the case with a modified form of *A. niger* NADP-glutamate dehydrogenase. For this reaction, the equilibrium constant (K_{eq}) numerically equals Γ (mass action ratio; see Chap. 9), at a concentration of NADP$^+$ where the enzyme shows no net forward or reverse reaction rate. A plot of net reaction rate versus Γ was used to determine the K_{eq} (Walvekar et al. 2014).

The first-order rate constants for the forward (V_{maxf}/K_{MS}) and reverse (V_{maxr}/K_{MP}) direction of a reversible enzyme-catalyzed reaction are not independent. However, they are related to the equilibrium constant (K_{eq}) of the overall reaction. Following conclusions may be drawn from the Haldane relationship:

(a) Both V_{maxf} and V_{maxr} may be written as their corresponding $k_{cat} \times [E_t]$ terms. Further, V_{max} depends on the enzyme concentration ($[E_t]$) and k_{cat} does not. Since the $[E_t]$ appears both in the numerator and the denominator, *the equilibrium constant (K_{eq}) is independent of enzyme concentration*.

(b) Haldane equation can be conveniently written as

$$\left(\frac{k_{catf}}{K_{MS}}\right) / \left(\frac{k_{catr}}{K_{MP}}\right) = K_{eq}$$

The equilibrium constant is then nothing but the ratio of k_{cat}/K_M values of the forward and reverse reactions. The reaction equilibrium constant is a thermodynamic constant and cannot change merely because an enzyme is present. Therefore, *K_{eq} puts a constraint on what values the enzyme kinetic constants (both k_{cat} and K_M) could take*. This is yet another case where k_{cat}/K_M for an enzyme assumes importance.

(c) For any reversible enzyme reaction, the "Haldanes should match." Some form of irreversibility (like inactivation of the enzyme during reaction!) may not satisfy the Haldane relationship—because K_{eq} of the reaction cannot be different.

(d) The Haldane equation does not constrain the equilibrium constant for the internal equilibrium at the enzyme active site (i.e., "ES $\rightleftarrows$ EP"). When the forward and the reverse rates are equal ($k_{catf} \times [ES] = k_{catr} \times [EP]$), we obtain

$$K_{int} = \frac{[EP]_{eq}}{[ES]_{eq}} = \frac{k_{catf}}{k_{catr}}$$

Note that the K_{eq} in solution ($= [P]/[S]$) may not be the same as that on the enzyme ($= [EP]/[ES]$). The K_{eq} values for chemical reactions vary over a wide range. Whereas the K_{int} values for all enzyme reactions are closer to unity than the corresponding K_{eq} values for the same reaction. The K_{int} values are near unity implies $\Delta G°$ "for the reaction at enzyme active site" is close to zero (more on this in Chap. 39 and Table 39.5; Hassett et al. 1982).

Kinetic feasibility of isozymes: Effectively there are two constraints operating on the kinetic constants of an enzyme catalyst. One is the diffusion rate barrier (on k_{cat}/K_M value) and the other is thermodynamic—set by the Haldane relationship. The reaction K_{eq} dictates only that the ratio $(V_{maxf}/K_{MS})/(V_{maxr}/K_{MP})$ remains constant. Within these constraints, many numerical solutions are possible—giving the same K_{eq} but with different k_{cat} and K_M values. This is very interesting because enzymes with different V_{max} (or k_{cat}) and K_M values could be constructed for the same reaction—without violating the two constraints mentioned above. One such example is shown in the box below.

Haldane Relationship and Isozymes

Consider a reversible reaction "S $\rightleftarrows$ P" catalyzed by two distinct enzymes namely, Enz-I and Enz-II. Suppose their kinetic constants are as shown:

Enz-I:	$V_{maxf} = V_{maxr} = 100$	$K_{MS} = 5$ mM	$V_{maxf}/K_{MS} = 20$
		$K_{MP} = 5$ mM	$V_{maxr}/K_{MP} = 20$
Enz-II:	$V_{maxf} = V_{maxr} = 100$	$K_{MS} = 100$ mM	$V_{maxf}/K_{MS} = 1$
		$K_{MP} = ??$	

We infer the following from these data:

(a) Enz-I is relatively more efficient in the forward direction. Its V_{maxf}/K_{MS} is 20 times that of Enz-II.
(b) The K_{eq} for this reaction (S $\rightleftarrows$ P) may be calculated from the given data for Enz-I.

$$\frac{V_{maxf} \times K_{MP}}{V_{maxr} \times K_{MS}} = \frac{100 \times 5}{100 \times 5} = 1$$

(c) We now predict that the K_{MP} for Enz-II should be 100 mM! Why so? Because it is the only numerical solution that gives an equilibrium constant of unity (i.e., $K_{eq} = 1$).

$$K_{eq} = \frac{V_{maxf} \times K_{MP}}{V_{maxr} \times K_{MS}} = \frac{100 \times 100}{100 \times 100} = 1$$

Any other value (including a K_{MP} of 5 mM, like for Enz-I) will yield a different K_{eq}—recall that reaction equilibrium constant is a thermodynamic parameter that the catalyst cannot tinker with.

Multiple catalytic designs for the same reaction are possible—this is the basis for the existence of isozymes in nature. Isozymes will have compensated for differences in their forward kinetic constants (V_{maxf} and/or K_{MS}) by suitable adjustments in their reverse kinetic constants (V_{maxr} and/or K_{MP})—thereby resulting in an identical K_{eq} value. The isozyme with a lower V_{maxf}/K_{MS} could either have an appropriately lowered V_{maxr}, an elevated K_{MP} or both.

A consequence of multiple kinetic solutions for the same reaction is that isoforms of an enzyme catalyst are possible. Nature has exploited these solutions in the form of *isozymes* (or isoenzymes). Isozymes are multiple molecular forms that catalyze the same chemical reaction. Among others, lactate dehydrogenases (the heart and the muscle isoforms) and alcohol dehydrogenase (yeast ADH-I and ADH-II) are excellent examples. Bacteria elaborate on two distinct isoforms of threonine deaminase for biosynthesis (with higher affinity for Thr; low K_M) and catabolism (with lower affinity for Thr; high K_M). Isozymes play critical roles in cellular metabolic regulation. We will have more to say on this later (in Chap. 38).

Although the Haldane relationship places limits on the kinetic constants, a wide range of enzyme kinetic behavior is still allowed. As a result, we can expect an enzyme evolved to be a more effective catalyst for one direction of a reaction than the other. Indeed, such *one-way enzymes* are possible. For instance, the limiting rate of the forward reaction catalyzed by methionine adenosyltransferase is about 10^5 times greater than its reverse reaction rate. The enzyme efficiency can be improved for one direction, at the expense of the other, by optimizing k_{cat}/K_M values to suit prevailing concentrations of substrate and product. One-way enzymes also make physiological sense—they may never be required to catalyze the reaction in the reverse direction in vivo. There may be no evolutionary pressure to achieve catalytic perfection in that direction! If the active site is strictly complementary to the transition state, then the enzyme will be an optimized catalyst for both directions. Efficiency in one direction could however be preferentially improved by evolving an active site that binds either S or P relatively better, with nearly the same transition state. Finally, since $V_{maxf} = k_{catf} \times [E_t]$, any unfavorable k_{catf} changes (arising out of thermodynamic constraints) during catalyst design/evolution can be compensated by the system. A cell can maintain the desired V_{maxf} by increasing $[E_t]$ (increased cellular abundance!) despite having a lower k_{catf}. Therefore, the cellular concentrations of various isozymes (and enzymes in general!) are not necessarily maintained at the same level (Futcher et al. 1999).

14.5 Enzyme Kinetics of the Non-Michaelis–Menten Kind

Not necessarily all enzymes follow Michaelis–Menten kinetics, and we will summarize key examples of such variations here (Hill et al. 1977). We already note that enzymes displaying cooperativity (either positive or negative) and substrate inhibition are atypical—they cannot be described by the classical Michaelis–Menten treatment (Fig. 14.4). There are other plausible atypical cases besides these two examples. They include: (1) Self-catalyzing enzyme systems like zymogen activation by limited proteolysis (trypsinogen to trypsin)—where the substrate concentration cannot be maintained, and more enzyme is generated in the reaction. (2) The approximation of "$[S_t] \approx [S]$" fails when the K_M is less than the total enzyme concentration $[E_t]$, employed in an assay. This is also true for tight-binding inhibitors whose K_I is less than the enzyme concentration used. (3) The depletion of enzyme occurs due to time-dependent covalent irreversible inhibition, its action on suicide substrates, diffusional and crowding effects, interfacial enzyme catalysis, and/or a processive mechanism. Representative examples of such atypical cases are discussed in detail in many places in this book.

In conclusion, one need not force fit the Michaelis–Menten equation for every enzyme (Srinivasan 2021). We may fit the nonlinear data to a model to better visualize the system. With an increase in the number of parameters, we may get good fits to the data, but this overfitting may be devoid of any biological significance. We may recall John von Neumann—"With four parameters I can fit an elephant, and with five I can make him wiggle his trunk." Occam's razor is the way to go.

14.6 Use and Misuse of the Michaelis–Menten Equation

Apart from its historical importance, the Michaelis–Menten equation undoubtedly remains a very important tool in enzyme kinetics. It is the first useful approximation for any new enzyme to be studied. One must however be confident that the Michaelis–Menten equation is obeyed by that enzyme—for not all enzymes adhere to Michaelis–Menten kinetics. Careful analysis of kinetic measurements may indicate possible deviations, such as substrate inhibition. There is no need to force the Michaelis–Menten equation on every enzyme—after all data dictates!

At some level, all scientific representations are approximations. In this sense, the Michaelis–Menten equation is an outcome of explicit postulates and assumptions made in its derivation. The most elementary consideration is that all the $v \rightarrow$ [S] kinetic data used are initial velocities. The general applicability of the Michaelis–Menten model also requires that $[S_t] \approx [S]$ and that $[E_t] << [S_t]$. If any one or more of these stipulated conditions are not met, then the model is unsuitable for use. Other relevant modifications to the Michaelis–Menten equation become necessary before adopting such a kinetic model.

Both V_{max} (which is $k_{cat} \times [E_t]$) and K_M are intrinsic constants for a given enzyme. The V_{max} in principle may be adjusted by varying the total enzyme present, whereas k_{cat} is unique for that catalyst. The Michaelis constant (the K_M) is a constant defined according to the Michaelis–Menten model (and hyperbolic kinetics). Any interpretation of K_M becomes irrelevant the moment the Michaelis–Menten model is not obeyed. Even within these confines, the K_M provides only an apparent measure of substrate affinity. That implies "$K_M = K_S$" is not always true (especially, when k_{cat} is not negligibly small in comparison to k_{-1}) and should not be taken for granted without sufficient evidence.

Despite the divergent views on interpreting k_{cat}/K_M as "specificity constant," "performance constant" (Koshland Jr), "catalytic proficiency" (Wolfenden), or "rate of substrate capture" (Northrop), we note that k_{cat}/K_M is of great practical significance. When the magnitude of k_{cat}/K_M is comparable to the diffusion-limited on-rate constant, the steady-state approximation operates. For all lower values (i.e., $k_{cat}/K_M < k_1$), the equilibrium mechanism is more appropriate. Lastly, catalytic perfection may not be the only constraint worked on the enzyme by evolution! Regulation and cost efficiency may be two others, at the least.

References

Albery WJ, Knowles JR (1976) Evolution of enzyme function and the development of catalytic efficiency. Biochemistry 15:5631–5640

Baath JA et al (2022) Sabatier principle for rationalizing enzymatic hydrolysis of a synthetic polyester. J Am Chem Soc Au 2:1223–1231

Briggs GE, Haldane JBS (1925) A note on the kinetics of enzyme action. Biochem J 19:338–339

Choudhury R, Punekar NS (2009) *Aspergillus terreus* NADP-glutamate dehydrogenase is kinetically distinct from the allosteric enzyme of other Aspergilli. Mycol Res 113:1121–1126

Eisenthal R, Danson MJ, Hough DW (2007) Catalytic efficiency and k_{cat}/K_M: a useful comparator? Trends Biotechnol 25:247–249

Futcher B et al (1999) A sampling of the yeast proteome. Mol Cell Biol 19:7357–7368

Hasinoff BB (1984) Kinetics of carbonic anhydrase catalysis in solvents of increased viscosity: a partially diffusion-controlled reaction. Arch Biochem Biophys 233:676–681

Hassett A, Blättler W, Knowles JR (1982) Pyruvate kinase: is the mechanism of phospho transfer associative or dissociative? Biochemistry 21:6335–6340

Hill AV (1910) The possible effects of the aggregation of the molecules of haemoglobin on its dissociation curves. J Physiol 40:iv–vii

Hill CM, Waight RD, Bardsley WG (1977) Does any enzyme follow the Michaelis—Menten equation? Mol Cell Biochem 15:173–178

Kari J et al (2017) An inverse Michaelis–Menten approach for interfacial enzyme kinetics. ACS Catal 7:4904–4914

Koshland DE Jr (2002) The application and usefulness of the ratio kcat/KM. Bioorg Chem 30:211–213

Mellors A (1971) The Haldane relationship—enzymes & equilibrium. Biochem Educ 4:71

Michaelis L, Menten ML (1913) Die kinetic der invertinwirkung. Biochem Z 49:333–369

Northrop DB (1998) On the meaning of Km and V/K in enzyme kinetics. J Chem Educ 75:1153–1157

Radika K, Northrop DB (1984) Correlation of antibiotic resistance with Vmax/Km ratio of enzymatic modification of aminoglycosides by kanamycin acetyltransferase. Antimicrob Agents Chemother 25:479–482

Radzicka A, Wolfenden R (1995) A proficient enzyme. Science 267:90–93

Srinivasan B (2021) A guide to the Michaelis–Menten equation: steady state and beyond. FEBS J 289:6086–6098

Teipel JW, Hass GM, Hill RL (1968) The substrate specificity of fumarase. J Biol Chem 243:5684–5694

Walvekar AS, Choudhury R, Punekar NS (2014) Mixed disulfide formation at Cys141 leads to apparent unidirectional attenuation of *Aspergillus niger* NADP-glutamate dehydrogenase activity. PLoS One 9(7):e0101662

Further Reading

Gunawardena J (2012) Some lessons about models from Michaelis and Menten. Mol Biol Cell 23:517–519

Johnson KA (2013) A century of enzyme kinetic analysis, 1913 to 2013. FEBS Lett 587:2753–2766 part of Historical perspectives: special issue on the occasion of the centenary of the Michaelis Menten paper

Johnson KA, Goody RS (2011) The original Michaelis constant: translation of the 1913 Michaelis Menten paper. Biochemistry 50:8264–8269

Srinivasan B (2021) Explicit treatment of non-Michaelis-Menten and atypical kinetics in early drug discovery. ChemMedChem 16:899–918

Wolfenden R (2014) Massive thermal acceleration of the emergence of primordial chemistry, the incidence of spontaneous mutation, and the evolution of enzymes. J Biol Chem 289:30198–30204

More Complex Rate Expressions

15

The Henri–Michaelis–Menten equation is the simplest rate law for an isolated enzyme-catalyzed reaction under clearly specified conditions. Not all enzyme reactions are this simple. More complex mechanisms may involve (a) multiple intermediates/ complexes (not just one ES complex) and (b) more than one substrate and/or product. Indeed, the most common enzyme mechanisms found in metabolism are reactions with two substrates and two products. The remarkable success of Michaelis–Menten formalism over the last century has led to the extension of this classical approach to more complex systems.

15.1   Investigating Enzyme Mechanisms Through Kinetics

Kinetic description of a complex enzyme-catalyzed reaction provides insights into its mechanism. Traditionally, kinetic studies are closely associated with the investigations of enzyme mechanisms. They are primarily used to understand reaction mechanisms. The process of elucidating reaction mechanisms through kinetics is an excellent example of the *Scientific Method* in practice. The sequence of steps involved in this process of mechanism building is shown in the box below.

Mechanism Building: The Process
(a) Collate all the available data that describe the enzyme reaction in question
(b) Postulate a minimal mechanism that accounts for all the enzymes' behavior and accommodates available data
(c) Analyze the proposed mechanism by deriving a rate equation for it
(d) Predict distinctive kinetic outcomes from the analysis
(e) Test these predictions by performing critical experiments
(f) Accept or reject the proposed mechanism based on these results
(g) If rejected, suitably modify the proposed mechanism, and iterate

---

© The Author(s), under exclusive license to Springer Nature Singapore Pte
Ltd. 2025
N. S. Punekar, *ENZYMES: Catalysis, Kinetics and Mechanisms*,
https://doi.org/10.1007/978-981-97-8179-9_15

The usual general practice in arriving at any enzyme mechanism is to set up a tentative reaction scheme based on initial, often preliminary, evidence. Subsequently, test whether this describes the experimental results. If it does not, the initial scheme has to be modified or replaced. The process is repeated until some enzyme mechanism has been developed which accounts for all the experimental results—rates, complexes, substrates, and product stoichiometries. The provisional mechanism is further strengthened by any new supporting data. However, a single piece of contradictory evidence is enough to discredit the mechanism. The mechanism that survives this scrutiny is considered consistent with the experimental data. In this sense, even an attractive mechanism—which is always a hypothesis—can never be proven beyond doubt.

There is no panacea for coming up with the correct model for a complex enzyme mechanism. Model building exercises and mechanism validation are legitimate parts of most kinetic analyses. This is iterated till a satisfactory explanation (mechanism!) is in place. Other chemical and physical methods may help bolster the case—but *kinetic data should be the ultimate arbiter of mechanism*—because it reports on the behavior of an enzyme while in action. Rate equations are at the heart of any kinetic study. They capture the essence of a mechanism in a mathematical form. If an enzyme reaction is well described by a rate equation, then the data should fit that equation well. While proposing a kinetic mechanism and describing a rate equation for it, the following points need to be kept in mind.

- Keep the proposed mechanism as simple as possible. Complexity should not be assumed unnecessarily ("Pluralitas non est ponenda sine necessitate"—the Occam's razor, William of Occam) (Wildner 1999). According to this principle of parsimony—Hypotheses should not be multiplied beyond necessity ("Entia non sunt multiplicanda praeter necessitatem"). For example, it is meaningless to propose a complex mechanism and invoke many intermediates (ES complexes) without actual evidence. This simplicity paradigm of scientific inquiry was also invoked by Newton when he said—"We are to admit no more causes of natural things than such as are both true and sufficient to explain their appearances. To this purpose the philosophers say that Nature does nothing in vain, and *more is in vain when less will serve*; for nature is pleased with simplicity, and affects not the pomp of superfluous causes" (Newton's Principia).
- Process of elucidating enzyme mechanisms works best when alternative mechanisms are postulated. Start with multiple working hypotheses (alternative mechanisms). It is easier to disprove a mechanism (falsification strategy of Popper) than to prove a mechanism with certainty. Kinetics is of great value as it can be used to test and eliminate putative mechanism(s). Beautiful mechanisms need not necessarily be true. After all, the proposed mechanism is just a model, not data. Also, just because experimental data fit a particular rate equation is not proof that the assumptions made to derive that rate equation were correct.
- Different mechanisms may display the same kinetic behavior. Therefore, kinetic methods alone cannot be used to unambiguously identify a mechanism. For example, a mechanism with one intermediate (ES complex) with four rate

constants (Fig. 14.1) and a mechanism with two intermediate complexes (and six rate constants) result in the same general rate expression. Their V_{max} and K_M terms may however be composed of different individual rate constants. Steady-state kinetics (and Michaelis–Menten formalism) therefore cannot establish the nature and number of intermediates. For that, other methods/ techniques (such as pre-steady-state tools) are required.

15.2 Notations and Nomenclature in Enzyme Kinetics

Rate equations and reaction equilibria are two complementary bits of information describing the kinetic mechanism of an enzyme. Representing both these becomes a challenge with increasing mechanistic complexity. A major difficulty in following the enzymology literature is the variety of abbreviations and notations used by different research groups. Beginners are bound to be confused by the diversity of nomenclature. A comprehensive and uniform nomenclature was set forth by WW Cleland in three landmark papers published in *Biochimica Biophysica Acta* (in 1963) (Cleland 1989). Table 15.1 lists the most common notations in enzyme kinetics literature and these are used throughout this book.

An enzyme mechanism may involve several different enzyme forms and various steps of reaction equilibrium. These need to be properly represented for clear understanding. The mechanistic scheme indicating various enzyme forms and the rate constants for individual step(s) can be shown in different ways. Correctly displaying a complex mechanism, with a large number of enzyme forms and/or steps, poses a challenge. *Cleland notations* are often preferred for their simple yet clear presentation of enzyme mechanisms. In this depiction (a) the reaction sequence is written from left to right, (b) the enzyme surface is denoted by a horizontal line, (c) various enzyme forms are denoted below this horizontal line while the central complexes (where bond breaking/forming chemistry takes place) are given in

Table 15.1 Notations commonly used in enzyme kinetics

Item	Notation
Substrates	A, B, C, and D, in the order of their addition to the enzyme (S, for a single substrate)
Products	P, Q, and R, in the order of their release from the enzyme
Inhibitors	I, J, etc.
Enzyme forms	E, F, etc.
Forward rate constants	k_1, k_2, k_3, etc.
Reverse rate constants	k_{-1}, k_{-2}, k_{-3}, etc.
Michaelis constants	K_A, K_B, K_C, etc. (K_M, in general)
Dissociation constants	$K_{IA}, K_{IB}, K_{IC}, K_I, K_J, K_P$, etc. ($K_D$, in general)

brackets, (d) vertical downward arrows represent substrate addition to that enzyme form, and (e) upward arrows from the enzyme surface indicate product dissociation. Even though the binding and release arrows are single headed, these steps are viewed as reversible steps. For reactions in aqueous solutions, water is in large excess (55.5 M) and its concentration usually remains constant. Therefore, water is not explicitly shown (either as substrate or product) in these mechanisms.

Many enzyme reactions are freely reversible. In these cases, the reactants (substrates) become products and products become substrates for the reverse direction. For instance, NADP-glutamate dehydrogenase (EC 1.4.1.4) in the forward (reductive amination) reaction has three substrates (viz., NADPH, 2-oxoglutarate, and NH_3) and two products (viz., glutamate and $NADP^+$) (Fig. 15.1). However, the same enzyme is an example of two substrates and three products reaction in the reverse (oxidative deamination). A few more common enzyme equilibria are depicted in Fig. 15.1 according to Cleland notations. The first case (and also the simplest) corresponds to a single substrate–single product equilibrium, used earlier to derive the Michaelis–Menten equation (see Fig. 14.1). All others are more complex. The addition of multiple substrates (or release of more than one product) may occur in various ways as shown. In a *sequential mechanism*, all the substrates must be added to the enzyme before any product can leave. This addition may be either *ordered* (e.g., malate dehydrogenase and glutamine synthetase) or *random* (e.g., alcohol dehydrogenase). In a *ping pong mechanism* (also known as the substituted-enzyme mechanism or double-displacement reaction), the substrate addition sequence is broken by the release of one or more products (e.g., nucleoside-diphosphate kinase and acetyl CoA carboxylase).

Appropriate rate expressions can be derived for every mechanism shown in Fig. 15.1 (and for many others that may be proposed!). How to study these mechanisms with the help of suitable rate equations is discussed later (see Part III). Different methods to derive rate equations, starting with postulated enzyme equilibria, will be discussed next.

15.3 Deriving Rate Equations for Complex Equilibria

The first step in obtaining a suitable rate expression for any reaction is to set up appropriate equilibria with relevant steps and corresponding rate constants. Once the equilibria with various steps and enzyme forms are set up to represent the mechanism, the derivation of an appropriate rate equation is straightforward. Recall that *deriving the rate equation is an exercise in evaluating the concentration of the productive (ES) complex* (Chap. 14). It is necessary to obtain [ES] in terms of [S] at any given instance. As the ES form alone breaks down into products the velocity is proportional to [ES]. The fraction of total enzyme ([E_t]) in the ES form is the key.

Fraction of the total enzyme in the ES form

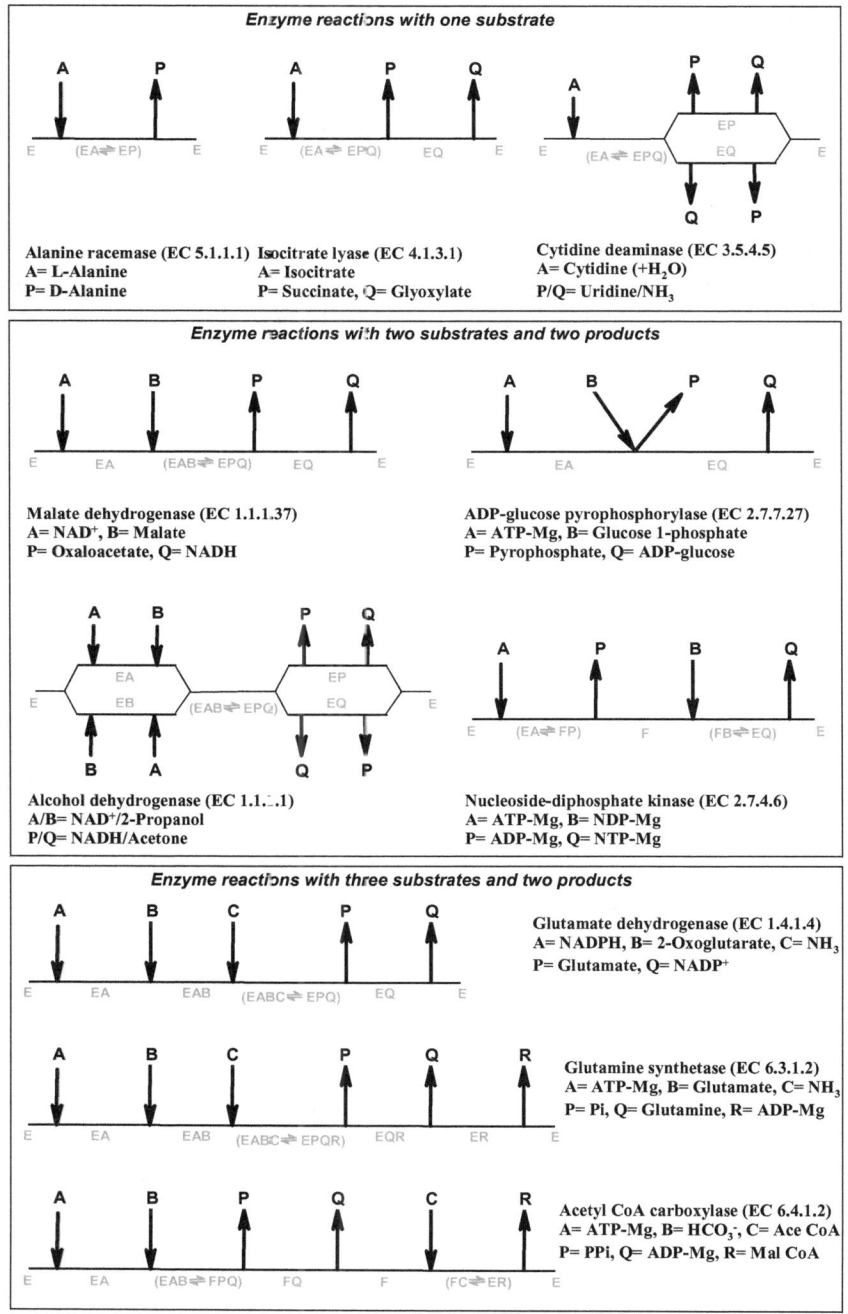

Fig. 15.1 Common enzyme mechanisms and their equilibria according to Cleland notations. All enzyme forms are in gray and central complexes are in brackets. Enzymes with their substrates/products and EC numbers are shown. Note that the same enzyme from a different organism, tissue, or organelle could have a different kinetic mechanism

$$f = [\text{ES}]/[\text{E}_t] = v/V_{\max}$$

Apart from the direct algebraic method (as originally used to derive the Michaelis–Menten equation), there are other ways of deriving rate equations for more complex equilibria. A few of these approaches (with relevant shortcuts and simplifications) are briefly described below.

Algebraic method: This method involves the following steps. (1) Set up proper equilibria for various reaction steps and enzyme forms, (2) make use of steady-state assumption and conservation equations, (3) evaluate the concentration of the ES complex, and finally, (4) present it in terms of [E_t]. As an example, we will derive the rate equation for the equilibria involving two enzyme forms (Fig. 14.1); however, we will also consider the reversible reaction, with [P] $\neq$ 0.

Rate Equation for the Equilibria Involving Two Enzyme Forms
Assuming steady state, we get

$$\frac{d[\text{ES}]}{dt} = k_1[\text{E}][\text{S}] + k_{-2}[\text{E}][\text{P}] - (k_{-1} + k_2)[\text{ES}] = 0$$

Rearranging for [E] in terms of [ES],

$$[\text{E}] = \left(\frac{k_{-1} + k_2}{k_1[\text{S}] + k_{-2}[\text{P}]} \right)[\text{ES}]$$

Substituting for [E] in the enzyme conservation equation ([E_t] = [E] + [ES]) and then solving for [E_t], we get

$$[\text{E}_t] = \left(\frac{k_{-1} + k_2}{k_1[\text{S}] + k_{-2}[\text{P}]} + 1 \right)[\text{ES}] \quad \text{and then} \quad [\text{ES}] = \frac{[\text{E}_t]}{\left(\frac{k_{-1}+k_2}{k_1[\text{S}]+k_{-2}[\text{P}]} + 1 \right)}$$

From the above two equations, we have both [E] and [E_t] expressed in terms of [ES]. Now consider the reaction velocity "v" expressed as substrate disappearance. This may be written as

$$v = -\frac{d[\text{S}]}{dt} = k_1[\text{E}][\text{S}] - k_{-1}[\text{ES}]$$

Substituting for [E] in terms of [ES] and then rearranging, we obtain

$$v = k_1[\text{S}] \left(\frac{k_{-1} + k_2}{k_1[\text{S}] + k_{-2}[\text{P}]} \right)[\text{ES}] - k_{-1}[\text{ES}]$$

(continued)

$$v = \left(\frac{k_1[\text{S}](k_{-1} + k_2)}{k_1[\text{S}] + k_{-2}[\text{P}]} - k_{-1} \right)[\text{ES}]$$

Substituting for [ES] in terms of $[\text{E}_t]$, we get

$$v = \frac{\left(\frac{k_1[\text{S}](k_{-1}+k_2)}{k_1[\text{S}]+k_{-2}[\text{P}]} - k_{-1} \right)}{\left(\frac{k_{-1}+k_2}{k_1[\text{S}]+k_{-2}[\text{P}]} + 1 \right)}[\text{E}_t]$$

Simplifying this equation we obtain

$$v = \left(\frac{k_1 k_2[\text{S}] - k_{-1}k_{-2}[\text{P}]}{k_{-1} + k_2 + k_1\text{S} + k_{-2}\text{P}} \right)[\text{E}_t]$$

Further simplification by dividing both the numerator and the denominator by $(k_{-1} + k_2)$, and then rearranging we get

$$v = \frac{\dfrac{k_2[\text{E}_t][\text{S}]}{\left(\frac{k_{-1}+k_2}{k_1} \right)} - \dfrac{k_{-1}[\text{E}_t][\text{P}]}{\left(\frac{k_{-1}+k_2}{k_{-2}} \right)}}{1 + \dfrac{[\text{S}]}{\left(\frac{k_{-1}+k_2}{k_1} \right)} + \dfrac{[\text{P}]}{\left(\frac{k_{-1}+k_2}{k_{-2}} \right)}}$$

$$v = \frac{\dfrac{V_{\text{maxf}}[\text{S}]}{K_{\text{MS}}} - \dfrac{V_{\text{maxr}}[\text{P}]}{K_{\text{MP}}}}{1 + \dfrac{[\text{S}]}{K_{\text{MS}}} + \dfrac{[\text{P}]}{K_{\text{MP}}}}$$

As mentioned before (see the Haldane relationship, Chap. 14), this rate equation is symmetric with respect to $\text{S} \rightarrow \text{P}$ and $\text{P} \rightarrow \text{S}$. It is a more general form of the Michaelis–Menten equation. If we put $[\text{P}] = 0$, then the equation collapses to the classic form of the Michaelis–Menten equation, as derived in the previous chapter.

The algebraic method, exemplified above, is good for a mechanism with few enzyme forms. The complexity of derivation increases with more enzyme forms and the algebra involved becomes daunting. One needs to solve "n" simultaneous equations for as many enzyme forms—to get their unknown concentrations. This is best achieved by solving the simultaneous algebraic equations by the *determinants* (of the nth order) method. While the method is time-consuming for complex enzyme mechanisms, it is quite useful in computer-assisted derivation of rate equations. Further details may be found in specialist books on enzyme kinetics.

King-Altman procedure: Complex equilibria with many enzyme forms are best handled by this method. The King-Altman method exploits the topological approach (King and Altman 1956). Various enzyme forms are set up with proper equilibria in the form of a figure. Care is taken to ensure that each enzyme form occurs only once in this figure. The fraction of $[\text{E}_t]$ present in each enzyme form is then evaluated using this representation. To do this, one lists all the possible patterns that interconnect all enzyme forms, but without forming closed loops. For example, for "n" enzyme species each pattern should contain "$n-1$" lines. A partition equation can now be written for each form—which defines the proportion of the enzyme in that form, in terms of individual rate constants and relevant concentrations. A partition equation for any enzyme form (E_n) can be written in terms of [E_t] generally as

$$[E_n] = \frac{D_n}{D_1 + D_2 + \ldots + D_n}[E_t],$$

where D_1 through D_n are numerators for respective enzyme forms while their sum ($\sum$) is the denominator. Thus, for each enzyme form there is an expression which when divided by the sum of all such expressions ($\sum$) gives the partition equation—describing the fraction of that enzyme form present in steady state. Suitable partition equations are then used to evaluate the rate in the forward direction. Derivation of the rate equation, for equilibria involving two enzyme forms (Fig. 14.1), by this approach is shown in the box below.

King-Altman Procedure for Equilibria with Two Enzyme Forms
The equilibria shown in Fig. 14.1 may be rewritten in the form of a figure as shown.

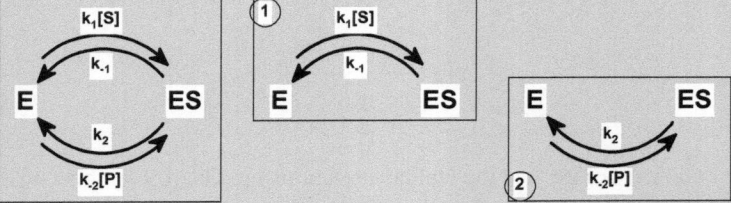

Two possible patterns that interconnect both enzyme forms, without forming closed loops, are shown in 1 and 2. With their help, partition equations corresponding to E and ES forms can now be written. Consider the formation of E for instance. It gets formed with a rate constant of k_{-1} (in box 1) and k_2 (in box 2). Accordingly, its partition equation may be written as

$$\frac{[E]}{[E_t]} = \frac{k_{-1} + k_2}{\Sigma}$$

Similarly, for ES we get

(continued)

$$\frac{[\text{ES}]}{[\text{E}_\text{t}]} = \frac{k_1[\text{S}] + k_{-2}[\text{P}]}{\Sigma}$$

From enzyme conservation equation ($[\text{E}_\text{t}] = [\text{E}] + [\text{ES}]$), we observe that Σ is the sum of all the numerator terms

$$\frac{[\text{E}]}{k_{-1} + k_2} = \frac{[\text{ES}]}{k_1[\text{S}] + k_{-2}[\text{P}]}$$

$$= \frac{[\text{E}_\text{t}]}{\Sigma} \quad \text{(and hence } \Sigma = k_{-1} + k_2 + k_1[\text{S}] + k_{-2}[\text{P}])$$

Furthermore, solving for $[\text{E}_\text{t}]$ in terms of $[\text{ES}]$ we get

$$[\text{E}_\text{t}] = \left(\frac{k_{-1} + k_2}{k_1[\text{S}] + k_{-2}[\text{P}]}\right)[\text{ES}] + [\text{ES}] = \left(1 + \frac{(k_{-1} + k_2)}{k_1[\text{S}] + k_{-2}[\text{P}]}\right)[\text{ES}]$$

On rearranging, this equation allows us to present $[\text{ES}]$ in terms of $[\text{E}_\text{t}]$ as

$$[\text{ES}] = \left(\frac{k_1[\text{S}] + k_{-2}[\text{P}]}{k_{-1} + k_2 + k_1[\text{S}] + k_{-2}[\text{P}]}\right)[\text{E}_\text{t}]$$

Note that the denominator (i.e., $k_{-1} + k_2 + k_1[\text{S}] + k_{-2}[\text{P}]$) of the above equation is the sum of all numerators (Σ) mentioned above. Next, we write an analogous expression for $[\text{E}]$ in terms of $[\text{E}_\text{t}]$ as

$$[\text{E}] = \left(\frac{k_{-1} + k_2}{k_{-1} + k_2 + k_1[\text{S}] + k_{-2}[\text{P}]}\right)[\text{E}_\text{t}]$$

Considering the reaction velocity "v" in terms of product formed, we write

$$v = \frac{d[\text{P}]}{dt} = k_2[\text{ES}] - k_{-2}[\text{E}][\text{P}]$$

Upon substituting for $[\text{E}]$ and $[\text{ES}]$ obtained from the above partition equations

$$v = k_2\left(\frac{k_1[\text{S}] + k_{-2}[\text{P}]}{k_{-1} + k_2 + k_1[\text{S}] + k_{-2}[\text{P}]}\right)[\text{E}_\text{t}] - k_{-2}\left(\frac{k_{-1} + k_2}{k_{-1} + k_2 + k_1[\text{S}] + k_{-2}[\text{P}]}\right)$$

$$\times [\text{E}_\text{t}][\text{P}]$$

Simplifying further

(continued)

$$v = \frac{k_1 k_2 [\text{S}][\text{E}_\text{t}] - k_{-1} k_{-2} [\text{P}][\text{E}_\text{t}]}{k_{-1} + k_2 + k_1 [\text{S}] + k_{-2} [\text{P}]}$$

Dividing both the numerator and the denominator by $(k_{-1} + k_2)$ and rearranging, we get

$$v = \frac{\frac{k_1}{k_{-1}+k_2} k_2 [\text{E}_\text{t}][\text{S}] - \frac{k_{-2}}{k_{-1}+k_2} k_{-1} [\text{E}_\text{t}][\text{P}]}{1 + \frac{k_1}{k_{-1}+k_2}[\text{S}] + \frac{k_{-2}}{k_{-1}+k_2}[\text{P}]}$$

By appropriate substitutions for $V_{\max}$ and K_M terms, the equation takes the following form:

$$v = \frac{\frac{V_{\text{maxf}}[\text{S}]}{K_{\text{MS}}} - \frac{V_{\text{maxr}}[\text{P}]}{K_{\text{MP}}}}{1 + \frac{[\text{S}]}{K_{\text{MS}}} + \frac{[\text{P}]}{K_{\text{MP}}}}$$

Notice that the above equation is identical to the one derived by the algebraic method. The King-Altman procedure is schematic in nature and one can write down the rate equation by inspecting patterns connecting different enzyme forms. It can be used for more complex schemes than the example described above. The procedure however becomes complicated with multi-substrate random mechanisms as it gives squared substrate terms. Also, if more than one closed loop is present, the precise number of unique patterns to be considered becomes nontrivial.

Net rate constant method: This method is a useful shortcut developed by Cleland (1975). It is ideal to derive rate equations for simple kinetic mechanisms without branched pathways. The protocol involves the following steps. (1) Set up appropriate equilibria with all enzyme forms, (2) represent steady-state flux at each step as unidirectional (net rate) constants (denoted by k' values) such that flux values in each step along with the distribution of enzyme species remain the same, (3) begin with an irreversible step in the original scheme, where the net rate constant and the real rate constant are the same ($k_\text{n}' = k_\text{n}$), and evaluate each net rate constant by going backwards, and finally (4) substitute net rate constants in a suitable equation to obtain the rate expression. How this procedure works is shown for a linear mechanism with three enzyme forms (see box below).

Net Rate Constant Method for Linear Equilibria

Consider the mechanism with three enzyme forms E, ES, and EP, as shown. Although the actual rate constants are different, for each step we substitute a net rate constant (k_n') such that $k_\text{n}' = k_\text{n} \times$ partition ratio. Note that k_n is the true forward rate constant for the step in question.

(continued)

$$E \underset{k_{-1}}{\overset{k_1[S]}{\rightleftharpoons}} ES \underset{k_{-2}}{\overset{k_2}{\rightleftharpoons}} EP \underset{k_{-3}[P]}{\overset{k_3}{\rightleftharpoons}} E$$

Corresponding Net-rate constants:

$$E \overset{k_1'}{\longrightarrow} ES \overset{k_2'}{\longrightarrow} EP \overset{k_3'}{\longrightarrow} E$$

If steady-state conditions operate, then by definition net rates for all the steps are equal. That is at steady state

$$v = k_1'[E] = k_2'[ES] = k_3'[EP]$$

Therefore,

$$\frac{[E]}{v} = \frac{1}{k_1'} \quad \text{and} \quad \frac{[ES]}{v} = \frac{1}{k_2'} \quad \text{and} \quad \frac{[EP]}{v} = \frac{1}{k_3'}$$

Since $[E_t] = [E] + [ES] + [EP]$, we know that

$$\frac{[E]}{v} + \frac{[EA]}{v} + \frac{[EP]}{v} = \frac{1}{k_1'} + \frac{1}{k_2'} + \frac{1}{k_3'} = \frac{[E_t]}{v}$$

And this can be suitably rearranged to get

$$v = \frac{[E_t]}{\frac{1}{k_1'} + \frac{1}{k_2'} + \frac{1}{k_3'}}$$

The concept of "net rate constant" is like "conductance" in electrical systems. Reciprocal of the net rate constant then becomes resistance and the sum of resistances (denominator term above) dictates what fraction of $[E_t]$ is in the productive from.

It now remains to plug in the values of individual net rate constants and simplify to obtain the rate expression. If we consider initial velocity conditions (i.e., $[P] = 0$), then the last step of the above linear mechanism becomes irreversible. And therefore $k_3' = k_3$. We now go backwards sequentially, to evaluate the net rate constant for the previous steps. For instance,

$$k_2' = k_2 \times \text{Partition ratio for EP}$$

The net rate constant thus is the real forward rate multiplied by the partition ratio for that enzyme form. For EP form,

(continued)

$$\text{Partition ratio} = \frac{\text{Rate of EP going forward}}{\text{Rate of EP going forward} + \text{Rate of EP returning to EA}}$$

This can be represented as

$$\text{Partition ratio for EP} = \frac{k_3'}{k_{-2} + k_3'} = \frac{k_3}{k_{-2} + k_3} \quad (\text{because } k_3' = k_3)$$

We now substitute this value of the partition ratio to obtain k_2' as shown below

$$k_2' = k_2 \times \frac{k_3}{k_{-2} + k_3} = \frac{k_2 k_3}{k_{-2} + k_3}$$

In a similar manner, k_1' (step previous to k_2' step) may now be evaluated. Finally, by substituting for k_2' from above and simplifying

$$k_1' = k_1[S] \times \frac{k_2'}{k_{-1} + k_2'} = \frac{k_1[S]\frac{k_2 k_3}{k_{-2}+k_3}}{k_{-1} + \frac{k_2 k_3}{k_{-2}+k_3}} = \frac{k_1 k_2 k_3[S]}{k_{-1}k_{-2} + k_{-1}k_3 + k_2 k_3}$$

This is how all the net rate constants (k_1' through k_3', in the given mechanism) are evaluated. These can now be substituted in the general form of the rate expression obtained earlier:

$$v = \frac{[E_t]}{\frac{1}{k_1'} + \frac{1}{k_2'} + \frac{1}{k_3'}}$$

The following equation is thus obtained:

$$v = \frac{[E_t]}{\frac{1}{\frac{k_1 k_2 k_3[S]}{k_{-1}k_{-2}+k_{-1}k_3+k_2 k_3}} + \frac{1}{\frac{k_2 k_3}{k_{-2}+k_3}} + \frac{1}{k_3}}$$

This equation can now be rearranged and simplified into the rate expression as shown below

$$v = \frac{[E_t]}{\frac{k_{-1}k_{-2}+k_{-1}k_3+k_2 k_3}{k_1 k_2 k_3[S]} + \frac{k_{-2}+k_3}{k_2 k_3} + \frac{1}{k_3}}$$

$$= \frac{k_1 k_2 k_3[E_t][S]}{k_{-1}k_{-2} + k_{-1}k_3 + k_2 k_3 + (k_2 + k_{-2} + k_3)k_1[S]}$$

(continued)

$$v = \frac{\left(\frac{k_2 k_3}{k_2 + k_{-2} + k_3}\right)[E_t][S]}{\left(\frac{k_{-1}k_{-2} + k_{-1}k_3 + k_2 k_3}{(k_2 + k_{-2} + k_3)k_1}\right) + [S]}$$

We recognize that the final form of the rate equation derived by net rate constant method resembles the typical Michaelis–Menten equation. Remarkably, the expression contains all the individual rate constants and thus allows us to obtain V_{max}/K_M and V_{max} in terms of these individual rate constants. The method therefore (a) is best suited for deriving rate expressions for isotope exchange, isotope partitioning, and positional isotope exchange studies and (b) shows good promise in interpreting isotope effects on V_{max}/K_M and V_{max} of the enzyme.

Another advantage of net rate constant method is that expressions for V_{max}/K_M or V_{max} may be obtained without deriving the entire rate equation. Consider V_{max}/K_M first. From the basics of Michaelis–Menten formalism we know that $v = (V_{max}/K_M) \times [S]$, at low [S]. By inspecting the above linear mechanism, we see that $k_1{}'$ is rate limiting at low [S]. And therefore,

$$v = k_1{}' \times [E_t] = \frac{k_1 k_2 k_3 [S]}{k_{-1}k_{-2} + k_{-1}k_3 + k_2 k_3} \times [E_t]$$

On comparison with the equation "$v = (V_{max}/K_M) \times [S]$" we can write

$$\frac{V_{max}}{K_M} = \frac{k_1 k_2 k_3 [E_t]}{k_{-1}k_{-2} + k_{-1}k_3 + k_2 k_3}$$

Similarly, now consider V_{max}. At saturating [S], $v = V_{max}$ and $k_1{}'$ can be neglected. Therefore,

$$v = \frac{[E_t]}{\frac{1}{k_2{}'} + \frac{1}{k_3{}'}} = \frac{[E_t]}{\frac{1}{\frac{k_2 k_3}{k_{-2} + k_3}} + \frac{1}{k_3}} = \frac{k_2 k_3 [E_t]}{k_2 + k_{-2} + k_3} = V_{max}$$

We can compare these expressions for V_{max}/K_M and V_{max} with the full rate expression above and identify the relevant terms contributing to them.

Other methods: There are a few other variations, in addition to the three methods described above, to derive a rate expression. A method described by Cha simplifies the rapid equilibrium segment containing many enzyme forms as though it were a single enzyme species (Cha 1968). A single lumped up rate constant is then used to represent this segment and a rate equation is derived. This is a useful tool when the random addition of substrates occurs in the mechanism—as it avoids squared terms. The assumption of rapid equilibrium is a useful simplification and need not actually be true for the method to work.

15.4   Enzyme Kinetics and Common Sense

It should be obvious from the general theme of this chapter that enzyme kinetic analysis provides valuable mechanistic insights. A minimal mathematical ability is required to meaningfully appreciate and use this tool. Although an added advantage, mathematical proficiency is not mandatory to apply kinetic methods to enzyme mechanisms. As ascribed to Einstein: *We should make things as simple as possible, but not simpler*. Derivation of a few rate equations was deliberately included to bring home this point. However, it is not a prerequisite to appreciate the subject matter of this book.

The kinetic methods described above allow us to derive and appreciate the connection between a mechanism and its corresponding rate equation. But the correctness of such an equation is only as valid as the assumptions made in deriving it. Some mechanisms may be quite complex and equations formidable. Nevertheless, doing kinetics can be fun so long as one understands what these equations mean. For most examples, the hard work (of deriving them!) has already been done; a lot can be accomplished by judicious use of these equations found in the literature. One need only develop a sense of discrimination and understand the conceptual meaning of the equation to be used. Appropriate use of equations found in the enzyme literature is as important as deriving new ones. As Cleland has stated: *All the mathematics in the world is no substitute for a reasonable amount of common sense*.

This commonsense approach coupled with elementary mathematical ability forms the basis of enzyme mechanisms described in this book. The emphasis will therefore be more on the conceptual framework of kinetic description and analysis.

References

Cha S (1968) A simple method for derivation of rate equations for enzyme-catalyzed reactions under the rapid equilibrium assumption or combined assumptions of equilibrium and steady state. J Biol Chem 243:820–825

Cleland WW (1963) The kinetics of enzyme-catalyzed reactions with two or more substrates or products: I. Nomenclature and rate equations. Biochim Biophys Acta 67:104–137. II. Inhibition: nomenclature and theory. Biochim Biophys Acta 67:173–187. III. Prediction of initial velocity and inhibition patterns by inspection. Biochim Biophys Acta 67:188–196

Cleland WW (1975) Partition analysis and concept of net rate constants as tools in enzyme kinetics. Biochemistry 14:3220–3224

Cleland WW (1989) Enzyme kinetics revisited: a commentary by W Wallace Cleland. Biochim Biophys Acta 1000:209–221

King EL, Altman C (1956) A schematic method of deriving the rate laws for enzyme-catalyzed reactions. J Phys Chem 60:1375–1378

Wildner M (1999) In memory of William of Occam. Lancet 354:1272

Enzyme Kinetic Data: Collection and Analysis

16

Michaelis–Menten formalism undoubtedly is a very important first approximation for any new enzyme to be studied. Although simple and elegant, it has only a limited range of applicability. Not all enzymes adhere to Michaelis–Menten kinetics and there are notable exceptions. Appropriate kinetic experimental design allows us to make this judgment. Collating good-quality kinetic data is the first task in enzyme characterization.

16.1 Obtaining Primary Data: Practical Aspects

A reliable and robust assay method is a prerequisite for obtaining enzyme data. Good kinetic practices also ensure that the best quality primary data are collected. Extensive coverage of both these aspects may be found in Chaps. 11 and 12. Additional considerations of importance in generating primary kinetic data are discussed below.

Reductionism in experimental design: The kinetic experimental design almost always takes a reductionist approach—varying one parameter at a time while keeping all others constant. The parameters that may be varied include [S], [P], pH, ionic strength, buffer species, activators, inhibitors, etc. Perhaps the most important and informative data set is the change in initial velocity versus substrate concentration (the $v \to [S]$ plot). Recall that even in a multi-substrate reaction, a series of bimolecular collisions take place to assemble the productive enzyme complex. Furthermore, such reactions can be treated as pseudo-uni-molecular with respect to one substrate by holding all others constant. In this sense, the primary data set is a $v \to [S]$ curve having 6–10 data points for every substrate (see below). Complete kinetic analysis of a bi-substrate reaction ($n = 2$) therefore requires approximately 10^2 independent assays to be performed. As the number of reactants and modifiers increases, the experimental data to be collected become enormous. Typically, the data volume increases as a power function of "n," where "n" is the number of reactants (substrate/product) and modifiers (activator/inhibitor) associated

© The Author(s), under exclusive license to Springer Nature Singapore Pte Ltd. 2025

N. S. Punekar, *ENZYMES: Catalysis, Kinetics and Mechanisms*, https://doi.org/10.1007/978-981-97-8179-9_16

Table 16.1 Variation of initial velocity with substrate concentration: original $v \rightarrow$ [S] data for arginase

[S] (Arginine, mM)	ΔA_{478}	[P] (Urea, mM)	v (μmol Urea $\times$ min$^{-1} \times$ mg^{-1})	[S] depleted (%)	$[\bar{S}]$ $(([S_i] + [S_f])/2)$ (mM)
6.3	0.173	0.86	15.7	13.7	5.8
12.5	0.329	1.65	30.1	13.2	11.7
14.2	0.348	1.74	31.7	12.2	13.3
16.2	0.402	2.01	36.7	12.4	15.2
20.0	0.413	2.06	37.7	10.3	19.0
35.0	0.634	3.17	57.9	9.1	33.4
50.0	0.650	3.25	59.4	6.5	48.4
100.0	0.842	4.21	76.9	4.2	97.9
150.0	0.975	4.88	89.0	3.3	147.6

Fig. 16.1 A v versus [S] plot of arginase data from Table 16.1. The line drawn through the data points is the nonlinear least-squares best fit to the Michaelis–Menten equation, representing a rectangular hyperbola

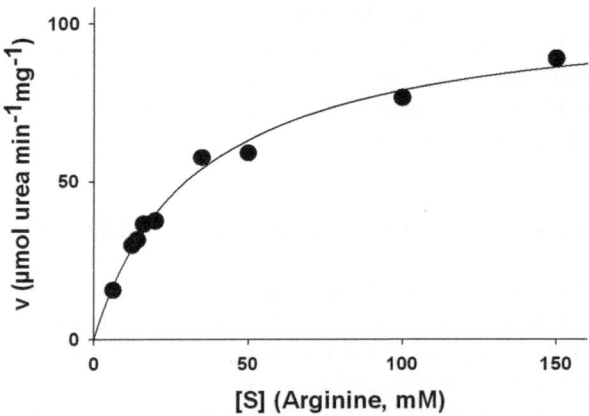

with that enzyme. For instance, *E. coli* glutamine synthetase is affected by at least eight reactants/modifiers. Accordingly, as many as 10^8 assays may be necessary to fully define its kinetics!

   Choice of substrate concentrations: A major objective in the kinetic profiling of any enzyme is to determine its kinetic constants V_{max} (and hence k_{cat}) and K_M. Initial velocity measurements made at different [S] are the original data required to achieve it. Table 16.1 provides a sample of $v \rightarrow$ [S] data for arginase (initial velocities obtained under steady-state conditions). We shall use these data to demonstrate various aspects of subsequent kinetic analysis. The first and most straightforward way of analyzing the original data is to plot a graph of $v \rightarrow$ [S] as shown in Fig. 16.1. Reasonable estimates of V_{max} and K_M may be obtained from such plots—provided the data cover a broad range of [S].

   Urea formed due to arginase action on L-arginine was estimated as ΔA_{478} by colorimetry (Archibald method). The original raw data are shown in black where [S] is an independent variable (decided by the experimenter) and ΔA_{478} (reflecting urea formed) is the dependent variable. All other parameters (shown in gray) are derived from this primary data set. For the majority of kinetic analysis [S] and v data (shown in bold face) are used.

The concentration of urea in the 200 μL reaction (after a 10 min assay) is calculated from a urea standard curve (slope, 0.16). Initial velocity (v) is calculated (see Chap. 2, Sect.2.6 for details) from here using the amount of enzyme protein (0.35 μg per 200 μL assay). The percent of [S] depleted is obtained as follows: for example, 2.06 mM of urea is formed stoichiometrically from 20 mM of arginine. This amounts to 10.3% of the substrate converted to the product. Accordingly, $[\bar{S}]$ will be calculated as [20 mM + (20.0 − 2.06) mM]/2 = 19 mM.

A broad range of [S] must be used to obtain reliable estimates of V_{max} and K_M. The two extreme cases [S] << K_M and [S] >> K_M, respectively, define V_{max}/K_M and V_{max} (Table 14.1) from this data set. One usually begins with the definition of V_{max}—the data representing high [S] (as will be discussed later). If only the lower [S] is covered, then the data will appear to be first order with respect to [S]. This is as defined by $v = (V_{max}/K_M) \times [S]^1$, the limiting case of the Michaelis–Menten equation (Fig. 14.2). On the other hand, measurements made only at higher [S] are biased toward V_{max}. From such data, V_{max} may be estimated but there is no way to determine K_M. Clearly, a range of substrate concentrations, between 0.1 and 10 K_M (0.33–3 K_M, at the least!), should be used to accurately determine the kinetic constants. The $v \to$ [S] graph around K_M is the region of maximum curvature for a rectangular hyperbola (Fig. 14.3). It is therefore wise to choose data points on both sides of K_M so that the curve is best defined!

Pilot experiments and iteration: A new enzyme study often begins with a reliable assay method but no prior knowledge of its kinetic constants. How to choose the appropriate [S] range then? It is common practice to conduct pilot experiments with data points spanning a broad range of [S]. From these preliminary data, a rough estimate of V_{max} and K_M is obtained. More reliable estimates can then be obtained by narrowing the [S] range between 0.33 K_M to 3 K_M (using a rough value of this K_M obtained from the pilot experiment). Finally, many data points are generated within this range to calculate V_{max} and K_M values. Obtaining a reliable and useful $v \to$ [S] data set for an enzyme is thus an iterative process. Within a couple of rounds, we should be able to reach the best range of substrate concentrations required.

Importance of measuring initial velocities: Precise measurement of initial velocity is of prime importance in kinetic analysis. Underestimation of initial velocity is a common problem when nonlinear time courses are observed. Data obtained using a "continuous assay" are therefore more reliable. One practical remedy against nonlinear time courses is to *use dilute enzyme solutions*. Working with suitably diluted enzymes also helps to (a) reduce rates to manageable levels, (b) conserve the precious enzyme, and (c) eliminate unwanted interactions, if any. Other difficulties in experimental measure of initial velocity at higher [S] may be due to reasons of limited solubility, interference in measurements, sensitivity of detection method, etc. An example of one such limitation and how to overcome it is shown in the box below.

Monitoring NADP-Glutamate Dehydrogenase Reaction Progress

The NADP-glutamate dehydrogenase reaction can in principle be continuously monitored by an increase in absorption at 340 nm. This is due to the reduction of $NADP^+$ to NADPH during the reaction. The enzyme (from *A. niger*) exhibits a K_M of about 10 µM for its substrate $NADP^+$. A maximal absorbance difference at 340 nm ($\varepsilon = 6220 \ M^{-1} \ cm^{-1}$) of 0.06 is obtained when all 10 µM $NADP^+$ is converted to NADPH. If only 10% substrate conversion is permissible, this value can be 0.006. The $\Delta A_{340 \ nm}$ values obtainable for $NADP^+$ concentrations below K_M (<10 µM) are even smaller! Attempts to achieve a larger $\Delta A_{340 \ nm}$ will surely lead to higher substrate depletion, nonlinear time course, and erroneous initial velocity. The poor sensitivity of the spectrophotometric assay, particularly in this example, makes it unsuitable for use. *Switching to a more sensitive detection method is a better option.* For instance, the fluorimetric estimation of the product (NADPH) allows precise initial velocity measurements, (a) even at $NADP^+$ concentrations below 10 µM and (b) with permissible (less than 10%) substrate depletion.

The elementary consideration in measuring the "[S] versus v" data is that true initial rates be recorded, at every substrate concentration tested. This, in practice, is however easier said than done—particularly at low [S] values. In the Michaelis–Menten formalism we assume that $[S_t] \approx [S]$. In practice, therefore, up to 5–10% depletion of $[S_t]$ may be tolerated over the assay period. This is because experimental errors (and variation) often contribute more than this substrate depletion effect. The depletion of the substrate is a significant problem at lower [S] ranges tested (Table 16.1). One should therefore ensure that less than 5–10% substrate is converted to product(s) while assaying the enzyme at the lowest [S] chosen. Any higher level of S $\rightarrow$ P conversion quickly results in deviation of rates from linearity—and underestimation of v. Since experimental errors are large at low [S], and it is desirable to limit substrate conversion to under 10%, many replicate measurements may be required at low [S] values.

Lee and Wilson suggested a modification to analyze kinetic data where significant substrate conversion (as high as 40% depletion) has occurred (Lee and Wilson 1971). Suppose the substrate concentration decreases from $[S_t]$ to $[S_f]$ by the end of the assay. Instead of using the initial substrate concentration added ([S] total, i.e., $[S_t]$) the arithmetic mean of $[S_t]$ and $[S_f]$ is recommended. The enzyme does not see $[S_t]$ throughout the assay period. Therefore, the arithmetic mean ($[\bar{S}] = ([S_t] + [S_f])/2$) is a more appropriate measure of substrate concentration in $v \rightarrow [S]$ plots. For instance, suppose the initial substrate concentration of 5.0 mM reduces to 4.0 mM (at the end of the assay) because of a 20% conversion. The effective average substrate concentration ($[\bar{S}]$) felt by the enzyme during the assay is 4.5 mM (and not 5 mM!). This difference is even larger when substrate depletion becomes higher (for instance, see the analysis of arginase data in Table 16.1). Finally, a word of caution is in order, however. The ***Lee–Wilson modification*** works well only when the substrate and/or product do not significantly inhibit the enzyme. This procedure is not suitable for enzymes where substrate inhibition is observed, or significant product inhibition occurs at low [P] levels.

Utility of the integrated form of the Michaelis–Menten equation: A complete time course (the reaction progress curve) is actually a more robust source of kinetic information. It allows us to characterize the rate behavior at different extents of substrate depletion and product accumulation. Kinetic analysis of a time course is possible with the integrated form of the corresponding rate equation. This approach is particularly well suited to reactions that can be monitored continuously. The *integrated rate equations are commonly used in chemical kinetics* but rarely in enzyme kinetics (of course apart from fast reaction kinetics—analysis of transients). The Michaelis–Menten equation can be integrated, and this may be written as shown

$$\frac{[S_t] - [S]}{t} = -K_M \times \frac{1}{t} \times \ln \frac{[S_t]}{[S]} + V_{max}$$

where $[S_t]$ and $[S]$ are substrate concentrations at time zero and time t, respectively. Therefore, the product formed after time t is ($[S_t]-[S]$). A plot of ($[S_t]-[S]$)/t versus $1/t \times \ln([S_t]/[S])$ should give a straight line with V_{max} as its intercept and K_M as its slope. In principle, with this integrated form of the equation, a single extended reaction time course analysis should suffice to obtain all the enzyme kinetic parameters (Schwert 1969; Moreno 1985). This also avoids mixing errors associated with initial-rate methods.

Then *why is it that initial-rate studies are popular in enzyme kinetics*? One reason is historical. Second, the integrated form (as shown above) does not incorporate the effects of product accumulation. When incorporated, however, the system quickly gets complicated—even for a single substrate-product example. The use of integrated rate equations does not provide immediately evident conclusions that are possible from initial rate measurements. In order to estimate kinetic parameters from integral rate curves or from initial rate measurements, initial substrate concentrations must be of the same order of magnitude as the K_M values. Therefore, it appears that integrated rate curves continue to be used as supplementary to initial velocity kinetics. Lastly, the initial velocity measurement method is advantageous in that individual variables like [S], [P], and [E_t] can be manipulated individually and at will.

16.2 Analyzing Data: The Basics

Investigating the kinetic properties of an enzyme implies learning how it responds to changes in the environment. The most common variable is [S] and it is essential to work over a wide range of [S] where the rate changes appreciably. Provided an enzyme obeys the Michaelis–Menten equation, good design of kinetic experiments requires that [S] values should extend on both sides of the K_M. Typically 3–5 data points below and an equal number above K_M are desirable. The data set for a well-represented $v \rightarrow$ [S] plot should thus have at least 6–10 well-spread points.

Variation, errors, and statistics: Whether Michaelis–Menten formalism operates or not, the $v \rightarrow$ [S] data for an enzyme represent a nonlinear relationship. Experimental data occupy only a segment of the rectangular hyperbola described by the equation (Fig. 14.3). Therefore, one must start the kinetic analysis with high-quality, original $v \rightarrow$ [S] data. Large errors are associated with measurements of v at lower [S]. This is because the initial velocity responds steeply in this [S] range. And many replicates may be required. This however requires care and attention, especially when working with unstable enzyme preparations. For instance, suppose we wish to vary one parameter ([S]) at four different values of the other (say the second substrate). For ten points per $v \rightarrow$ [S] data set, a total of 40 assays need to be performed. The enzyme may lose significant activity toward the end of this lengthy experiment. This should be checked of course. It is possible to pool original $v \rightarrow$ [S] data from separate experiments while evaluating kinetic constants (like V_{max} and K_M). But it is best to use a full set of $v \rightarrow$ [S] data generated in a single experiment—this minimizes "between experiments" variation. Repeat measurements should be performed to obtain a reliable data set because *any degree of sophisticated analysis will not transform bad data into good data!*

Errors cannot be avoided while obtaining kinetic data. It is important to appreciate the nature of these errors and their scatter. Due to error scattering, it becomes difficult to decide whether the measured data fit the assumed rate equation (such as the Michaelis–Menten equation and hyperbolic curve). Analysis of such data requires statistical tools, and regression methods in particular. This treatment is very helpful for nonlinear curves where systematic deviations are more difficult to detect by the eye. The correlation coefficient indicates the consistency of the data with the assumed model (and rate equation). Residual plots are used to measure the deviation of each value from the assumed function (as per the rate equation). Rigorous statistical analysis reduces the dangers of subjectivity in interpretation. Nevertheless, it is worth remembering that the accuracy of calculation cannot compensate for the lack of accuracy in collecting or recording data.

Best curve fitting can be generated by nonlinear least-squares fit of the data to the rate equation, such as the Michaelis–Menten equation. Many graphics programs are available today to perform nonlinear curve fitting. A few common examples are listed in Table 16.2. One should be reasonably familiar with the limitations of such programs, however. The two unknowns V_{max} and K_M are solved iteratively in such programs. Direct analysis of the untransformed data provides the most reliable estimates of V_{max} and K_M.

One objective of kinetic studies is to get an estimate of the intrinsic kinetic constants such as V_{max} (and hence k_{cat}) and K_M for the enzyme. Different means of analyzing enzyme kinetic data are discussed below.

Table 16.2 Software available to analyze enzyme kinetic data

Name	Software details
Cleland's Package	Suite of FORTRAN programs; *Methods in Enzymology* 63:103 (1979); Open source
SigrafW	Microsoft[R] Visual Basic Studio program; *Biochemistry and Molecular Biology Education* 33:399 (2005); Open source
Hyper and Median	Hyper.exe is a program for the analysis of enzyme kinetic data; http://homepage.ntlworld.com/john.easterby/abouthyp.html; Open source
Leonora	Steady-state Enzyme Kinetics by A. Cornish-Bowden; Supplement to *Analysis of Enzyme Kinetic Data*, Oxford University Press (1995)
DynaFit	BioKin, Ltd.; *Analytical Biochemistry* **237**, 260-(1996); http://www.biokin.com/; Open source/commercial
VisualEnzymics	Softzymics, Inc. 623 Brickhouse Road, Princeton, NJ 08540; http://www.softzymics com/; Commercial
SigmaPlot	Enzyme Kinetics Module; Systat Software Inc.; http://www.sigmaplot.com/; Commercial
EnzFitter	BIOSOFT, PO Box 1013, Great Shelford, Cambridge, CB22 5WQ GB—United Kingdom; http://www.biosoft.com/w/enzfitter.htm; Commercial

16.3 Plotting *v* Versus [S] Data

The v versus [S] plot: The most valuable insight on enzyme kinetic behavior is found in the original $v \rightarrow$ [S] plot (Fig. 16.1, for example). It is important to critically examine this plot before attempting to transform the original data into linear plot forms (e.g., Lineweaver–Burk plot, discussed later in this chapter). From the first look, the data may appear to follow a rectangular hyperbola. This can be easily checked as follows: obtain estimates of V_{max} and K_M from this data set as if the data fit a Michaelis–Menten equation; using these two constants, generate the rectangular hyperbola and compare this computed curve with the original curve. Within the limits of error, any systematic deviations become obvious by this comparison. Software exists to do this nonlinear curve fitting exercise (curve in Fig. 16.1 fitted to data in Table 16.1).

Reasonable estimates of V_{max} and K_M may be obtained manually from the original $v \rightarrow$ [S] plot—provided the data cover a broad range of [S]. One plots the data on graph paper and draws a curve by simply connecting the data points by straight lines. A horizontal line is drawn at the apparent plateau value of v and its point of intersection with the Y-axis then defines V_{max}. The point on the Y-axis where $v = V_{max}/2$ is then located. A horizontal line is drawn from this $V_{max}/2$ point to the point of intersection with the data curve. A vertical line from this intersection to the X-axis then defines the value of K_M (see Fig. 14.2, for instance).

Extraction of the two constants from a straightforward $v \rightarrow$ [S] plot is error-prone because of the nonlinear relationship. The V_{max} has to be obtained from the value of the asymptote to the X-axis. However, this involves geometrical extrapolation, which is difficult. In practical terms, it may also not be feasible to test higher

[S] values—substrate solubility being one major consideration. K_M is nothing but [S] at $V_{max}/2$, and therefore errors of estimation in V_{max} are carried over into the evaluation of K_M. The K_M directly determined from $v \rightarrow$ [S] plots are also subject to significant errors. Finally, a direct comparison of a group of hyperbolas obtained from different experiments is difficult. Several data analysis procedures and manipulations of the Michaelis–Menten equation therefore evolved and are in use over the years. Some of these important transformations are considered below.

Direct linear plot: This plot was suggested by Eisenthal and Cornish-Bowden (1974), where a series of "v–[S]" data pairs are directly plotted. For each "v–[S]" pair of data we can generate a straight line by marking v on the Y-axis (the V_{max} axis) and [S] on the negative side of the X-axis (the K_M axis). All these lines (n lines for as many "v–[S]" pairs of data!) must intersect at a point in the first quadrant with K_M and V_{max} as its coordinates. This result follows from the following transformation of the original Michaelis–Menten equation. Taking reciprocals on both sides,

$$\frac{1}{v} = \frac{K_M + [S]}{V_{max}[S]} \text{ and then on rearranging we obtain,}$$

$$\frac{V_{max}}{v} = \frac{K_M}{[S]} + 1$$

According to this equation, when $K_M = 0$ we get $V_{max} = v$, and again for $V_{max} = 0$ we obtain $K_M = -[S]$. Therefore, (a) we plot [S] on the negative side of the X-axis and (b) the point of intersection of all the lines has K_M and V_{max} as its coordinates. The Eisenthal–Cornish-Bowden direct linear plot of the data from Table 16.1 is plotted for example in Fig. 16.2. These lines in practice do not neatly converge but intersect over a range of values—a clear reflection of experimental errors. This in itself is very useful in revealing poor data points. If it is certain that (a) the enzyme obeys Michaelis–Menten formalism and (b) nonlinear curve fitting is not feasible, then the direct linear plot provides the best estimates of K_M and V_{max} from $v \rightarrow$ [S] data.

v versus log[S] plot: Apart from the simple $v \rightarrow$ [S] plot in the original paper, Michaelis and Menten also plotted their data as v versus log[S]. This plot is based on the rearrangement of the classical Michaelis–Menten equation as shown.

$$v = \frac{V_{max}[S]}{K_M + [S]} \text{ may be rearranged to } \frac{V_{max}}{v} = \frac{K_M}{[S]} + 1$$

Taking logarithms,

$$\log\left(\frac{V_{max}}{v} - 1\right) = \log K_M - \log[S]$$

The following variation of the above equation may appear more familiar:

Fig. 16.2 Direct linear plot
of Eisenthal and Cornish-
Bowden. The graph of
$v \rightarrow$ [S] data for arginase from
Table 16.1 is shown

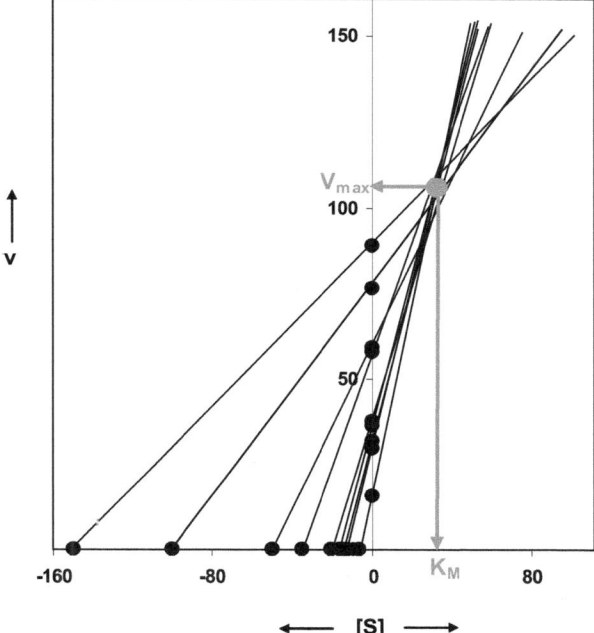

$$p[S] = pK_M + \log\left(\frac{V_{max} - v}{v}\right)$$

This is comparable to the dissociation of a weak electrolyte versus pH curve and takes a form very similar to the Henderson–Hasselbalch equation relating pH to pKa.

The advantage of a $v \rightarrow \log[S]$ plot is that the points corresponding to lower [S] are not crowded together. This semi-log plot is particularly useful in (a) comparing velocities over a large range of substrate concentration and (b) plotting initial velocity data for enzymes with vastly different K_M values, on a single graph. For instance, substrate affinities of different liver hexokinase isozymes and glucokinase span from μM to mM (K_M for glucose). Plotting this on the same X-axis would require a very long graph paper! A $v \rightarrow \log[S]$ plot, with fractional velocity (v/V_{max}) on the Y-axis, allows a convenient comparison on the same graph. The point of inflection (where $\log K_M = \log[S]$ as seen by putting $v = V_{max}/2$ in the above equation) provides a good estimate of K_M. It is only required to identify the midpoint (inflection point) of the curve to determine K_M (Fig. 16.3). However, it is better to analyze the $v \rightarrow \log[S]$ plot by nonlinear curve fitting tools, rather than manually.

The $v \rightarrow \log[S]$ plot is a useful method to diagnose the presence of cooperativity in enzyme kinetics. While such graphs are always sigmoid, the steepness of the curve at the inflection point provides a measure of cooperativity. The steepness (and hence cooperativity) is much more obvious in this semi-logarithmic plot than in the original $v \rightarrow [S]$ curve. Its steepness is also directly related to h—the Hill coefficient

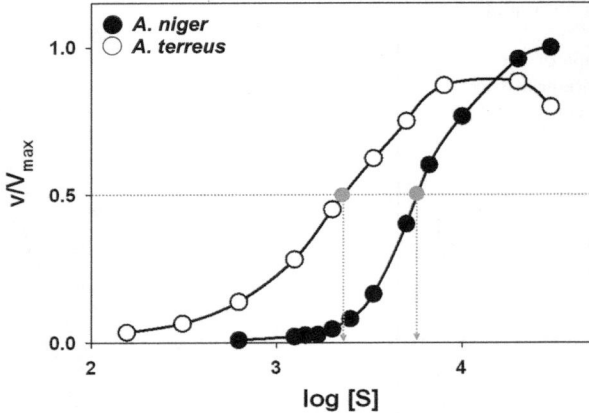

Fig. 16.3 The plot of $v \rightarrow \log[S]$. 2-Oxoglutarate (substrate) saturation of NADP-glutamate dehydrogenases from *Aspergillus niger* [●] and *Aspergillus terreus* [○] (redrawn from the original data in Fig. 14.4). Fractional velocity (v/V_{max} which is dimensionless) is plotted on the Y-axis and arrows point to the inflection point on the two curves. The curve for *A. niger* enzyme is steeper (higher h value)

(Chap. 14, Sect. 14.2.4). Such a graph for two NADP-glutamate dehydrogenases (with distinct h values) in Fig. 16.3 succinctly illustrates the point.

In fact, dose–response curves for enzyme inhibition also take a similar semi-logarithmic form. More on dose–response curve and the graphical determination of IC_{50} value (**I**nhibitor **C**oncentration giving 50% inhibition) is given in a later section (Fig. 21.8, Chap. 21). The effective range for binding in general extends over 2 logs of the ligand concentration. We may recall that acetic acid (pKa of 4.8) is mostly in the ionized state (CH_3COO^-) at pH 5.8 and is mostly in the unionized state at pH 3.8.

Hill plot: We noted earlier (Chap. 14, Sect. 14.2.4) that the Hill equation (and not the Michaelis–Menten equation) better describes the effect of cooperative interactions on the measured enzyme rate. The Hill coefficient h is a convenient and commonly used index of cooperativity. For Michaelian enzymes, h is one. It however takes other values (including nonintegers) for cooperative enzymes. The Hill coefficient is considered to represent the *minimum number* of interacting binding sites on an oligomeric enzyme (Hill 1910). Constant $K_{0.5}$ is similar (not the same!) to K_M but also contains terms related to the effect of substrate binding at one site to the binding at other sites. The Hill equation represents a nonlinear relationship between v and [S] of an enzyme. The $v \rightarrow$ [S] data can be directly fit to this equation (through nonlinear curve fitting protocols) to extract the three parameters—V_{max}, $K_{0.5}$, and h. Otherwise, the velocity data can be analyzed by using the linear form of the Hill equation shown below. Taking logarithms after rearranging the Hill equation we obtain

Fig. 16.4 The Hill plot. The
log[*v*/(V_{max}−*v*)] → log
[S] plots for (a) arginase
(*v* → [S] data from
Table 16.1) and (b) two
NADP-glutamate
dehydrogenases (data from
Fig. 14.4) are shown in the left
and right panels, respectively.
Note that in both cases axes
are marked in the logarithmic
scale. Only the data points in
the linear portion of the curves
are used to obtain the Hill
coefficient (***h***) and $K_{0.5}$

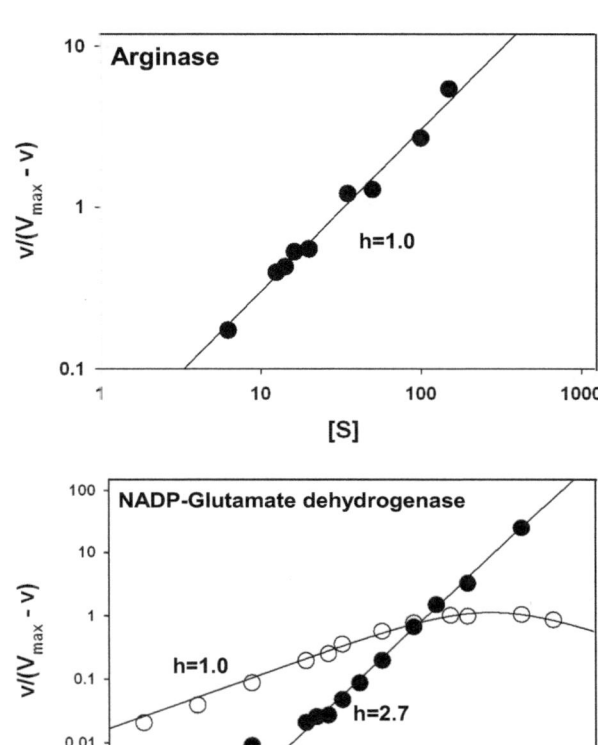

$$\log\left(\frac{v}{V_{max} - v}\right) = h \times \log[S] - \log K_{0.5}$$

A plot of log[*v*/(V_{max}−*v*)] as a function of log[S] should therefore yield a straight-line graph with a slope of ***h*** and a Y-axis intercept of −log$K_{0.5}$. Hill plots for three different enzymes are illustrated in Fig. 16.4.

A precise V_{max} value should be known beforehand to plot the log[*v*/(V_{max}−*v*)] → log[S] graph. This is not easily obtained because of the nonlinear *v* → [S] relation in the first place. If present, errors in V_{max} get carried further. Lastly, the linear region of this plot is the most meaningful portion of the curve (Fig. 16.4); and linearity prevails only over a limited region of substrate concentration (around [S] = $K_{0.5}$). For these reasons, it is desirable to determine V_{max}, $K_{0.5}$, and ***h*** from direct nonlinear curve fits to the Hill equation itself.

16.4   Linear Transforms of the Michaelis–Menten Equation

Accurate extraction of kinetic constants directly from $v \rightarrow$ [S] plots is error-prone because these graphs are nonlinear. There are several ways to transform the hyperbolic relation into a linear form. This is often attempted because linear relations are better amenable to extrapolations—and hence extraction of kinetic constants. The standard transforms of original $v \rightarrow$ [S] data for arginase (from Table 16.1) for linearization are listed in Table 16.3. These will be used to demonstrate various plots described subsequently.

Lineweaver–Burk plot: The Lineweaver–Burk plot (also known as the **double-reciprocal plot**) is a historically important, often-used linear transform of the Michaelis–Menten equation (Lineweaver and Burk 1934). This is done by taking reciprocals on both sides to obtain

$$\frac{1}{v} = \frac{K_M}{V_{max}} \frac{1}{[S]} + \frac{1}{V_{max}}$$

This equation describes a straight line and is of the form $y = mx + C$. A graph of $1/v \rightarrow 1/[S]$ should be linear with slope m $= K_M/V_{max}$ and intercept C $= 1/V_{max}$. When [S] $= \infty$ (and hence 1/[S] on the X-axis is zero), the Y-axis intercept should relate to maximal velocity and zero-order rate constant ($1/V_{max}$, as expected). While achieving [S] $= \infty$ may be nearly impossible in practice, extrapolation ([S] $= \infty$) is possible due to this linear transform and V_{max} is conveniently evaluated. The value of K_M can be obtained from the slope and intercept (dividing m by C) of such a plot. A representative double reciprocal plot (of the data from Table 16.3) is shown in Fig. 16.5.

The double-reciprocal plot overcomes the analysis difficulties due to the nonlinear (hyperbolic) $v \rightarrow$ [S] relation. The plot should however be used with much discretion. Casual evaluation of V_{max} and K_M from this plot can be flawed. The practical considerations associated with the Lineweaver–Burk plot can be serious and need attention. These aspects are described in the box below.

Table 16.3 A few standard transforms of original $v \rightarrow$ [S] data for arginase

[S] (Arginine, mM)	v ($\mu mol \times min^{-1} \times mg^{-1}$)	1/[S]	1/v	v/[S]	[S]/v
6.3	**15.7**	0.159	0.0637	2.492	0.401
12.5	**30.1**	0.080	0.0332	2.408	0.415
14.2	**31.7**	0.070	0.0315	2.232	0.448
16.2	**36.7**	0.062	0.0272	2.265	0.441
20.0	**37.7**	0.050	0.0265	1.885	0.531
35.0	**57.9**	0.029	0.0173	1.654	0.604
50.0	**59.4**	0.020	0.0168	1.188	0.842
100.0	**76.9**	0.010	0.0130	0.769	1.300
150.0	**89.0**	0.007	0.0112	0.593	1.685

The original $v \rightarrow$ [S] data from Table 16.1 are shown in bold

Fig. 16.5 The double-reciprocal plot. A graph of $1/v \rightarrow 1/[S]$ data for arginase from Table 16.3 is shown. The curved dotted lines (schematic and in gray) represent the nonlinear plots that may be obtained when the cooperativity of substrate binding is manifest

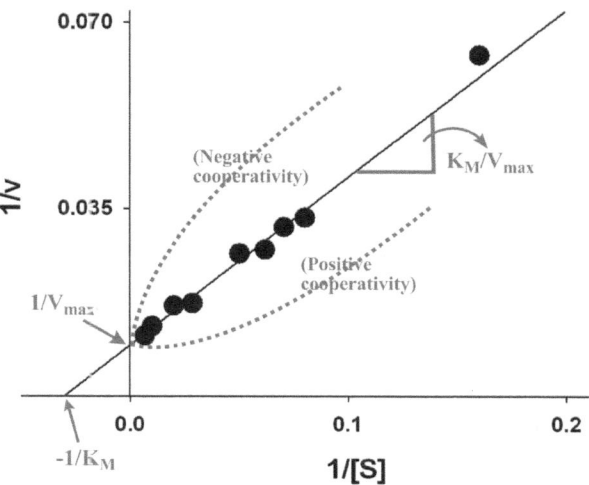

Practical Aspects of Double-Reciprocal Analysis
- Common practical problem encountered with the Lineweaver–Burk plot is the data spread. One should carefully choose [S] values such that they lead to evenly spaced points on the 1/[S] axis.
  With evenly spaced [S] values used for saturation, the data points as their reciprocals tend to cluster toward the Y-axis (see Fig. 16.5). A good experimental design and careful choice of initial [S] values can take care of this problem. One approach could be as follows: First, decide on the highest [S] to be used. Accordingly, prepare a 10X stock solution—on adding 0.1 mL of this to a reaction mixture (1.0 mL final volume), the desired highest [S] is obtained. Second, dilute the original 10X substrate solution into 1:2, 1:3, 1:4, 1:5, 1:6, 1:7, 1:8, etc. to generate working stock solutions of 1/2, 1/3, 1/4, 1/5, 1/6, 1/7, 1/8, etc. strengths. When these are used (by adding 0.1–1.0 mL reaction a decreasing concentration series of 0.500, 0.333, 0.250, 0.200, 0.167, 0.143, 0.125, etc. is obtained. On plotting their reciprocals (1/[S]), the data points will now be equally spaced at intervals of 1, 2, 3, 4, 5, 6, 7, 8, etc. For arginase example—starting with the highest concentration of 150 mM (Table 16.1)—data spread for an ideal Lineweaver–Burk plot would be to use 150 mM, 75 mM, 50 mM, 37.5 mM, 30 mM, 25 mM, 21.4 mM, 18.8 mM, 16.7 mM, 15 mM, etc. This range should also satisfy the additional important condition of data points on both sides of the K_M.
  Another approach is to exploit the rough estimate of K_M from a pilot experiment. Using this as a point in the middle, prepare a series of *relative* substrate concentrations of 1.0, 1.11, 1.25, 1.43, 1.67, 2.0, 2.5, 3.33, 5.0, and 10.0. Supposing the K_M is around 1.0 mM, then we can have the range

(continued)

as 0.5 mM,...0.1.0 mM,...0.5.0 mM. In this arrangement, the relative substrate concentration of "2" corresponds to 1.0 mM.

- Second important issue with the Lineweaver–Burk analysis is the way experimental errors are reflected in this plot. Errors are unevenly weighted in the form of reciprocals. Appropriate weighting to data points is best achieved by reliable curve fitting programs (Table 16.2) that account for the nonlinear error distribution. Small errors in lower v values lead to substantial errors in $1/v$, whereas similar errors in large v values lead to barely noticeable errors in $1/v$. A linear regression method cannot recognize this distortion because the errors themselves are nonlinear. This problem can be sorted by using suitable weighting factors—less importance to data points with large errors.

  Lineweaver–Burk analysis is possible even in the absence (nonavailability rather) of programs that incorporate nonlinear error distribution. The procedure involves the following steps: Original $v \rightarrow$ [S] data are first directly fit to the Michaelis–Menten equation. The K_M and V_{max} values so obtained are then plugged into the double-reciprocal equation (given above) to obtain a straight line. This straight-line fit is without the systematic errors arising from the improper weighting of data points. The trick therefore is to avoid the temptation of directly fitting a straight-line to $1/v \rightarrow 1/[S]$ values.

  Finally, it is best to obtain data of such good quality that the result is also obvious without statistical analysis. Quoting Henry Clay—*Statistics are no substitute for judgment.*

Despite its limitations (mentioned above), Lineweaver–Burk plots are of considerable value. Of the various approaches to linearize the Michaelis–Menten equation, only the *Lineweaver–Burk plot permits the individual display of v and [S] on the two axes*. In all others (see below) at least one coordinate is a composite of both v and [S]. The second big advantage of this plot is our ability to follow changes in the first-order and zero-order rate constants of an enzyme-catalyzed reaction simply by inspection. Recall that the reciprocal of its slope is V_{max}/K_M (first-order rate constant), while the reciprocal of its intercept is V_{max} (zero-order rate constant). *An increase in slope or intercept indicates a decreased first-order or zero-order rate constant, respectively.* We live in real time and reciprocal analysis (of this type) makes it difficult to grasp reality. Beginners in enzymology may find it awkward to appreciate the reciprocal relationship. Nevertheless, this information is particularly useful (a) in diagnosing the mechanistic details of multi-substrate enzymes and (b) in probing the mode of interaction between an enzyme and its inhibitor (see Chap. 17).

If and when Lineweaver–Burk plots of enzyme kinetic data are nonlinear then it is obvious that the original assumptions inherent in the Michaelis–Menten equation do not hold! Failure to measure true initial velocity, a common problem, should be quickly checked. Furthermore, the kinetic data should not be forced to fit a straight line. Instead, other kinetic models should be explored to address deviations from

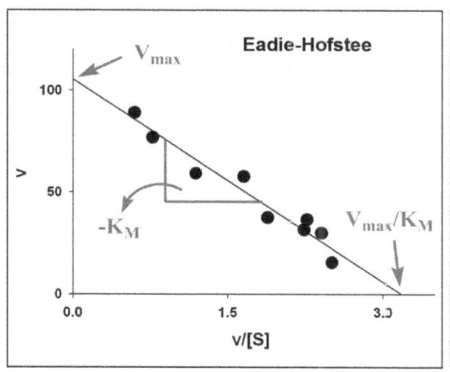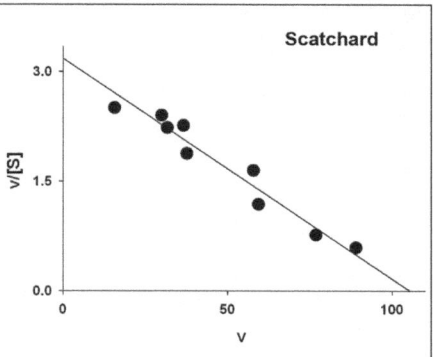

Fig. 16.6 The single-reciprocal plot acccrding to Eadie–Hofstee. Arginase data from Table 16.3 were graphed. Both the Eadie–Hofstee ($v \rightarrow v/[S]$; left panel) and Scatchard ($v/[S] \rightarrow v$; right panel) versions of the plot are shown

hyperbolic kinetics. Nonlinear Lineweaver–Burk plots result if multiple enzyme forms acting on the same substrate (isozymes with distinct kinetic characteristics) exist in the assay. With pure enzyme samples, however, curved double-reciprocal plots could mean any of the following: (a) substrate activation, (b) substrate inhibition, (c) multiple binding of substrate molecules, or (d) cooperativity of substrate binding. Some of these will be elaborated at appropriate places later. The curvature in the double-reciprocal graph is concave upwards for positive cooperativity and concave downwards for negatively cooperative enzymes (gray dotted lines schematically shown in Fig. 16.5). Departures from linearity are less obvious in a Lineweaver–Burk plot, but are better viewed in others like the Eadie–Hofstee plot and the Hanes–Woolf plot described below.

Eadie–Hofstee plot: This plot (Eadie 1942; Hofstee 1952) is one other way to transform the hyperbolic relation into a linear form for further analysis. The classical Michaelis–Menten equation can be rearranged by cross multiplying as shown:

$$v(K_M + [S]) = V_{max}[S]$$

Dividing both sides by [S] and rearranging, we obtain

$$v = -K_M \frac{v}{[S]} + V_{max}$$

This is again of the form $y = mx + C$ when v is plotted against $v/[S]$. A graph of $v \rightarrow v/[S]$ should be linear with negative slope m = K_M and the Y-axis intercept C = V_{max}. The arginase data from Table 16.3 are plotted in this form in Fig. 16.6, for example. This method of plotting original $v \rightarrow [S]$ data is a linearization through a single reciprocal; only [S] is in the reciprocal form. Since the X-axis represents a composite value ($v/[S]$), it is conceptually more difficult to appreciate the velocity changes as a function of [S] in this plot. Because of the different spread of data points

Fig. 16.7 The Woolf–Hanes single-reciprocal plot. Arginase data from Table 16.3 were graphed as the [S]/v → [S] plot

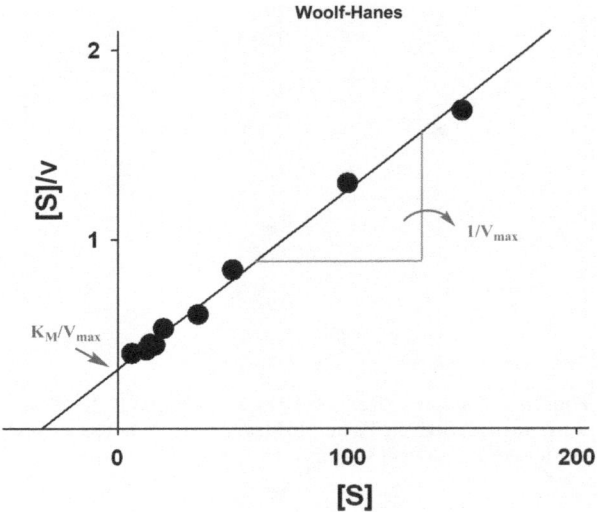

and error distribution, the Eadie–Hofstee plot is better suited to track departures from the typical Michaelis–Menten kinetics. The nonlinearity arising due to cooperativity is better viewed in this plot. However, actual quantitative analysis of cooperativity should be done through a Hill plot.

Another version of the Eadie–Hofstee plot is obtained when the two axes are interchanged (Fig. 16.6). A plot of v/[S] against v is equivalent to a Scatchard plot (normally used in binding analysis) (Scatchard 1949). The linear form of the Michaelis–Menten equation corresponding to this plot is

$$\frac{v}{[\text{S}]} = -\frac{1}{K_\text{M}}v + \frac{V_\text{max}}{K_\text{M}}$$

In this representation, the ratio of [S]$_\text{bound}$/[S]$_\text{free}$ (corresponding to v/[S]) is plotted against [S]$_\text{bound}$ (corresponding to v). Also, the K_M becomes the equivalent of K_D in Scatchard analysis for ligand binding.

Woolf–Hanes plot: The Woolf–Hanes plot is a single reciprocal analysis of v → [S] data where [S]/v is plotted against [S] (Haldane 1957). It is derived simply by multiplying the Lineweaver–Burk transformation throughout by [S]. The following linear transform of the Michaelis–Menten equation is thus obtained.

$$\frac{[\text{S}]}{v} = \frac{1}{V_\text{max}}[\text{S}] + \frac{K_\text{M}}{V_\text{max}}$$

This is again of the form $y = mx + C$ when [S]/v is plotted against [S]. The kinetic parameters are extracted from the slope ($1/V_\text{max}$) and intercept (K_M/V_max) of such a plot (Fig. 16.7). Since the Y-axis represents a composite value ([S]/v), the plot is conceptually more difficult to appreciate. This plot is not so commonly used in enzyme literature.

Table 16.4 Linear transforms of the Henri–Michaelis–Menten equation: a summary

Plot	Lineweaver–Burk (1934)	Eadie–Hofstee (1942)	Scatchard (1949)	Woolf–Hanes (before 1932)
Plot of $(Y \to X)$	$1/v \to 1/[S]$	$v \to v/[S]$	$v/[S] \to v$	$[S]/v \to [S]$
Y-axis intercept	$1/V_{max}$	V_{max}	V_{max}/K_M	K_M/V_{max}
X-axis intercept	$-1/K_M$	V_{max}/K_M	V_{max}	$-K_M$
Slope	K_M/V_{max}	$-K_M$	$-1/K_M$	$1/V_{max}$
Features	Individual display of v and [S] on two axes; Rate constants directly visualized from intercept and slope	View nonlinearity due to cooperativity and departure from hyperbolic kinetics	Same as Eadie–Hofstee but axes interchanged; ligand binding studies	Weighting of errors from original data is the least distorted

Note: V_{max} and V_{max}/K_M respectively correspond to zero-order and first-order rate constants in the Michaelis–Menten formalism

Salient features of the common linear transforms of the classical Henri–Michaelis–Menten equation are collated and compared in Table 16.4.

16.5 Summing Up

A very fundamental piece of enzyme kinetic insight is the primary *initial velocity* versus *substrate concentration* data set. The $v \to [S]$ data is a gold mine of kinetic information. Therefore, it is prudent to obtain reliable data in the first place—because any degree of sophisticated analysis will not transform bad data into good data! While collecting and analyzing enzyme kinetic data, the following key issues should be critically considered.

- Ensure that true initial velocities are measured at all the concentrations of substrate tested, particularly at lower [S] values.
- Use the untransformed data to figure out if it actually fits the Michaelis–Menten kinetics, or if this model is being imposed/forced on the data.
- A sufficiently broad range of substrate concentration should be tested to obtain the original $v \to [S]$ data. And ensure that the data are not biased to either low [S] or high [S].
- Use suitable and rigorous statistical analysis to account for experimental errors and data scatter.

Analysis of enzyme kinetic data must involve the relevant statistical analysis (Johnson 2019). Fancy statistical packages should not be used without clearly understanding what they can and cannot do! *Statistics should be used as a lamp post—to illuminate but not to lean on poor data.*

References

Eadie GS (1942) The inhibition of cholinesterase by physostigmine and prostigmine. J Biol Chem 146:85–93

Eisenthal R, Cornish-Bowden A (1974) The direct linear plot. Biochem J 139:715–720

Haldane JBS (1957) Graphical methods in enzyme chemistry. Nature 179:832

Hill AV (1910) The possible effects of the aggregation of the molecules of haemoglobin on its dissociation curves. J Physiol 40:iv–vii

Hofstee BHJ (1952) Specificity of esterases. J Biol Chem 199:357–364

Johnson KA (2019) New standards for collecting and fitting steady state kinetic. Beilstein J Org Chem 15:16–29

Lee H-J, Wilson IB (1971) Enzymic parameters: measurement of V and Km. Biochim Biophys Acta 242:519–522

Lineweaver H, Burk D (1934) The determination of enzyme dissociation constants. J Am Chem Soc 56:658–666

Moreno J (1985) The use of the integrated Michaelis-Menten equation in the determination of kinetic parameters. Biochem Educ 13:64–66

Scatchard G (1949) The attractions of proteins for small molecules and ions. Ann N Y Acad Sci 51:660–672

Schwert GW (1969) Use of integrated rate equations in estimating the kinetic constants of enzyme-catalyzed reactions. J Biol Chem 244:1278–1284

Further Reading

Bezerra RMF, Dias AA (2007) Utilization of integrated Michaelis-Menten equation to determine kinetic constants. Biochem Mol Biol Educ 35:145–150

Carrillo N, Ceccarelli EA, Roveri OA (2010) Usefulness of kinetic enzyme parameters in biotechnological practice. Biotechnol Genet Eng Rev 27:367–382

Part III

Elucidation of Kinetic Mechanisms

Approaches to Kinetic Mechanism: Overview

The order of the addition of substrates to and release of products from the enzyme active site, along with the establishment of relative rates of various events, defines the kinetic mechanism. These mechanisms fall into two broad groups: those where the full complement of substrates must assemble on the enzyme active site before the reaction occurs are termed "sequential mechanisms." In the other category, a product (s) is released between additions of two substrates and are called "ping-pong mechanisms." In this category, substitution on the enzyme active site groups occurs and hence is also known as the double displacement mechanism. The study of these mechanisms is best approached by rigorous experimental design where data are collected by systematically varying one parameter at a time. This method of reduction is in full display in enzyme kinetics. In fact, elucidating enzyme kinetic mechanisms offers the best example of how scientific hypotheses are tested. Elements of the scientific method in sequence include problem recognition → collation of available information → hypothesis building → experimentation → reasoning and deduction → refining the hypothesis. These steps are iterated to arrive at an enzyme mechanism. Finally, when experiments and measurements agree with the theory, truth is secured. In practical enzyme kinetics, this exercise translates into the following steps (Table 17.1) for the elucidation of mechanisms.

A very important aspect of kinetic mechanism elucidation is that a given set of data, at times, may fit/describe more than one unique kinetic scheme. For instance, all the available kinetic evidence could not distinguish between an S_N1 (carbonium ion formation) mechanism or covalent catalysis (involving the active site carboxylate). The S_N1 mechanism predicts retention of stereochemical configuration (at C1 of the glycosidic sugar) in the product, whereas covalent catalysis (via the acylal) passing through two Walden inversions also predicts retention. Often other methods (other than steady-state kinetics!) may have to be resorted to in such cases. More recent MALDI-TOF evidence for a lysozyme-acylal favors the covalent catalysis mechanism by lysozyme. Resolution of kinetic equivalence invariably requires more incisive experimentation to cleanly distinguish between different possibilities. If

N. S. Punekar, *ENZYMES: Catalysis, Kinetics and Mechanisms*, https://doi.org/10.1007/978-981-97-8179-9_17

Table 17.1 Steps to kinetic mechanism elucidation

1.	Write a minimal mechanism based on available information
2.	Experimentally obtain kinetic parameters/constants involved
3.	Build a probable mechanism through diagnostic experimentation involving:
(a)	A study of initial velocities
(b)	Use of different inhibitions (product, dead-end, substrate, alternate substrate)
(c)	Isotopic studies (both exchange analysis and isotope effects)
(d)	pH-dependence studies
This is a desirable order for experimentation but need not necessarily be rigid	
4.	Review the mechanism by reasoning and deduction
5.	Confirm and/or refine the mechanism by designing more experiments

Table 17.2 Nature of information obtained from experiments

Experimental approach	Nature of information obtained
1. Pre-steady-state kinetics	Detection of enzyme complexes/intermediates, rate-limiting k values
2. Variation of [S]; analysis of initial velocity patterns	Kinetic constants; sequence of complexes; binding order
3. Inhibition analyses; variation of [P], [I], etc.	Active site definition; sequence of complexes; binding order
4. Substrate/ product structures	Map of the active site and geometry
5. Isotope exchange study	Partial reactions; distinction between mechanisms
6. Isotope effects	Individual rate constants; chemical mechanism; TS structure
7. pH variation and kinetics	Relevant pKas for catalysis and/or binding; nature of functional groups

there are no unique testable differences between rival mechanisms, then it may never be possible to resolve the kinetic ambiguity (the so-called "black box").

17.1 Which Study Gives What Kind of Information?

All those ligand (substrate, inhibitor, activator, etc.) interactions that result in altered enzyme reaction rates may be exploited to understand the kinetic mechanism. Enzyme–ligand interactions that are kinetically silent are of no consequence to this study. For example, a molecule may bind to the enzyme without changing any of its kinetic properties. Such binding may be potentially useful in enzyme purification and/or stability studies but is useless in defining the kinetic mechanism. The nature of mechanistic information that can be gleaned from various kinetic studies is summarized in Table 17.2.

Of these, pre-steady state kinetics was introduced in an earlier section (Chap. 10). The remaining approaches form the subject matter of subsequent chapters and will be discussed in greater detail.

The K_M as an "apparent" dissociation constant (only when k_2 is very small compared to k_1; see Chap. 14) provides some measure of the strength of binding of S to E. Similarly, a kinetic dissociation constant for an inhibitor (namely K_I) may be obtained from steady-state kinetics. Complementary tools to define the strength of ligand binding, besides those gleaned from kinetic studies, are often necessary for a complete understanding of the enzyme mechanism. Binding constants (dissociation constants) for substrates, substrate analogs, inhibitors, activators, etc. may be obtained through several techniques. These include various spectroscopic techniques (UV–visible and fluorescence), isothermal titration calorimetry (ITC; Chap. 9), differential scanning calorimetry (DSC), surface plasmon resonance (SPR; to assess affinity and on/off rates), equilibrium dialysis, gel filtration chromatography, and kinetics of protection by ligands (Chap. 20). A comparison of kinetic dissociation constants with purely physical dissociation constants provides useful insights into binding synergism, antagonistic effects, etc.

17.2   Two Thumb Rules

There are two component activities to any kinetic study: one qualitative and the other quantitative. From a systematic analysis of the kinetic data by different plots and replots and slope and/or intercept changes, various kinetic constants like V_{max}, k_{cat}, K_M, K_{iA}, and K_P are evaluated. At the qualitative level, the inspection of slopes and intercepts of a double reciprocal plot is very informative. We are thus looking at the two limiting cases: sub-saturating (slope) and saturating (intercept) concentrations of the varied substrate. Any change in slope (which is K_M/V_{max}) points to a change in the first-order rate constant. An effect on the intercept ($1/V_{max}$) similarly reflects on the zero-order rate constant (see Chap. 14). The presence of an inhibitor brings about an increase in the magnitude of intercept, slope, or both. On the other hand, increasing substrate (other than the one whose saturation is being studied) concentration leads to lower intercept, slope, or both (Fig. 17.1).

A careful interpretation of slope and intercept changes is at the heart of understanding and postulating a kinetic mechanism. Let us therefore attempt to make some generalizations about how this can be done. We have seen earlier (derivation of rate equations) that along the reaction path few distinct, kinetically significant enzyme forms may occur. The total enzyme ($[E_t]$) is distributed into these forms depending on the extant equilibria (and/or steady state). A corresponding rate equation was derived from this description by evaluating the fraction of $[E_t]$ present as ES (the productive complex). Factors (such as substrate, inhibitor, activator, or pH) that perturb this equilibrium result in a redistribution of $[E_t]$ into various enzyme forms. Any consequent change in the concentration of ES leads to a change in reaction velocity. Intuitively, we can therefore predict how the slopes and intercepts of Lineweaver–Burk plots are affected by any substance (substrate, product, etc.) based on the distribution of enzyme forms and their equilibria. The converse of this exercise is of course of great practical value—we can set up appropriate equilibria from the slope and intercept effects caused by any substance. This is the

Fig. 17.1 Possible changes
in the slope and intercept of
double reciprocal plots. The
direction of change with
increasing concentrations of
(**a**) other substrate, activator,
etc. and (**b**) product, inhibitor,
etc. is indicated by respective
arrows

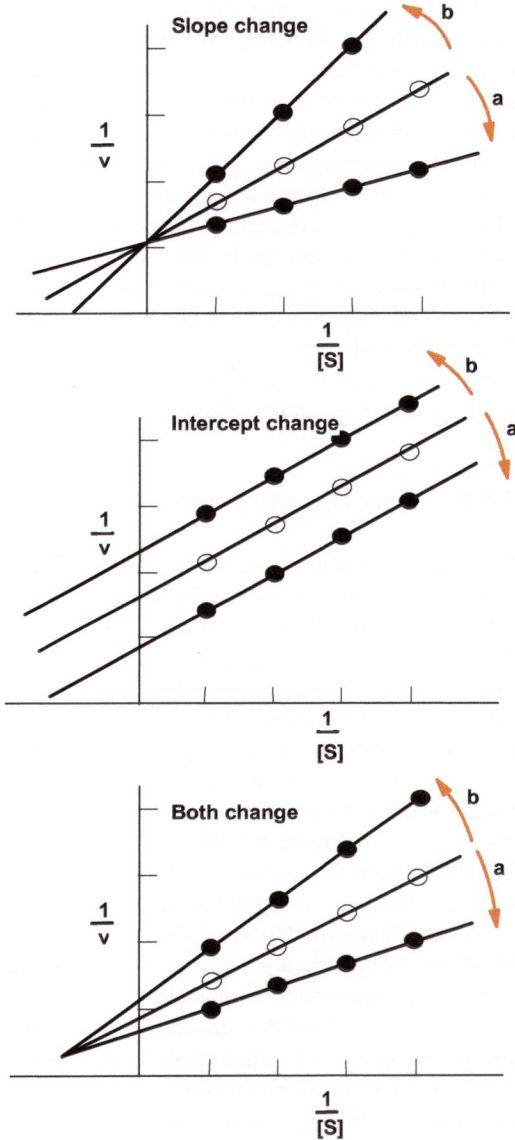

crux of the enzyme kinetic mechanism. From the expected redistribution of $[E_t]$ into different enzyme forms, two thumb rules may be stated here. These rules were framed by the famous enzymologist Cleland (1963, 1989), to predict slope and intercept effects for product and dead-end inhibitors. However, we generalize and extend them to any substance that binds on or more enzyme forms and perturbs the equilibrium (of $[E_t]$ distribution).

Rule I. A ligand (substrate, product, or inhibitor) affects the intercept (zero-order rate) of the double reciprocal plot when it combines reversibly with an enzyme form other than that with which the varied substrate combines.

Rule II. The slope (first-order rate) of the double reciprocal plot is affected when (a) the ligand and the varied substrate combine reversibly with the same enzyme form or (b) the ligand and varied substrate bind reversibly to two different enzyme forms that are connected by a series of reversible steps along the reaction path.

It goes without saying that once the slope and intercept effects have been independently predicted, they can be combined to generate the whole picture (pattern). Rule II.b requires that we know whether two enzyme forms, along the reaction path, are reversibly connected or not. Seeking such reversible connectivity with dead-end inhibitors (Chap. 19) is tricky since they do not form a part of the normal reaction sequence. A step becomes irreversible when it involves (1) the addition of the substrate at saturation ($[S] \to \infty$), (2) the release of the product under initial velocity conditions ($[P] = 0$), or (3) some irreversible chemical event (with a large negative ΔG) like CO_2 release, oxidation of an aldehyde to acid, or aromatization of a ring. Logical connections between (a) prediction of slope and intercept effects from a given kinetic scheme and (b) arriving at a kinetic mechanism from experimentally observed slope and intercept effects are better understood with suitable examples. A few case studies are therefore included in a concluding section of the elucidation of kinetic mechanism (Chap. 26).

References

Cleland WW (1963) The kinetics of enzyme-catalyzed reactions with two or more substrates or products: I. Nomenclature and rate equations. Biochim Biophys Acta 67:104–137. II. Inhibition: nomenclature and theory. Biochim Biophys Acta 67:173–187. III. Prediction of initial velocity and inhibition patterns by inspection. Biochim Biophys Acta 67:188–196

Cleland WW (1989) Enzyme kinetics revisited: a commentary by W Wallace Cleland. Biochim Biophys Acta 1000:209–221

One could in principle study the full course of an enzyme-catalyzed reaction until the equilibrium is reached. This is not practiced because (a) the equations soon become very complex to handle, (b) with reasonable progress of the reaction, inhibition by product(s) sets in, and (c) it is difficult to follow the many, simultaneously changing reactant concentrations. For these reasons, experiments are best conducted under steady-state and initial velocity conditions. Employing the method reduction approach—data are generated almost always by systematically changing one variable at a time.

The nature of the experiments conducted, and the information sought from the initial velocity data are as follows.

1. Monitor initial velocity "v" by varying the concentration of one substrate at different fixed concentrations of the others. If the enzyme reaction in question involves a single substrate, then there is not much information other than obtaining kinetic constants (V_{max} and K_M).
2. The $v \rightarrow$ [S] data are plotted in the double reciprocal format (Lineweaver–Burk plots), and the patterns are analyzed qualitatively.
3. If the double reciprocal plots are non-linear, then they suggest that the enzyme under study may exhibit (a) substrate inhibition—the curve being concave upward, or (b) cooperativity in its interaction with its substrates—concave upward for positive cooperative and concave downward for negative cooperative interactions (see Fig. 16.5). We will have more to say on these non-linear plots later.
4. Gradual changes in the slope and/or intercepts, as a function of the fixed substrate concentration, are noted. It may be recalled that changes in slope point to a change in V_{max}/K_M (which in turn reflects on the first-order rate constant). An effect on the intercept is similarly related to V_{max}, and the zero-order rate constant.
5. On quantitative analysis of slope and intercept changes, various kinetic constants are evaluated.

© The Author(s), under exclusive license to Springer Nature Singapore Pte
Ltd. 2025
N. S. Punekar, *ENZYMES: Catalysis, Kinetics and Mechanisms*,
https://doi.org/10.1007/978-981-97-8179-9_18

Several interesting variations of the initial velocity patterns are possible and are indeed observed. However, we will restrict ourselves to some of the more common patterns observed in such studies.

18.1   Intersecting Patterns

This is indicative of a sequential combination of the two substrates considered in the study. The equilibria representing such a general case are given below (Fig. 18.1).

Three individual situations commonly encountered could include (a) random, (b) preferred ordered, and (c) ordered addition of A and B. In an ordered sequential mechanism (the upper path; E → EA → EAB → Products) only EA is formed, whereas in the random case (both paths leading to EAB; E → EA → EAB → Products as well as E → EB → EAB → Products), both EA and EB are formed. A general equation derived for a two-substrate sequential case will look like:

$$v = \frac{V_{\max}[\text{A}][\text{B}]}{K_{\text{iA}}K_{\text{B}} + K_{\text{A}}[\text{B}] + K_{\text{B}}[\text{A}] + [\text{A}][\text{B}]}$$

This rate equation will be identical, and it cannot distinguish between the ordered versus random mechanism. Remember that so long as we reach the same EAB complex from the two routes, the equation will be symmetric. This is what is expected of any state function (path-independent property). Therefore, $K_{\text{iA}}K_{\text{B}} = K_{\text{iB}}K_{\text{A}}$ in the random mechanism. We note that K_{iA} is the kinetic dissociation constant of A from EA (and K_{iB} for the dissociation of B from EB) (Frieden 1957).

Determination/evaluation of kinetic constants and replots: On double reciprocal analysis, one obtains slope and intercept values at different fixed concentrations of B (Fig. 18.2). Similar plots can also be obtained with B as the varied substrate but at different fixed concentrations of A.

The slope and intercept values are further evaluated from the following relationships:

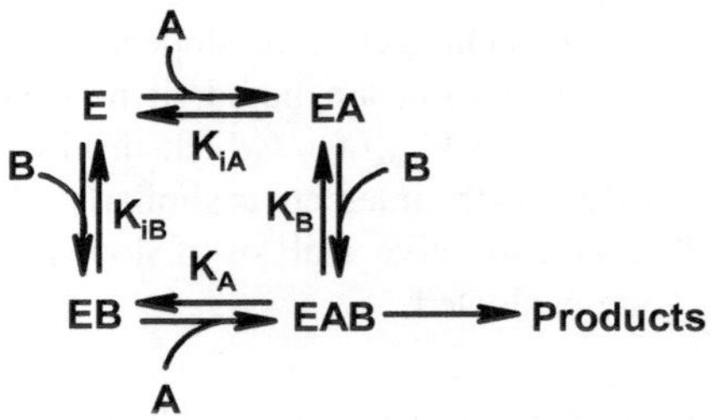

Fig. 18.1 Equilibria representing the sequential interaction of substrates in a bi-reactant mechanism

Fig. 18.2 Double reciprocal
plot for the sequential
mechanism with A as the
varied substrate. Replots of
slope → 1/[B] and
intercept → 1/[B] are shown
in the text

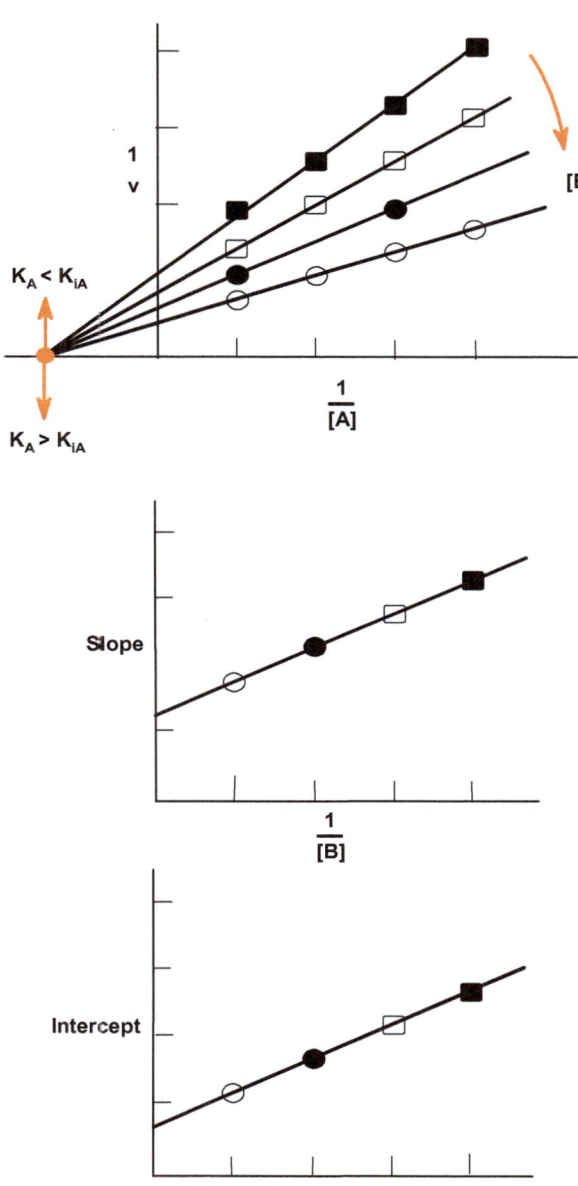

$$\text{Slope} = \frac{K_A}{V_{max}} + \frac{K_{iA}K_B}{V_{max}[B]}$$

$$\text{Intercept} = \frac{1}{V_{max}} + \frac{K_B}{V_{max}[B]}$$

A replot of "intercept → 1/[B]" is linear with (a) a reciprocal of the intercept giving V_{max} and (b) dividing slope by intercept gives K_B. Substituting these values in the "slope → 1/[B]" equation we can extract values of K_A and K_{iA}. Although useful, this is not of course the best way to evaluate these constants; one should use statistical fits to the entire data set. User-friendly software/programs are available for this purpose.

Interpretation: Intersecting Lineweaver–Burk patterns arise because both the slope (first-order rate) and the intercept (zero-order rate) change as a function of different fixed [B] values. Note that A and B bind two different enzyme forms at equilibrium, and hence both the slope and the intercepts change. These are indicative of a sequential mechanism, but we cannot distinguish various mechanistic subsets ranging from "random" to "ordered" by initial velocity pattern analysis alone. We need to perform other types of experiments to get there. These include (a) direct substrate binding experiments with radiolabels, e.g., LDH (from bovine heart muscle) binds NAD^+ (forming $E-NAD^+$) but not lactate (E-lactate does not form); it binds lactate only in the presence of NAD^+. Ordered addition of substrates is indicated—NAD^+ followed by lactate; (b) direct monitoring of binary and/or ternary complexes by MALDI-TOF; and (c) ordered versus random binding through product inhibition and isotope exchange studies (Chaps. 24 and 26).

Coordinates for the point of intersection depend upon the relative values of K_A and K_{iA}. The X-axis coordinate corresponds to $-1/K_{iA}$ (which is $-1/K_{iB}$ when analyzing for B) while the ordinate intercept value is $\frac{1}{v} = \frac{1}{V_{max}}\left(1 - \frac{K_A}{K_{iA}}\right)$ (which is $= \frac{1}{V_{max}}\left(1 - \frac{K_B}{K_{iB}}\right)$ when B is studied). How the expressions for these coordinates are obtained may be found in the appendix to this chapter.

The position of the cross-over point depends on the relative magnitudes of K_A and K_{iA} (the kinetic dissociation constant for A from EA). These are illustrated in Fig. 18.2. For example,

$K_A = K_{iA}$ then the lines intersect on the X-axis
$K_A < K_{iA}$ then the lines intersect above the X-axis
$K_A > K_{iA}$ then the lines intersect below the X-axis

There are examples (like Mg-ATP binding to creatine kinase) where $K_{iA} >> K_A$. This indicates a tighter binding of A to the enzyme when B is bound and is termed **synergistic** binding. Clearly, the K_{iA} (the kinetic dissociation constant for A from EA) may not be a simple dissociation constant but may contain additional rate constants—such as for enzyme conformational change—within it.

18.2 Parallel Patterns

Parallel initial velocity patterns are suggestive of a ping-pong (double displacement) mechanism. The equilibria representing such a general case are given (Fig. 18.3).

Two situations commonly encountered include (a) single-site ping-pong and (b) multi-site ping-pong mechanisms. In a two-substrate ping-pong mechanism, no ternary complex (EAB) is formed. A general equation derived for a two-substrate ping-pong case will look like

$$v = \frac{V_{\max}[A][B]}{K_A[B] + K_B[A] + [A][B]}$$

Notice that no K_{iA} term appears in the denominator of this rate equation (when compared with the expression for the sequential mechanism above). This is the same as putting $K_{iA} = 0$ and is consistent with the absence of EAB form of the enzyme, and hence an absence of slope effect in the double reciprocal plots. Now the question is—what may appear as a parallel pattern—is it really parallel? Here, we should quickly note that K_{iA} may have a very small value; and if so, it becomes quite difficult from the measured data to decide whether the slope is really changing or not. The lines may appear parallel but may actually intersect far away to the left of the origin—this is what happens with brain hexokinase with glucose as a substrate. We are thus relying on quantitative information to answer a qualitative question. While it is easy to conclude that a set of straight lines intersect (i.e., $K_{iA} \neq 0$) we need additional evidence to ensure that a pattern is genuinely parallel (i.e., $K_{iA} = 0$). For instance, D-amino acid oxidase showed "almost" parallel lines in initial velocity analysis. It was originally thought that the imino acid (oxidized amino acid that leads to pyruvate formation) is released from E-FADH$_2$ prior to O$_2$ binding. Subsequently, pyruvate was found to compete with alanine. This product inhibition pattern clearly showed the reaction to be sequential and not ping-pong. We will revisit the question "How parallel is parallel?" while dealing with reversible enzyme inhibition analysis (Chap. 21).

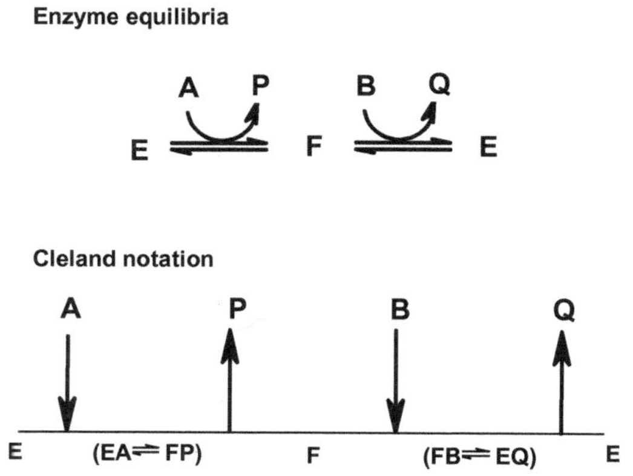

Fig. 18.3 Equilibria representing a bi-reactant ping-pong mechanism

The difficulty of concluding whether a set of lines is parallel (in double reciprocal plots) may be sorted by employing the Woolf–Hanes plot (see Chap. 16). For a ping-pong reaction, plots of [A]/v → [A] at various fixed [B] values will converge on the [A]/v axis. Such a convergence (and any deviation from the common point of interaction) is readily recognized.

Determination/evaluation of kinetic constants and replots: The double reciprocal form of the above equation will be

$$\frac{1}{v} = \frac{K_A}{V_{max}} \frac{1}{[A]} + \frac{1}{V_{max}}\left(1 + \frac{K_B}{[B]}\right)$$

Upon inspection and analysis, one obtains no change in slope (as expected with a parallel set of lines) while the intercept values at different fixed concentrations of B do change (Fig. 18.4).

Fig. 18.4 Double reciprocal plot for a ping-pong mechanism with A as the varied substrate. Replots of intercept → 1/[B] are shown as inset

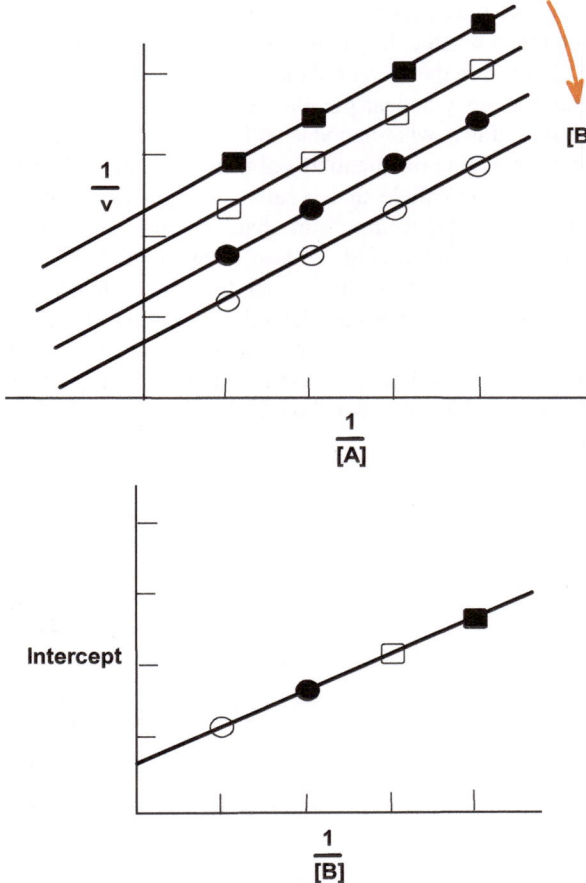

They are evaluated from the following relationships:

$$\text{Slope} = \frac{K_A}{V_{max}}$$

$$\text{Intercept} = \frac{1}{V_{max}} + \frac{K_B}{V_{max}[B]}$$

A replot of intercept → 1/[B] is linear with (a) reciprocal of the intercept giving V_{max} and (b) slope divided by intercept gives K_B. Substituting for V_{max} in the slope → 1/[B] equation we can assign a value for K_A. The best way of course is to evaluate these constants by suitable statistical fits to the entire data set.

Interpretation: Parallel patterns (in a two-substrate case) are normally indicative of a ping-pong mechanism. Consider the two substrate–two product classical ping-pong mechanism (Fig. 18.3). Employing the thumb rules listed earlier (see Chap. 17), the following predictions on the initial velocity patterns can be made (Table 18.1).

These predictions are based on an understanding of different enzyme forms and their equilibria. A binds to E and B binds to the F form of the enzyme; the two forms are not reversibly connected as product release is irreversible when [P] = 0. The arguments for the situation where [A] is varied are as follows:

(a) At low [A] we are looking at slope effect (first-order rate) and any level of [B] cannot affect the interaction of A with E; conversion of E → F is rate limiting and not F → E, hence no slope effect in double reciprocal plots.
(b) At A = ∞ we are looking at the intercept effect (zero-order rate) and all enzyme molecules are captured into the EA form; because F → E becomes rate limiting, any level of [B] affects the overall reaction rate and the intercept is affected.
(c) Summing up the results of a and b above, only the intercept (zero-order rate) of the double reciprocal plot changes as a function of different fixed [B] values; we obtain a parallel set of lines as a consequence.
(d) E and F enzyme forms get reversibly connected when finite levels of P are present. At low [A], now F can go back by binding with P or go forward by interacting with B. Therefore, varying [B] will affect the rate (and the slope). With A = ∞ situation, B affects the rate as before (in b above). The net result is an intersecting pattern.

Table 18.1 Expected initial velocity patterns for a two substrate ping-pong mechanism

Substrate varied	Fixed substrate at	Enzyme parameter affected		Initial velocity pattern
		Intercept (1/V_{max})	Slope (K_M/V_{max})	
A	B = ∞	Yes	No	Parallel
A	B = K_B	Yes	No	Parallel
B	A = ∞	Yes	No	Parallel
B	A = K_A	Yes	No	Parallel

B = ∞ implies saturating [B]; B = K_B implies sub-saturating [B]

Thus, a parallel pattern turns intersecting (and a slope effect is introduced!) if one of the products (say P) is deliberately included in the experiment. Such a change is diagnostic of P release interrupting the sequential addition of A and B (scheme for the ping-pong mechanism above). The addition and presence of P establish a reversible path between the two forms of the enzyme to which A and B bind, respectively. Notice that initial velocity measurements in the presence of added P can be made by estimating the other product Q.

It is not necessary that parallel patterns always mean a ping-pong mechanism. There are occasions where one obtains parallel patterns, but the mechanism is not ping-pong: (a) in a three-substrate fully ordered sequential mechanism, the intersecting A $\rightarrow$ C pattern is observed when [B] is sub-saturating. This pattern turns parallel with saturating levels of [B]. This is actually diagnostic of B being the middle substrate. The slope effect disappears because saturating B introduces an irreversible step (by forcing all the enzymes forward to the EAB form) in between the additions of A and C. Saccharopine dehydrogenase is an example of this kind with lysine as the middle substrate. A similar study with a three-substrate partly ordered sequential mechanism (first substrate ordered, e.g., ATP-citrate lyase) gives two parallel patterns. No parallel patterns are observed in the case of fully random or C-ordered three substrate mechanisms. Three substrate ping-pong mechanisms generally give at least two parallel patterns. A detailed analysis of all such cases may be found in Viola and Cleland (1982). (b) In a Theorell–Chance mechanism, A $\rightarrow$ B pattern appears parallel (see below).

18.3   Few Unique Variations

The relative magnitudes of various kinetic constants in a given mechanism sometimes lead to interesting examples. In such cases, the patterns appear distinct from the two typical cases described in A and B above. We will look at two unusual variations of initial velocity patterns here.

A two-substrate ordered mechanism where the addition of A is at equilibrium is called an equilibrium-ordered mechanism. As expected of a sequential mechanism both patterns are intersecting. However, two unusual features arise due to the equilibrium addition of A. (a) The slope replot of $1/v \rightarrow 1/A$ pattern passes through the origin indicating that $1/K_A$ tends to infinity (i.e., $K_A \approx 0$). Since A gets trapped on the enzyme to form the EAB complex at infinite [B], the corresponding double reciprocal plot is a horizontal line (with zero slope). (b) For the same reason, $1/v \rightarrow 1/B$ pattern intersects on the Y-axis (saturating [B]) giving no intercept effect. Both these features, diagnostic of an equilibrium ordered mechanism, also define the first substrate to add—that is A.

The second example is the Theorell–Chance mechanism. Here, the sequential ordered addition of A and B occurs while the central complexes (EAB $\leftrightarrow$ EPQ) are found in insignificant levels. At any sub-saturating level of [A], B takes it quickly forward to release P. Because of this apparent irreversibility, the slope effect is quite small and the A $\rightarrow$ B pattern looks nearly parallel. However, if we run the reaction in the slow reverse direction (with P and Q as substrates) the pattern definitely intersects. Alcohol dehydrogenase from horse liver follows the Theorell–Chance mechanism.

Because parallel patterns could appear outside of ping-pong mechanisms, further proof of the ping-pong mechanism will come from the demonstration of (a) relevant partial reactions and (b) the substituted form (the F form) of the enzyme. Partial reactions are best evidenced through isotope exchange studies (Chap. 24). The substituted (F form) form of the enzyme is often characterized by isolating and/or trapping it. We will have more to say on these forms with respect to their chemical mechanisms (Chap. 29 in Part IV).

Appendix

Coordinates for the point of intersection in the Lineweaver-Burk plots for sequential mechanism.

These coordinates can be readily evaluated from the general rate equation above by a bit of tedious algebra. Consider the double reciprocal form of the above equation:

$$\frac{1}{v} = \left[\frac{K_A}{V_{max}} \left(1 + \frac{K_{iA}K_B}{K_A[B]} \right) \right] \frac{1}{[A]} + \frac{1}{V_{max}} \left(1 + \frac{K_B}{[B]} \right)$$

At two different values of [B], i.e., $[B]_1$ and $[B]_2$, we obtain the same $1/v$ only at the cross-over point (point of intersection). We can thus equate the two rates and simplify the equation.

$$\frac{K_A}{V_{max}} \left(1 + \frac{K_{iA}K_B}{K_A[B]_1} \right) \frac{1}{[A]} + \frac{1}{V_{max}} \left(1 + \frac{K_B}{[B]_1} \right) = \frac{K_A}{V_{max}} \left(1 + \frac{K_{iA}K_B}{K_A[B]_2} \right) \frac{1}{[A]} + \frac{1}{V_{max}}$$
$$\times \left(1 + \frac{K_B}{[B]_2} \right)$$

And therefore $\left(\frac{K_{iA}}{[A]} + 1 \right) \left(\frac{1}{[B]_1} - \frac{1}{[B]_2} \right) = 0$

Since $\left(\frac{1}{[B]_1} - \frac{1}{[B]_2} \right)$ term cannot be zero by experimental design, we have $\frac{K_{iA}}{[A]} + 1 = 0$ and hence $\frac{1}{[A]} = -\frac{1}{K_{iA}}$. At the cross-over point, therefore, 1/[A] corresponds to $-1/K_{iA}$ (accordingly for B it is $-1/K_{iB}$). We can now substitute $-1/K_{iA}$ for 1/[A] in the equation above and simplifying:

$$\frac{1}{v} = -\left[\frac{K_A}{V_{max}} \left(1 + \frac{K_{iA}K_B}{K_A[B]} \right) \right] \frac{1}{K_{iA}} + \frac{1}{V_{max}} \left(1 + \frac{K_B}{[B]} \right)$$

$$\frac{1}{v} = \frac{1}{V_{max}} - \frac{K_A}{K_{iA}} \frac{1}{V_{max}}$$

$$\frac{1}{v} = \frac{1}{V_{max}} - \frac{K_A}{K_{iA}} \frac{1}{V_{max}} = \frac{1}{V_{max}}\left(1 - \frac{K_A}{K_{iA}}\right)$$

In a similar manner, we can obtain the ordinate intercept in the case of B.

References

Frieden C (1957) The calculation of an enzyme-substrate dissociation constant from the over-all initial velocity for reactions involving two substrates. J Am Chem Soc 79:1894–1896

Viola RE, Cleland WW (1982) Initial velocity analysis for *ter* reactant mechanisms. Methods Enzymol 87:353–366

Enzyme Inhibition Analyses

19

Enzymes are delicate protein catalysts with subtle conformational flexibilities. This makes them vulnerable and a number of environmental conditions and/or ligands could bring about a decline in the net catalytic activity. An enzyme may be irreversibly killed (inactivation by high temperature, extremes of pH, nonaqueous solvent, chemical modification, etc.) or inhibited by ligands that bind to it. Inhibitors are usually small molecular weight ligands that bring about a decrease in the rate of enzyme-catalyzed reaction. For a molecule to act as an inhibitor, it must physically interact with the enzyme. Interactions with the enzyme that do not affect its catalytic activity (that are kinetically silent!) are of no inhibitory consequence. For example, a molecule may bind to the enzyme without changing any of its kinetic properties. Such ligands may serve as potential baits in enzyme purification but are useless in the study of kinetic mechanisms.

A study of enzyme inhibition provides powerful insights into the reaction mechanism. The utility of such an inhibitor kinetic analysis with a detailed description is covered in the subsequent sections. We can classify inhibitors based on their chemical nature, and also the unique inhibitory features exhibited by them. The nature of enzyme inhibition may be reversible or irreversible. A few more common forms of inhibition and the relevant terminology are given in Table 19.1.

19.1 Reversible Versus Irreversible Inhibition

Reversible inhibitors are excellent tools to study enzyme kinetic mechanisms. It is important to establish the reversible nature of inhibition before embarking on its use to study enzyme mechanisms. A diagnostic test of reversibility is to physically separate E and I from their complex (EI) and show full recovery of the original enzyme activity. Dialysis, ultra-filtration, and gel filtration chromatography are useful techniques in separating enzyme molecules from their small molecular weight inhibitors. These techniques may not be able to differentiate between tight binding

© The Author(s), under exclusive license to Springer Nature Singapore Pte Ltd. 2025
N. S. Punekar, *ENZYMES: Catalysis, Kinetics and Mechanisms*,
https://doi.org/10.1007/978-981-97-8179-9_19

Table 19.1 Common
inhibitor types encountered

Enzyme inhibition category		Nomenclature
Irreversible	Nonspecific	Chemical modifications
	Active site directed	Affinity labels
		Suicide substrates
		Tight binding
Reversible	Site of binding	Isosteric
		Allosteric
	Extent of inhibition	Partial
		Complete
	Ligand types	Product
		Substrate
		Dead-end
	Kinetic features[a]	Competitive
		Uncompetitive
		Noncompetitive

[a] Each one of these may further be grouped as linear, hyperbolic, or parabolic with respect to their slope and/or intercept changes

inhibition and true irreversible inactivation. In such cases, one can look for the release of the original inhibitor molecule after denaturing the enzyme protein or it may be possible to remove the free inhibitor as it is released from EI (viz., by adsorption of the inhibitor on charcoal). Specific covalent inhibition usually involves two distinct steps—binding and inactivation. This can be tested by a "jump-dilution" approach. Both E and I are preincubated at saturating concentrations to allow complex formation first. This is followed by rapid dilution and assessment for recovery of enzyme activity. The EI complex formed will have two possible fates—either to dissociate back to E and I or to get inactivated irreversibly. The rate of enzyme recovery after jump-dilution provides an estimate of k_{off}, the dissociation rate constant.

Another simple approach to assess irreversibility is to study the effect of the inhibitor on the $v \rightarrow [E_t]$ curve for the enzyme (also refer to Chaps. 11 and 12, in Part II). Increasing concentrations of the enzyme are incubated in the absence or presence of a fixed concentration of the inhibitor. Subsequently, the enzyme activity remaining is measured in each case (Fig. 19.1). These are called diagnostic Ackermann–Potter plots (Ackermann and Potter 1949; Cha 1975).

An irreversible inhibitor would stoichiometrically inactivate (and depending on the rate of inactivation) and titrate out enzyme molecules. In theory, an irreversible interaction between enzyme and inhibitor may occur in many ways. One possibility is that the enzyme-inhibitor complexation is reversible in principle but seems irreversible due to tight binding. Another situation is when the enzyme reacts with an inhibitor and the modified form of the enzyme cannot be converted back into an active enzyme. In either case, the amount of enzyme inactivated will depend not only on the amount of inhibitor but also on the amount of enzyme present. A preincubation-dependent increase in inhibitor potency is a strong indication of either

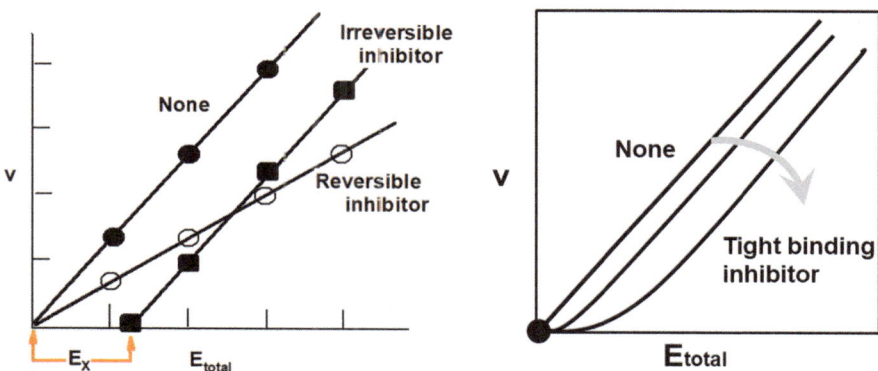

Fig. 19.1 Influence of an inhibitor on the enzyme concentration versus initial velocity curve (the Ackermann–Potter plot). Enzyme titrations by an irreversible inhibitor (left panel; E_x, the amount of enzyme titrated) and a tight binding inhibitor (right panel) are shown

reversible slow-binding or irreversible covalent inhibition. Regardless of the exact nature of the irreversibility, the active enzyme molecules remaining will be kinetically indistinguishable from the original enzyme molecules. Therefore, the curve will be parallel to the "control $v \rightarrow [E_t]$ curve" but will not pass through the origin. The point of intersection on the X-axis represents the amount of enzyme irreversibly inactivated by the concentration of the inhibitor used. On the other hand, in the presence of a reversible inhibitor, all the enzyme molecules will be active but are kinetically less efficient. Hence, the corresponding "$v \rightarrow [E_t]$ curve" will be of a lower slope but still passes through the origin. This is a simple and quick way to establish the reversible nature of an inhibitor. A word of caution—the enzyme-inhibitor incubation conditions should be carefully chosen; significant enzyme inactivation must occur in the given time of incubation with an irreversible inhibitor.

Two facts must be considered when studying a tight-binding inhibitor. First, the depletion of the free ligand by binding cannot be ignored. Second, both the association and dissociation reactions may be so slow that classical steady-state enzyme kinetics may not apply. With tight-binding inhibitors (under conditions where $[E_t] \gg K_I$), the curves in the Ackermann–Potter plots appear parallel but would be nonlinear in the early concentrations of the enzyme, and yet they intersect at the origin (Fig. 19.1).

19.2   Partial Versus Complete Inhibition

A simple inhibition experiment involves monitoring the rates of an enzyme-catalyzed reaction in the presence of increasing concentrations of the reversible inhibitor. The resultant $v \rightarrow [I]$ curve is nonlinear and asymptotic to the X-axis. At saturating values of a complete inhibitor, the enzyme activity tends to zero (Fig. 19.2a). On the other hand, the enzyme activity plateaus to a nonzero limiting

Fig. 19.2 Fractional inhibition analysis. The relative value of velocity is plotted in the $v \rightarrow [I]$ plot (panel **a**) and a double reciprocal plot of $1/i \rightarrow 1/[I]$ is shown in panel **b**

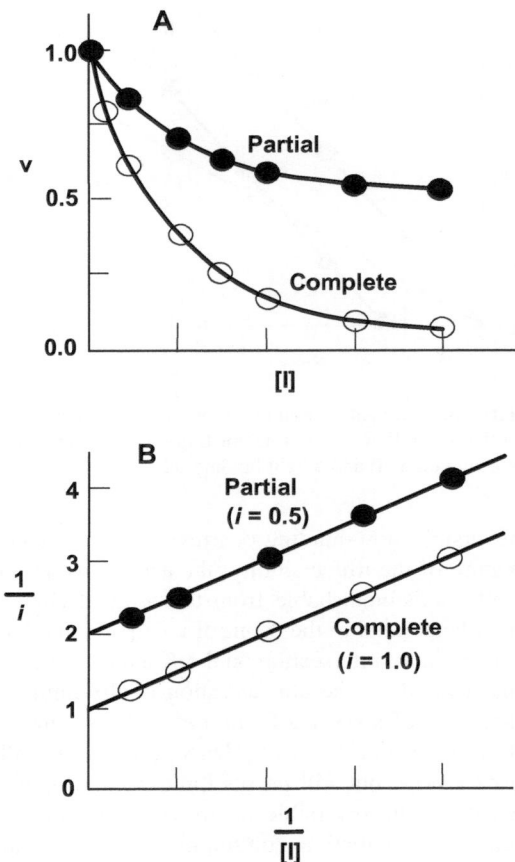

value by increasing the concentration of a partial inhibitor (Whiteley 2000a). However, due to the nonlinear, asymptotic nature of such curves and for reasons of experimental feasibility (unable to achieve a very high concentration of the inhibitor, in practice), it is difficult to determine the limiting value.

Fractional inhibition analysis is a convenient tool to distinguish between partial and complete inhibitors (Whiteley 2000b). For this, a relative quantity termed fractional inhibition (denoted "i") is defined as shown.

$$i = 1 - \frac{v_i}{v} = \frac{v - v_i}{v}$$

Here, v_i is the inhibited rate in the presence of the inhibitor. If the inhibition is complete, then i will take a value of unity (because at the saturating concentration of the inhibitor, v_i will be zero). For a partial inhibitor, however, v_i will never reach zero and i will be always less than one. The inhibition data are plotted as $1/i \rightarrow 1/[I]$ to obtain a linear plot (this is analogous to the Lineweaver–Burk treatment of v → [S] plot!). The Y-axis intercept of such a double reciprocal plot is unity for a complete inhibitor, whereas it will be greater than one for partial inhibitors (Fig. 19.2b).

19.3   Other Inhibitor Types

Inhibitors are often structurally related to the product(s) or substrate(s) of that enzyme. It is easy to appreciate that molecules structurally similar to a substrate/product can occupy the same space (pocket!) at the enzyme active site. Such inhibitors are called *isosteric* inhibitors. In some metabolic pathways, a terminal metabolite without any chemical analogy/reactivity to an earlier step is a powerful inhibitor of its own biosynthesis (Part VI, Chap. 38). Obviously, such structurally unrelated molecules cannot occupy the isosteric (orthosteric) enzyme active site; but inhibit by binding to a site distinct from the active site. Such inhibitors are called *allosteric* inhibitors and the site where they bind is called an allosteric site. The binding of an inhibitor at the allosteric site is communicated to the active site through the protein matrix—as a conformational change.

The product of the enzyme reaction can act as an inhibitor. *Product inhibition* may result due to the reversal of the forward reaction since it will be the substrate for the reaction backward. This mode of action will obviously be possible only if the full complement of products is present in the assay. In a multi-product reaction, however, the presence of a single product may lead to inhibited rates by playing musical chairs with substrates. Let us consider an example to illustrate these modes of inhibition. Lactate dehydrogenase catalyzes the following reversible reaction:

$$\text{Pyruvate} + \text{NADH} + \text{H}^+ \rightleftharpoons \text{Lactate} + \text{NAD}^+$$

In the presence of pyruvate and NADH, only the forward reaction occurs. If the assay also contains lactate and NAD^+, then the reverse reaction also becomes significant. The net forward rate will then be reduced (inhibition is seen) because of a certain backward reaction rate. In this sense product inhibition is a result of the reversal of the reaction. However, the presence of NAD^+ alone can inhibit the forward reaction. Since there is no reversal of the reaction possible (because lactate is missing from the assay!), inhibition by NAD^+ occurs because it can displace NADH from the enzyme active site and prevents the E.pyruvate.NADH ternary complex formation. Logically, because of their structural similarity, NAD^+ and NADH are expected to compete for the same active site pocket on the enzyme. We note that the E.pyruvate.NAD^+ complex, if at all formed, is not productive. Such a combination is termed a dead-end complex—implying that this enzyme form is not on the normal reaction path of the catalytic turnover.

The dead-end combination (exemplified with NAD^+ for lactate dehydrogenase reaction, above) may also occur with substrate/product analogs that are not substrates/products themselves for the enzyme. Such inhibitors are typical *dead-end* inhibitors; their complexes with the enzyme do not form part of the normal

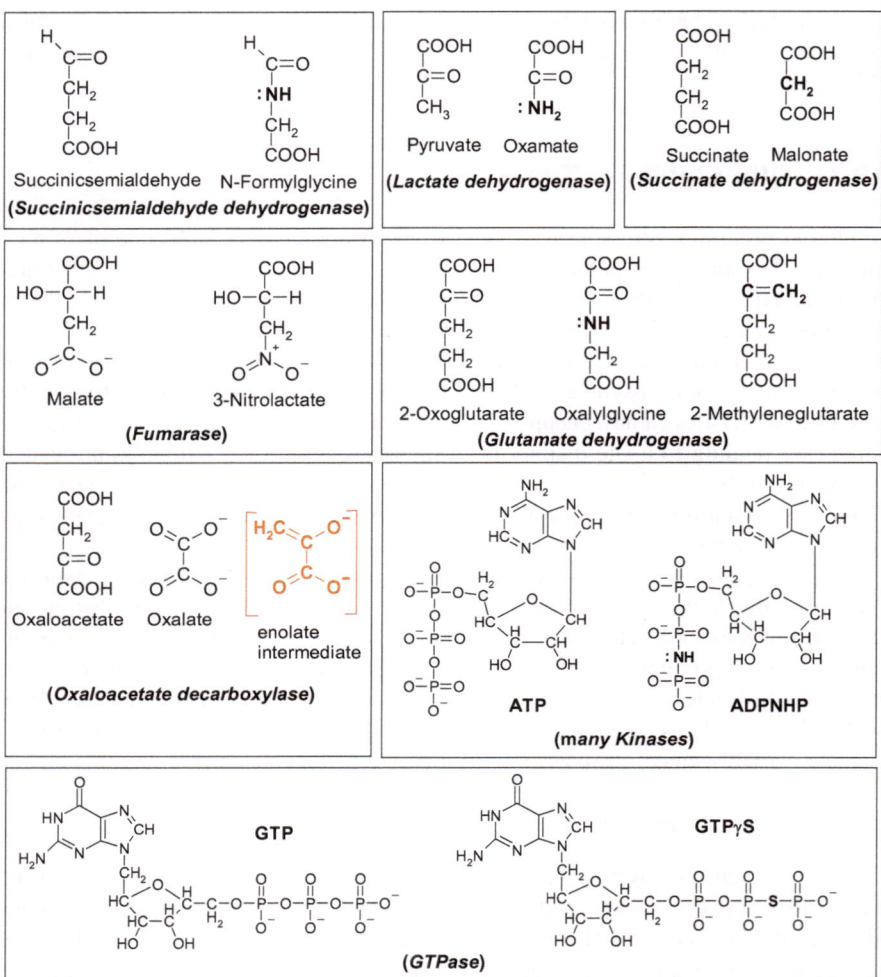

Fig. 19.3 Structures of a few dead-end inhibitors. The enzyme inhibited and the corresponding substrates that they mimic are also shown. Key structural differences are shown in bold

reaction sequence—hence, no catalysis occurs. Examples of some dead-end inhibitors and their target enzymes are shown in Fig. 19.3.

Dead-end inhibitors are excellent tools in the study of enzyme kinetic mechanisms. Besides initial velocity analysis and product inhibition (discussed above), one could use dead-end inhibitors to help deduce kinetic mechanisms. The thumb rules to predict dead-end inhibition patterns are similar to those employed for product inhibitions (Chap. 17) but with one exception. Being dead-end inhibitors, they cannot bring about the partial reversal of the reaction; hence they give more number of uncompetitive inhibition patterns.

Alternate products or substrates can be viewed as inhibitors of the enzyme reaction with their normal counterparts. In rare instances, the substrate itself acts as an inhibitor at higher concentrations. These cases and their relevance to the study of enzyme mechanisms will be discussed a little later (Chap. 22). Reversible inhibitors, especially the product and dead-end inhibitors, provide valuable insights into establishing enzyme kinetic mechanisms.

References

Ackermann WW, Potter VR (1949) Enzyme inhibition in relation to chemotherapy. Proc Soc Exp Biol Med 72:1–9

Cha S (1975) Tight-binding inhibitors—I: kinetic behavior. Biochem Pharmacol 24:2177–2185

Whiteley CG (2000a) Enzyme kinetics: partial and complete competitive inhibition. Biochem Educ 28:144–147

Whiteley CG (2000b) Mechanistic and kinetic studies of inhibition of enzymes. Cell Biochem Biophys 33:217–225

Further Reading

Rodriguez JG, Towns MH (2019) Analysis of student reasoning about Michaelis–Menten enzyme kinetics: mixed conceptions of enzyme inhibition. Chem Educ Res Pract 20:428–442

Irreversible inhibition of enzyme activity often results from covalent modification of the enzyme protein. Once the enzyme is covalently bound to an irreversible inhibitor it is permanently incapacitated. The inhibition is time-dependent and not freely reversible by procedures like dilution, dialysis, or gel filtration. Such inhibitors are often referred to as enzyme inactivators. Irreversible inhibition effectively decreases the concentration of the enzyme present—the net result being a reduced V_{max} (because $V_{max} = k_{cat} \times [E_t]$ and $[E_t]$ is actually reduced during irreversible inactivation)—while the K_M of the remaining active enzyme is unaltered. This is reminiscent of a reversible noncompetitive inhibition pattern where both V_{max} and V_{max}/K_M are affected (see Chap. 21). Therefore, for any new inhibitor, it is prudent to first establish whether that inhibitor is reversible or not! Without much hair-splitting on their nomenclature, we will consider three broad categories of enzyme inactivators in terms of their mechanism.

20.1 Chemical Modification Agents

A number of reagents are known in protein chemistry to modify specific types of amino acid side chains (de Gruyter et al. 2017). While chemical modification of proteins is an important tool, site-selective reactivity requires exquisite control over both chemo- and regioselectivity, under ambient, aqueous conditions. Such chemical probes were largely limited to reactions at nucleophilic cysteine and lysine residues. Various methods for achieving selective modification of both natural and unnatural amino acids are now available (Spicer and Davis 2014; Gunnoo and Madder 2016). Historically, these group-specific reagents have been extensively used to define residues that are essential for protein function, viz., enzyme active site residues. Table 20.1 lists some of the commonly used chemical modification reagents. They covalently modify the proteins and are irreversible. In some cases (like

Table 20.1 Chemical modification reagents as irreversible enzyme inhibitors

Amino acid (functional group)	Commonly used reagents
Arginine (guanidinium)	2,3-Butanedione, Phenylglyoxal
Cysteine (thiol)	Iodoacetamide, N-Ethylmaleimide (NEM), 4-Chloromercuribenzoate, Disulfides
Histidine (imidazole)	Diethylpyrocarbonate (DEPC)
Lysine (amino)	1-Fluoro-2,4-dinitrobenzene (FDNB), Trinitrobenzene sulfonic acid (TNBS)
Serine (hydroxyl)	Diisopropylfluorophosphate (DIFP), Phenylmethanesulfonyl fluoride (PMSF)
Tryptophan (indole)	N-Bromosuccinimide (NBS)

diethylpyrocarbonate), it may be possible to reverse this modification by using a stronger nucleophile (like hydroxylamine!).

By virtue of their ability to react with functionally important residues on the enzyme, these chemical modifiers act as irreversible inhibitors. They are useful in the identification of enzyme groups relevant for binding and/or catalysis. Chemical modification can become quite specific at times—especially when the residue in question enjoys a unique microenvironment in the protein (Eyzaguirre 1987). Histidine residues at the active site of ribonuclease A are susceptible to modification by diethylpyrocarbonate. Most kinases display a reactive arginine side chain at the active site to anchor substrate phosphate group(s). The case of super-reactivity of Ser-195 at the active site of chymotrypsin is well known. Ser-195 is the only residue titrated by diisopropylfluorophosphate among a total of 28 Ser residues found in the chymotrypsin sequence. Yet another example is the reactivity of Lys-126 (an anchor for substrate carboxylate) of bovine glutamate dehydrogenase to trinitrobenzene sulfonic acid. Such fortuitously selective chemical modification of active site residues was very valuable in defining them even before enzyme crystal structures were available. Although modern techniques like site-directed mutagenesis have largely supplanted this approach, chemical modification has its value in enzyme kinetic analysis. Protection experiments can provide a direct measure of the true dissociation constant for a substrate (see below). As this approach requires only catalytic amounts of the enzyme, it has been often applied to those cases where enough sample is not available for binding studies.

Despite its elegance and simplicity, the interpretation of chemical modification data requires caution. If a modification is actually occurring at an active site residue, then (a) there should be a stoichiometric relation between the extent of modification and extent of inactivation and (b) substrate/product/inhibitor that masks the residue on binding should protect against inactivation. By the same token, the caveats include (a) an essential residue titrated by this method may not be involved in binding, catalysis, or both! Irreversible inactivation may occur due to the modification of structurally important enzyme residues, (b) difficulty in pinpointing the role of a particular residue—the method cannot by itself distinguish between residues of equal reactivity. This has not however prevented the practical utility of such irreversible modifying reagents.

Interpretation of chemical modification data can be tightened by additional experiments. The number and position of amino acid residues modified can be tracked through radio-labeled reagents (e.g., ^{14}C-labeled iodoacetamide or N-ethylmaleimide). Isolation and sequence characterization of the labeled peptide can be part of routine protein chemistry. Chemical modification in combination with high-resolution mass spectrometry (like MALDI-TOF) is a powerful tool in recent times (Shuken 2023). A differential labeling strategy of comparing the inactivated enzyme, in the presence and absence of a suitable ligand (substrate or inhibitor), can further qualify the essentiality of that amino acid residue.

Chemical modification reagents irreversibly inactivate by reacting with groups on the enzyme surface. This happens without the prior formation of a specific noncovalent enzyme–inhibitor complex (Rakitzis 1984). Consequently, the inactivation follows a second-order reaction and is not a saturable phenomenon. We can treat the inactivation kinetics as shown below.

$$E + I \rightarrow EI^*$$

where EI* is inactive and the rate of inactivation (with k_1 as the second-order rate constant) given by

$$-\frac{d[E]}{dt} = \frac{d[EI^*]}{dt} = k_1[E][I]$$

When [I]>>[E], one can experimentally set up pseudo-first-order conditions (Chap. 8; Part II) with respect to fixed [E], then we have

$$-\frac{d[E]}{dt} = \frac{d[EI^*]}{dt} = k_{obs}[E], \quad \text{where } k_{obs} = k_1[I]^n$$

The following equation may be written for a general case where "n" the number of irreversible inhibitor molecules accounts for the inactivation and exponential decay of [E]

$$\log k_{obs} = \log k_1 + n \log[I]$$

Experimentally one follows the inactivation of a fixed amount of E at different [I] values. These data are first plotted as percent enzyme activity remaining versus time; and then also as log percent activity remaining versus time. From each straight line we obtain a set of k_{obs}—the pseudo-first-order rate constant values. A replot of $\log k_{obs} \rightarrow \log[I]$ should result in a straight line; the number of essential residues required to be modified for inactivation (stoichiometry from slope "n") and the second-order inactivation rate constant (k_I, from the intercept) are thus obtained. Typical results, for example, on N-ethylmaleimide inactivation of glutamine synthetase, are shown in Fig. 20.1 (Punekar et al. 1985). Besides the caveats mentioned

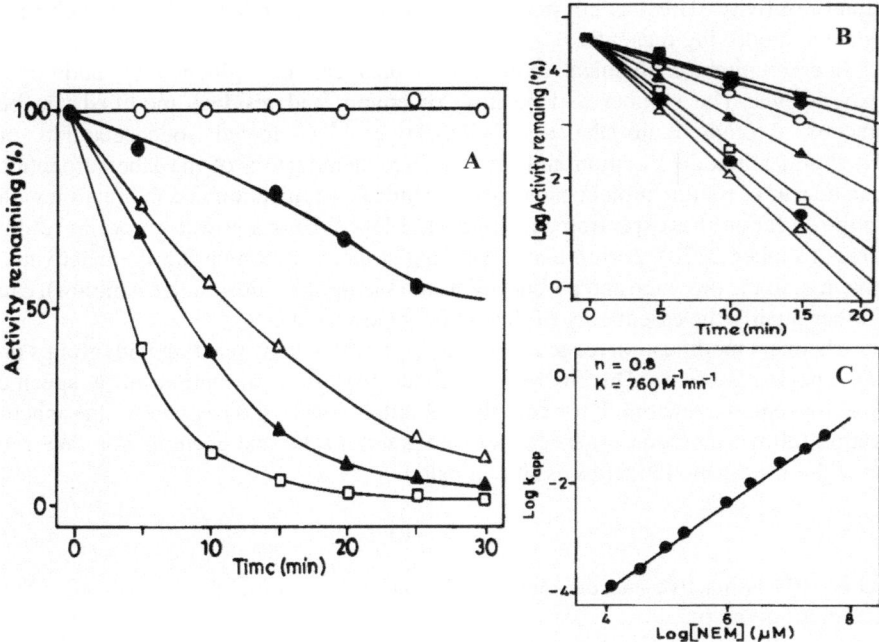

Fig. 20.1 N-Ethylmaleimide inactivation of glutamine synthetase. The time course of inactivation (**a**), semi-log plot of the same data (**b**), and a replot log $k_{obs} \rightarrow$ log[I] (**c**) are shown

earlier on, these experiments require that a large excess of [I] over [E] and pseudo-first-order conditions be satisfied.

The way to perform protection experiments is as follows. The concentration of the inactivating agent and enzyme is fixed to achieve a suitable maximal rate of inactivation (k_o). The decreasing inactivation rates (k_p values) are monitored as a function of increasing protecting ligand concentration [P]. Without belaboring the detailed derivation, the equation for kinetics of protection by a ligand against inactivation is as shown:

$$\log \frac{k_o - k_p}{k_p} = n \log[P] - \log K_P$$

Here, k_o and k_p are pseudo-first-order rate constants in the absence or presence of the protecting molecule P. A plot of $\log[(k_o - k_p)/k_p] \rightarrow$ log[P] provides the values of "n" and the dissociation constant K_P (Tian and Tsou 1982). Such an approach has been extensively used as an independent probe for establishing the binding of ligands to enzymes. Figure 20.2 depicts the protection afforded to glutamine synthetase activity by Mn^{2+} ions, against N-ethylmaleimide inactivation (Punekar et al. 1985). Conceptually, "n" gives the number of molecules of P required to interact with the enzyme to afford protection. The K_P obtained from this route is purely a physical

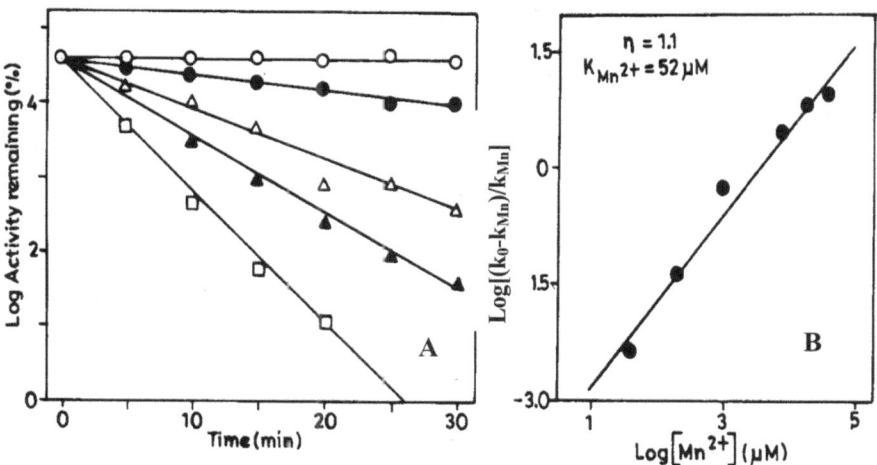

Fig. 20.2 Protection afforded by Mn^{2+} ions during N-ethylmaleimide inactivation of glutamine synthetase. Pseudo-first-order plots (**a**) and the corresponding replot (**b**) of $\log[(k_o - k_{Mn})/k_{Mn}] \rightarrow \log[Mn^{2+}]$ are shown

dissociation constant. It is distinct from those obtained by enzyme kinetic studies— (a) the K_M which is a composite of rate constants or (b) the K_I which is a kinetic parameter obtained in the presence of a full complement of substrates. Nevertheless, a comparison of these values gives useful insights into various ligand–enzyme interactions like synergistic binding.

20.2 Affinity Labels

Also known as active site-directed irreversible inhibitors, affinity labels combine the reactivity of chemical modification agents (see above) along with a structural moiety to provide specificity (Eyzaguirre 1987). They are covalent inactivators resembling the substrate but also contain a reactive functional group, such as an α-haloketone or a reactive ester. The best-known example of this category is the chymotrypsin inhibitor, N-4-toluenesulfonyl-L-phenylalanine chloromethylketone (TPCK) (Fig. 20.3). Affinity labels act as specific irreversible inhibitors—for instance, TPCK does not inactivate trypsin, whereas TLCK does.

Affinity labels differ from chemical modification agents in that a specific noncovalent EI complex is formed before the actual inactivation event. This mechanism, similar to the Michaelis–Menten formalism with Briggs–Haldane steady-state assumption (Chap. 14; Part II), may be represented as

$$E + I \rightleftarrows EI \rightarrow EI^*$$

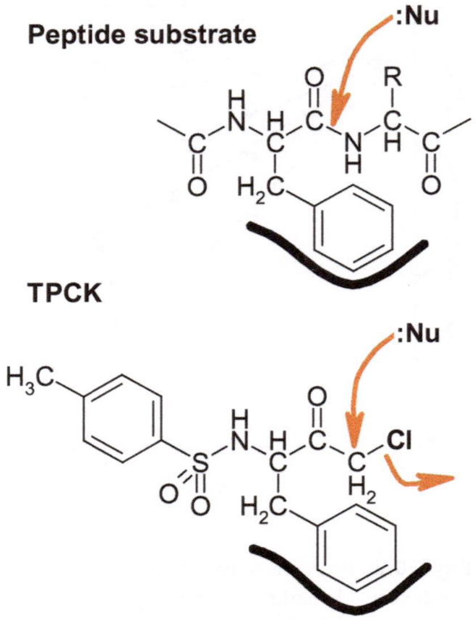

Fig. 20.3 Structure of N-4-toluenesulfonyl-L-phenylalanine chloromethylketone (TPCK) and the corresponding portion of chymotrypsin substrate. TLCK is trypsin-specific and contains L-lysine in place of L-phenylalanine

where EI represents the noncovalent enzyme–inhibitor complex, EI* is the irreversibly inhibited covalent adduct, and k_2 is the first-order rate constant for EI → EI* conversion. By analogy to the Michaelis–Menten equation we write

$$k_{\mathrm{obs}} = \frac{k_2[\mathrm{I}]}{K_{\mathrm{I}} + [\mathrm{I}]}$$

A plot of $k_{\mathrm{obs}} \rightarrow [\mathrm{I}]$ is hyperbolic. This is unlike the second-order rate for chemical modification agents where the $k_{\mathrm{obs}} \rightarrow [\mathrm{I}]$ plot is linear. As expected, saturation kinetics for the inactivation process is observed, and from a plot of $1/k_{\mathrm{obs}} \rightarrow 1/[\mathrm{I}]$ we can evaluate both k_2 and K_{I}.

The chemistry of affinity labels is very amenable, and given the substrate structural details, such labels can be synthesized for almost any enzyme. Enzyme literature is rich with such examples (De Cesco et al. 2017; Wu et al. 2019). Apart from their utility as inhibitors, they can be used to tag and identify key residues involved in the catalytic process of that enzyme. The incorporation of groups that are activated by light of specific wavelength makes affinity labels into useful photo-affinity labels.

20.3   Suicide Substrates

These are unreactive molecules normally recognized as substrates by the enzyme. But during the catalytic turnover the target enzyme converts them into reactive entities. These reactive species trap an essential catalytic group at the active site and kill the enzyme (Alston et al. 1983; Silverman 1995). At least a part of the

Fig. 20.4 A suicide substrate (the frog!) once into the catalytic cycle never gives up its interaction with the enzyme (the stork!)

enzyme catalytic action is required for the reactive species to be generated; then the substrate never gives up (Fig. 20.4). Therefore, such inhibitors are variously referred to as—k_{cat} inhibitors, Trojan horse inactivators, trap substrates, enzyme-activated irreversible inhibitors, etc. The reaction scheme for these mechanism-based inhibitors may be represented as shown:

$$E + I \rightleftharpoons EI \rightarrow EI^* \rightarrow E - I$$
$$\downarrow$$
$$E + P$$

where a noncovalent complex (EI) is first converted to an activated species (EI*); this can either inactivate the enzyme by trapping it or go through a complete catalytic cycle to release the corresponding product and the free enzyme. It is of course desirable that the partitioning (shown in bold) of EI* be more toward E–I than a full cycle of catalysis (and P release) to be an efficient suicide substrate.

Inactivation by suicide substrates is a first-order process and shows saturation kinetics with respect to inhibitor concentration. Since they interact at the enzyme active site, substrates competitively protect against such inhibitors. In the absence of the respective enzyme target, they are unreactive molecules, and hence nonspecific modifications of other proteins are minimized. This property makes them eminently suited as potential designer drugs. Penicillins (β-lactam antibiotics in general) act on their target enzyme in this mode—a penicilloyl enzyme is formed during catalysis

Table 20.2 Suicide substrates and their target enzymes

Suicide substrate	Enzyme acted upon	Application
Gabaculine	γ-Aminobutyrate transaminase	Anticonvulsant
5-Fluorodeoxyuridylate	Thymidylate synthase	Anticancer
Penicillin	DD-Transpeptidase	Antibacterial
Clavulanic acid	β-Lactamase	β-Lactam synergist
Exemestane	Aromatase	Breast cancer
Allopurinol	Xanthine oxidase	Gout
α-Difluoromethylornithine	Ornithine decarboxylase	Antiprotozoal

thereby blocking the active site Ser residue. A short, representative list of suicide substrates is given in Table 20.2.

20.4 Tight-Binding Inhibitors

Some ligands/inhibitors may bind the enzyme reversibly but very tightly. Such tight-binding inhibitors may be difficult to distinguish from truly irreversible inhibitors (also see Chap. 19). The tightness of interaction may appear irreversible although there is no covalent bond established to the enzyme. The noncovalent reversible but very tight interaction in the formation of the avidin–biotin complex (with a half-life of 2.5 years and an incredibly small dissociation constant of 10^{-13} M) are well known. Similarly, methotrexate is a high-affinity inhibitor of dihydrofolate reductase. Another example of such tight binding interaction is between purine nucleoside phosphorylase and immucillin-H (a TS inhibitor with an equilibrium dissociation constant of 23 pM). Besides other noncovalent forces, the strength for such interactions comes from well-directed hydrogen bonds; often some of these are low-barrier hydrogen bonds.

In some ways, this tight binding reflects the slow dissociation of the inhibitor. How does this tightness of binding come about? For a simple binding equilibrium,

$$E + A \rightleftharpoons EA$$

the equilibrium dissociation constant is given by $K_S = ([A][E])/[EA]$. This K_S may also be written as the ratio of the off and on rate constants (k_{off}/k_{on}). The larger the denominator (k_{on}), the smaller the K_S, the tighter the binding. However, there is an upper limit for k_{on}—the diffusion limit—of about 10^8 M^{-1} s^{-1} (see Chap. 14, Sect. 14.3.1). This implies that for a k_{off} value of unity, the best K_S achievable is 10^{-8} M ($= 1$ $s^{-1}/10^8$ M^{-1} s^{-1}, or about 10 nM). We have seen that many drugs, acting as enzyme inhibitors, bind much more tightly than this. Since this cannot happen only by increasing k_{on} (because diffusion puts an upper cap and limits it!) such tight binding inhibitors act so by decreasing the numerator (k_{off} value). Clearly, these inhibitors dissociate more slowly from their respective enzyme complexes (Morrison and Walsh 1988; Schloss 1988).

A major consequence of tight binding by an inhibitor is that steady-state kinetic analysis becomes complicated. It has been suggested that whenever $K_S/[E_t]$ is less than 1000, the steady state and related assumptions should be abandoned. For K_D values in the sub-nanomolar range, the assumption that $[I] >> [E_t]$ may no longer be valid. One needs to use tight-binding inhibitors at concentrations around (or lower than) that of the enzyme itself. In this sense, we cannot assume that $[I]_{free} \approx [I]_{total}$— and the inhibitor itself is titrated out. Thus, tight binding inhibitors provide a convenient means of accurately determining the proportion of active enzyme molecules in the given sample.

References

Alston TA, Porter DJT, Bright HJ (1983) Enzyme inhibition by nitro and nitroso compounds. Acc Chem Res 16:418–424

De Cesco S et al (2017) Covalent inhibitors design and discovery. Eur J Med Chem 138:96–114

de Gruyter JN, Malins LR, Baran PS (2017) Residue-specific peptide modification: a chemist's guide. Biochemistry 56:3863–3873

Eyzaguirre J (1987) Chemical modification of enzymes: active site studies. Ellis Horwood. isbn:0745800238, 9780745800233

Gunnoo SB, Madder A (2016) Chemical protein modification through cysteine. Chembiochem 17: 529–553

Morrison JF, Walsh CT (1988) The behaviour and significance of slow binding enzyme inhibitors. Adv Enzymol Relat Areas Mol Biol 61:201–301

Punekar NS, Vaidyanathan CS, Appaji Rao N (1985) Role of Mn[II] and Mg[II] in the catalysis and regulation of *Aspergillus niger* glutamine synthetase. Indian J Biochem Biophys 22:142–151

Rakitzis ET (1984) Kinetics of protein modification reactions. Biochem J 217:341–351

Schloss JV (1988) Significance of slow-binding enzyme inhibition and its relationship to reaction intermediate analogs. Acc Chem Res 21:348–353

Shuken SR (2023) An introduction to mass spectrometry-based proteomics. J Proteome Res 22: 2151–2171

Silverman RB (1995) Mechanism-based enzyme inactivators. Methods Enzymol 249:240–283

Spicer CD, Davis BG (2014) Selective chemical protein modification. Nat Commun 5:4740

Tian W-X, Tsou C-L (1982) Determination of the rate constant of enzyme modification by measuring the substrate reaction in the presence of the modifier. Biochemistry 21:1028–1032

Wu L et al (2019) An overview of activity-based probes for glycosidases. Curr Opin Chem Biol 53: 25–36

Reversible Inhibitions

21

Reversible inhibitors, especially the product and dead-end inhibitors, provide valuable insights to establish enzyme kinetic mechanisms. We have acknowledged earlier (Chap. 19) that reversible nature of inhibition has to be established before embarking on its use to study enzyme mechanisms. How these reversible inhibitors are employed in enzyme kinetic analyses are discussed in this chapter.

The nature of kinetic experiments conducted, and the information sought from reversible enzyme inhibition data are as follows:

1. Monitor initial velocity "v" by varying the concentration of one substrate at different fixed concentrations of the inhibitor. If the enzyme reaction in question involves more than one substrate, then the concentration of all other substrates (other than the one whose concentration is varied) is fixed.
2. The $v \rightarrow$ [S] data are plotted in the double reciprocal format (Lineweaver-Burk plots) to generate a series of curves—one for each fixed concentration of the inhibitor. These patterns are analyzed qualitatively.
3. Gradual changes in the slope and/or intercepts, as a function of the fixed inhibitor concentration, are noted. An inhibitor may affect the first-order rate constant (V_{max}/K_M which is reflected in slope changes) or the zero-order rate constant (V_{max} as reflected in intercept changes) or both.
4. On quantitative analysis of slope and intercept changes, various kinetic constants including K_I values are evaluated. Depending upon whether the slope/intercept increases as a linear function of [I] or not, the inhibition may also be classified as *linear*, *hyperbolic,* or *parabolic*.

Inhibition analyses through double reciprocal plots are most useful as their slope and/or intercept effects tell us directly about the effect of that inhibitor on the rate constants. At times the ***Dixon plot*** is a useful alternative (Dixon 1953). Like before, we monitor initial velocity "v" but by varying the concentration of the inhibitor at different fixed concentrations of the substrate in question. If the enzyme reaction

N. S. Punekar, *ENZYMES: Catalysis, Kinetics and Mechanisms*,
https://doi.org/10.1007/978-981-97-8179-9_21

involves more than one substrate, then the concentration of all substrates (other than the one whose concentration is varied) is fixed. The $1/v \rightarrow$ [I] data are plotted for every fixed concentration of the substrate. There are a few advantages of Dixon analysis. Since we are primarily varying [I], relatively high, fixed [S] values could be used. For the double reciprocal analysis, however, data at lower [S] values are very essential for accuracy of analysis (see Chap. 16; Part II). Dixon plots are diagnostic for nonlinear inhibitions (see below). A drawback of the Dixon plot patterns is, unlike the double reciprocal analysis, slope and intercept effects cannot be directly interpreted. It is also difficult to infer which of them is/are nonlinear. Finally, these plots are the best way to study the interaction of different inhibitors of the same enzyme through multiple inhibition analysis. This is particularly valuable in the analysis of enzymes regulated by multiple inhibitors such as glutamine synthetase (See Chap. 38; Part VI).

Three common reversible inhibition patterns (as double reciprocal plots) observed in reversible inhibition studies are described below. These concepts are nicely presented through cartoons by Tayyab (1990).

21.1 Competitive Inhibition

The scheme representing interaction equilibria for a competitive inhibitor with the enzyme is given below (Fig. 21.1).

A competitive inhibitor affects only the slope of a double reciprocal plot. Therefore, a series of lines that intersect on the Y-axis are obtained (Fig. 21.2). Using the equilibrium assumption (Chap. 15; Part II) a rate equation may be derived to represent this situation.

$$v = \frac{V_{\max}[S]}{K_M\left(1 + \frac{[I]}{K_I}\right) + [S]}$$

This equation is identical to classical Michaelis-Menten equation except that $(1 + [I]/K_I)$. term multiplies the K_M in the denominator. The equation and the equilibria do not make it explicit as to whether the inhibitor binds to the active site or a site elsewhere on the enzyme. Both isosteric and allosteric inhibitors behave kinetically identical. It is natural to expect that a competitive inhibitor structurally related to either a substrate or a product occupies the active site. Malonate and

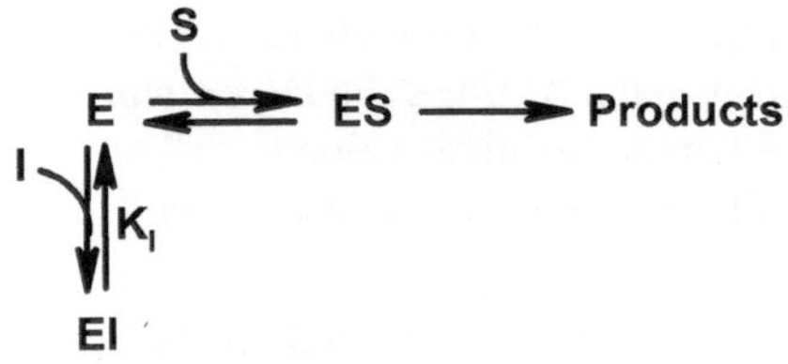

Fig. 21.1 Equilibria representing interaction of a competitive inhibitor with the enzyme

Fig. 21.2 Double reciprocal plots for the competitive inhibition of the enzyme with S as the varied substrate

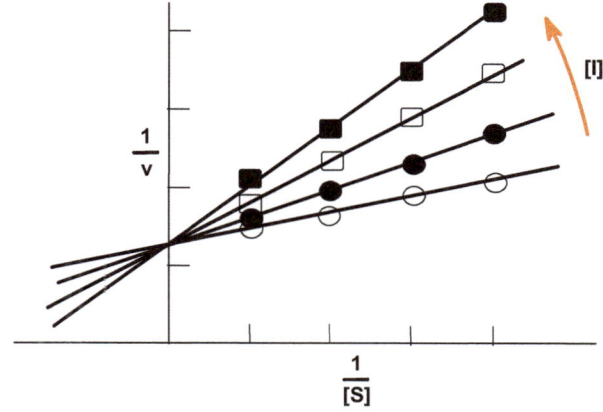

fumarate are well-characterized succinate dehydrogenase competitive inhibitors of this kind. In all other cases, therefore, additional data are required to establish where a competitive inhibitor binds on the enzyme.

Determination/evaluation of kinetic constants and replots: The double reciprocal form of the rate equation

$$\frac{1}{v} = \frac{K_M}{V_{max}}\left(1 + \frac{[I]}{K_I}\right)\frac{1}{[S]} + \frac{1}{V_{max}}$$

may be used to determine the K_I for the competitive inhibitor. Only the slope of the double reciprocal plot changes as a function of inhibitor concentration.

$$\text{Slope} = \frac{K_M}{V_{max}}\left(1 + \frac{[I]}{K_I}\right)$$

In a simple competitive inhibition as represented in the scheme above, the re-plot of slope → [I] yields a straight line; analysis of its slope and intercept (actually the intercept/slope of this line) gives the K_I. The slope re-plot may not be linear, however. While such nonlinear slope re-plots are diagnostic of inhibitor interaction, extracting a K_I from such analysis is not straightforward. The nonlinearity of inhibition is better visualized by Dixon analysis. These Dixon plots of $1/v$ → [I] are based on the rearranged format of the following double reciprocal form for competitive inhibition.

$$\frac{1}{v} = \frac{1}{V_{max}}\frac{K_M}{[S]K_I}[I] + \frac{1}{V_{max}}\left(1 + \frac{K_M}{[S]}\right)$$

Interpretation: Presence of a competitive inhibitor affects only the first-order rate constant (V_{max}/K_M; 1/slope of Lineweaver-Burk plot). The V_{max} is not affected and the lines intercept on the Y-axis (where [S] → ∞).

This is expected according to the equilibria described in the competition scheme (Fig. 21.1). While S and I compete for the free enzyme (form E), at infinite [S], all the enzyme molecules will be in the ES form—hence no inhibition by I occurs. At any finite level of [S], however, there is a proportion of E available for I to bind and inhibition results. These arguments do not assume anything about the equilibrium dynamics—leading to mutually exclusive binding of I and S to the enzyme form E. Clearly the kinetic consequence is the same whether (a) I displaces S at the active site (isosteric competitive) or (b) I binds elsewhere on E but changes its conformation such that S cannot access the active site (allosteric competitive). In any case, inhibition completely overcome by large excess of substrate is a hallmark of competitive inhibition. Finally, with competitive inhibitors that are nonlinear, slope re-plots may either be hyperbolic or parabolic. Such nonlinear competitive inhibition is indicative of more complex scheme of interaction between the inhibitor and the enzyme. For instance, a hyperbolic inhibition may result when a ternary complex of IES also forms, and this complex is active but is less productive than the ES complex. A parabolic competitive inhibition may occur due to multiple inhibitor molecules binding to the substrate binding site.

21.2 Uncompetitive Inhibition

The scheme representing interaction of an uncompetitive inhibitor with the enzyme is shown below (Fig. 21.3).

An inhibitor affecting only the intercept of a double reciprocal plot yields a series of lines that are parallel to each other (Fig. 21.4). Using the equilibrium assumption (Chap. 15; Part II), a rate equation may be derived to represent this situation.

$$v = \frac{V_{\max}[S]}{K_M + [S]\left(1 + \frac{[I]}{K_I}\right)}$$

This equation is identical to the Michaelis-Menten equation except that the [S] term in the denominator is multiplied by a factor $(1 + [I]/K_I)$. From the way the equilibria are represented, it is obvious that an uncompetitive inhibitor binds to ES complex but not to E (free enzyme). It is, however, conceptually a bit challenging

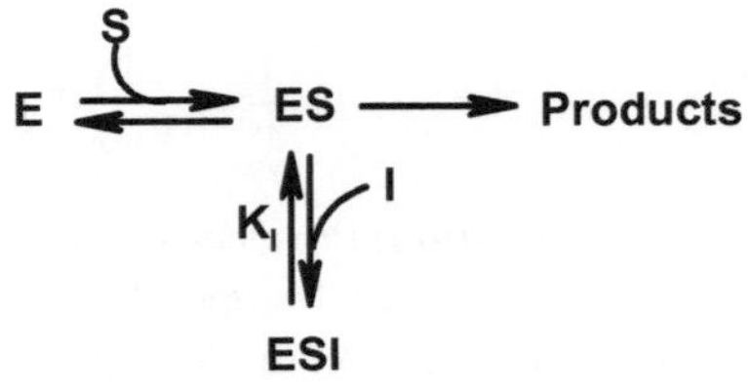

Fig. 21.3 Equilibria representing interaction of an uncompetitive inhibitor with the enzyme

Fig. 21.4 Double reciprocal plots for uncompetitive inhibition of the enzyme with S as the varied substrate

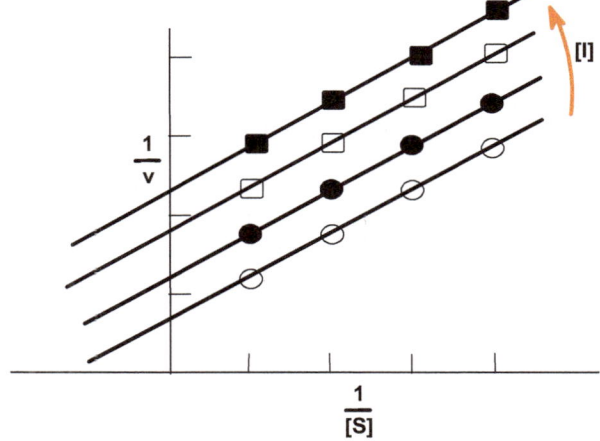

to imagine how an inhibitor can bind only the enzyme-substrate complex with a "single substrate" enzyme.

Determination/evaluation of kinetic constants and replots: The double reciprocal form of the rate equation

$$\frac{1}{v} = \frac{K_M}{V_{max}} \frac{1}{[S]} + \frac{1}{V_{max}} \left(1 + \frac{[I]}{K_I}\right)$$

could be used to determine the K_I for the uncompetitive inhibitor. The intercept of the double reciprocal plot changes with the inhibitor concentration according to the relation:

$$\text{Intercept} = \frac{1}{V_{max}} \left(1 + \frac{[I]}{K_I}\right)$$

The re-plot of intercept → [I] data for an uncompetitive inhibitor gives a straight line; analysis of its slope and intercept (actually the intercept/slope of this line) gives the K_I. If the intercept re-plot is non-linear, then extracting a K_I from such analysis is not straightforward; nonlinear curve fitting for the data could be resorted to. The non-linearity of inhibition is better visualized by Dixon analysis based on the following double reciprocal form of equation for uncompetitive inhibition:

$$\frac{1}{v} = \frac{1}{V_{max}} \frac{1}{K_I} [I] + \frac{1}{V_{max}} \left(1 + \frac{K_M}{[S]}\right)$$

Interpretation: An uncompetitive inhibitor affects only the zero-order rate constant (V_{max}; 1/intercept of the Lineweaver–Burk plot). Since V_{max}/K_M is not affected, a pattern with parallel set of lines (with no slope change) is obtained (Fig. 21.4).

According to the scheme for uncompetitive inhibition, at low [S], most of the enzyme will be in E; very little ES exists for I to bind. Hence, this inhibitor will not affect the slope (first-order rate constant). However, the Y-axis intercepts (where [S] → ∞) do change because all enzyme molecules will be in ES form and I binds to it. This also means uncompetitive inhibition (unlike competitive inhibition) cannot be overcome by increasing [S]. Intercept re-plots, whenever nonlinear, may either be hyperbolic or parabolic. Such nonlinear uncompetitive inhibition is indicative of more complex scheme of interaction between the inhibitor and the enzyme.

Inhibition of (a) arylsulfatase by hydrazine and (b) 5-enolpyruvylshikimate-3-phosphate synthase (EPSP synthase) by glyphosate are two interesting examples of uncompetitive inhibition. Uncompetitive inhibitors do not bind free enzyme but only the ES complex (Cornish-Bowden 1986). To visualize the physical picture of how this happens is quite a challenge—particularly with single-substrate enzyme reactions. The problem is how to imagine that the inhibitor has no affinity to (or binding site on) the free enzyme but one gets created in the ES form. A conformational change in the enzyme to reveal this site may need to be invoked. Alternately, the inhibitor may bind the enzyme bound substrate itself. Uncompetitive inhibitions, more common in multi-substrate enzyme mechanisms, are of diagnostic value in elucidating kinetic mechanisms. An uncompetitive dead-end inhibition by an analog of B, with A as the varied substrate, is diagnostic of an ordered mechanism.

21.3 Noncompetitive Inhibition

Interaction of a noncompetitive inhibitor with the enzyme may be represented by the following equilibria (Fig. 21.5).

A noncompetitive inhibitor affects both slope and intercept of a double reciprocal plot. Normally such a pattern shows a common point of intersection by a series of lines (Fig. 21.6). Using the equilibrium assumption (Chap. 15; Part II), a rate equation may be derived (see Appendix of this chapter) to represent this situation:

$$v = \frac{V_{max}[S]}{(K_M + [S])\left(1 + \frac{[I]}{K_I}\right)}$$

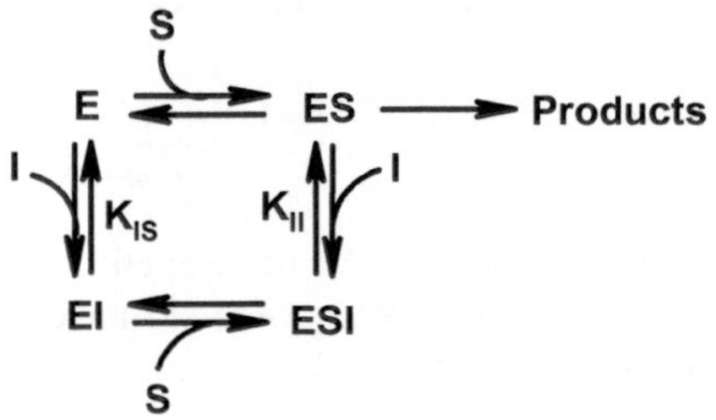

Fig. 21.5 Equilibria representing interaction of a noncompetitive inhibitor with the enzyme

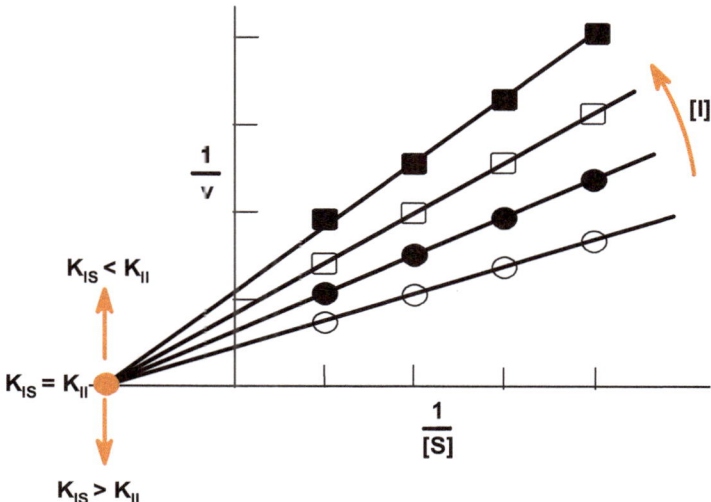

Fig. 21.6 Double reciprocal plots for noncompetitive inhibition of the enzyme with S as the varied substrate

Again, this equation is similar to the Michaelis-Menten form except that the entire denominator (K_M + [S]) is multiplied by a factor of (1 + [I]/K_I). It is obvious that a noncompetitive inhibitor combines the virtues of both a competitive (binds E) as well as an uncompetitive (binds ES) inhibition. Hence, both slope and intercept effects are, respectively, observed.

Determination/evaluation of kinetic constants and replots: The double reciprocal form of the rate equation

$$\frac{1}{v} = \frac{K_M}{V_{max}}\left(1 + \frac{[I]}{K_I}\right)\frac{1}{[S]} + \frac{1}{V_{max}}\left(1 + \frac{[I]}{K_I}\right)$$

could be used to determine the K_I for the noncompetitive inhibitor. From the way the noncompetitive inhibition equilibria are represented, it is not necessary that the inhibitor affinity to E and ES be identical. Therefore, the K_I obtained from the slope (termed more specifically as K_{IS}) and the intercept (termed K_{II}) may have different numerical values. In this sense, the slope and intercept of the double reciprocal plots change with the inhibitor concentration as shown below:

$$\text{Slope} = \frac{K_M}{V_{max}}\left(1 + \frac{[I]}{K_{IS}}\right)$$

$$\text{Intercept} = \frac{1}{V_{max}}\left(1 + \frac{[I]}{K_{II}}\right)$$

The two secondary plots (i.e., re-plots of slope → [I] data and intercept → [I] data) for a simple noncompetitive inhibitor give straight lines. The two sets of data may be now analyzed for K_I values, just the way we did before for competitive and uncompetitive inhibitors, respectively. It is possible that the slope re-plot, the intercept re-plot, or both of them may be nonlinear; such nonlinearity of inhibition is better visualized by Dixon analysis. A nonlinear curve fitting strategy will also be needed to obtain relevant K_I values.

Interpretation: A noncompetitive inhibitor affects both the zero-order rate constant (V_{max}; 1/intercept of the double reciprocal plot) and the first-order rate constant (V_{max}/K_M; 1/slope of the double reciprocal plot). According to the scheme for noncompetitive inhibition, the inhibitor binds an enzyme form both at low [S] (where form E predominates) and high [S] (where form ES predominates) conditions. Since Y-axis intercepts (where [S] → ∞) do change, noncompetitive inhibition (unlike competitive inhibition) cannot be overcome by increasing [S].

Noncompetitive inhibition (with few exceptions, see below) is manifest as an intersecting set of lines in the double reciprocal analysis. The coordinates of the point of intersection when evaluated (also refer to intersecting patterns in Chap. 18) are

Vertical (1/v) coordinate of the cross-over point $= \frac{1}{V_{max}} \left(1 - \frac{K_{IS}}{K_{II}} \right)$ and

Horizontal (1/[S]) coordinate of the cross-over point $= -\frac{K_{IS}}{K_{II}}$

The point of intersection is determined by the relative magnitudes of K_{IS} and K_{II}. If $K_{IS}/K_{II} = 1$, then the intersection is at the left of origin and on the X-axis at $1/v = 0$. This is a situation where $K_{IS} = K_{II}$ and is often known as pure noncompetitive inhibition. It is perfectly possible that $K_{IS} \neq K_{II}$ and such cases are at times referred to as *mixed inhibition*! Thus, depending on the relative values of K_{IS} and K_{II}, the point of intersection is located above, on, or below the X-axis (arrows in Fig. 21.6 indicate how the point of intersection moves).

Noncompetitive inhibition may also be viewed as a combination of two extremes, i.e., competitive (slope effect alone) and uncompetitive (intercept effect alone) (Fig. 21.7). When $K_{IS}/K_{II} \to 0$, it ends up being competitive (K_{II} is infinite; I has no affinity for ES) but with $K_{IS}/K_{II} \to \infty$ it simply collapses to being uncompetitive (K_{IS} is infinite; I has no affinity for E). However, if K_{IS} is very large (but not infinity!) then the lines in a noncompetitive pattern may appear parallel but actually intersect far away to the left of the origin. While it is easy to conclude that a set of lines intersect, we need to be cautious in concluding that a pattern is truly parallel.

Whenever intercept/slope re-plots are nonlinear, they may either be hyperbolic or parabolic. So long as either slope or intercept (or both) re-plots are linear, a crossover point is observed in the noncompetitive pattern. If no crossover point is observed but a series of "magic lines" are observed, then the inhibition is non-linear in both slope and intercept effects.

Fig. 21.7 Noncompetitive inhibition viewed as a combination of two extremes, namely, competitive (K_{IS}/K_{II} tends to zero) and uncompetitive (K_{IS}/K_{II} tends to infinity) cases. Only when $K_{IS} = K_{II}$ do the lines intersect on the X-axis

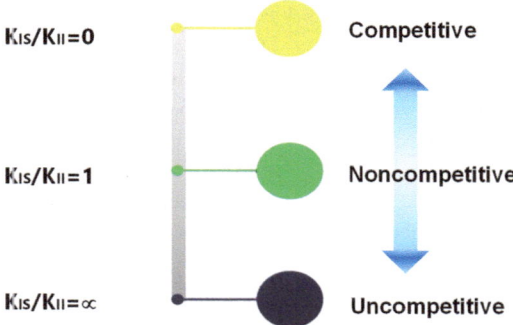

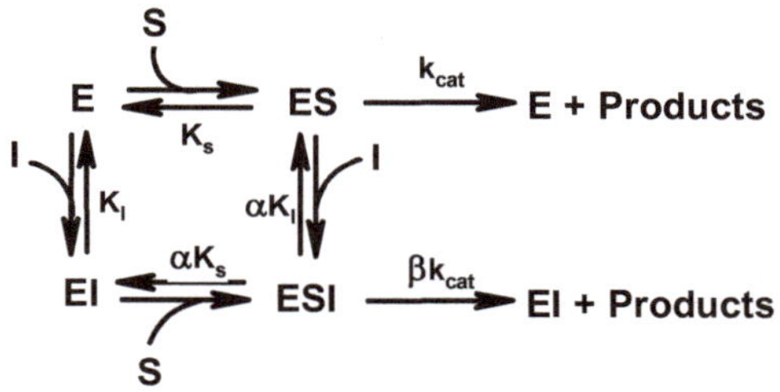

$K_{IS}/K_{II}=0$ Competitive

$K_{IS}/K_{II}=1$ Noncompetitive

$K_{IS}/K_{II}=\infty$ Uncompetitive

Fig. 21.8 A general equilibrium scheme for the interaction of an inhibitor with different enzyme forms

$$
\begin{array}{ccc}
& S & \\
E & \underset{K_s}{\rightleftharpoons} ES & \xrightarrow{k_{cat}} E + Products \\
I\,\Big\updownarrow K_I & \Big\updownarrow \alpha K_I\, I & \\
EI & \underset{\alpha K_s}{\rightleftharpoons} ESI & \xrightarrow{\beta k_{cat}} EI + Products \\
& S &
\end{array}
$$

21.4   Reversible Inhibition Equilibria: Another Viewpoint

The equilibria between the enzyme, substrate and the inhibitor may be treated by one other approach. It is useful to appreciate this kinetic representation as one also finds enzyme kinetic literature presented in this way. While different sets of nomenclature and/or representation are no doubt confusing, they need to be understood in order to fully savor the richness of enzyme kinetic literature. In this equilibrium treatment, dissociation constants for ES complex (K_S), EI complex (K_I), and k_{cat} are as shown in Fig. 21.8.

Two additional factors included in this scheme are: (1) α is the factor by which I changes the affinity of S to E and (2) β is the factor by which I alters the rate of product formation from ES complex. In the equilibrium scheme, both K_S and K_I are affected by the same α value by the presence of the other ligand. On thermodynamic grounds, $\Delta G°$ is a path-independent function and equilibrium constants (K_S, K_I, αK_S, and αK_I) are directly related to corresponding $\Delta G°$ values. As long as E goes to form the same ESI complex, by either route, the constant α multiplying K_S and K_I will be identical.

Significance of α and β values: The values of α and β provide useful information on the binding interactions and catalysis. Their relationship with the earlier classification of inhibitor types is summarized in Table 21.1.

Table 21.1 Kinetic significance of α and β values

Effect	Nature	Lines intersect	Parameter value	
Complete inhibition	Competitive	On Y-axis	$\alpha = 0$	$\beta = 0$
	Noncompetitive	On X-axis	$\alpha = 1$	$\beta = 0$
	Mixed	Above X-axis	$\alpha < 1$	$\beta = 0$
	Mixed	Below X-axis	$\alpha > 1$	$\beta = 0$
	Uncompetitive		$\alpha \gg 1$	$\beta = 0$
Partial inhibition			$\alpha = $ any	$0 < \beta < 1$
Activation			$\alpha = $ any	$\beta > 1$

This representation is particularly suited to describe enzyme activation and partial inhibition. A complete inhibitor will prevent the breakdown of ESI to products and therefore, $\beta = 0$. A partial block of this step is described by a value of β between 0 and 1. On the other hand, an activator will have a value of β greater than 1. We note that, depending on the value of α, partial inhibitors and activators can also be classified. The inhibition constants obtained from the slope (K_{IS}) and the intercept (K_{II}) effects for a noncompetitive inhibitor (see the earlier discussion on non-competitive inhibition and Fig. 21.6) are related to the α value as $K_{IS}/K_{II} = \alpha$. The corresponding rate equation will be

$$v = \frac{V_{max}[S]}{K_M\left(1 + \frac{[I]}{K_I}\right) + [S]\left(1 + \frac{[I]}{\alpha K_I}\right)}$$

For nonzero values of α other than unity, the lines do not intersect on the x-axis; these cases are denoted as mixed inhibitors in this treatment (Table 21.1).

21.5   IC$_{50}$ and Its Relation to K_I of an Inhibitor

The magnitude of K_I value for an inhibitor reflects its strength of interaction with the enzyme. The K_I being dissociation constant—smaller its value, more potent is the inhibitor. At times it is not feasible to rigorously determine the K_I value for an inhibitor. One other measure of relative inhibitory potency is the IC$_{50}$ value. This is the concentration of inhibitor required to achieve 50% inhibition of enzyme activity under a defined set of assay conditions. The enzyme reaction velocity, IC$_{50}$, and the inhibitor concentration are related as shown:

$$\frac{v_i}{v} = \frac{1}{1 + \frac{[I]}{IC_{50}}}$$

where v_i is the rate in the presence of inhibitor I and v is the control rate in the absence of I. The fractional activity (v_i/v) may then be plotted as a function of log[I]. The IC$_{50}$ can be graphically gleaned from such a dose-response curve (Fig. 21.9).

Fig. 21.9 Dose-response curve of enzyme activity as a function of inhibitor concentration. In this semi-log plot, fractional enzyme activity is plotted versus logarithm of inhibitor concentration

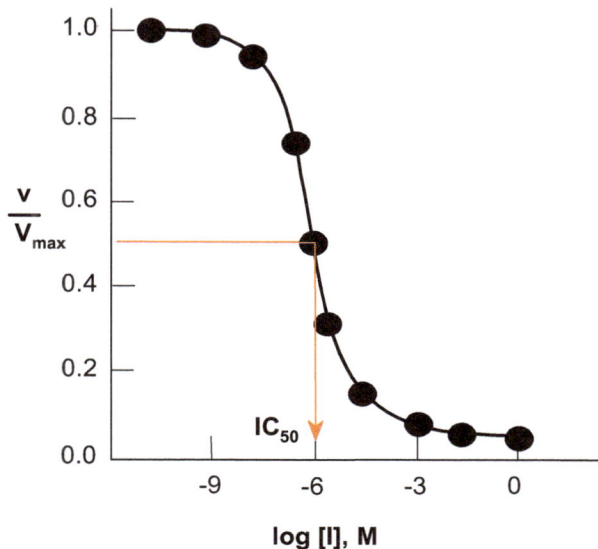

In practice, the inhibitor concentration is varied over several logs (at a single fixed [S]) to obtain estimates of IC$_{50}$. Presenting inhibitory potency of molecules as IC$_{50}$ values is very popular in pharmaceutical research. For instance, a series of potential inhibitors against a particular enzyme target may be screened and ranked quickly and conveniently according to their IC$_{50}$ values.

It is possible to relate IC$_{50}$ values of an inhibitor to its corresponding K_I provided the type of inhibition is known (Brandt et al. 1987). For example, the relationship between the K_I, K_M, [S], and IC$_{50}$ value can be derived for a competitive inhibitor as shown:

By definition, when [I] = IC$_{50}$, we have $v_i = v/2$.

Substituting respective velocity equations,

$$\frac{V_{\max}[S]}{K_M \left(1 + \frac{IC_{50}}{K_I}\right) + [S]} = \frac{V_{\max}[S]}{K_M + [S]} \frac{1}{2}$$

On simplifying we get,

$$\frac{K_M}{K_I} IC_{50} = K_M + [S] \text{ and therefore, } IC_{50} = K_I \left(1 + \frac{[S]}{K_M}\right)$$

This equation represents the Cheng-Prusoff relationship for a competitive inhibitor (Cheng and Prusoff 1973). It can be used to calculate K_I for an enzyme (at any given [S]) from the corresponding IC$_{50}$ value. Although the IC$_{50}$ determinations are popular for their experimental simplicity, two precautions are mandated.

Fig. 21.10 Changing IC_{50} values as a function of substrate concentration (dimensionless; relative to K_M) for three common modes of reversible inhibition. The plot is a schematic with an arbitrary K_I value and a representation of X-axis in log units around the K_M

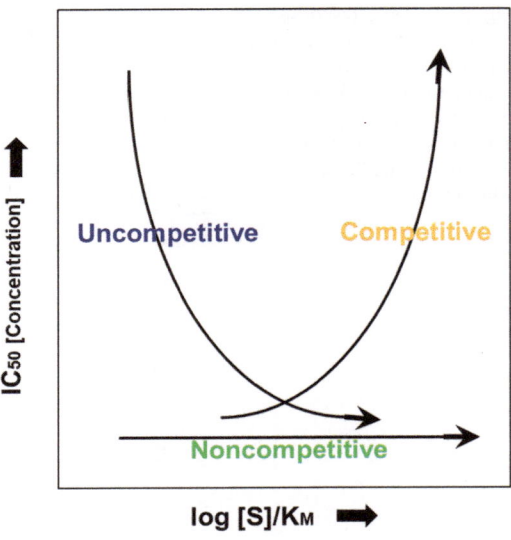

1. It is important to maintain and report the assay conditions along with the IC_{50} value. Otherwise, different IC_{50} values are not comparable. Among other things, the substrate concentration used in the assay influences the measured IC_{50} value (equation above).
2. Cheng-Prusoff relationships for different inhibitor types (e.g., competitive, uncompetitive, and noncompetitive) are not the same. The Cheng-Prusoff relationship is shown as a schematic graph for three types of inhibitors in Fig. 21.10. The plot of IC_{50} values with $[S]/K_M$ (a dimensionless quantity in its log form) remains qualitatively the same irrespective of the absolute value of K_I. The uncompetitive inhibitors are highly sensitive to small changes in $[S]$ below the K_M of the enzyme, whereas competitive inhibitors are highly sensitive to minor perturbations in $[S]$ above the K_M. For pure noncompetitive inhibitors ($\alpha = 1$), the potency remains unchanged as a function of $[S]$, while it increases or decreases nonlinearly for $\alpha < 1$ or $\alpha > 1$, respectively.

Therefore, IC_{50} values may be compared strictly between compounds exhibiting the same mode of inhibition. When in doubt, the IC_{50} values must be cross checked by performing a rigorous double reciprocal analysis. IC_{50} values and Cheng-Prusoff relationships are commonly used for high throughput inhibitor screening of a series of structurally related molecules (Srinivasan 2021). Such analysis is of immense practical value in quantitative structure activity relationship (QSAR) studies (see Chaps. 22, 26, and 36). In the final analysis, however, there is no substitute for a rigorously determined K_I value of an inhibitor.

Appendix

The rate equation for noncompetitive inhibition may be derived starting from an equilibrium assumption (hence K_S replaces K_M) and the scheme shown in Fig. 21.5. In this equilibrium E, ES, EI and ESI are the four different enzyme forms and ESI form is nonproductive. Defining any three equilibrium constants will automatically define the fourth (note: state function). In order to evaluate the concentration of the ES complex we use the following relations:

$$[ES] = \frac{[E][S]}{K_S}\ ; [EI] = \frac{[E][I]}{K_{IS}}\ ;\text{and} \ [ESI] = \frac{[ES][I]}{K_{II}} = \frac{[E][S][I]}{K_S K_{II}}$$

Evaluating the fraction of total enzyme ($[E_t]$) present in the ES form, we get

$$f = \frac{[ES]}{[E_t]} = \frac{[ES]}{[E] + [ES] + [EI] + [ESI]}$$

Substituting for the concentrations of all enzyme forms from the three equilibrium relationships in terms of [E],

$$f = \frac{\frac{[E][S]}{K_S}}{[E] + \frac{[E][S]}{K_S} + \frac{[E][I]}{K_{IS}} + \frac{[E][S][I]}{K_S K_{II}}} = \frac{\frac{[S]}{K_S}}{1 + \frac{[S]}{K_S} + \frac{[I]}{K_{IS}} + \frac{[S][I]}{K_S K_{II}}}$$

Simplifying (multiplying both the numerator and the denominator by K_S) further we obtain

$$f = \frac{[S]}{K_S + [S] + \frac{K_S}{K_{IS}}[I] + \frac{1}{K_{II}}[S][I]}.$$

This fraction ($f = [ES]/[E_t]$) may be now plugged in to obtain the rate equation:

$$v = f.V_{max} = \frac{V_{max}[S]}{K_S + [S] + \frac{K_S}{K_{IS}}[I] + \frac{1}{K_{II}}[S][I]}$$

This equation can be easily rearranged to the more familiar form:

$$v = \frac{V_{max}[S]}{K_S\left(1 + \frac{[I]}{K_{IS}}\right) + [S]\left(1 + \frac{[I]}{K_{II}}\right)}$$

Of course, if $K_{IS} = K_{II}$, then this simply becomes

$$v = \frac{V_{max}[S]}{(K_S + [S])\left(1 + \frac{[I]}{K_I}\right)}.$$

References

Brandt RB, Laux JE, Yates SW (1987) Calculation of inhibitor Ki and inhibitor type from the concentration of inhibitor for 50% inhibition for Michaelis-Menten enzymes. Biochem Med Metab Biol 37:344–349

Cheng Y-C, Prusoff WH (1973) Relationship between the inhibition constant (Ki) and the concentration of inhibitor which causes 50 percent inhibition (IC50) of an enzymatic reaction. Biochem Pharmacol 22:3099–3108

Cornish-Bowden A (1986) Why is uncompetitive inhibition so rare? A possible explanation, with implications for the design of drugs and pesticides. FEBS Lett 203:3–6

Dixon M (1953) The determination of enzyme inhibitor constants. Biochem J 55:170–171

Srinivasan B (2021) Explicit treatment of non-Michaelis-Menten and atypical kinetics in early drug discovery. ChemMedChem 16:899–918

Tayyab S (1990) Biochemistry through cartoons—understanding enzymes. Biochem Educ 18:42–43

Alternate Substrate (Product) Interactions

We have so far noted that a substrate molecule normally forms a productive complex at the enzyme active site. However, there are cases where substrate also interacts with the enzyme (or the ES complex) in a nonproductive fashion. If this interaction is *"kinetically silent,"* it will not show up in the routine steady state kinetic analysis. Other methods (like equilibrium dialysis, fluorescence difference spectroscopy, or MALDI-TOF) may, however, be able to detect such binding phenomena. Most often nonproductive interactions of substrate are not considered at all—except when they also interact with the same enzyme as activators or inhibitors.

22.1 Substrate Inhibition

A decrease in enzyme activity because of high substrate concentrations is termed substrate inhibition. Substrate inhibition may be observed due to one or more of the following reasons:

(a) Presence of a second set of low affinity binding sites for S; and when so bound, can lead to non-productive, inefficient enzyme forms.
(b) Unproductive binding of substrate by partial sub-site occupancy.
(c) Removal of an essential active site metal ion or cofactor by high [S].
(d) Presence of excess un-complexed substrate such as ATP; note that in most cases Mg-ATP is the true substrate. It is therefore important to use proper concentration ratios of ATP and Mg^{2+} (Chap. 30; Part IV).

A simple binding equilibrium (where excess substrate acts as an uncompetitive inhibitor) and the corresponding rate equation for substrate inhibition is shown below:

© The Author(s), under exclusive license to Springer Nature Singapore Pte Ltd. 2025
N. S. Punekar, *ENZYMES: Catalysis, Kinetics and Mechanisms*, https://doi.org/10.1007/978-981-97-8179-9_22

Fig. 22.1 Typical
$v \rightarrow$ [S] plots for substrate
inhibition of the enzyme.
Enzyme activity (plotted as
dimensionless v/V_{max}) as a
function of substrate
concentration (plotted as
dimensionless [S]/K_M) at four
different K_I/K_M values are
shown

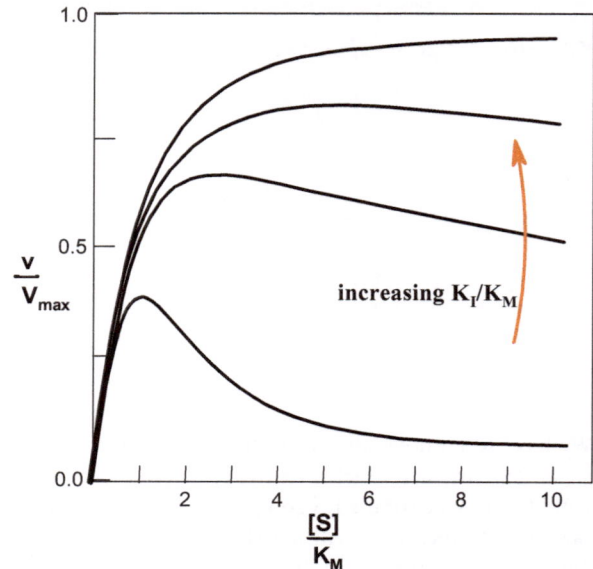

It is thus expected that an SES ternary complex does form. Because of the $[S]^2$
term, the $v \rightarrow$ [S] relationship is nonlinear even when double reciprocals are taken.
The nature of $v \rightarrow$ [S] plot (describing substrate inhibition) will depend on the
relative magnitudes of K_M and K_I; an optimum is apparent in these plots. This is
illustrated in Fig. 22.1.

$$v = \frac{V_{max}[S]}{K_M + [S]\left(1 + \frac{[S]}{K_I}\right)} = \frac{V_{max}[S]}{K_M + [S] + \frac{[S]^2}{K_I}}$$

Determination of kinetic constants and their significance: The double reciprocal
form of the rate equation for substrate inhibition is as shown:

$$\frac{1}{v} = \frac{K_M}{V_{max}}\frac{1}{[S]} + \frac{1}{V_{max}}\left(1 + \frac{[S]}{K_I}\right) = \frac{K_M}{V_{max}}\frac{1}{[S]} + \frac{1}{V_{max}} + \frac{1}{V_{max}}\frac{[S]}{K_I}$$

This equation may also be rearranged to

$$\frac{V_{max}}{v} = 1 + \frac{K_M}{[S]} + \frac{[S]}{K_I}$$

The [S]/K_I term becomes significant at high [S] values and a smaller K_I value—
leading the substrate inhibition to set in. At low [S] (and relatively large K_I values),
this equation collapses to a regular Michaelis-Menten eq. An estimate of K_M can be
made from $v \rightarrow$ [S] data at lower substrate concentrations by extrapolating the
apparent linear region of a normal double reciprocal plot (dashed line tangent in

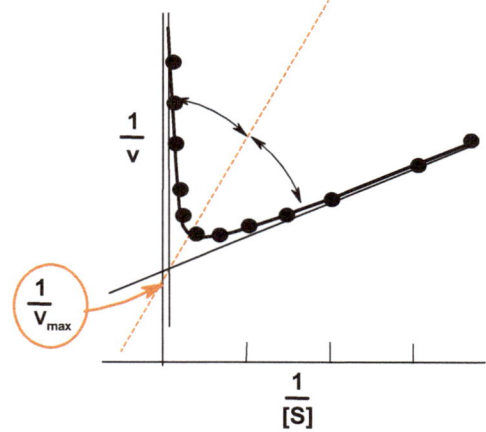

Fig. 22.2 Double reciprocal plot of $v \rightarrow$ [S] data for an enzyme showing substrate inhibition. Graphical estimates of V_{max} and K_M are indicated as extrapolations through broken lines

Fig. 22.2). Graphically the value of $1/V_{max}$ may be obtained by drawing a line bisecting the angle made by the two limbs of the $1/v \rightarrow 1/$[S] curve (the dotted line bisector; Fig. 22.2). The estimated V_{max} and K_M values may be subsequently substituted in the rate equation to obtain K_I. The best way to extract all the kinetic parameters (V_{max}, K_M, and K_I) is to fit the data to the equation. A simple graphical method to determine the kinetic parameters of substrate inhibition of complete and partial types is also available (Yoshino and Murakami 2015). This method consists of plotting experimental data as $v/(V_{max} - v)$ versus the reciprocals of the [S].

Substrate inhibitions are not usually important if [S] is kept relatively low—below its physiological levels. The evolutionary process normally eliminates dead-end combination of substrate that leads to substrate inhibition. However, if physiologically [S] does not reach inhibitory levels (for instance, aldehyde substrates that can be toxic to the cell), then dead-end combinations do persist. This often manifests as substrate inhibition in kinetic studies, particularly in the non-physiological direction. The phenomenon of substrate inhibition is estimated to occur in some 20% of enzymes. A partial list of enzymes that show substrate inhibition may be found in Reed et al. (2010). Substrate inhibition is often regarded as a non-physiological phenomenon. However, it may have biological value because of the following reasons:

(a) for many enzymes in vivo, substrate concentrations are to the right of their V_{max}, suggesting that such enzymes typically operate under substrate inhibition.
(b) some enzymes have specialized sites where a second substrate molecule can bind and act as an allosteric inhibitor (viz., ATP for phosphofructokinase). For them substrate inhibition is clearly a specially evolved feature.
(c) there is sufficient evidence to show that substrate inhibition plays critical regulatory roles in many metabolic pathways.

Without going into the kinetic origins, the possible biological functions of substrate inhibition with five enzymes, namely, tyrosine hydroxylase, acetylcholinesterase, phosphofructokinase, folate cycle enzymes, and DNA methyltransferase have been discussed (Reed et al. 2010). In each of these cases, substrate inhibition plays a distinctly different regulatory role. The classical substrate inhibition of phosphofructokinase ensures that resources are not wasted to make ATP when it is plentiful.

Substrate inhibition may be induced by an inhibitor in an ordered sequential mechanism. In most cases, such induced substrate inhibitions are partial (and not complete) because the inhibitor does escape from the EIB ternary complex at a reduced but finite rate. Inhibition of hexokinase by Mg-ATP is an example of this kind. Lyxose (an inhibitor resembling glucose, the first substrate) induces a competitive substrate inhibition by Mg-ATP. Such induced substrate inhibition occurs because much of the enzyme is trapped as E.Lyxose.Mg-ATP complex, from which lyxose cannot dissociate.

22.2 Use of Alternate Substrates in Enzyme Studies

Alternate products or substrates compete with the normal substrates for the same enzyme form(s). They may also be viewed as inhibitors of the normal reaction. Such an analysis does enrich our kinetic understanding of enzyme action. Glucose and galactose compete for the same form of hexokinase enzyme (Fig. 22.3). Therefore, galactose may be seen as a competitive inhibitor of glucose reaction with hexokinase and vice versa.

Two useful areas of application of alternate substrates (and alternate products—because they are substrates in the reverse direction) in the kinetic study of enzymes are enumerated below.

Information about the active site shape, geometry, and interactions: Enzymes are specific catalysts and hence, can accommodate a narrow range of substrate structures in their active sites. Conversely, testing different structural variants of the natural substrate for catalysis by the enzyme defines the active site. This was an attractive but simple option to probe active sites much before the use of X-ray structural data. Excellent insight into ovine brain glutamine synthetase active site

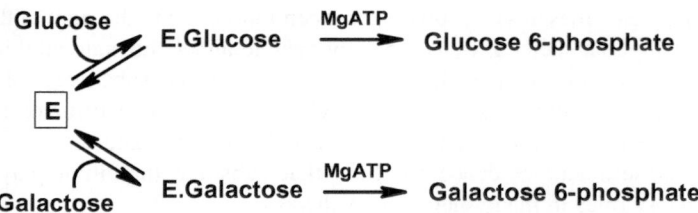

Fig. 22.3 Glucose and galactose compete for the same enzyme. Enzyme hexokinase is represented as E

Table 22.1 Interaction of different structural variants of L-glutamate with glutamine synthetase

Compound	Substrate activity[a] (%)	Competitive inhibition[a]
α-Aminomalonate	00	No
L-Aspartate	00	No
D-Aspartate	00	No
L-Glutamate	100	Yes
D-Glutamate	54	Yes
β-Glutamate	46	Yes
α-Methyl-L-glutamate	67	Yes
β-Methyl-D-glutamate (*threo*)	46	Yes
γ-Methyl-L-glutamate (*threo*)[b]	63	Yes
γ-Hydroxy-L-glutamate (*threo*)	89	Yes
γ-Hydroxy-D-glutamate (*threo*)	02	–
L-α-Aminoadipate	22	–
D-α-Aminoadipate	11	–

Adapted with permission from A. Meister, Adv Enzymol, 31:183–218. Copyright (Meister 1968) John Wiley & Sons Inc
[a] Both substrate activity and inhibition are scored against L-glutamate, the natural substrate
[b] Of the four possible isomers, only this is a substrate

was adduced by Meister and his colleagues (Meister 1968) by this approach (Table 22.1).

While some molecules may not be substrates they may interact at the active site and inhibit the enzyme. A quantitative structure-activity analysis for different substrates and non-substrates taken together succinctly defines the active site perimeters (Fig. 22.4). Molecular models were used to map the enzyme active site space occupied by L-glutamate. Arguments based on steric hindrance by various substitutions on the glutamate structure, the active site occupancy of this substrate was elegantly worked out. Accordingly, ovine brain glutamine synthetase binds L-glutamate in a fully extended conformation in which the carboxyl groups are as far apart as possible. The inter-carboxyl carbon distance of about 5 Å is required and the molecule is anchored at the active site through two carboxyl groups and an amino group.

It must be noted that glutamine synthetases from other sources possess a much higher degree of substrate specificity; many of the structures described above are not accepted while only L-glutamate serves as a substrate. On the other hand, L-phosphinothricin and L-methionine-DL-sulfoximine (Fig. 22.4) markedly inhibit all glutamine synthetases. The two potent tight binding inhibitors may be phosphorylated and resemble the γ-glutamyl phosphate intermediate of the enzyme reaction.

Several biologically relevant guanidinium group-containing compounds occur in nature. Life has solved the chemistry of guanidinium group hydrolysis by recruiting metal ions for activating water attack. The majority of these ureohydrolases exhibit an alkaline pH optimum (greater than pH 9.0) and contain a bimetallic Mn^{2+} cluster at the active site. While the apparatus to hydrolyze guanidinium group (to release

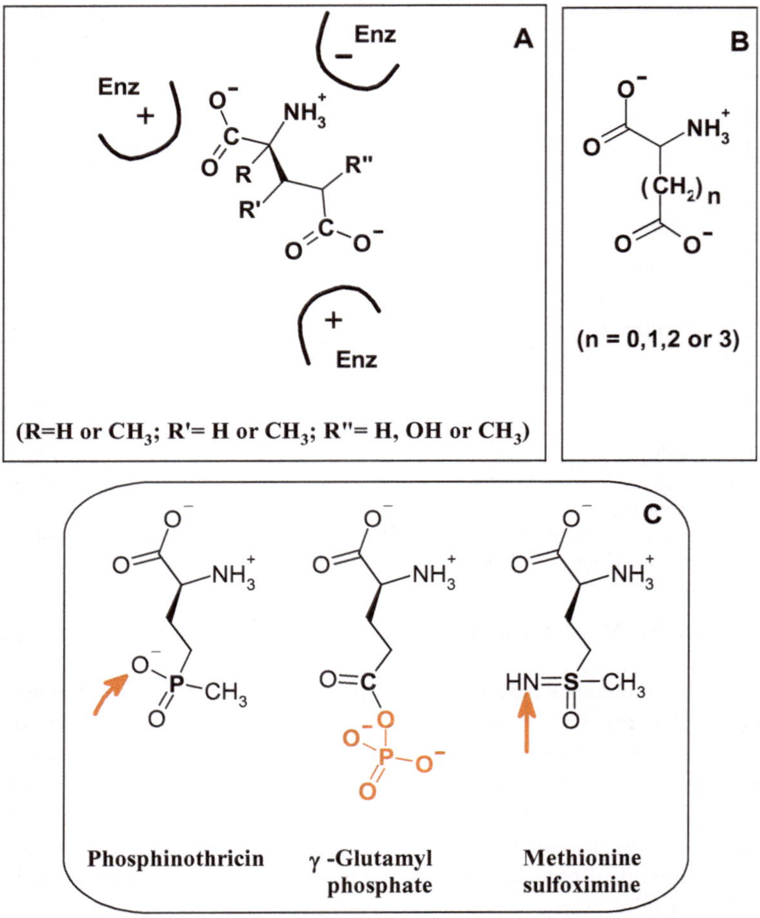

Fig. 22.4 Different substrate structural variants used to define glutamine synthetase active site geometry. (**a**) Systematic variations in glutamate structure at α, β, or γ carbon (corresponding to data in Table 22.1). (**b**) Variation of chain length separating the two carboxylates. (**c**) The γ-glutamyl phosphate enzyme bound intermediate. Arrows indicate the potential phosphorylation position on the two tight binding inhibitors, namely, L-phosphinothricin and L-methionine-DL-sulfoximine

urea) is more or less conserved, members of ureohydrolase superfamily display exquisite substrate specificity (Table 22.2). Indeed, they are classified based on this feature.

The *A. niger* 4-guanidinobutyrase is a Zn^{2+} enzyme and acts poorly on 3-guanidinopropionate. On the contrary, the *Candida parapsilosis* enzyme accepts both 4-guanidinobutyrase and 3-guanidinopropionate equally well as its substrate (Saragadam et al. 2019; Gaikwad et al. 2024). The substrate specificity of human arginase was extensively probed by site directed mutations. Arginase specificity could be changed (in the N130D variant), to recognize and accommodate agmatine

Table 22.2 Substrate structural requirements for three ureohydrolases[a]

Compound	Structure	Used as substrate by		
		Arginase	Agmatinase	4-Guanidino-butyrase
L-Arginine	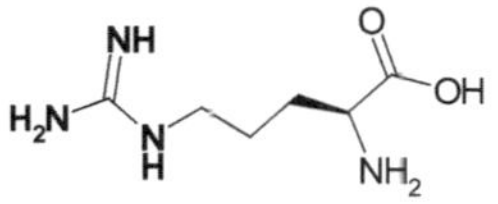	**Yes**	No	No
Agmatine	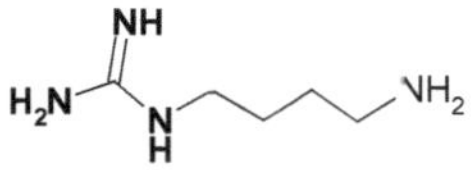	No	**Yes**	No
4-Guanidino-butyrate	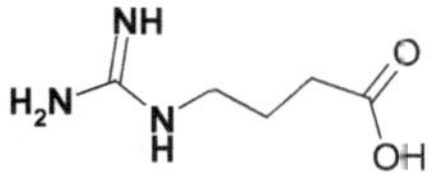	No	No	**Yes**
D-Arginine	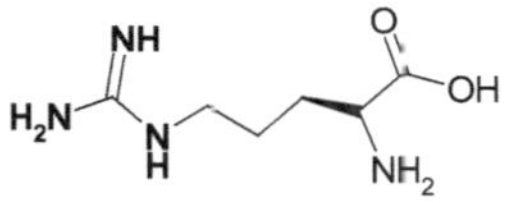	No	No	No
L-Homoarginine	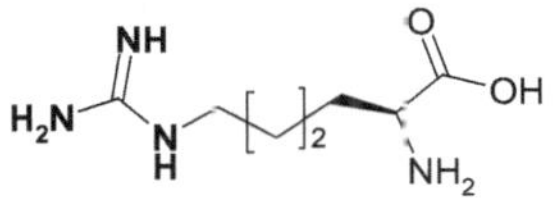	No	No	No

[a] Enzyme sources: arginase from *Aspergillus niger*, agmatinase from *Escherichia coli*, and 4-guanidinobutyrase from *Pseudomonas putida*

in place of arginine, thereby converting it into an agmatinase. Exploring the arginase active site with substrates, substrate analogues, and non-substrates has gained importance. Potent, selective arginase inhibitors (like NOHA) have found therapeutic application in channeling arginine pools into NO synthesis.

Most enzymes bind an otherwise conformationally flexible substrate, at their active sites, in a fixed geometry. As seen above, an extended L-glutamate conformation is frozen out at their respective active sites by glutamine synthetase and NADP-glutamate dehydrogenase. Alternate substrate structures with rigid geometry are useful in extracting such information. Phosphofructokinase phosphorylates the 1-position of D-fructose 6-phosphate in its furanose form. Because its two anomers (α-D-fructose 6-phosphate and β-D-fructose 6-phosphate) equilibrate very rapidly in water, it is hard to tell which anomer is the substrate. The β-anomeric structure may be frozen out as its corresponding 2,5-anhydro-D-mannitol-6-phosphate (Fig. 22.5). It is not a hemiacetal but an ether (−OH missing on C-2) and cannot undergo mutarotation; this is accepted as an alternate substrate by phosphofructokinase and not the corresponding α-isomer. The conclusion is inescapable that β-D-fructose 6-phosphate is the natural substrate for this enzyme. Similar studies with fructokinase showed that the β-furanose form of D-fructose is its substrate.

Fig. 22.5 Structural
similarity between α-D-
fructose 6-phosphate, β-D-
fructose 6-phosphate, and
2,5-anhydro-D-mannitol-6-
phosphate. Note that
2,5-anhydro-D-mannitol is an
ether; the −OH on C2 carbon
is missing and hence cannot
mutarotate

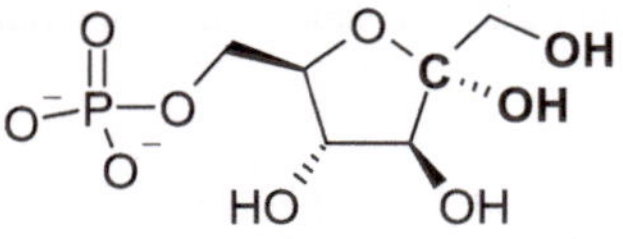

α-D-Fructose 6-phosphate

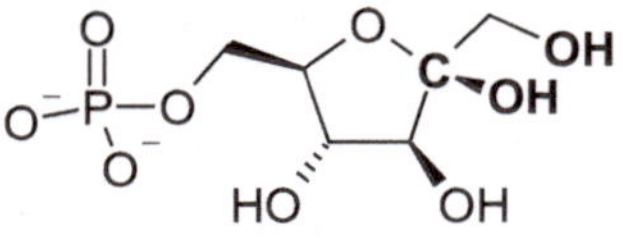

β-D-Fructose 6-phosphate

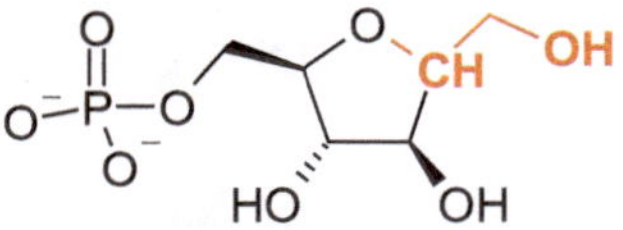

2,5-Anhydro-D-mannitol 6-phosphate

Substrate structures with fixed geometry may also be used to determine: (a) which conformational isomer of ATP is the substrate for a given enzyme and (b) whether the α-phosphate of ATP is ever coordinated to the divalent metal ion during reaction. Further treatment on the biochemistry of ATP and its reactivity may be found in a later section (Chap. 30).

Finally, the vast diversity of semisynthetic β-lactam structures (Chap. 36; Part V) provides practical examples of how the knowledge of their binding to the active sites of D-Ala-D-Ala carboxypeptidase (the target) and β-lactamase (that confers resistance) has been successfully exploited in new antibiotic discovery.

Understanding kinetic mechanism: Alternate substrates and substrate analogues provide useful kinetic information. We will discuss two of their possible applications here:

(a) Dead-end inhibition by an analogue of substrate B can be used as diagnostic of an ordered bi-reactant mechanism. This inhibition is uncompetitive with A as the varied substrate for the ordered case.
(b) A sticky substrate (high commitment to catalysis upon binding to E—see Chap. 24 for details) often leads to complex rate equations in the analysis of pH kinetics and isotope effects. A slow substrate with a K_M in the mM range will usually not be sticky. The binding of such slow alternate substrates will be in rapid equilibrium. Fructose 6-sulfate—being a mono-anion—binds 6-phosphofructokinase poorly. It therefore is a non-sticky, slow substrate when compared to fructose 6–phosphate (a natural substrate di-anion). Substrate affinity may also be lost by changing the reaction pH due to incorrect protonation states. For instance, creatine kinase loses affinity for creatine as the pH is decreased from 8.0 to 7.0; and it becomes non-sticky. In both these kinases, the

loss of substrate affinity also leads to a change in the kinetic mechanism. A normal random mechanism changes to equilibrium ordered (Chap. 18; see Sect. 18.3) with ATP-Mg binding before the other substrate.

References

Gaikwad SR, Punekar NS, Pathan EK (2024) Characterization of a novel 4-guanidinobutyrase from *Candida parapsilosis*. FEMS Yeast Res 24:foae003

Meister A (1968) The specificity of glutamine synthetase and its relationship to substrate conformation at the active site. Adv Enzymol Relat Areas Mol Biol 31:183–218

Reed MC, Lieb L, Nijhout HF (2010) The biological significance of substrate inhibition: a mechanism with diverse functions. Bioessays 32:422–429

Saragadam T, Kumar S, Punekar NS (2019) Characterization of 4-guanidinobutyrase from *Aspergillus niger*. Microbiology 155:396–410

Yoshino M, Murakami K (2015) Analysis of the substrate inhibition of complete and partial types. Springerplus 4:292

pH Studies with Enzymes

23

Enzyme catalyzed reactions involve one or more proton transfers. Acid-base chemistry permeates most of enzyme chemical mechanisms. Ionizable amino acid side chains of the enzyme protein are typically involved in such catalysis. Each ionizable group can be viewed as an acid and a conjugate base. A general exposure to acid-base chemistry and catalysis may be found in Chap. 28 of the book. The reader is well advised to read that background material to better appreciate the subject covered in this chapter.

Interactions of enzyme with its substrate (or any ligand for that matter) involve at least one or few groups whose correct ionization is necessary for its optimal function. The ionization state of acid-base groups on the enzyme, substrate, and inhibitor directly affects catalytic activity. These ionizable groups may affect the enzyme activity by influencing the binding or catalysis or both. Almost all protonation—deprotonation events occurring in an enzyme reaction—are catalyzed. The relevant acid-base groups on the enzyme may be identified with the help of techniques such as:

(a) Group specific chemical modifications (see Chap. 20)
(b) pH dependence of kinetic parameters
(c) Structure-activity correlations through physical techniques like X-ray, nuclear magnetic resonance (NMR) spectroscopy, etc.
(d) Site-directed mutation analysis (see Chap. 40; Part VI).

All the four approaches give useful complementary information; however, pH-dependence of enzyme kinetics gives the best insight as it reports on the reaction being catalyzed while it occurs. The pH dependence of kinetic parameters provides information on kinetic as well as chemical mechanism. The kinetic aspects will be the focus here while the other approaches pertaining to pH are dealt with in later sections (Chap. 28; Part IV).

N. S. Punekar, *ENZYMES: Catalysis, Kinetics and Mechanisms*, https://doi.org/10.1007/978-981-97-8179-9_23

23.1 Enzyme pH Optimum

Determination of activity as a function of pH is the first and simplest experiment one conducts to determine the effect of H^+ ions on the enzyme. In all enzymatic studies maintaining a well-defined pH (H^+ concentration) of the system is very crucial. Since a range of pH values (between 0 and 14 in an aqueous environment) are to be used, more than one kind of buffer ion may be required in the experiment. Effects on the enzyme due to switching of buffer species and ionic strength, if any, become important and have to be eliminated. This is achieved by the judicious use of suitable buffers (refer to Chap. 12; Part II). Most enzymes display a bell-shaped pH-activity curve with maximal activity around neutral pH. However, there are enzymes with a pH optimum in the acidic (such as pepsin) and alkaline (such as arginase) range as well (Fig. 23.1).

The decrease in activity on either side of pH optimum may result from (a) instability of the enzyme and/or (b) changes in the kinetic parameters of the enzyme due to pH. Enzyme stability over broad pH ranges is a desirable feature for industrial application. Information about the stability of the enzyme over the pH range studied is also necessary in designing correct kinetic studies.

To establish the pH stability profile, the enzyme is preincubated at different pH values. For each pH of preincubation, aliquots are withdrawn as a function of time and activity remaining is assayed. In this process, it must be ensured that the active enzyme remaining is stable after taking the preincubation sample into assay mixture. A plot of percent activity remaining versus pH (after a fixed time of preincubation) gives a fair idea about enzyme's pH stability. Meaningful pH dependence of enzyme activity is then sought within this range of pH defined for stability.

Experimental determination of pH optimum (plot of pH versus activity; Fig. 23.1) serves two purposes. It is of practical importance in enzyme assay optimization. Second, the ascending and descending limbs of such profiles give some idea about the range of pKas and hence possible ionizable groups involved.

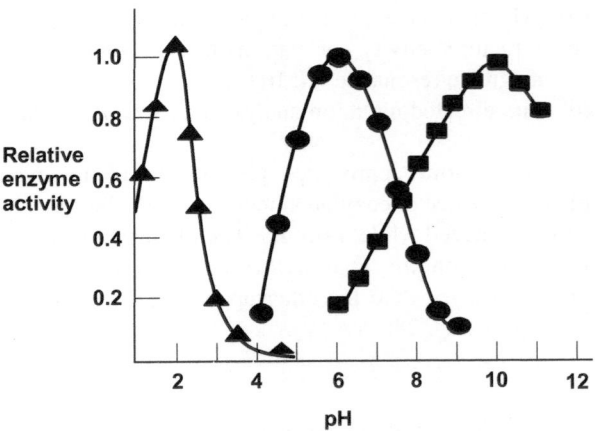

Fig. 23.1 pH-activity profile of an enzyme. Three representative enzyme examples shown are pepsin (▲), amylase (●), and arginase (■)

The pH behavior of an enzyme is a complex outcome of all the ionizable groups present on the enzyme and the substrate. A simplistic treatment of bell-shaped activity curve is to compare it to the ionization of a dibasic acid. Examples of enzymes in which at least two active site ionizable groups have been assigned for function are ribonuclease A (H12 and H119) and lysozyme (E35 and D52). Of many ionizable groups present on the enzyme and the substrate, two may be considered kinetically significant with the enzyme represented as HEH. The ascending limb defined by the ionization of the first proton (HEH $\rightleftarrows$ EH$^-$ + H$^+$) and the descending limb by the other (EH$^-$ $\rightleftarrows$ E^{2-} + H$^+$). The active enzyme species is the singly de-protonated form EH$^-$. In such a simplistic model representation, the pH behavior of V_{max}/K_M (the first-order rate constant) is given by the following equation:

$$\frac{V_{max}}{K_A} = \frac{\left(\frac{V_{max}}{K_A}\right)^0}{\frac{[H^+]}{K_{H1}} + 1 + \frac{K_{H2}}{[H^+]}}$$

where K_{H1} and K_{H2} are the acid dissociation constants for the two groups and $(V_{max}/K_M)^0$ is the pH-independent value when the two groups are correctly protonated. At lower pH values (i.e., high [H$^+$]), the $K_{H2}/[H^+]$ term in the denominator is insignificant while at higher pH values (i.e., low [H$^+$]), the $[H^+]/K_{H1}$ term becomes insignificant. Although much oversimplified, this equation does capture the essence of pH dependence of enzyme kinetic parameter (V_{max}/K_M in this case).

23.2   pH Kinetic Profiles

The best mechanistic information can be obtained by performing substrate saturation at different pH values. It must be borne in mind, however, that the kinetic mechanism may itself change with the change in pH of the reaction. For instance, the normal random mechanism exhibited by creatine kinase changes to an equilibrium ordered one (Chap. 18; Mg-ATP binding first) as the pH is decreased from 8.0 to 7.0. Such effects have to be checked beforehand. The pH-kinetic experiments are conducted so that effects of ionic strength and changing buffer species are properly controlled. With these data, one can simultaneously determine the effects of pH on the kinetic constants such as V_{max}, V_{max}/K_M, and the K_M. Similarly, one can determine the pH dependence of (a) K_I for a competitive inhibitor and (b) $K_{activator}$ for an activator. The variation in kinetic parameters with pH is best plotted as log-log plots; this makes sense as pH is a logarithmic scale ($-\log[H^+]$). While the horizontal axis is always pH, the Y-axis could be $\log V_{max}$, $\log V_{max}/K_M$, pK_I ($-\log K_I$), p$K_{activator}$ ($-\log K_{activator}$), p$K_{metal\ ion}$ ($-\log K_{metal\ ion}$), etc. Profiles typically schematized to indicate such log-log plots are shown in Fig. 23.2.

Interpretation of pH profiles: Any singly ionizable group can access either a protonated or a deprotonated state. As fraction of a proton cannot be transferred, one encounters line segments with zero or unit slopes. Rarely, ionization of two protons

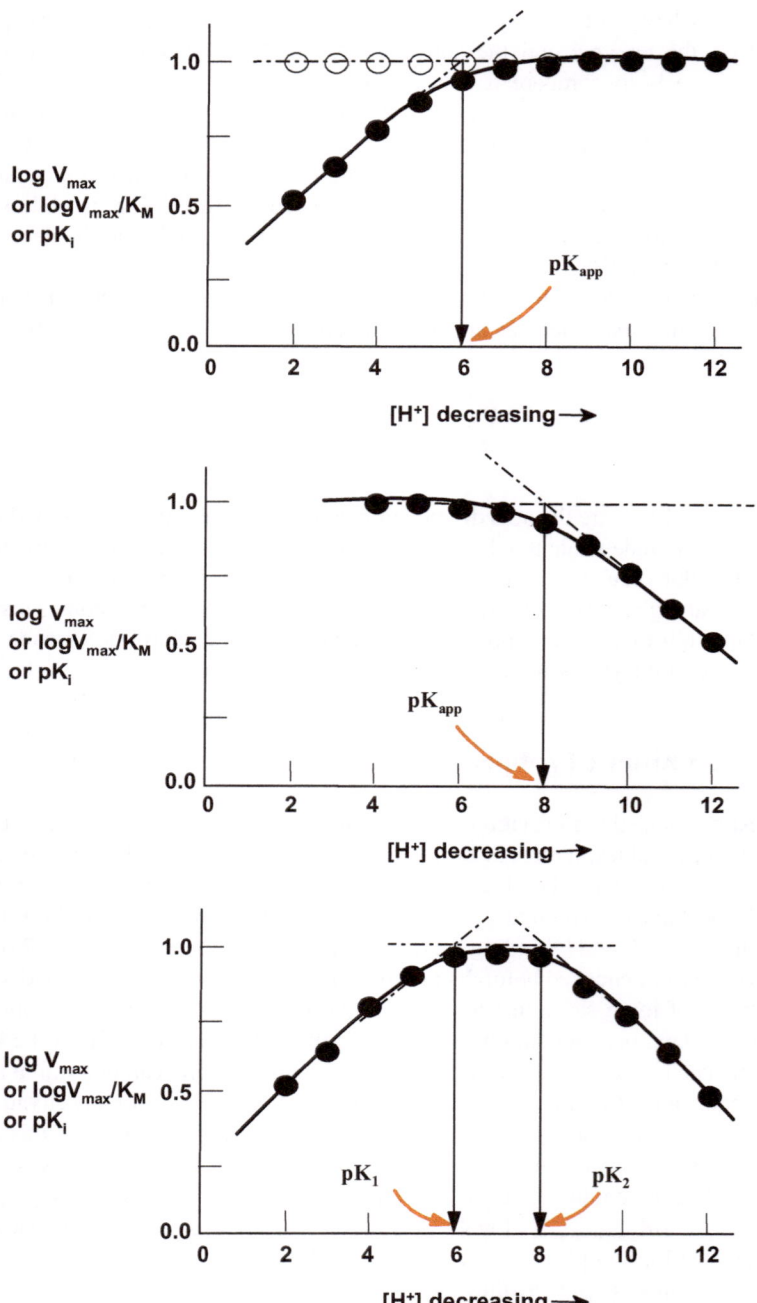

Fig. 23.2 Possible pH versus kinetic constant profiles encountered for an enzyme. Protonation of the group results in decreased activity/binding (top panel), de-protonation of the group results in decreased activity/binding (middle panel), and combination of the two—indicating titration of two ionizable groups (bottom panel). In each case, the kinetic constant (V_{max}, V_{max}/K_M, or K_I) is plotted

at a single pH may however appear as lines with a slope of 2. A curve with a unit positive slope followed by a plateau (i.e., zero slope; Fig. 23.2 top panel) indicates that the protonation of that group results in decreased activity/binding. The point of intersection (extrapolated to pH axis) gives the pKa of the group involved. At higher pH values a group may lose a proton and this may lead to decreased activity/binding (line segment with zero slope followed by one with slope −1; Fig. 23.2 middle panel). The presence of both an ascending and a descending limb with a plateau in the middle (Fig. 23.2 bottom panel) represents an enzyme example with two kinetically significant group protonations. Going from the left to right, deprotonation of the first group increases activity/binding while the deprotonation of the second group decreases activity/binding. We have already come across an equation describing such a pH behavior of V_{max}/K_M above. The observed pKa values of the two groups give a glimpse of their possible identity.

If one encounters two ionizable groups with their pKa values not sufficiently apart (more than 2 pH units), then it is difficult to distinguish the two and identify which one is protonated. We can further imagine an unusual situation: going up the pH scale, a group behaving like that in Fig. 23.2 middle panel may occur before a second group with a profile like that in Fig. 23.2 top panel. This is an example of a **reverse protonation** state and the plateau in Fig. 23.2 bottom panel is suppressed (to much less than the true value). For an active enzyme, therefore, the group with lower pKa is required in the protonated state while the group with higher pKa must be in the deprotonated form. For example, fumarase catalysis requires an active site carboxylate in the deprotonated state and a protonated imidazole. Such reverse protonation cases are not uncommon. It is, however, conceivable that the catalyst may be locked into an inactive state—possibly because of the proximity of two oppositely charged groups (they may form a salt bridge). Mother Nature may have therefore selected suitable pairs of ionizable groups to avoid such thermodynamic wells; lysozyme active site recruits two carboxylates—alternately acting as a general base and general acid for catalysis.

We make a few general observations on the pKas and acid-base groups obtained from pH kinetics studies. Groups that titrate in the $\log V_{max} \rightarrow$ pH profiles are mainly implicated in the catalytic step and their ionization occurs in the ES complex (saturating concentration of all reactants). Since V_{max}/K_M signifies first-order rate constant and interaction of S with E, $\log V_{max}/K_M \rightarrow$ pH data reflect on essential ionizable groups on the free substrate and the enzyme form with which it combines. Such groups therefore are important for substrate binding as well as catalysis. This interpretation is further simplified if ionization of acid-base group(s) on the substrate molecule is accounted for, or they do not exist (like in glucose). An example of this is glucose isomerase (actually a xylose isomerase from *Actinoplanes* spp.) and its site directed mutant (H54Q) form. This imidazole group (pKa around pH 6.0) of His54 is

---

Fig. 23.2 (continued) on the Y-axis as its log value. Open circles (in top panel) indicate the profile when that group does not titrate

Fig. 23.3 Interaction of glutamate dehydrogenase with the γ-COOH of 2-oxoglutarate through its active site Lys amino group. Analogous active site interactions with oxalylglycine (competitive inhibitor, middle) and 2-oxovalerate (alternate substrate, right) are also shown

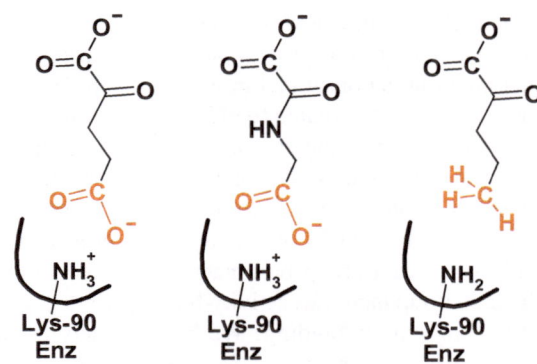

titrated in the $\log V_{max} \rightarrow$ pH profile of the native glucose isomerase but not in the H54Q mutant (the profile with open circles, Fig. 23.2 top panel).

The Michaelis constant—K_M for the substrate—is not an independent kinetic parameter but may be viewed as a ratio of V_{max} and V_{max}/K_M. Therefore, its pH dependence is a composite of $\log V_{max} \rightarrow$ pH and $\log V_{max}/K_M \rightarrow$ pH plots. The $pK_M \rightarrow$ pH profile is rarely plotted. Such profiles are useful indicators of acid-base groups involved in binding (found either on the enzyme or on the substrate) provided K_M approximates K_S (Chap. 14; Part II). The pH dependence of K_I for a competitive inhibitor is the simplest pH profile to interpret. In these $pK_I \rightarrow$ pH profiles ionizable groups responsible for binding alone are titrated. The K_I being equilibrium dissociation constant, the pKa values obtained from such plots are actual pKa values. They can be meaningfully compared with the group pKas obtained from $\log V_{max} \rightarrow$ pH and $\log V_{max}/K_M \rightarrow$ pH profiles. pH studies on bovine glutamate dehydrogenase provide an excellent example for such analysis (Rife and Cleland 1980). Glutamate dehydrogenase titrates an ionizable group around pH 8.0—both in the pK_M profile for 2-oxoglutarate (substrate) and pK_I profile for oxalylglycine (competitive inhibitor). However, this group is not seen in the pK_M profile for 2-oxovalerate—an alternate substrate but missing the γ-COOH (Fig. 23.3).

Since –COOH group is fully deprotonated at this alkaline pH, a positively charged group on the enzyme is clearly implicated. Deprotonation of this enzyme Lys side chain adversely affects 2-oxoglutarate binding but not that of 2-oxovalerate; indeed, this Lys-NH$_2$ must stay unprotonated for 2-oxovalerate binding. The involvement of this group (K90) anchoring the γ-COOH of 2-oxoglutarate was clearly borne out much later, by X-ray structural data on this enzyme.

Finally, pK_I and $pK_{activator}$ profiles provide information about the ligands provided by the enzyme that binds the metal ion. The pKa for the ionization of metal ion-bound water may also be seen in such profiles.

23.3 Identifying Groups Seen in pH Profiles

X-Ray, NMR, and chemical modification information on active site functional groups are useful but are limited by the fact that (1) they represent one of the many protein conformations and (2) even when active site pictures with a substrate or inhibitor bound are available, one may be looking at nonproductive complexes. An analog/inhibitor may bind differently from the way a productive substrate interacts at the active site. Site-directed mutations are useful in ruling out important roles for a particular group, but mere loss of activity may result from structural changes and not necessarily due to specific role in catalysis.

The pKa values of functionally important acid-base groups are usually significantly shifted from corresponding pKa for groups on free amino acids. Such perturbation of normal pKas is due to unique active site environment—created by the protein—and shielding of these groups from the bulk aqueous phase (Knowles and Jencks 1976). Nevertheless, comparison of experimentally determined enzyme pKa values with reported pKas of amino acid side-chain functional groups can help in identifying their chemical nature. Regardless of our knowledge on the exact nature of the acid-base group, its role in catalysis and/or binding can often be assigned from pH-kinetics profiles. Two different techniques may be useful in identifying the functional group whose pKa is observed in enzyme pH profiles.

Inspection of their enthalpy of ionization (ΔH_{ion}): The temperature dependence of equilibrium ionization constant is different for different acid-base groups (Table 23.1). These may be broadly categorized into three groups: groups with low ΔH_{ion} (carboxylate and phosphate), groups with moderate ΔH_{ion} (imidazole, thiol, and phenolic OH), and groups with high ΔH_{ion} (amino, metal bound water, and guanidinium). In practice, pH profiles are generated at different temperatures and the pKa values so obtained are plotted against respective $1/T$ values. The slope of such a plot gives $\Delta H_{ion}/2.303R$. Typical ΔH_{ion} values of 6 kcal/mol and 12 kcal/mol corresponding to a ΔpKa (between 0 and 25 °C) of 0.4 and 0.8 pH units, respectively. A potential identification of groups from their characteristic pKa and ΔH_{ion} is thus feasible. A word of caution: unusual ΔH_{ion} values may be obtained when the ionization of the group is coupled with protein conformational change (Knowles and Jencks 1976). In such cases, experimentally observed ΔH_{ion} may include a component for that as well.

Through solvent perturbation: According to the Bronsted definition of acids and bases, any species of a functional group that has a tendency to lose a proton is an acid. This definition eminently suits our understanding of the role of acid-base groups at the enzyme active site. We can classify all ionizable functionalities on

Table 23.1 Range of ΔH_{ion} values for ionizable groups on the enzyme

Group	pKa range (at 298 K)	ΔH_{ion} (kcal mol^{-1})
Carboxylate, phosphate	2–5	0 ± 1.5
Imidazolium, thiol, phenolic OH	5–10	6 ± 1.5
Amine, metal-bound water, guanidinium	8–12	12 ± 1.5

Table 23.2 Different Bronsted acid groups on the enzyme

Acid type	Functional group
Neutral acids ($X - H \rightleftarrows X^- + H^+$)	Carboxylate, phosphate, thiol, phenolic OH, metal-bound water
Cationic acids ($X - H^+ \rightleftarrows X: + H^+$)	Imidazolium, amine, guanidinium

the enzyme (including side chains of amino acid residues) into two groups: cationic acids and neutral acids (Table 23.2).

A clear difference between the two Bronsted acid classes is in the net charge generated upon dissociation. Two charges are generated (one proton and a −ve ion) on dissociation of a neutral acid while there is one +ve charged species on either side of the dissociation equilibrium for a cationic acid (Table 23.2). This has implications to their differential behavior upon solvent perturbation (altered dielectric constant, etc.). The pKa values of neutral acids are elevated by adding water miscible organic solvents like different alcohols, formamide, dioxane, dimethyl sulfoxide, or N, N-dimethylformamide. Cationic acid pKas are largely unaffected by such solvent addition. In practice, pH profiles are generated with and without the solvent and the relevant group pKas are compared. The groups whose pKas are elevated in the presence of an added solvent are all neutral acids. The two histidine residues (H12 and H119) at the active site of ribonuclease A were identified by such a study. There are two limitations of solvent perturbation approach. Groups that are not exposed to solvent will not be affected or accounted for. Second, since 30–50% of the solvent may have to be used, it becomes important to ensure that the enzyme is not inactivated by such addition.

References

Knowles JR, Jencks WP (1976) The intrinsic pKA-values of functional groups in enzymes: improper deductions from the pH-dependence of steady-state parameter. CRC Crit Rev Biochem 4:165–173

Rife JE, Cleland WW (1980) Determination of the chemical mechanism of glutamate dehydrogenase from pH studies. Biochemistry 19:2328–2333

Isotopes in Enzymology

Isotopes are atoms of an element that contain different numbers of neutrons in their atomic nucleus. They have the same atomic number, i.e., same number of protons (of course an equal number of electrons) but differ in their atomic mass due to the neutron numbers. For example, the three isotopes of hydrogen have one proton and one electron each but contain zero (hydrogen), one (deuterium), or two (tritium) neutrons. Chemical properties of any element are determined by its electronic configuration (and atomic number) and therefore, all the isotopes of that element react similarly. Isotopes provide an excellent tool for the study of reactions because they are isoelectronic and isosteric. This feature makes isotopes eminently suitable as tracers in enzyme research. Both radioactive and stable isotopes commonly used in enzymology are listed in Table 24.1. While remarkably similar in their chemical reactivity, their mass differences inflict subtle but definite changes in the rate of bond forming/breaking events involving them. These "isotope effects" —valuable in understanding enzyme mechanisms—are discussed in a subsequent section (Chap. 25).

Substrates/products labeled with radioactive isotopes are used in enzyme assays. Stable isotopes are less commonly used to monitor enzyme reactions. Their detection is either inherently less sensitive or requires involved instrumentation like NMR spectroscopy or mass spectrometry (Wang et al. 2021). Such non-radioactive methods are not easily amenable for routine assays. NMR active labels (like ^{15}N and ^{13}C) are of value in solution dynamic studies of enzymes. These heavy isotopes (^{13}C in particular) are very valuable also in the study of metabolism (metabolic flux measurements and *Metabolomics*). Coupled with accurate mass measurements, substrates appropriately labeled with heavy isotopes provide the necessary data for kinetic isotope effect analysis.

© The Author(s), under exclusive license to Springer Nature Singapore Pte Ltd. 2025
N. S. Punekar, *ENZYMES: Catalysis, Kinetics and Mechanisms*, https://doi.org/10.1007/978-981-97-8179-9_24

Table 24.1 Isotopes commonly used in enzyme studies

Element (most abundant form)	Radioactive isotope			NMR active isotopes	Heavy isotopes
	Isotope	β-Emission (MeV)	Half-life		
Hydrogen (^{1}H)	^{3}H	0.018	12.3 years	^{1}H, ^{2}H	^{2}H
Carbon (^{12}C)	^{14}C	0.154	5700 years	^{13}C	^{13}C
Nitrogen (^{14}N)	–	–	–	^{15}N	^{15}N
Oxygen (^{16}O)	–	–	–	^{17}O	^{17}O, ^{18}O
Phosphorus (^{31}P)	^{32}P	1.718	14.3 days	^{31}P	–
Sulfur (^{32}S)	^{35}S	0.167	87.1 days	–	–

24.1   Enzyme Assays with a Radio-labeled Substrate

Radioisotope measurements are particularly resorted to when other simpler methods of assay like colorimetry, spectrophotometry, fluorimetry, or polarography are not feasible. A substrate bearing one or more radioisotopes is used in enzyme kinetic measurements so that the product formed is radioactive. Almost always enzyme assays with radio-labeled substrates are fixed time, end-point assays—the reaction is stopped, and using an appropriate technique, labeled product is separated from the remaining substrate. A chromatographic step is often employed for this purpose. For example, galactokinase reaction is monitored by using ^{14}C-galactose as substrate; the labeled galactose-1-phosphate formed is resolved from unreacted ^{14}C-galactose by a suitable ion exchange resin and its radioactivity counted. Similarly, ^{14}CO$_2$ released from 1-^{14}C-L-glutamate is captured in a base and counted to assay glutamate decarboxylase (Kumar et al. 2000). The amount of radioactivity in the product, as a function of time, gives the measure of reaction progress. While using radiolabels to monitor reaction rates, the following considerations are important:

1. The position of the radiolabel in the substrate should be carefully chosen. Atomic positions away from and not involved in bond breaking/bond forming steps must be used. Such remote labeling ensures that no "isotope effects" are introduced.
2. Sensitivity of radiotracer detection requires that a good post-reaction separation of labeled substrate from the product formed be achieved. This is critical in obtaining satisfactory blanks and controls.
3. In enzyme kinetic studies, the labeled substrate is usually mixed with "cold" (unlabeled) substrate to achieve required substrate concentration without having to use high quantities of radioactivity. The specific radioactivity is so adjusted that minimal amount of the radiolabel will provide good signal-over-background readings.

Commonly used radioisotopes are β-emitters, and the radioactive decay process follows first-order kinetics (Chap. 8; Part II). Each radioisotope is associated with a characteristic half-life (Table 24.1). The standard unit for radioactivity is the Curie

(Ci)—the quantity of any substance that decays at a rate of 2.22×10^{12} disintegrations per minute (dpm). The proportion of radio-labeled molecules in the given substrate sample is expressed as specific radioactivity. Convenient units for specific radioactivity are dpm/µmol and µCi/mmol.

In practice, one cannot directly measure radioactivity in a given sample because the efficiency of counting the number of disintegrations (in a scintillation counter) is never 100%. Therefore, the counts per minute (cpm) are measured, and using a factor for counting efficiency, the sample radioactivity can be calculated in dpm units. Going from cpm to dpm to specific radioactivity, one can easily convert the data into conventional enzyme velocity units.

24.2 Isotope Partitioning

Apart from their use as tools in enzyme assays (as mentioned above), radiolabels are employed to probe the enzyme reaction mechanism. Exchange of label from the product to its cognate substrate can be followed even while the reaction is actually proceeding in the forward direction. For example, in a reaction catalyzed by glucose-6-phosphatase, the reverse flow of label from ^{14}C-glucose back to glucose-6-phosphate was monitored although the net forward reaction was occurring. Such experiments provide insights into the order of product release.

Radioisotopes are also used to probe the fate of a substrate molecule sitting on the enzyme as EA binary complex. Recall the earlier Michaelis-Menten formalism (Chap. 14; Part II) employed to derive the rate equation. We can now imagine the same enzyme-substrate complex but with a radio-labeled substrate (A^*) (Fig. 24.1). Once formed, this EA^* has two possible fates: A^* gets converted to product (catalysis and product release with a forward rate constant k_{cat}) or A^* is released from the complex (with a rate constant k_{-1}) even before catalysis can occur.

With suitable experimental design it should be possible to measure the ratio of two rate constants—k_{cat}/k_{-1}. This ratio—without units—is a measure of *commitment-to-catalysis*; it is also variously referred to as *partition coefficient* (also see Chap. 15; Part II) or *stickiness ratio*. This ratio is negligible when k_{cat} is very small compared to k_{-1}—the commitment-to-catalysis is very low—the EA complex has greater tendency to dissociate rather than to convert A to P. If k_{cat} is much greater than k_{-1}, then commitment-to-catalysis is very high—the substrate is very sticky and all the substrate that binds E goes on to form product. It should however be clear that

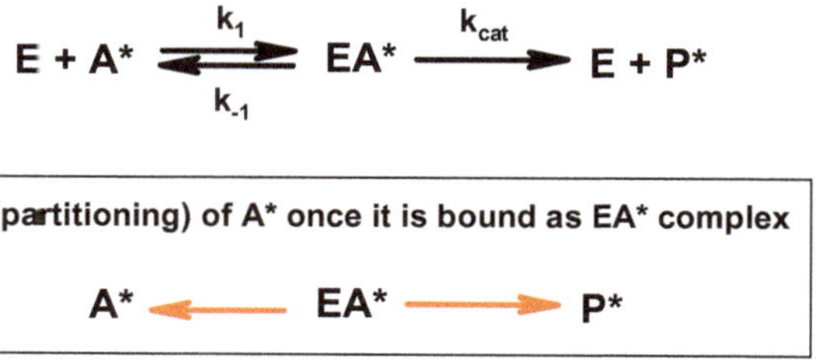

Fig. 24.1 Equilibria showing the fate of enzyme-bound isotopically labeled substrate

stickiness (given by k_{cat}/k_{-1}) and affinity (the association constant, k_1/k_{-1}) are not the same.

The commitment-to-catalysis can be experimentally measured by the substrate trapping procedure (called Rose experiment). A schematic of this elegant but powerful protocol is shown below.

1. Generate EA^* complex in a small volume (say 20 μL) by mixing known amount of enzyme and labeled substrate.
2. Dilute this with rapid mixing into a cocktail containing other reaction components also containing >1000-fold excess of unlabeled A.
3. Incubate briefly (10–15 s) for the enzyme to go through several catalytic turnovers and permit enzyme bound A^* to react.
4. Quench the reaction (usually with acid) and measure radiolabel in A^* and P^*.
5. Steps 1 through 4 are repeated at several concentrations of the other substrate (other than A, say B). The label trapped as P^*, at several concentrations of B, are recorded. Suitable controls are taken each time.

From a double reciprocal analysis of the isotope trapping data ($1/P^* \rightarrow 1/[B]$) we obtain (a) maximal labeled P^* formed in the first turnover and (b) the apparent K_M for B, for the trapping process. The commitment-to-catalysis is determined at saturating [B]. Such an isotope trapping strategy was first described for yeast hexokinase by Irving Rose, to show that glucose was very sticky. Isotope partitioning analysis of inosine phosphorylase (reaction shown below) provides an excellent example of this procedure.

Inosine + Phosphate⇌Hypoxanthine + Ribose-1-phosphate

Of the 30 μM of enzyme-[8-^{14}C]inosine complex, almost 20 μM was trapped and recovered as [8-^{14}C]hypoxanthine at saturating phosphate concentration. The remaining 10 μM of the EA^* complex dissociated back to release the bound substrate, [8-^{14}C]inosine. A commitment-to-catalysis of about 2 was calculated from these data ($k_{cat}/k_{-1} = 20$ μM/10 μM = 2).

There are practical limits as to when an isotope partition study is feasible. For single substrate enzymes, for example, the reaction begins as soon as enzyme and substrate are mixed. Measuring partition coefficient (stickiness) then requires rapid mix, chemical quench approaches—this is equipment and technique intensive (Chap. 10; Part II). A simpler alternative in such cases is to measure stickiness by monitoring micro-viscosity effects on the V_{max}/K_M of the enzyme (refer to Chap. 14; Part II, and Chap. 39; Part VI, relating to significance of k_{cat}/K_M).

One other variation in isotope partitioning experiment is positional isotope exchange (PIX). This method measures the rate of internal isotope exchange within a substrate molecule. PIX rate is usually expressed as a ratio to the overall reaction rate. The PIX analysis of argininosuccinate lyase reaction is a good example of this approach.

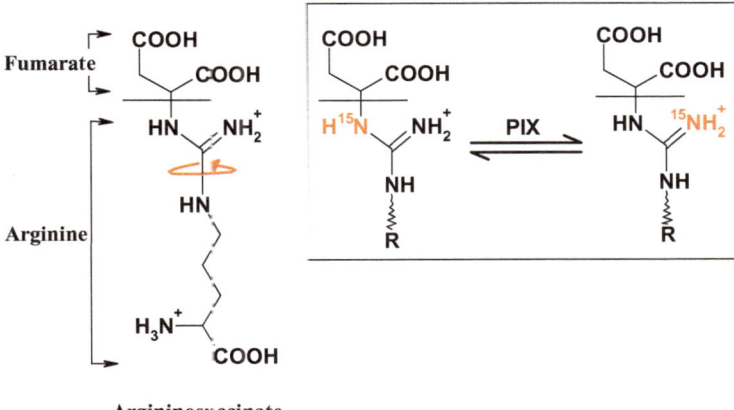

Fig. 24.2 Positional isotope exchange (PIX) in argininosuccinate lyase reaction. Gray line shows the C–N bond cleaved by the enzyme and the cyclic arrow indicates a free rotation around that bond at the enzyme active site

$$\text{Argininosuccinate} \rightleftarrows \text{Fumarate} + \text{Arginine}$$

The PIX rate was measured by ^{15}N NMR spectroscopy while its overall reaction was progressing toward arginine formation. The enzyme mobilizes the bond between the bridge nitrogen and the second carbon of the dicarboxylic acid (Fig. 24.2).

The bridge to non-bridge ^{15}N exchange rate, as a function of fumarate concentration was hyperbolic. It could thus be concluded that (a) the guanidinium group of arginine can freely rotate in the active site and (b) fumarate leaves much faster than arginine from the enzyme surface. Middlefort and Rose used PIX to establish the formation of glutamyl-phosphate as an intermediate in the glutamine synthetase reaction (Rose 2006). A PIX of the β-γ bridge oxygen of ATP-Mg provided crucial mechanistic evidence for γ-glutamyl phosphate as an intermediate in glutamine synthetase reaction (Chap. 35; Part IV).

24.3   Isotope Exchanges at Equilibrium

In the previous sections we saw the utility of isotopes (radioisotopes in particular) in elucidating enzyme mechanisms. They were examples of when the net chemical reaction was occurring in one direction. Isotope exchange kinetics is also possible in a system at equilibrium, and they are useful for defining kinetic mechanisms. Data from carefully executed isotope exchange studies are a powerful supplement to steady-state kinetic analysis (Boyer 1978; Rose 1995). More importantly, they provide excellent evidence in discriminating ordered and random sequential mechanisms. We recall that, in some cases, steady-state kinetic data give quantitative information, and this may have to be used to answer qualitative questions. For example, the presence of a K_{iA} term with finite value is used to conclude against parallel initial velocity patterns (Chap. 18). However, when these values are extremely small, a clear-cut decision becomes difficult. In such cases, isotope exchange data give unambiguous yes-no answers (see below).

We can assemble substrates and products in their equilibrium concentrations (K_{eq} = [products]/[substrates]); but a net flow of matter (either substrates $\rightarrow$ products or products $\rightarrow$ substrates) will not occur if the activation energy barrier is very high—the system is said to be at *static equilibrium*. Addition of an enzyme catalyst—opens the channel so to speak—and makes it dynamic. In a *dynamic equilibrium* there is both forward and backward flux, but no net reaction takes place because the two rates match (Chap. 9; Part II). The isotope exchange between reactant-product pairs in the presence of the enzyme is the first evidence of the dynamic nature of equilibrium. Glucose isomerase provides a simple example of this concept (Chap. 9). Similar label transfer experiments can also be performed for reactions with multiple substrates/products.

Nature of experiments conducted: Equilibrium isotope exchanges are conducted with the help of a suitably labeled substrate or a product. Most often such experiments are analytical—used to decide reaction types (ping-pong, ordered, or random) and not to obtain kinetic constants. Nevertheless, following points are worth considering in the design of a clean isotope exchange study:

(a) Label should be incorporated into a remote position of the reactant such that isotope effects are negligible. The label should serve only as a marker and should not lead to a different mechanism altogether.

(b) The reaction mixture at equilibrium is so assembled that a very small quantity of the tracer with high specific radioactivity is introduced. The concentration of the labeled component added should be insignificant and not perturb the equilibrium set up.

(c) Apart from the purity of the radio-labeled reactant, it is critical to ensure functional purity of the enzyme. High sensitivity of tracer detection makes the interference by contaminating enzyme activities problematic. ATP $\leftrightarrow$ ADP exchange rate for a kinase (such as hexokinase) is wrongly estimated if the enzyme sample contains another ATPase as a contaminant. It must be ensured that additional activities, if any, exhibited by the enzyme sample are due to the same enzyme. For instance, the γ-glutamyl transferase activity is catalyzed by the same glutamine synthetase active site where glutamine synthesis occurs. One can ascertain that two different activities are due to the same enzyme (Chap. 13; Part II), if they co-purify to a constant ratio of specific activities during various stages of purification.

(d) Since isotope exchanges are set up with all the reactants (substrates and products) present, abortive dead-end complexes (See Chap. 19 and later in Chap. 26, Fig. 26.3) may be formed. Their formation influences and severely impedes the interpretation of exchange data.

24.4 Partial Reactions and Ping-Pong Mechanism

Ping-pong mechanisms involve double displacements, and a substituted form of the enzyme (denoted as F form) occurs during the catalytic cycle (Chap. 18). This implies that atom/group transfer on to the enzyme from a particular substrate can occur even in the absence of rest of the substrates. Because of this partial reaction, isotope exchange can be demonstrated between a reactant and product in the absence of other reactants and products. In fact, an early and brilliant example of this work was by H.A. Barker and colleagues on bacterial sucrose phosphorylase. Sucrose phosphorylase catalyzes the following reaction:

$$\text{Sucrose} + \text{Phosphate} \rightleftharpoons \text{Fructose} + \text{Glucose-1-phosphate}$$

The isotope exchanges arising out of its two partial reactions are

(a) Glucose-1-phosphate + [^{32}P]Phosphate $\rightleftharpoons$ Glucose-1-[^{32}P]
phosphate + Phosphate
(b) Sucrose + [^{14}C]Fructose $\rightleftharpoons$ [^{14}C]Sucrose + Fructose

Notice that sucrose is not required for the first exchange and phosphate is not required for the second exchange. Such isotope exchanges are best evidence of ping-pong mechanism and glucosyl-enzyme as an essential covalent intermediate is indicated in this case. The following kinetic scheme (Fig. 24.3) adequately accounts for the observed exchanges.

The enzyme form "F" is the glucosyl-enzyme intermediate of sucrose phosphorylase. Either sucrose or glucose-1-phosphate can charge the enzyme with glucose to form the glucosyl enzyme. This glucosyl group can be picked up either by [^{32}P] phosphate (B $\leftrightarrow$ Q exchange, "a" above) or by [^{14}C]fructose (A $\leftrightarrow$ P exchange, "b" above) from the medium.

Other examples involving partial reactions include enzymes that transfer amino groups (such as aspartate transaminase), acyl groups (such as 3-oxoacid CoA transferase, serine transacetylase and transpeptidase), carboxyl group (like transcarboxylase), a three carbon unit (a transaldolase), a two carbon unit

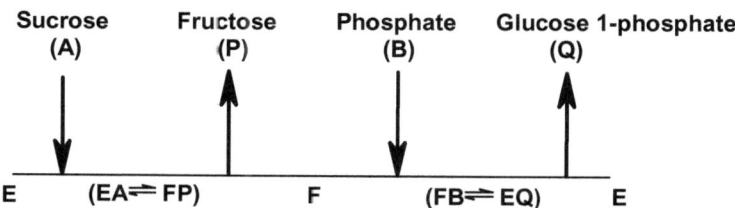

Fig. 24.3 Kinetic scheme for sucrose phosphorylase. The two enzyme forms E (free enzyme) and F (glucosyl-enzyme) are shown

(a transketolase), phosphate group transfer (nucleotide diphosphokinase), phosphate group migration (phosphoglyceromutase), two electron transfer (like in methylenetetrahydrofolate reductase), etc. As expected, in all these cases isotope exchanges between the first substrate and the product occur in the absence of other substrate and/or product. The different F-forms for these reactions may be referred to in a later section (Chap. 29; Part IV).

24.5 Sequential Mechanisms

The full complement of substrates needs to assemble at the enzyme active site for any chemistry to take place in a sequential mechanism. Therefore, as expected, no partial reactions are possible. In such situations, isotope exchanges are possible only when all reactants and products are present. Accordingly, the design of experiments is somewhat involved so as to maintain equilibrium condition.

Typically, in a two substrate-two product enzyme reaction of the type,

$$A + B \rightleftarrows P + Q$$

the the $A \leftrightarrow Q$ isotope exchange rate is monitored as a function of increasing concentrations of B–P pair. The concentrations of B and P are simultaneously raised to maintain the [B]/[P] ratio and hence, the position of equilibrium $K_{eq} = ([P][Q]/[A][B])$. For the converse experiment, $B \leftrightarrow P$ isotope exchange is measured as [A] and [Q] are simultaneously raised, while maintaining the [A]/[Q] ratio. For other possible combination of exchanges (namely, $A \leftrightarrow P$ and $B \leftrightarrow Q$) to occur, atoms/groups must be transferred between the respective substrate-product partners. With hexokinase reaction for example, there is nothing common between ADP and glucose and a study of ADP $\leftrightarrow$ glucose exchange is not possible. Therefore, which exchanges are experimentally observable depends on the reaction chemistry under consideration.

Ordered mechanism: The addition of substrates and release of products being ordered, only the outer pair—A and Q—can interact (respectively bind or leave) with the E form of the enzyme.

$$A + E \rightleftarrows EA + B \rightleftarrows (EAB \rightleftharpoons EPQ) \rightleftarrows EQ + P \rightleftarrows E + Q$$

Increasing the concentrations of B-P pair (but maintaining the [B]/[P] ratio) will initially increase the rate of $A \leftrightarrow Q$ isotope exchange (hyperbolic). At higher concentrations, however, most of the enzyme gets locked up into central complexes and very little E form will be available for interaction. Therefore, $A \leftrightarrow Q$ isotope exchange rate is inhibited as the concentration of B-P pair is raised. In a fully ordered mechanism, this inhibition can drive the $A \leftrightarrow Q$ exchange rate to zero. This profile for $A \leftrightarrow Q$ exchange rate is shown in Fig. 24.4 (panel I). The exchange profile for the inner pair—B and P—is quite different, however. Increasing concentrations of the A–Q pair facilitates $B \leftrightarrow P$ isotope exchange by driving more and more enzyme into

Fig. 24.4 Equilibrium
isotope exchange profiles
observed with enzymes.
Schematic isotope exchange
data for complete inhibition
(panel I), partial inhibition
(panel II), and hyperbolic
(panel III) patterns of
exchange is shown

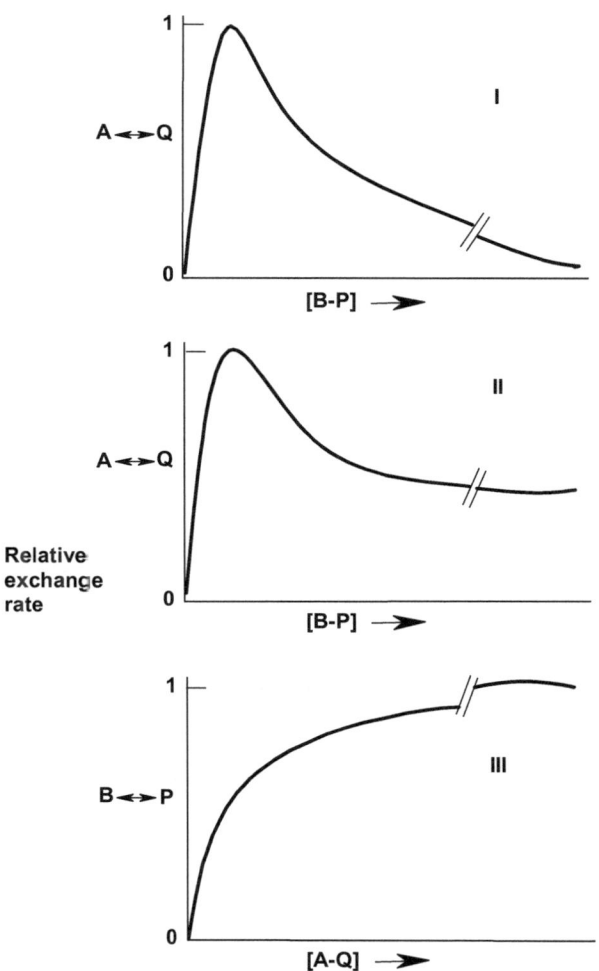

central complexes. Ultimately the B $\leftrightarrow$ P exchange rate reaches a plateau and a
hyperbolic profile results (Fig. 24.4, panel III).

From the rates of exchange between cognate pairs of reactants, Silverstein and
Boyer (1978) showed that lactate dehydrogenase follows an ordered mechanism
with the coenzymes (NAD^+ and NADH) forming the outer Michaelis complexes.
Malate dehydrogenase also shows such exchange patterns and is an example of fully
ordered mechanism (Fig. 24.5).

The $NAD^+ \leftrightarrow$ NADH exchange rate increased and eventually fell to zero when
[malate-oxaloacetate] was raised; whereas malate $\leftrightarrow$ oxaloacetate exchange rate
followed hyperbolic pattern and reached a plateau with increasing [NAD^+–
NADH]. The substrate inhibition of exchange rate is typical for the outer pair and

Fig. 24.5 Kinetic scheme for malate dehydrogenase. The coenzymes form the outer substrate-product pair

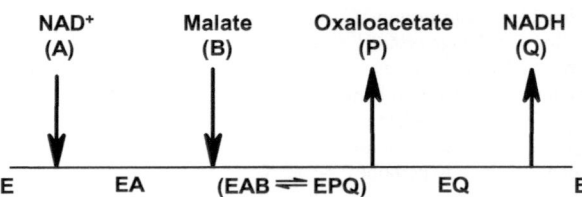

thus we conclude that NADH is the first substrate to add while NAD^+ is the last product to leave the enzyme.

While Theorell-Chance mechanism is an example of an ordered mechanism (but without significant central complexes) it is not possible to pile up central complexes. Therefore, no substrate inhibition (by alcohol-aldehyde pair) of $NAD^+ \leftrightarrow NADH$ exchange rate occurs with liver alcohol dehydrogenase.

Random mechanism: As random addition of substrates (and release of products) can occur, no fixed inner and outer pairs can be defined. The E form of the enzyme can bind to all the reactants. Because of this the enzyme is not locked up into central complexes and the $A \leftrightarrow Q$ isotope exchange rate is not inhibited as the concentration of B–P pair is raised. In effect the rates of exchange between both cognate pairs of reactants ($A \leftrightarrow Q$ as well as $B \leftrightarrow P$) follow hyperbolic kinetics (Fig. 24.4, panel III). The three isotope exchanges possible with hexokinase reaction—glucose $\leftrightarrow$ glucose-1-phosphate, ATP $\leftrightarrow$ glucose-1-phosphate and ADP $\leftrightarrow$ ATP—are hyperbolic and show no substrate inhibition. Other examples of sequential random mechanisms where such exchange data are available include creatine kinase and yeast alcohol dehydrogenase.

Formation of a ternary complex in random mechanisms can follow both the routes, namely, $E \rightarrow EA \rightarrow EAB$ or $E \rightarrow EB \rightarrow EAB$ (see Chap. 18). It is however not necessary that these alternative paths are used equally by an enzyme. If one of the routes is taken most of the time, then such a random mechanism is termed as preferred ordered sequential mechanism. In such cases (say $E \rightarrow EA \rightarrow EAB$ is preferred), one observes an inhibition of $A\leftrightarrow Q$ isotope exchange. However, even at very high concentrations of B–P pair, the $A \leftrightarrow Q$ exchange does not go to zero but reaches a limiting value (Fig. 24.4, panel II). This partial inhibition of exchange is characteristic of a ***preferred-order*** in the random mechanism and the relative magnitude of inhibition gives the extent to which the two paths are followed. In fact, isotope exchange is a very sensitive tool to detect minor reaction pathways.

In summary (a) isotope exchange in the absence of the second (other) substrate is diagnostic of a ping-pong mechanism, (b) isotope exchanges at equilibrium are possible for sequential mechanisms only when the full complement of reactants (all substrates and all products) are present, and (c) substrate inhibition of $A \leftrightarrow Q$ exchange is observed by raising B-P levels only in an ordered mechanism.

References

Boyer PD (1978) Isotope exchange probes and enzyme mechanisms. Acc Chem Res 11:218–224

Kumar S et al (2000) Metabolic fate of glutamate and evaluation of flux through the 4-aminobutyrate (GABA) shunt in *Aspergillus niger*. Biotechnol Bioeng 67:575–584

Rose IA (1995) Isotopic strategies for the study of enzymes. Protein Sci 4:1430–1433

Rose IA (2006) Mechanistic inferences from stereochemistry. J Biol Chem 281:6117–6119

Wang L, Amelung W, Willbold S (2021) ^{18}O isotope labeling combined with ^{31}P nuclear magnetic resonance spectroscopy for accurate quantification of hydrolysable phosphorus species in environmental samples. Anal Chem 93:2018–2025

Isotope Effects in Enzymology

So far, we have seen how chemically identical behavior of isotopes are exploited in elucidating enzyme mechanisms. Despite their remarkable chemical similarity, isotopic substitution does affect reaction rates that directly involve them. Isotopic substitution can influence the equilibrium position, affect reaction rate, or both. The former (affecting the equilibrium constant) is termed *equilibrium isotope effect* while the latter (affecting the rate constant) is termed *kinetic isotope effect* (KIE). Commonly encountered types of isotope effects in enzyme study are given with examples below.

Equilibrium isotope effect: Consider the following exchange reaction where D^+ is exchanged with two hydrogen atoms (as H^+ ions) of water.

$$2D^+ + H_2O \rightleftarrows D_2O + 2H^+$$

The equilibrium constant for this reaction is 8.2. The D-O bond is stiffer (shorter by 0.04 Å and stronger) than the H-O bond because of the heavy deuterium atom. Thermodynamic stability of D_2O is relatively higher (hence has a lower zero-point energy) than H_2O. Therefore, at equilibrium, there will be much more D_2O than H_2O. For the same reasons, D_2O as a solvent can affect the ionization of an acid group. The dissociation of acetic acid in D_2O is relatively less favored than in H_2O.

$$CH_3COOD \rightleftarrows CH_3COO^- + D^+$$

The acid dissociation constant (itself an equilibrium constant!) is given by the ratio of the two rate constants, k_1/k_{-1}. We can consider k_{-1} (reverse) as a simple bimolecular collision rate constant. As this process is purely diffusion controlled, it is unaffected by the isotopic substitution. However, as k_1 (forward) involves breaking of the CH_3COO-D bond, it is affected by isotopic substitution and is higher for CH_3COO-H. As a direct consequence of this, pKa of acetic acid is raised in D_2O (and therefore, pD = pH + 0.4). These two simple examples describe how an

N. S. Punekar, *ENZYMES: Catalysis, Kinetics and Mechanisms*, https://doi.org/10.1007/978-981-97-8179-9_25

equilibrium isotope effect comes about. It should be mentioned that the equilibrium isotope effect is presented as a non-unity ratio of the equilibrium constants ($K_{\text{light isotope}}/K_{\text{heavy isotope}}$).

Solvent isotope effect: Isotopic substitution of solvent protons by the heavy isotope (such as in D_2O) can affect the ionization of an acid group and more importantly the rate of the enzyme reaction itself. Such solvent isotope effects are particularly important when solvent protons participate directly or are exchanged via the ionizable groups on the enzyme/substrate during catalysis. Solvent isotope effects are often useful in distinguishing between nucleophilic versus general base catalysis (Chap. 4; Part I and Chap. 29; Part IV).

Kinetic isotope effect: The KIEs reflect changes in the vibrational frequencies of reactants as they pass through the rate-determining transition states to form products. Because they directly report on the kinetic reaction path, the KIE is perhaps the most powerful tool available for a mechanistic enzymologist. A KIE is usually written as a ratio of rate constants for the light and heavy isotopic reactants. For example, the isotope effect for a C-H bond versus C-D bond cleavage may be written as k_{C-H}/k_{C-D} (or simply k_H/k_D). When the isotopic substitution is at the reaction center (and directly participates in bond-breaking/making events), then the observed KIE is termed as primary isotope effect (1°KIE). Consider the following schematic example:

$$X - C_\alpha - C_\beta - C_\gamma -$$

The effect due to isotopic substitution at X—in the X-C_α bond-breaking event— is a ***primary KIE***. Secondary kinetic isotope effect (2°KIE) arises when the isotopic substitution is further removed from the scene of action. Accordingly, they are denoted as ***α-secondary*** or ***β-secondary*** KIEs, etc.—where α and β denote the position of the isotopic substitution relative to the atom undergoing bond cleavage. KIEs typically decrease in magnitude as the point of isotopic substitution lies further from the reaction center. Most enzyme KIE studies are confined to primary and α-secondary effects—expectedly these are the ones giving most useful information on the enzyme mechanisms.

Normal versus *inverse kinetic isotope effect*: An isotopic substitution by a heavier atom (like H by D) makes that bond stiffer and stronger. If this bond is to be broken during reaction, then that isotope will have less restrained bonding environment in the transition state when compared to the reactant state. In such cases, the reactant bearing the heavy isotope reacts less rapidly and a normal isotope effect is observed—the KIE value is above unity. An inverse isotope effect (and KIE value below unity) results in the case where isotope in question experiences a more restricted bonding environment in the transition state. It is obvious that such information provides valuable insights into the nature of the transition state itself.

25.1 Magnitude of the Observed Isotope Effect

As full theory of the origin of KIEs is quite complex and involves many factors, we will consider a simple approximation to evaluate it from first principles. The elementary theory for the case of C-H bond versus C-D bond cleavage step is considered below. In this treatment, the C-H and C-D bonds have very little difference in terms of their electronic, translational, and rotational properties. Their vibrational motion is considered harmonic and that one of these stretching modes becomes the bond breaking event in the transition state. The vibrational frequencies (as seen in the infra-red region) representing their respective zero-point energies are distinct for the two bonds; this is the major factor contributing to KIEs because the energy level of the transition state for both the reactions is approximately same. These conditions are schematically represented in Fig. 25.1.

The zero-point energy (ZPE) for a C-X bond (considered as a harmonic oscillator) is given by

$$ZPE = \frac{1}{2}h\nu = \frac{1}{2}hc\bar{\nu}$$

where h = Plank's constant (6.64×10^{-34} J.s), c = velocity of light (3×10^{10} cm.s^{-1}) and $\bar{\nu}$ = vibrational frequency in cm^{-1}. Based on these definitions, we could write the ZPE for the C-H and C-D bonds as follows:

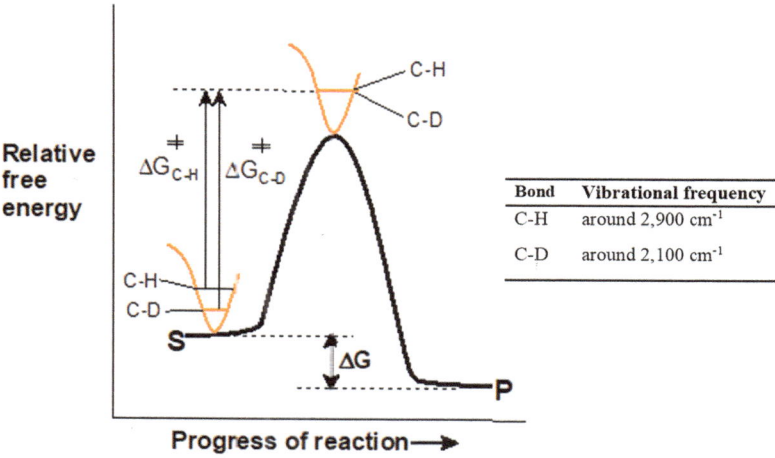

Bond	Vibrational frequency
C-H	around 2,900 cm^{-1}
C-D	around 2,100 cm^{-1}

Fig. 25.1 Reaction coordinate diagram for the C-H bond cleavage in an exergonic reaction. The heavier isotopomer (C-D) lies at lower energy as expected for a shorter bond length. However, the transition state for both C-H and C-D bond cleavages are of similar energy. Therefore, ΔZPE significantly contributes to the difference between $\Delta G^{\ddagger}$ for C-H and C-D bonds. The accompanying table gives the vibrational frequencies (observed in infra-red region) for the C-H and C-D bonds

$$ZPE_{C-H} = \frac{1}{2} hc\bar{v}_{C-H} \text{ and } ZPE_{C-D} = \frac{1}{2} hc\bar{v}_{C-D}$$

and hence, the energy difference between the two bonds as

$$\Delta ZPE = \frac{1}{2} hc\bar{v}_{C-H} - \frac{1}{2} hc\bar{v}_{C-D}$$

From Hook's law, the vibrational frequency is given by:

$$\bar{v} = \frac{1}{2\pi c} \sqrt{\frac{\kappa}{\mu}}$$

where κ is the force constant of the bond and μ is reduced mass ($= \frac{m_1 m_2}{m_1 + m_2}$). The difference between the C-H and C-D bonds thus boils down to differences in their reduced masses. Therefore,

$$\bar{v}_{C-H} = \frac{1}{2\pi c} \sqrt{\frac{\kappa}{\mu_{C-H}}} \text{ and } \bar{v}_{C-D} = \frac{1}{2\pi c} \sqrt{\frac{\kappa}{\mu_{C-D}}}$$

We thus obtain

$$\frac{\bar{v}_{C-H}}{\bar{v}_{C-D}} = \sqrt{\frac{\mu_{C-D}}{\mu_{C-H}}} \text{ and } \bar{v}_{C-D} = \bar{v}_{C-H} \sqrt{\frac{\mu_{C-H}}{\mu_{C-D}}}$$

and the ZPE difference between the two bonds is obtained by substituting these values in the equation above.

$$\Delta ZPE = \frac{1}{2} hc\bar{v}_{C-H} - \frac{1}{2} hc\bar{v}_{C-H} \sqrt{\frac{\mu_{C-H}}{\mu_{C-D}}}$$
$$= \frac{1}{2} hc\bar{v}_{C-H} \left(1 - \sqrt{\frac{\mu_{C-H}}{\mu_{C-D}}}\right)$$

We can evaluate ΔZPE from this equation (in fact, for any two isotopic pairs) by substituting respective parameters. Typical $\bar{v}_{C-H}$ is obtained from IR frequency data (2900 cm^{-1}, from the table in Fig. 25.1) and the respective reduced masses are calculated [where $\mu_{C-H} = (12 \times 1)/(12 + 1) = 0.923$ and ($\mu_{C-D} = (12 \times 2)/(12 + 2) = 1.714$]. Therefore,

$$\Delta ZPE = \frac{1}{2} \times 6.64 \times 10^{-34} \times 3 \times 10^{10} \times 2900 \left(1 - \sqrt{\frac{0.923}{1.714}}\right) = 7955 \times 10^{-24} J$$

This corresponds to 4789 J.mol^{-1} (when multiplied by the Avogadro number, 6.02×10^{23} mol^{-1}). Since the reaction rate constant is given by $k = (kT/h)e^{-\frac{\Delta G^{\ddagger}}{RT}}$, we can find the ratio,

$$\frac{k_{C\text{-}H}}{k_{C\text{-}D}} = \frac{e^{-\frac{\Delta G^{\ddagger}_{C\text{-}H}}{RT}}}{e^{-\frac{\Delta G^{\ddagger}_{C\text{-}D}}{RT}}} = e^{\frac{\Delta G^{\ddagger}_{C\text{-}D} - \Delta G^{\ddagger}_{C\text{-}H}}{RT}} = e^{\frac{\Delta ZPE}{RT}}$$

At 25 °C, this value is around 7.0 ($= e^{\frac{4789}{8.314 \times 298}}$). When a deuterium isotope effect is fully manifest, one obtains a $k_{C\text{-}H}/k_{C\text{-}D}$ value of around 7.0. Most often this value ranges from 2 to 15 and if no isotope effect is observed, then it will be unity. As a rule of thumb, an observed $k_{C\text{-}H}/k_{C\text{-}D}$ value between (a) 4 and 7 is indicative of a symmetric TS for that C-H bond cleavage event and (b) 1 and 4 means an asymmetric TS (for cleavage of this bond) or it is a secondary effect (2°KIE with no bond cleavage).

The magnitude of any KIE thus depends on the following factors that influence the bond-breaking/forming events at the transition state:

(a) The actual mass difference (in terms of reduced mass, μ) due to the isotopic substitution. We have noted that the reduced mass for C-D bond is almost double than that for C-H bond. For other isotopic substitutions, this difference is much smaller and consequently the KIE is smaller. For example, the reduced mass for ^{12}C-^{12}C bond is 6.00 [$\mu_{C\text{-}C} = (12 \times 12)/(12 + 12)$] while for ^{12}C-^{13}C bond, it is just 6.24 [$\mu_{C\text{-}C} = (12 \times 13)/(12 + 13)$]. Therefore, a maximum KIE of 1.06 may be observed for ^{12}C-^{13}C bond cleavage. The expected KIEs for other examples are $k_{12C}/k_{14C} = 1.09\text{--}1.15$; $k_{14N}/k_{15N} = 1.04$; and $k_{O16}/k_{O18} = 1.08$.
(b) The force constant—that is how tightly the atom is held. This has a bearing on the bond order (and in turn bond length) associated with that bond.
(c) Whether any other step, other than due to the isotopic substitution, is more rate-limiting. If another step is significant in determining the reaction rate, then the anticipated isotope effect may be either small or not observed at all. Indeed, such information can be actually used in elucidating enzyme reaction mechanisms.

25.2 Experimental Approaches to Measure Isotope Effects

It is obvious, from the theory on how the isotope effects are manifest, that very sensitive and accurate methods are required to measure them. Three different experimental approaches may be taken to measure KIEs.

Direct comparison: A common strategy is to synthesize many different substrate molecules, each with a specific position isotopically labeled. Then, the various kinetic constants (V_{max} and V_{max}/K_M) can be measured for the normal as well as the suitably isotope-labeled substrate. These rate constants can be directly compared to note the isotope effects on different kinetic parameters. In the case of lactate dehydrogenase, for example, the hydrogen on the C2 (asymmetric) carbon of lactate (Fig. 25.2a) can be substituted by deuterium.

Using the two kinds of substrates, the deuterium isotope effect on V_{max} and V_{max}/K_M can be directly measured (Fig. 25.2b, above). This approach of direct comparison is excellent as it measures all the required kinetic parameters in one

Fig. 25.2 (a) Lactate structure with two different isotopic substitutions on its C2 carbon and (b) Schematic double reciprocal plots when these two isotopomers are used separately as lactate dehydrogenase substrates

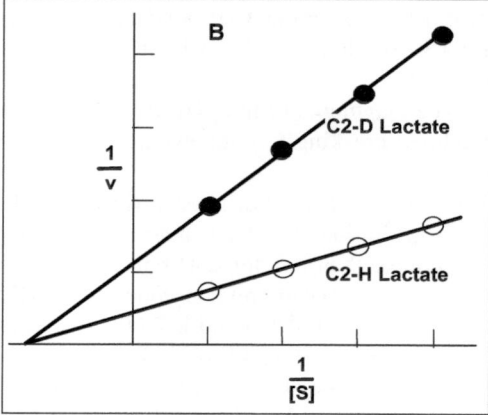

go. However, the method suffers in that it requires very high degree of label substitution. Impurities in the labeled substrate affect the measurement of V_{max} effect whereas accurate measurement of substrate concentration is crucial as it influences the KIE on V_{max}/K_M. Apart from the factors listed before, in practice, the extent of isotopic labeling possible at the given position (of substrate structure) also determines the actually measured effect. 90% label enrichment at the C2-H of lactate (by C2-D) gives only 90% of the maximal KIE. While high degree of isotopic enrichment at one position may be possible with deuterium, this is almost always not feasible for other isotopes.

Equilibrium perturbation: In this approach, labeled substrate (S^*) and unlabeled product are mixed at a calculated equilibrium ratio in the presence of the enzyme. One observes a temporary displacement of this equilibrium with time as $S^* \rightarrow P$ conversion is slow compared to that from $S \rightarrow P$. The system thus takes time to reach equilibration of the label on both sides; and from this time transient, the KIE can be obtained. This method is useful only for reversible reactions. Further, high label substitution in the substrate is required and temperature maintenance is crucial for it affects the equilibrium.

Internal competition method: This method exploits the fact that S and S^* (labeled S) compete with each other to form ES complex and subsequently for turnover. There are different ways one can set this competition, but we will exemplify this with glucose-6-phosphate dehydrogenase reaction.

$$\text{Glucose-1-phosphate} + \text{NAD}^+ \rightarrow \text{6-Phosphogluconate} + \text{NADH}$$

The natural abundance of ^{13}C at C1 of glucose-6-phosphate is 1.1%. If there is discrimination by the enzyme (say ^{13}C glucose-6-phosphate is slowly converted), then with time the substrate remaining is enriched with ^{13}C at C1 of glucose-6-phosphate; its abundance rises above 1.1%. Simultaneously, the abundance of ^{12}C in the product (6-phosphogluconate) increases. The C1 of 6-phosphogluconate can be quantitatively converted into carbon dioxide by oxidation (enzymatic or chemical method). The corresponding enrichment of ^{12}C in carbon dioxide (the ratio of ^{12}C/^{13}C in the CO_2 gas) can be directly measured in an isotope ratio mass spectrometer. This protocol can be used to measure the KIE due to ^{13}C at the C1 position, for glucose-6-phosphate dehydrogenase reaction.

Use of an additional *remote label* in the reactant, apart from the atom expected to experience KIE, has made this method much more versatile.

25.3 Applications of KIEs in Enzymology

The value of KIE studies in enzymology can be better appreciated through specific examples. Rather than giving a mere list of these applications, we shall simply present them as selected case studies. Therefore, in no way such a treatment can be exhaustive but can only be representative. For more details, the reader may wish to look up detailed literature on this enzyme frontier (Cleland 2003, 2005; Klinman 2014).

Elucidating kinetic mechanism: Enzymes achieve catalytic power by setting up multiple steps, often of comparable energetic barriers, along the reaction coordinate. The isotope sensitive chemical step may be buried between other rate limiting enzymatic steps like a rate-limiting product release, a rate-limiting enzyme conformational change, etc. A large magnitude of the observed primary isotope effect on V_{max} is indicative of the fact that the isotopic substitution is part of a major rate-limiting step. A full deuterium isotope effect of 7.0 on V_{max} suggests that bond cleavage to that hydrogen determines the overall rate of the reaction; no isotope effect indicates that some other step is rate-limiting. Often a single step is not solely rate-limiting and hence deuterium isotope effects on V_{max} are in the range 1.5–2.0; the KIEs are between 2.0 and 4.0 for hydrolytic reactions in water. Physical binding steps are insensitive to isotope effects; hence, no isotope effect on the K_M is expected whenever K_M equals K_S. As K_M is a complex of rate constants, effect of isotopic substitution on any one of its contributing rate constants will manifest as a KIE on K_M. As expected, only when the K_M is altered by heavy isotopic substitution, the KIEs on V_{max} and V_{max}/K_M are different. From such studies, it should be possible to tease out some of the individual rate constants along the enzyme reaction scheme.

There are a few examples of how KIEs are influenced by the presence of allosteric regulators. The mammalian glutamate dehydrogenase is allosterically activated by ADP and inhibited by GTP. The presence of ADP increases the deuterium isotope effects on V_{max} and V_{max}/K_M from 1.05 to 1.3. AMP nucleosidase is activated by

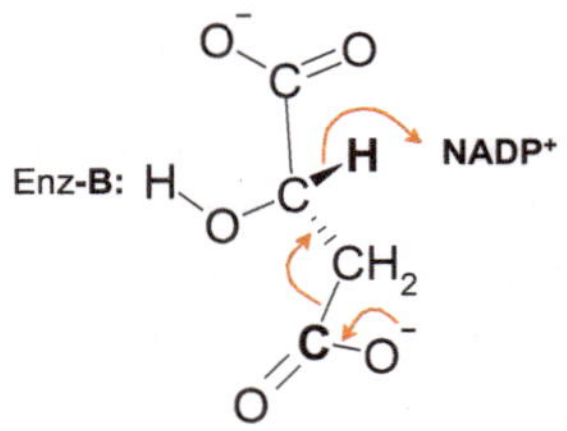

Fig. 25.3 Representation of malic enzyme reaction mechanism where hydride transfer and decarboxylation occur in concert. The atoms of malate where isotopic substitutions are used are shown in bold

Mg-ATP and more bond order remains in the leaving group (a C1′-N9 bond order of 0.16 versus 0.21), in the presence of this activator.

Deciding chemical mechanism: KIEs provide valuable information to the chemical mechanism of the enzyme reaction under consideration. As enzyme chemical mechanisms are covered in detail later (in Part IV), only a reference to the applications of isotope effects is made here. A major use of isotope effects is to decide whether the given reaction follows a concerted or stepwise mechanism. We will take two examples to demonstrate this concept.

Malic enzyme: The reaction catalyzed by malic enzyme is as shown:

$$\text{Malate} + \text{NADP}^+ \rightleftharpoons \text{Pyruvate} + \text{CO}_2 + \text{NADPH}$$

Mechanistically, this oxidative decarboxylation of malate involves two bond breaking events: the cleavage of the C2-H bond with concomitant hydride transfer to NADP$^+$ and the cleavage of C3-C4 bond leading to CO_2 release (Fig. 25.3).

Respective isotopic substitutions, namely, C2-H by C2-D and C3-C4 by C3-^{13}C4 give expected primary isotope effects indicating that both these bond cleavages contribute towards determining the overall rate of this reaction. We may wish to know whether the two bond cleavage events occur simultaneously (concerted) or follow one after the other (sequential). This can be verified by a ***double kinetic isotope effect*** study. Here we monitor the change in observed ^{13}C KIE due to a deuterium substitution (at C2). With malic enzyme, a ^{13}C KIE (for C3-^{13}C4) of 1.0302 was found. However, when the substrate C2-H was replaced by C2-D, this ^{13}C KIE decreased to 1.0250. This is indicative of the fact that hydride transfer and decarboxylation occur in different steps of the kinetic mechanism. If they were to occur in the same step (i.e., if concerted), then replacement of C2-H by C2-D should have made that step more rate-limiting and increased the size of the observed ^{13}C KIE. As the mechanism is stepwise, the C2-H to C2-D substitution made some other step rate-limiting and hence, decreased the size of the observed ^{13}C KIE. In the reaction sequence therefore, the ^{13}C-sensitive step comes later than the deuterium sensitive step—the hydride transfer occurs first and is followed by decarboxylation (Fig. 25.4).

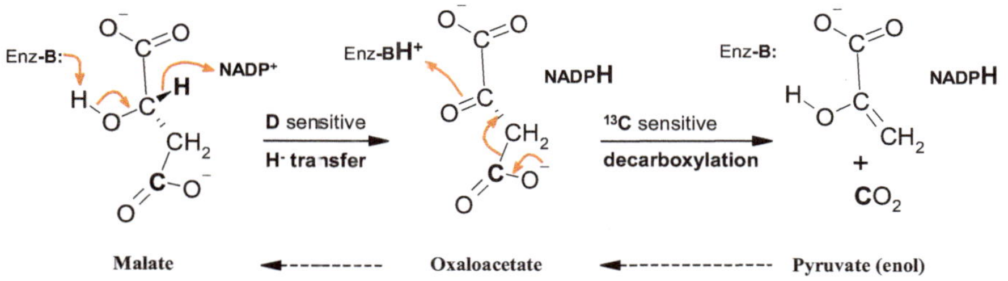

Fig. 25.4 Deuterium and ^{13}C-sensitive steps in the reaction mechanism of malic enzyme. The dotted arrows indicate the sequence of events in the direction of reverse reaction

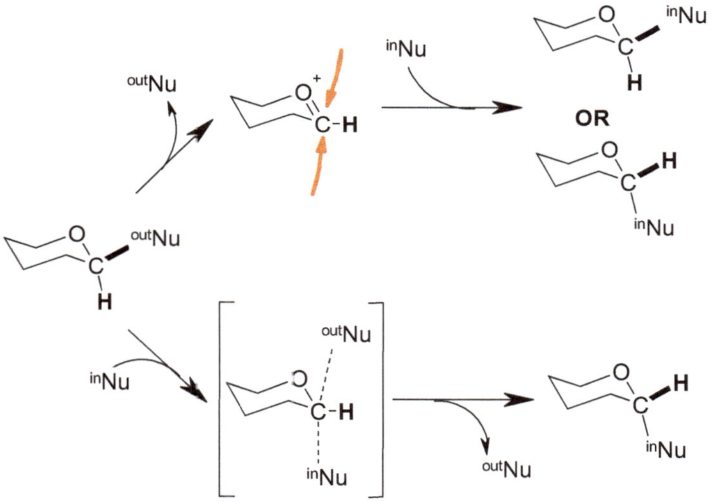

Fig. 25.5 Mechanism of glycosyl transfer: The two possible paths for this reaction are a stepwise mechanism (via an oxycarbonium; upper route) or a S_N2 mechanism (second-order nucleophilic substitution; lower route). The atoms where α-secondary KIE is measured are in bold

This means, on the enzyme, malate is first oxidized to oxaloacetate and then decarboxylated. It goes with the same intuition that such double isotope effects are not symmetric—in the reverse direction (reductive carboxylation of pyruvate) ^{13}C-sensitive step arrives before the hydride transfer.

Glycosyltransferases: Secondary isotope effects are very useful in determining the enzyme chemical mechanism in many cases. Two possibilities exist in the glycosyltransferase chemistry (Fig. 25.5). One can visualize an S_N1 reaction (a stepwise mechanism involving first the formation of an oxycarbonium intermediate, the upper path) or a S_N2 reaction (second-order nucleophilic substitution, the lower path). A detailed treatment on nucleophiles and nucleophilicity may be found in a later section (Chap. 29; Part IV).

In the S_N1 mechanism, because the outgoing nucleophile (^{out}Nu:) leaves first, the C1 carbon of the sugar changes from a tetrahedral (sp^3) arrangement to a flattened sp^2 configuration. This is subsequently attacked by the incoming nucleophile (^{in}Nu:). During reaction, the hybridization state of glycosidic C1 goes from $sp^3 \rightarrow sp^2 \rightarrow sp^3$. The S_N2 mechanism (as shown) always results in an inversion of configuration at C1 (the Walden inversion). Also, the C1 atom simultaneously experiences the effects of incoming and outgoing nucleophiles. Clearly by comparison, the bond order changes around the C1 atom are different for S_N1 and S_N2 mechanisms. This is exploited, through α-secondary isotope effects, to distinguish the two reactions. The α-secondary KIE can be measured by an isotopic substitution in the substrate C1-H (to C1-D). Typically, such an effect for S_N1 reaction (oxycarbonium ion mechanism) is larger (between 1.07 and 1.13) than that observed for S_N2 reaction (1.00 $\pm$ 0.06). The observed α-secondary isotope effect (a $k_H/k_D = 1.11$) for lysozyme catalyzed hydrolysis is best accommodated by an oxycarbonium ion mechanism. In this sense, isotope effects serve as a guide in choosing the most likely mechanism.

Understanding enzyme transition state: By definition, the transition state (TS) is the highest energy point on the lowest energy path between reactants and products. We have seen earlier (Chap. 4; Part I) that one of the major reasons enzymes catalyze reactions is by binding/stabilizing the TS in preference to either the substrate or the product. Rate accelerations are attributed to tightness of TS binding by an enzyme.

Enzymatic transition states are dynamic entities with lifetimes defying direct physical/experimental observation. Early work on analysis of TS and reaction mechanism relied on the introduction of various chemical substituents near/around the reaction center and monitoring their effects on the reaction rates. This information was interpreted through linear free energy relationships—like Hammett equation and Bronsted relation—to arrive at mechanistic details about the reaction and its TS. Such a chemical approach through systematic use of structural homologues is not suitable for the study of enzyme TS. The substrate selectivity/specificity of an enzyme severely limits the number of structural variants that can be employed.

Isotopic substitutions lead to more subtle changes and are eminently suited to probe the active site chemistry. In this background, kinetic isotope effects have the potential to provide direct information on the enzymatic TS. However, this path has been less traversed by researchers and for very few enzyme reactions. A major problem is that the intrinsic KIE for the chemical step, whose TS we wish to understand, may be difficult to access. Enzymes achieve catalytic power by setting up multiple steps, often of comparable energetic barriers, along the reaction coordinate. The isotope sensitive chemical step may be buried between other rate limiting enzymatic steps which could be rate-limiting product release, rate-limiting enzyme conformational change, etc. It would therefore be necessary to understand and uncover or account for them in order to obtain relevant intrinsic KIE from the observed KIE. For instance, a high commitment-to-catalysis (Chap. 24) results in underestimation of intrinsic KIE. Over the last three decades, these difficulties have been surmounted, and it is now possible to approach the TS of almost any enzyme with some preparation.

Once the intrinsic KIEs are available for an enzyme reaction, then the TS structure can be deduced in the usual physical organic chemistry sense. The experimental steps involved in TS analysis through KIEs are as follows:

(a) Synthesize substrates with appropriate isotopic labels at every position around the reaction center.
(b) Accurately measure the KIEs using these substrates. Correct them to get intrinsic values.
(c) A truncated TS is computed by fixing bond lengths and bond angles to match the observed KIEs.
(d) From this partial structure, generate the complete TS structure by optimization through semi-empirical methods (best fit to data by trial and error) and computational enzymology.

In effect, isotopic substitution at different positions in the substrate structure reports (via KIEs) on what really happens there during reaction. If an atom remains in the same binding environment, both in the substrate and in the TS, then there will be no KIE observed. Atoms that become vibrationally less constrained in the TS give normal KIE ($k_{light}/k_{heavy} > 1$) with the heavier isotopic substrate reacting more slowly. Conversely, atoms more constrained at the TS cause inverse KIE ($k_{light}/k_{heavy} < 1$) with the heavy isotope-labeled substrate reacting more rapidly. A qualitative picture of the TS can be constructed based on all such observed KIEs. Taken together, KIEs and computational chemistry provide a conceptually complete picture of the TS.

Arriving at the nature of enzyme TS complex is challenging, as typically positions of more than 10,000 atoms would have to be determined. A two-pronged approach is needed to decipher the features of enzyme transition-states: (a) measuring KIEs (as mentioned above) and (b) performing computational quantum chemistry. The KIEs give information about the geometry (the shape of the electron cloud surrounding the atoms—the van der Waals surface of the TS) and the electrostatic charge distribution. The two together lead us to the atomic structure of the TS. Computational chemistry is then used to sift through many possible TSs to find the one that matches the experimentally observed KIEs. The best fit structure contains the information about both geometry and electrostatic charge—the complete description of the TS.

Apart from a clear understanding of the reaction chemistry involved, TS analysis has a practical value. Knowledge of the TS for an enzymatic reaction provides information to design stable analogs as TS inhibitors (Schramm, 1998, 2011). A comparison of molecular electrostatic potential surface of the substrate with the TS is possible. It may then be feasible to design molecules bearing electrostatic potential surfaces like the TS. They can be synthesized and tested for their potential for enzyme inhibition. TS-like structures for several N-ribosyltransferases were defined through the observed KIEs and computational chemistry. For example, a potent inhibitor of purine nucleoside phosphorylase (9-deazainosine iminoribitol with a K_I of 20 pM) was achieved by this approach (Fig. 25.6). 5'-Methylthioadenosine/S-

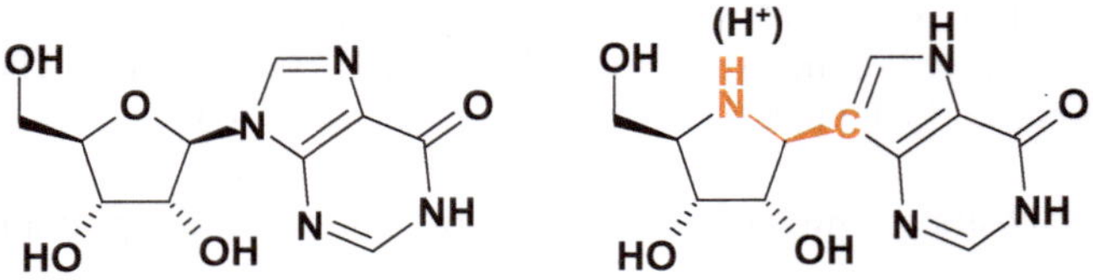

Inosine (K_M = 15 μM) **9-Deazainosine iminoribitol (K_I = 20 pM)**

Fig. 25.6 Structures of the substrate (inosine) and the synthesized TS inhibitor of purine nucleoside phosphorylase

adenosylhomocysteine nucleosidase catalyzes the hydrolytic cleavage of adenine from methyl-thioadenosine. Informed by TS analysis inhibitors both early and late dissociative transition state analogs were designed for this enzyme (Thomas et al., 2012). Late TS analogs have an extra atom to mimic the longer distance between the two atoms undergoing bond cleavage. Such strategies have practical value in the design of enzyme inhibitors as drug molecules (see Chap. 36 for more details). It must be noted that TS complex structures are not necessarily identical for homologous enzymes.

Drug design using TS analysis is in its infancy but has much to promise. This approach differs from the two traditional methods, namely, structure-based drug design and screening of chemical and/or natural product libraries. These two are general in that they can be used against most pharmaceutical targets of interest—including ion channels, receptors, and enzymes. However, the TS analysis approach to drug discovery is through a rational design and limited to enzyme targets. But then, enzyme catalysis is at the heart of life processes.

References

Cleland WW (2003) The use of isotope effects to determine enzyme mechanisms. J Biol Chem 278: 51975–51984

Cleland WW (2005) The use of isotope effects to determine enzyme mechanisms. Arch Biochem Biophys 433:2–12

Klinman JP (2014) The power of integrating kinetic isotope effects into the formalism of the Michaelis–Menten equation. FEBS J 281:489–497

Schramm VL (1998) Enzymatic transition states and transition state analog design. Annu Rev Biochem 67:693–720

Schramm VL (2011) Enzymatic transition states, transition state analogs, dynamics, thermodynamics, and lifetimes. Annu Rev Biochem 80:703–732

Thomas K et al (2012) Femtomolar inhibitors bind to 5′-methylthioadenosine nucleosidases with favorable enthalpy and entropy. Biochemistry 51:7541–7550

Further Reading

Klinman JP, Kohen A (2013) Hydrogen tunneling links protein dynamics to enzyme catalysis. Annu Rev Biochem 82:471–496

Schramm VL (2012) Freezing time: targeting the briefest moment in chemistry may lead to an exceptionally strong new class of drugs. Scientist 26:5

We conclude our forays into enzyme kinetic mechanisms by summarizing three practical aspects. Some examples of relating mechanisms with steady state kinetic data are detailed. Secondly, a general scheme that one should follow in orderly collection of kinetic data is given. This chapter (in Part III) is an attempt to bring home the point that enzyme kinetic study is not just "blue sky" research—but has real practical value.

26.1 How to Relate Mechanisms with Steady State Kinetic Data

A product inhibits the enzyme catalyzed reaction by virtue of its binding to one or more enzyme forms. The product inhibition patterns offer useful inputs in deciding the kinetic mechanism of a multi-substrate and/or multi-product enzyme. In a bi-reactant mechanism, information on the slope and intercept effects due to a product is normally obtained in a systematic manner. For example, concentration of one substrate (say A) is varied at different fixed levels of product inhibitor (say P). This experiment itself should be performed at two distinct fixed concentrations of B—once with saturating (noted as $B = \infty$) and again with sub-saturating (noted as $B = K_B$). As P is the inhibitory product used in this case, the reaction rates may be followed by formation of Q. For Q as the inhibitor, we need to monitor P, however. Thus, there are eight different product inhibition patterns possible for a bi-reactant mechanism.

We recall here that saturation with a fixed substrate (a) may result in an irreversible step in the mechanism (see Chap. 17) and (b) leads to no inhibition by the product that competes for the same enzyme form. Armed with an understanding of equilibria, irreversible steps, how slope (V_{max}/K_M; 1/slope of Lineweaver–Burk plot) and intercept (V_{max}; 1/intercept of Lineweaver–Burk plot) are affected and the thumb rules (described in Chap. 17), we can predict various product inhibition patterns for any given mechanism. *Converse of this is what we do in practice*—experimentally

N. S. Punekar, *ENZYMES: Catalysis, Kinetics and Mechanisms*, https://doi.org/10.1007/978-981-97-8179-9_26

obtain various slope/intercept effects and from these data arrive at the enzyme kinetic mechanism. The expected product inhibition patterns for three common bi-reactant mechanisms are given below.

26.1.1 Ordered Mechanism

Lactate dehydrogenase is an example of an ordered bi-substrate reaction. The substrate product pair of $NAD^+/NADH$ is the outer pair and binds the free enzyme according to the scheme (Fig. 26.1):

The predicted slope/intercept effects and the product inhibition patterns are given in Table 26.1.

26.1.2 Random Mechanism

Hexokinase is an example of random bi-substrate reaction. The substrates and products bind various enzyme forms according to the scheme (Fig. 26.2).

The predicted slope/intercept effects and the product inhibition patterns are given in Table 26.2.

The prediction of product inhibition patterns (for both ordered and random mechanisms) shown above is made with the understanding that the EBQ dead-end complex is possible and is formed. Here Q is not only acting as a product inhibitor

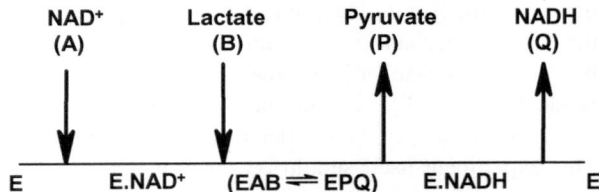

Fig. 26.1 Ordered bi-substrate reaction of lactate dehydrogenase

Table 26.1 Product inhibition patterns for an ordered bi-reactant mechanism

Product inhibitor	Substrate varied	Fixed substrate at	Enzyme parameter affected		Inhibition pattern
			Intercept $(1/V_{max})$	Slope (K_M/V_{max})	
P	A	$B = \infty$	Yes	No	Uncompetitive
P	A	$B = K_B$	Yes	Yes	Noncompetitive
P	B	$A = \infty$	Yes	Yes	Noncompetitive
P	B	$A = K_A$	Yes	Yes	Noncompetitive
Q	A	$B = \infty$	No	Yes	Competitive
Q	A	$B = K_B$	No	Yes	Competitive
Q	B	$A = \infty$	No	No	No inhibition
Q	B	$A = K_A$	Yes	Yes	Noncompetitive

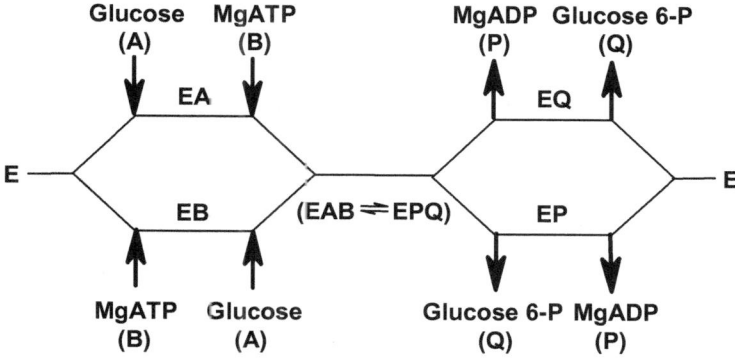

Fig. 26.2 Random bi-substrate reaction of hexokinase

Table 26.2 Product inhibition patterns for a random bi-reactant mechanism

Product inhibitor	Substrate varied	Fixed substrate	Enzyme parameter affected		Inhibition pattern
			Intercept $(1/V_{max})$	Slope (K_M/V_{max})	
P	A	$B = \infty$	No	No	No inhibition
P	A	$B = K_B$	No	Yes	Competitive
P	B	$A = \infty$	No	No	No inhibition
P	B	$A = K_A$	No	Yes	Competitive
Q	A	$B = \infty$	No	Yes	Competitive
Q	A	$B = K_B$	No	Yes	Competitive
Q	B	$A = \infty$	No	No	No inhibition
Q	B	$A = K_A$	Yes	Yes	Noncompetitive

but is also combining with EB in a dead-end fashion. We can rationalize occurrence of such complexes from the physical picture of the enzyme active site and how substrates and products occupy their respective places there. Consider a kinase reaction where; the γ-Ⓟ from ATP is transferred to an acceptor R-OH. This could be hexokinase where acceptor R-OH is nothing but glucose.

$$\text{Ado-}ⓅⓅⓅ + \text{HO-R} \rightleftarrows \text{Ado-}ⓅⓅ + Ⓟ\text{O-R}$$

Besides their respective single occupancy on the enzyme (EA, EB, EP, and EQ) the following four ternary complexes (Fig. 26.3) could be anticipated in principle.

Two of these (complexes 3 and 4 in Fig. 26.3) are dead-end complexes. The EAP complex (complex 3) is expected to be observed as one Ⓟ (the γ-Ⓟ is missing). However, the formation of complex 4 (the EBQ complex) depends on whether the extra piece (Ⓟ group here) can be accommodated at the active site or not. We now can generalize this to other enzyme examples: (a) EBQ should form for smaller groups (like hydride from NADH), (b) EBQ may form with reduced affinity for groups like acetyl or phosphoryl, and (c) not at all for larger ones like glycosyl or

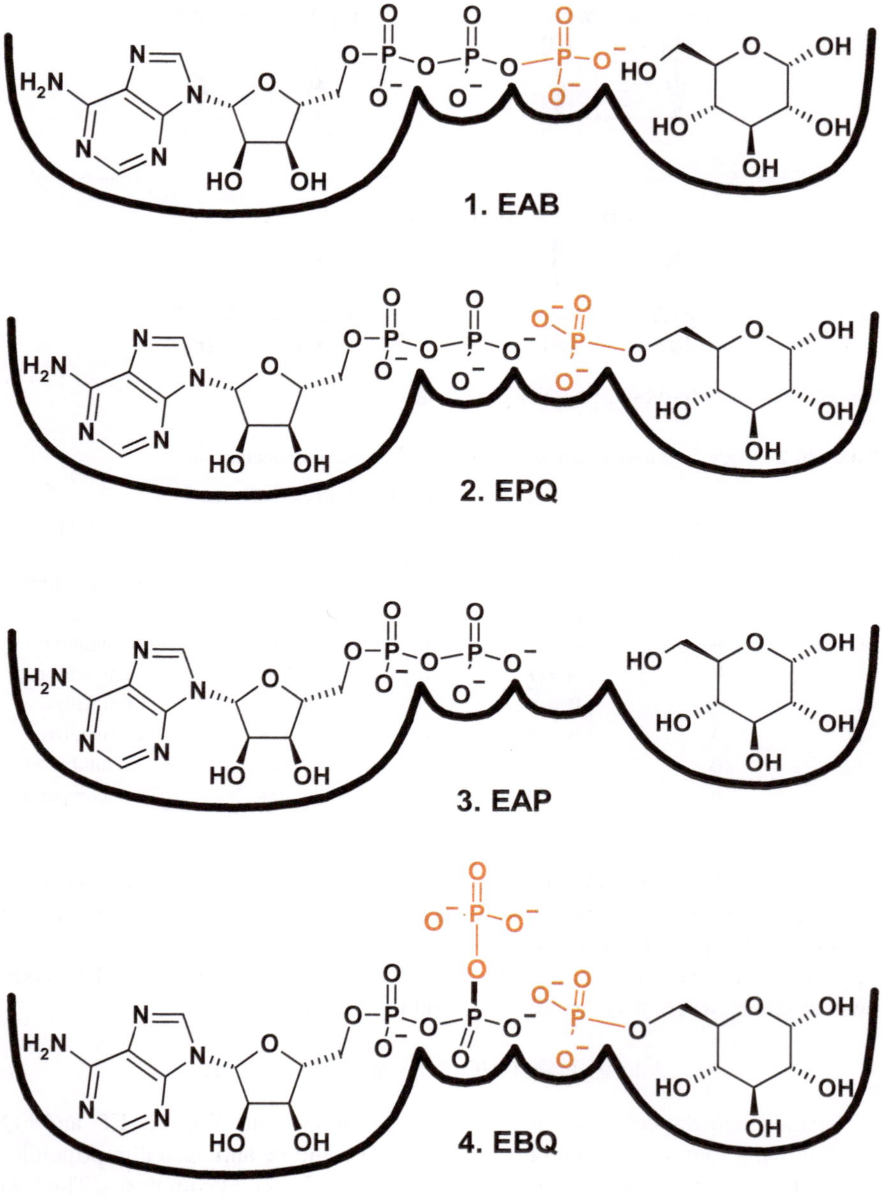

Fig. 26.3 Different ternary complexes possible with the enzyme in a kinase mechanism. The scheme for hexokinase with glucose is simplified by not showing Mg^{2+} in Mg-ATP. The four complexes are: (1) EAB (E.Mg-ATP.glucose), (2) EPQ (E.Mg-ADP.glucose 6-phosphate), (3) EAP (E.Mg-ADP.glucose), and (4) EBQ (E.Mg-ATP.glucose 6-phosphate)

adenosyl group. The above predictions (Tables 26.1 and 26.2) will obviously change if dead-end complexes other than EAP (like EBQ) are also formed.

26.1.3 Ping-Pong Mechanism

4-Aminobutyrate transaminase (GABA transaminase) is an example of ping-pong bi-substrate reaction. GABA and succinic semialdehyde form one substrate-product pair for partial reaction that defines the ***ping*** part and 2-oxoglutarate and L-glutamate form the other pair (defining the ***pong*** part). Schematically this may be shown as in Fig. 26.4.

The predicted slope/intercept effects and the product inhibition patterns are given in Table 26.3.

These product inhibition patterns are predicted for a ping-pong mechanism involving a single active site—both the half reactions occurring in the same site. However, with multi-site ping-pong mechanism, the product inhibition patterns are distinct. Transcarboxylase is an interesting example of a two-site ping-pong mechanism involving a biotin shuttle.

$$\text{Oxaloacetate} + \text{Propionyl CoA} \rightleftharpoons \text{Pyruvate} + \text{Methylamlonyl CoA}$$

Reactions on the transcarboxylase enzyme surface with two sites is shown in Fig. 26.5. Since the thioesters (propionyl CoA and methylmalonyl CoA) combine at

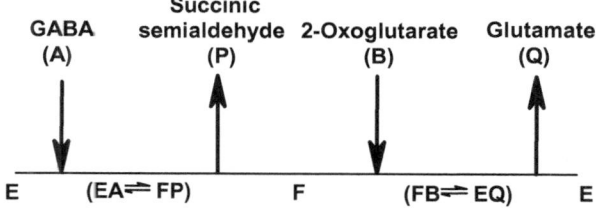

Fig. 26.4 Ping-pong bi-bi mechanism of GABA transaminase. The two enzyme forms E (pyridoxal phosphate-enzyme) and F (pyridoxamine phosphate-enzyme) are shown

Table 26.3 Product inhibition patterns for a ping-pong bi-reactant mechanism

Product inhibitor	Substrate varied	Fixed substrate	Enzyme parameter affected		Inhibition pattern
			Intercept $(1/V_{max})$	Slope (K_M/V_{max})	
P	A	B = ∞	No	No	No inhibition
P	A	B = K_B	Yes	Yes	Noncompetitive
P	B	A = ∞	No	Yes	Competitive
P	B	A = K_A	No	Yes	Competitive
Q	A	B = ∞	No	Yes	Competitive
Q	A	B = K_B	No	Yes	Competitive
Q	B	A = ∞	No	No	No inhibition
Q	B	A = K_A	Yes	Yes	Noncompetitive

Fig. 26.5 Schematic representation of the transcarboxylase reaction. Oxaloacetate (A) and pyruvate (P) interact with site S1 while the thioester pair (propionyl CoA and methylmalonyl CoA, shown as B and Q, respectively) interacts at the S2 site. The swinging arm with biotin (X) shuttles the carboxyl group from S1 to S2

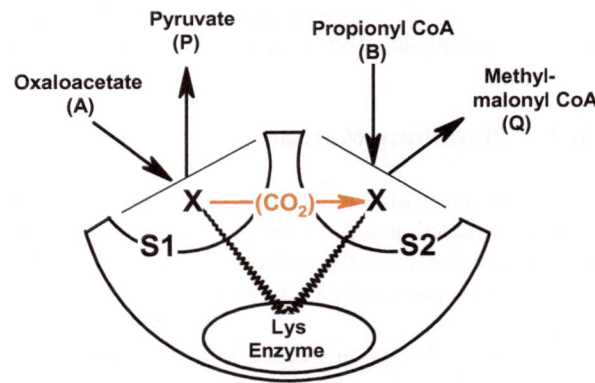

one site, they compete with each other while oxaloacetate and pyruvate compete for the other site. Because of this, the product inhibition patterns are reverse of the patterns for the classical ping-pong mechanism.

In multi-site ping pong mechanisms, the group to be transferred is moved from one site to the other. This movement may involve a swinging arm attached to the enzyme such as biotin (transcarboxylase), lipoic acid (pyruvate dehydrogenase complex), phospho-pantetheine (fatty acid synthase) or a protein channel connecting the two sites (glutamate synthase).

26.2   Assigning Kinetic Mechanisms: An Action Plan

Various lines of kinetic experimentation (presented in earlier sections) provide different bits of information on the overall enzyme mechanism. The tools range from steady state kinetics to isotopic analysis to tracking enzyme intermediates. Obviously, no single line of practical enquiry provides all the details needed to set down the complete mechanistic scheme. It is therefore essential to decide when to undertake which type of experiment(s)—studying a new enzyme requires an action plan. A logical sequence of collecting data is thus an important part of this experimental strategy. An organized decision tree capturing this mental process is outlined below (Fig. 26.6).

It is much more demanding to distinguish amongst the possible sequential mechanisms (Leskovac et al. 2004). Isotope exchange study is powerful enough to tell the subtle differences between various sequential schemes. A parallel initial velocity pattern quickly leads one to ping-pong mechanism but the caveat of "how parallel is parallel" requires considerable care. The apparently parallel initial velocity pattern for brain hexokinase (with glucose as substrate) became unambiguously intersecting with fructose as alternate substrate. Collecting confirmatory evidence adequately settles such issues.

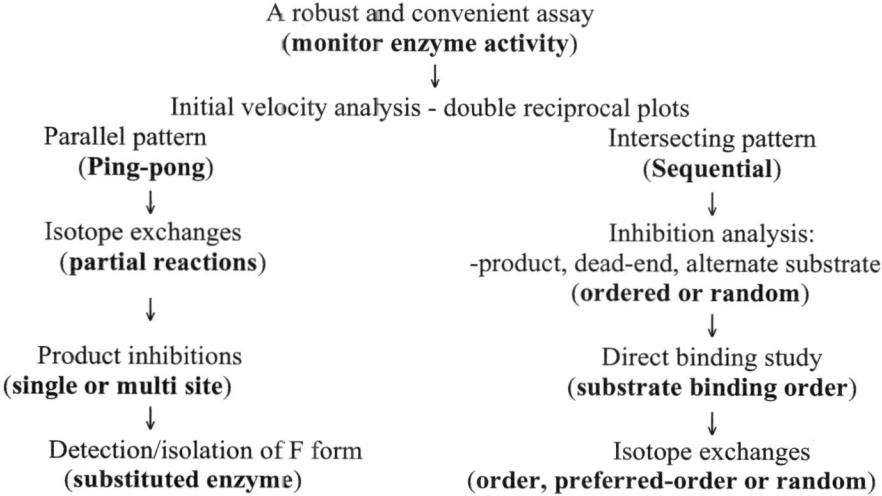

Fig. 26.6 Flow chart outlining the mental process leading to enzyme kinetic mechanism from experiments. Conclusions reached after each data generated are given in parenthesis in bold

26.3 Practical Relevance of Enzyme Kinetics

Apart from the intellectual satisfaction of having understood the inner workings of a remarkable catalyst, there are many direct benefits of a detailed kinetic analysis. This aspect of enzyme kinetic mechanism is often underappreciated by many. We will enumerate three important areas where the knowledge of enzyme kinetics directly becomes relevant.

26.3.1 Affinity Chromatography and Protein Purification

Moderate affinities encountered with substrates, products, or their analogs hinder their use as baits/ligands for affinity purification. The intrinsic ligand affinity may be further compromised upon chemical cross-linking for immobilization on the matrix. Transition state analogs offer useful affinity ligands; they often exhibit tighter binding and therefore, are promising candidates. They can also be used as eluents for conventional substrate affinity chromatography.

Creatine kinase was conveniently purified with one such strategy. The transition state of creatine kinase involves a flat phosphate intermediate (see Fig. 30.1 in Chap. 30). Nitrate ion (NO_3^-) mimics this flat phosphate moiety. This knowledge was used to purify creatine kinase by immobilizing it on ADP column. The enzyme binds tightly to this column in the presence of creatine and NO_3^-—a tight transition state complex is formed. The bound enzyme could be released selectively by adding free ADP to compete with the column bound ADP.

Secondly, prior knowledge of kinetic mechanism permits judicious development of binding/elution strategies in enzyme purification. Enzyme interaction with a substrate analog affinity matrix can be strengthened by suitably including another substrate in the binding buffer. This may be achieved on the matrix itself, either by exploiting a synergistic binding or through kinetically locking the ternary (EAB) complex. For instance, glutamate dehydrogenase binds poorly to NAD-Sepharose; inclusion of glutarate (substrate mimic and a dead-end inhibitor) in the developing buffer retards the enzyme significantly on this column. Bound enzyme protein may be released simply by excluding the second ligand (e.g., glutarate) from the elution medium. Enzyme mechanism and substrate analogs can be used to arrive at best conditions for retardation/binding/elution from substrate analog affinity matrices. Such kinetic lock-in strategies hold much promise in large scale enzyme purifications.

26.3.2 Dissection of Metabolism

Specific enzyme inhibition serves as a very valuable tool in teasing out metabolism. Competitive inhibition of succinate dehydrogenase by malonate is a classic example in the discovery of citric acid cycle. Attempts to define the glycolytic sequence and mitochondrial electron transport chain also made use of specific inhibitors. Inhibition of an enzyme in vivo may be seen as a metabolic cross-over at that step of metabolic pathway.

Starting from kinetic properties of individual enzymes it may be possible to reconstruct the characteristics of an entire pathway. Capturing metabolic complexity and structure through such a bottom-up approach is one of the objectives of "Systems Biology." Finally, the kinetic characterization of the key enzymes around a metabolic branch point provides some indications of relative in vivo flux to competing pathways. Everything else being equal, the enzyme with lower K_M for the common substrate (metabolite) dictates the pathway direction (refer to Chap. 39). For example, arginase and nitric oxide synthase compete for the cellular pool of arginine; the knowledge of their inhibition kinetics was exploited to design specific inhibitors and augment flux through nitric oxide synthase.

The disadvantage of a competitive inhibitor is that higher concentrations of the substrate can nullify its effect. Non-competitive inhibitors could be more effective as they cannot be overwhelmed by more substrate.

26.3.3 Enzyme Targeted Drugs in Medicine

Enzymes are catalysts that make and break specific covalent chemical bonds. They bind as well as catalyze—while other protein classes including receptors, transporters, antibodies and even DNA only bind, thereby offering unique features

Table 26.4 Examples of enzyme targeted therapy

Enzyme target	Drug example	Mode of inhibition	End use
Dihydropteroate synthase	Sulfanilamide	Competitive	Antibacterial
Dihydrofolate reductase	Methotrexate	Tight binding	Leukemia
Cyclooxygenase	Ibuprofen	Tight binding	Anti-inflammatory
HMG CoA reductase	Atorvastatin	Substrate analogue	Hypercholesteremia
Xanthine oxidase	Allopurinol	Mechanism based	Gout
α-Amylase	Acarbose	Transition state	Diabetes
Angiotensin converting enzyme	Captopril	Transition state	Hypertension
HIV retropepsin	Saquinavir	Transition state	AIDS
Alanine racemase	D-Cycloserine	Covalent PLP-adduct	Tuberculosis
Triacylglycerol lipase	Orlistat	Covalent adduct	Obesity
D-Ala-D-Ala carboxypeptidase	β-Lactams	Covalent adduct	Antibacterial

for drug design. Several enzyme targets have been exploited and the corresponding inhibitors are in the market as drugs (Alexander et al. 2017). Many of the drugs are natural compounds whose molecular basis of action was clarified *post facto*. With excellent understanding of enzyme action over time, rational drug design has arrived. Experimental access to transition state features of a target enzyme is an excellent opportunity to design potent drugs. We have already come across some cases of tight binding and suicide enzyme inhibitors in use (Chap. 19). Few more representative examples of successful drugs, their target enzyme, and the nature of inhibition are given in Table 26.4. Enzyme activators have not had much practical success so far. Activators of glucokinase may yet find applications in the future diabetes therapy.

Captopril was the first rationally designed enzyme-targeted drug. It may be viewed as a transition state inhibitor of angiotensin converting enzyme (ACE; peptidyl dipeptidase A). Similarly, HIV retropepsin inhibitors were developed through detailed kinetic knowledge and a variety of structure-assisted drug design techniques. Saquinavir contains a hydroxyethylamine isostere moiety and functions as a transition state analog. It is quite clear from the current state of art that rational drug design through a thorough understanding of enzyme kinetics, transition state, and active site structure is here to stay. It has overtaken the traditional serendipity-based screening approach in pharmaceutical industry (De Cesco et al. 2017).

Finally, enzyme targets are also valuable in other applications. Few organo-phosphorus compounds inhibiting acetyl cholinesterase are in the market as insecticides/pesticides. Three successful herbicides act via potent inhibition of their respective target plant enzymes. Phosphinothricin is activated on phosphorylation by glutamine synthetase to a tight binding inhibitor which mimics its transition state. N-Phosphonomethyl glycine (glyphosate) acts as an herbicide by preventing aromatic amino acid biosynthesis in plants (Boocock and Coggins 1983). It acts as an uncompetitive inhibitor and is thought to resemble the transition state of

5-enolpyruvylshikimate-3-phosphate synthase (EPSP synthase) belonging to the shikimate pathway. Branched chain amino acid biosynthesis is blocked by sulfonylureas. Sulfonylurea herbicides inhibit in a time-dependent manner, bear no resemblance to acetolactate synthase (the target enzyme in plants) substrates, and are non-competitive inhibitors. The first two of the above herbicides are interesting molecules in that they contain a direct C-P bind in them. In all the three cases, genes expressing inhibitor resistance have been isolated. These herbicides in conjunction with their resistance genes have found direct use in developing genetically modified crops.

References

Alexander SPH et al (2017) The concise guide to pharmacology 2017/18: enzymes. Br J Pharmacol 174:S272–S359
Boocock M, Coggins JR (1983) Kinetics of S-enolpyruvylshikimate-3-phosphate synthase inhibition by glyphosate. FEBS Lett 154:127–133
De Cesco S et al (2017) Covalent inhibitors design and discovery. Eur J Med Chem 138:96–114
Leskovac V et al (2004) A general method for the analysis of random bi-substrate enzyme mechanisms. J Ind Microbiol Biotechnol 31:155–160

Part IV

Chemical Mechanisms and Catalysis

Chemical Reactivity and Molecular Interactions

27

Every step of metabolism is in essence a chemical reaction. The vast majority of these reactions in an organism are catalyzed. Enzymes act as catalysts to facilitate the bond forming and/or breaking steps of these chemical transformations. Enzyme catalyzed or not, chemical reactions proceed mainly through the formation and cleavage of chemical bonds. In some cases, the catalyst itself participates covalently in the overall chemical scheme. A reaction is therefore best understood through an appreciation of chemical reactivity and how molecules interact with each other. The nature of chemical bonds and associated molecular interactions is reviewed in this chapter. A thorough understanding of these basic chemical mechanisms is essential to appreciate how enzymes facilitate chemical reactions.

27.1 Atoms, Molecules, and Chemical Bonding

All the elements in the periodic table (except for inert gases!) display various degrees of reactivity. Consequently, they occur in nature as compounds. The reactivity of an element is the reflection of its atomic structure and electronic configuration (Table 27.1). Hydrogen, carbon, nitrogen, oxygen, phosphorus, and sulfur dominate the reactions in biochemistry and hence, are the elements we most often encounter in enzyme chemistry. The valence electrons of these atoms participate in chemical reactions; they belong to the s and p orbitals of highest energy level. For instance, carbon has four electrons in its 2s and 2p orbitals—defining its valency as four.

A chemical bond may be defined as the force that holds the atoms together within a molecule. The valence electrons are available for the formation of covalent bonds. When two atoms with appropriate electronic configuration approach each other at close enough range they can enter into bond formation. Bonding between atoms leads to molecules.

© The Author(s), under exclusive license to Springer Nature Singapore Pte Ltd. 2025
N. S. Punekar, *ENZYMES: Catalysis, Kinetics and Mechanisms*,
https://doi.org/10.1007/978-981-97-8179-9_27

Table 27.1 Electronic structure of elements encountered in enzyme mechanisms

Element	Atomic number	Radius (Å)	Electronic configuration	Valency (most relevant)	Electro-negativity (Pauling scale)
H	1	1.2	$1s^1$	1	2.20
C	6	2.0	$1s^2,2s^22p^2$	4	2.55
N	7	1.5	$1s^2,2s^22p^3$	3	3.04
O	8	1.4	$1s^2,2s^22p^4$	2	3.44
P	15	1.9	$1s^2,2s^22p^6,3s^23p^3$	5	2.19
S	16	1.9	$1s^2,2s^22p^6,3s^23p^4$	2	2.58

27.1.1 Covalent Bonds

Among the different explanations of chemical bonding, Molecular Orbital Theory has found a wide acceptance. According to this theory, when two atoms approach each other for bonding, their respective valence atomic orbitals (one from each atom) can combine to form two molecular orbitals. The distribution of valence electrons among these molecular orbitals (the bonding and antibonding molecular orbitals) is called the electronic configuration of the molecule. Valence electrons from atoms entering into bond formation are distributed between bonding and antibonding (shown with an asterisk) orbitals—according to their order of energy levels.

$$\sigma 1s < \sigma^* 1s < \sigma 2s < \sigma^* 2s < \pi 2p, \text{etc.}$$

Since bonding orbitals occur at a lower potential energy, electron occupancy in them stabilizes the bond. Electrons occupying the antibonding orbitals destabilize the bond. As long as there are more valence electrons in the bonding orbital than in the antibonding orbital, the bond (and the molecule) is stable. We then define the bond order as follows:

Bond order
    = ½ (number of bonding electrons − number of antibonding electrons)

For instance, if the bonding molecular orbitals contain a pair of electrons more than the antibonding molecular orbitals, then the bond order is one. Integral bond order (**n**) values of 1, 2, or 3 correspond to single, double, or triple covalent bonds, respectively. A single bond thus means that the bonding atoms share a pair of electrons between them. Fractional bond orders are possible and may be encountered in a resonance stabilized molecule (see below).

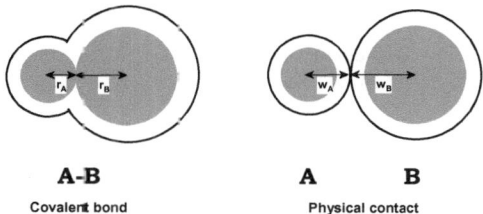

A-B   **A   B**

Covalent bond   Physical contact

Fig. 27.1   Covalent radius, van der Waals radius, and bond length. In a covalent diatomic molecule A–B, the bond length is the sum of the two covalent radii ($R = r_A + r_B$). This distance is shorter than the sum of their respective van der Waals radii ($= w_A + w_B$)

Table 27.2   Average covalent bond lengths relevant to enzyme chemistry

Bond type	Bond length (Å)	Bond type	Bond length (Å)
C–H	1.14	C–O	1.43
C–C	1.54	C=O	1.24
C=C	1.34	C–N	1.51
C≡C	1.20	C=N	1.32

Two different atoms can approach each other to the extent permitted by their van der Waals radii (Fig. 27.1). This considers the overall size of the two atoms including their valence shells. Any closer approach is sterically not feasible. However, when the two atoms enter into covalent bonding their valence atomic orbitals overlap—then the two atoms come closer than their van der Waals radii can allow. The equilibrium inter-nuclear separation distance of the covalently bonded atoms is called bond length. The bond length is the sum of the covalent radii of the two participating atoms (Fig. 27.1). The two atoms participating in a covalent bond cannot approach closer than the bond length due to repulsive forces between them. However, their separation by a distance longer than the bond length leads to weakening of the covalent bond—due to less-than-optimal overlap of their atomic orbitals.

An empirical relation (according to Linus Pauling) between bond order (**n**) and bond length (R) may be given as shown.

$$R_n = R_1 - 0.3 \times \log \mathbf{n}$$

With increasing bond order, the bond length decreases. Accordingly, a carbon–carbon double bond is shorter than the corresponding single bond. A few bond lengths relevant in biochemical reactions are listed in Table 27.2. Bond lengths in a molecule can be measured by spectroscopy, X-ray, and electron diffraction methods.

While covalent bonds are quite stable (bond enthalpy of about 90 kcal × mol^{-1}; Table 27.3, below), electrons from these bonds can get displaced—leading to reactivity of a particular molecule. The atoms with larger electro-negativity (Table 27.1) tend to pull more electron density toward them. A covalent bond may be affected and/or polarized due to different electro-negativities of the participating

Table 27.3 Strengths of
covalent and non-covalent
chemical bonds

Bond type	Bond strength[a] ($kcal \times mol^{-1}$)
van der Waals attraction	0.1 (per atom)
Hydrogen bond	1–3
Ionic	3–80
Covalent	90

[a]Bond strength is given as the energy required for breaking
it. Hydrogen bonds and ionic interactions are weakened in an aqueous environment as water competes in such interactions.
(kJ = 0.24 kcal)

atoms and their neighbors. These influences are variously ascribed to *inductive*
effect, *electromeric* effect, *hyper-conjugation*, and *resonance*. Some molecules
cannot be represented by a single covalent structure. Instead, they are better shown
as equivalent to a combination of two or more simple structures. Such resonance
structures tend to stabilize the overall molecular structure by delocalization of
electron density. Examples include the peptide bond, the aromatic benzene nucleus,
and the carboxylate group. One measurable outcome of resonance stabilization is the
unusual bond lengths (and bond orders). The peptide bond has a partial (40%)
double bond character. The C–N bond length of a peptide bond is shorter than a
typical C–N bond but longer than the C=N bond (Table 27.2). The carbon–carbon
bond length between the adjacent carbon atoms of benzene is 1.45 Å, which is
intermediate between the expected lengths for single and double bonds. Similarly,
the negative charge on the carboxylate group is equally shared between the two
oxygen atoms—both the oxygens and the C–O bonds are equivalent.

When two atoms sharing electrons in a covalent bond are of equal electro-
negativity, the resulting bond is nonpolar. A carbon–carbon bond is one such
example. However, when the two participating atoms are of different electro-
negativity, the covalent bond is polarized. Covalent bonds acquire degrees of polar
character depending on the electro-negativities of the bonding atoms. Proportionately more electron density resides with the more electronegative partner of the
covalent bond. For example, the carbon of a C–O covalent bond will carry a $\delta+$
charge while the oxygen will carry a corresponding $\delta-$ charge.

27.1.2 Directional Property of Covalent Bonds

Biochemical reactions often involve establishing or cleaving covalent bonds at the
carbon atom in a molecule. The electronic configuration of carbon atom either in its
ground state ($1s^2, 2s^2 2p^2$) or in the excited state ($1s^2, 2s^1 2p_x^1 2p_y^1 2p_z^1$) suggests that
the four valence electrons are not identical and hence, the four bonds should not be
equivalent. However, the four single bonds around a tetravalent carbon (such as in
methane) are equivalent. This is due to the hybridization of 2s and 2p orbitals of
carbon. The similar energies of the 2s and 2p orbitals can interact to form hybrid
orbitals of equivalent energy and shape. In fact, three types of hybrid orbitals
(Fig. 27.2) are possible with carbon: (a) Combination of one 2s orbital with three

Hybridization	Orbitals and Geometry	Bond angle	Covalent bonds	Shape and example
sp³ carbon; tetrahedral	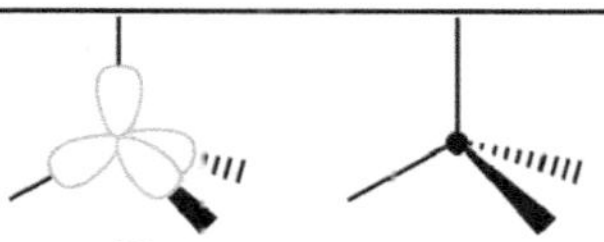	109.5°	4 σ bonds	Tetrahedral; Methane
sp² carbon; trigonal	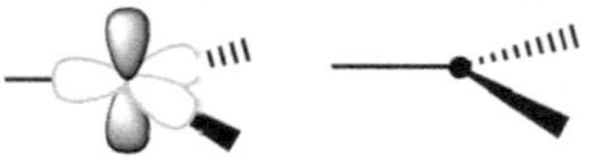	120.0°	3 σ bonds and 1 π bond	Planar; Ethylene
sp¹ carbon; linear	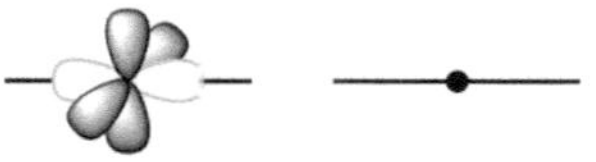	180.0°	2 σ bonds and 2 π bonds	Linear: Acetylene
sp³ nitrogen; tetrahedral	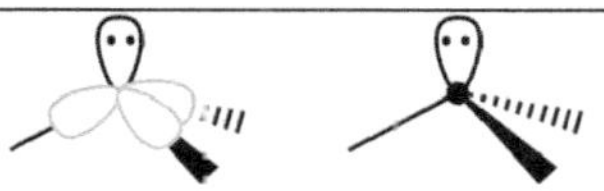	107.3°	3 σ bonds (1 lone pair)	Trigonal pyramid; Ammonia
sp³ oxygen; tetrahedral		104.5°	2 σ bonds (2 lone pair)	Bent; Water

Fig. 27.2 Types of hybridization leading to directionality of covalent bonds. A covalent σ bond is formed due to maximal overlap of participating orbitals. A lateral overlap of p orbitals (dumbbell shaped, shaded grey) result in a π bond. A double bond is made of one σ bond and one π bond while a triple bond consists of one σ bond and two π bonds

2p orbitals yields four sp³ orbitals. The four sp³ orbitals are equivalent and can form four σ bonds along the apices of a regular tetrahedron. In this tetrahedral geometry, the four bonds on the carbon are equally separated from each other. (b) It is possible that one s and two p orbitals mix to give three new hybrid orbitals. In this sp² hybridization, the new hybrid orbitals allow for three trigonal planar σ bonds to be formed. The two lobes of the remaining p orbital are perpendicular to this plane (Fig. 27.2) and are capable of a π bond through a lateral overlap. A double bond at carbon is thus in effect a combination of a σ bond and a π bond. (c) When a single p orbital combines with the 2s orbital, two equivalent sp orbitals result. With this linear hybridization, the remaining two p orbitals are placed perpendicular to each other. They can enter into π bonds through lateral overlaps. A triple bond at carbon is therefore a combination of one σ bond and two π bonds.

Different hybridization modes of s and p orbitals clearly accounts for the directional property of covalent bonds around carbon. From the second row of the periodic table, nitrogen and oxygen are the other two important elements in enzyme reactions. Both can exist in sp³ or sp² hybridization states (Fig. 27.2). The lone pair of nitrogen occupies one of the hybrid orbitals whereas oxygen has two such orbitals bearing a pair of electrons each. The bond angles in sp^3 hybridized nitrogen and oxygen are smaller than those in carbon due to lone pair repulsions.

27.1.3 Non-covalent Interactions and Intermolecular Forces

Apart from the strong covalent bonds discussed above, molecules can interact with each other through different non-covalent forces. They include van der Waals interactions, hydrogen bonding, and ionic interactions. These are generally weaker than the covalent bonds (Table 27.3). Both ionic and hydrogen bonds are further weakened in aqueous environments. These three weak attractive forces are very important in enzyme catalysis, particularly due to their readily reversible nature.

The fact that noble gases can be liquefied suggests that even they display molecular interactions. Weak van der Waals interactions do occur between all molecules. These are contributed by (a) dipole–dipole interactions, (b) dipole–induced dipole interactions, and (c) London dispersion forces. Although weak, many of them together can produce significant cooperative interactions. They are important in protein structure, hydrophobic recognition, and enzyme catalysis.

A hydrogen atom bonded to an electronegative atom (like N or O) can interact with another electronegative atom (like N or O) bearing a lone pair of electrons. This weak charge interaction is called a hydrogen bond (Herschlag and Pinney 2018). The atom to which the hydrogen is covalently bonded is referred to as the hydrogen bond donor and the other electronegative atom is called the hydrogen bond acceptor. In a hydrogen bond (shown as dotted line) of the type ">N–H·····O=C<", N is the donor and O is the acceptor. Typically, the length of a hydrogen bond is longer than the corresponding covalent bond. Strength of a hydrogen bond depends on distance, direction (angle), and the nature of the participating electronegative atoms. The more nonlinear a hydrogen bond (bend at the H atom!) is, the weaker it gets—bond directionality matters. While Cl and N are of comparable electro-negativity, chlorine cannot form a hydrogen bond due to its larger size. Hydrogen bonds are central to the structure and the catalytic apparatus of an enzyme molecule. A very strong hydrogen bond results when the H atom is equally shared and strongly bonded to both the donor and acceptor atoms. Such low barrier hydrogen bonds (LBHBs) do occur and form part of the catalytic strategy of many enzymes (Borshchevskiy et al. 2022) (also see Chap. 4).

That quintessential scientist, JBS Haldane once famously said "even the Pope is 70% water"! Water is the universal protic solvent and most life processes have evolved around the unique properties of water (Table 27.4). It is the smallest and most abundant molecule in a cell. But for its ordered hydrogen bonded network, water would not be in liquid form under ambient conditions. All biological (and biochemical) processes (and reactions) are either directly or indirectly under the influence of characteristic features of water (Chaplin 2001, 2006; Frenkel-Pinter et al. 2021; Nordén 2019). The biological catalysis has primarily/predominantly originated in an aqueous environment is consistent with this.

Table 27.4 Properties of water

Property	H_2O	D_2O
Molecular mass	18.015	20.028
Melting point	273.0 K	276.8 K
Boiling point	373.0 K	374.4 K
Density at 298.0 K	1.000	1.106
Maximum density at	277.0 K	284.2 K
Viscosity (centipoise)	0.89	1.11
Dielectric constant	78.39	78.06
O–H/O–D bond length	0.958 Å	0.918 Å
[H$^+$] measure (as $-\log$ [H$^+$])	pH	pD ($=$ pH $+$ 0.41)

Fig. 27.3 (a) Hydrogen bonding and the structure of water. Dashed lines represent hydrogen bonds and all the bond lengths are in Å. (b) Dipole moment of water molecule

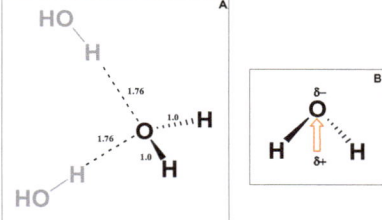

Hydrogen bonds play a major role in the structure and solvent properties of water. The unusually high boiling point, melting point, and density behavior are the manifestation of strong intermolecular hydrogen bonding in water (Fig. 27.3). Because of difference in the electronegativity of O and H, water also has a large dipole moment. A substitution of H by D (heavy isotope of hydrogen) significantly changes the solvent properties of water. Some of these differences are listed in Table 27.4. Most acids are 3–5 times weaker in D_2O than in H_2O. The shorter O–D bond (by about 0.04 Å or 0.004 nm) leads to changes in polarizability and solvent structure. Because of this, hydrogen bonds and hydrophobic interactions are also affected. Even proton transfer rates may be affected leading to what is known as solvent isotope effects.

Two oppositely charged atoms/groups attract each other—these electrostatic forces are called ionic interactions or salt bridges. The strength of this Coulombic attraction depends directly on the two charges involved (Q_1 and Q_2) and is inversely proportional to the square of the distance (r) between them.

$$F = \frac{Q_1 \times Q_2}{r^2 \times D},$$

where D is the dielectric constant.

Ionic interactions also depend on the dielectric constant (D) of the medium. They are nearly as strong as a covalent bond in vacuum but are greatly weakened by the aqueous environment (with typical bond energies in water of the order 3–5 kcal $\times$ mol^{-1}). Therefore, salt bridges are stronger in the hydrophobic interior of a protein than on the solvent exposed surface. Charged ligands often form salt bridges with their charged counterparts at the enzyme active site. For instance, negatively charged phosphate group is bound through positively charged active site guanidinium group (of arginine residue).

27.2   Chemical Reaction Mechanisms

Living beings depend on a myriad of chemical reactions—the sum total of metabolism. These chemical transformations are almost invariably brought about by enzyme catalysts. Many biochemical reactions may appear quite complex but are not. Their apparent complexities are largely due to the variety of reactant structures involved. However, at the mechanistic level, these reactions are simply the elementary reactions of organic chemistry. Relevant description of some of these reaction types and the principles of underlying chemical mechanisms will follow.

27.2.1 Cleaving and Forming Covalent Bonds

Enzymes utilize many of the same mechanisms that are well known to chemists—for the synthesis and degradation of organic compounds. There are two possible ways of forming and cleaving a carbon–carbon single bond. Upon *homolytic cleavage*, the participating carbon atoms depart with one electron each from the C–C bond (of the two electrons shared by them). This type of fragmentation results in two radicals (Fig. 27.4). Two carbon atoms (with one unpaired electron each) come together to establish a covalent bond—in the reverse of homolytic fission. Free radicals are

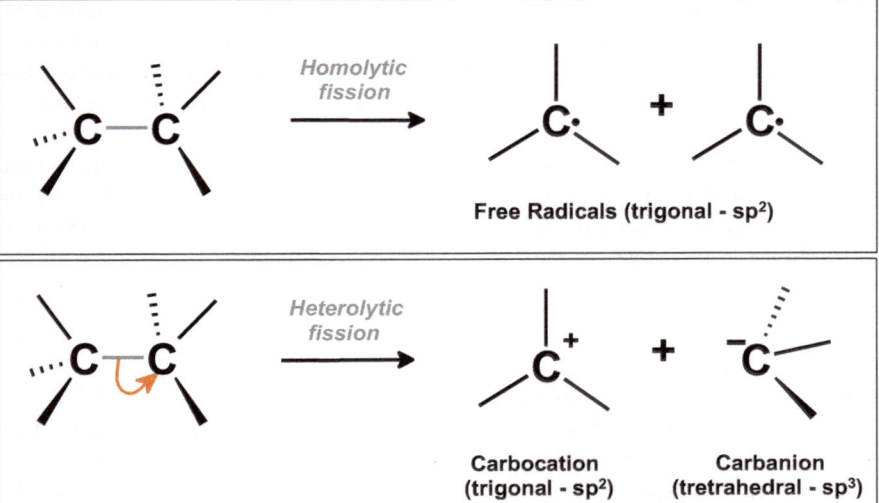

Fig. 27.4 Homolytic versus heterolytic fission of a carbon–carbon covalent bond

Fig. 27.5 Stabilization of carbocations and carbanions. The carbonyl group can act as a device to present carbocations (top) and carbanions (bottom) for reaction. The carbanion shown in effect is also an enolate

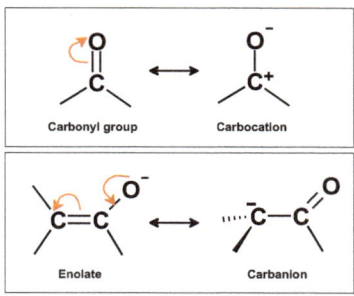

generally very reactive and unstable species. The larger the number of alkyl substituents on the carbon carrying the unpaired electron, more stable is that free radical (due to hyperconjugation). The carbon atom (bearing the lone electron in a free radical) is sp^2 hybridized (trigonal) with the third p orbital bearing the lone electron. Although not common, enzymes catalyzing reactions with free radical reaction intermediates are known. Ribonucleotide reductase, involved in the biosynthesis of deoxyribonucleotide precursors of DNA, is an important example of this type.

The second mode of C–C bond cleavage is the heterolytic cleavage (Fig. 27.4). Fragmentation by *heterolytic cleavage* generates a pair of oppositely charged ionic species. The species carrying the positively charged carbon atom is *carbocation*. The other product of heterolytic cleavage is the *carbanion*, with a negatively charged carbon atom. The combination of a carbocation with a carbanion leads to the formation of a covalent C–C bond—which is the exact reverse of a heterolytic cleavage. Both forms of the charged carbon species, namely, carbocations and carbanions, are extremely unstable. The carbon bearing the positive charge in the carbocation is sp^2 hybridized (trigonal) with an empty unhybridized p orbital. More the number of alkyl substituents on the carbon bearing the positive charge, more stable is the carbocation. The order of stability for carbanions is the exact reverse of carbocations. Furthermore, carbanion carbon is sp^3 hybridized with the lone pair occupying one of the vertices of a tetrahedron.

Enzymes make use of the ionic mechanism in most C–C bond cleavage and formation events. Since both carbocations and carbanions are unstable, how are they generated and stabilized during reaction? A carbocation is generated/stabilized in the context of the overall molecular structure. The positive charge may be distributed and stabilized through resonance. For instance, the carbonyl carbon ($>C=O$) can be a suitable carbocation for reaction. It has a large dipole moment with significant negative charge on its oxygen atom while the carbon atom bears an equal amount of positive charge. In effect the carbonyl group is a resonance hybrid of charged and uncharged structures, imparting a carbocation character to the carbon atom (Fig. 27.5). Similarly, any molecular arrangement that stabilizes the negative charge on carbon, in principle supplies a carbanion. Quite often the carbanion is attached to a functional group that allows the negative charge to be delocalized by resonance. Distribution of negative charge over several atoms, in addition to the carbon atom in question, stabilizes the carbanion. An adjacent β-carbonyl group (and not an α-carbonyl group!) often provides such an apparatus (Fig. 27.5). Due to such resonance stabilization, the reactivity of the bond β to the carbonyl becomes ~10^{33} times greater than a typical hydrocarbon bond. This unique reactivity underlies every C–C bond-breaking reaction of central metabolism (Rabinowitz and Vastag 2012). Resonance delocalization of negative charge allows the carbon atom to react as a carbanion—without letting full formal negative charge to develop on that carbon. Chemistry involving enolates (such as with pyruvate) are excellent examples of this concept.

The unequal heterolytic cleavage of a covalent bond brings us to yet another important definition of electron-rich and electron deficient groups/centers. A ***nucleophile*** (literally *nucleus lover*) is a species that forms a covalent bond to its reaction partner (the ***electrophile***—*electron lover*) by donating both bonding electrons. All molecules or ions with a free pair of electrons can act as nucleophiles. Because nucleophiles donate electrons, they are by definition ***Lewis bases*** (see Chap. 28). A nucleophile is thus an electron-rich chemical reactant that is attracted by electron-deficient compounds. Examples of nucleophiles are anions such as COO^-, or a compound with a lone pair of electrons such as $-NH_2$ (amine), NH_3 (ammonia), or H_2O (water). In the same sense, carbanions are nucleophiles and carbocations are electrophiles. The chemistry of formation of the C–C bond thus involves the nucleophilic attack by carbanion on to a carbocation. The carbonyl group in a molecule, for instance, often sets up the electrophilic (carbocation) or nucleophilic (carbanion) center required for many enzyme chemistries. A detailed discussion on nucleophiles, nucleophilic chemistry, and its role in catalysis is available in a subsequent chapter (Chap. 29).

27.2.2 Logic of Pushing Electrons and Moving Bonds

To appreciate how enzymes work, familiarity with two languages is necessary. One is the kinetic/thermodynamic description of reactions—that formed the significant thrust of earlier two sections (both Part II and Part III). Second is the description of the reaction mechanism—which involves moving electrons and making/breaking of bonds. The first is physical and the second is chemical (largely organic chemistry!). Most organic reactions are polar in nature, where nucleophile and electrophiles participate. Electron flow is the key to molecular reactivity—nucleophiles donating electrons to the electrophiles. Curly arrows (shown in color in each reaction) are used to describe reaction mechanisms. A ***curly arrow represents the movement of a pair of electrons*** with the result that a bond is formed between a nucleophile and an electrophile. A simple example would be the association/dissociation equilibrium of a Bronsted acid. The lone pair electrons on the carboxylate oxygen of acetate are transferred to empty 1s orbital of H^+ (Fig. 27.6). Exact opposite of this electron pair movement occurs during dissociation of the acid—which is a bond cleaving event.

Charge is conserved in each step of a reaction—***net charge cannot be created or destroyed***. If we start with neutral molecules (like acetic acid) and make a cation (like

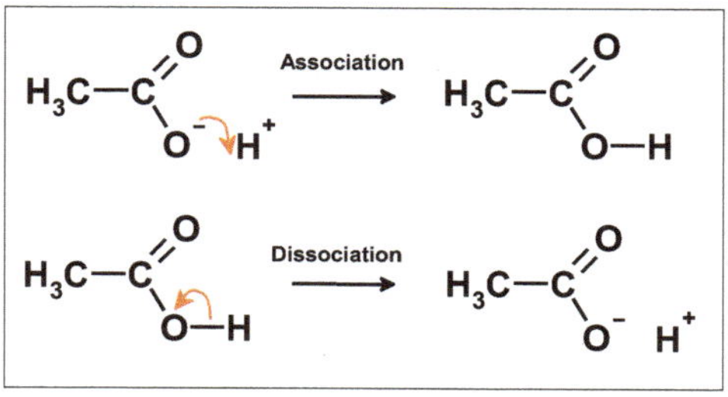

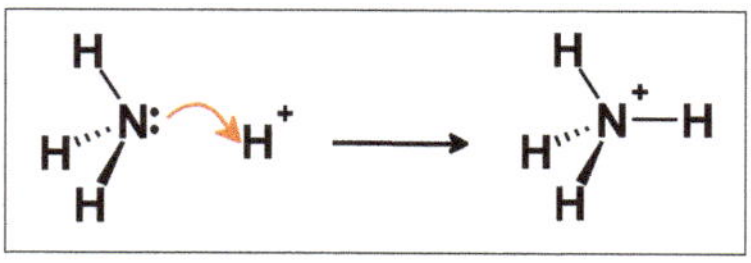

Fig. 27.6 Curly arrows are used to show the movement of electron pair. Ionization equilibrium of acetic acid (top) and protonation of ammonia (bottom) are shown as examples

H^+), we must make a corresponding anion (like acetate). Protonation of ammonia is yet another example of this conservation of charge concept (Fig. 27.6). There is one net positive charge on either side (H^+ the reactant and NH_4^+ the product). Curly arrows are also used to show movement of electrons within molecules. Enolization is an example where two lone pair shifts (two curly arrows; Fig. 27.5) occur.

Curved arrows are vital to understand reaction mechanisms. Curly arrows are important conceptual representation of electron pair movements in a chemical mechanism. Such arrows always represent the *movement of electrons and not atoms*. The tail of the arrow shows the source of electron pair—usually from a lone pair, a π bond or a σ bond. The arrowhead indicates the ultimate destination of the electron pair—often to an electronegative atom that can harbor a negative charge. Oxygen of a carbonyl group is one such atom (Fig. 27.5).

A tricky question is to know where to begin the first arrow. It is usual to start pushing electrons (first curly arrow!) from the nucleophile, anion, or a lone pair. But some mechanisms are better understood as electron pulling—generally by a reagent (electrophile) such as a cation, an acid, or a Lewis acid. All the curly arrows in a given mechanism move in the same direction. However, we may (a) either draw the first arrow from a nucleophile (electron pushing) or (b) end the first arrow into an electrophile (electron pulling).

The conventions and concepts used to write proper chemical mechanisms are summarized in the box below.

Guidelines to a Chemical Mechanism
In writing and understanding chemical reaction mechanisms the following broad guidelines operate:

(continued)

- Identify the nucleophilic and electrophilic atoms taking part in the reaction.
- Decide whether the mechanism involves electron pushing (arrow to begin from the nucleophile) or electron pulling (beginning at the electrophile).
- Mark the lone pair of electrons on the nucleophilic atom.
- Draw curly arrow(s) to show the flow of electrons from an electron-rich center to an electron-deficient center.
- Multiple curly arrows always move in the same direction—they never meet head on or end up in a single atom.
- C, N, and O atoms can have a maximum of eight electrons in the outer valence shell ($2s, 2p_x, 2p_y, 2p_z$) while H has two ($1s$). Carbon has a valency of 4.
- If you make a new bond to uncharged H, C, N, or O you must break one of the existing bonds in the same step.
- Define the charges clearly and ensure that overall charge is conserved (before and after) in the mechanism.

Drawing curly arrows and writing reaction mechanisms is like learning swimming. Once you have mastered the skill (of course with some practice!), it is difficult to forget or make mistakes.

27.3   Stereochemical Course of Reaction

Majority of biomolecules are chiral compounds. The presence of a carbon atom bonded to four different groups (***abcd***) leads to chirality (asymmetry) and such atoms are termed ***stereo-centers or chiral centers***. Consider alanine for example. It is chiral because the α-carbon is bonded to four different groups, namely, –COOH, –NH$_2$, –CH$_3$, and –H. There can be two distinct three-dimensional arrangements (***configuration***) of the four groups around the α-carbon. The two nonsuperimposable mirror-image forms are called enantiomers—the L- and D-forms of alanine. In modern nomenclature, the two are designated as 2S-alanine and 2R-alanine, respectively. The L- and D-forms of alanine are identical in their physical properties except for their interaction with *plane-polarized* light. Hence, such structural isomers are also commonly called ***optical isomers***. It is possible that a molecule may have more than one chiral carbon in it. For instance, threonine has two asymmetric centers—one at the α-carbon and the other at β-carbon.

Glycine, unlike alanine, does not have a chiral α-carbon and is optically inactive. Apart from –COOH and –NH$_2$ groups, the other two substituents on its α-carbon are hydrogen atoms. However, these two –H atoms are stereochemically distinct. Carbon centers that are surrounded by ***aabc*** groups are known as ***prochiral*** centers. For instance, the α-carbon of glycine or the C-1 of ethanol is prochiral. In general, a molecule is said to be prochiral if it can be converted from achiral (such as ***aabc***) to

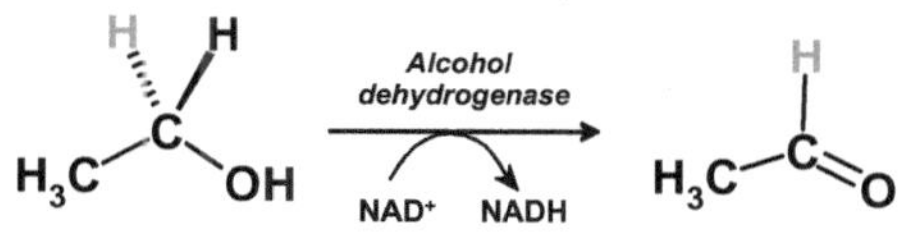

Fig. 27.7 Stereochemistry of alcohol dehydrogenase reaction. Only the *proR*-hydrogen of ethanol is selectively removed by the enzyme. The C-1 of ethanol is an example of prochiral carbon with ***aabc*** arrangement of groups around it. The *proR*-hydrogen is shown in black, and *proS*-hydrogen is in gray

chiral (***abcd***) in a single chemical step. Numerous biological reactions involve prochiral compounds. Nevertheless, the two identical substituents are selectively recognized by enzymes (Ogston hypothesis!). The two H atoms attached to C-1 of ethanol are distinguished by alcohol dehydrogenase. In the chiral active site environment, ethanol is held in a fixed orientation—so that only *proR*-hydrogen is removed (Fig. 27.7). Other examples of such chiral discrimination at a prochiral center include aconitase (acting on citrate) and fumarase (acting on fumarate to form L-malate). Such discrimination is possible because by themselves enzymes are chiral catalysts. Enzyme active sites almost always provide a chiral environment and act in a stereo-specific manner. When this is not the case, then a non-enzymatic step may be involved.

The stereochemical course of an enzyme reaction may be determined by suitable isotopic labeling of substrate. The fate of this label is subsequently monitored to deduce the reaction path. For instance, the deuterium label in *proR* position of ethanol alone is removed by alcohol dehydrogenase—*proS* hydrogen is retained in the product acetaldehyde (Fig. 27.7). In principle, one can construct chiral methyl groups with three different isotopes (H, D, and T) of hydrogen. Similar stereochemical strategy is used to analyze the stereochemistry of phosphoryl transfer reactions. Notably, phosphates contain three apparently identical oxygen substituents, and they can be labeled with ^{16}O, ^{17}O, and ^{13}O, the three isotopes of oxygen (we will have more on phosphate chemistry in a subsequent chapter).

It is important to know the exact chirality relationship between the reactant and the product. The absolute configuration of a compound (e.g., the reaction product) can be assigned by (a) converting it chemically into a compound of known stereochemistry and then measuring its optical activity, (b) complexing it with a known chiral reagent and then determining its relative configuration by NMR spectroscopy or X-ray crystallography, and (c) using an enzyme of known chiral specificity. For a more detailed treatment on stereochemistry, the reader is encouraged to consult elementary texts on organic chemistry.

27.4   Common Organic Reaction Types

Enzyme reactions involve breaking/making covalent bonds in their substrates. Such events involving covalent bonds are greatly influenced by the surrounding chemical environment and functional groups. A *functional group* is a group of atoms within a molecule that has a characteristic chemical behavior, such as a carboxylate (–COOH) group in acetic acid or a thiol (–SH) in cysteine. A functional group that makes off with a pair of electrons (of the σ bond that is broken) is called an outgoing nucleophile or a *leaving group*. In a given reaction, we may come across an incoming nucleophile (^{in}Nu:) that replaces an outgoing nucleophile (^{out}Nu:). Although a vast variety exists, we will restrict ourselves to more common organic reaction types encountered in enzyme reactions here.

27.4.1 Nucleophilic Displacements

In a nucleophilic substitution (S_N type) reaction, one nucleophile (the leaving group) is substituted by another on a saturated sp^3-hybridized carbon atom. In an **S_N1** (Substitution, Nucleophilic, **1**st order) reaction, the substrate undergoes a spontaneous dissociation to generate a carbocation intermediate. After this rate-determining first-order event, the carbocation reacts with the substituting nucleophile to form the product. Thus, S_N1 reactions occur in two steps and usually take place with tertiary or allylic carbon of the substrate. Conversion of geranyl diphosphate to geraniol is an example of S_N1 reaction (Fig. 27.8).

An **S_N2** (Substitution, Nucleophilic, **2**nd order) reaction takes place in a single step—where the incoming nucleophile attacks the electrophilic carbon with its electron pair. Because the incoming and outgoing nucleophiles are on opposite sides (180° apart), the stereochemistry at the reacting center is inverted during an **S_N2** reaction. Methylation of norepinephrine to epinephrine by *S*-adenosylmethionine is one such reaction. The amine nitrogen of norepinephrine (incoming nucleophile) displaces *S*-adenosylhomocysteine (outgoing nucleophile or the leaving group) on the electrophilic methyl carbon atom (Fig. 27.8). Other important examples of nucleophilic displacement include acyl, phosphoryl, and glycosyl transfer reactions. We will revisit few of them as specific cases in the later chapters.

27.4.2 Elimination Reactions

Reactions involving elimination of HX to yield an alkene are important in many biochemical pathways. Particularly common are dehydration (removal of H_2O) and deamination (removal of NH_3) reactions. An *elimination* reaction in the reverse direction simply describes an *addition* reaction. Addition and elimination reactions generally occur adjacent (at α, β position) to a carbonyl group. The substrates are normally thioesters, carboxylic acids, ketones, or aldehydes. Mechanistically

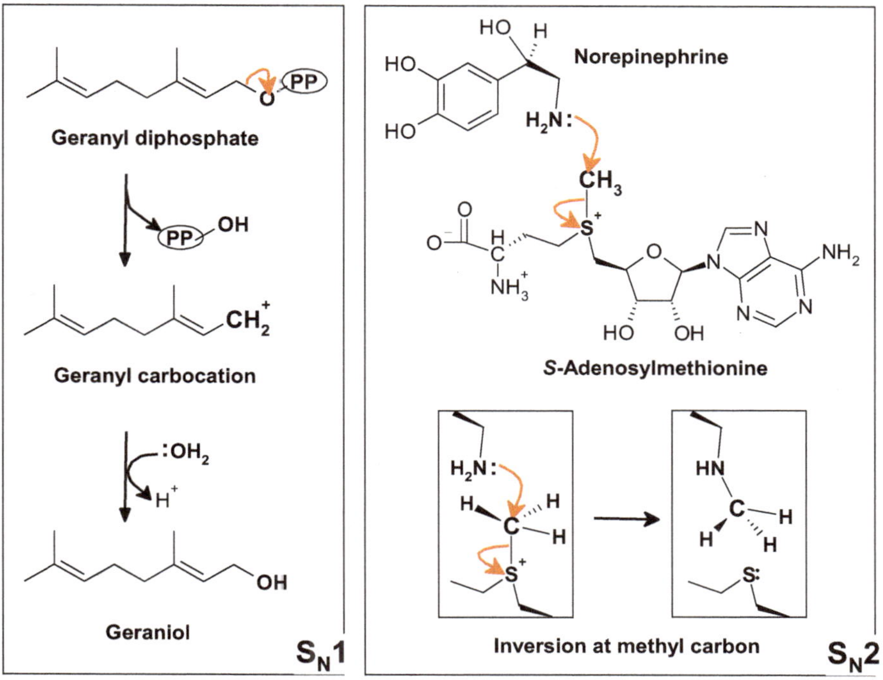

Fig. 27.8 Nucleophilic substitution reaction mechanisms. Examples of S_N1 (geraniol synthase; left panel) and S_N2 (phenylethanolamine N-methyltransferase; right panel) type reactions are shown

elimination reactions are more complex and may be classified into E1, E2, or E1cB reaction types. The elimination of HX involves the cleavage of a C–H bond and the cleavage of C–X bond. The timing of C–H and C–X bond cleavages in the reactant determines the type of elimination reaction (Fig. 27.9). In **E1** (Elimination **first** order) mechanism, first the C–X bond breaks to generate a carbocation. This carbocation then undergoes base abstraction of H⁺ (C–H bond break) to form the double bond. The **E1** reactions, found in organic chemistry, are rarely encountered in enzymology. In the **E1cB** (Elimination 1st order, conjugate Base assisted) reaction, the C–H bond cleavage occurs first to give a carbanion intermediate. This carbanion in a subsequent step loses X⁻ to give the alkene. Eliminations with **E1cB** mechanism are quite common in biological chemistry—the carbanion intermediate being stabilized by the functional residues at the enzyme active site. A single-step **E2** (Elimination 2nd order) mechanism is followed when C–X and C–H bond cleavages are concerted, i.e., occur simultaneously (Fig. 27.9). When **E2** mechanism operates, the H and X are eliminated from opposite faces of the molecule (*anti* elimination). With **E1cB** reaction, however, reaction stereochemistry may be *anti* or *syn* (same side)—depending on the active site geometry.

A well-documented case of **E1cB** mechanism is aspartase reaction (Fig. 27.9). Aspartases from different organisms show high sequence homology, and this

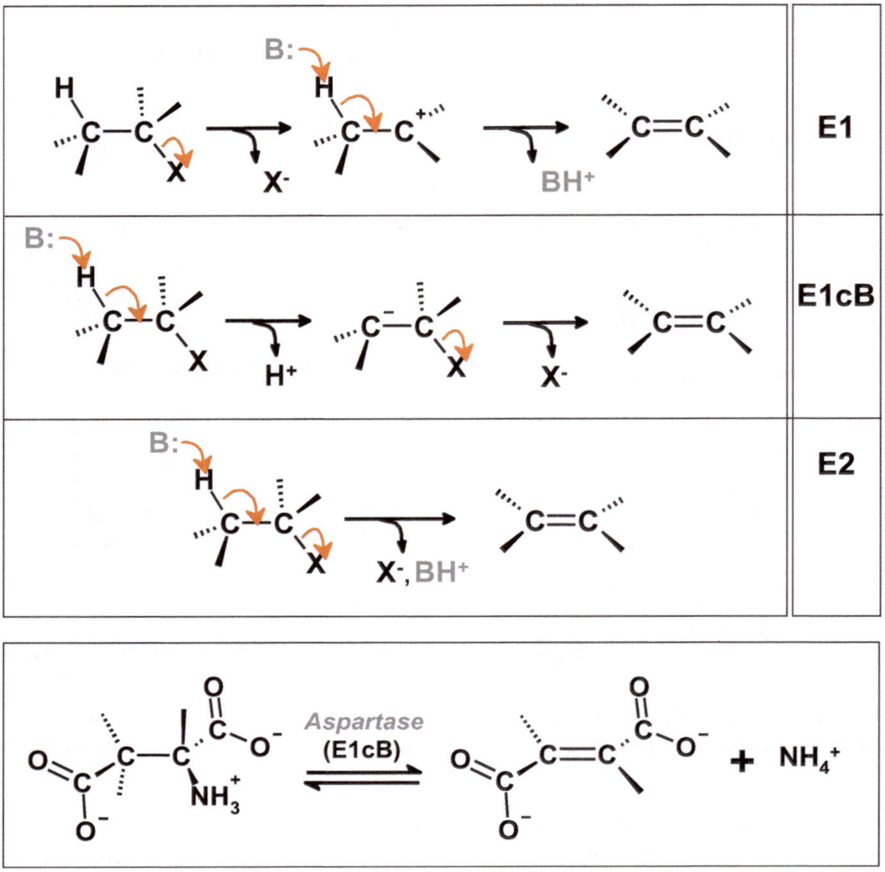

Fig. 27.9 Three possible elimination reaction mechanisms. While the **E1** reaction goes through a carbocation intermediate, the **E1cB** mechanism involves a carbanion intermediate. In the **E2** mechanism, breaking of C–H and C–X bonds is simultaneous. Aspartase reaction (shown below) follows **E1cB** mechanism where the C–N bond breaks to eliminate ammonium

homology extends to functionally related enzymes such as the class II fumarases, the argininosuccinate lyase, and adenylosuccinate lyase. Other examples include dehydratases like 3-dehydroquinate dehydratase (shikimate pathway) and β-hydroxyacyl ACP dehydratase (fatty acid biosynthesis). Since addition/elimination reactions are reversible, a dehydratase performs hydration in the opposite direction. For instance, *trans*-2-enoyl CoA hydratase adds water across a double bond during fatty acid oxidation.

The dehydration of a β-hydroxycarboxylic acid by aconitase (citrate → *cis*-aconitate → isocitrate) is likely an example of **E2** elimination (Fig. 27.9).

27.4.3 Carbon–Carbon Bond Formation

Reactions that lead to formation or cleavage of C–C bonds are central to metabolic logic. They are the key steps in building (synthesis) and degradation (catabolism) of diverse cellular metabolites. Some of the most fundamental reactions of this type involve carbon dioxide as substrate (carboxylation) or product (decarboxylation). These reactions will be discussed a little later (Chap. 32). A few less widely distributed C–C bond formation reactions include steps in terpene synthesis (via carbocation intermediate) and lignin biosynthesis (via free radical intermediate). Besides carboxylation/decarboxylation, two other C–C bond forming reactions are of great importance. Significantly both are carbonyl condensation reactions. A carbonyl condensation results in bond formation between the carbonyl carbon of one partner and the α carbon of the other carbonyl partner. This reactivity is because the α hydrogen of a carbonyl compound is weakly acidic and susceptible to base capture. The enolate so formed (in its carbanion form; Fig. 27.5) functions as a nucleophile.

The *aldol reaction* is the condensation of two carbonyl compounds (aldehyde or ketone) via an enolate intermediate. It yields a β-hydroxy carbonyl compound from two molecules of aldehyde or ketone. This is an example of nucleophilic addition reaction. One molecule reacts with base to generate a nucleophilic enolate, which then attacks the carbonyl carbon of the second molecule (Fig. 27.10). Reverse of aldol condensation is also possible and constitutes a C–C bond cleavage reaction. The enzymes which catalyze aldol reactions are known as aldolases. Fructose-1,6-bisphosphate aldolase reaction from glycolysis is a prototype. Other examples include citrate synthase (aldol condensation) and ATP citrate lyase (aldol cleavage). A few amino acids can also undergo aldol-type cleavage. The carbanion formed from such C–C bond breaks is stabilized through a cofactor like pyridoxal phosphate (e.g., serine hydroxymethyltransferase).

Carboxylic esters can react to form β-keto-esters, via an ester enolate intermediate. Such reactions are known as *Claisen condensation* reactions. Claisen ester condensation is more difficult than an aldol reaction because the C–H bond adjacent to an ester is significantly less acidic than the proton next to a ketone. The carbonyl group of a thioester is more ketone-like and better suited for Claisen condensation. Therefore, we often encounter thioesters of coenzyme A (CoA) in biological reactions. In principle, one molecule of ester reacts with base to give a nucleophilic enolate ion. This enolate adds to the second molecule in a nucleophilic acyl substitution reaction (Fig. 27.10). The initial alkoxide expels the leaving group (thiolate is a better leaving group—hence thioesters!) to regenerate a carbonyl group and form the β-keto ester product. There are many examples of Claisen reactions involving acetyl CoA in biological systems (Zhou et al. 2020). These include condensation reactions in the biosynthesis and assembly of fatty acids, polyketides, and steroids. Two acetyl CoA units condense to form 3-ketobutyryl CoA, which in turn condenses with another molecule of acetyl CoA to give hydroxymethylglutaryl CoA (HMG CoA—onward to cholesterol biosynthesis).

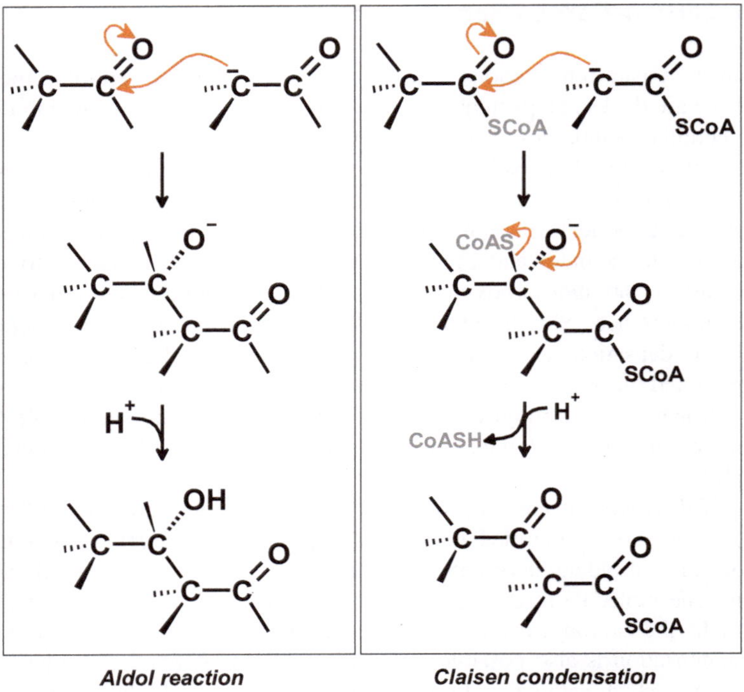

Aldol reaction   **Claisen condensation**

Fig. 27.10 Biochemically important carbon–carbon bond forming reactions. Aldol reaction (left panel) yields a β-hydroxy carbonyl compound from two molecules of aldehyde or ketone. The key step is the nucleophilic addition of enolate (carbanion) to the other >C=O group. In Claisen condensation (right panel), two molecules of an ester combine to yield a β-keto-ester. The key event here is the nucleophilic acyl substitution by the enolate (carbanion). Example of a thioester (acetyl CoA) condensation is shown

Claisen condensation reaction is reversible; a β-keto ester can be cleaved by a suitable base to yield two ester molecules. It is worth noting that an essential reversal of Claisen condensation reaction occurs in fatty acid catabolism. For example, thiolytic cleavage of β-ketoacyl CoA each time releases one molecule of acetyl CoA.

27.5   Summing Up

The variety of chemical reactions catalyzed by enzymes is vast. Indeed, some of the chemistry—like the formation of a C–P bond—was novel even to organic chemists. The nature of the reaction catalyzed by enzymes forms the basis of EC classification. The six general reaction categories include oxidation-reduction reactions, group transfers, hydrolysis, isomerizations, and synthetic steps. At the mechanistic level, nucleophiles, nucleophilic attack, and general acid-base catalyzed proton transfers permeate most of the bioorganic chemistry and enzymology. A basic understanding of their reactivity is essential to appreciate enzyme chemical mechanisms. These aspects will be elaborated in the two subsequent chapters.

References

Borshchevskiy V et al (2022) True-atomic-resolution insights into the structure and functional role of linear chains and low-barrier hydrogen bonds in proteins. Nat Struct Mol Biol 29:440–450

Chaplin M (2006) Do we underestimate the importance of water in cell biology? Nat Rev Mol Cell Biol 7:861–866

Chaplin MF (2001) Water: its importance to life. Biochem Mol Biol Educ 29:54–59

Frenkel-Pinter M et al (2021) Water and life: the medium is the message. J Mol Evol 89:2–11

Herschlag D, Pinney MM (2018) Hydrogen bonds: simple after all? Biochemistry 57:3338–3352

Nordén B (2019) Role of water for life. Mol Front J 3:1–17

Rabinowitz JD, Vastag L (2012) Teaching the design principles of metabolism. Nat Chem Biol 8: 497–501

Zhou S et al (2020) Coenzyme A thioester-mediated carbon chain elongation as a paintbrush to draw colorful chemical compounds. Biotechnol Adv 43:107575

Further Reading

Clayden J, Greeves N, Warren S (2014) Organic chemistry, 2nd edn. Oxford University Press, Oxford

McMurry JE, Begley TP (2015) The organic chemistry of biological pathways, 2nd edn. WH Freeman, New York

Acid–Base Chemistry and Catalysis

28

Acids and bases are enormously important in enzyme chemistry. A thorough knowledge of acid–base chemistry is crucial to understand reaction mechanisms and catalysis. According to the Bronsted definition of acids and bases, any substance (or a functional group) that has a tendency to lose a proton is an acid. Correspondingly, a base will then be a proton (H^+) acceptor. This definition, as we will see below, eminently suits our understanding of the role of acid–base groups at the enzyme active site. However, Lewis provided a broader definition of acids and bases. Accordingly, a Lewis acid is a substance that accepts an electron pair from a base while a Lewis base is a substance that donates an electron pair to an acid. Lewis acids are involved in enzyme-catalyzed processes as cofactors. Metal cations such as Mg^{2+}, Mn^{2+}, Zn^{2+}, and iron–sulfur clusters are Lewis acids. One way of visualizing this acidity is to consider water molecule coordinated to Zn^{2+}, for instance. Because a lone pair of oxygen is donated to Zn^{2+} (the Lewis acid), the O–H bond of bound water is better polarized and therefore more readily loses a proton (i.e., H^+). While the definition of base is practically same in the two definitions, the concept of Lewis acids is much broader and goes beyond just the H^+ donors.

28.1 Acids and Bases

28.1.1 Bronsted Acid/Base and Context of Water

The Bronsted–Lowry concept of acids and bases best describes the proton transfers during enzyme catalysis—reactions occurring in water. Every Bronsted acid has a conjugate base associated with it—while every base has a conjugate acid form. Consider the following proton transfer reaction:

N. S. Punekar, *ENZYMES: Catalysis, Kinetics and Mechanisms*,
https://doi.org/10.1007/978-981-97-8179-9_28

$$AH + B \rightarrow A^- + BH^+$$

The acid AH upon losing H^+ ion becomes its conjugate base A^-. Similarly, the base B by accepting a proton becomes its conjugate acid BH^+. Being the universal protic solvent, water can behave as an acid or as a base. It can accept a proton from an acid or donate a proton to a base. Therefore, it is no surprise that naked H^+ species never occur in aqueous solutions. They invariably are found as solvated hydronium ions (H_3O^+; and we will always mean H_3O^+ when we talk of protons in water). The equilibrium extent of these reactions depends on the relative strengths of the acids and bases involved.

$$AH + H_2O \rightleftharpoons A^- + H_3O^+$$

$$B + H_2O \rightleftharpoons BH^+ + OH^-$$

With a strong acid, water acts as a base and becomes protonated to H_3O^+. A base on the other hand would deprotonate water to give hydroxide ion (OH^-)—here, water is acting as an acid. Such compounds that act either as an acid or a base are called *amphoteric*.

Acids differ in their ability to donate protons. They can be simply compared by measuring the ease with which they transfer a proton to the solvent—water. The strength of an acid in an aqueous solution is expressed by its acid dissociation constant K_a, as defined in the box below. The stronger the acid larger is its K_a.

Acid Dissociation Constant

$$K_a = \frac{[H_3O^+] \times [A^-]}{[AH]}$$

This equation can be rearranged to

$$[H_3O^+] = \frac{K_a \times [AH]}{[A^-]}$$

Taking logarithms on both sides and changing the sign,

$$-\log[H_3O^+] = -\log K_a - \log \frac{[AH]}{[A^-]} = -\log K_a + \log \frac{[A^-]}{[AH]}$$

$$pH = pK_a + \log \frac{[A^-]}{[AH]}$$

This equation is known as the ***Henderson–Hasselbalch equation*** (Po and Senozan 2001). When the concentrations of acid and its conjugate base are

(continued)

equal, the value of $[A^-]/[AH]$ is 1.0. Therefore, the log $([A^-]/[AH])$ is zero and at this point pH is exactly equal to pK_a. Thus, the pK_a of a Bronsted acid is the pH at which it is half dissociated to its conjugate base.

The Two pK_as of Water:

The pK_a for H_3O^+ may be calculated from the following acid dissociation equilibrium:

$$H_3O^+ \rightleftharpoons H_2O + H^+$$

Accordingly, $K_a = ([H_2O] \times [H^+])/[H_3O^+]$. However, another molecule of water accepts this proton to form H_3O^+. Therefore, $K_a = [H_2O] = 55.56$ M and pK_a for H_3O^+ will then be $-\log(55.56) = -1.74$.

The pK_a for the ionization of H_2O will follow from its acid dissociation equilibrium:

$$H_2O \rightleftharpoons OH^- + H^+$$

The K_a for H_2O will be $= ([OH^-] \times [H^+])/[H_2O]$. This can be evaluated by plugging in the values for the three concentrations. Therefore, $K_a = (1.0 \times 10^{-14})/55.56 = 1.80 \times 10^{-16}$ and then $-\log(1.80 \times 10^{-16}) = 15.74$.

The two pK_as of water should not be confused with the pH of water. The pK_a of H_2O is 15.74 and the pK_a of H_3O^+ is -1.74. The pH of pure water at 25 °C is 7.0 and this is not its pK_a!

To compare a wide range of acidities (acid strengths), the negative logarithm of K_a (i.e., $-\log K_a$) is more convenient. Obviously, a stronger acid will then have a smaller pK_a. It also follows from the Henderson–Hasselbalch equation that when $[AH] = [A^-]$, we get $pH = pK_a$. By mapping their pK_a values on to the pH scale in water, different Bronsted acids/bases can be compared. It also follows that the ionization of a strong acid generates a weak conjugate base, and the protonated form of a strong base is a weak conjugate acid. Finally, when two bases compete for H^+, the stronger base wins the hand of a proton.

Besides being both a proton donor and a proton acceptor, water dissociates to produce H^+ and OH^-. The ionization of water due to different states of de-protonation is as shown:

$$H_3O^+ \xrightarrow{pK_a = -1.74} H_2O \xrightarrow{pK_a = 15.74} HO^- \xrightarrow{pK_a = 21.00} O^{2-}$$

Of these, the first two conjugate acid/base forms of water (namely, H_3O^+/H_2O with a pK_a of -1.74 and H_2O/OH^- with a pK_a of 15.74) are responsible for its amphoteric nature in aqueous solutions. They also define the limits of pH scale in aqueous solutions (Silverstein and Heller 2017). The concentration of H^+ ions (same as H_3O^+) in pure water is 10^{-7} M. In terms of pH scale (negative logarithm of $[H^+]$), this corresponds to a pH of 7.0. In fact, $pH = -\log ([H^+]/1 \text{ M})$ implies that pH is a unitless quantity (for standard state is considered as 1 M). From the concept of pH and the ionic product of water (K_w), the pH scale in aqueous solution spans from zero to 14. While a pH of 7.0 is neutral (i.e., equal number of H^+ and OH^- ions), acidic solutions have a pH of <7.0 and lower the pH more acidic the solution. Similarly, higher the pH more basic is the solution.

28.1.2   Factors Affecting Bronsted Acid/Base Strength

Broadly three factors influence the strength of a Bronsted acid (or a base).

First, the chemical structure defines the reactivity of a molecule. The number of chlorine atom substitutions on the methyl group of acetic acid influences the acidity. Trichloroacetic acid is a much stronger acid than acetic acid. Similarly, the acidic nature of the phenolic–OH increases in the order phenol $< p$-nitrophenol $<$ picric acid (2,4,6-trinitrophenol). With respect to basicity, aniline is a weaker base than ethylamine, for instance. As a rule, more stable the conjugate base (e.g., $F^- > OH^- > NH_2^-$) the stronger the acid (HF $> H_2O > NH_3$). Also, more electronegative the element on which the negative charge resides the more stable is the conjugate base.

Second, the medium by virtue of its solvation effects greatly influences the strength of an acid/base. Dissociation of HCl in gas phase is strongly endothermic— and it does not dissociate. The same molecule readily ionizes in water and is a very strong acid. This is due to favorable solvation of the dissociated charged species. Consider acetic acid as a solvent in place of water. Since acetic acid is more acidic than water, HCl is a weaker acid when acetic acid is the solvent. Similarly, ammonia is more basic than water and therefore acetic acid (normally a weak acid) behaves as a strong acid in ammonia. The converse is true for the Bronsted bases. In solvents more acidic than water, all bases behave as stronger bases and in solvents less acidic than water they appear as weaker bases. When a base stronger than the hydroxide ion is added to water, it will be converted into its conjugate acid with equimolar release of OH^- ions. A natural outcome of this solvent influence is the *leveling effect* observed in water. All acids with pK_a values below -2.0 (note: pK_a of H_3O^+ is -1.74) appear as equally strong acids. This is because, in water, such acids readily protonate the solvent and generate H_3O^+ equivalents. Therefore, there cannot be a stronger proton donor than H_3O^+. In the same way, all bases with pK_as >16.0 (note: pK_a of OH^- is 15.74) appear equally strong because they are virtually completely protonated by the solvent—water. The solvent properties of water reflect its O–H bond strength. A substitution of H by D makes it more rigid and hence most acids are 3–5 times weaker in D_2O than in H_2O.

Lastly, the microenvironment of a Bronsted acid/base greatly influences its ionization and pK_a. Consider ionization of the two carboxylate groups of glutamic acid in water. The pK_a of the α-COOH is 2.19 while that of its γ-COOH is 4.25. The amino group (α-NH$_2$) is a good proton acceptor and thus facilitates the ionization of the adjacent α-COOH—its pK_a is lowered. The γ-COOH experiences no such effect and hence shows a pK_a similar to that of acetic acid (pK_a of 4.76). Similarly, the

Fig. 28.1 Perturbation of normal pK_a values due to polar and nonpolar microenvironmental effects. Hydrophobic pocket is shown in gray

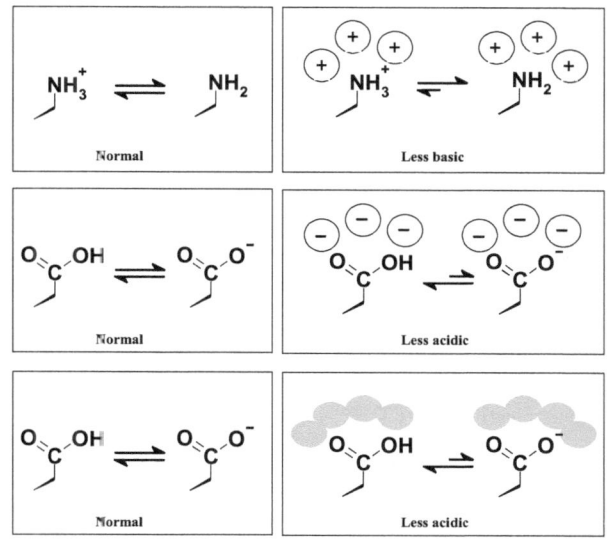

α-NH$_2$ becomes less basic (pK_a of 9.67) than the isolated amino group of a primary amine (e.g., ethylamine has a pK_a of 10.75). In general, the pK_a of a group would be altered if it is involved in a salt bridge formation with an oppositely charged residue or is surrounded by like charges (Fig. 28.1). Ionization (and separation of charges) is not favored in a hydrophobic environment. The nonaqueous hydrophobic microenvironment (often found at enzyme active sites) destabilizes the charged species and shifts the equilibrium towards a thermodynamically more favorable unionized state. For instance, acetic acid is a weak acid in benzene than in water; pK_a of the –COOH group is elevated in a hydrophobic environment.

All the three factors have been brought to bear by Nature in fine-tuning pK_a values—in order to optimize acid–base catalysis at enzyme active sites.

28.1.3 Ionizable Groups Relevant to Enzyme Structure and Function

The pH dependence of enzyme activity is a reflection of the acid–base groups involved in substrate binding and catalysis. Ionizable amino acid side chains of the enzyme protein are typically involved in such catalysis (see Chap. 23). Typical pK_a values of such acid–base groups that populate the 0–14 pH range in water, are listed in Table 28.1. Very few acid–base groups other than those of the polypeptide contribute to enzyme function. Water bound to a metal ion (cofactor) on the enzyme can also ionize with characteristic pK_a of around 8.0–9.0. The actual value of course will depend on the nature of the metal ion (such as Mg^{2+}, Mn^{2+}, Zn^{2+}, or Co^{2+}) and the other ligands it is coordinated to.

The relevance of the acid–base properties of functional groups must be in the context of water and the physiological pH of around 7.0. Enzyme active site bases

Table 28.1 pK_a values of Bronsted acid–base groups found on enzymes

Functional group	Structural form Acid⇌Base	Typical pK_a (Range)	Acts at pH 7.0 as
Carboxylate			Base
α-COOH (C-terminal)		3.4 (2.0–7.0)	
β-COOH (Asp)		3.9 (2.0–7.0)	
γ-COOH (Glu)		4.3 (2.0–7.0)	
Imidazole (His)		6.5 (6.0–8.0)	Acid or base
α-Amino (N-terminal)		7.5 (6.0–8.0)	Acid
ε-Amino (Lys)		10.5 (6.0–10.5)	Acid
Sulfhydryl (Cys)		8.3 (7.5–9.0)	Acid
Hydroxyl (Tyr)		10.0 (8.5–10.5)	Acid
Guanidinium (Arg)		>12.0	Positive charge
Hydroxyl (Ser, Thr)		13.5 (9.0–13.5)	Acid (?)
Me^{2+}-H_2O (Mg^{2+}, Mn^{2+}, Zn^{2+}etc)		8.0 (7.0–9.0)	Acid

must therefore be deprotonated around pH 7.0 but have pK_a values just below 7.0. The imidazole side chain of histidine residue is one such ideal acid–base group. It is a versatile reagent for enzymatic acid–base chemistry. Many of the pK_as listed in Table 28.1 can be shifted by the microenvironment in the vicinity of the enzyme active site (Grimsley et al. 2009). Because of such perturbations, it is often difficult to unambiguously assign an acid–base group to experimentally observed pK_as. For instance, (a) the active site lysine –NH_2 of acetoacetate decarboxylase exhibits an unusually low pK_a of 6.0 (Ishikita 2010), (b) the pK_a of the active site glutamate – COOH of lysozyme is attenuated to 6.0 from its typical pK_a of 4.3, and (c) papain has

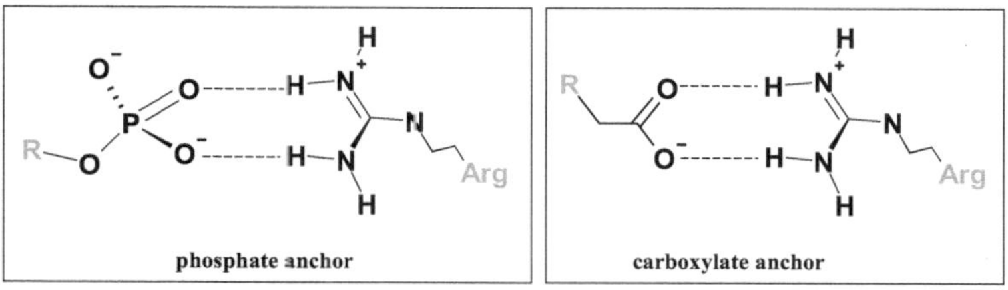

Fig. 28.2 The guanidinium group of arginine residue provides an excellent site to strongly bind carboxylate and phosphate groups through a cyclic, bifunctional, hydrogen-bonded ion pair (gray lines)

a histidine whose imidazole side chain pK_a is 3.4. Factors that influence the pK_a of such groups include the neighboring charged groups and shielding from bulk solvent by the surrounding hydrophobic residues (e.g., see Fig. 28.1).

Proton transfer from a group with lower pK_a to a group of higher pK_a (and not its reverse) is thermodynamically favored. When it is fully transferred to the hydrogen bond acceptor atom, we call it a proton transfer. In a normal hydrogen bond, the H atom is usually found on the hydrogen bond donor, but is weakly bonded to the hydrogen bond acceptor atom. Low barrier hydrogen bonds (LBHB) are short, very strong hydrogen bonds—the H atom equally shared between two electronegative atoms of nearly equal pK_as. The pK_as of the two participating groups have to match for a normal hydrogen bond to become LBHB. Such transitions are demonstrated in many enzymes. This pK_a matching, albeit ephemeral, is associated with catalytic events (Kahyaoglu et al. 1997). Thus, LBHBs represent one aspect of transient pK_a perturbation of acid/base groups that participate in a hydrogen bond.

Some of the groups (included in Table 28.1) have pK_as far from 7.0 and are unlikely to contribute to enzyme catalysis by acting as acid/base groups (Holliday et al. 2009). These include the guanidinium group of Arg and hydroxyl groups of Ser and Thr. Their participation cannot be categorically ruled out, as their pK_as could be sufficiently perturbed through microenvironmental effects. The active site serine – OH of chymotrypsin functions at about 5 pH units below its pK_a—around 8.0. However, it acts as a nucleophile during catalysis (see next chapter for details). Guanidinium group of arginine has a very high pK_a (>12.0); around neutral pH, where most enzymes function, this group remains fully protonated. It thus cannot be a good acid/base but presents a permanent positive charge. The guanidinium group provides a site through which carboxylate or phosphate can be bound as a cyclic, bi-dentate, hydrogen-bonded ion pair (Fig. 28.2). Arginine side chain is thus ideally suited to hold substrates in position at the active site. There is growing evidence that guanidinium group of arginine residue may act as a specific anchor for carboxylate (e.g., carboxypeptidase A) and phosphate (e.g., hexokinase) groups in many enzymes.

28.2   General Acid–Base Catalysis

Rates of many reactions are accelerated in the presence of an acid or base. In a simplest form, H^+ or OH^- may directly participate (act as the acid or the base) in catalysis. When H^+ (or H_3O^+) directly acts as the catalytic acid group, then it is called *specific acid catalysis*. Similarly, when the catalytic base is OH^-, it is termed *specific base catalysis*. Other components of an aqueous solution (other than H^+ or OH^-) including buffer species may act as Bronsted acid/base catalyst. Reaction rate acceleration due to a Bronsted acid (other than H^+) is referred to as *general acid catalysis*. Participation of a Bronsted base (other than OH^-) similarly defines the *general base catalysis*.

Hydrolysis of esters is catalyzed by both acids and bases. This reaction is an excellent example of specific acid/base catalysis in water. The contribution of specific acid catalysis decreases with increasing pH while that of specific base catalysis increases with increasing pH. This is expected as $[H^+]$ decreases with increasing pH while $[OH^-]$ is increasing. Both $[H^+]$ and $[OH^-]$ are very low at pH = 7.0. Accordingly, the contribution by specific acid/base catalysis is minimal and the experimentally observed hydrolysis rate constant (k_{Obs}) is the lowest at pH 7.0 (Fig. 28.3).

Contributions of Specific and General Acid Catalysis
Consider an acid-catalyzed hydrolysis of a reactant (S).

$$S \quad \underset{k_{Obs}}{\overset{H_2O,\ H^+,\ AH}{\longrightarrow}} \quad \text{Products}$$

The hydrolysis rate of S can be written as

(continued)

Fig. 28.3 Influence of pH on the rate of ester hydrolysis in water. The logarithm of k_{Obs} is plotted on Y-axis as pH (on the X-axis) is a logarithmic scale

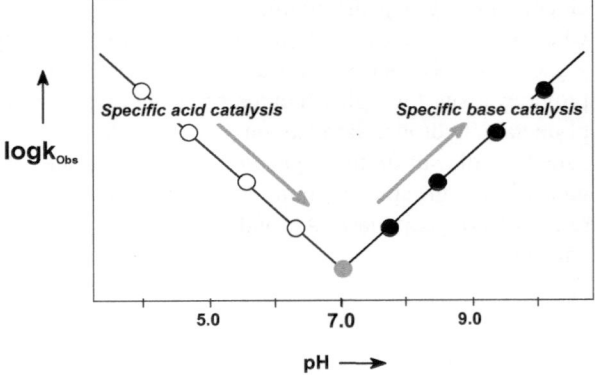

Fig. 28.4 Effect of buffer (general acid) concentration on the rate of an acid-catalyzed reaction. By simultaneously adjusting [AH] and [A⁻], the pH is held constant (according to Henderson–Hasselbalch equation) while molarity of the buffer is varied. General acid catalysis will almost always contain the specific acid catalysis component in it (intercept = $k_0 + k_H \times$ [H⁺])

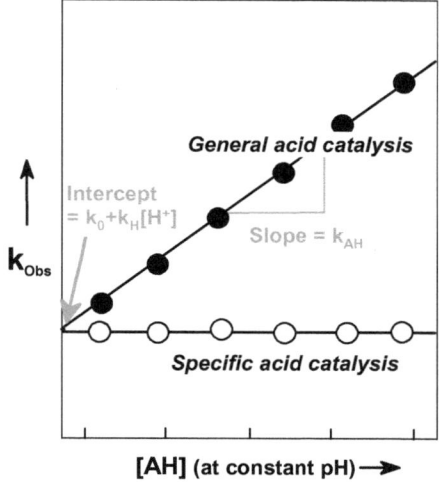

[AH] (at constant pH) ⟶

$$-\frac{d[S]}{dt} = k_{Obs} \times [S]$$

where k_{Obs} is the experimentally observed pseudo-first-order rate constant for hydrolysis. This acid catalysis rate constant in turn has the following components:

$$k_{Obs} = k_0 + k_H \times [H^+] + k_{AH} \times [AH]$$

where

k_0 = first-order rate constant for uncatalyzed reaction

k_H = second-order rate constant for H⁺ catalyzed reaction (specific acid catalysis) and

k_{AH} = second-order rate constant for the reaction catalyzed by AH (general acid catalysis).

For **specific acid catalysis** $k_{AH} = 0$ and therefore $k_{Obs} = k_0 + k_H \times [H^+]$. At a constant pH, this term itself will be a constant and independent of [AH]. The plot of k_{Obs} against [AH] at constant pH results in a linear plot with slope zero (Fig. 28.4). In practice, k_0 and k_H are better estimated by using a very strong acid alone (which is fully ionized, i.e., [AH] = 0 and so is its contribution to catalysis). Since [H⁺] is very small, specific acid catalysis constant can be safely ignored at or above neutral pH.

With **general acid catalysis** however, the component of specific acid catalysis will always be there. The linear plot of k_{Obs} versus

(continued)

[AH] (at constant pH) will show a gradient (slope $= k_{AH}$) and the intercept of this plot equals "$k_0 + k_H \times [H^+]$" (Fig. 28.4).

Note: The above analysis focuses on contributions to the reaction rate by specific and general acid catalysis. Catalysis by base can also be treated similarly. But this is deliberately not shown to avoid repetition of the concept.

A general acid (or base) catalysis always comes with the specific acid (or base) catalysis component. Consider a reaction catalyzed by the general acid AH. Since AH dissociates into A^- and H^+ in water, both AH (a general acid) and H^+ (a specific acid) species are available for participation in catalysis. The two contributions can be distinguished by the effects of pH and the acid group concentration (i.e., [AH]) on the rate of chemical reaction. Dependence of the reaction rate on the overall concentration of H^+ ions indicates that specific acid catalysis is occurring. We can also conduct the reaction at various buffer strengths but keeping the pH (i.e., [H^+]) constant. In such cases, the ionic strength difference can be overcome by suitable salt addition. A linear dependence of reaction rate on the buffer concentration (but the ratio [A^-]/[AH] remaining same) indicates general acid catalysis (Fig. 28.4). Such a rate dependence on buffer concentration is not observed with specific acid catalysis.

28.2.1 Effectiveness of a General Acid Catalyst

The effectiveness of a general acid catalyst depends on its acid strength. This is given by the empirical relation known as the **Brønsted relationship** (Chap. 8). According to this relation:

$$\log k_{AH} = \alpha \times \log K_{AH} + C$$

where

k_{AH} = rate constant of the catalytic step
K_{AH} = dissociation constant (K_a) of the acid AH
α = Brønsted parameter and
C = a constant

The Bronsted parameter α indicates the sensitivity of the catalytic step for changes in acid strength (pK_a) of AH. A plot of $\log k_{AH}$ against $\log K_{AH}$ (this is nothing but "$-pK_a$") gives a slope $= \alpha$. Conceptually, the Brønsted parameter (also known as Bronsted coefficient, normally ranges between 0 and 1) indicates to what extent a proton is transferred from the acid (AH) to the substrate in the transition state. When $\alpha = 1$, every change in acid strength fully affects catalysis. The proton is (almost) completely transferred to the substrate in the transition state (see acid

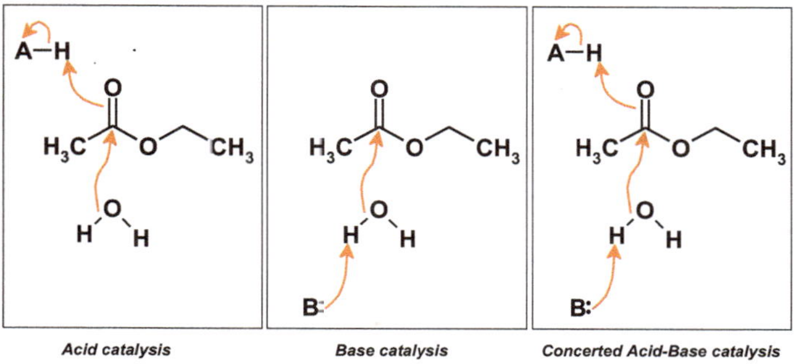

| Acid catalysis | Base catalysis | Concerted Acid-Base catalysis |

Fig. 28.5 Acid–base-catalyzed hydrolysis of ethyl acetate. Ester hydrolysis may be facilitated by the acid AH (acid catalysis), the base B: (base catalysis) or both (in concert). In water, when AH = H⁺, it is specific acid catalysis and when B: = OH⁻, it is specific base catalysis. Concerted acid–base catalysis involves simultaneous presence and action of a general acid (AH) and a general base (B:). Such an arrangement of groups is best provided at enzyme active sites shielded from the bulk aqueous medium

catalysis; Fig. 28.5). However, when $\alpha = 0$, the reaction is insensitive to changes in acid strength—all acids catalyze the reaction equally strongly (because $\alpha \times \log K_{AH} = 0$ and $\log k_{AH}$ equals the constant C). Here, the proton is hardly transferred in the transition state of the reaction. Lastly, when $\alpha = 0.5$, the proton is transferred halfway between the acid anion A⁻ and the substrate in the transition state—indicating a symmetrical TS for the reaction.

There is also a Brønsted relation for general base catalysis which may be similarly written as: $\log k_{B:} = -\beta \times \log K_{BH^+} + C$. Here, the coefficient β has the same meaning as α for general acid catalysis but has a negative sign as we are using the acid dissociation constant for the conjugate acid of the base (B:).

The Hammett equation $(\log(k_x/k_0) = \rho \times \sigma_X)$ is yet another empirical relationship like Bronsted relation. It is more general and may be used to assess the effect of structure (such as any group X) on reactivity; here, k_x and k_0 are rate constants for the reactant structure with substituent X and the standard (where X = H), respectively. The Hammett constant ρ (analogous to Bronsted coefficient, α) measures the sensitivity of the reaction to electronic effects (Fersht and Kirby 1967). Generally, a positive ρ value means more electrons in the TS than in the reactant and a negative ρ value means fewer electrons in the TS. More details on linear free energy relationships (like Bronsted relationship and Hammett equation) and their utility may be found in specialized texts (Kirsch 1972).

28.2.2 Ester Hydrolysis: An Example

Hydrolysis of carboxylate esters is frequently encountered both in organic chemistry and enzymology. Both share acid–base catalysis as a common reaction feature. It is therefore instructive to make a comparison of the two in terms of mechanism of acid–base catalysis. Consider the hydrolysis of ethyl acetate for instance (Fig. 28.5). The mechanism of ester hydrolysis involves the formation of a transition state with (a) further polarization of the carbonyl group and (b) partial charge transfer between the ester and water molecule. Such a transition state can be stabilized through proton transfer by an acidic group (AH) to the carbonyl oxygen of the ester. Alternatively, the same transition state may be stabilized by a base (B:) accepting a proton from the attacking water molecule.

Ethyl acetate hydrolysis in water is subject to specific acid (where AH is H^+ and $k_{Obs} = k_0 + k_H \times [H^+]$) or specific base (where B: is OH^- and $k_{Obs} = k_0 + k_{OH} \times [OH^-]$) catalysis. Therefore, k_{Obs} will not contain contributions due to general acid or general base catalysis (k_{AH} or $k_{B:}$ terms, respectively; see box above). Ester hydrolysis at enzyme active sites is different, however. Since both $[H^+]$ and $[OH^-]$ are quite small at physiological (near neutral) pH, the contribution of specific acid/base catalysis is marginal. Active sites of esterases (such as lipases, cutinases, and acetyl cholinesterase) contain general acid–base groups for catalysis. Simultaneous and appropriate positioning of both a general acid and a general base greatly facilitates the reaction (Fig. 28.5). Enzyme-catalyzed ester hydrolysis is an excellent example of general acid–base catalysis.

28.3   Summing Up

That "enzyme catalyzed reactions involve one or more proton transfers" is an understatement. General acid–base chemistry permeates most of enzyme chemical mechanisms. General acids and bases, respectively, will function only below or above their pK_a values (see Chap. 4). Ionizable amino acid side chains of the enzyme protein are typically involved in such catalysis. Each ionizable group can be viewed as an acid and also a conjugate base. Around neutral pH, as is evident from Table 28.1, the carboxylate groups (of C-terminal, aspartate/glutamate side chain) are present in the deprotonated form and act as bases. Imidazole group of histidine can act either as an acid or as a base under physiological conditions. The protonated amino group (of lysine), phenolic OH (of tyrosine), and thiol (of cysteine) all can function as general acids around neutral pH.

Some of the acid–base chemistries that take place at the enzyme active site seem almost impossible—particularly because the available acid/base groups on an enzyme are of moderate pK_a values. However, enzymes can carry out bifunctional (at times multifunctional!) catalysis. Protonation of the substrate molecule occurs at one location at the same time as deprotonating it in another region (concerted acid–base catalysis; Fig. 28.5). Such simultaneous H^+ donation/abstraction events make it possible to deprotonate substrate groups with apparently very high pK_a. Ketosteroid isomerase is one such example: a proton is abstracted from a C–H bond adjacent to a keto group by simultaneous protonation of $>C=O$ to form an enol. Such *1,3-prototropic shifts* (where a proton is moved from the first atom to the third) are quite commonly observed during enzyme catalysis. These proton transfer events may involve a single or two different acid–base groups. When the same acid–base group shuttles a proton from one atom to another, that proton is not easily lost to the solvent. This can be checked by incorporating a suitable isotope (e.g., tritium) label in the substrate.

Finally, a Bronsted base is a species that accepts a proton—forms a bond with H^+. This is equivalent to the base making a nucleophilic attack on to a proton. The Bronsted base could attack an electrophile other than H^+, in principle. If this happens, then we call that same base—a nucleophile. How such nucleophilic (and electrophilic) reactions contribute to catalysis forms the subject of next chapter.

References

Fersht AR, Kirby AJ (1967) Structure and mechanism in intramolecular catalysis. The hydrolysis of substituted aspirins. J Am Chem Soc 89:4853–4857 and 4857–4863

Grimsley GR, Scholtz JM, Pace CN (2009) A summary of the measured pK values of the ionizable groups in folded proteins. Protein Sci 18:247–251

Holliday GL, Mitchell JBO, Thornton JM (2009) Understanding the functional roles of amino acid residues in enzyme catalysis. J Mol Biol 390:560–577

Ishikita H (2010) Origin of the pKa shift of the catalytic lysine in acetoacetate decarboxylase. FEBS Lett 584:3464–3468

Kahyaoglu A et al (1997) Low barrier hydrogen bond is absent in the catalytic triads in the ground state but is present in a transition-state complex in the prolyl oligopeptidase family of serine proteases. J Biol Chem 272:25547–25554

Kirsch JF (1972) Linear free energy relationships in enzymology, Chapter 8. In: Chapman NB et al (eds) Advances in linear free energy relationships. Plenum Publishing Company Ltd., pp 369–400

Po HN, Senozan HM (2001) The Henderson-Hasselbalch equation: its history and limitations. J Chem Educ 78:1499–1503

Silverstein TP, Heller ST (2017) pKa values in the undergraduate curriculum: what is the real pKa of water? J Chem Educ 94:690–695

Nucleophilic Catalysis and Covalent Reaction Intermediates

29

The concept of nucleophiles and electrophiles is closely related to bases and acids of the acid–base chemistry. Electrophiles are essentially same as Lewis acids while the nucleophiles are equivalent to Lewis bases. In practice, however, acid–base concept involves electron donation to H^+ and the terms "electrophile" and "nucleophile" are normally used to indicate electron donation to a carbon atom. Nucleophiles contain a pair of electrons in a high-energy filled orbital that they can donate to an electrophile. As a complement, electrophiles are species with an empty atomic orbital of lower energy that can accept a lone pair (from a nucleophile). Nucleophiles can be neutral (like the amino group) or negatively charged (like $-S^-$, the thiolate ion) while the electrophiles can be neutral or positively charged (like the carbocation or a divalent metal ion). If the electrophile is a proton, then the nucleophile in question is a base, by definition (see previous chapter).

29.1   Nucleophiles and Electrophiles Available on the Enzyme

Enzymes as proteins have a range of nucleophilic groups available to them. Some R groups of amino acid residues are excellent nucleophiles. In fact, most ionizable groups (see Table 28.1) available on the enzyme can in principle act as nucleophiles in their deprotonated state. Enzyme active site acid/base groups that exist in the deprotonated form around pH 7.0 are potential nucleophiles for catalysis. Some of the more commonly encountered nucleophiles are the carboxylate ($-COO^-$ of Asp, Glu, and the C-terminus), the amino ($-NH_2$ of Lys and the N-terminus), the imidazole (of His), the thiolate ($-S^-$ of Cys), the phenolic group ($-ArO^-$ of Tyr), and the alkoxide ($-O^-$ of Ser and Thr). The reactivity of these nucleophilic groups may be further modified by the active site micro-environment. Metal ions (like Zn^{2+}) enhance the reactivity of bound water molecule making it a better nucleophile than bulk water. Clearly, there is no scarcity of functional groups on the enzyme surface for nucleophilic catalysis. These side chains attack electrophilic portions of

N. S. Punekar, *ENZYMES: Catalysis, Kinetics and Mechanisms*, https://doi.org/10.1007/978-981-97-8179-9_29

substrates to form a covalent bond between enzyme and substrate. We will address these covalent reaction intermediates a little later. The identity of the actual nucleophile (of the enzyme) involved in the catalytic mechanism may be ascertained by a combination of tools like chemical modification, pH dependence of enzyme kinetics, X-ray structural data, site-directed mutation studies, etc.

The availability of electrophilic groups on a protein is a different matter. Polypeptide enzymes are very poorly endowed with good electrophilic reagents/groups for catalysis. Therefore, nature has recruited many small molecules to fill this need. Several coenzymes form covalent adducts with substrates. These covalent intermediates in turn generate new electrophilic groups and they function as electron sinks during catalysis. Since much of enzyme chemistry is carbanion chemistry, coenzymes (like pyridoxal phosphate and thiamine pyrophosphate) function to stabilize them as their electrophilic adducts. Apart from divalent cations, organic molecules serve this purpose as cofactors and prosthetic groups. Table 29.1 lists some of the more commonly encountered nonprotein components employed by enzymes for chemistry. The list includes both electrophilic and redox reagents. A detailed discussion on their role in redox chemistry (Chap. 31) and electrophilic catalysis (Chap. 33) may be found in later chapters.

29.1.1 Nucleophilicity Versus Basicity

A nucleophile is a Lewis base that uses an available electron pair to bond to (electrophilic) carbon. All nucleophiles are also bases—as they can abstract a proton. Then are basicity and nucleophilicity interchangeable? Not really. However, one factor that correlates well with nucleophilicity is basicity—a strong base is usually a strong nucleophile. The phenoxide of 4-nitrophenol (a strong acid whose conjugate base is weak) is a poor nucleophile in comparison to that of phenol (relatively weak acid). This correlation is *not* obeyed when two nucleophiles that attack through different atoms are compared. For instance, thiophenate ($C_6H_6-S^-$) is a stronger nucleophile than phenoxide ($C_6H_6-O^-$) by four orders of magnitude but is a weaker base (Table 29.2). Similarly, alkoxide ions are strong bases but not very good nucleophiles while their thiolate analogs are weak bases but good nucleophiles. The small and electronegative oxygen keeps its nonbonding electron pairs close to itself whereas sulfur is lot larger and less electronegative than oxygen. Thus, sulfur lone pairs are more easily available to form a bond with an electrophile reflecting on the *greater nucleophilicity of sulfur relative to oxygen*.

Clearly, nucleophilicity and basicity are related but also they differ in the following way. Base strength depends on the position of the equilibrium for that base to accept a proton from water. Nucleophile strength, however, is based on relative rates of reaction with a common electrophile. A good nucleophile is one that rapidly forms a new bond with carbon. Nucleophilicity is a kinetic property while basicity a thermodynamic property (pK_a, the proton ionization equilibrium). Nucleophilicity order is structure dependent as well as solvent dependent. Many factors determine the nucleophilic power of a functional group. Among others, these include

Table 29.1 Coenzymes and their functions in enzyme catalysis

Coenzyme/Cofactor	Structure	Enzyme example (function)	Vitamin
L-Ascorbic acid		Prolyl hydroxylase (Redox)	Vitamin C
Biotin		Pyruvate carboxylase (Electrophile)	Biotin
Tetrahydrofolate (FH$_4$)		Serine hydroxymethyltransferase (Redox and electrophile)	Folic acid

(continued)

Table 29.1 (continued)

Coenzyme/Cofactor	Structure	Enzyme example (function)	Vitamin
Nicotinamide adenine dinucleotide (NADH and NADPH)		Lactate dehydrogenase (Redox)	Niacin
Coenzyme A (CoA)		Citrate synthase (Acyl activation; good leaving group)	Pantothenic acid
Pyridoxal phosphate (PLP)		GABA transaminase (Electrophile)	Pyridoxal

Cofactor	Structure	Enzyme example	Vitamin
Flavin adenine dinucleotide (FAD) and Flavin mononucleotide (FMN)		D-Amino acid oxidase (Redox)	Riboflavin
Thiamine pyrophosphate (TPP)		Transketolase (Electrophile)	Thiamine

(1) The groups/atoms relevant to chemistry are marked gray in each structure. (2) $NADP^+$ differs from NAD^+ in having a phosphate group on 2′ OH (shown as (P)). (3) R= H for FMN and R= AMP attached via its phosphate for FAD. (4) Other cofactors like lipoic acid, cyanocobalamin (vitamin B_{12}), nickel (I) hydrocorphin coenzyme F-430, and phylloquinone (vitamin K) also perform important functions but are less frequently encountered. While not included here for the sake of brevity, they will be referred to in relevant chapters later

Table 29.2 Nucleophilicity and basicity are related but different

Nucleophile	Structural form	Nucleophilicity (n)	pK_a
Phenol (Phenolate)		5.75	10.0
4-Nitrophenol (4-Nitrophenoxide)		(<5.75)	7.2
Thiophenol (Thiophenoxide)		9.92	6.6

The nucleophilicity parameter "n" is from the *Swain–Scott relationship*: $\log(k/k_0) = s \times n$ wherein "s" is a sensitivity parameter, k = rate constant of the reaction with that nucleophile and k_0 = rate constant of the reaction with standard nucleophile.

(a) the strength of the carbon–nucleophile bond, (b) solvation energy of the nucleophile, (c) steric hindrance if any, and (d) the electro-negativity and polarizability of the nucleophilic atom. Certain nucleophiles have two adjacent electronegative atoms (such as NH_2OH, NH_2NH_2, and HOO^-, etc.). Because of this α-effect, they are more reactive than expected from their pK_as alone.

While a well-defined nucleophilicity scale is elusive, few empirical equations (also see Table 29.2) have attempted to quantify it. According to ***Edwards equation***,

$$\text{Nucleophilic power} = \log \frac{k}{k_0} = \alpha\mathbf{P} + \beta(\mathbf{pKa} + \mathbf{1.74})$$

where,

k = rate constant of the reaction with that nucleophile
k_0 = rate constant of the reaction with standard nucleophile (water)
$\mathbf{P}$ = polarizability, related to refractive index ratio (Nu/water) and α and β are constants dependent on the reaction.

The Edwards equation relates nucleophilic power to basicity (the pK_a term). While basicity is closely related to nucleophilicity this is not the full story! A change in the nucleophilic atom can dramatically affect nucleophilicity. This atom change is reflected through the polarizability factor (P in the equation). For example, O is less polarizable than S; the value of P is large for a soft nucleophile like $–S^-$. On the other hand, pK_a of a hard nucleophile (like $–O^-$) is larger. In general, and particularly at physiological (near neutral) pH, $–S^-$ (thiolate) is therefore a better nucleophile than $–O^-$ (an alkoxide or a phenoxide). The thiolate anion of cysteine (pK_a of ~8.0; Table 28.1) exists in appreciable concentrations at physiological pH ranges. This anion ($–S^-$; due to its electronic and polarizability properties) is 10–100 times more nucleophilic than normal oxygen or nitrogen bases of comparable pK_a values.

29.1.2 Concept of a Good Leaving Group

Another aspect related to the concept of nucleophilicity is the notion of a better leaving group. Leaving groups are the fragments that retain the electrons in a heterolytic bond cleavage. Since they keep the electron pair, leaving groups are quite often nucleophiles (and bases). Weaker bases are more stable with the extra pair of electrons and therefore make better leaving groups. Furthermore, the effectiveness of a leaving group increases with the group's *energetic stability after it has left*. Thus, a weak base is a better leaving group than a strong base. Similarly, a molecule that is neutral after leaving is generally a better leaving group than one that is negatively charged after leaving. A good leaving group can be recognized as being the conjugate base of a strong acid. It thus makes perfect sense that $R–S^-$ (thiolate; the conjugate base of a stronger acid) is therefore a better leaving group than $R–O^-$ (alkoxide; the conjugate base of a weaker acid). This may be one good reason why nature chose thiols in many enzyme chemistries.

Nature has repeatedly exploited the "better leaving group" feature in its synthetic designs. An otherwise energetically difficult reaction is driven forward by its departure. Often a high-energy intermediate is built with a good leaving group for this purpose. A number of reactions coupled to ATP hydrolysis (see Chap. 30) work on this principle.

Nucleophilic acyl substitution reactions are an important class of reactions catalyzed by enzymes. These include acyl transfer chemistry of amide, ester, and thioester bonds. The reaction involves substitution of the leaving group ($^{out}Nu:$) bonded to the carbonyl carbon by an attacking nucleophile ($^{in}Nu:$).

$$R\text{-}CO\text{-}^{out}Nu : +^{in}Nu : \rightleftarrows R\text{-}CO\text{-}^{in}Nu : +^{out}Nu :$$

The greater the stability of the carbonyl compound, the less reactive it is. Consequently, we find that amides are the least reactive because of resonance stabilization. There is much resonance in esters ($R–CO–OR'$) than in thioesters ($R–CO–SR'$). Thioesters therefore have little or no double bond character in their $C–S$ bond while esters have appreciable double bond character. In terms of reactivity, these compounds may be ranked as Amide < Ester < Thioester < Acyl phosphate. In addition, breaking up is easy with esters and thioesters as they provide better leaving groups ($R–O^-$ and $R–S^-$, respectively). Such "good leaving group" chemistry is indeed exploited in enzyme reactions involving peptide bond hydrolysis (e.g., subtilisin and papain) and transpeptidation (e.g., intein-mediated protein splicing).

29.2 Nucleophilic (Covalent) Catalysis

Nucleophilic catalysis, whenever recruited by an enzyme, makes important contributions to its catalytic power (Chap. 4). In nucleophilic catalysis, the catalyst reacts with an electrophilic center of the reactant to form a covalent intermediate in

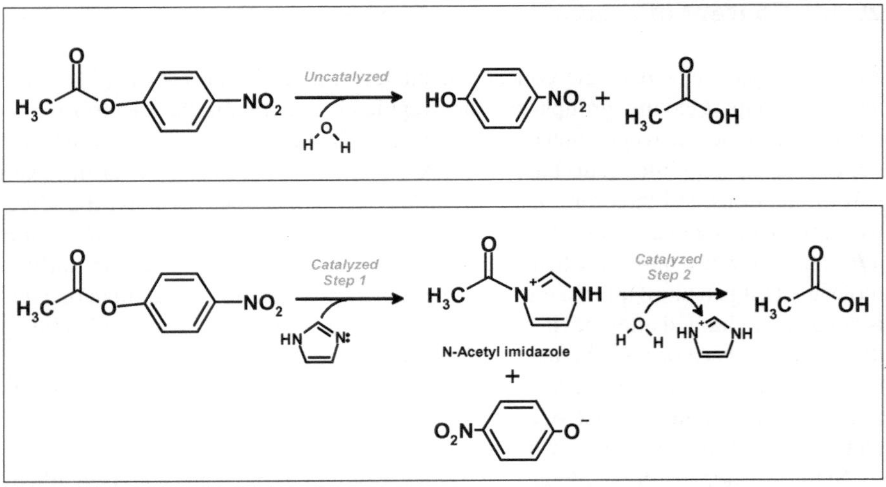

Fig. 29.1 Imidazole-catalyzed hydrolysis of 4-nitrophenyl acetate. Both acetylation of imidazole (Step 1) and hydrolysis of acetyl imidazole (Step 2) are faster than the direct interaction of 4-nitrophenyl acetate with water (uncatalyzed reaction). Imidazole is a true nucleophilic catalyst since it is recovered intact after the reaction and acetyl imidazole forms during catalysis

the reaction mechanism. Therefore, it is sometimes also referred to as *covalent catalysis*. Catalysis involves lowering the energy of activation for that reaction. One way to do this is to change the reaction mechanism in ways which introduce new steps with lower activation energy. Typically, the original reaction is broken down into two or more steps. The catalytic nucleophile first forms a covalent intermediate with the reactant. The original nucleophile then attacks this intermediate to displace the catalytic nucleophile—in a nucleophilic substitution reaction. In the final analysis, the catalytic nucleophile does not end up in the product but is regenerated. Hydrolysis of 4-nitrophenyl acetate catalyzed by imidazole (Fig. 29.1) is an excellent example of this mode of catalysis.

Hydrolysis of acetic anhydride by pyridine is another such example. Yet another interesting nonenzymatic model for nucleophilic catalysis is the decarboxylation of acetoacetic acid catalyzed by aniline. Here, the decarboxylation reaction is facilitated by the formation of an aniline–acetoacetate covalent adduct (the imine intermediate).

Criteria for Nucleophilic Catalysis
The following conditions have to be met in order to ascertain that nucleophilic (covalent) catalysis is involved. The relevant covalent intermediate:

- must be detected, isolated, chemically characterized and shown to be present during the reaction.

(continued)

- should be kinetically competent—its reactivity rate should be faster than the overall reaction rate.
- if demonstrated with model reactions, then it is crucial to show that the same mechanism is operating in the actual reaction.

In addition to these lines of direct evidence, one can obtain much indirect support through kinetic and other data. These will be discussed in some detail later in this chapter.

29.2.1 Measures of a Good Nucleophilic Catalyst

The features that make a good catalytic nucleophile include (a) the catalytic nucleophile should be a better nucleophile (react faster) than the reactant and more nucleophilic than the solvent, (b) it must be a better leaving group in the covalent intermediate, and (c) the covalent intermediate should be thermodynamically less stable than the final product under the reaction conditions. According to these requirements, the catalyst must be both *a very effective nucleophile and a good leaving group*. All these criteria are well satisfied by imidazole in the example shown in Fig. 29.1. Imidazole is a better nucleophile than water in attacking the carbonyl carbon of 4-nitrophenyl acetate. The covalent intermediate (N-acetyl imidazole) was isolated and is more susceptible to water attack than the reactant, 4-nitrophenyl acetate. The N-acetyl imidazole formed is less stable than the reaction products; otherwise, it would not be kinetically competent and hence would accumulate. It reacts with water several orders of magnitude faster than 4-nitrophenyl acetate. This would not be possible if imidazole was a poor leaving group. In summary, imidazole makes 4-nitrophenyl acetate hydrolysis a two-step event. It thus provides a lower energy ($\Delta G^{\neq}$) reaction path—the hallmark of a catalyst.

29.2.2 Nucleophilic Catalysis Versus General Base Catalysis

Almost invariably nucleophiles can also act as good general bases. They can act directly (nucleophilic attack) or by abstracting a proton (general base) from the solvent (water) or the substrate. Then how are we to distinguish between the two? Nucleophilic attack leads to a new bond established between the nucleophile and the carbon atom. Detection of this *covalent intermediate is proof for nucleophilic catalysis*. The intermediate must of course satisfy the criteria laid out (see box above). When a covalent intermediate is unstable, it may sometimes be possible to show its existence by trapping it chemically. Detecting or trapping a covalent intermediate may not necessarily be easy or trivial. Our inability to track/trap such an intermediate, however, does not constitute as proof against nucleophilic catalysis. As the golden rule of the scientific method goes—*absence of evidence is not evidence of absence!* (after Carl Sagan). As mentioned above, the intermediate may be very unstable or short-lived. We recall that the mechanism of lysozyme was revised when the covalent adduct between Asp52 (acting as a nucleophile and not a base) and C1 of the substrate glycoside was detected (recently by electrospray ionization mass spectrometry—ESI-MS).

Catalysis by Nucleophile or Base?

Since a nucleophile can also act as a base (and vice versa), it is tricky to decide which role the catalytic group is playing. For instance, an active site carboxylate group (of Glu) is thought to function as a base (in thermolysin) or a nucleophile (in carboxypeptidase A) in amide bond hydrolysis. How to tell them apart is illustrated with the help of two cases involving hydrolysis of 4-nitrophenyl acetate.

Catalysis by Acetate:

$$\text{4-Nitrophenyl acetate} \quad \overset{H_2O,\ CH_3COO^-}{\underset{k_{Obs}}{\longrightarrow}} \quad \text{4-Nitrophenol + Acetic acid}$$

The hydrolysis of 4-nitrophenyl acetate by acetate anion may be viewed both as a general base catalysis and/or nucleophilic catalysis (Fig. 29.2). The two are kinetically identical and the same rate expression ($k_{Obs} = k_0 + k_{Acetate} \times [CH_3COO^-]$) fits them both. However, mechanistically the two are different. Acetic anhydride is an obligate, covalent intermediate with nucleophilic catalysis. It can be trapped by reacting it with aniline to form acetanilide. Since acetanilide is not formed—no acetic anhydride is formed—nucleophilic catalysis may be ruled out (with caution, of course). In the case of acetate-catalyzed 4-nitrophenyl acetate hydrolysis, evidence points to general base catalysis (Fig. 29.2; top left box).

Catalysis by Imidazole:

$$\text{4-Nitrophenyl acetate} \quad \overset{H_2O,\ Imidazole}{\underset{k_{Obs}}{\longrightarrow}} \quad \text{4-Nitrophenol + Acetic acid}$$

Imidazole can be a base as well as a nucleophile. Both general base catalysis and nucleophilic catalysis lead to the same rate expression ($k_{Obs} = k_0 + k_{Imidazole} \times [Imidazole]$). While there may be contribution by imidazole to general base catalysis, nucleophilic catalysis does occur. The covalent intermediate—*N*-acetyl imidazole—is detected spectroscopically and is trapped by aniline as acetanilide (Fig. 29.3). What is more, the rate of 4-nitrophenol formation shows an initial burst phase followed by steady state. This indicates step 1 is faster than step 2 for this reaction (see Fig. 29.1).

Corroborating evidence for nucleophilic catalysis may be obtained from additional experiments.

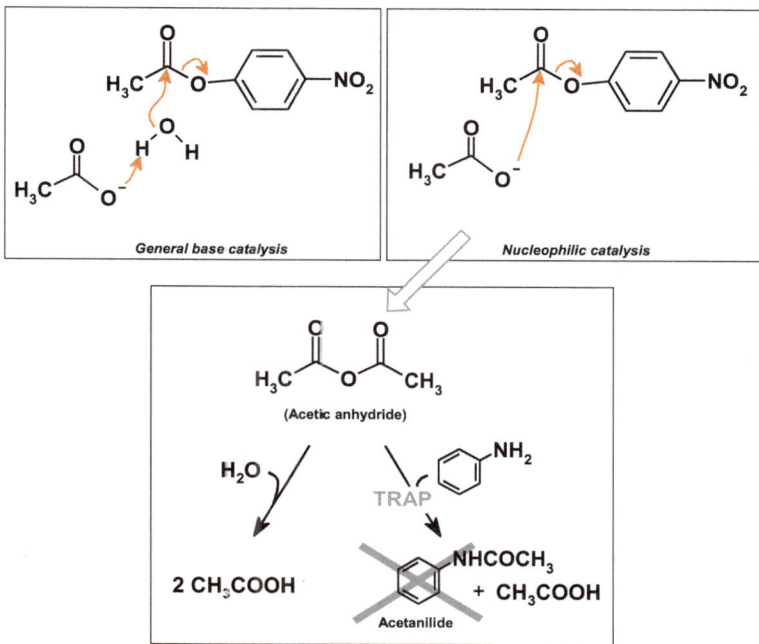

Fig. 29.2 Acetate-catalyzed hydrolysis of 4-nitrophenyl acetate. Acetanilide is not trapped during the catalyzed reaction implying that no acetic anhydride is formed. Nucleophilic catalysis (top right) may thus be ruled out in favor of general base catalysis by acetate (top left)

1. Solvent isotope effects (see Chap. 25) are often useful in distinguishing between nucleophilic versus general base catalysis. One conducts the reaction in water and D_2O to measure the deuterium kinetic isotope effect on the rate. General base catalysis involves proton abstraction steps and the cleavage of an O–H (O–D) bond; Substantial isotope effect in D_2O may therefore be observed. No such O–H bond weakening is expected for a nucleophilic attack. Therefore, nucleophilic catalysis will not show a solvent isotope effect. Whereas absence of significant solvent deuterium kinetic isotope effect supports nucleophilic catalysis, the data needs to be interpreted with caution. Solvent isotope effects (in D_2O) may also arise when "base catalyzed attack of water" is the rate-determining step. In some cases, the solvent isotope effect may be masked (or is borderline) due to other rate-determining events, thereby making the interpretation difficult.

2. Basicity and nucleophilicity are related but different. This is clearly reflected in the Bronsted relation. A linear Bronsted plot (good correlation) implies a general base involvement. Groups with different polarizability but the same pK_a can be compared. Strong deviations in the Bronsted plots (and large β values) thus suggest the involvement of nucleophilic catalysis. Similarly, steric hindrance is not so important for general base catalysis (proton transfer) but is critical for a nucleophilic attack.

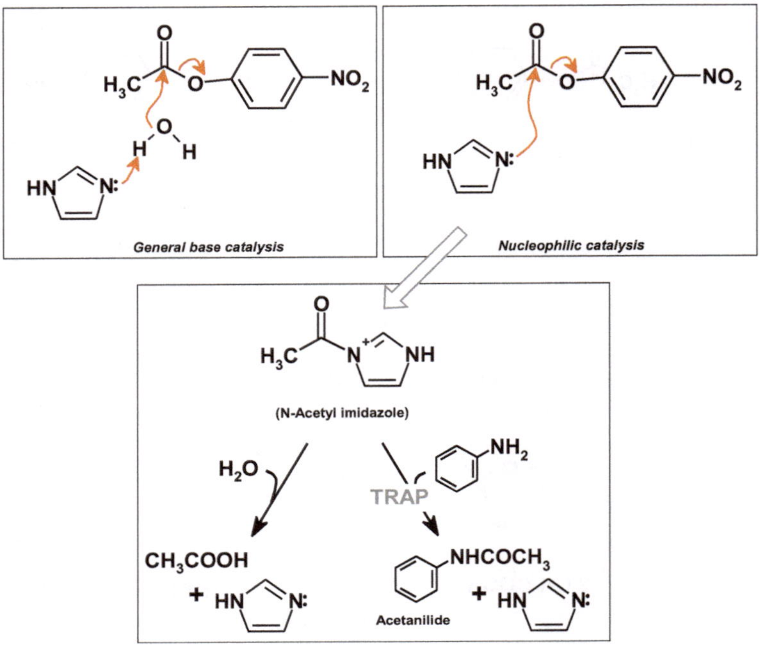

Fig. 29.3 Imidazole-catalyzed hydrolysis of 4-nitrophenyl acetate. Acetanilide is trapped in the presence of aniline; thus *N*-acetyl imidazole is formed (and detected by UV spectroscopy) during the catalyzed reaction. The reaction definitely occurs by nucleophilic catalysis (top right) but some extent of general base catalysis may not be ruled out (top left)

3. Competition by a nucleophile (same or similar to the leaving group) slows down the nucleophilic catalysis. This is because the extra nucleophile addition drives the equilibrium backward to form the covalent intermediate. However, if the added nucleophile acts as a general base catalyst, then the reaction rate is further accelerated.

These approaches are well suited to probe and analyze nonenzymatic models of nucleophilic catalysis (Fersht and Kirby 1967). They differ from each other in the simplicity of approach and the strength of evidence in subsequent data interpretation. For these very reasons, their utility in analyzing enzyme mechanisms may be limited.

29.3   Covalent Reaction Intermediates

Electrophiles have a positively polarized, electron poor atom that can accept an electron pair from a nucleophile. On the other hand, nucleophiles are electron rich and can donate a pair of electrons to an electrophile. In a large majority of reactions, this complementarity leads to nucleophile donating an electron pair to electrophile, with the formation of a covalent bond. Recall the formation of N-acetyl imidazole (Fig. 29.3) during imidazole-catalyzed hydrolysis of 4-nitrophenyl acetate. Covalent reaction intermediate(s) is thus a feature of nucleophilic attack during a reaction. Enzyme catalysis is no exception to this rule. Reversible non-covalent binding of substrate(s) to the enzyme is a precondition for catalysis. In some enzyme reactions, however, one or more discrete covalent intermediates are formed after these binding events. The reactive covalent intermediates assist catalysis by—(a) constraining the reactants within the active site (entropic contribution), (b) providing better leaving group (nucleophile) options, and (c) moving them on a leash between different subsites.

How Covalent Reaction Intermediates are Formed?
Enzyme-bound covalent intermediates can be formed in three different ways.

- an enzyme nucleophile attacks the substrate electrophilic center to form the covalent bond. The acyl-enzyme intermediate of chymotrypsin is a classic example of this kind. The enzyme Ser195 bonds to the carbonyl carbon of the scissile peptide bond and releases the amino group (of the first product).
- a substrate nucleophilic group attacks an electrophilic center on the enzyme. The substrate amino acid (via its amino group) attacks the carbonyl carbon of enzyme-bound pyridoxal phosphate (PLP) forming a covalent adduct (the Schiff's base). In transaminases, the amino group is held on to the enzyme (as pyridoxamine phosphate).
- a nucleophilic group of one substrate may attack the other substrate to generate a covalent intermediate physically enclosed in the enzyme active site. For instance, γ-COO$^-$ of glutamate makes a nucleophilic attack on the γ-phosphate of ATP to form enzyme-bound γ-glutamyl phosphate intermediate in glutamine synthetase. A closed active site environment serves to protect and direct such reactive intermediates to desired chemistry. At no stage in the reaction, the substrate (or portion of it) is covalently attached to glutamine synthetase. In contrast, a portion of the substrate is covalently held on to the enzyme in the first two cases.

There are many instances where the substrate (or a part of it) is covalently held on to the enzyme (few examples are listed in Table 29.3). These covalent enzyme adducts may arise by the attack of an enzyme group (either nucleophilic or electrophilic) on to the substrate. Such reactions involving transfer of groups may be generally represented as follows:

$$\text{E} + \text{A-X} \rightleftharpoons \text{E-X} + \text{A} \quad \text{followed by} \quad \text{E-X} + \text{B} \rightleftharpoons \text{E} + \text{B-X}$$

Table 29.3 Examples of reaction intermediates covalently linked to enzyme

Covalently linked to	Intermediate	Enzyme examples
Enzyme provides the nucleophile		
Serine (–OH)	*O*-Acyl enzyme	Acetylcholinesterase, Chymotrypsin
Cysteine (–SH)	*S*-Acyl enzyme	Papain, Glyceraldehyde-3-phosphate dehydrogenase, Glutamate synthase
Serine (–OH)	*O*-Phospho enzyme	Alkaline phosphatase, Phosphoglucomutase; Phosphodiesterase (via Thr-OH)
Histidine (-imidazole)	N^1-Phospho enzyme	Glucose-6-phosphatase, Nucleoside-bisphosphate kinase, Succinyl CoA synthetase
Tyrosine (–OH)	*O*-Sulfo enzyme	Arylsulfate sulfotransferase
Lysine (–NH$_2$)	Imine adduct (Schiff's base)	Fructose-1,6-bisphosphate aldolase, Acetoacetate decarboxylase, Transaldolase
Lysine (–NH$_2$)	AMP-enzyme	DNA ligase (NAD$^+$)
Glutamate (–COOH?)	Glycosyl enzyme	Sucrose phosphorylase, β-Galactosidase
Enzyme provides the electrophile		
Pyruvoyl group	Imine adduct (Schiff's base)	Histidine decarboxylase (bacterial)
Pyridoxal phosphate	Aldimine (Schiff's base)	Glutamate decarboxylase, L-Alanine aminotransferase
Biotin	*N*-Carboxy-biotin	Acetyl CoA carboxylase, Transcarboxylase, Pyruvate carboxylase
Thiazolium ring of Thiamine pyrophosphate	Hydroxyethyl TPP; 1,2-Dihydroxyethyl TPP	Pyruvate decarboxylase; Transketolase

There is a net transfer of group "-**X**" from one substrate to the other via the enzyme-bound covalent intermediate (**E-X**). Often such reactions follow ping-pong kinetics with the substituted enzyme (**E-X**) representing the "**F**" form of the enzyme (see Chaps. 18 and 24). Most common examples of group transfer reactions involve acylation, phosphorylation, or glycosylation of an enzyme nucleophile. These groups are subsequently transferred from the enzyme covalent adduct to another incoming nucleophile. However, hydrolysis ensues whenever the second nucleophile is water.

29.4 Detecting Intermediates and Establishing Their Catalytic Competence

A major objective in understanding how enzymes function is to look at all the reaction steps. Regardless of whether the covalent intermediates are formed between—(a) enzyme and substrate or (b) between two substrate molecules—they provide important mechanistic clues. Proving the existence of transient and/or covalent intermediates and showing that they indeed participate in the catalytic process (i.e., their kinetic and chemical competence) requires multiple lines of evidence (Frey 2015). We will look at each of these briefly, with suitable examples.

29.4.1 Steady-State Kinetics

This approach provides relatively limited information about covalent intermediates and their kinetic competency. Often the initial rate equation is identical for different chemical mechanisms (see box above—"catalysis by nucleophile or base?"). In two substrate ping-pong mechanism, the active site retains a portion of the first substrate. The "F" form of the enzyme is obtained when the product of the first substrate departs. Ping-pong kinetics (parallel lines obtained in initial velocity analysis; Chap. 18) of this type is indicative of a covalent enzyme intermediate. The phosphorylated form of nucleoside–bisphosphate kinase and pyridoxamine phosphate form of a transaminase are well-known examples (Table 29.3).

Burst-phase kinetics (chymotrypsin catalysis with a poor substrate like 4-nitrophenyl acetate being a classic example) is suggestive of a covalent intermediate participation. Rapid release of 4-nitrophenol (colored yellow) is proportional to the active enzyme present and forms the initial burst phase. This is followed by linear, slower steady-state because the acyl–enzyme intermediate breaks down slowly (Fig. 29.4). The initial burst would be seen only if the second deacylation step is slow. With good substrates (like amides) this is not the case, and bursts are hardly visible.

Enzyme-bound intermediates were also inferred from burst kinetics in catalysis by alkaline phosphatase, glyceraldehyde-3-phosphate dehydrogenase, and aminoacyl-tRNA synthetase.

29.4.2 Isotope Exchange Studies

Ping-pong mechanisms involve double displacement and a substituted form (the "F" form) of the enzyme occurs during the catalytic cycle. Group transfer from an appropriate substrate can occur even in the absence of the other(s). Corresponding isotope exchanges can therefore be detected (Chap. 24). Consider L-alanine transaminase reaction, for example.

$$\text{L-Alanine} + \text{2-Oxoglutarate} \rightleftarrows \text{Pyruvate} + \text{L-Glutamate}$$
$$\quad\text{(A)}\qquad\qquad\text{(B)}\qquad\qquad\text{(P)}\qquad\qquad\text{(Q)}$$

Upon incubating labeled pyruvate (product P) and L-alanine (substrate A), label exchange ensues in the presence of the enzyme. That is, L-alanine becomes labeled with time and a partial reaction, even in the absence of B or Q, is thus detected. Most likely explanation is that the amino group is held on to the enzyme (covalent pyridoxalamine phosphate intermediate) while pyruvate can freely enter/exit the

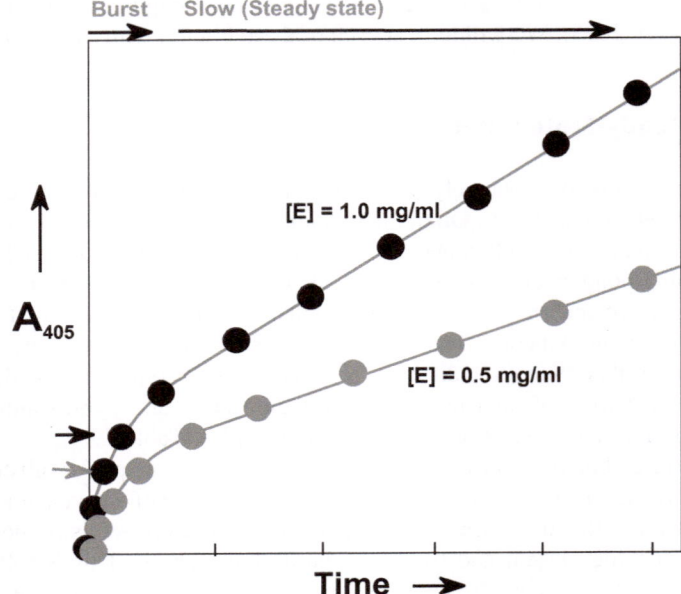

Fig. 29.4 Burst-phase kinetics observed with hydrolysis of 4-nitrophenyl acetate by chymotrypsin. Formation of 4-nitrophenol is monitored as increase in A_{405} with time. The extent (its amplitude—gray arrows on Y-axis) and the slope of burst are proportional to the total catalytically active enzyme. This property can indeed be used to determine the concentration of active enzyme present in a given preparation

active site to react. With caution, such partial exchanges constitute an operational test for the covalent reaction intermediate.

The mechanism of glutamine synthetase offers a different example of reaction intermediate. In this three-substrate sequential mechanism, γ-glutamyl phosphate is formed at the active site. Nucleophilic attack by ammonia to this intermediate displaces phosphate to form glutamine (Fig. 29.5). The covalent intermediate (-γ-glutamyl phosphate) is inferred by the positional isotope exchange (PIX; for a detailed treatment see Chap. 24) study. The ^{18}O-labeled β,γ bridge oxygen of ATP exchanges with non-bridge β-phosphate oxygens—for this exchange to occur

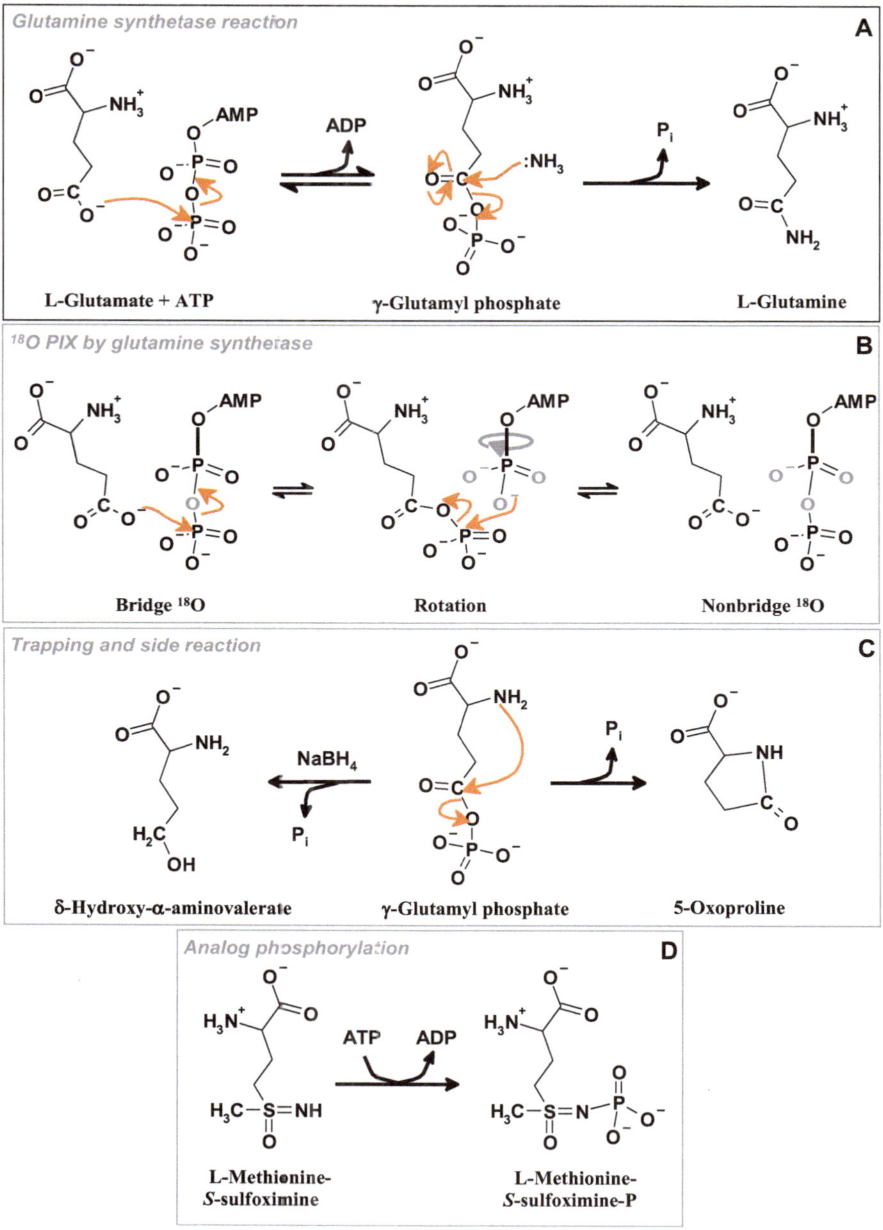

Fig. 29.5 Glutamine synthetase catalysis proceeds via γ-glutamyl phosphate as the obligate covalent reaction intermediate. (**a**) Glutamine synthetase reaction. (**b**) Scrambling of ^{18}O-label evidenced by PIX. (**c**) Trapping γ-glutamyl phosphate by NaBH$_4$ reduction and 5-oxproline forming side reaction. (**d**) Phosphorylation of L-methionine-S-sulfoximine

glutamate must be present. However, this ^{18}O-label scrambling can occur in the absence of ammonia. This PIX data is consistent with glutamate-dependent reversible mobilization of the ATP γ-phosphate group.

29.4.3 Inference from Analogs and Side Reactions

Geometric analogs of substrate–substrate covalent intermediates (formed at the active site) are expected to achieve tight binding. Examples of this kind include analogs of γ-glutamyl phosphate (for glutamine synthetase) and aminoacyl adenylate (for aminoacyl t-RNA synthetase). Here the enhanced binding affinity is taken to indicate the similarity of the intermediate analog to the true covalent intermediate. Such evidence, however, is suggestive but not decisive.

The same active site (of many enzymes) may also exhibit side reactions. Analyzing these side reactions often provides clues to the covalent intermediates formed. Again, glutamine synthetase offers an excellent example of this concept. Apart from ^{18}O-label exchange (β,γ bridge to non-bridge PIX in ATP), the enzyme is capable of the following additional, side reactions: (1) Synthesis of pyroglutamate (5-oxoproline) from ATP and L-glutamate in ammonia-depleted conditions, (2) Formation of ATP by the enzyme when incubated with synthetic acyl phosphate and ADP, and (3) ATP-dependent covalent phosphorylation of L-methionine-S-sulfoximine (an irreversible inhibitor of glutamine synthetase) on its sulfoximine nitrogen. All these side activities are consistent with the formation of γ-glutamyl phosphate—an activated covalent intermediate—during the normal reaction of glutamine synthetase (Fig. 29.5).

29.4.4 Direct Observation and/or Trapping

Direct observation of covalent intermediates may be possible whenever they are sufficiently stable, and they possess readily detectable spectral properties. It is practically difficult to observe/isolate any intermediates for enzyme reactions with their usual turnover numbers ($>1 \ s^{-1}$). Either one resorts to fast reaction kinetic methods (Chap. 10) or attempts to slow down the rate of their breakdown by reducing the temperature (thermal trapping). In any case, the intermediates are best detected when they possess coenzyme chromophores like NAD$^+$, FAD, pyridoxal phosphate, and cobalamin. Electron spin resonance (ESR; for radical intermediates), nuclear magnetic resonance (NMR; for structural information), X-ray crystallography (for structures and distances of "frozen" intermediates), ESI-MS are some of the specialized but powerful tools that provide information on intermediates. The α-imino acid intermediates, postulated in the reductive amination reactions, are often elusive (Paczelt et al. 2023). An enzyme-bound iminoglutarate reaction intermediate was detected in the crystal structure of NADP-glutamate dehydrogenase (Prem Prakash et al. 2018). An intermediate covalently bound to the enzyme was detected and the lysozyme catalytic mechanism was finalized (Kirby 2001). An

Table 29.4 Enzyme reaction intermediates trapped by borohydride reduction

Enzyme example	Intermediate	Intermediate trapped as
Fructose-1,6-bisphosphate aldolase	Imine adduct (Schiff's base)	N^e-Dihydroxyisopropyl derivative of active site lysine
Histidine decarboxylase (bacterial)	Imine adduct (Dehydroalanine)	N^α-Carboxyethyl derivative of histidine and histamine (formed from enzyme pyruvoyl group and α-amino group of substrate)
Lactate racemase	Lactyl thiolester	Lactaldehyde (formed by reducing thioester of enzyme cysteine)
Glutamine synthetase	γ-Glutamyl phosphate	δ-Hydroxy-α-aminovalerate ($-CO\sim OP$ reduced to $-CH_2OH$)

In the first three examples, the intermediate is covalently bound to the enzyme. In the last case, the intermediate is non-covalently held at the active site

inherently unstable putative hemiketal phosphate intermediate, noncovalently bound to the enzyme 3-deoxy-D-manno-2-octulosonate-8-phosphate synthase, was visualized by the application of time-resolved electrospray ionization (ESI) mass spectrometry. The enzyme reaction was directly monitored on a short millisecond timescale, in the ESI time-of-flight (ESI-TOF) mass spectrometer (Li et al. 2003).

Reactive intermediates when present may be trapped by suitable chemical reagents. Identifying covalent intermediates by chemical trapping is a time-tested tool for the enzyme chemist. Although very powerful, trapping has two limitations—it is a destructive method and what is trapped may be an artifact of the procedures employed. Methods to trap intermediates include (a) acid/base treatment to stabilize the ephemeral intermediate, (b) intercept the reactive species with nucleophiles like hydroxylamine to form a stable derivative, (c) reduce the intermediate chemically with sodium borohydride or lithium aluminum hydride, or (d) oxidation to detect thiol esters and vicinal diol intermediates. We have already seen an example of an activated carbonyl intermediate (e.g., N-acetyl imidazole in Fig. 29.3) captured by aniline (acting as intercepting nucleophile). Chemical trapping of intermediates by $NaBH_4$ is very valuable in probing enzyme reaction mechanisms (Table 29.4). Such trapping studies lend implicit experimental support to the proposed reaction mechanism.

29.4.5 Stereochemical Evidence

Examining the stereochemical course of a reaction is yet another approach to probe the formation of a covalent intermediate (Tanner 2002; Rose 2006). It lends direct support to the formation of covalent intermediates, if any. We note that an S_N2 reaction always involves an inversion of configuration (Fig. 27.8). If two discrete S_N2 steps occur one after the other, then overall retention of configuration is seen—due to two consecutive inversions (one from the donor to the enzyme and the other from the enzyme to the acceptor). If no covalent intermediate is formed during a

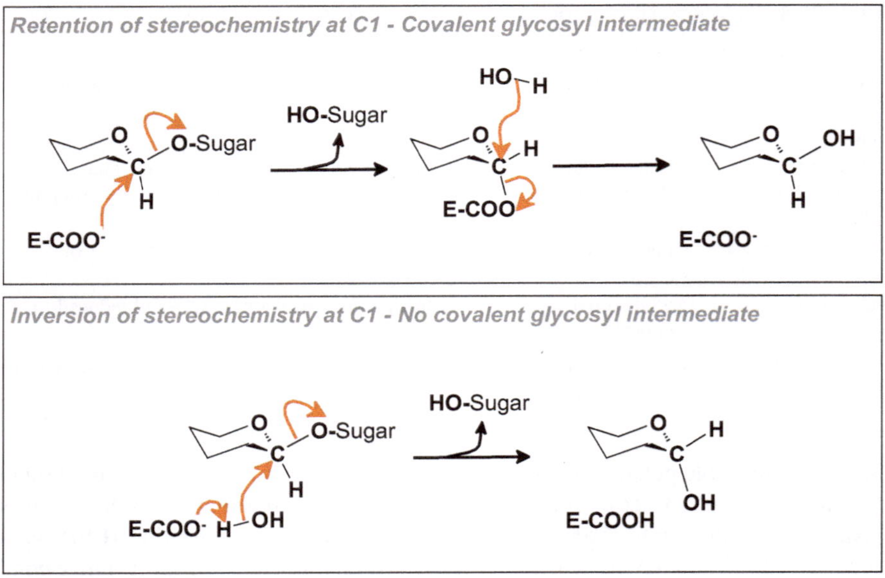

Fig. 29.6   Stereochemistry at the C-1 carbon of glycosidases. Formation of covalent intermediate is associated with retention of stereochemistry (top panel) while its absence results in inversion (bottom panel) at the glycosidic carbon. Only relevant structural details are shown for the sake of clarity

reaction cycle, then the product should show inverted configuration with respect to the substrate. A product with an inversion thus indicates (a) that covalent intermediate may not form during the reaction and (b) a direct in-line transfer between the substrates.

Glycosidases provide elegant examples to demonstrate this concept. In case of lysozyme, hydrolysis of the glycosidic linkage occurs with retention of stereochemistry at the glycosidic carbon. An enzyme nucleophile (Asp52-COO$^-$) covalently attaches to the glycosidic carbon while the departing sugar leaves from the other side (Kirby 2001). Upon subsequent attack of water, the product is formed with ***retention of stereochemistry*** at the glycosidic center (Fig. 29.6). Similar retention of stereochemistry was observed with sucrose phosphorylase. The two enzymes are thus retaining glycoside transferases. Evidence of retention of stereochemistry implies the presence of a covalent intermediate during reaction.

There are also glycosidases (e.g., some cellulases) where the ***inversion of stereochemistry*** occurs at the glycosidic center. A covalent intermediate may not form in such cases (Fig. 29.6). These enzymes also contain an active site carboxylate (–COO$^-$) but its role is different. X-ray crystallography data indicates that it is not near enough to form a covalent bond. Instead, it assists in the direct nucleophilic attack of water.

A final word of caution on the interpretation of stereochemical evidence is in order. We have seen earlier that steady-state kinetics can never prove a mechanism—it can only rule out alternative pathways and narrow down the choice. Similarly, stereochemical criteria by themselves cannot solve a reaction mechanism. Consider this. Inversion of configuration can arise from a single nucleophilic displacement reaction—but it can also arise from three, five, or any odd number of successive displacements. Likewise, retention of configuration implies not necessarily just two successive displacements but any even number. Therefore, a direct observation of a competent intermediate constitutes an unambiguous proof of its existence.

29.4.6 Catalytic Competence of an Intermediate

It is always a major challenge to conclusively establish the identity of a chemical reaction intermediate (Purich 2002). May be, existence of the postulated species is supported by a small subset of experiments listed above. Multiple lines of evidence are required to make a strong inference on its existence (metaphorically, a minimum of three legs are required for a stool to be stable and stand!). Whenever feasible it should be isolated, synthesized, and characterized for the two criteria. An intermediate once proposed (through direct and indirect experimental reasoning) should still satisfy the twin criteria of chemical competence and catalytic competence.

The postulated intermediate is *chemically competent* only if it is converted by the enzyme to go to products. It should also serve to go back to the substrate in the case of reversible enzyme reactions. γ-Glutamyl phosphate satisfies this standard and qualifies as a chemically competent intermediate of glutamine synthetase reaction (Fig. 29.5). Intermediates that are artifacts of the analysis itself most likely do not satisfy this condition.

The proposed (maybe isolated) intermediate should react to form products (in either direction) at rates at least as fast as the rate of overall reaction. Further, not one but all the actual intermediates in the mechanism of an enzyme must be kinetically competent. The *kinetic competence* implies that the intermediate has the capacity to be formed and discharged at rates equal to or greater than the overall rate of the normally occurring reaction. The acyl-enzyme of chymotrypsin is a true covalent intermediate during its catalytic cycle (Fig. 29.4). For a good substrate, acylation and deacylation rates are comparable with no net accumulation of the acyl-enzyme intermediate. With a poor substrate like 4-nitrophenyl acetate, however, acyl-chymotrypsin accumulates significantly as the deacylation step becomes rate-limiting. It is important to note that correspondingly the overall rate of the reaction itself slows down with the poor substrate still satisfying the role of acyl-enzyme as a kinetically competent intermediate. Phosphoglucomutase is another example where the phospho-enzyme intermediate was shown to be both kinetically and chemically competent.

In conclusion, any exceptions to the two criteria of catalytic competence rules out that species as an intermediate on the main pathway of an enzymatic mechanism.

29.5 Summing Up

Nucleophilic catalysis is an important weapon in the armory of enzymes. While not all enzymes employ covalent intermediates during their catalytic cycle, the actual number is quite large. By dividing the overall reaction into a suite of partial reactions, covalent catalysis achieves an energetically easy path. Enzymes participate in covalent catalysis—often by themselves becoming covalent partners. For this, they employ a range of nucleophilic groups provided by their amino acid side chains or cofactors/prosthetic groups.

Nature has carefully chosen active site nucleophiles for their reactivity (Ribeiro et al. 2023). They may be further modulated by the microenvironment effects of the active site. Consider subtilisin for example. The imidazole side chain of histidine (with a pK_a around 7.0) is an effective base at neutral pH. By correlation, its unprotonated form should be the most effective nucleophile. A stronger base like the Ser-O^- however makes a better nucleophile. But its concentration in the bulk aqueous phase (at pH 7.0) will be very small. The active site microenvironment ensures that it is generated and stabilized. While both are essential active site residues for subtilisin, nature has chosen Ser-OH as the nucleophile and His (imidazole) as the general base for catalysis. Subtilisin like all other serine proteases bears Ser-OH as its active site nucleophile.

When compared to the native form, the Ser→Ala mutant is a very poor catalyst. This identifies active site Ser-OH as a valuable nucleophilic tool. The residual activity of the Ser→Ala mutant of subtilisin then, by default, must use OH^- for the initial attack. Papain—a cysteine protease—bears a Cys-SH as its active site nucleophile. But the Ser→Cys mutant of subtilisin (called thio-subtilisin) is inactive. This is surprising in that generally $–S^-$ (thiolate) is a better nucleophile than $–O^-$ (alkoxide). Clearly modulation of reactivity, nucleophilicity and geometry at the active site are important.

Covalent reaction intermediates are the direct manifestation of nucleophilic catalysis. Because of their short-lived nature, it is a challenge to track them down. A solid proof of their existence therefore requires multi-pronged data—chemical, kinetic, spectroscopic, and stereochemical. Ultimate proof of their involvement in the enzyme mechanism must ensure that they are kinetically and chemically competent.

References

Fersht AR, Kirby AJ (1967) Structure and mechanism in intramolecular catalysis. The hydrolysis of substituted aspirins. J Am Chem Soc 89:4857–4863

Frey PA (2015) Transient intermediates in enzymology. J Biol Chem 290:10610–10626

Kirby AJ (2001) The lysozyme mechanism sorted– after 50 years. Nat Struct Biol 8:737–739

Li Z et al (2003) A snapshot of enzyme catalysis using electrospray ionization mass spectrometry. J Am Chem Soc 125:9938–9939

Paczelt V et al (2023) Glycine imine—the elusive α-imino acid intermediate in the reductive amination of glyoxylic acid. Angew Chem Int Ed 62:e202218548

Prakash P, Punekar NS, Bhaumik P (2018) Structural basis for the catalytic mechanism and α-ketoglutarate cooperativity of glutamate dehydrogenase. J Biol Chem 293:6241–6258

Purich DL (2002) Covalent enzyme-substrate compounds: detection and catalytic competence. Methods Enzymol 354:1–27

Ribeiro AJM et al (2023) A global analysis of function and conservation of catalytic residues in enzymes. J Biol Chem 295:314–324

Rose IA (2006) Mechanistic inferences from stereochemistry. J Biol Chem 281:6117–6119

Tanner ME (2002) Understanding Nature's strategies for enzyme-catalysed racemization and epimerization. Acc Chem Res 35:237–246

Phosphoryl Group Chemistry and Importance of ATP

30

Reactions involving transfer of phosphoryl groups are central to the metabolism of all living beings. This chapter brings out the unique features that make phosphate and its derivatives ideal candidates for driving metabolism. Chemistry at the phosphorus atom is almost always catalyzed by a suitable enzyme. Reaction mechanisms pertaining to these enzymes are presented. Aspects of high-energy compounds like ATP and their role in group transfer reactions are highlighted.

30.1 Why Nature Chose Phosphates?

Phosphoric acid esters and anhydrides are cardinal players in metabolism and enzyme chemistry (Westheimer 1987). They are found uniformly in the pathways of all biomolecules—nucleic acids, proteins, carbohydrates, and lipids. A few important phosphate compounds representing a range of linkages are listed in Table 30.1.

Phosphoric acid, its esters, and anhydrides are particularly selected by biological systems on the following counts. First, even as its diester (read genetic material like DNA!) phosphate retains one negative charge and is thus noticeably stable to nucleophilic attack. Consider this with 55.5 M of water as a reasonable nucleophile around. Second, the permanent negative charge serves to retain phosphate compounds inside—as they cannot cross the phospholipids' bilayer without assistance. Third, negative charges on phosphates are excellent specificity/recognition entities of phosphorylated substrates in binding to enzymes. Lastly, phosphate (and at times pyrophosphate) is usually a good leaving group in many nucleophilic displacement reactions.

Phosphoric acid is tribasic acid. Its successive ionization constants differ by factors of $>10^5$. The three pK_as of phosphoric acid are well spaced—$pK_{a1} = 2.12$, $pK_{a2} = 7.21$, and $pK_{a3} = 12.32$. Not many other polyanionic compounds are endowed with this feature. Therefore, phosphoric anhydrides exhibit many favorable

N. S. Punekar, *ENZYMES: Catalysis, Kinetics and Mechanisms*, https://doi.org/10.1007/978-981-97-8179-9_30

Table 30.1 Phosphate derivatives in metabolism

Nature of linkage to phosphoric acid	Examples
Monoester	Glucose-6-phosphate, Dihydroxyacetone phosphate
Diester	DNA, RNA, and Phospholipids
Enolic ester	Phosphoenolpyruvate
Amide	Phosphocreatine
Anhydride	ATP, Pyrophosphate and Acetyl phosphate
Pyrophosphate ester	Isopentenyl pyrophosphate, 5-Phosphoribosyl-1-pyrophosphate

properties. They are protected by these negative charges from rapid attack of water and other nucleophiles. Thus, even though thermodynamically unstable, they are kinetically quite stable in aqueous environment. This remarkable combination of thermodynamic instability and kinetic stability makes them stand out as potentially *energy-rich* compounds (like ATP, see below). They can drive uphill chemical reactions in the presence of a suitable catalyst. Nature prefers phosphate esters because they are stable, yet they can be attacked and cleaved by enzymic hydrolysis.

Arsenate ($HAsO_4^{2-}$) and phosphate (HPO_4^{2-}) have striking similarities with nearly identical pK_a values, similarly charged oxygen atoms, and close thermochemical radii. Therefore, discriminating phosphate from arsenate is a paramount challenge for enzymes. Then why not arsenate? Arsenate esters, unlike phosphate esters, are notoriously unstable in aqueous solutions—upon water attack they rapidly decompose to alcohol and arsenate (Elias et al. 2012; Tawfik and Viola 2011). Rapid hydrolysis of the corresponding high-energy arsenate esters (arsenolysis) leads to wasteful "futile cycles" and therefore decouples metabolism.

30.2   Chemical Mechanisms at the Phosphoryl Group

Phosphoryl group ($-PO_3^-$) transfers are ubiquitous in intermediary metabolism. And they are invariably enzyme-catalyzed reactions (Knowles 1980; Cleland and Hengge 2006). Every thermodynamically uphill step of metabolism involves a displacement at the phosphorous atom of a phosphoric monoester or anhydride. Mechanistically phosphoryl group transfer reactions can be studied at three levels: (a) whether a phospho-enzyme is formed during the reaction, (b) which is the rate-limiting step and the nature of the transition state, and (c) whether the displacement at phosphorus atom is associative or dissociative in nature.

30.2.1 Bond Cleavage at Phosphorus Atom

In all phosphoryl and pyrophosphoryl group transfers, the phosphorus atom reacts as an electrophilic center. It is therefore handed over from one nucleophile to the other.

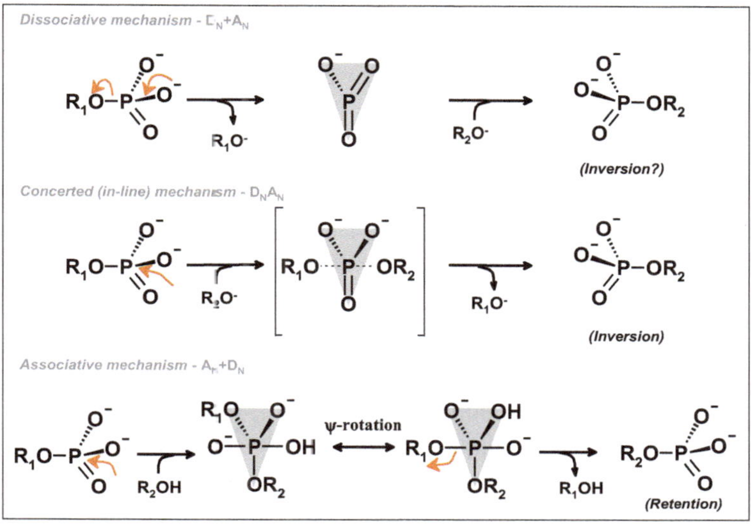

Fig. 30.1 Mechanisms of phosphoryl group transfer. The trigonal plane defining the phosphoryl group is shown as gray triangle (1) Dissociative pathway goes through an unstable metaphosphate intermediate. While a racemic product is expected in solution, the spatial arrangements of substrates at the enzyme active site govern the stereochemical outcome. (2) The concerted (in-line) reaction between the two substrates leads to inversion. (3) The associative (adjacent) mechanism always leads to retention of configuration. The pathway goes through pseudo (ψ)-rotation at the penta-coordinate intermediate stage. Subsequently, the leaving group (R_1OH) always leaves from an apical position of the trigonal bipyramid

All enzymatic phosphoryl transfers proceed with the cleavage of the phosphorus–oxygen bond and a nucleophile forms a bond to phosphorus. During phosphoryl ester hydrolysis for instance, the oxygen of water appears in phosphate—implying a nucleophilic attack by "O" of water on P. This is easily demonstrated by performing hydrolysis in ^{18}O-labeled water. Whenever phosphorylation serves to activate a group, the situation is different. The acyl group transfer and glycosyl group transfer reactions are illustrative. In both these cases the carbon–oxygen bond is cleaved, and a nucleophile forms a bond to carbon. We will discuss the *activation of groups* and the concept of a *good leaving group* later on.

Phosphoryl group-transfer reactions broadly fall into three types (Fig. 30.1). These differ in the mechanism of bond cleavage/formation at the phosphorus atom.

1. In the ***dissociative mechanism,*** a meta-phosphate (PO_3^-) intermediate is formed prior to attack by the incoming nucleophile (Fig. 30.1). The outgoing nucleophile dissociates in the first step (D_N) while the incoming nucleophile attacks the meta-phosphate in the second (A_N). In the IUPAC nomenclature, it is denoted as $D_N + A_N$ mechanism and is analogous to the S_N1 reaction mechanism in carbon chemistry (see Chap. 27). Evidence (through kinetic isotope effects; Chap. 25) for

a meta-phosphate-like transition state (loose TS) was obtained for the bovine protein tyrosine phosphatase. Similarly, stabilized meta-phosphate entity has been experimentally observed with fructose-1,6-bisphosphatase and *Lactococcus lactis* β-phosphoglucomutase.

2. The ***concerted mechanism*** proceeds via a penta-coordinate transition state. In this case (which is S_N2 like and is also known as D_NA_N mechanism) the attacking nucleophile enters opposite the leaving group. In this ***in-line mechanism,*** no reaction intermediate is involved. Adenylate kinase is an example of this type. Its mechanism involves an inversion at the P atom; this can be tested using a chiral phosphate ester and examining the stereochemical outcome of the overall reaction (see below).

3. The third possibility is of an ***associative mechanism***. Here the incoming nucleophile attacks first to give a penta-coordinated phosphorane intermediate (Fig. 30.1). It is therefore shown as $A_N + D_N$ mechanism. The penta-covalent intermediate is of trigonal bipyramid geometry; the five substituents therefore are found either at *equatorial* or at *apical* position. In this ***adjacent mechanism,*** the nucleophile enters on the same side as the leaving group. Since groups can enter or leave only from the apical position, the trigonal bipyramid formed has to rearrange. This movement—termed ***pseudorotation***—brings the originally equatorial leaving group to apical position for expulsion. A well-characterized example of associative mechanism is bovine pancreatic ribonuclease A. The 2′-OH of ribose sugar attacks the phospho-diester via an associative mechanism to form a divalent transition state stabilized by Lys-41.

We know that tetravalent carbon compounds are stable. But phosphorus can form stable trivalent (planar), tetravalent (tetrahedral), and pentavalent (a trigonal bipyramid with three equatorial and two axial bonds to P) compounds. This has a bearing on the relevant bond orders to P atom, possible during reaction. Concerted mechanism of phosphoryl transfer reactions straddles the two outer limits of an associative (five bonds to P) to a fully dissociative (three bonds to P) transition state. The character of the transition state thus ranges from (a) being associative with the sum of the axial bonds between one and two, (b) to S_N2 with the sum of the axial bonds equal to one, (c) to dissociative with the sum of the axial bonds less than one. As a general rule, at least with nonenzymatic phosphoryl transfer reactions, the trend is loose transition states for monoesters, a synchronous reaction for diesters, and a tight transition state for triesters. As the phosphate is more esterified (like in phosphodiesters and phosphotriesters), these stabilized phosphoesters require additional bond order from the nucleophile to achieve transition state. In reality, progressively more associative mechanisms occur on a continuum. This also depends on whether a good leaving group is found on the P atom (such as a concerted mechanism) or not (such as in a fully associative mechanism).

30.2.2 Stereochemistry of Phosphoryl Transfer

Stereochemical course of a reaction provides excellent clues to the reaction mechanism. This is true as well with the chemical mechanisms at P atom. Phosphates contain three apparently identical oxygen substituents. And it is convenient that three isotopes of oxygen are available—^{16}O, ^{17}O, and ^{18}O. Chiral phosphate esters can be prepared to act as enzyme substrates. Upon reaction, chiral phosphate ester products may be generated (Fig. 30.1). Absolute configuration of such products can be analyzed, albeit with some chemical (technical) skill and spectroscopy. In the final outcome, we can clarify whether the enzymatic reaction proceeds with retention or inversion of configuration at the P atom.

Defining the stereochemical course for phosphate (when compared to its ester) is a bit tricky. Phosphate has four equivalent oxygen atoms while only three isotopes of oxygen are available to mark them. This limitation can be overcome by using sulfur as the fourth substituent (Fig. 30.2). Fortunately, many enzymes do accept substrates with corresponding thiophosphate groups. Analyzing the configuration of $[^{16}O, ^{17}O, ^{18}O]$-thiophosphate product formed is all that is needed.

Stereochemical evidence is the most diagnostic of mechanistic pathway of phosphoryl group transfer. This is particularly true in cases where a phosphorylated enzyme intermediate cannot be isolated and characterized. A single displacement at the phosphorus generally results in inversion of stereochemistry. Reactions with no phosphoryl enzyme intermediate (a direct *in-line* attack leading to phosphoryl transfer) usually go with inversion. Phosphokinases proceed with inversion at P suggesting a direct transfer between the two substrates. Phosphomutases—viewed as internal kinases—proceed with retention. This results from two inversions because of a double displacement involving a phosphoryl enzyme intermediate. In enzymes

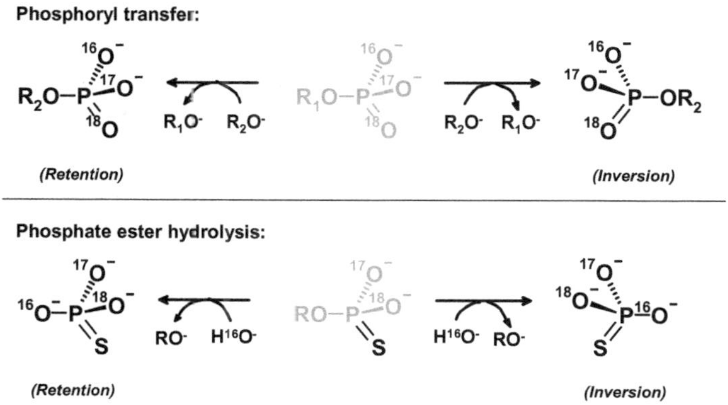

Fig. 30.2 Stereochemistry at the phosphorus atom. The phosphoryl group has the same tetrahedral geometry as the saturated carbon compound. The three oxygen isotopes (^{16}O, ^{17}O, and ^{18}O) along with the unique R group on the fourth oxygen define the stereochemistry of phosphoryl transfer. For phosphate ester hydrolysis, however, S atom of thiophosphate serves as the fourth substituent to define the stereochemistry of phosphate release

such as alkaline phosphatase—where a phosphoryl enzyme form exists—the reaction ends up with retention. The presence of a phospho-enzyme intermediate (for *E. coli* alkaline phosphatase) was demonstrated by incorporation of ^{32}P from the labeled substrate on to the enzyme Ser residue. The associative (adjacent) mechanism always leads to retention as the configuration is retained in each step.

Phosphoryl Transfer Mechanism: Single or Double Displacement?
The transfer of a phosphoryl group (from the donor substrate to the acceptor substrate) in principle can occur in two different ways. (1) The phosphorylium group may be directly transferred from the donor substrate to the acceptor substrate—a *single displacement* route. (2) The phosphoryl group is first transferred to a suitable group on the enzyme (phosphoenzyme covalent intermediate is formed) and is then transferred from the enzyme to the acceptor substrate in the second step—a *double displacement* route.

Examples to illustrate the two different modes of phosphoryl transfer are adenylate kinase and nucleoside bisphosphate kinase. The chemical reactions catalyzed by the two enzymes are apparently similar.

Adenylate kinase:

$$AMP + Mg\text{-}ATP \rightleftarrows ADP + Mg\text{-}ADP$$

Nucleoside bisphosphate kinase:

$$Mg\text{-}GDP + Mg\text{-}ATP \rightleftarrows Mg\text{-}GTP + Mg\text{-}ADP$$

However, adenylate kinase follows a single displacement mechanism while nucleoside bisphosphate kinase operates through a double displacement mechanism. A kinetically competent phospho-enzyme (see Table 29.4) has been demonstrated for the latter enzyme. The types of experimental evidence to support and contrast these two mechanisms are tabulated below.

Single Displacement (Adenylate kinase)	Double Displacement (Nucleoside bisphosphate kinase)
• Phospho-enzyme intermediate not formed	• Phospho-enzyme intermediate is formed
• Stereochemical inversion at P center	• Stereochemical retention at P center
• Sequential kinetic mechanism	• Ping-pong kinetic mechanism
• No partial exchange reactions occur	• Partial exchange reactions are observed

30.3   Adenosine Triphosphate: Structure Relates to Function

The most common phosphoryl group donor in metabolism is adenosine triphosphate—commonly abbreviated as ATP (Ramasarma 1998). It was identified as a derivative of adenosine with three phosphates by Fiske and SubbaRow in 1929. ATP is a derivative of adenosine-5′-phosphate with two more phosphates attached to its 5′-phosphate via anhydride linkages. Todd et al confirmed this by chemical synthesis in 1949. As noted above, and in contrast to carboxylic anhydrides, phosphoric anhydride groups in ATP are protected by their negative charges from rapid attack by water (and other nucleophiles). Fritz Lipmann's observation was prescient in ascribing the kinetic stability of ATP to the negative charges in ATP. This makes hydrolysis of ATP thermodynamically favorable (large $-\Delta G°$) but kinetically unfavorable (large $\Delta G^{\neq}$)—a virtue exploited by nature to use ATP as an ideal free energy currency. ATP is universally conserved as the principal energy currency. One reason suggested for this choice may be that its formation is chemically favored in aqueous solution under mild prebiotic conditions (Hon et al. 2022). However, there are examples of atypical glycolysis where phosphofructokinase uses pyrophosphate instead of ATP. The reduced thermodynamic driving force of this reaction could be increased by replacing the pyrophosphate-requiring native phosphofructokinase with one that uses ATP (Pinna et al. 2022).

More recently, ATP has been assigned another cellular role as a hydrotrope. Hydrotropes are small molecules that solubilize hydrophobic molecules in aqueous solutions. They are amphiphilic molecules and differ from classical surfactants with little aggregation features. As a natural cellular hydrotrope, ATP may act to keep proteins soluble (Patel et al. 2017; Rice and Rosen 2017).

30.3.1 ATP Is a High-Energy Compound

Combination of kinetic stability and thermodynamic instability imparts ATP its *energy-rich* nature. Hydrolysis of ATP can therefore be coupled to drive uphill chemical reactions in the presence of a suitable enzyme catalyst. *ATP provides energy by group transfer and not by simple hydrolysis.* The term *energy rich* implies that (a) it has a high phosphate group transfer potential and (b) on hydrolysis of its phosphoric anhydride bonds sufficient ΔG (free energy) is available for the formation of other bonds. The $\Delta G°$ for ATP hydrolysis (ATP $\rightleftarrows$ ADP + ⑪) is negative and very large (around 7.0 kcal/mol, at 25 °C and pH 7.0). This corresponds to an equilibrium constant for hydrolysis of about 140,000 M!

What makes ATP a *high-energy compound* and confers it *high phosphate group transfer potential*? Several factors that contribute include (a) Electrostatic repulsions between neighboring negative charges of ATP, (b) Relative bond energies of the reactants and products, and (c) Better solvation and relative resonance stabilization of the products (ADP + ⑪) of hydrolysis. Other compounds (listed above ATP in Table 30.2) may also be recognized as high-energy compounds on similar grounds. Since negative charges on ATP are a function of phosphate group ionization, pH and

Table 30.2 $\Delta G°$ of hydrolysis for important phospho-compounds

Phospho-compound	Structure[a]	$-\Delta G°$ (kcal/mol) (25 °C, pH 7.0)
Phosphoenolpyruvate		14.8
1,3-Bisphosphoglycerate		11.8
Acetyl phosphate		10.3
Phosphocreatine		10.3
Inorganic pyrophosphate (Ⓟ–Ⓟ)		8.0

Compound	Structure	ΔG°
ATP ($\rightarrow$ AMP + $\circled{P}$-$\circled{P}$)		7.7
ATP ($\rightarrow$ **ADP** + $\circled{P}$)		**7.3**
Glucose-1-phosphate		4.8
Glucose-6-phosphate		3.3

[a] The ΔG° of hydrolysis for *high-energy* bond (shown in gray) is listed in the last column (1.0 kcal = 4.184 kJ). Note that glucose-1-phosphate and glucose-6-phosphate are phosphate esters that do not contain a *high-energy* bond. Whenever abbreviated, the phosphoryl group is shown as "$\circled{P}$" throughout this book

ionic strength have a significant effect on $\Delta G°$ for ATP hydrolysis (Alberty 1969). ATP invariably occurs as a complex with divalent metal ions—Mg^{2+} in particular (see below). For these reasons, the likely $\Delta G°$ for ATP hydrolysis in vivo may be as high as 12.0 kcal/mol (and not 7.3 kcal/mol).

Like ATP, several phosphorylated compounds can transfer their phosphoryl groups to water. The associated free energy changes ($\Delta G°$ of hydrolysis) are listed in Table 30.2. Larger the negative $\Delta G°$ for a phospho-compound, greater is its phosphoryl group transfer potential. This sets up a nice hierarchy of energy transfer in coupled reactions. Under standard conditions, for instance, compounds above ATP (in Table 30.2) can transfer phosphoryl group to ADP. And those below cannot. Phosphoenolpyruvate has the highest negative $\Delta G°$; this is because the enol formed quickly coverts to the keto form making the reaction further exergonic. Phospho-enolpyruvate has a higher phosphoryl transfer potential than ATP; it can therefore phosphorylate ADP to ATP.

30.3.2 Complexes of ATP with Metal Ions

Metabolically relevant ATP occurs as a divalent metal ion complex. Most often this metal ion is Mg^{2+}. The shielding of ATP negative charges (Chap. 4) is important for access to nucleophiles. Mg-ATP^{2-} is the true substrate of most enzymes that are generally described as *ATP-dependent*. In reality, there is nothing like a pure solution of Mg-ATP^{2-}. An equimolar mixture of ATP and $MgCl_2$ at pH 7.0 contains the following species at different concentrations (in their decreasing order!): Mg-ATP^{2-}, ATP^{4-}, $HATP^{3-}$, Mg^{2+}, Cl^-, $MgHATP^-$, Mg_2-ATP, and $MgCl^+$. Moreover, their proportions vary with (a) the total [ATP] and [$MgCl_2$], (b) the pH and buffering species present, and (c) the ionic strength (Alberty 1969). For these reasons, studying interactions of Mg-ATP^{2-} with an enzyme obviously requires much care. Otherwise, phenomena like spurious cooperativity (Fig. 30.3) may occur (Punekar et al. 1985). To avoid such kinetic artifacts, it is important to accurately evaluate the true concentration of Mg-ATP^{2-} at a given total [ATP] and [$MgCl_2$]. The stability constants for many of the metal complexes of biochemical interest have been measured. It is thus possible to calculate the concentration of any complex given the concentrations of free components. There are simple computer programs to do such iterative calculations.

Three common experimental designs are found in literature while deciding [ATP] and [$MgCl_2$] to fix Mg-ATP^{2-} concentration. It does not help to either a) vary [ATP] and [$MgCl_2$] at equimolar ratios or b) fix [$MgCl_2$] at a very high value (say 10 mM) while [ATP] is varied. The third and the best option is to keep the total [$MgCl_2$] in constant excess over the total [ATP]. One final factor to consider is the variation in ionic strength as the component ($MgCl_2$ and ATP) concentrations are increased. It is desirable to maintain the ionic strength constant. A value of about 0.15 M may be appropriate since it approximates the ionic strength in vivo.

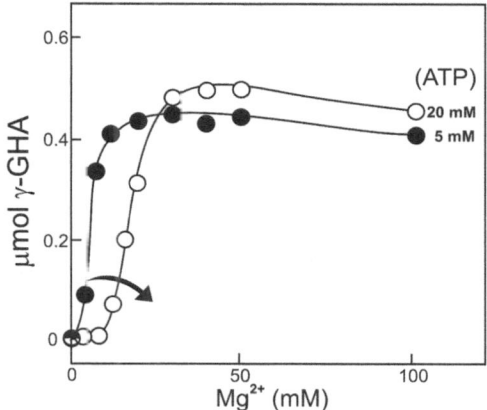

Fig. 30.3 Mg^{2+} saturation profile of glutamine synthetase. Mg-ATP is the true substrate of this enzyme (see Table 30.3 later, for the reaction). A small fraction of total ATP is present as Mg-ATP at lower $[Mg^{2+}]$. The higher the initial [ATP] used, this curve is shifted further to the right (arrow). This apparent sigmoidicity is thus a manifestation of equilibrium between free [ATP], free $[Mg^{2+}]$, and their various complexes and does not reflect cooperative enzyme kinetics

A method of continuous variation may also be used to arrive at the best ratio of [ATP] and $[MgCl_2]$. This protocol—called the *Job Plot*—provides useful information in optimizing the assay under a standard set of conditions. Mole fractions of [ATP] and $[MgCl_2]$ are varied such that the total molarity remains constant. If the enzyme prefers a 1:1 complex, then the enzyme activity will be maximal at mole fraction of 0.5 (as shown in Fig. 30.4). However, if the active species is a 1:2 complex (of ATP:Mg^{2+}), then the enzyme will be most active at mole fraction of 0.33.

Free ATP in solution (when not in a complex with divalent metal ion) takes a linear extended conformation. It is a flexible molecule because P–O single bonds enjoy many degrees of freedom. However, ATP conformation is frozen as a metal complex. ATP assumes specific folded forms, when so complexed. Different divalent metal ions interact differently with ATP; variations include interactions with α-, β-, and γ-phosphates and N7 of adenine (Fig. 30.5). This has a bearing on which ATP-metal ion complex can serve as substrate for a given enzyme.

Both tri-dentate and bi-dentate (Mg^{2+} bonded to O atoms of β- and γ-phosphate) complexes of Mg-ATP are formed in solution. Furthermore, β,γ-bi-dentate Mg-ATP exists as a rapidly equilibrating mixture of Λ and Δ screw-sense isomers (Fig. 30.6). These two isomers have opposite CD spectra. It is tricky to determine which of the three Mg-ATP forms the true substrate for a given enzyme. This can be attempted in the following ways:

1. ATP coordination complexes of ATP with Cr^{3+}, Co^{3+}, and Rh^{3+} are inert in water—their stable screw-sense isomers can be separated as pure Λ and Δ forms

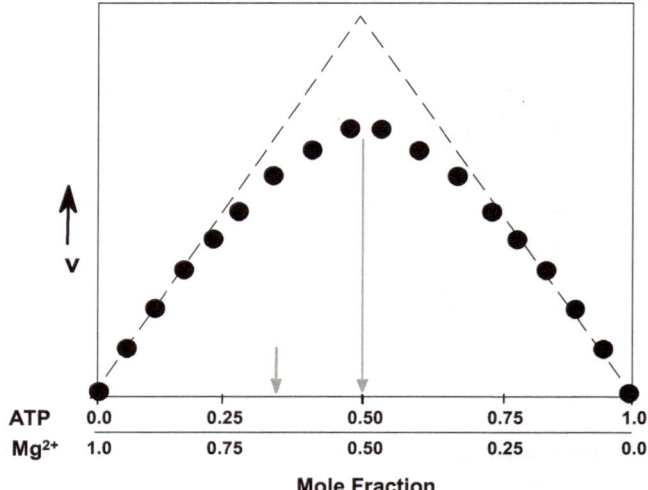

Fig. 30.4 Job plot for an enzyme requiring Mg-ATP as substrate. The maximal enzyme activity is at the mole fraction of 0.5 (1:1 ratio of ATP:Mg^{2+}; the long vertical gray arrow). The peak activity would move to the left for an optimal mole fraction of 0.33 (1:2 ratio of ATP:Mg^{2+}; short gray arrow)

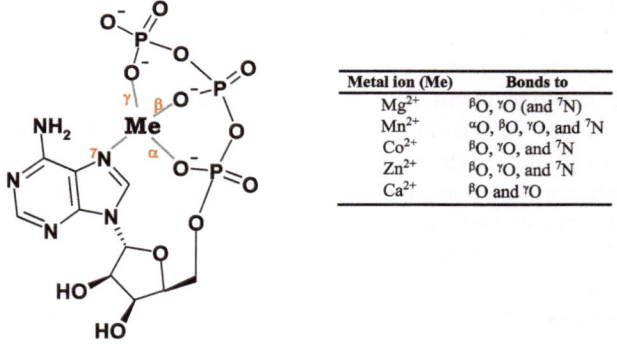

Metal ion (Me)	Bonds to
Mg^{2+}	$^{\beta}O$, $^{\gamma}O$ (and ^{7}N)
Mn^{2+}	$^{\alpha}O$, $^{\beta}O$, $^{\gamma}O$, and ^{7}N
Co^{2+}	$^{\beta}O$, $^{\gamma}O$, and ^{7}N
Zn^{2+}	$^{\beta}O$, $^{\gamma}O$, and ^{7}N
Ca^{2+}	$^{\beta}O$ and $^{\gamma}O$

Fig. 30.5 Each divalent metal ion interacts differently with ATP. Different combinations of phosphate oxygens ($^{\alpha}O$, $^{\beta}O$, and $^{\gamma}O$) and/or ^{7}N of adenine coordinate to the metal ion are shown in the accompanying table

and tested as possible substrates. For instance, hexokinase uses the Λ isomer of β,γ-bi-dentate Cr-ATP while adenylate kinase prefers the Δ isomer.

2. Chiral sulfur substituted ATP complexes differently with Mg^{2+} and Cd^{2+}. Mg^{2+} prefers to coordinate to O over S by a factor of 31,000, whereas Cd^{2+} prefers S

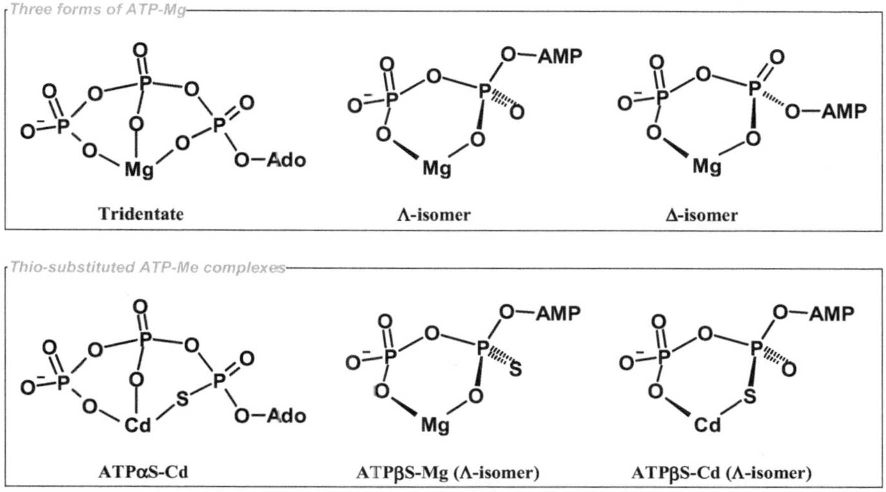

Fig. 30.6 Various forms of Me-ATP complexes in solution. The screw-sense isomers (Λ and Δ forms) of Mg-ATP are in rapid equilibrium. The α-phosphate of ATPαS is coordinated to Cd through the S atom. The β-phosphate of ATPβS becomes chiral as one O is replaced by S atom. The screw-sense isomers formed with Mg and Cd are reversed when the other stereoisomer of ATPβS is used. The adenosine moiety is abbreviated as Ado

over O by a factor of 60. And when sulfur replaces one of the two non-bridge oxygens of β-phosphate, the resulting ATPβS becomes chiral (Fig. 30.6). Therefore, with ATPβS the two metal ions form distinct (predominant) screw-sense complexes. Specific preference of an enzyme for Mg^{2+} or Cd^{2+} complex of ATPβS is thus indicative of its screw-sense isomer specificity. The two Λ screw-sense complexes shown in Fig. 30.6 are substrates for yeast hexokinase, for example.

3. ATPαS is useful in deciding whether the α-phosphate is coordinated to the metal ion during reaction. If α-phosphate coordination is significant for enzyme catalysis, then a reversal in the α-S isomer specificity is expected when Mg^{2+} is replaced by Cd^{2+}. The Mg^{2+} to Cd^{2+} switch affects α-S isomer specificity of creatine kinase but not of hexokinase. Thus, β,γ-bi-dentate Mg-ATP is the substrate for hexokinase whereas it is the tri-dentate Mg-ATP for creatine kinase.

30.3.3 ATP Binding to Enzymes

Negative charges of ATP tend to protect it from the attack by incoming nucleophiles. Then how is it handled at the enzyme active site? ATP exists in vivo largely as a complex with Mg^{2+}. Typically, Mg^{2+} neutralizes two of the negative charges. Ion pair interactions with active site Arg (guanidinium group) and Lys (ammonium group) residues contribute to binding (also hydrogen bonds) and further negative charge neutralization. Together such interactions generally define the productive

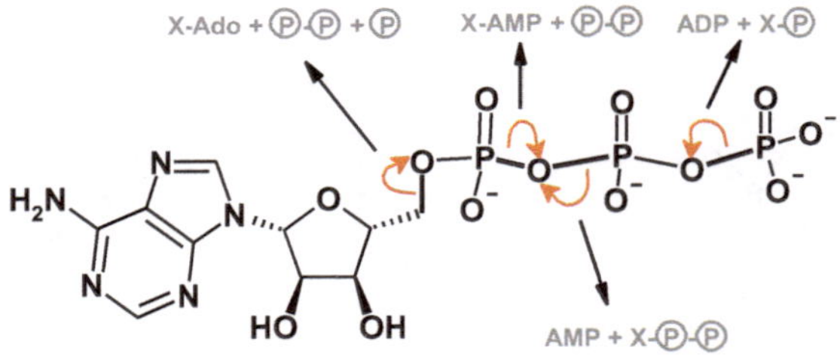

Fig. 30.7 Different modes of ATP cleavage. The electron movements (gray arrows) indicate the bond cleaved when a nucleophile (X:) attacks one of the P or C atoms of ATP. Some representative enzyme examples of these may be found in Table 30.3

binding of ATP to the enzyme active site. While many subtle variations on this theme are possible, there is a degree of active site conservation to accommodate ATP. The sequence popularly known as **Walker motif**-A (also called P loop) is thought to be the nucleotide binding site in many proteins. It consists of the sequence $A/GX_4GKT/S$ (in one-letter amino acid code) and is flanked by a β-strand and an α-helix. Walker motif loops around the tri-phosphate moiety of ATP. Walker motif-B on the other hand, consists of X_4D (where X is almost exclusively a hydrophobic residue), occurs at the end of a β-strand and interacts with the Mg^{2+} ion coordinated to the tri-phosphate moiety of ATP.

30.3.4 Different Modes of ATP Cleavage

ATP is a versatile molecule serving as a free energy source as well as a donor of its constituent groups namely, phosphate (Ⓟ), pyrophosphate (Ⓟ-Ⓟ), and AMP. Accordingly, ATP hydrolysis may occur at its γ-, β-, or α-phosphate group (Fig. 30.7).

The ATP α-bond is cleaved in nucleotidyltransferase reactions, the γ-bond is cleaved in kinase (phosphotransferase) reactions. Both α- and γ-bonds are mobilized in S-adenosylmethionine synthesis (Frey and Magnusson 2003; Struck et al. 2012). Cleavage at the β-bond is rare. Both α- and β-bond cleavages are thought to proceed through an associative transition state—as it is crowded at the P atom. With its two negative charges and because it is a good leaving group, a dissociative transition state is expected when the γ-bond of ATP is cleaved. Representative examples of all these modes of ATP cleavage are listed in Table 30.3.

30.4 Investing Group Transfer Potential to Create Good Leaving Groups

The **group transfer potential** of a compound is defined as the negative of free energy of its hydrolysis. The transfer of a group (such as γ-phosphate from ATP) to water (hydrolysis) provides a standard for use in comparing such reactions. The more

Table 30.3 Different modes of ATP cleavage in metabolism

ATP cleavage pattern	Enzyme and reaction	Comment[a]
ATP → ADP + Ⓟ	*Glutamine synthetase* Glutamate + NH_3 + ATP → Glutamine + ADP + Ⓟ	Bond between β-γ phosphates is split
ATP + X → ADP + X-Ⓟ	*Hexokinase* Glucose + ATP → Glucose 6-Ⓟ + ADP	The γ-phosphate is transferred to acceptor
ATP → AMP + Ⓟ-Ⓟ	*Aminoacyl-tRNA synthetase* Amino acid + ATP + tRNA → Aminoacyl-tRNA + AMP + Ⓟ-Ⓟ	Bond between α-β phosphates is split
ATP + X → AMP + X-Ⓟ-Ⓟ	*5-Phosphoribosyl-1-pyrophosphate synthetase* Ribose-5-Ⓟ + ATP → 5-Phosphoribosyl-1-pyrophosphate + AMP	Bond between α-β phosphates split with pyrophosphate transfer to acceptor
ATP + X → X-AMP + Ⓟ-Ⓟ	*FAD synthetase* FMN + ATP → FAD + Ⓟ-Ⓟ	Bond between α-β phosphates split with AMP transfer to acceptor
ATP + X → X-Adenosine + Ⓟ-Ⓟ + Ⓟ	*S-Adenosyl-methionine synthetase* Methionine + ATP → S-Adenosyl-methionine + Ⓟ-Ⓟ + Ⓟ	Bonds at α and γ phosphates split with adenosine transfer to acceptor

[a]Refer Fig. 30.7 to visualize which bond(s) of ATP are mobilized

positive the group transfer potential, the greater is the tendency to transfer the group to an acceptor. (This concept is like how we define standard reduction potential and rank reducing agents). Thus, a comparison of group transfer potentials provides the means for establishing the direction of group transfer among various donor–acceptor molecules. Recall that since phosphoenolpyruvate has higher phosphoryl transfer potential, it can phosphorylate ADP to ATP (Table 30.2). The concept of group transfer potential is not just limited to phosphoryl transfers but may be extended to others like acyl group transfers (Perler 1998). Some relevant examples of compounds with high group transfer potential are listed in Table 30.4.

Leaving groups in a reaction are quite often nucleophiles (see Chap. 29). And a *good leaving group* is one that is stable in solution. Generally, lower the base strength, the greater the ease of expulsion. In keeping with this, the leaving group order is acetate (pK_a, 4.76) > phosphate (pK_a, 7.00) > hydroxyl (pK_a, 15.8). Alkoxides (RO^-; pK_a, 16.0) and amide ions (RNH^-; pK_a, 30) are difficult to expel because they are strongly basic. They need to be protonated to leave from a reaction center—enzymes provide suitable acid groups to donate protons and help them leave. More importantly, a poor leaving group must be activated for expulsion. This is done through energy-rich compounds that exhibit high group transfer potential (Table 30.4). A major objective of investing phosphate transfer potential of ATP into different recipient substrates is thus to achieve better leaving group chemistry for the reaction. Tagging poor leaving groups with different parts of ATP serves to activate them. For instance, methyl transfer from methionine is difficult because basic thiolate anion has to be expelled. But when methionine is activated as *S*-adenosyl-methionine (see Table 30.3), a facile leaving group is created. This is *S*-adenosyl-homocysteine—a neutral, nonbasic sulfur compound. In this sense, *high group transfer potential is used to craft a good leaving group*. We will elaborate this concept below by describing acyl-group activation, frequently encountered in metabolism.

30.4.1 Activating Carboxylate for Acyl Group Transfer

Nucleophiles need to approach and attack the carboxylate carbon to form various products. Important examples include ester synthesis (R–OH as the incoming nucleophile), peptide/amide bond formation (R–NH$_2$ as the incoming nucleophile),

Table 30.4 Compounds with high group transfer potential

Group transferred	High-energy compound	Examples
Acetyl (acyl) group	Oxygen-ester	Aminoacyl tRNA, Acylcarnitine
	Thioester	Acyl CoA, Acyl carrier protein
	Mixed anhydride	Acetyl phosphate
Phosphoryl group[a]	Acid anhydride	ATP, Ⓟ-Ⓟ
	Phosphoramidate (P–N bond)	Phosphocreatine

[a]More examples of high phosphoryl group transfer potential may be found in Table 30.2

and reduction of carboxylate to aldehyde (hydride as the incoming nucleophile). But the free carboxylate ($-COO^-$) group in water is not reactive towards a nucleophilic attack. This is because the incoming nucleophile has to displace O^{2-} from the carbonyl carbon (Fig. 30.8). The O^{2-} species is a very poor leaving group making the event thermodynamically unfavorable. Therefore, a modifying group must be built and tagged to the leaving carboxylate oxygen. Upon activation, this oxygen first becomes part of a good leaving group (Perler 1998). Subsequently, the incoming nucleophile displaces this group to establish a bond with the carbonyl carbon (Fig. 30.8).

Different representative examples of acyl activation in enzyme reactions are summarized in Table 30.5 below.

30.5 Summing Up

Inorganic phosphate, phosphorylated metabolites, and high-energy phosphate compounds play pivotal yet universal roles in cellular metabolism. Phosphoric acid, a tri-basic acid with unique ionization properties, is well exploited by nature. Phosphorylation imparts charge and polarity to compounds. Such tags prevent them from escaping out of the cellular compartments.

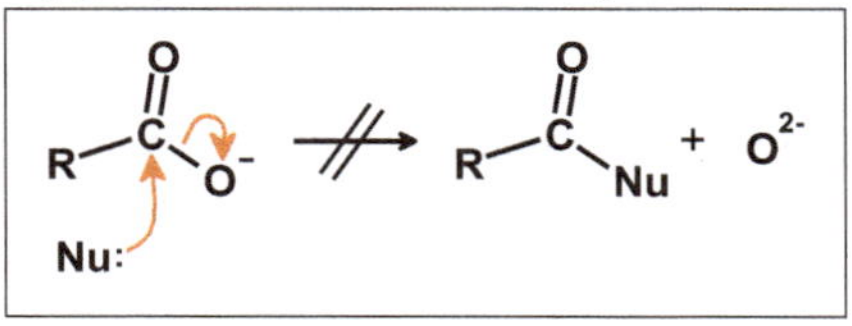

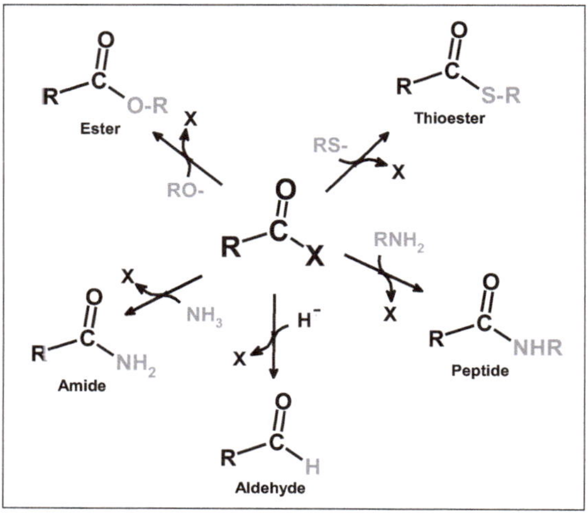

Fig. 30.8 Activating carboxylate for acyl group transfer. Since O^{2-} is a very poor leaving group, the carboxylate is first tagged with a better leaving group (shown as $-X$) like Ⓟ or AMP. This can be easily displaced by the incoming nucleophile

Table 30.5 Enzyme reaction mechanisms involving acyl activation

Activated acyl group	Nucleophile which accepts the acyl group	Enzyme example
Acyl phosphate—*Phosphate as leaving group*		
γ-Glutamyl phosphate	NH$_3$	Glutamine synthetase
γ-Glutamyl phosphate	NH$_2$ of cysteine	γ-Glutamylcysteine synthetase (GSH biosynthesis)
γ-Glutamyl phosphate	Hydride (H$^-$)	γ-Glutamyl phosphate reductase
Aspartyl-β-phosphate	Hydride (H$^-$)	Aspartate β-semialdehyde dehydrogenase
1,3-Bisphosphoglycerate[a]	Enzyme-Cys-SH	Glyceraldehyde-3-phosphate dehydrogenase
Acyl adenylate—*AMP as leaving group*		
Fatty acyl-AMP	CoA-SH	Fatty acyl-CoA synthetase
Amino acyl-AMP[b]	NH$_2$ of amino acid	Non-ribosomal peptide synthetase
Amino acyl-AMP	tRNA–CCA–OH	Amino acyl-tRNA synthetase

[a]The acyl group is first transferred to enzyme-Cys-SH. This thioester is later reduced to aldehyde
[b]In the non-ribosomal peptide synthetase, the amino acyl group is first transferred to a thiol (Cys-SH and 4′-phosphopantetheine-SH) on the enzyme. This thioester is then attacked by the next incoming amino acid to form the peptide bond

Enzyme substrates are often derivatized with phosphate, adenosine, or AMP. Presence of an additional group like this provides molecular features for better recognition/discrimination at the enzyme active site. They in turn contribute to binding energy that can be exploited for catalysis.

The phosphate group has the same tetrahedral geometry as that of a saturated carbon center. Substitution by four different groups on a P atom generates a chiral center.

Phosphorus atom provides an electrophilic center in most reactions involving nucleophilic attack. Whenever group transfer occurs, phoshorylium (phosphoryl group, PO$_3^-$) is transferred and not the phosphate (PO$_4^{3-}$).

Group transfers are best achieved in a reaction with good leaving groups. High-energy compounds are used to modify the reactants so that a better leaving group is created. Phosphate, pyrophosphate, and AMP are examples of leaving groups derived from ATP.

In the final analysis, the ester/amide bond synthesis is an example of dehydration reaction. And ATP, being an effective anhydride, acts as a remarkable dehydrating agent. In this sense, high-energy compound nature of ATP is employed to drive uphill reactions. ATP provides chemical energy by group transfers and not by simple hydrolysis.

References

Alberty RA (1969) Standard Gibbs free energy, enthalpy, and entropy changes as a function of pH and pMg for several reactions involving adenosine phosphates. J Biol Chem 244:3290–3302

Cleland WW, Hengge AC (2006) Enzymatic mechanisms of phosphate and sulfate transfer. Chem Rev 106:3252–3278

Elias M et al (2012) The molecular basis of phosphate discrimination in arsenate-rich environments. Nature 491:134–137

Frey PA, Magnusson OT (2003) S-Adenosylmethionine: a wolf in sheep's clothing, or a rich man's adenosylcobalamin? Chem Rev 103:2129–2148

Hon S et al (2022) Increasing the thermodynamic driving force of the phosphofructokinase reaction in *Clostridium thermocellum*. Appl Environ Microbiol 88:e0125822.

Knowles JR (1980) Enzyme-catalyzed phosphoryl transfer reactions. Annu Rev Biochem 49:877–919

Patel A et al (2017) ATP as a biological hydrotrope. Science 356:753–756

Perler FB (1998) Breaking up is easy with esters. Nat Struct Biol 5:249–252

Pinna S et al (2022) A prebiotic basis for ATP as the universal energy currency. PLoS Biol 20: e3001437

Punekar NS, Vaidyanathan CS, Appaji Rao N (1985) Role of Mn[II] and Mg[II] in the catalysis and regulation of *Aspergillus niger* glutamine synthetase. Indian J Biochem Biophys 22:142–151

Ramasarma T (1998) A profile of adenosine triphosphate. Curr Sci 74:953–966

Rice AM, Rosen MK (2017) ATP controls the crowd. Science 356:701–702

Struck A-W et al (2012) S-Adenosyl-methionine-dependent methyltransferases: highly versatile enzymes in biocatalysis, biosynthesis and other biotechnological applications. Chembiochem 13:2642–2655

Tawfik DS, Viola RE (2011) Arsenate replacing phosphate - alternative life chemistries and ion promiscuity. Biochemistry 50:1128–1134

Westheimer FH (1987) Why nature chose phosphates. Science 235:1173–1178

Enzymatic Oxidation-Reduction Reactions 31

Life is interposed between two energy levels of the electron.
—Szent Gyorgyi

The driving force for all life as we know it is derived from reduction-oxidation (redox) reactions. Most biological oxidations are often coupled to cellular energy production. Typically, carbon compounds (such as carbohydrates) are oxidized to carbon dioxide while oxygen is reduced to water. Enzymes play a significant role in connecting the series of redox reactions, ultimately involving oxygen. In mitochondrial electron transport chain, electrons are passed from NADH along a series of electron acceptors/donors (oxidants and reductants) to O_2. Molecular oxygen is the final oxidant (terminal electron acceptor) of aerobic metabolism. Biological reductions—on the other hand—are employed to store energy in chemical forms for later use. In photosynthetic organisms, the reduction of carbon dioxide (to carbohydrates) is powered by sunlight while water is oxidized (to oxygen). This broad canvas of redox reactions serves to drive pumps, maintain concentration gradients across membranes, and generate metabolites that have high group transfer potential and/or are energy rich. Not surprisingly, oxidoreductases form a significant group (EC 1.x.x.x) of well represented enzymes (Chap. 7).

31.1 What Are Oxidation-Reduction Reactions?

Oxidation-reduction reactions involve the transfer of electrons between two chemical species. A compound which loses electron(s) is oxidized while the compound that gains electron(s) is reduced. A compound that donates its electron(s) is called a **reductant** or a **reducing agent**. Conversely, the electron accepting molecule is the **oxidizing agent** or **oxidant**. The oxidation and reduction events must occur together—that is—an oxidant has to be present to accept the electron(s) from a reductant. In other words, for a molecule to be reduced, some other molecule must

N. S. Punekar, *ENZYMES: Catalysis, Kinetics and Mechanisms*, https://doi.org/10.1007/978-981-97-8179-9_31

get oxidized. In this sense, redox reactions may be compared to proton transfers in acid-base chemistry. An acid can lose/donate a proton only when a base accepts it. No oxidant can gain electrons without another substance (reductant) losing electrons. A complete oxidation-reduction reaction is thus a combination of two half reactions (*redox couples*). When the two half reactions are combined, the component with greater tendency to gain electrons (the oxidant) gets reduced at the expense of the other (the reductant—which loses electrons).

31.1.1 Reduction Potential: Measure of Tendency to Lose Electrons

We rank acid strengths based on their pK_as; the lower the pK_a, the stronger the acid (Chap. 28). Similarly, the tendency to gain or lose electrons may be used to rank order compounds. The strength of an oxidizing agent can be measured electrochemically by dissolving it in water and measuring the voltage required to reduce it. This in fact defines a scale of *standard reduction potentials* (or *redox potentials*) for each oxidant/reductant. Table 31.1 contains a selective list of redox couples and their standard reduction potentials. A large positive standard reduction potential indicates high electron affinity of that compound, and that it is a strong oxidant. Its conjugate reductant is a poor electron donor and a weak reducing agent. For example, the conjugate redox pair of $\frac{1}{2}O_2/H_2O$ has the highest positive standard reduction potential (+0.816 V). This makes O_2 the strongest available oxidizing agent and

Table 31.1 Standard reduction potentials for few biologically relevant redox couples

Redox couples (shown in the direction of reduction)	E'° (Volts)
$\frac{1}{2}O_2 + 2H^+ + 2e^- \rightarrow H_2O$	+0.816
$O_2 + 2H^+ + 2e^- \rightarrow H_2O_2$	+0.300
Cytochrome c (Fe^{3+}) + $1e^- \rightarrow$ Cytochrome c (Fe^{2+})	+0.250
Dehydroascorbate + $2H^+ + 2e^- \rightarrow$ Ascorbate	+0.060
$2H^+ + 2e^- \rightarrow H_2$	**0.000^a**
Oxaloacetate + $2H^+ + 2e^- \rightarrow$ Malate	−0.175
FAD + $2H^+ + 2e^- \rightarrow FADH_2$	$−0.180^b$
Pyruvate + $2H^+ + 2e^- \rightarrow$ Lactate	−0.190
Cystine + $2H^+ + 2e^- \rightarrow$ 2 Cysteine	−0.220
GSSG + $2H^+ + 2e^- \rightarrow$ 2 GSH	−0.230
Lipoate + $2H^+ + 2e^- \rightarrow$ Dihydrolipoate	−0.290
$NAD^+ + 2H^+ + 2e^- \rightarrow$ NADH + H^+	−0.320
$NADP^+ + 2H^+ + 2e^- \rightarrow$ NADPH + H^+	−0.320
$2H^+ + 2e^- \rightarrow H_2$	−0.421

All the E'° values are for standard conditions of unit activity (1 M concentration) at 25 °C and pH 7.0 ([H^+] = 10^{-7} M); Gases are at 1 atmospheric pressure
[a]By convention, the standard hydrogen electrode at pH 0 ([H^+] = 1 M) has an E° of zero. However, at pH 7.0 its measured E'° value is −0.421 V
[b]This E'° value is for free FAD/$FADH_2$ redox couple. Depending upon the active site microenvironment in a given flavoprotein, this value varies from −0.450 V to +0.150 V

H_2O (its conjugate reductant) the weakest reducing agent. At the other end of the spectrum, the $2H^+/H_2$ redox pair (-0.421 V) has a large negative standard reduction potential.

31.1.2 Reduction Potentials and Reaction Thermodynamics

Reduction potentials measure the affinity for electrons. The standard reduction potential is measured in volts and denoted as E°. The hydrogen electrode (representing the $2H^+/H_2$ redox pair) is assigned an arbitrary E° value of 0.00 V (Table 31.1) and all others are ranked relative to it. Many redox couples of biological interests involve protons. Standard reduction potentials for biological redox couples are thus more meaningful when expressed at pH 7.0 (near physiological conditions and not at pH = 0). Therefore, such reduction potentials (defined at pH 7.0 and 25 °C) are shown as E'° (and not as E°). Accordingly, when $[H^+]$ is 10^{-7} M (i.e., pH is 7.0), the E'° of $2H^+/H_2$ redox pair (the hydrogen electrode) becomes -0.421 V.

The reduction potential (E) not only depends on the chemical nature of a given redox couple (E'° as listed in Table 31.1) but also on their relative concentrations (activities). This relationship is described by the *Nernst equation*:

$$E = E^\circ + \frac{RT}{nF} \ln \frac{[\text{electron acceptor}]}{[\text{electron donor}]}$$

where n is the number of electrons transferred per molecule, and F is the Faraday constant (23.063 kcal $\times$ V^{-1} $\times$ mol^{-1}). The actual E value thus depends on the concentrations of oxidized and reduced species of the redox couple. We note that the Nernst equation and the free energy relationship between ΔG, ΔG° and the composition of the reaction mixture ($\Delta G = \Delta G^\circ + RT \ln \Gamma$; see Chap. 9) are analogous.

Electrons spontaneously flow from a compound with a lower reduction potential to that with a higher reduction potential. Therefore, under standard conditions, they are transferred from reduced component of any conjugate redox pair in Table 31.1 to the oxidized component of any conjugate redox pair above it. By convention, the tendency of electron flow (indicated by $\Delta E'^\circ$) for any oxidation-reduction reaction between two redox couples may be calculated as follows:

$$\Delta E'^\circ = E'^\circ_{\text{electron acceptor}} - E'^\circ_{\text{electron donor}}$$

For example, electrons can spontaneously flow from H_2 (-0.421 V) to O_2 ($+0.816$ V)—i.e.—the process is thermodynamically feasible ($\Delta E'^\circ = +1.237$ V). The more positive the $\Delta E'^\circ$ is, greater is the tendency for electron flow between the two redox couples. In this sense, ΔE for a redox reaction and its corresponding ΔG are related by:

$$\Delta G^\circ = -n \times F \times \Delta E'^\circ \quad (\text{and also, } \Delta G = -n \times F \times \Delta E)$$

where again, n is the number of electrons transferred per mole (equivalents per mole) in the redox reaction and F is the Faraday constant (23.063 kcal$\times$V^{-1} $\times$ mol^{-1}). A positive ΔE therefore means that ΔG is negative, and the reaction is spontaneous (i.e., thermodynamically favorable; Chap. 9). By analogy, *the criterion of spontaneity for a given oxidation-reduction reaction is ΔE (and not $\Delta E'^\circ$)*. All these concepts are illustrated with the help of lactate dehydrogenase reaction in the box below.

Redox Chemistry of Lactate Dehydrogenase Reaction

The following reversible reaction is catalyzed by lactate dehydrogenase:

$$\textbf{Pyruvate} + \textbf{NADH} + \textbf{H}^+ \rightleftarrows \textbf{Lactate} + \textbf{NAD}^+$$

Viewed from left to right, pyruvate is reduced to lactate while NADH is oxidized to NAD$^+$ in this reaction. In the reverse direction (from right to left), NAD$^+$ is reduced to NADH at the expense of lactate oxidation. *Oxidation-reduction reactions are thus always coupled*. For convenience, however, we can describe them as two half reactions (redox couples), both written in the direction of reduction as shown below:

$$\text{Reduction of pyruvate}: \quad \text{Pyruvate} + 2\text{H}^+ + 2\text{e}^- \rightarrow \text{Lactate}$$

$$\text{Reduction of NAD}^+: \quad \text{NAD}^+ + 2\text{H}^+ + 2\text{e}^- \rightarrow \text{NADH} + \text{H}^+$$

The E'° values for these two redox couples, under standard conditions, may be obtained from Table 31.1. The electrons flow from NAD$^+$/NADH couple (-0.320 V) to pyruvate/lactate couple (-0.190 V) because the latter is at higher positive standard reduction potential. This means pyruvate/lactate half reaction goes as shown, while the NAD$^+$/NADH couple undergoes oxidation. The NADH is the reducing agent, and pyruvate is the oxidizing agent. The $\Delta E'^\circ$ for this reaction may be accordingly calculated.

$$
\begin{aligned}
\Delta E'^\circ &= E'^\circ{}_{\text{pyruvate/lactate}} - E'^\circ{}_{\text{NAD/NADH}} \\
&= \text{-}0.190 \text{ V} - (\text{-}0.320 \text{ V}) \\
&= +0.130 \text{ V}
\end{aligned}
$$

The ΔG° for the reaction may now be obtained, as the two are related. It is a two-electron reduction ($n = 2$) and therefore,

$$\Delta G^\circ = \text{-}n \times F \times \Delta E'^\circ = \text{-}2 \times 23.063 \times 0.130 = \text{-}5.996 \text{ kcal} \times \text{mol}^{-1}$$

(continued)

Since the $\Delta G°$ is negative (-5.996 kcal×mol^{-1}), ***reduction of pyruvate by NADH is a spontaneous process under standard conditions*** (at 25 °C and pH 7.0 and with pyruvate, lactate, NAD$^+$, and NADH all at 1.0 M).

Lactate dehydrogenase (rat skeletal muscle) is exclusively cytosolic. The approximate concentrations of lactate (5.0 mM), pyruvate (0.5 mM), NAD$^+$ (0.5 mM), and NADH (0.001 mM) in this compartment are reported. Reduction potentials for the two redox couples under these conditions (at 25 °C) may now be calculated as shown.

$$E_{\text{Pyr/Lac}} = E'^{\circ} + \frac{RT}{nF} \ln \frac{[\text{Pyruvate}]}{[\text{Lactate}]} = -0.190\,\text{V} + \left(\frac{2.303 \times 1.987 \times 298}{2 \times 23063} \log \frac{[0.5]}{[5.0]} \right) \text{V}$$

$$= \text{-}0.190\,\text{V} + 0.0296 \times \log{(0.1)}\,\text{V}$$

$$= \text{-}0.190\,\text{V} \text{-} 0.0296\,\text{V} = \text{-}0.2196\,\text{V}$$

Similarly,

$$E_{\text{NAD/NADH}} = E'^{\circ} + \frac{RT}{nF} \ln \frac{[\text{NAD}^+]}{[\text{NADH}]}$$

$$= -0.320\,\text{V} - \left(\frac{2.303 \times 1.987 \times 298}{2 \times 23063} \log \frac{[0.5]}{[0.001]} \right) \text{V}$$

$$= \text{-}0.320\,\text{V} + 0.0296 \times \log{(500)}\,\text{V}$$

$$= \text{-}0.320\,\text{V} + 0.0799\,\text{V} = \text{-}0.240\,\text{V}$$

The NAD$^+$/NADH couple (-0.240 V) has a more negative reduction potential than that of pyruvate/lactate couple (-0.2169 V). Accordingly, NADH reduces pyruvate but the ΔG for this reduction ($n = 2$) is different. This may now be calculated as before.

$$\Delta G = \text{-}n \times F \times \Delta E = \text{-}2 \times 23.063 \times (\text{-}0.2196\,\text{V} + 0.240\,\text{V})$$

$$= \text{-}0.941\,\text{kcal} \times \text{mol}^{-1}$$

This ΔG value is different from $\Delta G°$ (-5.996 kcal×mol^{-1}; the standard free energy change) for this redox reaction—it indicates the actual free energy change at the physiological concentrations of the redox couples prevailing in the muscle.

If the pyruvate/lactate ratio in the liver is 1:100, then $E_{\text{Pyr/Lac}}$ will be -0.249 V. This is more negative than that of NAD$^+$/NADH couple (-0.240 V) and hence, lactate now reduces NAD$^+$ (ΔG for the reaction as written above will be $+0.415$ kcal×mol^{-1}). The redox reaction actually goes from right to left in the liver. It may be reiterated that *the criterion of spontaneity for a reaction is ΔE (or ΔG) and not $\Delta E'^{\circ}$ (or $\Delta G°$).* Reversible feature of lactate dehydrogenase reaction forms the basis of a functional Cori's cycle.

31.1.3 Effect of pH on Standard Reduction Potentials

We noted that the standard hydrogen electrode at pH 0 ($[H^+] = 1$ M) has an E° of zero by convention (Table 31.1). And at pH 7.0, its measured E'° value is -0.421 V. In general, pH does influence the E° value of all redox couples that involve H^+ ions in their half reactions. The $NAD^+/NADH$ couple is a common example of this kind. As long as the pH of the reaction is maintained (recall that at standard condition, pH is 7.0) its E'° value is -0.320 V. At any other pH, the reduction potential of $NAD^+/$ NADH couple will change. We can calculate the effect of pH on E by suitably including $[H^+]$ term in the Nernst equation. For unit increase in pH, the E becomes more negative by 0.059 V (for one electron reduction). For $NAD^+/NADH$ couple (where $n = 2$), however, the ΔE will be 0.0295 V for each pH unit change.

31.2   How Enzymes Influence Redox Reaction Rates

We noted above that reduction of pyruvate by NADH is a spontaneous process under standard conditions. However, when pyruvate and NADH are mixed together, no reaction occurs. While this reaction is thermodynamically allowed (that is spontaneous with ΔG negative), there exists a kinetic barrier. This is where enzymes come into picture. As with any other reaction, an enzyme can facilitate and catalyze redox reactions. ***Overcoming kinetic barriers of redox reactions*** is their foremost feature.

At times, the reduction potential difference between two reactants (redox couples) may be large—making the reaction difficult. Redox couples with intermediate reduction potentials—*mediators*—often facilitate such reactions. They act as "go between" the two reactants. Obviously, such mediators themselves are redox-active groups/compounds. Except for the thiols (reversible oxidation-reduction of cysteine⇌cystine; Table 31.1), there is not much on the polypeptide chain for an enzyme to offer. Further, when a substrate is oxidized, electrons have to be moved out. There are no obvious electrophilic groups in the amino acid side chains of proteins. Thus, redox enzymes without exception show obligate requirements for either (a) an organic coenzyme to act as an electron acceptor or (b) a redox-active transition metal to pass the parcel of electrons to another acceptor molecule. Several non-protein coenzymes, cofactors, and prosthetic groups are recruited for this purpose. This is the second feature through which enzymes contribute to redox catalysis. More common examples of these are listed in Table 31.2.

Enzymes can modulate redox potentials of bound substrates and/or cofactors. This is done by virtue of their ability to create unique micro-environments—the third important feature of redox enzymes. Two examples of ***redox modulation*** illustrate this point well. The E'° value of bound $FAD/FADH_2$ couple varies from -0.450 V to $+0.150$ V, depending upon the flavoprotein active site (Table 31.1). When the active site preferentially binds FAD over $FADH_2$, then the E'° of this redox couple becomes more negative. Conversely, if $FADH_2$ is favored over FAD, then E'° tends to become more positive. These conclusions simply follow from the Nernst equation (see box above—Redox chemistry of lactate dehydrogenase reaction—for a similar

Table 31.2 Cofactors participating in enzymatic oxidation-reduction reactions

Cofactor/Prosthetic group (Enzyme example)	Reduced form → Oxidized form[a]
Nicotinamide adenine dinucleotides (Lactate dehydrogenase)	NAD(P)H NAD(P)$^+$
Biopterin (Phenylalanine hydroxylase)	5,6,7,8-Tetrahydrobiopterin 7,8-Dihydrobiopterin
Glutathione (Glutathione peroxidase)	2 GSH GSSG
Lipoamide (Dihydrolipoyl dehydrogenase)	Dihydrolipoate Lipoate

(continued)

Table 31.2 (continued)

Cofactor/Prosthetic group (Enzyme example)	Reduced form → Oxidized form[a]
Flavins (FAD and FMN) (Succinate dehydrogenase, Glutathione reductase)	1,5-Dihydroflavin Flavin semiquinone Oxidized flavin
Ascorbic acid (Vitamin C) (Prolyl hydroxylase; 2-ketoglutarate decarboxylating)	Reduced Ascorbate radical Oxidized
Iron-sulfur clusters (Fe-S)$_n$ (where $n = 2$ or 4) (Dihydroxylating dioxygenase; Ferridoxins)	(Fe-S)$_2$ cluster
Heme (Cytochrome oxidase)	Fe^{2+} Fe^{3+}
Transition metal ions (like Cu, Fe, and Mo) (Laccase; Catechol dioxygenase)	Cu^{1+} → Cu^{2+}; Fe^{2+} → Fe^{3+}

[a]Number of electrons transferred in oxidation step(s) is shown. While proton transfers may accompany oxidation, this inventory is not explicitly shown. Groups/atoms relevant to oxidation are marked in gray. See Table 29.1 for details of R groups in NAD(P), FAD, and FMN. Amino acid side chains, water, O$_2$, etc., provide the fifth and sixth ligands (X and Y) to heme iron

calculation). Cytochromes provide yet another example where enzyme protein brings about a shift in the reduction potential by differential binding. They all contain heme—an iron atom stuck between four N atoms of a porphyrin. But the fifth and the sixth ligands donated by the polypeptide greatly influence the reduction potential of the bound heme. This forms the basis for the "bucket brigade" of cytochromes lining up the electron transport chains. Cytochrome a_3 (+0.385 V) of cytochrome oxidase can easily steal an electron from cytochrome c (+0.235 V) in the mitochondria.

31.3 Mechanisms and Modes of Electron Transfer

The carbon encountered in an enzyme substrate exists in a range of oxidation states. For example, various oxidation states of a one carbon compound are shown below (Fig. 31.1). Considering their electro-negativities (H < C < O), the notional number of electrons present on carbon decreases by two in each step. More reduced compounds are richer in H than O and conversely, more oxidized compounds are richer in O than H.

Biological oxidation reaction may be broadly viewed as: (a) removal of electrons from a substrate molecule or (b) its direct combination with oxygen, where oxygen atom accepts the electron(s). We will look at reactions involving molecular oxygen a little later. A substrate molecule is oxidized when electrons are transferred from it to an acceptor. Such transfers occur in multiples of *single electron currency—the reducing equivalent*. In practice this may be through $1e^-$ transfers or $2e^-$ transfers. A vast majority of organic substrate oxidations are $2e^-$ transfer events. The two electrons may be transferred in a single step, or they may be moved one at a time (two $1e^-$ transfer steps). Single electron transfers generally have high energy barriers and require stabilization of a radical. *If two $1e^-$ steps are involved—then a free radical intermediate must form*. Free radicals may be detected by their typical EPR signals. Few structures such as flavins and ascorbic acid (Table 31.2) help stabilize such reactive species. Other chemical apparatus available for enzymes to do this are quinoid structures of some coenzymes (vitamins E and K and coenzyme Q), and redox-active transition metals (like Fe, Cu, and Mo). *Quinoproteins* are a recently characterized group of enzymes (quinoproteins, copper-quinoproteins, and quinohemoproteins) whose catalytic mechanisms involve free radical intermediates

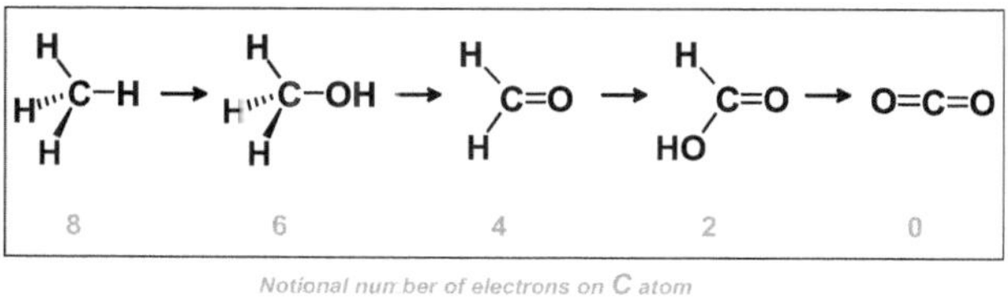

Notional number of electrons on C atom

Fig. 31.1 Various oxidation states of one-carbon compounds. The oxidation state of carbon increases from left (methane) to right (carbon dioxide)

on quinone-containing prosthetic groups. Pyrrolo-quinoline quinone (PQQ) is one such prosthetic group found in a few bacterial dehydrogenases and oxidases (Duine 1999; Klinman 1996). PQQ is considered the "third coenzyme" following nicotinamide and flavin. More recently, structure and activity data provide direct evidence for the existence of eukaryotic PQQ-dependent enzymes (Takeda et al. 2019). Bioinformatic studies suggest a widespread occurrence of quinoproteins in eukaryotes as well as prokaryotes.

The more common $2e^-$ transfers can occur by two distinct mechanisms: (a) Enzymatic dehydrogenations offer the first example where hydride (H^-, a proton with two electrons) transfer may be the oxidation step. Reactions of nicotinamide cofactors (NAD and NADP) apparently go through hydride transfer step(s). (b) In the second option, a substrate proton is first abstracted. A substrate carbanion is therefore expected to form—at least transiently. In a subsequent step, $2e^-$ transfer from the carbanion to an acceptor takes place. Some of the flavoenzymes operate through this mechanism.

31.4   Pterine and Folate Cofactors

Pterines and folic acid derivatives (Table 31.2), in their many forms perform diverse roles in nature, ranging from pigments (butterfly colors) to cofactors for numerous redox and one-carbon transfer reactions. The unique chemistry of the pteridine heterocycle is responsible for this extraordinary diversity of function. The pteridine ring system resembles the isoalloxazine ring of the flavine coenzymes (Table 31.2). Pterines are important redox cofactors (and include folic acid, tetrahydrobiopterin, and molybdopterin). The active form of biopterin is 5,6,7,8-tetrahydrobiopterin. This is oxidized to 7,8-dihydropterin during hydroxylation of the phenylalanine aromatic ring by phenylalanine hydroxylase (see Table 31.4 later in this chapter). A similar role is played by biopterin during the formation of 3,4-dihydroxypheylalanine from tyrosine. Folate differs structurally from biopterin in that the substituent on C6 of the pteridine ring (Table 31.2) consists of p-aminobenzoic acid further linked to polyglutamic acid. Tetrahydrofolate (THF) is a versatile cofactor that participates in redox chemistry and also functions to transfer "one-carbon" units in several oxidation states (Table 31.3). The "one-carbon" units are covalently attached to THF at its N^5, N^{10}, or both N^5 and N^{10} positions (Matthews and Drummond 1990). The one-carbon units enter the THF pool from serine—by the action of serine hydroxy methyltransferase (see Table 33.4)—at N^5, N^{10}-methylene-tetrahydrofolate. The other one-carbon (methyl group) transfer chemistries involve S-adenosyl-L-methionine, coenzyme B_{12}, and the nickel (I) hydrocorphin coenzyme F-430 (of methane forming step in methanogenic archaea; Thauer 2019).

The importance of folate as a cofactor is obvious as it participates in the biosynthesis of several amino acids and nucleotides (Benkovic 1980). Both 7,8-dihydropterin and 7,8-dihydrofolate are reduced back to their active tetrahydro-states by dihydrofolate reductase. This enzyme is a key target for cancer

Table 31.3 Enzyme reactions of different "one-carbon" tetrahydrofolate derivatives

Tetrahydrofolate (THF) derivative[a]	Enzyme reaction
N^5-Methyl-THF	Homocysteine methyltransferase
N^5,N^{10}-methylene-THF	Serine hydroxymethyltransferase, thymidylate synthase
N^5-Formyl-THF, N^{10}-Formyl-THF	Formylmethionyl-tRNA synthase, GAR transformylase, AICAR transformylase
N^5-Formimino-THF	Glutamate formiminotransferase
N^5,N^{10}-Methenyl-THF	N^5,N^{10}-Methenyl-THF cyclohydrolase

[a]The oxidation states of "one-carbon" units correspond to methanol (methyl-THF), formaldehyde (methylene-THF), and formate (formyl-THF, formimino-THF, and methenyl-THF) as shown in Fig. 31.1

chemotherapy (inhibited by methotrexate; see Chap. 20, Irreversible inhibitions). Similarly, sulfanilamide (a sulfa drug and a structural analog of p-aminobenzoic acid) is a medically valuable antibacterial agent (Table 26.4).

Nicotinamide and riboflavin are by far the most common redox cofactors used by enzymes in nature. We will describe the salient features and associated chemical mechanisms for them in some detail below. The support for their proposed mechanisms has come from actual enzyme reactions as well as from corresponding model reactions.

31.5 Nicotinamide Cofactors

Nicotinamide cofactors frequently participate in enzymatic redox reactions. They take part in oxidations and reductions via their pyridine ring (of the nicotinamide), and hence are sometimes also termed pyridine nucleotides. Change in absorbance (at 340 nm) associated with their oxidation-reduction is a convenient way of monitoring these enzyme reactions (Chap. 11). Nicotinamide adenine dinucleotide (NAD^+) was the first enzyme cofactor to be discovered, by Harden and Young, in the conversion of glucose to ethanol by yeast. Subsequently, nicotinamide adenine dinucleotide phosphate ($NADP^+$) was identified (by Warburg and Christian in 1934) as the redox cofactor responsible for glucose-6-phosphate oxidation in erythrocytes. Total chiral synthesis of NAD^+ was accomplished in 1957 by Todd's group. $NADP^+$ differs from NAD^+ in having an extra phosphate group (see Table 29.1 for structures). $NADP^+$ sports the additional phosphate group on its $2'$ OH of adenosine moiety. Despite this structural difference, both $NAD^+/NADH$ and $NADP^+/NADPH$ redox pairs have identical reduction potentials (-0.320 V, Table 31.1). Physiologically, however, $NAD^+/NADH$ redox couple functions largely as an electron acceptor (oxidizing agent) and is preferred in catabolic reactions. $NADP^+/NADPH$ couple on the other hand is the choice reductant (electron donor) in biosynthetic steps of metabolism.

Enzymes display a conserved structural motif to bind NAD^+—called the Rossmann fold. Active sites of many dehydrogenases bind NAD^+ in an extended conformation. In the bound state, the orientation of nicotinamide group can be either *anti* (away from; as in malate dehydrogenase) or *syn* (toward; as in glyceraldehyde-3-phosphate dehydrogenase) position with respect to its N-glycosidic bond. In most enzymes, the pyridine nucleotide cofactor is bound reversibly (more like a substrate). However, UDP-galactose epimerase is known to contain stoichiometric, tightly bound NAD^+ at its active site.

NAD(P)H is a powerful biological reducing agent but is stable in air. Reduced pyridine nucleotides (both NADH and NADPH) do not react with oxygen—in this sense, they are distinct from reduced flavins which do (see below). The NAD^+/NADH couple is a common participant in all dehydrogenases, wherein $>CH–XH$ is oxidized to $>C = X$ (where X is N or O). As with some reductase reactions, this redox couple also is used to reduce C=C double bonds. The redox chemistry with NAD^+/NADH couple is believed to proceed with a *hydride transfer* step. Oxidation of the C–H bond is often shown as its heterolytic cleavage—accompanied by a hydride transfer to C-4 position of NAD^+. Failure to equilibrate this transferable hydrogen with solvent (water) protons is suggestive of this mechanism. However, this does not constitute a final proof of hydride transfer —because some enzymes can exclude water from their active site during catalysis. When a hydride from NAD(P)H is transferred to carbon atom, acid-base dissociation of the resultant C–H bond is exceedingly slow. But whenever the hydride is transferred to a hetero atom (which is electronegative), then it exchanges rapidly with solvent protons. Isotopic labeling studies coupled with observed deuterium kinetic isotope effects have led to the general acceptance of a hydride transfer mechanism. Such a mechanism for lactate dehydrogenase reaction is depicted in Fig. 31.2.

The pyridine ring is puckered and non-planar in the reduced state (NADH or NADPH), whereas it is planar when oxidized. The C-4 of NADH has two non-equivalent hydrogens (labeled A and B); either one can move out as a hydride during its oxidation. As shown in Fig. 31.2, lactate dehydrogenase is A-side specific. Technically, this means that the H from *proR* position of C-4 of NADH is selectively transferred to pyruvate by this enzyme. In the reverse direction, the hydride is transferred to the *re* face of NAD^+ (which is planar). With respect to their choice of C-4 hydrogen on pyridine nucleotides (whether it is NAD or NADP), dehydrogenases may be either A-side specific or B-side specific. For instance, glyceraldehyde-3-phosphate dehydrogenase is B-side specific—it picks up the H from *proS* position of C-4 of NADH (this corresponds to *si* face of NAD^+, in the reverse reaction). The trans-hydrogenase of animal mitochondria—transfers hydride between NADH and $NADP^+$—is *proR* specific (A-side) for one and *proS* specific (B-side) for the other substrate.

Finally, it is worth noting that pyridine nucleotides (and NAD^+ in particular) have important non-redox roles in metabolism as well. It is the substrate for poly (ADP-ribose) polymerase (PARP), *E. coli* DNA ligase, and toxins of diphtheria and cholera. In all these cases, ADP-ribose of NAD^+ is transferred to different acceptors.

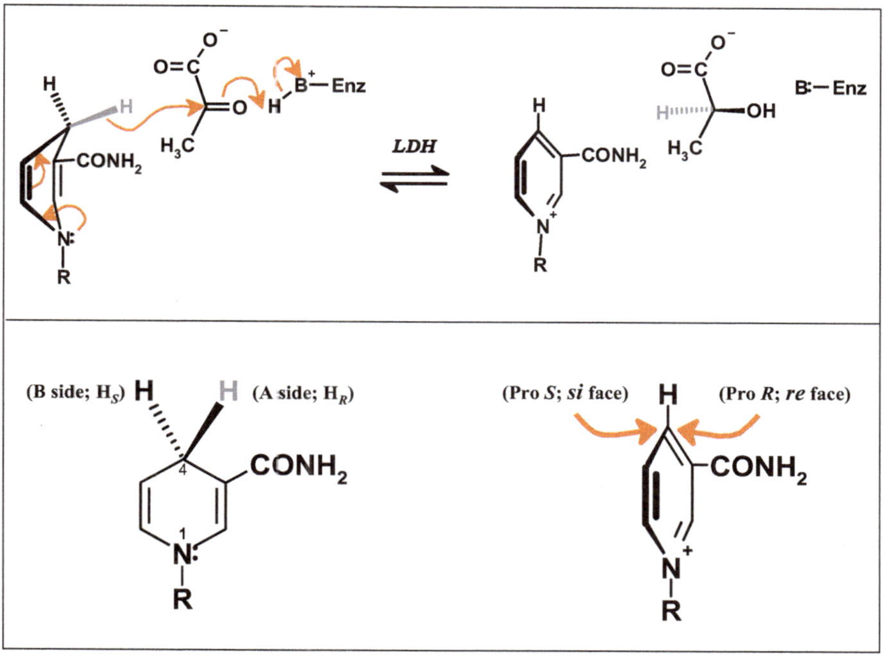

Fig. 31.2 Hydride transfer in the lactate dehydrogenase (LDH) reaction mechanism. This enzyme reaction is specific for the C-4 *R* hydrogen (bottom panel: H in gray, A-side specificity) of NADH. In the reverse direction, it transfers the hydride to C-4 *proR* position (*re* face) of NAD⁺

31.6 Flavins and Flavoenzymes

The two common flavin coenzymes (FMN and FAD) are chemically modified versions of riboflavin (vitamin B2). The ribitol side chain when phosphorylated is FMN or when attached through a diphosphate to adenosine gives FAD (see Table 29.1 for structures). Riboflavin itself was first isolated from eggs and its structure was elucidated in 1935. Extensively conjugated, tricyclic isoalloxazine ring system of riboflavin imparts it (and its cofactors) the bright yellow color. More importantly, this isoalloxazine ring system forms the redox-active structure of FMN and FAD. The oxidized cofactor absorbs in the visible region with one peak around 450 nm (Fig. 31.3). This 450 nm peak is absent in the corresponding $2e^-$ reduced form (1,5-dihydroflavin). The oxidized form of flavin is planar while the 1,5-dihydroflavin is bent with the two outer ring planes forming an angle (of ~30°) along the N5–N10 axis. One can visualize enzyme active sites selectively stabilizing one of these two forms—a possible basis for shifting the reduction potential of the flavin cofactor. For instance, D-amino acid oxidase binds FAD about 10^7-fold tighter than FADH₂ (1,5-dihydroflavin). Unlike with nicotinamide coenzymes, flavin cofactors are much more tightly bound to flavoenzymes (K_D ranging from 10^{-7} M

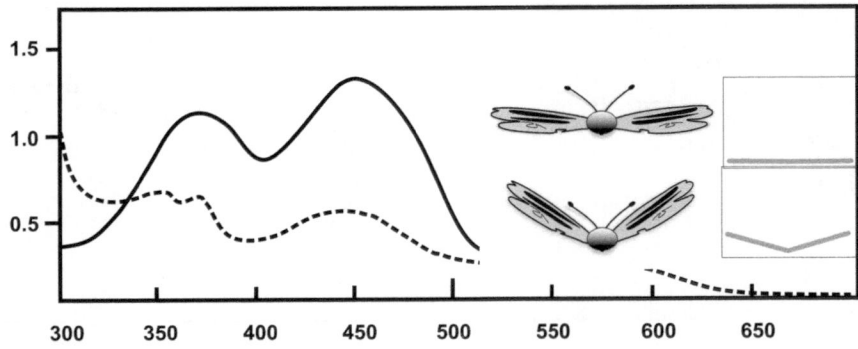

Fig. 31.3 Absorption spectra of reduced and oxidized FAD. Spectral properties, particularly those of the reduced flavin, are greatly influenced by their micro-environment at the active site. Representative spectrum of reduced FAD in an enzyme bound form is shown; $FADH_2$ (1,5-dihydroflavin) in free solution and in air, is unstable. The corresponding spectral features of FMN and $FMNH_2$ are almost identical

to 10^{-14} M). In some cases, the flavin (through a methyl C on its C-8) is covalently bound to the apoenzyme (through His residue of succinate dehydrogenase, for example).

FAD (as also FMN) can exist in three distinct oxidation states: FAD, FADH (semiquinone radical), and $FADH_2$ (1,5-dihydro FAD). Since flavin semiquinone is reasonably stable (at least in enzyme bound form), it can be a significant intermediate in flavoenzyme catalysis. Flavins can participate both in $1e^-$ and $2e^-$ transfer reactions. They are crucial as adapters/mediators between $2e^-$ oxidations (of organic compounds) and $1e^-$ oxidations (like in the electron transport chain). This feature also allows them to react with molecular oxygen (see below). On the contrary, pyridine nucleotides are always restricted to $2e^-$ (hydride) transfer reactions. Interestingly, 5-deazaflavin (where the flavin N-5 is replaced by a C atom) behaves more like a nicotinamide cofactor—capable of only $2e^-$ transfers. It is considered a "nicotinamide in a flavin's clothing." Factor F_{420} ($E'^\circ = -0.360$ V) found in methanogenic bacteria is an example of naturally occurring 5-deazaflavin. It serves as an essential catabolic cofactor (for low potential hydride transfers) in methanogenic, sulfate-reducing, and likely methanotrophic archaea. The F_o variant is exclusively used as a light-harvesting chromophore for DNA photolyases across the three domains of life (Greening et al. 2016).

Flavoenzymes may be classified according to the nature of: (a) electron acceptors that accept electrons from the reduced flavin and (b) electron donors that transfer electrons to the oxidized flavin nucleus (Fraaije and Mattevi 2000; Joosten and van Berkel 2007; Mattevi 2006). Different substrates like alcohol (D-lactate dehydrogenase), aldehyde (glucose oxidase), amine (D-amino acid oxidase), C–C bond (acyl CoA dehydrogenase), and thiols (glutathione reductase) may donate electrons to reduce the enzyme-bound flavin. When we analyze the molecules that receive electrons from a dihydroflavin (enzyme bound $FADH_2$, for instance), two broad categories may be observed. One group using molecular oxygen (O_2) as the electron acceptor includes oxidases, monooxygenases, dioxygenases, and metallo-flavoenzymes. We shall address this group in the next section. The second category includes all flavoproteins—such as succinate dehydrogenase and glutathione reductase—that do not use O_2 as electron acceptor. These dehydrogenase mechanisms represent a major class of reactions wherein flavin coenzymes take part. Varied experimental evidence (from model reactions and enzyme examples) supports different roles for its participation in these reactions. Possible mechanisms for such substrate dehydrogenations are:

(a) Direct hydride transfer from the substrate to oxidized flavin (to its N-5).
(b) Nucleophilic attack by substrate hetero-atom (other than C) at C4a (the bridge C between C-4 and N-5) of oxidized flavin.
(c) Nucleophilic attack by substrate carbanion at N-5 of oxidized flavin.
(d) A radical mechanism involving $1e^-$ transfers.

Glutathione reductase is a well characterized enzyme among the flavoprotein dehydrogenases. It regenerates glutathione (GSH) from its oxidized disulfide (GS-SG) by the following reaction:

$$GS\text{-}SG + NADPH + H^+ \rightleftarrows 2\ GSH + NADP^+$$

Based on their reduction potentials (Table 31.1), equilibrium position favors the reduction of GS-SG (oxidized glutathione; the disulfide) to GSH (reduced; the thiol) at the expense of NADPH. The hydride from NADPH is first transferred to the bound flavin (to its N-5) of glutathione reductase. Bound $FADH_2$ further transfers the two electrons to reduce a disulfide—of two cysteine residues at the enzyme active site. These cysteine thiols reduce GS-SG to two molecules of GSH. In turn, the active site disulfide is regenerated for the next catalytic cycle (Fig. 31.4). Other flavoenzymes with a similar chemical mechanism (but with different disulfide substrates) include dihydrolipoamide dehydrogenase, thioredoxin reductase, and trypanthione reductase.

Like glutathione reductase, many other flavoenzymes utilize NAD(P)H to reduce their bound flavin coenzyme. However, some of them eventually transfer these electrons to molecular oxygen. These will be the subject of our next topic.

31.7 Reactions Involving Molecular Oxygen

Molecular oxygen (O_2, also known as dioxygen) is the ultimate electron acceptor in aerobic metabolism. An oxygen atom has eight electrons with the configuration $1s^2, 2s^2 2p^4$ (Table 27.1). However, in nature, it exists as a diatomic molecule with a double bond (bond order is 2) between the two O atoms. Interestingly enough, it exists as a di-radical in the ground state (O_2 has a triplet ground state). Its two unpaired electrons in the valence orbitals have parallel spins, making it diamagnetic.

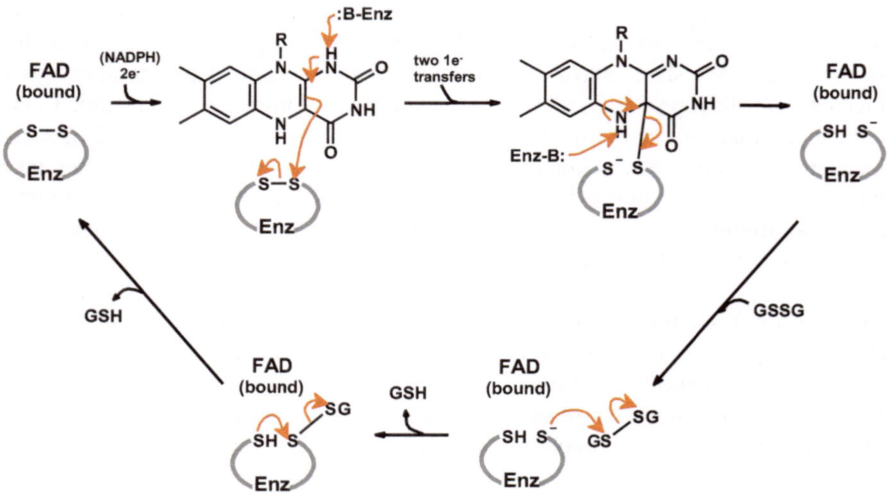

Fig. 31.4 Glutathione reductase reaction mechanism. Reduction of the enzyme disulfide requires the transfer of $2e^-$ from NADPH via the bound FAD (top reactions). The two active site cysteine thiols in turn reduce GSSG (bottom reactions). All proton transfer steps are not explicitly shown for the sake of clarity

The low reactivity of O_2 is closely related to its triplet ground state, with the following molecular orbital description:

$$(2s\sigma)^2 \ (2s\sigma^*)^2 \ (2p\sigma)^2 \ (2p\pi)^4 \ (2p\pi^*)^{1,1}$$

The peculiar electronic configuration of O_2, according to Pauli's exclusion principle, dictates that it can accept only unpaired electrons. Therefore, electrons must be transferred to O_2 one at a time. Such single electron transfers are in obvious contrast to the transfer of electrons in pairs seen in most redox reactions, discussed above. Special cofactors are required to transfer electrons from a two-electron donor to one-electron acceptor (and vice versa). FAD is one such cofactor that can participate in both $1e^-$ and $2e^-$ redox reactions. In general, the successively reduced products of molecular oxygen include the following. Some of them are very reactive or generate more reactive species.

$$O_2 \quad \xrightarrow{e^-} \quad O_2^{-\bullet} \quad \xrightarrow{e^-, 2H^+} \quad H_2O_2 \quad \xrightarrow{2e^-, 2H^+} \quad 2H_2O$$
$$\text{(Dioxygen)} \qquad \text{(Superoxide)} \qquad \text{(Hydrogen peroxide)} \qquad \text{(Water)}$$

Then, how do enzymes activate O_2 without hurting themselves? A predominantly rate-limiting first electron transfer to O_2 is followed up with subsequent electron and proton transfers occurring in rapid steps. The active site geometry further delimits/ controls the reactivity of oxygen to generate the desired regio—and stereochemical products— while minimizing deleterious side reactions (Klinman 2007). Redox

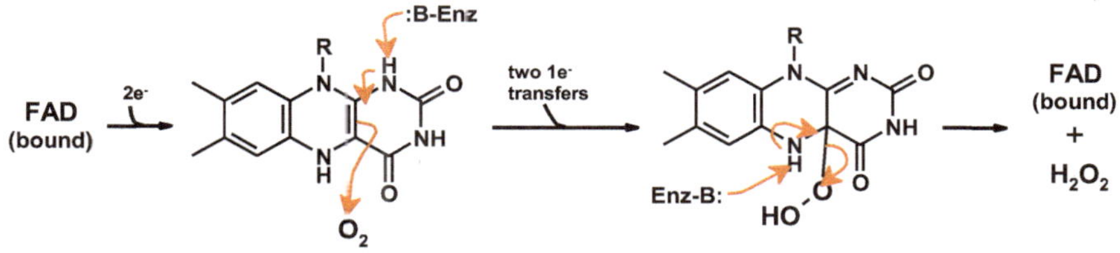

Fig. 31.5 The FAD peroxy adduct of D-amino acid oxidase and formation of hydrogen peroxide

biology is the fundamental aspect of aerobic life—being the terminal electron acceptor, they cannot avoid oxygen (Halliwell 2006). Nature has evolved antioxidant defenses so that aerobes can survive oxygen. So, life is a balance between these two processes. Besides small molecular antioxidants, a battery of enzymes is available for defense against the reactive oxygen species. These include enzymes that remove oxidants or oxidant precursors viz., superoxide dismutases, peroxiredoxins, thioredoxin/thioredoxin reductase, GSH peroxidase isoforms, and catalases. Nevertheless, there is some collateral damage to biological macromolecules, including proteins (Hawkins and Davies 2019). Exposure of biological molecules to oxidants is inevitable and commonplace.

Reactions of molecular oxygen usually occur at a prosthetic group on the enzyme (Malmstrom 1982). The reactive N5–C4a locus of isoalloxazine moiety is responsible for the extraordinary catalytic versatility of the flavins. Dioxygen reacts with a fully reduced flavin (such as bound $FADH_2$) to yield a flavin-peroxide. The C4a of the isoalloxazine is where this peroxy adduct forms (Fig. 31.5). The final fate of such peroxides is decided by the nature of the active site of that particular enzyme. Recently, a flavin cofactor with a distinct nitrone moiety (the flavin-N5-oxide) was reported as an oxygenating species (Saleem-Batcha and Teufel 2018). The redox active metal center of an enzyme may also react with molecular oxygen. The metal center may (a) transfer electrons to bound O_2 or (b) activate the organic substrate so that it can react with O_2, also bound at the active site.

Enzymatic insertion of atoms from dioxygen into organic substrates was first demonstrated, independently by Hayashi and Mason, in the 1950s. In their pioneering work, oxygen isotopes (^{18}O in particular) were exploited. Since then, more than 200 enzymes are known to use molecular oxygen as one of their substrates. These O_2 utilizing enzymes are classified into *Oxidases* and *Oxygenases*. In an oxidase-catalyzed reaction, O_2 functions only as an electron acceptor. In D-amino acid oxidase for instance, bound $FADH_2$ reduces molecular oxygen to hydrogen peroxide.

In oxygenase catalyzed reactions, however, one or both the atoms of O_2 are incorporated into organic substrates. Oxygenases in turn may be (a) *Dioxygenases*, which catalyze the insertion of both atoms of O_2 into the organic substrate or (b) *Monooxygenases* (mixed function oxidases), which catalyze the insertion of

Table 31.4 Enzyme active sites for molecular oxygen to react

Prosthetic group	Examples
Heme iron	Cytochrome oxidase, Cytochrome P_{450}, Catalase
Non-heme iron	Mono-oxygenases and Di-oxygenases, Prolyl hydroxylase (2-oxoglutarate dependent), Superoxide dismutase
Copper (mono-metallic)	Amine oxidase, Lytic polysaccharide monooxygenase
Copper (bi-metallic)	Laccase, Oxidases, Superoxide dismutase
Manganese	Photo-system II (oxygen-evolving complex), Superoxide dismutase
Flavin (FAD or FMN)	D-Amino acid oxidase, Oxygenases, Oxidases
Biopterin	Phenylalanine hydroxylase

one atom of O_2 into the organic substrate while the other one is reduced to water (Bugg 2003).

Participation of enzyme R-groups (amino acid side chains) alone is not enough to catalyze the reactions involving O_2. All enzymes that activate/reduce O_2 are conjugated proteins; molecular oxygen is activated by flavin, heme, or metal cofactor containing enzymes. Associated cofactors/prosthetic groups at the reaction centers include flavins, heme, iron (Fe^{3+}/Fe^{2+}), or copper (Cu^{2+}/Cu^+). For instance, lytic polysaccharide monooxygenases are mono copper enzymes, where the copper is bound in a characteristic histidine-brace (an apparatus for their remarkable oxidative power). The copper cofactor is first reduced by an external reductant, after which the enzyme reacts with either O_2 or H_2O_2. The powerful oxygen species generated can then hydroxylate the C1 or the C4 carbon in the scissile glycosidic bond (Eijsink et al. 2019). The structural, mechanistic, and chemical basis of flavoprotein monooxygenases catalysis is now well studied (Toplak et al. 2021; Phintha and Chaiyen 2023). They participate in reactions such as hydroxylation, epoxidation, dehalogenation, halogenation, heteroatom oxygenation, Baeyer–Villiger oxidation, α-hydroxylation of ketones, non-oxidative carbon-hetero bond cleavage, and light emission. This unmatched versatility in chemistry is a result of extensive fine-tuning and regiospecific functionalization of the flavin cofactor, achieved by unique architecture of the substrate-binding active sites. Non-heme iron prosthetic groups are common in oxygenases and some oxidases. Phenylalanine hydroxylase uses tetrahydro-biopterin along with Fe^{2+} for its activity. Table 31.4 lists the prosthetic groups exploited by enzymes acting on molecular oxygen. However, the list will be different, and gets expanded, when the reactants are either superoxide or hydrogen peroxide, instead of O_2.

31.7.1 Metal Ions in Redox Catalysis

Metal ions contribute to enzyme catalysis by shielding/stabilizing charges and by enhancing the nucleophilicity of water (Chap. 4). Multivalent metal ions can act as super (Lewis) acids. In addition, redox-active metal ions directly participate in electron transfer chemistry. Reduction potential of a redox-active metal center can be modulated, and it depends on the nature of ligands coordinating the metal ion. Metal ions can be classified as hard acids, soft acids, and borderline cases. Hard acids prefer a hard ligand, while soft acids prefer softer ligands (Irving-Williams stability series and Pearson's Hard-Soft-Acid-Base classification). For instance, Fe^{3+} (hard acid) prefers hard ligands (with O) more than Fe^{2+} (borderline case, preferring N, S containing ligands). Clearly, ligand preference versus the actual ligands present, influence the reduction potential (and metal reactivity) of a redox-active metal. This forms the basis of redox manipulation at active sites—enzymes offer appropriate side chain residues to modulate the metal center chemistry. A more detailed treatment of this subject may be found in specialized books on bio-inorganic chemistry.

Binding of metal cofactor (regardless of whether it is redox-active or not) to the apoenzyme can be followed by one or more techniques. Metal ion binding may be tested by: (a) kinetic competition with other metal ions and use of chelators, (b) gel-filtration, equilibrium dialysis and/or ultra-centrifugation, (c) difference spectral titrations (with UV-visible, fluorescence, optical rotatory dispersion, or circular dichroism), and (d) resonance techniques (like nuclear magnetic resonance, proton relaxation rates, electron spin resonance, or electron paramagnetic resonance). The actual ligands coordinating the metal center may be directly visualized by X-ray structure analysis. Depending on their strength of metal binding, enzymes may be grouped into *metallo-enzymes* and *metal-activated enzymes*. This distinction, however, is purely arbitrary and is based on the magnitude of the binding constant—when the binding is strong ($K_D < 10^{-8}$ M), it is a metalloenzyme but when the metal is weakly bound ($K_D > 10^{-8}$ M), it is a metal-activated enzyme. In practice, however, we see a continuum of metal ion-binding constants and it is difficult to classify borderline cases.

Metal ion requiring enzymes may be distinguished into three groups, depending on who donates the coordinating ligands to the metal. The metal ion is bound to the enzyme via the substrate (E-S-M) in a substrate bridge complex. Instead, the enzyme may independently bind the substrate and the metal—the enzyme bridge complex (M-E-S). Third, we can visualize a metal bridge complex, where enzyme makes contact with both the metal ion and the substrate, individually and in combination. These different binding types can be experimentally verified by the set of analytical techniques listed above.

31.8   Summing Up

Oxidation reduction reactions provide the driving force to all biological reactions. A reaction with positive reduction potential ΔE (and hence negative ΔG) is thermodynamically feasible. Like with all other reactions, enzymes hasten redox reaction rates by lowering kinetic barrier. When required, they recruit cofactors and help in tuning their redox potential, facilitate electron transfers, modulate oxygen reactivity, and exert control over the nature of redox substrate used.

Pyridine nucleotides and flavin coenzymes are two frequently encountered molecules in redox enzyme chemistry. They serve two distinct purposes. NAD

(P) is adept at $2e^-$ transfers (hydride transfers) and is stable in the O_2 environment. FAD (and FMN) is able to participate in both $1e^-$ and $2e^-$ transfers—hence, act as step-down or step-up adapters in redox reactions. Reduced flavin can therefore react with molecular oxygen.

Not all oxidation-reduction reactions occur at the carbon atom. Flavin-dependent N-hydroxylating enzymes catalyze reactions to generate N–O functional groups such as N-hydroxy, oxazine, isoxazolidine, nitro, nitrone, oxime, or N-nitroso, and azoxy units (Mügge et al. 2020). Similarly, flavin-dependent S-monooxygenases are also known (Valentino et al. 2020). Other examples include nitrate reductase and sulfate reductase. Nitrogenase—containing flavin, molybdenum and iron-sulfur center—performs a multi-electron reduction of nitrogen ($N_2 + 6H^+ + 6e^- \rightarrow 2NH_3$). Multi-component enzyme systems like dihydroxylating dioxygenases even contain a mini-electron transport chain (NADH $\rightarrow$ FAD $\rightarrow$ FeS cluster $\rightarrow$ Non-heme Fe) where electrons ultimately flow to substrates.

It is not surprising that nature has evolved a range of enzymes, coenzymes, and cofactors to perform redox reactions—as they fuel all the carbon-based life processes. Many radical and redox chemistries pose frontiers in enzymatic reaction mechanism, where reaction intermediates and pathways need investigation.

References

Benkovic SJ (1980) On the mechanism of action of folate- and biopterin-requiring enzymes. Annu Rev Biochem 49:227–252

Bugg TDH (2003) Dioxygenase enzymes: catalytic mechanisms and chemical models. Tetrahedron 59:7075–7101

Duine JA (1999) The PQQ story. J Biosci Bioeng 88:231–236

Eijsink VGH et al (2019) On the functional characterization of lytic polysaccharide monooxygenases (LPMOs). Biotechnol Biofuels 12:58

Fraaije MW, Mattevi A (2000) Flavoenzymes: diverse catalysts with recurrent features. Trends Biochem Sci 25:126–132

Greening C et al (2016) Physiology, biochemistry, and applications of F_{420}- and F_o-dependent redox reactions. Microbiol Mol Biol Rev 80:451–493

Halliwell B (2006) Reactive species and antioxidants. Redox biology is a fundamental theme of aerobic life. Plant Physiol 141:312–322

Hawkins CL, Davies MJ (2019) Detection, identification, and quantification of oxidative protein modifications. J Biol Chem 294:19683–19708

Joosten V, van Berkel WJH (2007) Flavoenzymes. Curr Opin Chem Biol 11:195–202

Klinman J (2007) How do enzymes activate oxygen without inactivating themselves? Acc Chem Res 40:325–333

Klinman JP (1996) New quinocofactors in eukaryotes. J Biol Chem 271:27189–27192

Malmstrom BG (1982) Enzymology of oxygen. Annu Rev Biochem 51:21–59

Mattevi A (2006) To be or not to be an oxidase: challenging the oxygen reactivity of flavoenzymes. Trends Biochem Sci 31:276–228

Matthews RG, Drummond JT (1990) Providing one-carbon units for biological methylations: mechanistic studies on serine hydroxymethyltransferase, methylene tetrahydrofolate reductase, and methyltetrahydrofolate-homocysteine methyltrasferase. Chem Rev 90:1275–1290

Mügge C et al (2020) Flavin-dependent N-hydroxylating enzymes: distribution and application. Appl Microbiol Biotechnol 104:6481–6499

Phintha A, Chaiyen P (2023) Unifying and versatile features of flavin-dependent monooxygenases: diverse catalysis by a common C4a-(hydro)peroxyflavin. J Biol Chem 299:105413

Saleem-Batcha R, Teufel R (2018) Insights into the enzymatic formation, chemical features, and biological role of the flavin-N5-oxide. Curr Opin Chem Biol 47:47–53

Takeda K et al (2019) Fungal PQQ-dependent dehydrogenases and their potential in biocatalysis. Curr Opin Chem Biol 49:113–121

Thauer RK (2019) Methyl(alkyl)-coenzyme M reductases: nickel F-430-containing enzymes involved in anaerobic methane formation and in anaerobic oxidation of methane or of short chain alkanes. Biochemistry 58:5198–5220

Toplak M, Matthews A, Teufel T (2021) The devil is in the details: the chemical basis and mechanistic versatility of flavoprotein monooxygenases. Arch Biochem Biophys 698:108732

Valentino H et al (2020) Structure and function of a flavin-dependent S-monooxygenase from garlic (*Allium sativum*). J Biol Chem 295:11042–11055

Carboxylations and Decarboxylations

32

Except in science fiction, all life as we know it today is based on carbon chemistry. Living beings either assimilate the required carbon from already made organic compounds (most heterotrophs) or fix inorganic carbon dioxide to produce the organic compounds (most autotrophs). Carbon dioxide is the end-product of respiration in all domains of life. Plants, algae, and cyanobacteria fix CO_2 during photosynthesis while some others do it by using inorganic compounds (lithoautotrophs) (Berg et al. 2010; Berg 2011). In this sense, carbon dioxide is the substrate (for carboxylation) or the product (of decarboxylation) of many enzymatic reactions in biology. A carboxylating enzyme usually links either CO_2 or HCO_3^- with an organic acceptor molecule. Enzymatic carboxylations are physiologically significant routes for CO_2 assimilation; CO_2 is necessarily reduced in any chemical carbon fixation reaction. Enzymes that fix carbon dioxide either use an external reductant or oxidize the substrate itself. And the reverse of carboxylation is decarboxylation. Carboxylations and decarboxylations are an important class of enzymatic reactions that make and break carbon-carbon bonds. These reactions have assumed special importance in the light of carbon economy and carbon capture (Bierbaumer et al. 2023; Matthews 2023).

32.1 Reactions and Reactivity of CO_2

Carbon dioxide (CO_2) represents only 0.036% of the atmospheric gases and can diffuse through cell membranes. Its hydration product bicarbonate (HCO_3^-), however, cannot. Solubility of carbon dioxide in water, and at physiological pH, is low. It promptly gets hydrated in water, and the equilibrium favors the bicarbonate form with an apparent pK_a (for [HCO_3^-/CO_2]) of 6.3. Under slightly alkaline conditions, [HCO_3^-] is therefore much higher than that of dissolved [CO_2)]; this makes usage of HCO_3^- advantageous. In fact, this is one reason why a decarboxylation event (enzymatic or not) often becomes an irreversible step of a reaction. By the same

N. S. Punekar, *ENZYMES: Catalysis, Kinetics and Mechanisms*, https://doi.org/10.1007/978-981-97-8179-9_32

Table 32.1 Carboxylases use either CO_2 or HCO_3^- as substrate

Enzymes[a] acting on	
Carbon dioxide (CO_2)	Bicarbonate (HCO_3^-)
Ribulose-1,5-bisphosphate carboxylase-oxygenase (RubisCO)	Carbamyl phosphate synthetase
PEP carboxykinase	PEP carboxylase
PEP carboxytransphosphorylase	Pyruvate carboxylase
Pyruvate synthase	Acetyl CoA carboxylase

[a]Each one of them additionally requires cofactors like divalent metal ion, biotin, or thiamine pyrophosphate. Bicarbonate is usually activated to its carbonic-phosphoric mixed anhydride (carboxy phosphate) for reaction; "X" corresponds to phosphate or N-1 of biotin. See specific examples for details

token, carboxylations are uphill reactions (endergonic) and require input of energy—with HCO_3^- in particular. No wonder that HCO_3^- is activated by phosphorylation to its mixed anhydride carboxy phosphate (see biotin-dependent carboxylases and Fig. 32.2 later).

The reversible hydration of carbon dioxide to bicarbonate is spontaneous but slow (takes several seconds). It can be accelerated by carbonic anhydrase (EC 4.2.1.1; typically, a Zn^{2+}-metalloenzyme).

$$CO_2 + H_2O \rightleftharpoons HCO_3^- + H^+$$

Carbonic anhydrase is an ancient enzyme ubiquitous among organisms from all three domains of life (Smith et al. 1999). It is recruited to facilitate efficient fixation of CO_2 during photosynthesis; and is a component of carboxysomes, the carbon-concentrating compartments from blue-green algae (Scott et al. 2024). The α- and β-carboxysomes of cyanobacteria contain ribulose-1,5-bisphosphate carboxylase/oxygenase (Rubisco) and carbonic anhydrase. Since HCO_3^- can diffuse through the proteinaceous carboxysome shell but CO_2 cannot, carbonic anhydrase generates high concentrations of CO_2 for carbon fixation by Rubisco (Wang et al. 2019). Carbonic anhydrase plays a physiologically significant role (as demonstrated in yeast) in generating HCO_3^- for the carboxylation reactions catalyzed by pyruvate carboxylase, acetyl CoA carboxylase, and carbamyl phosphate synthetase (Aguilera et al. 2005).

Both carbon dioxide (CO_2) and bicarbonate (HCO_3^-) are essential molecules in various physiological processes. The two forms have profound differences in geometry and charge (linear neutral CO_2 versus trigonal planar anion HCO_3^-). Due to these differences in geometry and charge, enzymes specifically use either CO_2 or HCO_3^- (Table 32.1). The carbon atom of CO_2 is a reasonable electrophile. A nucleophilic (carbanion) species is best suited to attack this carbon. Molecules undergoing carboxylation possess structural features that stabilize a carbanion for this attack. Many carboxylation reactions occur by the attack of a substrate carbanion on to the carbon atom of CO_2. Bicarbonate is not as good an electrophile as CO_2. The HCO_3^- anion must be activated to a more electrophilic species for reaction. This is achieved by metal ion coordination, dehydration at the enzyme active site, or by covalent activation. A carboxylase is thus capable of using either CO_2 or HCO_3^-, and not both. Table 32.1 lists a few examples from these two groups of enzymes.

How do we know whether a carboxylase uses CO_2 or HCO_3^- as its substrate? The chemical (non-enzymatic) equilibrium between CO_2 and HCO_3^- in water is reached over several seconds. This time window provides a good kinetic clue. Consider a carboxylase specific for CO_2 as its substrate. When this enzyme reaction is started with CO_2, initial rates are faster. However, with time, the effective $[CO_2]$ decreases due to its hydration to HCO_3^-. This in turn leads to a decreased enzymatic rate and a lower steady state rate is attained. The net result is an observed burst in the early part of the carboxylase time course. In summary, if a carboxylase shows initial burst kinetics with CO_2, then it uses CO_2 as substrate. For HCO_3^- as the substrate a lag in the time course is observed. A carboxylase requiring HCO_3^- as substrate behaves exactly the opposite—a burst with HCO_3^- and a lag with CO_2 in its kinetics. Careful investigation of lag and burst in carboxylation kinetics is thus a useful tool to distinguish whether that carboxylase uses CO_2 or HCO_3^- as its substrate. We finally note that the lag/burst kinetics should be abolished by including carbonic anhydrase (to rapidly establish $CO_2 \rightleftharpoons HCO_3^-$ equilibrium)—to confirm our results.

Carboxylation reactions entail the attack of a suitable carbanion to capture either CO_2 or HCO_3^-. Carboxylases may be classified according to (a) the nature of the attacking carbanion and (b) the cofactor recruited to stabilize the carbanion and/or activate carbon dioxide. Specific mechanistic details available for some well understood carboxylases are described later. Carboxylation reactions of pyruvate and PEP are central to carbon metabolism, and these are discussed next.

32.2 Carboxylation Chemistry with Pyruvate and Phosphoenolpyruvate

Pyruvate is a center piece of carbon metabolism. It is either the substrate or the product in several critical reactions (Table 32.2). A large majority of them are carboxylation or decarboxylation reactions.

Pyruvate predominantly exists as a keto acid while enolpyruvate quickly converts to the keto form in aqueous solution (Fig. 32.1, also see Fig. 27.5). Enolpyruvate offers a resonance stabilized carbanion at its C-3 carbon. This carbanionic carbon is where carboxylation of pyruvate (or PEP, its phosphorylated enol) occurs. The reaction also requires biotin as a cofactor (see below).

Phosphoenolpyruvate (PEP)—the enolase reaction product of glycolysis—participates in a few key biosynthetic steps. PEP is the pyruvoyl donor in two reactions, namely, those of 5-enolpyruvylshikimate-3-phosphate synthase

Table 32.2 Metabolic steps involving pyruvate as substrate or product

Reaction (shown as physiologically relevant)	Enzyme; cofactor(s)
Decarboxylations:	
Pyruvate → Acetaldehyde + CO_2	Pyruvate decarboxylase; TPP
2 Pyruvate → α-Acetolactate + CO_2	α-Acetolactate synthase; TPP
Pyruvate + CoASH + NAD^+ → Acetyl CoA + NADH + H^+ + CO_2	Pyruvate dehydrogenase complex; TPP, FAD, Lipoate
Pyruvate + O_2 → Acetate + CO_2	Pyruvate oxidase; TPP, FAD
Malate + NAD^+ ⇌ Pyruvate + NADH + H^+ + CO_2	Malic enzyme; Mn^{2+}
Oxaloacetate → Pyruvate + CO_2	Oxaloacetate decarboxylase; Mn^{2+}
Carboxylations:	
Pyruvate + HCO_3^- + ATP → Oxaloacetate + ADP + ⑫	Pyruvate carboxylase; Biotin
Acetyl CoA + CO_2 + $2e^-$ ⇌ Pyruvate + CoASH	Pyruvate synthase; Ferredoxin, TPP
Others:	
Pyruvate + CoASH ⇌ Acetyl CoA + $HCOO^-$	Pyruvate-formate lyase
Pyruvate + NADH + H^+ → Lactate + NAD^+	Lactate dehydrogenase
Pyruvate + L-Glutamate ⇌ L-Alanine + 2-Oxoglutarate	Glutamate-pyruvate transaminase; PLP
PEP + ADP → Pyruvate + ATP	Pyruvate kinase

Cofactor abbreviations are: *TPP* thiamine pyrophosphate, *FAD* flavin adenine dinucleotide, *PLP* pyridoxal phosphate. Some of these reactions are unique to anaerobic metabolism (pyruvate-formate lyase) and autotrophic carbon fixation mechanisms of archaea (pyruvate synthase and other reactions of reductive TCA cycle)

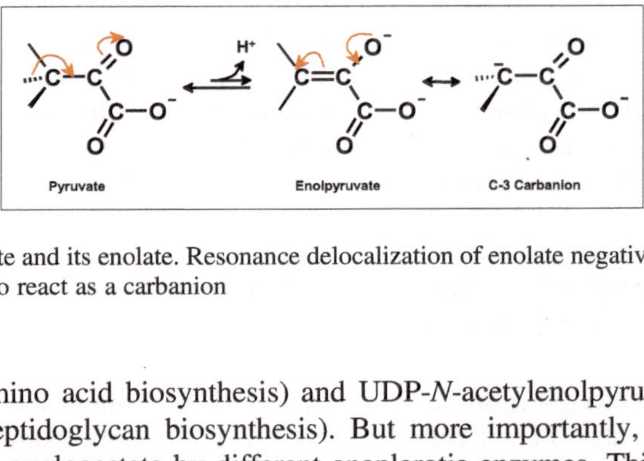

Fig. 32.1 Pyruvate and its enolate. Resonance delocalization of enolate negative charge allows its C-3 carbon atom to react as a carbanion

(of aromatic amino acid biosynthesis) and UDP-*N*-acetylenolpyruvylglucosamine synthase (of peptidoglycan biosynthesis). But more importantly, PEP is readily carboxylated to oxaloacetate by different anaplerotic enzymes. This carboxylation is invariably accompanied by transfer of the high energy phosphoryl group of PEP to an acceptor; nature of phosphoryl acceptor may, however, vary.

The three well known reactions that carboxylate PEP (box below and Table 32.1) seem to involve enol form of pyruvate as the common reactive intermediate. All of them have a divalent metal ion (Mn^{2+} or Mg^{2+}) requirement. The C-3 carbanion of enolpyruvate attacks either CO_2 (PEP carboxykinase and PEP carboxytransphosphorylase) or activated bicarbonate (PEP carboxylase; enolpyruvate attacks the carbonyl phosphate formed first).

Enzymes that Carboxylate PEP

PEP carboxylase (from plants and bacteria)

$$PEP + H_2O + HCO_3^- \rightleftharpoons Oxaloacetate + ⓟ$$

PEP carboxykinase (from fungi, plants, and mammals)

$$PEP + GDP + CO_2 \rightleftharpoons Oxaloacetate + GTP$$

PEP carboxytransphosphorylase (*Propionobacterium* sp.)

$$PEP + ⓟ + CO_2 \rightarrow Oxaloacetate + ⓟ\text{-}ⓟ$$

32.3   Cofactor Assisted Carboxylations

Carboxylation of PEP is facilitated by the associated phosphoryl group transfer. Similarly, all other carboxylations require an input of energy (to activate HCO_3^- in particular) and a suitable cofactor (Table 32.2). Prominent cofactors used in these reactions include biotin, vitamin K, and divalent metal ions. The vitamin K-dependent carboxylase (of prothrombin) is a unique example from blood clotting cascade. The enzyme generates a carbanion on C-4 of the glutamyl residue by abstracting the *proS* hydrogen. This carbanion captures CO_2 to form γ-carboxy-glutamate residues on prothrombin. Autotrophic CO_2 fixation mechanisms of some bacteria (such as pyruvate synthase) are rare examples where thiamine pyrophosphate (TPP) participates in a carboxylation reaction. Here the hydroxyethyl carbanion on TPP (of pyruvate synthase) makes a nucleophilic approach to CO_2. We note here that normally TPP is the cofactor in decarboxylation reactions (discussed later).

Biotin-dependent carboxylases form a physiologically significant group. Ribulose-1,5-bisphosphate carboxylase-oxygenase (RubisCO) is the first and single most important step of photosynthetic CO_2 assimilation. We turn now to these two examples in some detail.

32.3.1 Biotin-Dependent Carboxylases

Biotin is a cofactor in many crucial carboxylation reactions. It is therefore an important water-soluble vitamin whose deficiency causes dermatitis. First isolated from egg yolk in 1936, it binds tightly with avidin (a protein from egg white). Avidin-biotin complex is one of the strongest non-covalent interactions known (K_D of 10^{-15} M and a $t_{1/2}$ of 2.5 years). Most biotin-dependent enzymes are therefore inhibited when incubated with avidin. Biotin is a bicyclic ring with a substituted urea as a functional group for catalytic function (Table 29.1). The cofactor is covalently

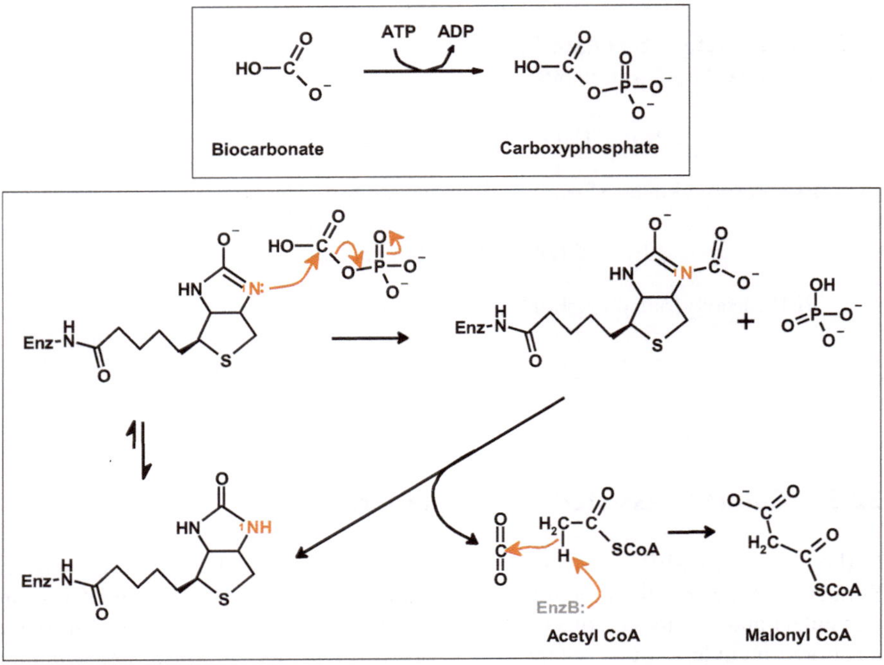

Fig. 32.2 Biotin-dependent carboxylation of acetyl CoA

attached to the ε-amino group of an active site lysine. This charging occurs post-translationally, at the expense of energy (ATP → AMP + $\circled{P}$-$\circled{P}$) and is akin the charging of lipoamide on to transacetylase.

Biotin-dependent carboxylases include enzymes that carboxylate acetyl CoA, propionyl CoA, pyruvate, and urea. More recently, 2-oxoglutarate carboxylase was identified as a component of reductive TCA cycle (Aoshima 2007). This new member of biotin carboxylases is similar to pyruvate carboxylase. Biocarbonate is the substrate in all these reactions and ATP is required to activate it. One atom of ^{18}O from isotopically labeled HCO_3^- ends up in the phosphate formed from ATP (Knowles 1989). This is consistent with the transfer of ATP γ-phosphate to generate *carboxy phosphate* intermediate (Fig. 32.2), at the enzyme active site. It is believed that carboxy phosphate is attacked by N^1 (in its deprotonated state) of biotin to form a carboxy-biotin intermediate. This intermediate (or CO_2 generated from it) is attacked by the substrate carbanion (nucleophile) to form the carboxylated product and regenerate biotin. The entire mechanistic sequence (for acetyl CoA carboxylase) is shown in Fig. 32.2.

Mechanistic evidence was gathered by several elegant isotope exchange experiments and trapping the relevant intermediates. Enzyme bound carboxy-biotin was trapped by diazomethane and stabilized as its methyl derivative (Biotin>N^1-$COOCH_3$). The kinetic and chemical competence of carboxy-biotin was also shown.

The carboxylated enzyme was capable of (a) carboxylating the acceptor substrate and (b) synthesizing ATP from ADP and Ⓟ. Their rates were comparable to the overall carboxylation reaction rates. The partial exchanges observed (see box below) support a ping-pong kinetic mechanism. All the biotin-dependent carboxylases studied so far proceed with retention of configuration at the carbon atom being carboxylated.

Exchange Reactions Observed with Acetyl CoA Carboxylase

Acetyl CoA carboxylase (and other biotin-dependent carboxylases) is a three-component enzyme:

1. Biotin is covalently bound to the small biotin carboxyl carrier protein (BCCP).
2. Biotin carboxylase charges HCO_3^- on to form carboxy-biotin; it is responsible for the HCO_3^- dependent ADP-ATP exchange.

$$[^{14}C]ADP + ATP \rightleftarrows ADP + [^{14}C]ATP$$

3. Carboxyl transferase transfers the active "CO_2" from carboxy-biotin to the acceptor carbanion (acetyl CoA); this is responsible for the acetyl CoA-malonyl CoA exchange.

$$[^{14}C]Acetyl\ CoA + Malonyl\ CoA \rightleftarrows Acetyl\ CoA + [^{14}C]Malonyl\ CoA$$

Transcarboxylase is an interesting variation of the biotin-dependent carboxylases. This enzyme (from *Propionibacterium shermanii*) is not a carboxylase but simply transfers the "CO_2" from a donor to an acceptor via the enzyme bound biotin.

$$Propionyl\ CoA + Oxaloacetate \rightleftarrows Pyruvate + S\text{-Methylmalonyl CoA}$$

Transcarboxylase is an example of two sites, ping-pong kinetics where the biotin on a flexible swinging arm ferries "CO_2" between two separate active site regions—the donor site and the acceptor site (see Fig. 26.5).

At least three instances of carboxylation at the N atom may be noted. Carbamyl phosphate synthetase is an example of ammonia carboxylation that occurs without the need for biotin. The ammonia N makes a direct nucleophilic attack on carboxy phosphate (enzyme bound) to form carbamate—which subsequently gets phosphorylated to carbamyl phosphate. However, the $-NH_2$ group of urea is normally an unreactive nucleophile. It is thought that such N is deprotonated before its attack on carboxy phosphate. Both the carboxylation at N^1 of biotin and that of urea (by urea carboxylase leading to allophanate) are examples of such *N*-carboxylations.

32.3.2 RubisCO

Ribulose-1,5-bisphosphate carboxylase-oxygenase (abbreviated as RubisCO) is the enzyme responsible for fixing atmospheric CO_2 by green plants and photosynthetic bacteria. RubisCO is the most abundant protein in the world. It neither carboxylates an energy-rich molecule (like PEP) nor uses ATP for activation. RubisCO is one of the abundant proteins localized in the chloroplasts and contains Cu^{2+} and requires Mg^{2+} or Mn^{2+} for its carboxylation activity. Carboxylation of ribulose-1,5-bisphosphate is associated with its irreversible cleavage to two molecules of 3-phosphoglycerate. This reaction forms the first step of Calvin–Benson cycle in plants.

Many labeling experiments have defined the chemical mechanism of RubisCO (Fig. 32.3). These include the following: (a) Radioactivity from $^{14}CO_2$ ends up in the C-1 carboxylate of one half of the 3-phosphoglycerate molecules formed, (b) ^{3}H from the solvent is incorporated at C-2 of the same 3-phosphoglycerate molecule that has got $^{14}CO_2$, and (c) ^{3}H on the C-3 of ribulose-1,5-bisphosphate rapidly

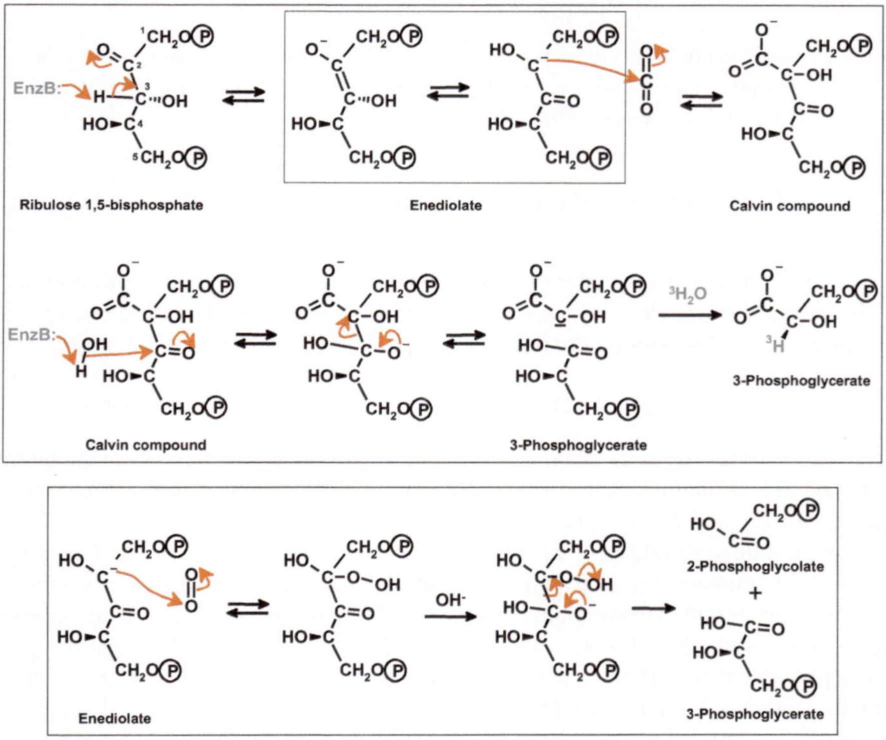

Fig. 32.3 Carboxylation reaction mechanism of ribulose-1,5-bisphosphate carboxylase-oxygenase. The bottom panel shows the oxygenase reaction catalyzed by the same enzyme. The carbon originating from CO_2 is in bold

exchanges with solvent protons in the presence of this enzyme. A six-carbon intermediate compound (2-carboxy-3-keto derivative of ribulose-1,5-bisphosphate) was postulated by Calvin and Benson. Molecules structurally related to this postulated intermediate are excellent reversible inhibitors of RubisCO. We have come across one such inhibitor earlier in 2′-carboxy-D-arabinitol 1,5-bisphosphate (see Fig. 4.9).

RubisCO reaction mechanism begins by proton abstraction from C-3 of ribulose-1,5-bisphosphate (which accounts for tritium exchange with solvent) and enolization of C-2 ketone. The resulting enediol now can display carbanion character at its C-2 and attacks CO_2 to form the carboxylated intermediate (Calvin compound). Nucleophilic attack by water at C-3 and the C–C bond cleavage (between C-2 and C-3) produces two 3-phosphoglycerate molecules. The 3-phosphoglycerate that bears the newly fixed carbon is the one whose C-2 carbanion picks up a solvent proton.

RubisCO is unable to strictly discriminate between CO_2 and O_2 for its substrate (Griffiths 2006). The consequence is the process of photorespiration observed in plants. Apart from its carboxylation reaction, RubisCO is capable of an oxygenase activity. RubisCO functions as a monooxygenase (in the absence of CO_2) and splits ribulose-1,5-bisphosphate into 2-phosphoglycolate and 3-phosphoglycerate. One atom of oxygen from $^{18}O_2$ ends up in the carboxylate group of 2-phosphogycolate. This supports the mechanism, where the enediol (its C-2 carbanion) attacks O_2 (Fig. 32.3).

32.3.3 Autotrophic CO_2/HCO_3^- Assimilation Pathways

Reductive pentose phosphate cycle (Calvin–Benson cycle) along with RubisCO is undoubtedly of singular importance in the autotrophic CO_2 fixation in nature. However, to date, six autotrophic CO_2 fixation mechanisms (Table 32.3) are known (Erb 2011). All of them of course carry out carboxylation steps either with CO_2 or HCO_3^- or both. The observed diversity reflects the variety of the organisms and the ecological niches existing in nature (Berg 2011; Scott et al. 2024).

32.4   Decarboxylation Reactions

Carboxylations and decarboxylations are complementary reactions, yet essential components of the global carbon cycle. Mechanistically, decarboxylations are the reverse of carboxylations—but with a difference. While either CO_2 or HCO_3^- is used in a carboxylation step, *decarboxylation invariably results in CO_2 as its product*. It is a different matter, however, that CO_2 so formed quickly gets hydrated (by carbonic anhydrase) to HCO_3^- in the aqueous setting. This makes decarboxylations irreversible and thermodynamically downhill. An organic compound undergoes decarboxylation by losing its carboxylate group by releasing CO_2. Typically, this reaction goes through a transition state where a carbanion develops on the carbon atom losing the -COOH group (Fig. 32.4). Decarboxylation rates can be

Table 32.3 Autotrophic CO_2/HCO_3^- assimilation mechanisms

Assimilation pathway[a]	Carboxylases used	Occurrence
Calvin–Benson cycle (reductive pentose phosphate cycle)	RubisCO (and PEP carboxylase)	Plants, algae, cyanobacteria, and many aerobic/facultative aerobic proteobacteria
Reductive citric acid cycle (rTCA cycle)	Pyruvate synthase (Fd), 2-Oxoglutarate synthase (Fd), Isocitrate dehydrogenase, PEP carboxylase	Green sulfur bacteria, anaerobic or microaerobic bacteria
Reductive acetyl CoA pathway (requires strict anoxic conditions)	CO Dehydrogenase/Acetyl CoA synthase, Pyruvate synthase (Fd), 2-Oxoglutarate synthase (Fd)	Prokaryotes like acetogenic bacteria, methanogenic archaea, psychrophiles, hyperthermophiles
3-Hydroxypropionate cycle	Acetyl CoA/Propionyl CoA carboxylases (Biotin)	Green non-sulfur phototrophs
3-Hydroxypropionate/ 4-Hydroxybutyrate cycle	Acetyl CoA/Propionyl CoA carboxylases (Biotin), Pyruvate synthase (Fd) and PEP carboxylase	(Hyper)Thermophilic organisms, *Crenarchaeota*
Dicarboxylate/ hydroxybutyrate cycle	Pyruvate:Ferredoxin oxidoreductase, phosphoenolpyruvate carboxylase	Thermoproteales and Desulfurococcales

[a]Depending on the environment and nutritional state some organisms may display more than one assimilation pathway (Erb 2011; Bierbaumer et al. 2023)

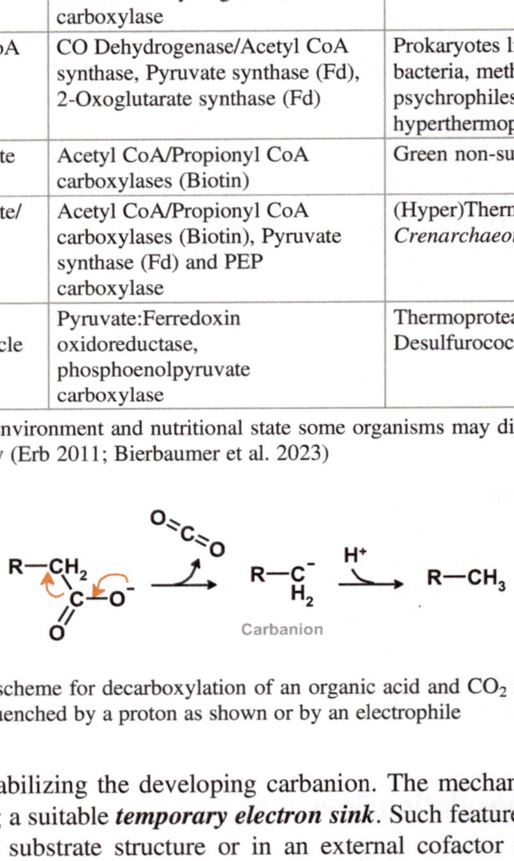

Fig. 32.4 General scheme for decarboxylation of an organic acid and CO_2 release. The product carbanion may be quenched by a proton as shown or by an electrophile

accelerated by stabilizing the developing carbanion. The mechanism(s) to do this include providing a suitable ***temporary electron sink***. Such features may be present either within the substrate structure or in an external cofactor recruited for this purpose. We will look at these possibilities in some detail.

Most common organic acids undergoing decarboxylations also contain additional functional groups. For instance, these may be α-amino acids (all the 20 and more), α-keto acids (pyruvate, 2-oxoglutarate, and 2-keto acids of branched chain amino acids), β-keto acids (oxaloacetate and acetoacetate), βγ-unsaturated acids (*cis*-aconitate), or β-hydroxy acids (malate, isocitrate and 6-phosphogluconate). An incipient carbanion forms during the decarboxylation of all these acids (Fig. 32.4). A β-keto acid substrate offers the possibility of enolization, this in turn can function as a useful electron sink. The β-keto group therefore assists in decarboxylation by delocalizing negative charge. An α-keto group cannot be used in this manner. Both α-amino acids and α-keto acids lack a built-in structural feature to stabilize this developing negative charge. External electron sinks like PLP (α-amino acid decarboxylations; Chap. 33) or TPP (α-keto acid decarboxylations) are required/ recruited to help decarboxylate them. While we will deal with these a little later, decarboxylation mechanisms for β-keto acids are discussed first.

32.4.1 β-Keto Acid Decarboxylation Mechanisms

The β-keto group provides natural assistance in decarboxylating β-keto acids. This is contingent upon how easily it can be enolized. Low basicity of the β-carbonyl group sets a high energy barrier and needs further assistance. Nevertheless β-keto group is exploited to support the developing carbanion in different ways. First, it is made a better electron sink by protonating the developing enolate form. For instance, non-enzymatic decarboxylation of acetoacetic acid is pH dependent and is 50 times faster in its acid form than the corresponding acetoacetate ion. Second, divalent metal ions may assist in stabilizing the negative charge. Mn^{2+} is very well able to accept the developing enolate ion during oxaloacetate decarboxylation (Fig. 32.5; also see Fig. 4.6); Indeed, oxaloacetate decarboxylase requires Mn^{2+} for activity. Oxalate ($^{-}OOC–COO^{-}$) mimicking the initial enolate product is an

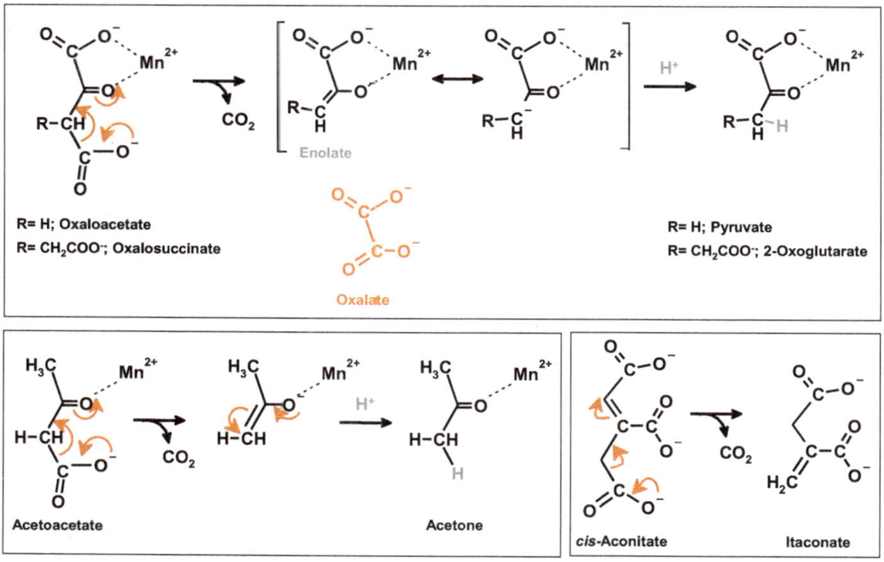

Fig. 32.5 Decarboxylation of β-keto acids. Among the carboxylate groups, the carboxylate that has a suitably positioned β-keto moiety is lost as CO_2. Mn^{2+} functions to accept the developing negative charge. Proposed decarboxylation of *cis*-aconitate by exploiting appropriately located double bond is also shown

inhibitor of this enzyme. This is in keeping with the proposed oxaloacetate decarboxylase chemical mechanism.

Enzymatic decarboxylation of a β-hydroxy acid shares mechanistic similarities with that of a β-keto acid. The β-hydroxy moiety cannot function as a useful electron sink. Therefore, it is converted to a β-keto acid prior to its decarboxylation. Well characterized examples of this kind include malic enzyme, isocitrate dehydrogenase, and 6-phosphogluconate dehydrogenase. Both malic enzyme and isocitrate dehydrogenase require bound Mn^{2+} for activity. They also require $NAD(P)^+$ which accepts the reducing equivalents ($2e^-$) and oxidizes the β-hydroxy acid. The two enzyme reactions appear to proceed with the generation of a β-keto acid intermediate. Analysis of kinetic isotope effects, at least with malic enzyme, supports a mechanism where malate oxidation precedes its decarboxylation (see Fig. 25.4). Oxalosuccinate is the proposed intermediate of isocitrate dehydrogenase reaction (see box below). Recently a loss of function mutant of isocitrate dehydrogenase was implicated in cancer metabolism—it simply reduces 2-oxoglutarate to 2-hydroxyglutarate (see Chap. 39). Although a β-keto acid intermediate seems chemically reasonable, its actual demonstration can be tricky. It may be tightly bound to the active site—not accessible for trapping reagents—or may be ephemeral.

Malic Enzyme:

$$Malate \xrightarrow{-2e^-} [Oxaloacetate] \rightarrow Pyruvate + CO_2$$

Isocitrate Dehydrogenase:

$$Isocitrate \xrightarrow{-2e^-} [Oxalosuccinate] \rightarrow 2\text{-}Oxoglutarate + CO_2$$

Oxaloacetate (as also oxalosuccinate) is both an α-keto acid and a β-keto acid. Only the carboxylate that is β to the keto group (previously the hydroxyl group) is lost as CO_2. Interestingly, *cis*-aconitate decarboxylase might exploit the suitably located double bond to facilitate decarboxylation and form itaconate.

32.4.2 Decarboxylation of Acetoacetate

As expected for a β-keto acid, both non-enzymatic and enzymatic decarboxylation of acetoacetic acid is promoted by Mn^{2+}. The divalent metal ion acts as a super-acid catalyst, polarizes the β-keto group, and makes it a better electron sink. Acetoacetate decarboxylases requiring Mn^{2+} for catalysis (Fig. 32.5) are less common. Instead, there is an efficient and interesting way to use β-keto moiety as an electron sink. This

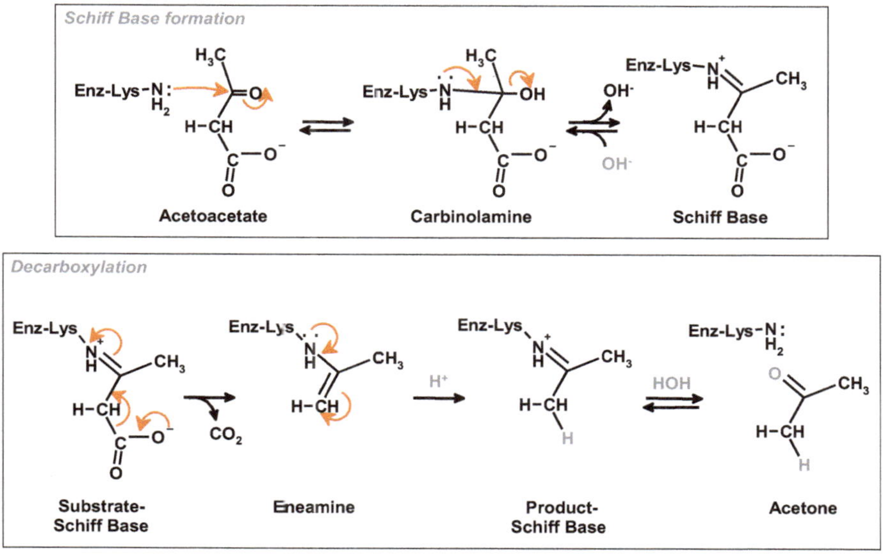

Fig. 32.6 Acetoacetate decarboxylase reaction mechanism. Formation of Schiff base between the amino group and a carbonyl is a pH dependent reversible process. During this process, carbonyl oxygen gets equilibrated with the oxygen of water. The reversible protonation of enamine on the enzyme by solvent protons (shown in grey in bottom panel) is responsible for observed deuterium exchange in D_2O. Note that acetone is symmetric, and all its methyl hydrogens are equivalent for this exchange

more effective strategy employs a protonated Schiff base form. While it is difficult to protonate the oxygen of β-carbonyl group (poor base), corresponding imine nitrogen (Schiff base) is readily protonated. This ***cationic imine*** is an excellent sink to stabilize the carbanion formed during decarboxylation. *The formation of imines at adjacent carbonyl groups is a general mechanism for catalysis when carbanions are generated during the reaction.* No wonder that amines are effective catalysts in decarboxylating β-keto acids. For instance, aniline (pK_a of 4.8) can accelerate acetoacetate (pK_a of 3.7) decarboxylation rate maximally at pH 4.2. The "aniline-acetoacetate complex" breaks down much faster than the acid alone. A well-documented enzyme example is acetoacetate decarboxylase from *Clostridium acetobutylicum*. In its reaction mechanism (Fig. 32.6), active site lysine $-NH_2$ forms a Schiff base with acetoacetate. This cationic imine acts as an electron sink to stabilize the carbanion formed at C-2 in the transition state.

The proposed acetoacetate decarboxylase chemical mechanism is supported by several lines of experimental evidence: (a) sodium borohydride ($NaBH_4$) inactivates this enzyme in the presence of acetoacetate, (b) a lysine residue is radio-labeled by $NaBH_4$ reduction in the presence of 3-[^{14}C]-acetoacetate (or [^{14}C]-acetone), (c) the ε-NH_2 of this active site lysine has an unusually low pK_a, (d) the enzyme incorporates ^{18}O from $H_2{}^{18}O$ into the product acetone, and e) the enzyme catalyzes the exchange of deuterium from D_2O into acetone to produce CD_3COCD_3. In sum, these data justify the *formation of an initial substrate imine intermediate and of an enamine of acetone* in the acetoacetate decarboxylase reaction mechanism. In contrast, an imine intermediate does not form in the case of Mn^{2+}—requiring acetoacetate decarboxylase (Fig. 32.5). Accordingly, $NaBH_4$ inactivation and ^{18}O exchange are also not observed with this Mn^{2+}-dependent catalysis.

32.5 Thiamine Pyrophosphate and α-Keto Acid Decarboxylations

Unlike β-keto acids, α-keto acids are difficult to decarboxylate. The α-carbonyl group of the α-keto acid cannot be used as an electron sink, and it cannot stabilize the developing carbanion. Hence, all α-keto acid decarboxylations need cofactor assistance. Nature has found an elegant solution to this chemical problem by recruiting thiamine pyrophosphate (TPP) as an external electron sink. The thiazolium ring of TPP physically participates in decarboxylation chemistry (Frank et al. 2007; Jordan 2003). The following chemical features make thiazolium (and hence TPP) an ideal cofactor:

1. The C-2 proton of the thiazolium ring is very acidic (dissociates with a $t_{1/2}$ of 2 min at pH 5.0). This is because (a) the C-2 carbon is sandwiched between two electronegative atoms (N and S) and (b) in its active conformation, proximal amine N on pyrimidine ring helps this deprotonation.

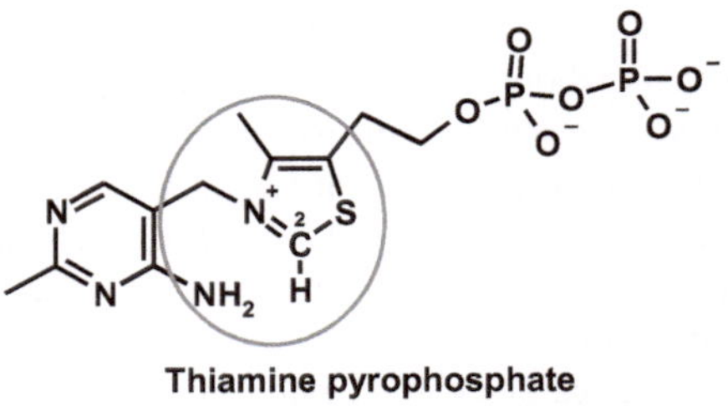

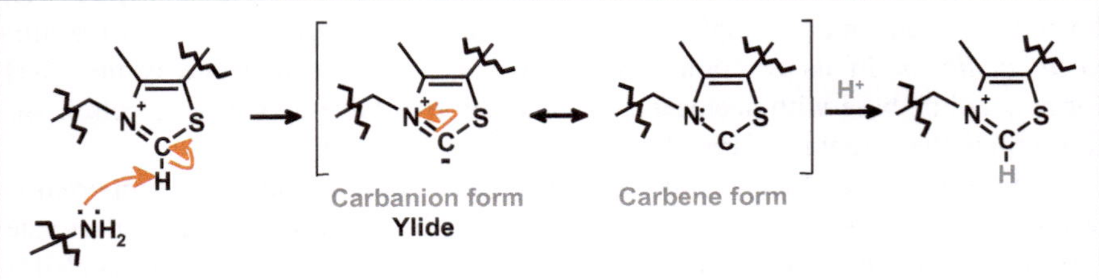

Thiamine pyrophosphate

Fig. 32.7 Thiamine pyrophosphate and its thiazolium ring. The C-2 proton is quite acidic and can exchange with solvent water. Along with the neighboring positive charge on N, TPP carbanion is an ylide

2. The carbanion at C-2 (Fig. 32.7) initiates the nucleophilic attack to the substrate carbonyl. The five-member ring thiazolium (with $-N=C-S-$) is best suited for the purpose (both on thermodynamic and kinetic grounds) whereas imidazolium (with $-N=C-N-$) and oxazolium (with $-N=C-O-$) are not. The carbanion at C-2 with the neighboring positively charged N atom is actually an **ylide**—a neutral dipolar molecule with formal positive and negative charges on adjacent atoms.

3. The thiazolium carbon-nitrogen double bond (and its cationic imine) acts as an electron sink to stabilize the substrate carbanion. This arrangement is akin to the cationic imine of Schiff base used in decarboxylating acetoacetate (Fig. 32.6). Thus, TPP furnishes the required electron sink for α-keto acid decarboxylations.

Decarboxylation of pyruvate is an important example of TPP-assisted electrophilic catalysis. There are many variants of pyruvate decarboxylation. In every case, the ylide carbanion (of TPP) first attacks the keto group of pyruvate and forms a lactyl adduct. Enzyme active site electrostatics (Fig. 4.3) contributes to expulsion of CO_2. Decarboxylation of the adduct results in the carbanion of hydroxyethyl TPP (HETPP; Fig. 32.8). HETPP is a stabilized carbanion since the "$>C=N^+<$" moiety of TPP acts as an electron sink—by forming an enamine intermediate. The fate of HETPP carbanion depends on the type of reaction catalyzed by that particular enzyme. All these possibilities are shown in Fig. 32.8.

Many reaction intermediates in TPP catalysis were recently observed by crystallography. For instance, pyruvate oxidase enzyme forms bound with 2-lactyl TPP (or its stable phosphonate analog), enamine of TPP, and 2-acetyl TPP provide snapshots of its chemical mechanism.

Oxidative decarboxylation of pyruvate is central to aerobic energy metabolism. This is done by a multi-enzyme complex of three distinct activities: E1—pyruvate decarboxylase, E2—transacetylase, and E3—dihydrolipoyl dehydrogenase.

Partial Reactions of Pyruvate Dehydrogenase Complex

$$Pyruvate + E1\text{-}TPP \rightarrow E1\text{-}HETPP + CO_2$$

$$E1\text{-}HETPP + E2\text{-}Lipoate + CoASH \rightarrow Acetyl\ CoA + E1\text{-}TPP + E2\text{-}Dihydrolipoate$$

$$E2\text{-}Dihydrolipoate + NAD^+ \rightarrow E2\text{-}Lipoate + NADH + H^+$$

Overall reaction stoichiometry:

$$Pyruvate + NAD^+ + CoASH \rightarrow Acetyl\ CoA + NADH + CO_2$$

The acetyl-lipoamide thioester formed by the first enzyme (E1) is used in the next step to generate acetyl CoA. The third activity (E3, with FAD as the redox device)

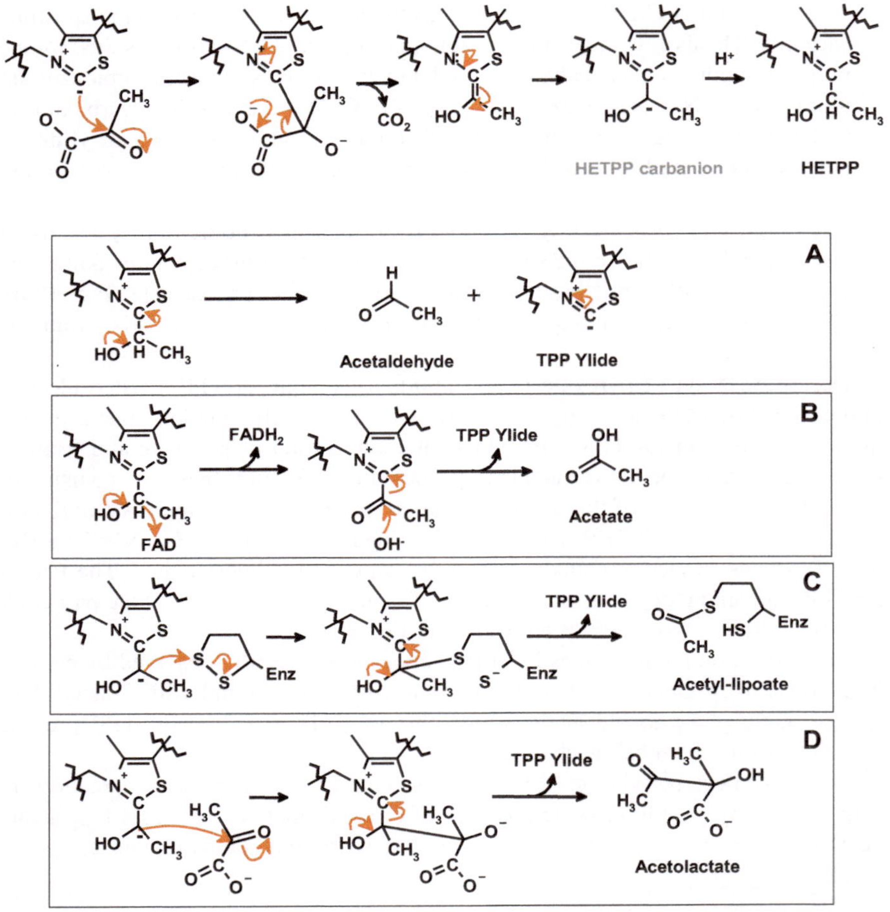

Fig. 32.8 Different decarboxylation reactions of pyruvate involving TPP. Formation of hydroxyethyl TPP (HETPP) carbanion and its subsequent fates are shown. (**a**) Protonation of HETPP and regeneration of coenzyme ylide gives acetaldehyde (non-oxidative decarboxylation; yeast pyruvate decarboxylase). (**b**) Oxidation of HETPP to acetyl-TPP followed by hydrolysis yields acetate (oxidative decarboxylation; *E. coli* pyruvate oxidase). (**c**) When HETPP carbanion reacts with a disulfide (lipoamide), with subsequent coenzyme ylide release, acetyl thioester is formed (oxidative decarboxylation; pyruvate dehydrogenase complex). (**d**) Attack of HETPP carbanion on keto carbon of another pyruvate molecule leads to α-acetolactate (condensation; α-acetolactate synthase)

oxidizes dihydrolipoate back by reducing NAD^+. The mechanism of this dehydrogenase resembles that of glutathione reductase described in the previous chapter (Fig. 31.4). The metabolic significance of pyruvate dehydrogenase complex is obvious from the range of five cofactors used in its chemistry—TPP, FAD, NAD^+, CoASH, and lipoate. Two other α-keto acid dehydrogenase complexes,

Ribulose-5-Ⓟ + Ribose-5-Ⓟ ⇌ Glyceraldehyde-3-Ⓟ + Sedoheptulose-7-Ⓟ

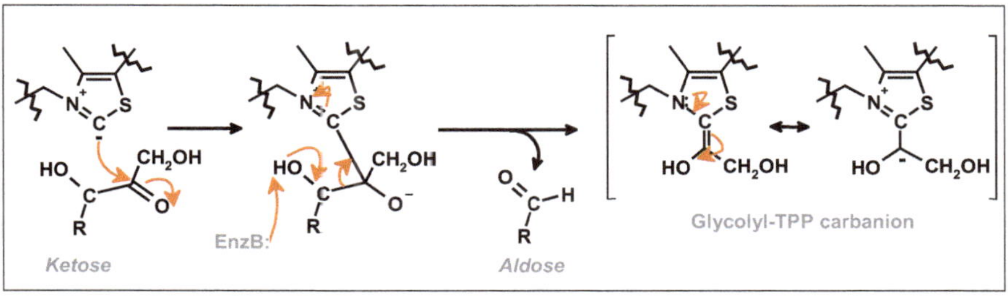

Fig. 32.9 Transketolase mechanism involving a glycolyl-TPP carbanion. Reaction for the synthesis of sedoheptulose-7-Ⓟ is shown for example. Glyceraldehyde-3-Ⓟ is released from ribulose-5-Ⓟ while the glycolyl group is retained on the TPP. In a reversal of these steps, ribose-5-Ⓟ accepts the glycolyl group to form sedoheptulose-7-Ⓟ

mechanistically very similar to pyruvate dehydrogenase complex, are 2-oxoglutarate dehydrogenase complex of Krebs cycle (2-oxoglutarate→Succinyl CoA) and the branched chain keto acid dehydrogenase complex of valine, isoleucine, and leucine catabolism.

Electrophilic catalysis and stabilization of substrate carbanion are hallmarks of TPP-dependent decarboxylations. For the same reasons, TPP is an ideal cofactor for carbon-carbon bond formation chemistry as well. The two-carbon transfers are catalyzed by the family of TPP-dependent transketolases. In a transketolase reaction, the TPP carbanion (thiazolium ylide) attacks the carbonyl carbon of a ketose sugar (Fig. 32.9). Instead of a decarboxylation, a carbon-carbon bond cleavage occurs in this adduct and an aldose (two carbons short) is released as the first product. In an exact reversal of this reaction sequence, the glycolyl-TPP carbanion now attacks an aldose (same or a different one) and the 2-hydroxyacetyl group is transferred.

The 2-hydroxyacetyl (also called the α-ketol) group transfers are critically important reactions of pentose phosphate pathway and the Calvin–Benson cycle. These are *two-carbon transfers* brought about by TPP-dependent transketolases. Transaldolases, on the other hand, carry out *transfer of three-carbon fragments* (dihydroxyacetone equivalents). Between them, transaldolases and transketolases move one-carbon equivalents (CH_2O) around and bring each carbon of every sugar into the metabolic pool. Indeed, the two activities together contribute to build glucose (from six individually fixed CO_2 molecules) and also generate a variety of aldoses and ketoses required for cellular metabolism.

32.6 Summing Up

Carbon dioxide is the substrate for carboxylation and product of decarboxylation in a number of biological reactions.

Both CO_2 and HCO_3^- could serve as carboxylation substrates. In carboxylation, a suitable carbanion captures either one of these species. But HCO_3^- is not as good an electrophile as CO_2. Prominent cofactors used in these reactions include divalent metal ions, biotin, and vitamin K. Besides, input of energy (in terms of phosphoryl group transfer) may be required to activate HCO_3^-.

Decarboxylations invariably result in the release of CO_2 as the product. Since, carbon dioxide is rapidly hydrated to HCO_3^- in water, decarboxylation steps are irreversible and thermodynamically downhill. A carbanion develops when –COOH is lost as carbon dioxide. Stabilizing such a carbanion transition state is an essential trick of enzyme catalysis. The developing negative charge on the C atom could be handled through suitably placed temporary electron sinks. Such sinks may be found on the substrate itself (β-keto group), on the enzyme (Schiff base through Lys-NH_2), or on the cofactor (a divalent metal ion, TPP, or pyridoxal phosphate).

Carboxylation reactions lead to carbon capture (fixing atmospheric CO_2) and decarboxylations release CO_2 (an end-product of respiration). The two are complementary in preserving the atmospheric CO_2 balance.

References

Aguilera J et al (2005) Carbonic anhydrase (Nce103p): an essential biosynthetic enzyme for growth of *Saccharomyces cerevisiae* at atmospheric carbon dioxide pressure. Biochem J 391:311–316

Aoshima M (2007) Novel enzyme reactions related to the tricarboxylic acid cycle: phylogenetic/functional implications and biotechnological applications. Appl Microbiol Biotechnol 75:249–255

Berg IA (2011) Ecological aspects of the distribution of different autotrophic CO_2 fixation pathways. Appl Environ Microbiol 77:1925–1936

Berg IA et al (2010) Autotrophic carbon fixation in archaea. Nat Rev Microbiol 8:447–460

Bierbaumer S et al (2023) Enzymatic conversion of CO_2: from natural to artificial utilization. Chem Rev 123:5702–5754

Erb TJ (2011) Carboxylases in natural and synthetic microbial pathways. Appl Environ Microbiol 77:8466–8477

Frank RAW, Leeper FJ, Luisi BF (2007) Structure, mechanism, and catalytic duality of thiamine dependent enzymes. Cell Mol Life Sci 64:892–905

Griffiths H (2006) Designs on Rubisco. Nature 441:940–941

Jordan F (2003) Current mechanistic understanding of thiamin-dependent enzymatic reactions. Nat Prod Rep 20:184–201

Knowles JR (1989) The mechanism of biotin-dependent enzymes. Annu Rev Biochem 58:195–222

Matthews ML (2023) Engineering photosynthesis, nature's carbon capture machine. PLoS Biol 21: e3002183

Scott KM, Payne RR, Gahramanova A (2024) Widespread dissolved inorganic carbon-modifying toolkits in genomes of autotrophic Bacteria and Archaea and how they are likely to bridge supply from the environment to demand by autotrophic pathways. Appl Environ Microbiol 90: e01557–e01523

Smith KS et al (1999) Carbonic anhydrase is an ancient enzyme widespread in prokaryotes. Proc Natl Acad Sci USA 96:15184–15189

Wang H et al (2019) Rubisco condensate formation by CcmM in β-carboxysome biogenesis. Nature 566:131–135

Electrophilic Catalysis and Amino Acid Transformations

Side chains of many amino acid residues are known to participate in nucleophilic catalysis by forming covalent enzyme-substrate intermediates. Enzymes have a choice of many nucleophilic groups but have little to offer in terms of good electrophiles. Therefore, a few small molecules (cofactors and prosthetic groups) are recruited by nature to complement an apoenzyme—resulting in a functional holoenzyme. These small molecules act as temporary electron sinks during catalysis by forming covalent adducts with substrates. Much of enzyme chemistry is carbanion chemistry. The abstraction of a proton or decarboxylation from an sp^3 carbon leaves behind a carbanion that is not so stable. Developing carbanions may be stabilized—hence effecting rate accelerations—by suitably placing temporary electron sinks. Coenzymes like pyridoxal phosphate and thiamine pyrophosphate function to stabilize them via their electrophilic adducts. We already listed more commonly encountered electrophilic reagents in Table 29.1. Decarboxylations involving carboxylic acids (other than amino acids) were covered previously (Chap. 32). Amino acid transformations including decarboxylation offer a different chemical challenge and are described in this chapter.

Covalent intermediates are as common to electrophilic as they are for nucleophilic catalysis. Proving the existence of covalent intermediates (like Schiff bases) and showing that they indeed participate in the catalytic process (i.e., their kinetic and chemical competence) requires multiple lines of evidence. We noted in Chap. 29 with regard to nucleophilic catalysis that these may include data from steady-state kinetics, isotope exchange studies, inference from analogs and side reactions, stereochemical evidence, and direct observation and/or trapping. These same tools may be used to understand electrophilic catalysis as well. Sodium borohydride inactivation of an enzyme (in the presence of appropriate substrate) provides diagnostic evidence for an imine (Schiff base) intermediate. Intermediate formation (such as a Schiff base) also makes the overall reaction proceed in a stepwise manner.

A covalent bond is established between the enzyme and the substrate in one of two ways: (a) an enzyme nucleophile may attack an electron deficient center on the

© The Author(s), under exclusive license to Springer Nature Singapore Pte Ltd. 2025
N. S. Punekar, *ENZYMES: Catalysis, Kinetics and Mechanisms*, https://doi.org/10.1007/978-981-97-8179-9_33

substrate or (b) an enzyme bound electrophile may be attacked by the electron rich center of the substrate. With decarboxylases, aldolases, and trans-aldolases, the substrate provides the carbonyl component, while the enzyme provides the amine. In pyridoxal 5′-phosphate (PLP) enzymes, however, it is the substrate that provides the amine component, and the coenzyme provides the carbonyl component. In both cases, the cationic imine—in a conjugated system—is excellent for extensive charge delocalization. These Schiff base systems act as temporary electron sinks to stabilize carbanion intermediates that develop during enzyme catalysis.

Schiff Base Chemistry A suitably positioned Schiff base is a simple yet effective electrophile and a good tool to delocalize/hold electrons. The formation of a Schiff base requires a carbonyl group and an amino function (Fig. 33.1). This is possible either with a "substrate ($>C=O$) and enzyme (NH_2)" (see aldolase in Table 29.3 and acetoacetate decarboxylase in Fig. 32.6) or "substrate (NH_2) and enzyme ($>C=O$)" (as in the case of a PLP-dependent transaminase, see below). Carbonyl groups are very rare on enzymes; more often an enzyme contributes lysine amino group that reacts with the substrate carbonyl to form a Schiff base.

We will first describe protein electrophiles that are derived from amino acid side chains on enzymes.

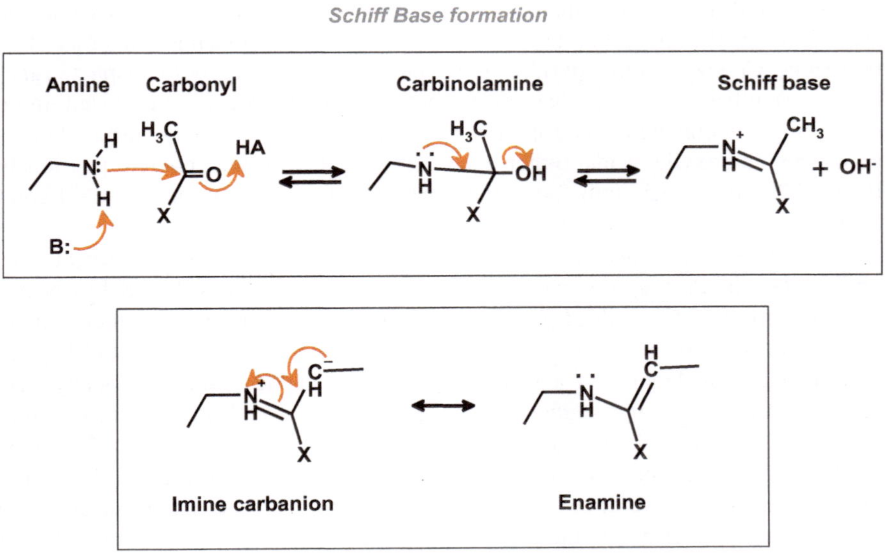

Schiff Base formation

Fig. 33.1 Schiff base formation requires a carbonyl group and an amine function. X in the carbonyl compound can be H (for aldehydes) or a group with C (for ketones). This reversible reaction is affected by pH because the Schiff base can be protonated. It is the protonated Schiff base that acts as an electron sink with the lone pair momentarily residing on the N atom. A carbanion adjacent to the protonated imine (bottom panel) is stabilized by the formation of enamine

33.1 Protein Electrophiles

Amino acid side chains have very little to offer as electrophilic groups. Apart from protons, as the most frequent electrophiles, enzymes also use metal ions and organic cofactors as electrophiles. Despite the fact that there is paucity of electrophilic groups in the side chains of proteogenic amino acids, electrophilic catalysis by enzymes is not uncommon. Then, where do the required enzyme electrophiles come from? There are two possibilities, and both are exploited in enzyme design by nature: (a) post-translation modification/conversion of a few amino acid side chains into electrophiles for catalysis (Atkins and Gesteland 2002; Cooke et al. 2009; van Poelje and Snell 1990) or (b) the recruitment of an electrophilic cofactor (like PLP) derived from vitamins (Percudani and Peracchi 2003; Phillips 2015). We will discuss the electrophiles generated from amino acid side chains first. The post-translational modification of the polypeptide chain may result in a catalytically functional electrophile. These examples are listed in Table 33.1 below.

33.1.1 Enzymes with Dehydroalanine and 4-Methylideneimidazole-5-One (MIO)

Enzymes belonging to Lyase Class (EC 4.x.x.x) utilize a range of prosthetic groups including thiamine pyrophosphate (TPP), iron-sulfur clusters, pyridoxal 5'--phosphate (PLP), biotin, or peptide-dehydroalanine (actually MIO moiety). Amino acid lyases form important and interesting members of this group. The elimination of ammonia from an L-amino acid is a chemical challenge. It requires the abstraction of a β-proton in the face of a positively charged α-ammonium group. In L-aspartate ammonia lyase (Fig. 27.9), the presence of the substrate β-carboxyl (electron-withdrawing) group facilitates this process. Therefore, the enzyme does not require any extra help in ammonia elimination. With histidine ammonia lyase (HAL) and phenylalanine ammonia lyase (PAL), however, this is much more difficult, as the β-proton in these amino acids is considerably less acidic. Members of this lyase group have recruited an essential electrophilic group—which was believed to be dehydroalanine (Fig. 33.2)—post-translationally derived from a serine residue. However, recently solved X-ray structures support the presence of 4-methylideneimidazole-5-one (MIO; synonym, 3,5-Dihydro-5-methylidene-4H-imidazol-4-one) as the functional group instead. MIO is formed by the autocatalytic cyclization of the inner Ala-Ser-Gly tripeptide motif on these enzymes. Both the structure and the process of MIO cyclization are mechanistically similar to the well-characterized fluorophore of green fluorescence protein.

MIO may be regarded as a modified dehydroalanine with much-enhanced electrophilicity. It helps increase the acidity of the abstractable β-proton. MIO is a very strong electrophile because (a) delocalization of the N lone pair into α,β-unsaturated carbonyl system is blocked and (b) upon addition of a nucleophile, the imidazolone becomes aromatic. The α-amine of the substrate (phenylalanine ammonia lyase)

Table 33.1 Electrophiles derived from amino acid side chains on enzymes

Amino acid residue(s) converted to	Nature of modification	Electrophile	Enzyme example
Lysine → Pyrrolysine	Pre-translational; dedicated codon usage	*(chemical structure)*	Methylamine methyltransferase
Glutamate + Tyrosine → PQQ (Pyrroloquinoline quinone)	Post-translational; free cofactor	*(chemical structure)*	Bacterial glucose dehydrogenase and methanol dehydrogenase
Serine → Pyruvate	Post-translational; peptide bound	*(chemical structure)*	Pyruvoyl enzymes like histidine decarboxylase
(Alanine → Dehydroalanine) -Ala-Ser-Gly- →-MIO-(4-methylideneimidazole-5-one)	Post-translational; peptide bound	*(chemical structure)*	Phenylalanine ammonia lyase and histidine ammonia lyase

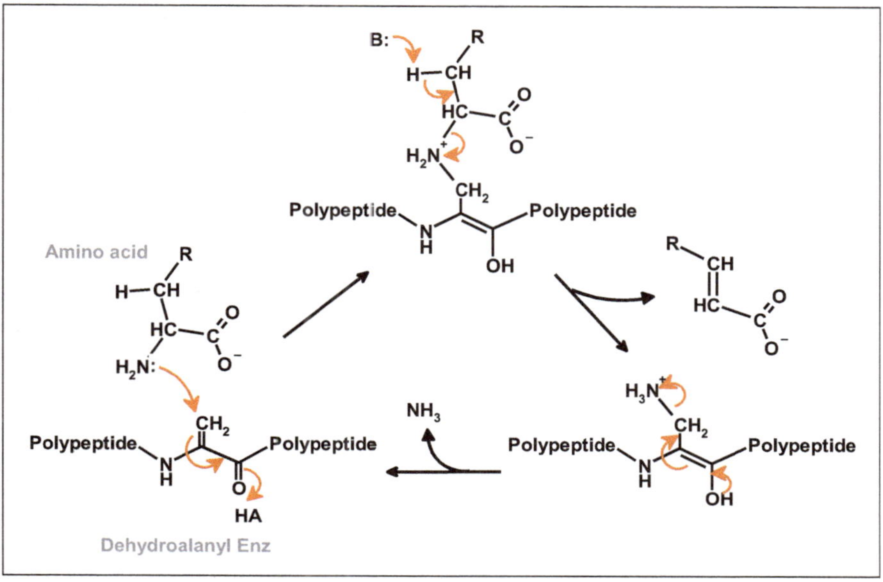

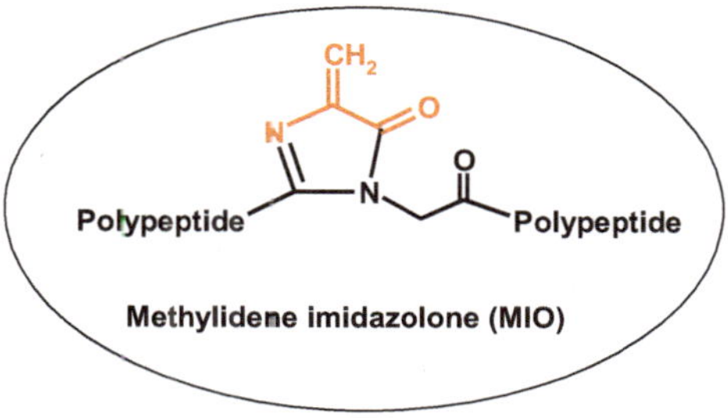

Fig. 33.2 The mechanism of dehydroalanine-dependent ammonia elimination. The substrate specificities of histidine ammonia lyase (R = imidazole) and phenylalanine ammonia lyase (R = phenyl) are different. MIO (structure shown below) may be regarded as a modified dehydroalanine of enhanced nucleophilicity

forms a covalent adduct with the exocyclic alkene of the MIO prosthetic group (Fig. 33.2). The bound amino acid is then deprotonated at the benzylic position by an enzymatic base (E1cB mechanism). Cinnamic acid is released by the lyase upon the loss of ammonia. During the catalytic cycle, N^3 atom of MIO changes its hybridization state from $sp^3 \rightarrow sp^2$ and then back. Based on the high electrophilicity of this prosthetic group and active site geometry, an alternative mechanism has also been proposed. According to this proposal, a Friedel—Crafts reaction-type attack at the substrate aryl side chain occurs.

33.1.2 Pyruvoyl-Dependent Enzymes

Decarboxylations are most common among the many metabolic transformations involving amino acids. Some important products of decarboxylation include γ-aminobutyrate, histamine, serotonin, dopamine, putrescine, and 3-(S-adenosyl) propylamine. Amino acid decarboxylases need to stabilize the developing carbanion during the course of catalysis. This is typically achieved by pyridoxal 5′-phosphate (PLP) as the electrophilic cofactor. PLP mechanisms will be the subject of a later section in this chapter (see below). However, several decarboxylases are known that do not use PLP but instead require pyruvate as a covalently bound prosthetic group. Table 33.2 provides a summary of pyruvoyl-dependent enzymes, along with their distribution and metabolic significance.

A combination of approaches is used to identify the presence of the pyruvoyl group on these enzymes. These include the following:

1. Cyanide, hydroxylamine, phenylhydrazine, and sodium borohydride ($NaBH_4$) inhibit pyruvoyl-dependent enzymes by reacting with the catalytically essential carbonyl group.
2. Pyruvoyl-dependent enzymes can be distinguished from those containing PLP by spectral analysis. While PLP absorbs (and fluoresces), the pyruvoyl group has negligible absorbance above 300 nm.
3. Covalently bound pyruvoyl group can be released by mild hydrolysis of the enzyme protein; it can later be reduced to lactate—by lactate dehydrogenase.
4. The pyruvoyl group can be distinguished from dehydroalanine (discussed earlier) by reduction with [^{3}H]-$NaBH_4$. The pyruvoyl group is obtained on subsequent hydrolysis as [^{3}H]-lactate while dehydroalanine residues yield [^{3}H]-alanine.

Table 33.2 Pyruvoyl-dependent enzymes and their metabolic significance

Decarboxylase	Metabolic significance	Reported from
Histidine → Histamine	Unknown	*Lactobacilli*
Aspartate → β-Alanine	Precursor of pantothenate, coenzyme A, acyl carrier protein, etc.	*E. coli*
4′-Phospho-pantothenoylcysteine → 4′-Phospho-pantotheine	Precursor of coenzyme A and acyl carrier protein	*E. coli*, mammalian liver
S-Adenosyl-methionine → 3-(S-adenosyl) propylamine	Polyamine precursor	*E. coli*, yeast, mammalian liver
Phosphatidylserine → Phosphatidylethanolamine	Phospholipid of the membrane	*E. coli*, yeast, mammalian liver

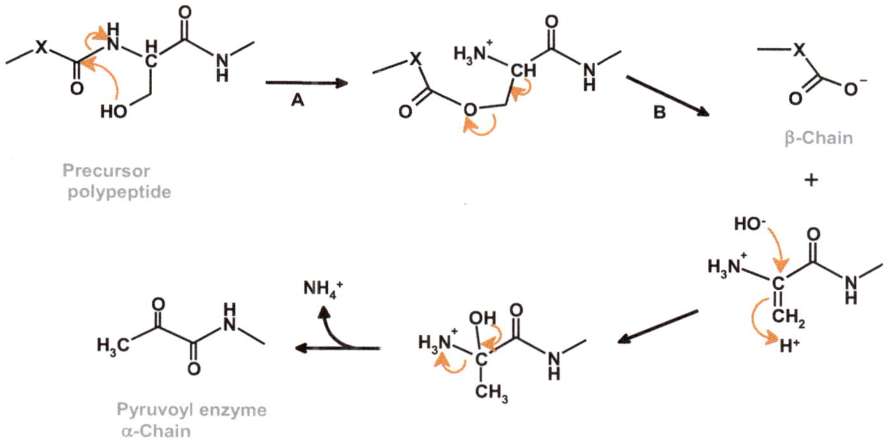

Fig. 33.3 Post-translational formation of pyruvoyl group. Pyruvate is always found at the N-terminus of α chain. It is formed by serinolysis (step A) and α,β-elimination (step B) events in the proenzyme

The pyruvoyl group on the enzyme is derived post-translationally. In all cases examined so far, a precursor polypeptide (proenzyme) is processed at a specific X–Ser bond to give two chains: a β chain with X at its –COOH terminus and the α chain with a pyruvoyl moiety in amide linkage at its N-terminus. The pyruvoyl moiety thus arises from an internal Ser residue. The mechanism of pyruvate formation by the serinolysis of proenzyme is shown in Fig. 33.3.

Histidine decarboxylase of *Lactobacillus* (a Gram-positive bacterium) is the best characterized pyruvoyl enzyme. Its decarboxylation mechanism is well understood, and possibly all other pyruvoyl-dependent decarboxylases act by a similar mechanism. The decarboxylation reaction can be visualized as a two-stage process: (a) labilization of the –COO⁻ group with loss of CO_2 and (b) its replacement by a proton. During catalysis, the pyruvoyl group forms a Schiff base (Fig. 33.4) with the substrate amine while the protein component provides the necessary binding pocket and groups required for general acid-base chemistry. A hydrophobic carboxylate binding pocket promotes the energetically favorable loss of negative charge by decarboxylation. The decarboxylation of pyruvoyl-amino acid Schiff base generates a resonance-stabilized carbanion. Proton addition to the azomethine carbon (in the imine carbanion in Fig. 33.4) gives the Schiff base of amine product with enzyme. Resolution of this Schiff base results in the free amine and regenerates the pyruvoyl enzyme for another catalytic cycle. Both the Schiff base intermediates (with the amino acid substrate and the product amine) can be trapped as acid-stable secondary amines. Further, their reduction by $NaBH_4$ covalently tags the protein.

The overall chemical mechanism of the pyruvoyl enzyme is analogous to the PLP-dependent amino acid decarboxylation (described in the next section; see Fig. 33.7). Histidine is the only amino acid so far known to be decarboxylated by both types of decarboxylases—one with pyruvoyl prosthetic group and the other

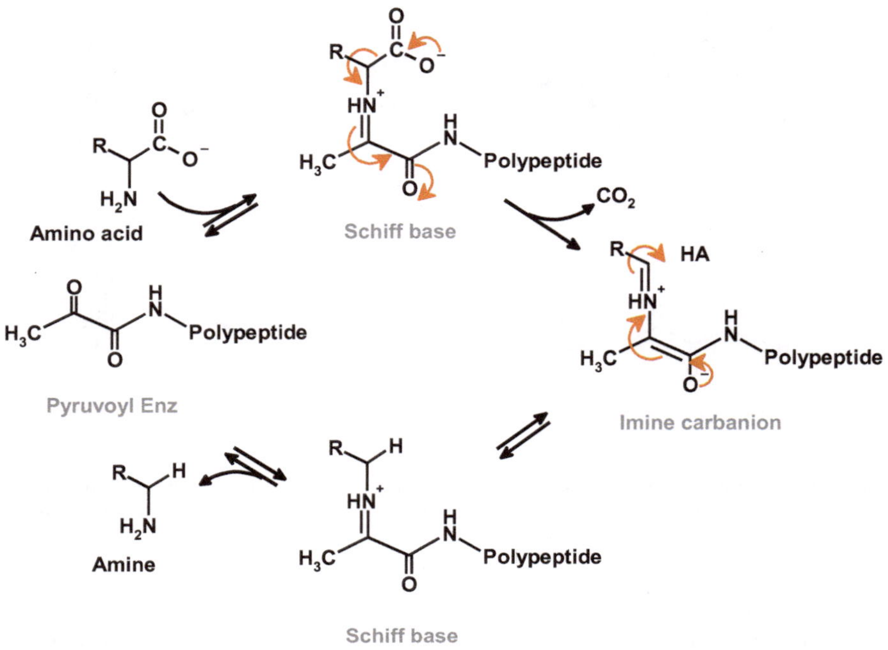

Fig. 33.4 The mechanism of amino acid decarboxylation by pyruvoyl-dependent enzymes

with PLP. While both act by Schiff base mechanism, they differ in their catalytic apparatus and distribution in the tree of life. Nature has clearly invented two excellent solutions to the same chemical problem.

33.2   Reactions Involving Pyridoxal Phosphate (PLP)

Proteogenic amino acids have exclusively L-configuration. Being predominant N-containing metabolites, they are often used as precursors of other cellular nitrogenous products. Apart from the MIO and pyruvoyl chemistry discussed above, breakdown and transformation of amino acids comes in two more forms:

(a) Oxidation of amino group by NAD(P)$^+$ (dehydrogenase) and flavin-dependent (oxidase) enzymes. These are already mentioned in Chap. 31.
(b) PLP-dependent reactions at the α-, β-, and γ-positions (with respect to the amino group) of amino acids. These wide ranges of reactions catalyzed by PLP-dependent enzymes are the subject of this section.

   Pyridoxal phosphate (PLP) is a crucial cofactor participating in most amino acid transformations. PLP (Fig. 33.5) is derived from vitamin B6 (pyridoxine) upon phosphorylation. The importance of PLP lies in its ability to provide an excellent

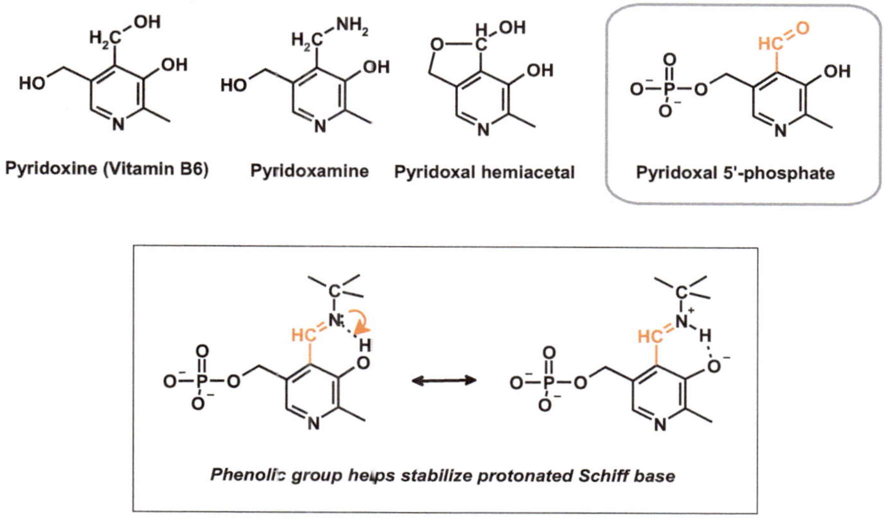

Fig. 33.5 Structures of pyridoxine, pyridoxamine, pyridoxal hemiacetal, and pyridoxal phosphate. The aldehyde group is masked in free pyridoxal as it forms hemiacetal. Phosphorylation of the –CH$_2$OH group ensures that the PLP aldehyde does not form a hemiacetal and is free to react with amino groups. The phenolic –OH on the pyridine ring is suitably positioned to stabilize the protonated PLP Schiff base (right, box below) during reaction. The positively charged N atom tends to withdraw electrons from C$_\alpha$ of the amino acid—thereby weakening its binding to –H, –R, and –COO$^-$

electron sink (for electrophilic catalysis). Carbanion intermediates that develop during enzyme catalysis are stabilized as covalent derivatives of PLP. The bound amino acid becomes a part of the extended conjugation with the pyridine ring and the lone pair of electrons on the ring nitrogen. Glycogen phosphorylase is an exception, where PLP does not function as an electrophilic apparatus. Instead, PLP performs a structural role with no apparent catalytic function. The phosphate group of PLP may also participate in a proton shuttle.

33.2.1 PLP Forms a Schiff Base

Pyridoxal in solution exists as an internal hemi-acetal (Fig. 33.5). In the PLP form, however, the aldehyde group is free, and the phenolic O$^-$ is hydrogen bonded to protonated imine. The PLP-aldehyde in most enzymes is bound to an active site lysine (through its ε-amino group) as a Schiff base. The PLP-bound holoenzyme, in many cases, can be converted into apoenzyme by dialysis in the presence of free cysteine (substituted cysteine-aldimine leaves the enzyme active site, taking away PLP with it). PLP and its various derivatives exhibit unique absorption and/or fluorescence spectra (Table 33.3). Much of the PLP chemistry is well understood through exquisite spectral properties of this cofactor.

Table 33.3 Spectral properties of pyridoxal phosphate and its derivatives

Coenzyme form of PLP	Absorption peak at	Fluorescence peaks (Excitation: Emission)
Free PLP	330 nm and 388 nm	Reduced PLP-Schiff base, λ_{ex}323 nm and λ_{em}390 nm
Lys (ε-amino)-PLP aldimine (Resting enzyme)	430 nm	–
Amino acid-PLP aldimine (the ES complex)	430 nm	–
Amino acid-PLP quinonoid intermediate	490 nm (detected by stopped flow kinetics)	
Amino acid-PLP ketimine intermediate	340 nm (detected by stopped flow kinetics)	
Enzyme bound PMP	330 nm	

The spectral data taken mostly from those reported for aspartate transaminase

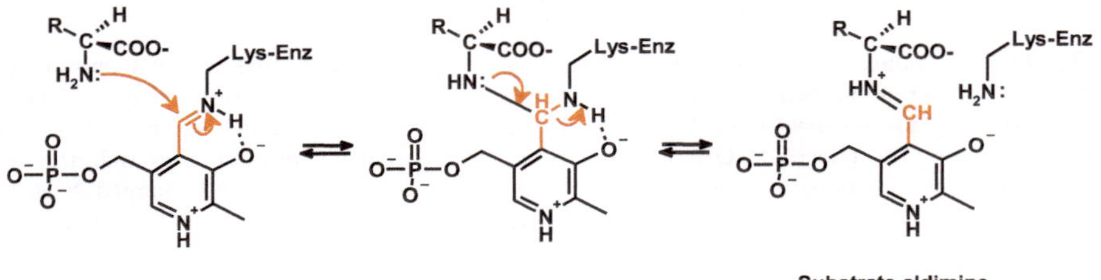

Fig. 33.6 Formation of substrate aldimine intermediate on the enzyme. Lysine (ε-NH$_2$) from the enzyme active site Schiff base is displaced by the substrate amine. The resulting PLP-substrate Schiff base (aldimine) is the starting point for various chemical routes

The enzyme active site lysine (in Schiff base with PLP) is displaced by the substrate amino group, when the enzyme binds its amino acid substrate. The substrate aldimine intermediate (Fig. 33.6) is the starting point for all the diverse reactions involving PLP chemistry. Its formation sets the stage for subsequent bond-breaking and bond-forming steps. If all the chemistry occurs on the PLP, what then is the role of the apoenzyme? The protein component stabilizes a particular substrate imine form—by selective protonation of the appropriate carbanion. Typically, PLP catalysis involves the following sequence of events:

1. Formation of substrate Schiff base (initial imine).
2. Chemical changes via relevant carbanions.
3. Formation of a product imine.
4. Hydrolysis of the product imine.

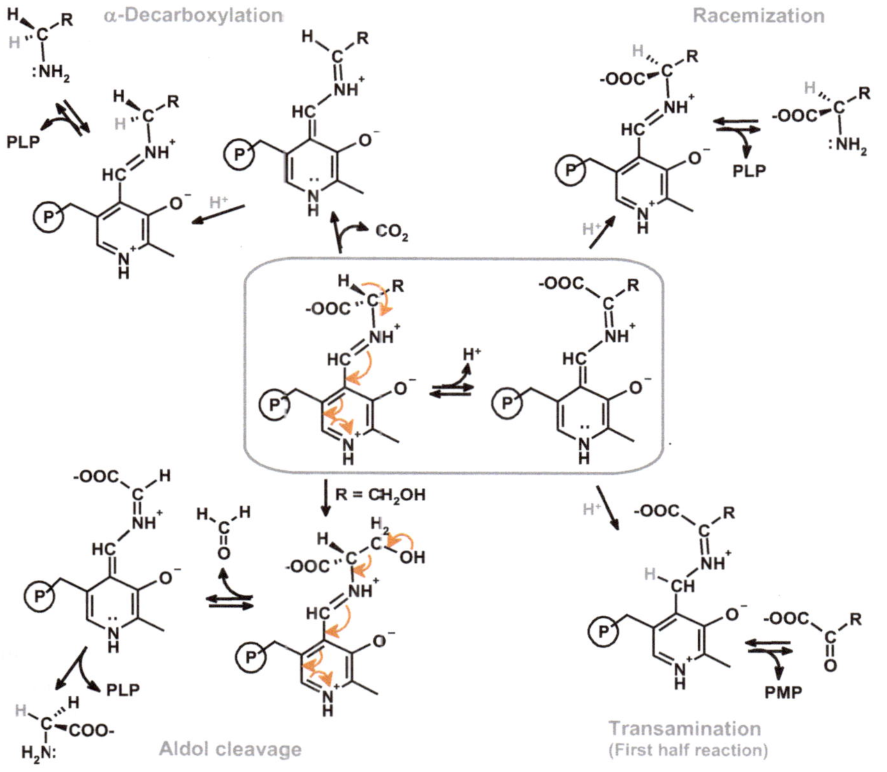

Fig. 33.7 Reactions occurring at the α-carbon of an amino acid. These reactions originate either from the aldimine or the quinonoid intermediate (boxed in gray). The quinonoid intermediate results when the C_α–H bond is mobilized. Enzyme active site holds the substrate in such a way that the bond around C_α to be broken is held out of plane of the pyridine ring

As noted above, the first step of amino acid chemistry with PLP is the formation of Schiff base. What happens subsequently depends on which bond around the α-carbon of the amino acid substrate is mobilized. One of the three bonds of the α-carbon is oriented such that it sticks out of plane (of pyridine ring of PLP) for cleavage. The substrate α-carbon is thus brought into the extended conjugate double bond system of the cofactor; this permits the resonance stabilization of developing carbanion. The chemical control obviously rests with the active site environment offered by the apoenzyme. Chemical mechanisms for the stabilization and protonation of various carbanion species are depicted in Fig. 33.7. Three most common reactions at the amino acid α-carbon are: (a) racemization, (b) transamination, and (c) decarboxylation.

33.2.2 Reactions at Amino Acid α-Carbon

Formation of the aldimine adduct makes the amino acid α-proton very acidic. This aldimine proton can be easily abstracted, and the resulting carbanion is stabilized through resonance—the pyridine ring acting as an electron sink—generating the quinonoid (ketimine) species. Donating the proton to the opposite face (of α-carbon which is sp^2 hybridized) results in net inversion—an example of ***racemization*** of α-amino acid. The racemized product is detached from the PLP by simple reversal of each step—with active site lysine ε-amino group reforming its imine linkage with PLP in the end. Proton abstraction/donation at the α-carbon may be carried out by a single active site base or two distinct groups. These two possibilities may be distinguished by whether deuterium label on the α-carbon is retained in the product or not. A very small group of *PLP-independent racemases* also produce D-amino acids. For instance, specific racemases for proline, glutamate, and aspartate are known. These enzymes operate by a two-base mechanism: the α-proton is removed from one face by an active site base, and a proton is donated by a protonated base on the other face. In these PLP-independent racemases, the developing carbanion on the α-carbon is stabilized due to the neighboring $-COO^-$, possibly through a strong "low barrier" hydrogen bond to an enzyme residue.

A second transformation at the α-center of an amino acid is ***transamination***— the conversion of amino acid to its corresponding keto acid. If the transamination involves an isolated amino group (like in GABA), then the product will be an aldehyde. The overall reaction for a transamination involves two half reactions (Fig. 33.8). The first half reaction produces the PMP (pyridoxamine) form of the

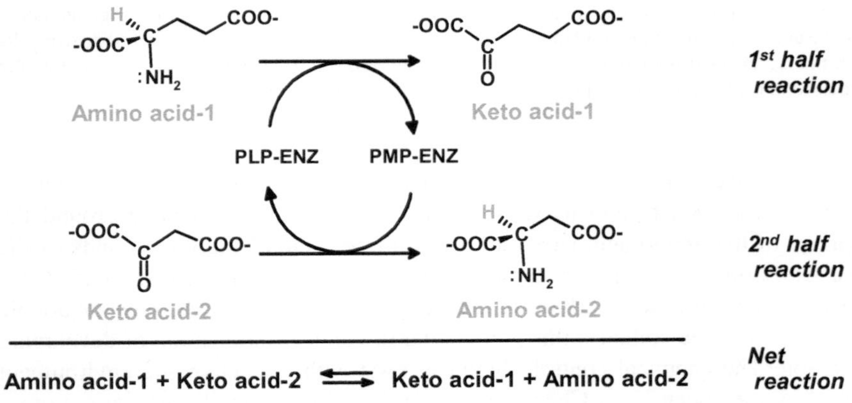

Fig. 33.8 Transamination of an amino acid comprises two half reactions. The first half reaction may be viewed as the oxidation of the amino acid-1 to keto acid-1, while in the second half, keto acid-2 is reduced to amino acid-2. The reactions shown here are for aspartate transaminase; amino acid-1 is aspartate, and the corresponding keto acid-1 is oxaloacetate. Glutamate (amino acid-2) and 2-oxoglutarate (keto acid-2) form the other substrate-product pair. In the case of alanine transaminase, however, L-alanine (amino acid-1) and pyruvate (keto acid-1) will be involved

enzyme. This is when the second keto acid substrate comes in and binds. A reversal of all the steps (by tracing the reverse path of first half reaction) regenerates the PLP enzyme back and releases the second amino acid product. The following experimental evidences support the depicted alanine transaminase reaction mechanism: (a) Kinetic data best fit a ping pong (Bi Bi) mechanism with the PMP-enzyme as the "F" form, (b) unique spectral signature of enzyme bound PMP is observed, (c) enzyme interacts with aspartate to stoichiometrically release oxaloacetate, in the absence of second substrate (2-oxoglutarate), and (d) The PMP-enzyme on incubation with keto acid stoichiometrically forms the corresponding amino acid. Because of the two partial reactions, relevant isotope exchanges (see Chap. 24) are also detected.

Aspartate transaminase (serum glutamate-oxoglutarate transaminase, SGOT) and serum glutamate-pyruvate transaminase (SGPT) are the two clinically relevant and well-studied amino acid transaminases.

The PLP adduct of an α-amino acid can also undergo **decarboxylation**. Here PLP acts as a $2e^-$ sink to facilitate the decarboxylation event. After the cleavage of the C_α-COO^- bond, protonation of the α-carbon and product release from PLP regenerates the primary amine. Besides pyruvoyl-dependent enzymes (see Table 33.2 and the previous section A. Protein electrophiles—*Pyruvoyl-dependent enzymes*), these PLP enzymes constitute a major class of amino acid decarboxylases. Examples of PLP-dependent decarboxylation products include GABA, histamine, dopamine, and 5-hydroxytryptamine (serotonin).

33.2.3 Reactions at Amino Acid β- and γ-Carbons

We have so far looked at PLP-dependent reactions that occur at the α-carbon—the C atom to which the substrate amino group is attached. A small, but important set of α-amino acid reactions occur at the β- and γ-positions as well (Fig. 33.9). Mechanistically, all these take off from the quinonoid intermediate.

Interestingly, in some of the reactions (particularly those at γ-carbon), PLP acts as a $4e^-$ sink (first $2e^-$ on the ring N and the next two on the adjacent iminium N) rather than a $2e^-$ sink ($2e^-$ on the ring N). The reaction mechanism at γ-carbon begins with aldimine adduct and leads to the following sequence of events:

Aldimine adduct → abstract C_α-H → Quinonoid intermediate (ring N as $2e^-$ sink) → abstract C_β-H → stabilize β-carbanion (adjacent iminium as $2e^-$ sink) → β,γ-unsaturated imine intermediate → reaction of γ-elimination/ γ-replacement

Examples of enzymes that exploit PLP as $4e^-$ sink include γ-cystathionase, L-methionine γ-lyase, cystathionine synthase, threonine β-epimerase, and threonine synthase.

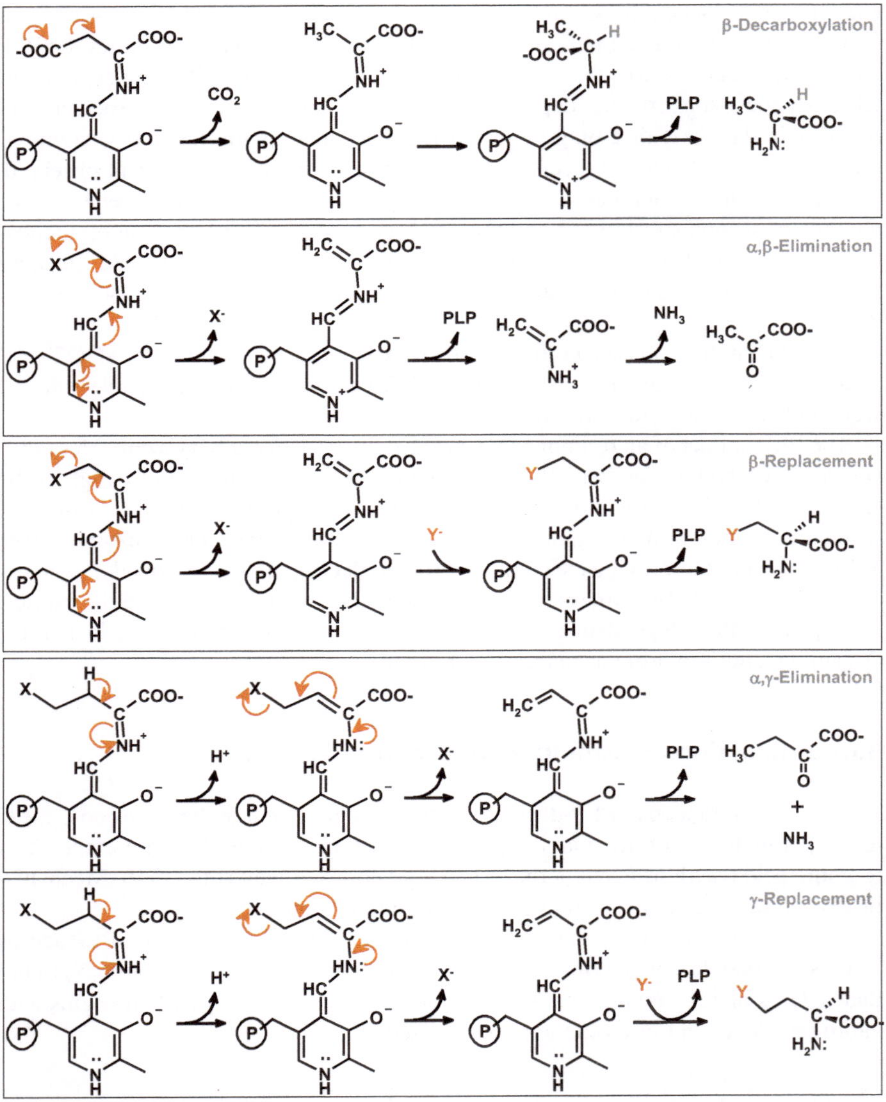

Fig. 33.9 Reactions occurring at β- and γ-carbon of an amino acid. All these reactions originate from the quinonoid intermediate (see Fig. 33.7 for details). Ability of the enzyme active site to quench a suitable carbanion is the key to these reactions. A second proton from the substrate is abstracted by the enzyme (with PLP acting as a $4e^-$ sink) in case of γ-elimination/replacement reactions

Table 33.4 is a compilation of various chemical events associated with reactions that are known to occur at C_α, C_β, and C_γ atoms of the amino acid substrate during PLP catalysis.

Table 33.4 Summary of various chemical events in PLP catalysis

Amino acid reaction at	Events 	Example
The α-Carbon		
Racemization	C_α-H removed/added	Alanine racemase
Transamination	Protonation of aldimine at PLP aldehydic C	Glutamate-oxaloacetate transaminase
Decarboxylation	Cleavage of C_α-COO$^-$ bond but C_α-H not mobilized	Glutamate decarboxylase
α,β-Aldolytic cleavage	C_α-H not mobilized	Serine hydroxymethyltransferase
The β-Carbon		
Decarboxylation	C_α-H mobilized	Aspartate β-decarboxylase
Elimination	C_α-H mobilized and 2e$^-$ oxidation[a] at C_α	Tryptophanase
Replacement	C_α-H mobilized and replacement of group on C_β	Tryptophan synthase
The γ-Carbon		
Elimination	C_α-H mobilized followed by the C_β-H to stabilize C_β-carbanion; elimination from C_γ- to form β,γ-unsaturated imine and 2e$^-$ oxidation[a] at C_α	γ-Cystathionase
Replacement	C_α-H mobilized followed by the C_β-H to stabilize C_β-carbanion; replacement of group on C_γ	Cystathionine synthase

[a]2e$^-$ oxidation at C_α invariably results in α-keto acid as one of the products

Selective cleavage of a specific bond of the PLP external aldimine results in a carbanionic intermediate. From here, the different reaction pathways diverge, leading to multiple activities: transamination, decarboxylation, racemization, elimination, and synthesis (Table 33.4). However, carbanionic intermediates are also known to react with electrophiles and oxidizing agents—one of them being O₂. Some PLP-dependent decarboxylases are able to consume molecular oxygen, transforming an amino acid into a carbonyl compound. One such enzyme able to perform the side paracatalytic reaction is the PLP-dependent *E. coli* glutamate decarboxylase; the oxygenase side reaction (at 0.1% rate of the main decarboxylation) produces succinic semialdehyde instead of γ-aminobutyric acid. Now, there is evidence to show that a small number of PLP-dependent enzymes employ molecular oxygen as a co-substrate (Hoffarth et al. 2020; Bisello et al. 2020). They are not para-catalytic and belong to the bacterial and fungal kingdoms—found in organisms synthesizing bioactive compounds.

33.3   Pyridoxamine Phosphate (PMP) Enzymes: Amino- and Deoxy-Sugar Biosynthesis

The versatility of PLP/PMP-dependent enzymes to catalyze a diverse array of chemical reactions is attributed to fine-tuning of the cofactor-substrate interactions in the active site. The vitamin B6-dependent enzymes involved in deoxy sugar biosynthesis may carry out transamination, deoxygenation, and even aldolase reactions to generate a variety of unique structures. The deoxy amino sugars—so produced—are components of many valuable natural products. Like in all other PLP-dependent transamination, sugar aminotransferases transfer an amino group from an amino acid donor to a ketone bearing acceptor. The NDP-activated keto sugar is the amino acceptor, whereas the amino donor is either L-glutamate or L-glutamine—or rarely L-aspartate. The position of the amino transfer is dictated by the enzyme specificity and the location of the substrate's keto group (Romo and Liu 2011). Of course, the keto group is generated prior to transamination, by standard dehydrogenation or isomerization reactions. The commonly encountered deoxy amino sugars are aminated at the C-3 and C-4 positions (Fig. 33.10).

While homologous to proteins in the PLP-containing aspartate aminotransferase superfamily, the sugar dehydrases contain a bound PMP moiety. Interestingly, some contain a [2Fe-2S] cluster in addition to PMP. A functional iron–sulfur cluster

Fig. 33.10 Reactions leading to amination of deoxy sugar occur on the NDP derivative of that sugar. The NDP moiety could be TDP, CDP, or GDP depending on the specificity of the enzyme

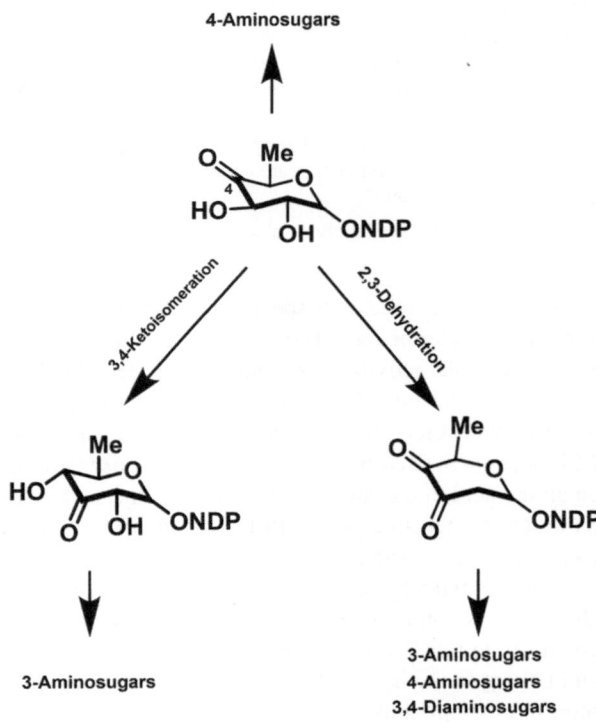

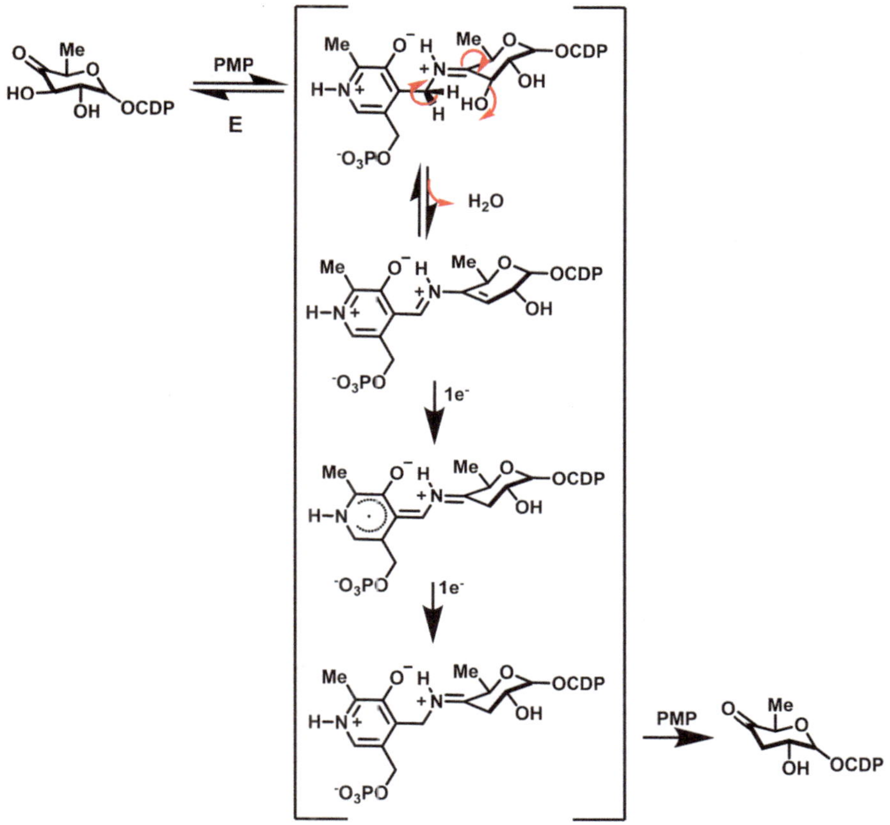

Fig. 33.11 The PMP-dependent dehydratase and an iron–sulfur containing flavodoxin-NADH reductase combine to produce the deoxy sugar. The example is that of CDP-6-deoxy-L threo-D-glycero-4-hexulose 3-dehydratase and its reductase, in the biosynthesis of CDP-L-ascarylose

(an obligatory one electron transfer cofactor) implies that a radical intermediate (the glucoseen) must form in the reduction process. Also, they are distinct from typical aminotransferases in that the enzyme lacks the highly conserved Schiff-base-forming lysine but has a histidine residue instead. Therefore, the cofactor is non-covalently held at the enzyme active site. Also, the Lys to His substitution may have transformed a normal PLP-dependent aminotransferase into a unique PMP-dependent catalyst that has now become a dehydrase (Smith et al. 2008; Romo and Liu 2011). Thus, PMP appears to have a dual function of being responsible for the anion-induced dehydration reaction and also for being an integral part of the subsequent redox (two one-electron transfer) process (Fig. 33.11).

In another route to deoxy sugars, the enzyme bound PLP is first transaminated to the PMP form with L-glutamate as the amino donor. This PMP forms a Schiff base with the keto sugar and the resulting adduct undergoes a 1,4-dehydration to

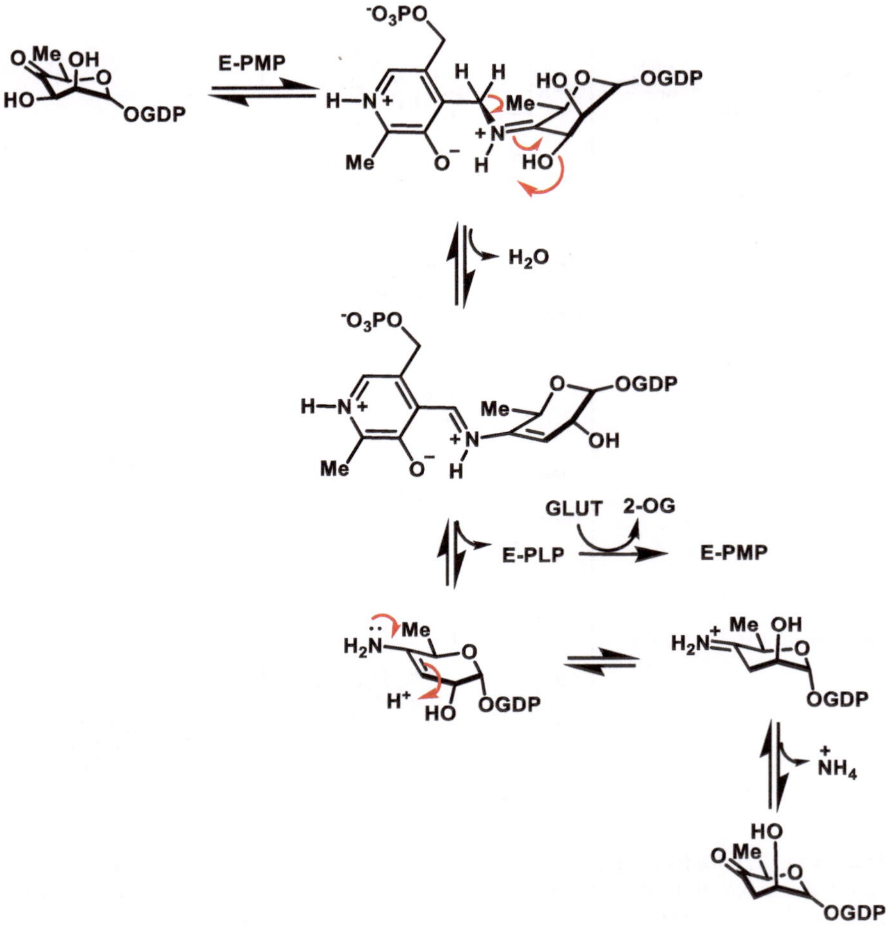

Fig. 33.12 The GDP-4-dehydro-6-deoxy-α-D-mannose 3-dehydratase (EC 4.2.1.168) is involved in β-L-colitose biosynthesis. The last two steps are non-enzymatic and spontaneous (E-PMP: PMP form of the enzyme; GLUT: glutamate; 2OG: 2-oxoglutarate)

eliminate the 3-OH group. The PLP is restored upon product hydrolysis and an unstable enamine intermediate is released (Cook et al. 2006; Romo and Liu 2011). This enamine intermediate tautomerizes to an imine, which spontaneously hydrolyses to release ammonia and the final product. An example of an enzyme involved in β-L-colitose biosynthesis is shown in Fig. 33.12.

Besides the unique mechanistic enzymology of B6 described here, many amino and deoxy sugars are components of bacterial cell membranes, flagella, toxins, and diverse secondary metabolites. As potential drug targets, these enzymes are important.

33.4 Summing Up

Stabilizing a carbanion transition state is often crucial for enzyme catalysis. The developing negative charge on a C atom could be handled through suitably placed temporary electron sinks. The sink may be found on the substrate itself (a β-keto group directly or its Schiff base through Lys-NH$_2$), on the enzyme (pyruvoyl group and 4-methylideneimidazole-5-one) or on the cofactor (a divalent metal ion, thiamine pyrophosphate (see Chap. 32), or pyridoxal phosphate). PLP is a versatile electrophile chosen by nature. While PLP-chemistry is more or less synonymous with reactions of amino acid metabolism, there are a small number of amino acid chemistries where protein electrophiles (like pyruvoyl group and 4-methylideneimidazole-5-one) have been recruited.

PLP-dependent reactions play a crucial yet irreplaceable role; they function to recycle nitrogen content of the cellular amino acid pool through catabolism, biosynthesis, and interconversions. All these enzymes/reactions make use of PLP cofactor as a temporary electron sink. Typically, the amino group of the substrate amino acid forms a Schiff base to initiate the proceedings. The bond to be broken (around the C$_\alpha$) sticks out of plane, as defined by the pyridine ring. For instance, racemases and transaminases bind their substrate such that the amino acid C$_\alpha$-H sticks out and it is mobilized. It is the C$_\alpha$-COO$^-$ group that is held out of plane by a decarboxylase. Reactions at the C$_\beta$ and C$_\gamma$ of amino acid may be less common but are equally important.

The non-enzymatic analogs of transamination and reductive amination reactions that convert keto acids into amino acids are of considerable interest in prebiotic evolution. They do indeed occur readily without catalysts but under high temperatures (Mayer and Moran 2023). However, in the absence of enzymes they are slow and require primordial catalysts such as protons or metal ions.

References

Atkins JF, Gesteland R (2002) The 22nd amino acid. Science 296:1409–1410

Bisello G et al (2020) Oxygen reactivity with pyridoxal 5′-phosphate enzymes: biochemical implications and functional relevance. Amino Acids 52:1089–1105

Cook PD, Thoden JB, Holden HM (2005) The structure of GDP-4-keto-6-deoxy-D-mannose-3 dehydratase: a unique coenzyme B6-dependent enzyme. Protein Sci 15:2093–2106

Cooke HA, Christianson CV, Bruner SD (2009) Structure and chemistry of 4-methylideneimidazole-5-one containing enzymes. Curr Opin Chem Biol 13:460–468

Hoffarth RE, Rothchild KW, Ryan KS (2020) Emergence of oxygen and pyridoxal phosphate-dependent reactions. FEBS J 287:1403–1428

Mayer RJ, Moran J (2023) Mechanism and catalysis of nonenzymatic analogs of amino acid biosynthesis. Adv Phys Org Chem 57:1–39

Percudani R, Peracchi A (2003) A genomic overview of pyridoxal-phosphate dependent enzymes. EMBO Rep 4:850–854

Phillips RS (2015) Chemistry and diversity of pyridoxal-5′-phosphate dependent enzymes. Biochim Biophys Acta 1854:1167–1174

van Poelje PD, Snell EE (1990) Pyruvoyl-dependent enzymes. Annu Rev Biochem 59:29–59
Romo AJ, Liu H (2011) Mechanisms and structures of vitamin B6-dependent enzymes involved in
    deoxy sugar biosynthesis. Biochim Biophys Acta 1814:1534–1547
Smith P et al (2008) Structure and mutagenic conversion of E1 dehydrase: at the crossroads of
    dehydration, amino transfer, and epimerization. Biochemistry 47:6329–6341

Further Reading

Hayashi H (1995) Pyridoxal enzymes: mechanistic diversity and uniformity. J Biochem 118:
    463–473
John RA (1995) Pyridoxal phosphate-dependent enzymes. Biochim Biophys Acta 1248:81–96

Free Radicals and Radical Enzymology

34

For a long time, it was held that free radicals are bad actors and described as reactive oxygen or nitrogen species (ROS, RNS) such as $O_2^{\cdot-}$, $HO^{\cdot}$, and $NO^{\cdot}$. Redox biology is the fundamental aspect of aerobic life. Being the terminal electron acceptor, aerobes cannot avoid oxygen and its reduction. Uncontrolled reduction of O_2 and its reduced species can be deleterious to critical to life biomolecules like DNA. It is now well accepted that many essential biochemical reactions also involve organic radicals. Reactions catalyzed by metalloenzymes like cytochrome P450, ribonucleotide reductase, and the adenosylcobalamin (AdoCbl or B12) enzymes employ organic radical intermediates. Generally, radical chemistry with enzymes is resorted by nature, for reactions that would be difficult or impossible to catalyze by polar mechanisms. Most often such reactions involve an H-atom abstraction (not a proton) from a recalcitrant C–H bond. The radical S-adenosylmethionine (radical SAM; rSAM) enzymes are the more recent superfamily of such enzymes with over half a million members—widely spread across the tree of life. Information explosion through analysis of sequenced genomes suggests that $>500,000$ enzymes/proteins exploit radicals for their chemistry.

As we have seen in an earlier overview chapter (Chap. 27, Chemical Reactivity and Molecular Interactions), the emphasis has been on reactions that involve heterolytic cleavage of covalent bonds leading to generation of nucleophiles, electrophiles, two electron, and proton transfers. However, this paradigm had to be modified significantly—with the advent of organic free radical chemistry involving enzyme reactions. Homolytic cleavage of covalent bonds produces free radicals—groups bearing unpaired electrons. Ribonucleotide reductases were the first enzymes shown to harbor "free radicals" (Reichard and Ehrenberg 1983). These ubiquitous enzymes are found throughout the nature and catalyze one of the most central reactions in all of biology—providing precursors (deoxyribonucleotides) for DNA synthesis. Yet another key biological process of photosynthesis exploits free radicals in the oxygen-evolving complex of photosystem II.

© The Author(s), under exclusive license to Springer Nature Singapore Pte Ltd. 2025
N. S. Punekar, *ENZYMES: Catalysis, Kinetics and Mechanisms*,
https://doi.org/10.1007/978-981-97-8179-9_34

Despite an early beginning, the progress in the field of radical enzymology was slow. This was because we were dealing with free radicals—which are ephemeral and highly reactive species. Their fleeting lifetimes make it difficult to observe them on the usual time scales. Advent of powerful and fast detection techniques like ENDOR (electron–nuclear double resonance), EPR (electron paramagnetic resonance), ESEEM (electron spin–echo envelope modulation), EXAFS (X-ray absorption fine structure), NRVS (nuclear resonant vibrational spectroscopy), XAS (X-ray absorption spectroscopy), XANES (X-ray absorption near edge spectroscopy), RFQ (rapid freeze-quench), Cryo-EM (cryo-electron microscopy), and availability of expressed proteins, site-directed incorporation of unnatural amino acids—using orthogonal tRNA synthetase technology—have allowed a detailed dissection of the free radical chemistry on the enzyme (Stubbe and Nocera 2021). Whenever proton transfers are coupled to electron transfers (PCET, Proton-Coupled Electron Transfer viz., in type I ribonucleotide reductases, see below), free radical transport across the protein matrix becomes crucial. The proton resting mass is $\sim$2000 times larger than that of the electron, hence a proton transfer is fundamentally limited to short distances; small changes in proton distance induced by conformational changes can significantly impact on radical transport. Accordingly, protein conformational changes can influence the radical enzymatic rates.

From a slow start with ribonucleotide reductase—particularly the B12 kind—the field of radical enzymology has picked up rapidly in recent years. The free radical based reactions are often tempered and assisted by cofactors such as flavins (FMN, FAD), thiamin pyrophosphate (TPP), and pyridoxal phosphate (PLP/PMP), with exemplars in both primary and secondary metabolism. The two major sources of free radicals in enzyme reactions are adenosylcobalamin (AdoCbl or coenzyme B12) (Banerjee and Ragsdale 2003) and rSAM (radical SAM; SAM with $[4Fe-4S]^{1+}$ cluster) (Broderick et al. 2014) (Fig. 34.1). Both initiate free radical chemistry by providing the $5'$-dAdo$^{\bullet}$ radical first; subsequent steps are, however, unique to the particular enzyme in question.

These two reservoirs of free radicals eventually produce stable and transient radicals of tyrosine (Y$^{\bullet}$), glycine (G$^{\bullet}$), cysteine (C$^{\bullet}$), and tryptophan (W$^{\bullet}$) residues within the enzyme proteins (Table 34.1). These radical species (with an unpaired electron) have been characterized using EPR spectroscopy and with specifically isotope-labeled proteins. The transient protein radical so generated interacts directly with the substrate to perform chemistry. Some enzyme free radicals, however, are well-stabilized and come with have half-lives of several hours. Their stability is due to the protected location within the protein, buried away from the solvent. A select summary of known radical enzymes is given in the table below.

Besides the "critical to life" ribonucleotide reductases, "radical enzymology" encompasses many facets of metabolism. These include methylations, dehydrogenations, oxidations, carbon-carbon bond rearrangements, sulfur insertion reactions, and many critical cofactors (organic as well as complex metal clusters). We will crisply summarize the two major radical enzyme chemistries in the following paragraphs. Mechanistic details of many radical reactions continue to be a work in progress.

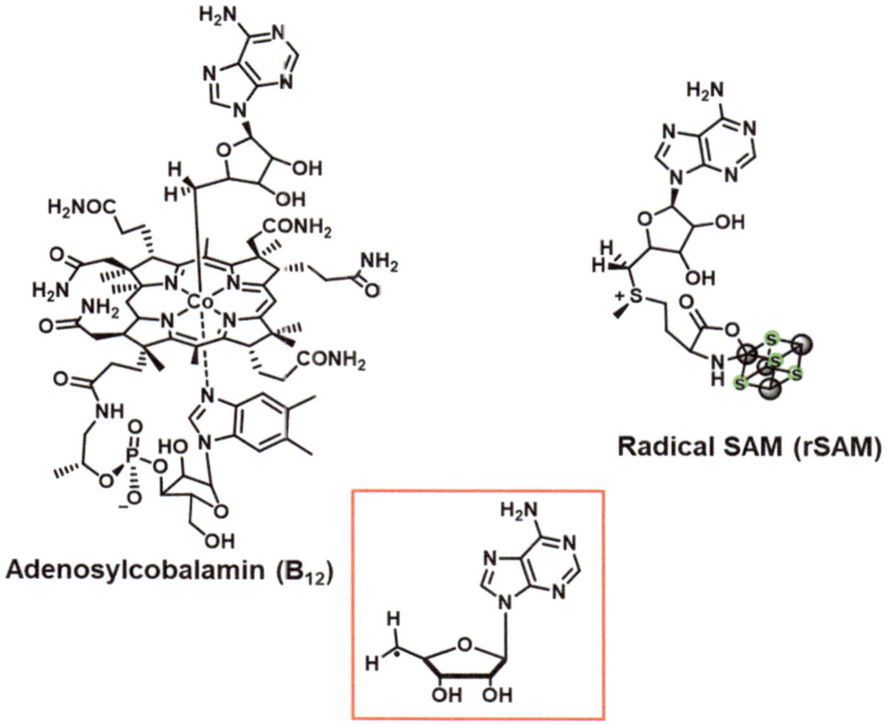

Fig. 34.1 The structures of adenosylcobalamin (AdoCbl or coenzyme B12) and rSAM (SAM with [4Fe-4S]$^{1+}$ cluster). The free radical 5′-dAdo$^•$ generated by these two free radical reservoirs is shown in the box

34.1 The B12 Enzymes

The AdoCbl-dependent enzymes are classified based on the specifics of the migrating group and the receiving carbon (Brown 2005). The Class I enzymes are mutases, catalyzing carbon skeleton rearrangements in which the migrating group is a carbon fragment. The C–C bond is cleaved, and the receiving carbon has two hydrogens. These enzymes include glutamate mutase (EC 5.4.99.1) and methylmalonyl CoA mutase (EC 5.4.99.2). The Class II enzymes are eliminases, which catalyze the migration and subsequent elimination of a hydroxyl or amino group so that a C–O or C–N bond is cleaved. Ribonucleotide reductase (EC 1.17.4.2) belongs here, and it catalyzes the reduction of NTPs to dNTPs. The Class III enzymes are the aminomutases, which catalyze the migration of an amino group to an adjacent carbon so that a C–N bond is cleaved. These enzymes additionally require pyridoxal phosphate (PLP) for reaction (viz., β-lysine-5,6-aminomutase; EC 5.4.3.3) (Banerjee 2003; Wu et al. 2011).

Table 34.1 Enzymes and associated cofactors in generating free radicals

Enzyme	Cofactor(s)	Radical(s)
Ribonucleotide reductase Type I; aerobic, *E. coli*	Diferric cluster[a]	tyrosine (Y•), cysteine (C•)
Ribonucleotide reductase Type II; *Lactobacillus leichmannii*	B12 (AdoCbl)	5'-dAdo•, cysteine (C•)
Glutamate mutase	B12 (AdoCbl)	5'-dAdo•
Methylmalonyl CoA mutase	B12 (AdoCbl)	5'-dAdo•
Glycyl radical enzyme activating enzymes	SAM, [4Fe-4S] cluster	5'-dAdo•
Ribonucleotide reductase Type III; anaerobic, *E. coli*	SAM, [4Fe-4S] cluster (of glycyl activating enzyme)	glycine (G•), cysteine (C•)
Pyruvate formate lyase	SAM, [4Fe-4S] cluster (of glycyl activating enzyme)	glycine (G•), cysteine (C•)
Lysine 2,3-aminomutase	SAM, [4Fe-4S] cluster, PLP	5'-dAdo•, PLP•
ThiC in TPP synthesis	SAM, [4Fe-4S] cluster	5'-dAdo•
Biotin synthase	SAM, [4Fe-4S] cluster	5'-dAdo•
Lipoyl synthase	SAM, [4Fe-4S] cluster	5'-dAdo•
Cytochrome c peroxidase	Heme/Fe	oxyferryl/tryptophan cation radical
Prostaglandin H synthase	Heme/Fe	tyrosine (Y•)

[a]Ribonucleotide reductases with other metal cluster compositions are also known from other organisms

B12 structure contains an adenosyl moiety and has a relatively weak Co–C bond linking the adenosyl moiety to the rest of the cofactor. In all B12-dependent enzymes, the 5'-deoxyadenosyl radical (5'-dAdo•) is first formed by homolysis of the cobalt-carbon bond of the coenzyme. This 5'-dAdo• radical reacts very rapidly to abstract hydrogen, either from the substrate (viz., glutamate mutase) or an enzyme cysteine residue (viz., ribonucleotide reductase). Like all B12-dependent enzymes, the initial homolysis of the coenzyme generates the 5'-dAdo• in glutamate mutase (Huhta et al. 2002; Wu et al. 2011). This abstracts the migrating hydrogen atom from the substrate to form a glutamate radical (from C-4). Rearrangement of this substrate radical occurs via its fragmentation to give acrylate and glycyl radical as intermediates (Fig. 34.2). The two recombine to give the methylaspartyl radical (and subsequently methylaspartate).

Regardless of the radical initiating cofactor, the formation of a thiyl radical (cysteine; C•) is common to all ribonucleotide reductases and that C• initiates nucleotide reduction chemistry (Stubbe and van der Donk 1998; Stubbe and Nocera 2021). The chemical mechanism for B12-dependent ribonucleotide reductase was worked out extensively through biochemical studies including the use of mechanism-based inhibitors, site-directed mutagenesis, and X-ray crystallography. Catalysis is initiated by H-atom abstraction from the 3' position of the ribose moiety by the thiyl radical (Fig. 34.3). Subsequent elimination of the 2'-OH group as water occurs simultaneously with deprotonation of the 3'-OH. Reduction of the

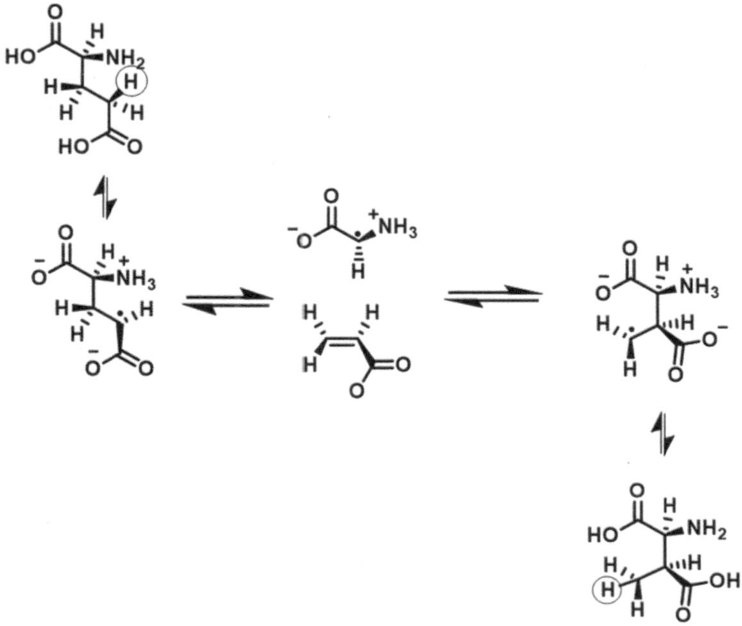

Fig. 34.2 The mechanism of glutamate mutase. The 5′-dAdo˙ radical formed from adenosylcobalamin (AdoCbl or coenzyme B12) abstracts H-atom (circled) from glutamate to initiate the radical chemistry

intermediate α-keto radical occurs via oxidation of two cysteines. This generates a 3′-deoxynucleotide radical, which re-abstracts the originally removed H-atom to regenerate the thiyl radical. The protein disulfide is reduced back to the two thiols by thioredoxin reductase (using NADPH as the reductant) to initiate the next round of catalysis.

34.2   Radical S-Adenosyl-L-Methionine Enzymes

These enzymes are found across the three kingdoms and catalyze an amazingly diverse set of reactions. The four enzymes viz., pyruvate formate lyase (PFL) and its associated activating enzyme (PFL-AE), anaerobic ribonucleotide reductase, lysine 2,3-aminomutase, and biotin synthase, provided early pointers to a new type of radical reactions. These enzymes initiate radical reactions using a fundamental new mechanism of catalysis—different from the B12 enzymes. The novel biological cofactor (rSAM) consists of a $[4Fe-4S]^{1+}$ cluster and S-adenosylmethionine (SAM). The unique iron of this $[4Fe-4S]^{1+}$ cluster is coordinated by the amino and carboxylate moieties of SAM, forming a classical five-member chelate ring (Fig. 34.1). This distinctive structural feature, along with its unique protein fold, is common to every radical SAM enzyme examined using X-ray crystallographic or

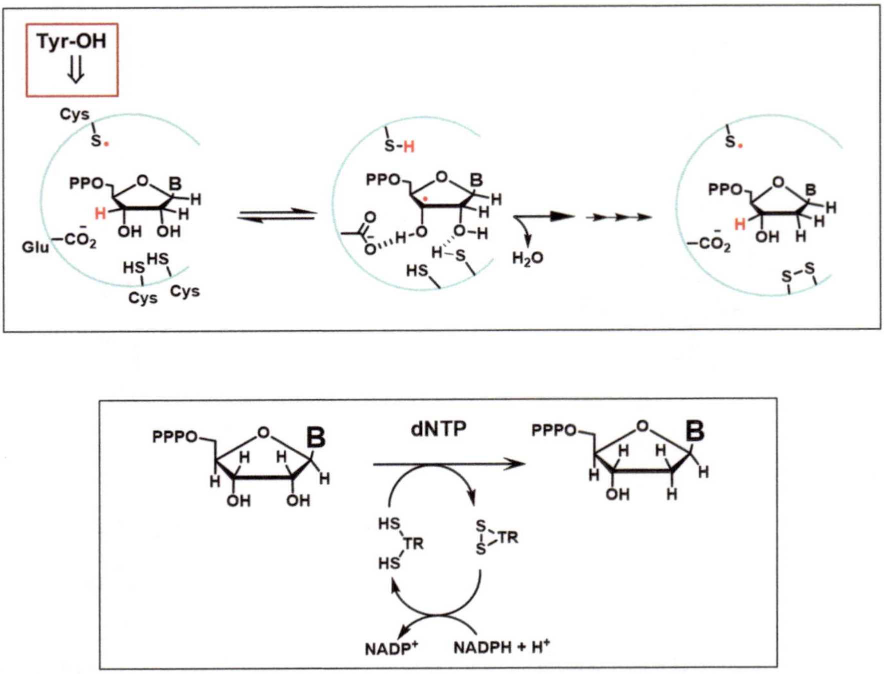

Fig. 34.3 The mechanism of type I ribonucleotide reductase—a schematic. The Tyr (Y•) radical generates Cys (C•) radical and initiates the nucleotide reduction chemistry. The C• abstracts H-atom (in red) from the 3′C of ribose (of NDP) to form the substrate radical. This rearranges through series of steps to deoxyribose of the substrate, leaving the enzyme oxidized (as its disulfide). The lower panel shows the overall ribonucleotide reductase step of a type II enzyme which acts in dNTP synthesis. The reduction of enzyme disulfide formed is reduced back to thiols by thioredoxin reductase

ENDOR spectroscopic methods. These enzymes initiate a radical reaction via the generation of a 5′-deoxyadenosyl radical (5′-dAdo•) intermediate (Broderick et al. 2014).

Although the rSAM superfamily (radicalsam.org) was recognized through bioinformatics in 2001, pyruvate formate lyase (PFL) gave the earliest clues to this class. The activation of PFL involves the generation of a stable protein radical. This activation requires the presence of iron, SAM, and an activating enzyme. The glycyl radical (G•) on PFL is located on a specific glycine residue and is one of the early stable protein radicals characterized. The activating enzyme (AE) in fact is the first radical SAM enzyme shown to contain a catalytically essential iron—sulfur cluster, and to use SAM for activation. Besides PFL, enzymes that are known to function through a catalytic glycyl radical are the anaerobic ribonucleotide reductase, B12-independent glycerol dehydratase among others. All these glycyl radical enzymes contain a stable, catalytically essential glycyl radical in their active state. The glycyl radicals are generated by glycyl radical enzyme activating enzymes

Fig. 34.4 The structure of Ω intermediate in the rSAM reaction mechanism. This intermediate contains an Fe–C5′–adenosyl bond

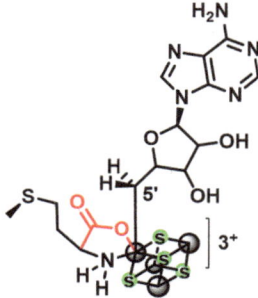

(AEs) —which are radical SAM enzymes. These AEs function either as distinct enzymatic entities (e.g., in the case of PFL) or as intimate subunits of the glycyl radical enzymes that they activate (e.g., with anaerobic ribonucleotide reductase).

To date, lysine 2,3-aminomutase is one of the best understood radical SAM enzymes, with extensive spectroscopic, biochemical, and structural information. The enzyme converts L-α-lysine to L-β-lysine. Its action is dependent on the Fe-S cluster, PLP, SAM, and anaerobic conditions. The Fe-S clusters of most radical SAM enzymes are air (O_2) sensitive and require anaerobic conditions for isolation and handling. This poses difficulties in preparing and assaying fully active radical SAM enzymes (with their true turnover numbers). The rSAM cofactor closely mimics the role for B12 in the adenosylcobalamin-dependent radical reactions. Accordingly, the radical SAM enzymes may produce the same intermediate, namely, the 5′-deoxyadenosyl radical (5′-dAdo•), which the AdoCbl enzymes generate. The parallels in reactivity for these two disparate cofactors led to calling SAM as a "poor man's adenosylcobalamin." Coenzyme B12 and SAM both contain adenosyl moiety, but SAM does not have that relatively weak Co–C bond. The involvement of $[4Fe-4S]^{1+}$ cluster coordinated to SAM within the active site clearly underscores the possibility of an Fe–C bond (Horitani et al. 2016). An organometallic intermediate Ω, exhibiting an Fe–C5′–adenosyl bond, was identified in all rSAM enzymes studied so far (Fig. 34.4). The Ω intermediate liberates 5′-dAdo• through homolysis of the Fe–C5′ bond, in analogy to Co–C5′ bond homolysis in B12. Due to the simple elegance of SAM as a radical precursor, and the exquisite control of 5′-dAdo• reactivity in these enzymes, the rSAM enzymes are pervasive in the tree of life. Despite the importance of the AdoCbl cofactor, however, there are fewer than 20 enzymes known to use AdoCbl to initiate radical reactions.

The sulfonium sulfur of SAM is at the center of its refined chemistry. It can undergo three modes of regioselective cleavages. SAM is a common methyl donor for numerous biochemical reactions. The bond between the sulfonium sulfur and the methyl carbon is cleaved heterolytically (via a nucleophilic mechanism) to generate CH_3^+, for the methyl group donation. The other two S–C bonds of the SAM sulfonium group can also undergo heterolytic cleavage in biochemical reactions, although that is not very common. The homolytic cleavage of the S–C(methyl) bond is not observed (as the $CH_3^•$ radical is very reactive and unstable) while only a couple

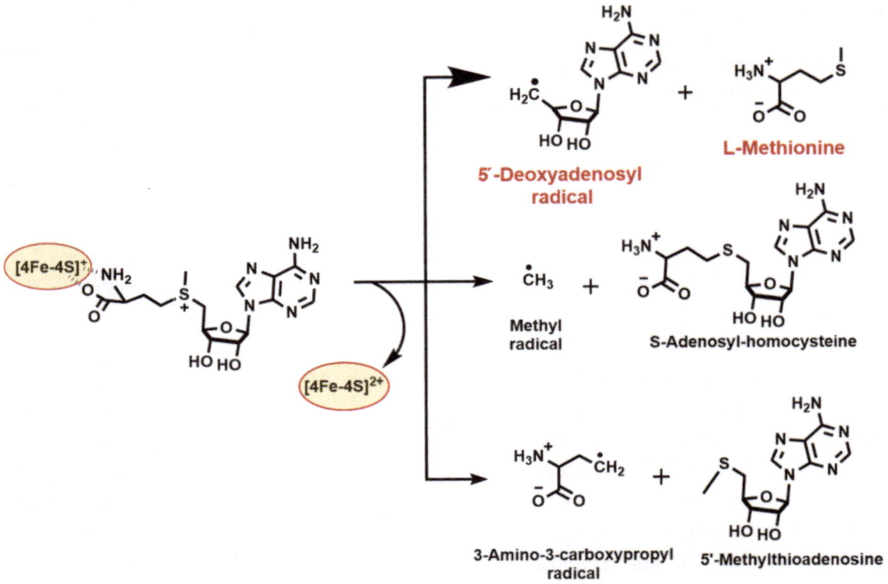

Fig. 34.5 Three different possible modes of homolytic cleavage of the S–C bonds in rSAM. The cleavage of SAM to generate methionine and the 5′-dAdo˙ is a reductive cleavage event with [4Fe-4S]⁺ going to [4Fe-4S]²⁺. The homolytic cleavage of the S–C(5′) bond (top reaction) leading to the formation of 5′-dAdo˙ radical is the most represented example

of enzymes are known to generate a radical by cleaving the third S–C bond (S–C(γ)). In the most rSAM reactions, it is the homolysis of the S–C(5′) bond that generates methionine and the 5′-dAdo˙ radical (Fig. 34.5; top arrow). Both the coordination chemistry and the protein environment dictate the cleavage of the S–C(5′) bond and the subsequent selective reactivity (of the 5′-dAdo˙ formed) toward product formation (Broderick et al. 2023). The chemical mechanism of rSAM goes through three distinct and sequential intermediates prior to substrate transformation: (a) The reductive cleavage of SAM generates the organometallic intermediate Ω (with a Fe–C5′ bond), (b) cleavage of the Fe–C5′ bond of Ω liberates "free" 5′-dAdo˙, and finally, (c) the 5′-dAdo˙ radical generates a substrate radical upon H-atom abstraction from the substrate. The substrate radical then undergoes specific transformations characteristic of that enzyme.

The cleavage of SAM to generate methionine and the 5′-dAdo˙ is a reductive cleavage event, requiring the input of one electron. This reductive cleavage of SAM occurs in most radical SAM enzymes in vitro even in the absence of the substrate—forming methionine and 5′-dAdoH. The 5′-dAdoH possibly results from quenching of the 5′-dAdo˙ radical by solvent or protein. The rSAM enzymes may use SAM as a substrate (then 5′-dAdoH is one of the products) or catalytically (the substrate/product radical regenerates SAM back) (Broderick et al. 2014). The 5′-dAdo˙ radical abstracts an H-atom from the active site glycine residue of PFL polypeptide to generate the glycyl radical (G˙) (Fig. 34.6).

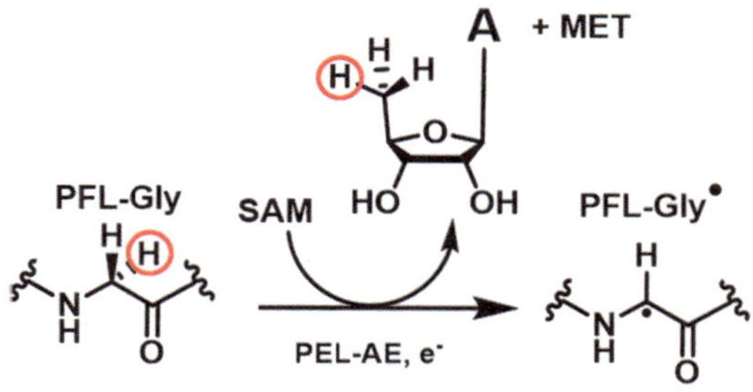

Fig. 34.6 The generation of a glycyl radical (G$^{\bullet}$) at the active site of PFL by the activating enzyme PFL-AE (which employs rSAM chemistry). PFL with glycyl radical (PFL-G$^{\bullet}$) is the active form of this enzyme

While many rSAM superfamily members exploit the catalytic power of 5′-dAdo$^{\bullet}$ radical, there are others that bind auxiliary cofactors and extend the catalytic repertoire of SAM. Interestingly, some rSAM enzymes use AdoCbl to facilitate challenging methylation and radical rearrangement reactions (Bridwell-Rabb et al. 2022). Two important enzymes catalyzing sulfur insertion reactions are: (1) lipoyl synthase which catalyzes the insertion of two sulfur atoms into C–H bonds of an octanoyl moiety to generate the lipoyl cofactor (Cronan 2024) and (2) biotin synthase catalyzing thiazole ring formation in the final step of biotin biosynthesis. Both these enzymes recruit auxiliary Fe–S clusters besides the rSAM cofactor.

34.3 Summing Up

The AdoCbl (B12) and radical SAM are the two important biological free radical reservoirs. Both contribute to the only route in all of biology to make deoxynucleotides and hence, DNA. Despite the crucial role played by the AdoCbl cofactor, there are very few enzymes known to use this mechanism to initiate radical reactions. On the other hand, radical SAM enzymes are pervasive in all three kingdoms of life. They play critical roles in numerous biosynthetic pathways including antibiotic production, posttranslational modifications, synthesis of protein cofactors, and in the synthesis of the nonprotein ligands of some of the most complex biological metal clusters known. It seems that radical SAM enzymes (with a cofactor consisting of SAM coordinated to $[4Fe-4S]^{1+}$ cluster) provide avenues for a wide variety of difficult transformations. Besides, better understanding of free radical processes can make it possible to engineer and tame fleeting radical intermediates for asymmetric catalysis (Zhou et al. 2021).

References

Banerjee R (2003) Radical carbon skeleton rearrangements: catalysis by coenzyme B12-dependent mutases. Chem Rev 103:2083–2094

Banerjee R, Ragsdale SW (2003) The many faces of vitamin B12: catalysis by cobalamin-dependent enzymes. Annu Rev Biochem 72:209–247

Bridwell-Rabb J, Li B, Drennan CL (2022) Cobalamin-dependent radical S-adenosyl methionine enzymes: capitalizing on old motifs for new functions. ACS Bio Med Chem Au 2:173–186

Broderick JB, Broderick WE, Hoffman BM (2023) Radical SAM enzymes: nature's choice for radical reactions. FEBS Lett 597:92–101

Broderick JB et al (2014) Radical S-adenosylmethionine enzymes. Chem Rev 114:4229–4317

Brown KL (2005) Chemistry and enzymology of vitamin B12. Chem Rev 105:2075–2149

Cronan JE (2024) Lipoic acid attachment to proteins: stimulating new developments. Microbiol Mol Biol Rev 88:e00005-24

Horitani M et al (2016) Radical SAM catalysis via an organometallic intermediate with an Fe-[5′-C]-deoxyadenosyl bond. Science 352:822–825

Huhta MS et al (2002) A novel reaction between adenosylcobalamin and 2-methyleneglutarate catalyzed by glutamate mutase. Biochemistry 41:3200–3206

Reichard P, Ehrenberg A (1983) Ribonucleotide reductase – a radical enzyme. Science 221:514–519

Stubbe J, Nocera DG (2021) Radicals in biology: your life is in their hands. J Am Chem Soc 143:13463–13472

Stubbe J, van der Donk WA (1998) Protein radicals in enzyme catalysis. Chem Rev 98:705–762

Wu B et al (2011) Aminomutases: mechanistic diversity, biotechnological applications and future perspectives. Trends Biotechnol 29:352–362

Zhou Q et al (2021) Stereodivergent atom-transfer radical cyclization by engineered cytochromes P450. Science 374:1612–1616

Integrating Kinetic and Chemical Mechanisms: A Synthesis

35

Enzymes are chemical catalysts par excellence. Mechanistic understanding of enzyme catalysis therefore emphasizes two key features—the kinetic pathway and the chemical route taken to achieve it. Elucidation of kinetic mechanisms was elaborated in several chapters earlier (in Part III). Chemical tools and cofactor reactivity exploited by these catalysts formed the major focus of this section (Part IV). Assimilating the kinetic and the chemical line of inquiry completes the comprehension of enzyme function in vitro. The origins of enzyme catalytic power (see Chap. 4) make better sense through this fusion. There may be multiple chemical solutions for catalyzing a particular reaction—for instance—peptide bond hydrolysis (Chap. 5). A chosen chemical mechanism may place constraints as to what kind of kinetic schemes are feasible. Besides, different kinetic mechanisms may exist for the same enzyme catalyzed reaction. We will attempt to bring together these two broad lines of evidence in this chapter through examples.

35.1 Competence of the Proposed Reaction Intermediate

When a reactant is converted to product, one or more bonds are broken and/or formed. This is true for an enzyme catalyzed reaction as well. One or more transition states and intermediates occur along the reaction path. The transition states differ from intermediates in terms of their lifetime of existence. A *transition state* is ephemeral and occurs only at the top of a potential energy peak on the reaction coordinate (see Chap. 3). As bond vibration modes are translated into bond breaking/forming events in the transition state, the time frames for the existence of transition state are less than the time required for intra-molecular vibrations ($<10^{-12}$ s). For these reasons, the transition state does not have a finite lifetime. In contrast, an intermediate in a reaction persists longer than the time required for inter-/intra-molecular bond vibrations. All the bonds holding the atoms together are fully established and the *intermediate* does not have any partial bonds (that are in the

N. S. Punekar, *ENZYMES: Catalysis, Kinetics and Mechanisms*, https://doi.org/10.1007/978-981-97-8179-9_35

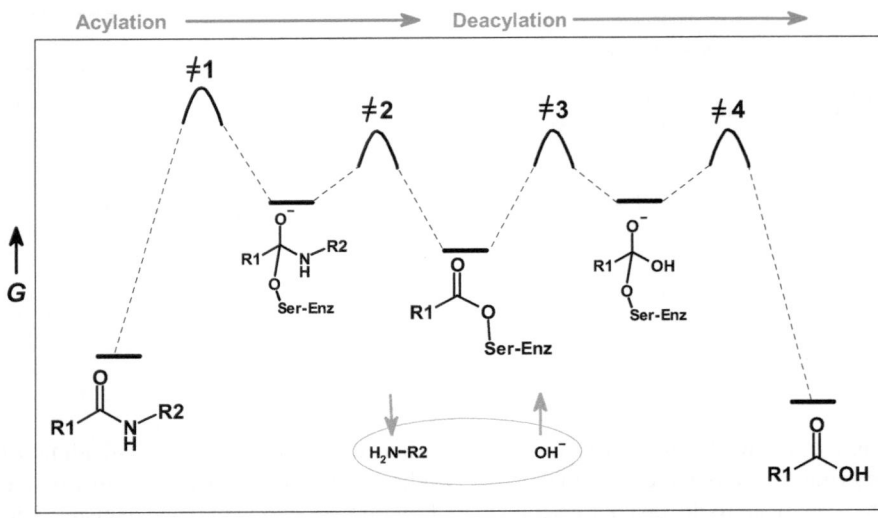

Fig. 35.1 Reaction coordinate for peptide bond hydrolysis by a serine protease. Four transition states ($\neq 1$ through $\neq 4$) and three intermediates are shown—the acyl-enzyme intermediate in the middle is flanked by two tetrahedral intermediates. For a good substrate, acylation and deacylation rates are comparable with no net accumulation of the acyl-enzyme intermediate. With a poor substrate like 4-nitrophenyl acetate ($K_M = 20 \ \mu M$ and $k_{cat} = 77 \ s^{-1}$), however, acyl-enzyme accumulates significantly as the deacylation step becomes rate-limiting (acylation rate $= 37 \ s^{-1}$ and deacylation rate $= 1.3 \times 10^{-4} \ s^{-1}$)

process of forming or breaking). Reaction intermediates are located at potential energy wells on the reaction coordinate—such that their energy levels lie between those of transition state and the substrate or the product. A schematic reaction coordinate for the serine protease (such as chymotrypsin) catalyzed peptide bond hydrolysis is depicted in Fig. 35.1. Because intermediates have finite lifetimes, it is often possible to observe and/or trap them.

The intermediates formed during enzyme catalyzed reactions may be non-covalent or covalent. We have seen before (Chaps. 10 and 29) that intermediates can be detected and/or trapped. Much more needs to be done besides demonstrating their actual existence. If an intermediate is indeed on the S → P reaction coordinate, then one may be able to demonstrate its kinetic and catalytic competence. An intermediate—proposed through direct and/or indirect experimental reasoning—must satisfy both these features. The following criteria are employed to verify the presence and participation of postulated intermediate:

- Detect and isolate the proposed intermediate (see Chap. 29).
- Determine its structure. If unstable, analyze the breakdown products for indicative clues.
- Look for chemical precedence and thermodynamic basis for such intermediates.

- Synthesize relevant model compounds and examine their chemical reactivity. Test if they are converted by the enzyme to products.
- A postulated intermediate is **chemically competent** only if it is converted by the enzyme to the product (and should form the substrate in the case of reversible enzyme reactions).
- Only those intermediates that form and collapse at rates equal to or greater than the overall enzymatic reaction rate are considered **kinetically competent**. This should be true for all the postulated intermediates in an enzyme mechanism.
- A compound that is kinetically not competent cannot be an intermediate in the reaction coordinate.

Intermediates like γ-glutamyl phosphate (glutamine synthetase), acyl-enzyme (chymotrypsin) and phospho-enzyme (phosphoglucomutase) do satisfy the above criteria of catalytic competence. The concepts of kinetic and chemical competence lead to complete understanding of an enzyme mechanism. We will now describe several well studied cases of such mechanistic syntheses.

35.2   Glutamine Synthetase

Glutamine synthetase (glutamate-ammonia ligase) catalyzes a reaction with three reactants and three products.

$$\text{L-Glutamate} + NH_4^+ + \text{Mg-ATP} \rightarrow \text{L-Glutamine} + \text{Mg-ADP} + ⓟ$$

In this three-substrate sequential mechanism, γ-glutamyl phosphate is formed at the active site. Nucleophilic attack by ammonia to this intermediate displaces phosphate to form glutamine (Fig. 29.5). The kinetic and chemical competence of γ-glutamyl phosphate was established through the following multiple lines of evidence.

1. Attempts to synthesize and directly test the reactivity of γ-glutamyl phosphate have not been successful due to its instability. This compound spontaneously cyclizes to pyrrolidone carboxylate (also known as pyroglutamate or 5-oxoproline) by loss of phosphate (a good leaving group). However, cis-1-amino-1,3-dicarboxycyclohexane (a glutamate mimic) is a substrate for glutamine synthetase that cannot cyclize γ-glutamyl phosphate. This compound is phosphorylated by the enzyme using Mg-ATP, in the absence of ammonia. The analogous acyl-phosphate (Fig. 35.2) can be isolated and identified. Second, β-glutamate (3-aminoglutarate) is a substrate for mammalian glutamine synthetase, but its acyl-phosphate is relatively stable and can be prepared. The enzyme drives the formation of Mg-ATP when incubated with Mg-ADP and synthetic β-glutamate phosphate (Fig. 35.2). The acyl-phosphate is thus *chemically competent* to form the substrate (ATP), in the reverse reaction.

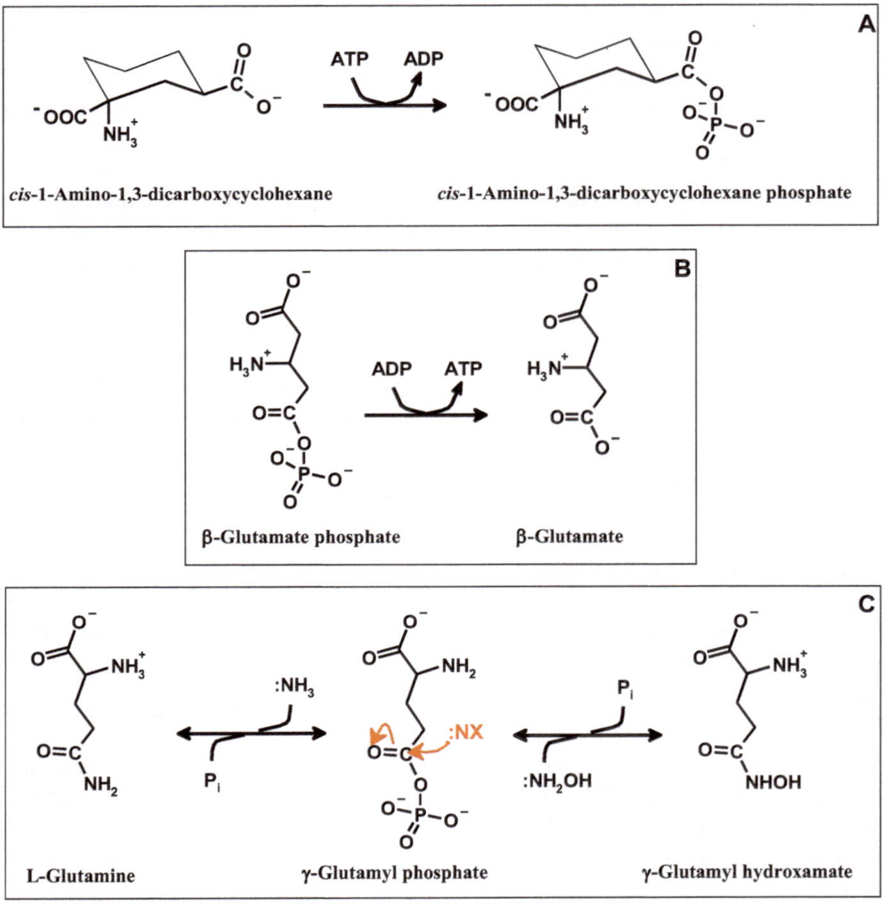

Fig. 35.2 Glutamine synthetase catalysis proceeds via enzyme bound γ-glutamyl phosphate as the obligate covalent reaction intermediate. (**a**) The enzyme phosphorylates *cis*-1-amino-1,3-dicarboxycyclohexane and the carboxyl-phosphate formed is stable. (**b**) ATP can be synthesized from ADP and synthetic β-glutamate phosphate. (**c**) The enzyme bound, activated γ-carboxylate of glutamate can be intercepted by **NX** (either NH_2OH or NH_3); this forms the basis of γ-glutamyl transferase activity of all glutamine synthetases

2. Several indirect clues also support the chemical intermediacy of γ-glutamyl phosphate (Fig. 29.5). Activated form of glutamate (such as γ-glutamyl phosphate) cyclizes to 5-oxoproline, and this is the product formed by glutamine synthetase in the absence of ammonia. Only an activated carboxylate group is reduced by sodium borohydride to aldehyde and then to alcohol; borohydride does not reduce a free carboxylate, γ-amide of glutamine or 5-oxoproline. The enzyme bound γ-glutamyl phosphate is reduced to δ-hydroxy-α-aminovalerate by borohydride demonstrating that the γ-carboxylate of glutamate is indeed

activated by glutamine synthetase. The enzyme mistakes the sulfoximine moiety of L-methionine-S-sulfoximine for the γ-carboxylate of glutamate and phosphorylates it. The L-methionine-S-sulfoximine N-phosphate along with Mg-ADP forms an extremely tight complex with glutamine synthetase. Finally, the glutamate carboxyl oxygen (^{18}O label) ends up in inorganic phosphate upon glutamine synthesis, suggesting the existence of an acyl-phosphate intermediate.

3. In the presence of Mg-ADP and phosphate or arsenate (AsO_4^{3-} is a structural analog of PO_4^{3-}), glutamine synthetase also catalyzes the γ-glutamyl group transfer.

$$\text{L-Glutamine} - NH_2OH \rightleftharpoons \gamma\text{-Glutamyl hydroxamate} + NH_3$$

This reaction is also consistent with the existence of an activated γ-carboxylate of glutamate during the reaction. NH_2OH in place of NH_3 can intercept this derivative to form γ-glutamyl hydroxamate; either of them can generate a tetrahedral intermediate (Fig. 35.2) during the enzymatic reaction. The γ-glutamyl transferase reaction follows a ping pong kinetic mechanism with glutamine as the leading substrate and NH_3 as the first product. The activated γ-carboxylate of glutamate bound to the enzyme (the F form) is then quenched by NH_2OH to release the second product.

4. The formation of γ-glutamyl phosphate is also inferred from the positional isotope exchange (PIX; for a detailed treatment see Chap. 24) study. The ^{18}O-labeled β,γ bridge oxygen of ATP exchanges with non-bridge β-phosphate oxygens: the γ-phosphate of ATP is transferred to γ-carboxylate of glutamate, the β-phosphate then can interchange the bridge ^{18}O to the two of its non-bridge ^{16}O atoms (Fig. 29.5). This exchange requires the presence of glutamate but can occur in the absence of ammonia; PIX occurs as the reaction cannot proceed beyond the formation of γ-glutamyl phosphate. In the presence of ammonia, however, the PIX rate decreases while glutamine formation is favored. Further, PIX rate in the absence of ammonia is equal to the initial velocity of complete reaction (i.e., glutamine synthesis). This demonstrates the *kinetic competence* of γ-glutamyl phosphate.

5. The steady-state kinetic data points to a sequential ter–ter reaction mechanism. Since the formation of γ-glutamyl phosphate is known to require only glutamate and Mg-ATP, isotope exchange between Mg-ATP and Mg-ADP is expected (when glutamate is present) in the absence of ammonia. This does not occur, possibly because the active site does not close for reaction unless all three substrates assemble. The observed equilibrium isotope exchanges support a random mechanism. But as both glutamate-glutamine (with ^{14}C label) and ammonia-glutamine (with ^{15}N label) exchange rates are much larger than that of Mg-ATP and Mg-ADP (with ^{14}C label) exchange, the microscopic association/dissociation rates for the nucleotides are slower than those for the amino acids.

γ-Glutamyl phosphate satisfies the kinetic criteria and also qualifies as a chemically competent intermediate of glutamine synthetase reaction. Intermediates that are artifacts of the analysis itself most likely do not stand such rigorous evaluation.

35.3   Glutamate Dehydrogenase

Glutamate dehydrogenase follows a sequential ter–ter reaction mechanism.

$$2\text{-Oxoglutarate} + NH_4^+ + NAD(P)H \rightleftharpoons L\text{-Glutamate} + H_2O + NAD(P)^+$$

The reductive amination of 2-oxoglutarate proceeds through an enzyme bound 2-iminoglutarate intermediate (Fig. 4.3). The following lines of evidence support this proposition.

1. Reduction of Δ^1-pyrroline-2-carboxylate (a cyclic-α-imino acid) by NADPH is catalyzed by glutamate dehydrogenase.
2. A primary deuterium isotope effect with α-deutero-L-glutamate is demonstrated in the first observable step of the reverse reaction. Kinetic evidence exists that ammonia is released from the enzyme in a step following the hydrogen transfer step of the reverse reaction.
3. Sodium borohydride reduction of 2-oxoglutarate in the presence of ammonia and enzyme leads to the production of an excess of L-glutamate over the D-enantiomer.
4. The active site electrostatics permits glutamate dehydrogenase to discriminate between iminium and carbonyl groups; the interaction decreases in the order— iminium ion ($>C=NH_2^+$) > 2-methyleneglutarate ($>C=CH_2$) > 2-oxoglutarate ($>C=O^\delta$), possibly due to charge repulsion (Choudhury and Punekar 2007). Spectral and structural evidence exists for the presence of this 2-iminoglutarate intermediate on the enzyme (Prakash et al. 2018).

Two mechanisms were proposed for the formation of the 2-iminoglutarate intermediate: (a) nucleophilic attack of ammonia on a covalently bound Schiff base in the "E-NADPH-2-Oxoglutarate" ternary complex and (b) reaction of ammonia with carbonyl group of 2-oxoglutarate in this ternary complex. The rates of carbonyl oxygen exchange (with water; Fig. 35.3) in the ternary complex must be widely different for the two mechanisms (much faster exchange if a Schiff base is involved). The "sequential" nature of the kinetic mechanism is borne out by the fact that no such ^{18}O exchange occurs in the absence of NADPH. However, this ^{18}O exchange with the solvent could be followed in the ternary complex. When measured, the loss of label from the $>C=^{18}O$ containing ternary complex is at least 10^5 times slower (and hence *kinetically not competent*) than the rate of reductive amination reaction. This result also provides evidence against the involvement of an enzyme bound Schiff base in the mechanism.

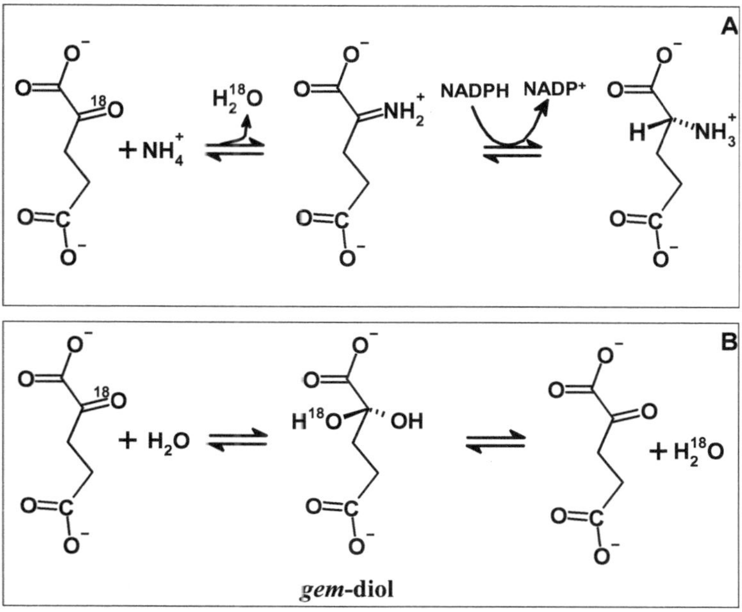

Fig. 35.3 Possible exchange of carbonyl ^{18}O with the solvent in glutamate dehydrogenase reaction. (**a**) Enzyme catalysis proceeds via an enzyme bound 2-iminoglutarate intermediate. (**b**) The much slower ^{18}O exchange due to formation of 2-oxoglutarate *gem*-diol disqualifies this species from being a catalytically competent intermediate

In the absence of ammonia, 2-oxoglutarate is not reduced by glutamate dehydrogenase. The rate of 2-oxoglutarate carbonyl ^{18}O exchange with water—due to *gem*-diol formation (Fig. 35.3) —is orders of magnitude slower than the overall enzyme reaction rate. Also, the *gem*-diol of 2-oxoglutarate is not a substrate for this enzyme. The 2-oxoglutarate *gem*-diol is therefore not on the reaction path of this enzyme. The enzyme strongly favors ammonia over water and the active site discriminates against the carbonyl reaction with water.

35.4   Disaccharide Phosphorylases

Phosphorylases are enzymes that reversibly phosphorolyze glycosides—to produce sugar 1-phosphates—with strict substrate specificities. This catabolic pathway—involving the phosphorolysis and direct production of phosphorylated sugars without consuming ATP—is energetically efficient. Among the glycoside hydrolase families, GH94 group is primarily comprised of phosphorylases that catalyze reversible phosphorolysis of β-D-glucosides to form α-D-glucose 1-phosphate (αGlc1P) with inversion of the anomeric configuration. We will consider three such enzyme activities and contrast their mechanistic details with sucrose phosphorylase—a retaining phosphorylase (Duodoroff et al. 1947).

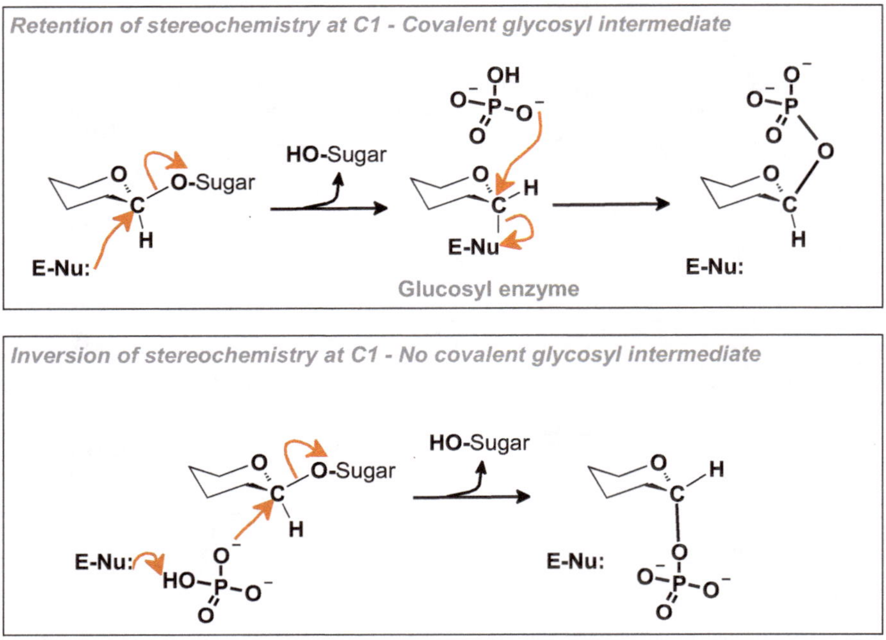

Fig. 35.4 Stereochemistry at the C-1 carbon of disaccharide phosphorylases. Formation of covalent intermediate is associated with retention of stereochemistry (top panel) while its absence results in inversion (bottom panel) at the glycosidic carbon. Only relevant structural details (of the sugar structures) are shown for the sake of clarity

Table 35.1 Stereochemical outcome of disaccharide phosphorylase action

Phosphorylase substrate	C-1 configuration of substrate	C-1 configuration of Glc1P[a]	Outcome
Maltose + Ⓟ ⇄ Glc1P + Glucose	α	β	Inversion
Cellobiose + Ⓟ ⇄ Glc1P + Glucose	β	α	Inversion
Cellobionate + Ⓟ ⇄ Glc1P + Gluconate	β	α	Inversion
Sucrose + Ⓟ ⇄ Glc1P + Fructose	α	α	Retention

[a]Glc1P: α- or β- D-glucose 1-phosphate

1. The glycosidic bridge oxygen (if ^{18}O labeled) of the four respective disaccharide substrates will not be found in the product—glucose 1-phosphate (Fig. 35.4). Glucosyl transfer with the cleavage of C-1 and bridge O bond in the substrate glycoside is indicated.
2. Stereochemical outcome of phosphorolysis by the four disaccharide phosphorylases is summarized in the table below (Table 35.1).

The phosphorylation of maltose, cellobiose, and cellobionate by the corresponding enzymes are examples with inversion of the anomeric configuration.

This implies a direct S_N2 displacement by oxygen of inorganic phosphate (Ⓟ in the table above) at the C-1 of the glycoside (Fig. 35.4) and suggests a ternary-complex mechanism. Sucrose phosphorylase on the other hand works with retention of stereochemistry. This result means either (a) two S_N2 displacements or (b) an S_N1 (carbonium ion) mechanism possibly with only one side attack. Additional data (see below) permits us to discriminate between these two possibilities.

3. The first three enzymes (listed in Table 35.1) do not catalyze partial exchanges, which is again consistent with a ternary-complex mechanism for them. Sucrose phosphorylase does catalyze partial exchanges, however (see Fig. 24.3). Sucrose is not required for the exchange of label between glucose 1-phosphate and [^{32}P] phosphate; and phosphate is not required for the second exchange between sucrose and [^{14}C] fructose. Such isotope exchanges are best evidence of ping-pong mechanism (with a glucosyl-enzyme as the covalent intermediate).

4. The glucosyl-enzyme form of sucrose phosphorylase is labile, but this bond is stable at acidic pH. Thus, the [^{14}C] glucose bound enzyme can be isolated by incubating the enzyme with [^{14}C] glucose containing sucrose and quenching it at pH 3.0. Under similar treatment, the enzyme does not get labeled when incubated with [^{14}C] glucose, [^{14}C] fructose, or [^{14}C] fructose containing sucrose. Only the glucosyl moiety derived from sucrose is covalently linked to sucrose phosphorylase during catalysis; this glucosyl-enzyme is the F form in the ping-pong kinetics.

5. We noted above (see γ-glutamyl transferase activity of glutamine synthetase) that arsenate (AsO_4^{3-}) is a structural analog of phosphate (PO_4^{3-}). Arsenate esters—unlike phosphate esters are unstable in aqueous solutions—upon water attack, they rapidly decompose to alcohol and arsenate. Glucose 1-phosphate in the absence of arsenate is stable in water. Only sucrose phosphorylase (and not the other three enzymes in Table 35.1) catalyzes the arsenate dependent hydrolysis (arsenolysis) of glucose 1-phosphate. Arsenolysis may be interpreted as follows: glucosyl form of sucrose phosphorylase can be formed when glucose 1-phosphate is present; this covalent enzyme intermediate is attacked by arsenate to give the unstable glucose 1-arsenate; the arsenate ester of glucose breaks down non-enzymatically to glucose and arsenate in water.

Glucose 1-phosphate + Enzyme $\rightleftarrows$ Glucosyl enzyme + Ⓟ
Glucosyl enzyme + AsO_4^{3-} $\rightleftarrows$ Glucose 1-arsenate + Enzyme
Glucose 1-arsenate + $H_2O \rightarrow$ Glucose + AsO_4^{3-}
Net Reaction: Glucose 1-phosphate + $H_2O \rightarrow$ Glucose + Ⓟ

The other three phosphorylases do not show arsenolysis—the result is consistent with the absence of glucosyl-enzyme in their sequential mechanism.

6. The inverting phosphorylases (such as those acting on maltose, cellobiose, and cellobionate) follow a sequential BiBi mechanism. In each case, double reciprocal plots of the initial velocities against various initial concentrations of

respective disaccharide and Ⓟ give a series of lines intersecting at a point. In contrast, sucrose phosphorylase involves a double displacement (ping-pong BiBi) mechanism—with fructose released prior to Ⓟ addition. As expected, two competitive and two noncompetitive product inhibitions are observed with sucrose phosphorylase.

We may summarize—from the mechanistic details of disaccharide phosphorylases discussed above—that a reaction may be catalyzed by different chemical and kinetic strategies. However, a glucosyl-enzyme intermediate is compatible only with ping-pong mechanism (see Fig. 24.3) and not a sequential BiBi mechanism.

35.5 Acyl Transferases

Acyl transfers are very common in metabolism and regulation. An acyl group is accepted by nucleophiles such as amines, alcohols, thiols, phosphates, and carboxylates. Acylation of alcohols represent an important group of enzymatic reactions among them. Mechanistically, acyl transfer to alcohol can occur in one of the two modes (Fig. 35.5): (a) direct nucleophilic attack by the alcohol on the acyl donor via a tetrahedral intermediate (or transition state) to form products or (b) formation of an acyl-enzyme intermediate involving an enzyme nucleophile and subsequent transfer of acyl group to the alcohol. We will now compare the two mechanisms.

35.5.1 Direct Nucleophilic Attack

Both the substrates must occupy the active site before the alcohol attacks the acyl donor. Therefore, such enzymes are predicted to follow a sequential kinetic mechanism. And this is observed experimentally (double reciprocal plots of the initial velocities against various initial concentrations of respective substrates give a series of lines intersecting at a point; Fig. 18.3). Examples include carnitine acetyltransferase (rapid equilibrium random kinetics), choline acetyltransferase (Theorell–Chance kinetics) and serine acetyltransferase (steady-state ordered at pH 7.5). A conserved histidine residue is found in the active site of all such enzymes. The imidazole side chain acts as a general base to remove the alcoholic proton so as to prepare it for the ensuing nucleophilic attack (Fig. 35.5). The initial attack on the acyl thioester (of acetyl CoA) by serine –OH is most likely the rate determining step in serine acetyltransferase; the general base accepting the alcoholic proton in the ternary complex sets up serine for this attack (Johnson et al. 2005). Solvent kinetic isotope effects on V_{max} and V_{max}/K_{Serine} are linearly dependent on percent of D_2O present—a result consistent with single proton in flight in the rate determining step. If multiple proton transfers were involved, then this dependence would be nonlinear.

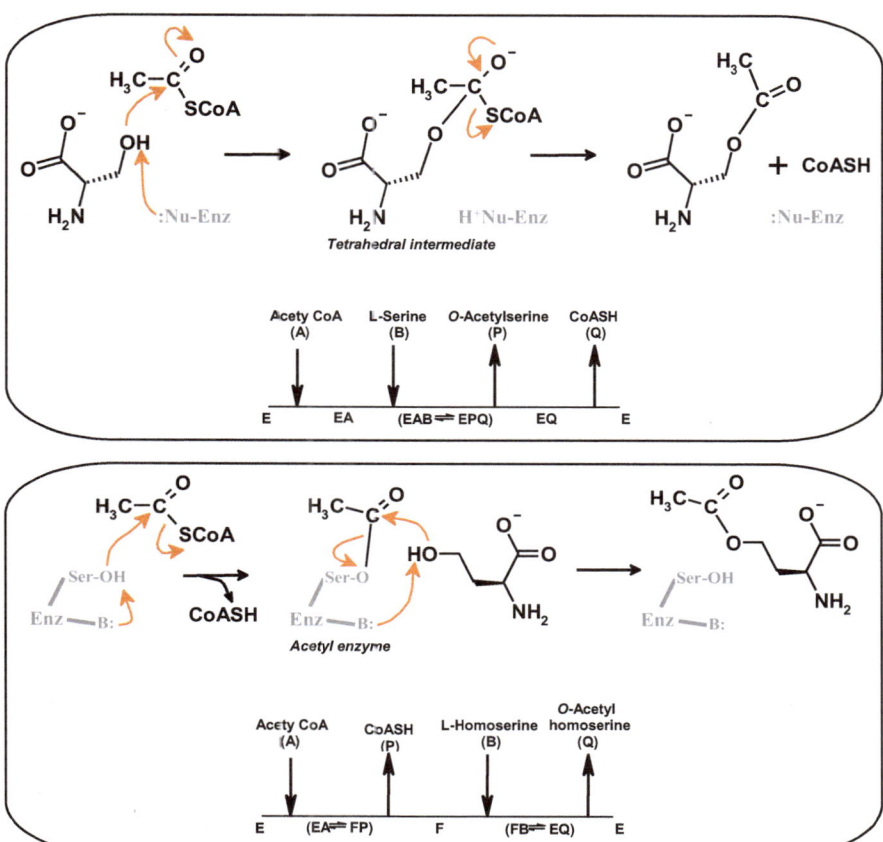

Fig. 35.5 Two possible mechanistic modes of acyl transfer to alcohol. Sequential mechanism (top panel): Direct nucleophilic attack by serine –OH on acetyl CoA—serine acetyltransferase. The Enz-Nu: is a conserved His residue. Ping-pong mechanism (bottom panel): Acyl-enzyme formation by the transfer of acetyl group first to an active site nucleophile (possibly a Ser-OH or Cys-SH). Homoserine attacks the acyl-enzyme (the F form) and receives the acetyl group in the second half reaction—homoserine acetyltransferase

35.5.2 Acyl-Enzyme Formation

Acyltransferases of this class display the predicted ping-pong kinetics. A ternary complex of the two substrates and the enzyme does not occur; the double reciprocal plots of the initial velocities give a parallel pattern (Fig. 18.4), typical for double displacement mechanism. The acyl group is transferred to an active site nucleophile during the first half of reaction. The acyl acceptor substrate receives it in the second half of the reaction. Depending on the enzyme, either a serine or a cysteine residue is the catalytic nucleophile at the active site. Examples of enzymes that go through an acyl-enzyme intermediate include homoserine succinyltransferase (active site Cys) and homoserine acetyltransferase (active site Ser?). Most lipases contain a catalytic serine that is acylated during catalysis. When water is made limiting (non-aqueous conditions), lipases can facilitate transacylation—a property gainfully exploited for trans-esterification reactions in the industry. In this situation, the ping-pong BiBi mechanism and the acyl-enzyme intermediate of lipases become obvious.

Both sequential and ping-pong kinetics are mechanistic possibilities for acyltransferase catalysis. But an acyl-enzyme intermediate forms only with ping-pong kinetics.

35.6   Chymotrypsin

Among the serine proteases chymotrypsin is a well-studied example. Its chemical and kinetic mechanism was established through a series of experimental evidence.

1. Incubation of chymotrypsin with diisopropyl fluorophosphate (DFP) leads to inhibition of the enzyme. This inhibition can be reversed by strong nucleophiles. When ^{32}P-labeled DFP is incubated with chymotrypsin a radioactive protein is obtained. Upon complete inhibition, one mole of the radiolabel is bound per mole of the enzyme. O-Diisopropyl phosphoryl serine is obtained on acid-hydrolysis of this labeled enzyme. The serine modified by diisopropyl fluorophosphate is at position 195 and is the only one among the 29 Ser residues found in chymotrypsin. The "free" amino acid serine does not react with DFP. The enzyme active site contains a serine residue, and this Ser-OH is the nucleophile. A catalytically critical Ser residue (serving as a nucleophile) is also found at the active site of many esterases and lipases.
2. Reagents modifying the amino acid histidine also inhibit the enzyme. The pH-activity studies reveal that a group with pK_a of 6.6 is important in catalysis. Active site directed irreversible inhibitors like N-4-toluenesulfonyl-L-phenylalanine chloromethylketone (TPCK) (Fig. 20.3) covalently modify His-57 residue.
3. The side-group of Ser-195 is the true catalytic nucleophile; its nucleophilic reactivity depends on a histidine side-group. X-Ray diffraction studies on crystalline chymotrypsin show that Ser-195 is in hydrogen bonding distance to the imidazole side chain of His-57. It is also seen that Asp-102 in turn hydrogen bonds to the active site His-57 side chain (Craik et al. 1987).
4. p-Nitrophenyl acetate is a substrate (albeit a poor one) of chymotrypsin and when acted upon yields an acetyl enzyme intermediate. The acyl-enzyme of chymotrypsin is a true covalent intermediate during its catalytic cycle. With a poor substrate like 4-nitrophenyl acetate, however, acyl-chymotrypsin accumulates significantly as the deacylation step becomes rate-limiting. This manifests as the observed burst kinetics (Fig. 29.4). O-Acetyl serine can be isolated by careful hydrolysis of this acetyl enzyme. It is worth noting that the acyl-enzyme, despite its accumulation, is still a kinetically competent intermediate; this is because the overall reaction rate itself slows down with this poor substrate. For a good

substrate, acylation and deacylation rates are comparable with no net accumulation of the acyl-enzyme intermediate (Fig. 35.1).

An acyl-enzyme intermediate of chymotrypsin was observed through UV spectroscopy, by using *trans*-cinnamoyl esters as substrates. This acyl-ester was both chemically competent as a reaction intermediate and kinetically competent (as it was converted to product at a rate at least as fast as the overall reaction rate).

Serpins present themselves as the proteinaceous suicide substrates of serine proteases. They act by generating a kinetically incompetent acyl-enzyme thereby titrating out the active enzyme. The trapping of transpeptidase (in bacterial cell wall biosynthesis) by β-lactam antibiotics, is a similar example. The intermediate penicilloyl enzyme is a dead-end as it is not kinetically competent.

5. Chymotrypsin reaction involves ordered product release; the amino product is released first before the acyl-enzyme is attacked by water. A UniBi kinetic mechanism is described by not considering water as a reactant (55.5 M; pseudo-first order kinetics). Reactions like transpeptidation and/ or peptide synthesis can be demonstrated when the second substrate is a nucleophile other than water (non-aqueous enzymology). As expected, a ping-pong BiBi mechanism and the acyl-enzyme intermediate of chymotrypsin catalysis become obvious here.

6. Site-directed mutagenesis was used to assess the contribution of Ser-195 toward trypsin catalysis. The kinetic mechanism of S195A mutant may not follow the ping-pong BiBi mechanism as no acyl-enzyme can form. The S195A mutant retains the ability to hydrolyze peptide bonds (at 10^3–10^4-fold above the uncatalyzed rates); this significant residual rate is attributed to the stabilization of oxyanion intermediate by the serine protease active site.

35.7 Aldolases and Transaldolases

The aldolytic cleavage of fructose 1,6-bisphosphate is a key reaction of glycolysis. This reaction is catalyzed by fructose 1,6-bisphosphate aldolase. The same enzyme is also responsible for the catalysis of the reverse reaction (retro-aldol condensation) to form fructose 1,6-bisphosphate during gluconeogenesis (Fig. 35.6).

We have noted earlier (Chap. 4) that—during catalysis—the stabilization of carbanion on C-3 of dihydroxyacetone phosphate may occur in one of the two ways: (a) a Schiff base formed with enzyme Lys-NH_2 (Class I aldolase) or (b) stabilizing the charge on carbonyl oxygen by the active site Zn^{2+} (Class II aldolase; the yeast enzyme has one Zn^{2+} per active site). Of the two, Class I aldolases have been extensively characterized for their mechanism of action (Samland and Sprenger 2006).

The collection of kinetic and chemical evidence in support of the Schiff base mechanism (Class I fructose 1,6-bisphosphate aldolase) is listed below.

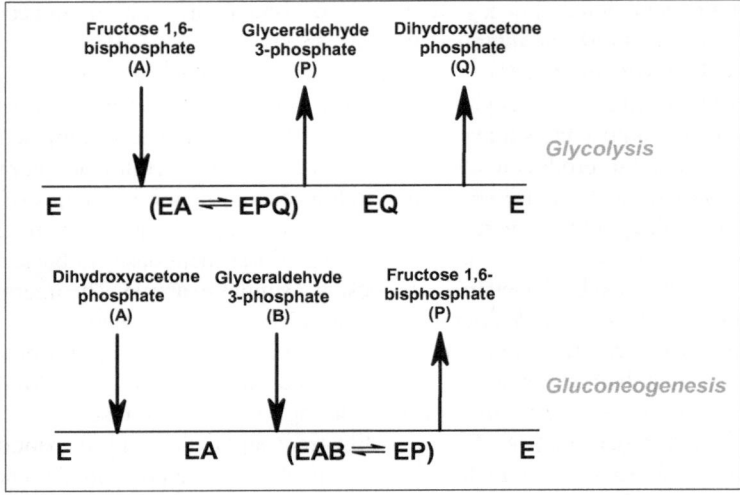

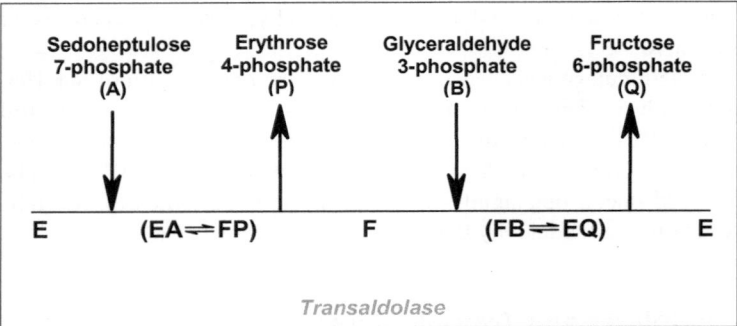

Fig. 35.6 Aldolase and transaldolase reactions. Cleland notation for fructose 1,6-bisphosphate aldolase (both directions; top panel) and transaldolase (bottom panel) are shown

1. Reduction by sodium borohydride ($NaBH_4$) in the presence of either dihydroxy-acetone phosphate or fructose 1,6-bisphosphate, but not in the presence of glyceraldehyde 3-phosphate, irreversibly inactivates the enzyme. If ^{14}C-dihy-droxyacetone phosphate was used in this reaction, then the radiolabel gets incorporated into the protein. Digestion of this modified protein provides evidence that the covalent intermediate trapped was the imine (cationic form of Schiff base; Fig. 35.7) between the modified lysine and dihydroxyacetone phosphate (Table 29.4). In a similar experiment, trapping of the lysine adduct was also demonstrated for acetoacetate decarboxylase reaction.

2. Active site labels display—(a) saturation behavior with respect to the rates of inactivation, (b) competition of this inactivation by corresponding substrate or

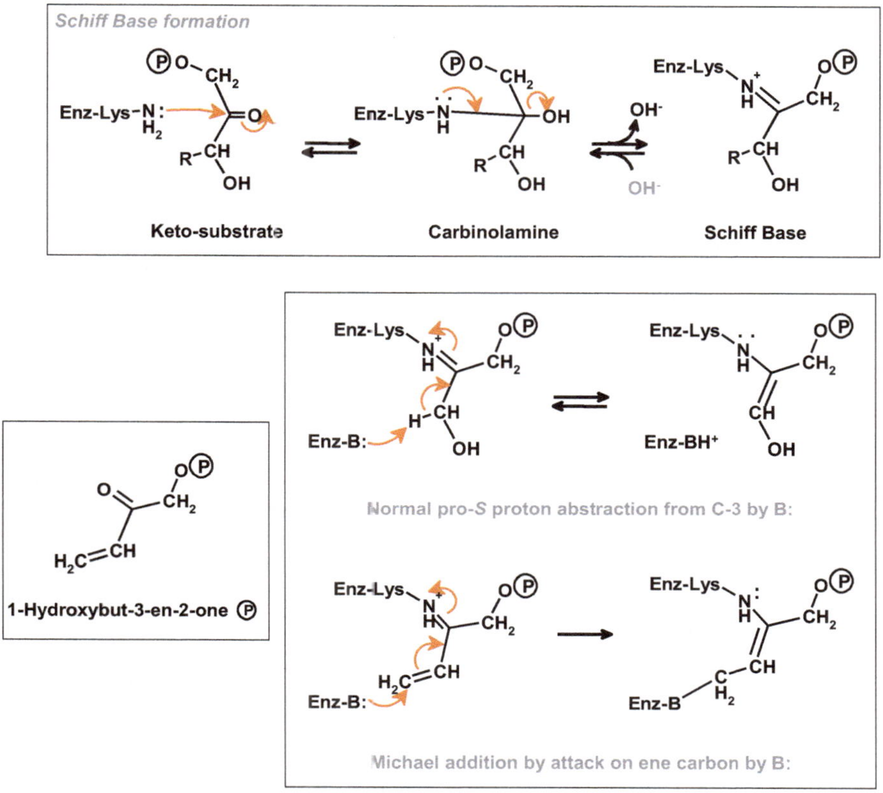

Fig. 35.7 Class I aldolase reaction mechanism. Reversible formation of Schiff base with keto-substrate (top panel); the substrate is dihydroxyacetone phosphate when R = H. The abstraction of pro-S hydrogen from C-3 by an enzyme base is contrasted with Michael addition of the same group as a nucleophile to 1-hydroxybut-3-en-2-one phosphate (an active site label) (bottom panel)

competitive inhibitor, and (c) the K_D values measured through the inactivation rates are similar to their respective K_M and K_I values. All these three aspects were tested with 1-hydroxybut-3-en-2-one phosphate as an active site label for fructose 1,6-bisphosphate aldolase.

1-Hydroxybut-3-en-2-one phosphate (active site label) binds to aldolase before covalently modifying it (by Michael addition; Fig. 35.7). From the saturation curve with pseudo-first order inactivation rate constants a K_D of 99 µM was observed. The irreversible inactivation by 1-hydroxybut-3-en-2-one phosphate is competitively inhibited by the substrate dihydroxyacetone phosphate. A dihydroxyacetone phosphate dissociation constant of 1.4 µM (calculated from such protection experiments) is close enough to the kinetically obtained K_I (4.5 µM) for this substrate.

3. Incubation with the enzyme facilitates the release of ^{18}O from the carbonyl group of dihydroxyacetone phosphate (but not from that of glyceraldehyde

3-phosphate) into the solvent. This is consistent with the formation of Schiff base between the enzyme and dihydroxyacetone phosphate. For the same mechanistic reason, acetoacetate decarboxylase also catalyzes the exchange of ^{18}O from $H_2^{18}O$ into acetone (its decarboxylation product).

4. The tritium exchange between ^3H-labeled dihydroxyacetone phosphate ($\circledP OCH_2\text{-}CO\text{-}C^3H_2\text{-}OH$) with water is catalyzed by the enzyme in the absence of glyceraldehyde 3-phosphate. This is an example of isotope exchange in the absence of one or more reactants—suggestive of a covalent intermediate with the enzyme that is responsible for the observed exchange. Further, the rate of this tritium exchange at equilibrium falls sharply with increasing concentration of glyceraldehyde 3-phosphate. This inhibition of exchange is indicative of glyceraldehyde 3-phosphate release before dihydroxyacetone phosphate (ordered product release). Accordingly, in the reverse reaction, dihydroxyacetone phosphate goes on to the enzyme first.

5. The two hydrogens on the pro-chiral C-3 of dihydroxyacetone phosphate are not equivalent. Hence only one of them (pro-S hydrogen) is available for exchange. One may contrast this situation with (a) the enolization of pyruvate, by pyruvate kinase reaction and (b) the exchange of deuterium from D_2O into acetone to produce CD_3COCD_3 by acetoacetate decarboxylase. The three methyl hydrogens of pyruvate and the six methyl hydrogens of acetone, respectively, are torsio-symmetric. Therefore, all the methyl hydrogens of pyruvate and acetone (note that the two methyl groups on acetone are equivalent) are exchangeable.

6. The C-3 of dihydroxyacetone phosphate is a pro-chiral center. Only its pro-S hydrogen is removed by a base on the enzyme active site. Its subsequent exchange with water by acid-base reaction is responsible for the tritium exchange mentioned in evidence 4 above. This exchange rate establishes only a lower limit to the rate at which the labeled intermediate itself is formed (Fig. 35.8). Consistent with this abstraction of pro-S hydrogen, the aldolase chemistry occurs with overall retention of configuration at C-3. The enzyme facilitates the stereo-specific attack by dihydroxyacetone phosphate imine on to only one face of the bound aldehyde (Fig. 35.8).

35.7.1 Transaldolase

While two mechanistically distinct aldolases (the Class I with Schiff base and the Class II with Zn^{2+}) are known, transaldolases with only Schiff base mechanism are reported. The best studied transaldolase is an important member of the pentose phosphate pathway (and the Calvin–Benson cycle). It catalyzes the transfer of three carbon (dihydroxyacetone) units, between the aldose acceptor molecules. The bound dihydroxyacetone phosphate enamine in transaldolase is held sufficiently long at the active site without imine hydrolysis. This feature allows the erythrose 4-phosphate to diffuse out of the active site, other aldose substrate (glyceraldehyde

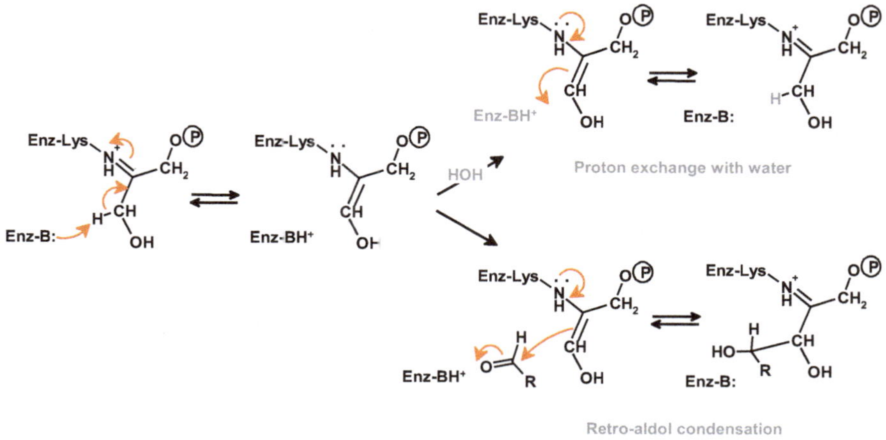

Fig. 35.8 Class I aldolase reaction mechanism. Subsequent to the abstraction of proton from C-3 of dihydroxyacetone phosphate, the enzyme bound intermediate suffers one of the two fates. It either exchanges the proton with water (upper route) or is captured by the aldehyde substrate to form the retro-aldol condensation product (lower route). Fructose 1,6-bisphosphate is formed when this RCHO is glyceraldehyde 3-phosphate

3-phosphate) to diffuse in and react with the bound dihydroxyacetone phosphate. In the retro-aldol condensation step (Fig. 35.8), one aldehyde (R_1CHO) leaves and is replaced by another (R_2CHO). Clearly, ping-pong BiBi kinetic mechanism ensues with the dihydroxyacetone phosphate adduct (Schiff base) on enzyme Lys-NH_2 as the F form.

The mechanistic difference between transaldolase and aldolase reaction is only that the release of dihydroxyacetone phosphate from the E-DHAP complex is much more rapid in the aldolase. The essential mechanistic similarity between the two enzymes is evident when FBP aldolase was able to carry out transaldolation upon limited proteolysis (removal of 3–4 amino acid residues from the carboxyl-terminus by carboxypeptidase A treatment). The limited proteolysis resulted in an aldolase whose proton abstraction ability becomes rate determining—it holds on to the bound dihydroxyacetone phosphate for a longer interval.

We may summarize from the above that aldolase catalysis can be achieved either by deploying a Schiff base (as in Class I aldolases) or a Zn^{2+} (as in Class II aldolases). For mechanistically similar reasons, acetoacetate decarboxylase also comes in two forms: one employing a Schiff base and the other an active site Mn^{2+} (Figs. 32.5 and 32.6). As expected, the Schiff base acetoacetate decarboxylase displays the relevant proton and ^{18}O exchanges with water. These are mechanistic features consistent with a carbanion stabilized during catalysis by both aldolase and acetoacetate decarboxylase. Finally, it appears that only with a Schiff base mechanism, there is sufficient time and scope for one aldehyde to leave and the other to enter the active site. This is consistent with the fact that transaldolases with

Schiff base mechanism alone are known and no Class II transaldolases have been reported so far (Samland and Sprenger 2006).

35.8 Ribonuclease A

Bovine pancreatic ribonuclease A (RNase A) is a compact enzyme with a polypeptide chain of 124 amino acids. It is an efficient catalyst in the hydrolysis of RNA phosphodiester bonds (Raines 1998).

1. The single polypeptide of RNase A contains four histidine residues. Incubation with iodoacetate leads to enzyme inhibition. The inhibition by iodoacetate is prevented if substrate or substrate analogs are present during incubation. Iodoacetate modifies an essential histidine required for RNase A activity.
2. When ^{14}C-iodoacetate is incubated with RNase A, a radioactive protein is obtained. One mole of ^{14}C-iodoacetate is bound per mole of enzyme upon complete inhibition. Complete hydrolysis of the inhibited enzyme with acid releases carboxymethyl histidine. This histidine was found to be His-119 upon sequence analysis; a minor product was His-12. Both His-119 and His-12 are important for RNase A catalysis and are spatially located close to each other. X-Ray data also support this.
3. The pH-activity studies revealed two general acid-base catalytic groups at the active site. Their cationic acid nature was indicated based on organic solvent effects (Chap. 23, Table 23.2). The pH titration of ^{1}H chemical shifts (by NMR) for His-119 and His-12 match the pH-activity profile of RNase A. Incorporating fluorohistidine at these two positions gives an enzyme with lower pH optimum—expected with the chemical reactivity differences between His and fluoroHis side chains.
4. Limited proteolysis by subtilisin cuts RNase A polypeptide into two fragments, S-peptide (residues 1–20) and S-protein (residues 21–124). The two separate fragments are inactive but their 1:1 mixture combines to give full activity. Both His-119 and His-12 are at the active site and are located close to each other.
5. RNase A cleaves RNA chain after pyrimidine residues. It also hydrolyzes monomeric cyclic-2′,3′-phosphates of UMP and CMP. The product is always a 3′-phosphate. The alkaline hydrolysis of RNA also proceeds through a cyclic-2′,3′-phosphate intermediate but here both the 2′- and 3′-phosphate are formed. While cyclic-2′,3′-phosphate is the true intermediate (and is kinetically competent) in the RNase A catalysis, the enzyme controls the reaction specificity to open up the cyclic-2′,3′-phosphate only as a 3′-phosphate. One may also expect that this portion of the substrate (the 5′ fragment) may leave the active site later than the 3′ fragment of the RNA.
6. Uridine vanadate is a potent competitive inhibitor of RNase A (Fig. 35.9). The fact that its structure mimics the cyclic-2′,3′-phosphate intermediate in the enzyme catalyzed reaction makes it a powerful inhibitor.

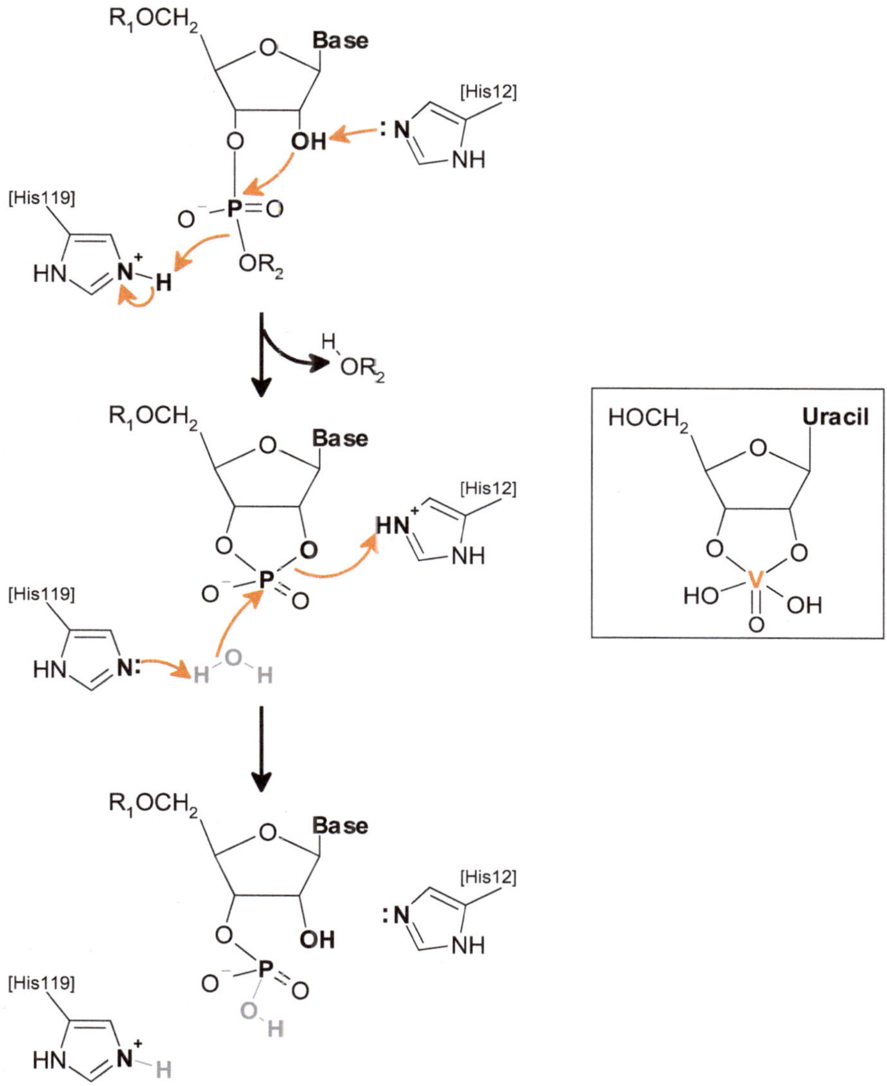

Fig. 35.9 Reaction mechanism of RNase A. His-12 imidazole group (base) abstracts the proton from 2′-OH while His-119 imidazole group (acid) protonates the leaving group ⁻OR₂ (the 3′ portion of the RNA substrate). The 2′,3′-cyclic phosphate is opened up by water where the roles of His-12 and His-119 are reversed. The enzyme specificity controls this step so that only the 3′-phosphate results. The structure of uridine vanadate (a potent inhibitor) is shown in the box

A reasonable idea of the active site and the reaction mechanism of RNase A may be arrived at from the above data. This is shown in the figure below (Fig. 35.9).

35.9   Summing Up

All enzyme mechanisms go through a transition state (or many transition states) — ephemeral species without finite lifetime—found at the top of a potential energy peak on the reaction coordinate. Intermediates, on the other hand, persist and are located at potential energy wells on the reaction coordinate; they can either be non-covalent or covalent complexes of the enzyme. Genuine intermediates must satisfy the twin criteria of kinetic and chemical competence. It may be possible to trap the reaction intermediates in certain situations.

For many chemical reactions, both sequential and ping-pong kinetics are feasible mechanistic solutions in enzyme catalysis. Such examples were described above with acyltransferases and disaccharide phosphorylases. Among redox reactions, those involving NAD(P) follow a sequential mechanism (e.g., alcohol dehydrogenase and lactate dehydrogenase) whereas ones with FAD follow ping-pong kinetics (e.g., glucose oxidase and methylenetetrahydrofolate reductase). Enzyme covalent intermediates are invariably associated with ping-pong reaction kinetics and not with sequential mechanisms. The well-accepted Phillips mechanism for lysozyme reaction had to be suitably modified once the covalent glycosyl-enzyme intermediate was demonstrated.

Both random and ordered addition of substrates and/or release of products are equally feasible kinetic solutions for a sequential mechanism. Alcohol dehydrogenases with either sequential-ordered or sequential-random mechanism have evolved and are reported in the literature. A change from one type of sequential mechanism to another may occur—for example, because of pH change. Creatine kinase exhibits equilibrium-ordered kinetics at pH 7.0 with Mg-ATP adding before creatine; but the back reaction shows a random addition of Mg-ADP and phosphocreatine. However, the kinetic mechanism is random in both directions at pH 8.0. This change in mechanism is because enzyme loses affinity for creatine at pH 7.0— as the active site general base becomes protonated. Consequently, Mg-ATP adds first and is at equilibrium. A pH-dependent change of mechanism is also reported for serine acetyltransferase (Fig. 35.5). The enzyme follows a sequential mechanism with acetyl CoA adding first and then Serine. However, it is equilibrium-ordered at pH 6.5 and steady-state ordered (with nearly parallel initial velocity pattern; this was mistaken for a ping-pong mechanism since $K_{\text{Acetyl CoA}} > K_{\text{IAcetyl CoA}}$) at pH 7.5. Again, the shift in mechanism is due to change in reactant affinity at different pH values.

Site-directed mutagenesis enables enzymologists to selectively replace active site residues and ask some really interesting mechanistic questions. However, it remains a distinct possibility that mutant enzymes might follow a different reaction pathway. This is what was noticed with triosephosphate isomerase and serine proteases.

Subtilisin is the most studied bacterial serine protease with active site catalytic residues S221, H64, and D32. This triad synergistically accelerates amide bond hydrolysis, by contributing a factor of $\sim 2 \times 10^6$ to the total catalytic rate enhancement of 10^9–10^{10}. Similar results were obtained for trypsin (with the active site catalytic triad of S195, H57, and D102). The residual activity, in the absence of catalytic triad, results from the transition state stabilization at the active site. For the S221A mutant, the reaction cannot proceed by the usual serine acyl-enzyme intermediate (ping-pong mechanism). Instead, direct attack of water on the scissile peptide bond may occur to form a single tetrahedral intermediate that collapses to the products. With the catalytic nucleophile missing (in the S221A mutant), the mechanism changes to a sequential one. While a mechanistic change from an acyl-enzyme to a direct one with tetrahedral intermediate may be possible, the converse is not because the essential nucleophile would be missing.

An active site apparatus with an oxyanion hole and favoring a tetrahedral reaction intermediate (after water attack) is a common mechanistic feature of proteases, esterases and carbonic anhydrases. It is therefore no wonder that a weak esterase activity is displayed by chymotrypsin, pepsin, carboxypeptidase A, and carbonic anhydrase. On similar grounds, some phosphotransferases exhibit phosphatase activity. Such promiscuity exposes the underlying common mechanistic features of enzyme action.

References

Choudhury R, Punekar NS (2007) Competitive inhibition of glutamate dehydrogenase reaction. FEBS Lett 581:2733–2273

Craik CC et al (1987) The catalytic role of active site aspartic acid in serine proteases. Science 237: 909–913

Duodoroff M, Barker HA, Hassid WZ (1947) Studies with bacterial sucrose phosphorylase. I. The mechanism of action of sucrose phosphorylase as a glucose-transferring enzyme (transglucosidase). J Biol Chem 168:725–732

Johnson CM, Roderick SL, Cook PF (2005) The serine acetyltransferase reaction: acetyl transfer from an acylpantothenyl donor to an alcohol. Arch Biochem Biophys 433:85–95

Prakash P, Punekar NS, Bhaumik P (2018) Structural basis for the catalytic mechanism and α-ketoglutarate cooperativity of glutamate dehydrogenase. J Biol Chem 293:6241–6258

Raines RT (1998) Ribonuclease A. Chem Rev 98:1045–1065

Samland AK, Sprenger GA (2006) Microbial aldolases as C–C bonding enzymes – unknown treasures and new developments. Appl Microbiol Biotechnol 71:253–264

Part V

Exploiting Enzymes

Enzyme Inhibitor Design: Drug Discovery

36

Two specific classes of biologic macromolecules are the major targets of modern drug therapy: enzymes and receptors. The features that make enzymes attractive drug targets are the same as catalysis, specificity, and regulation (Chap. 3). These are reflected in the drug discovery process as: potency, selectivity, and specificity, and diversity/novelty of mechanisms Broadly, every enzyme in the metabolism is a potential target for manipulation. Its catalytic activity could be attenuated through the action of inhibitors or by targeting the expression of the active enzyme (by gene knockout or RNA interference). Enhancing/activation of enzyme activity may also be possible through overexpression or the use of specific activators. Our emphasis here is to look for ways and means to control the activity of enzyme protein already present, and not to go through genetic manipulations. Enzymes constitute the targets for almost half the drugs (including herbicides, pesticides, etc.) available in the market. Many of the top 100 drugs in use worldwide are enzyme inhibitors of one or the other kind. Although a bit dated, an excellent perspective on the mechanistic basis of enzyme-targeted drugs by Robertson (2005) provides a good starting point. This review also comes with a supplementary table containing a complete list of more than 300 drugs and their enzyme targets. Also, reviews on rational design of enzyme inhibitors as potential drugs by McLeish and Kenyon (2003) and Copeland et al. (2007) give a good account of this topic.

36.1 Enzyme Activators as Drugs?

While enzyme inhibitors have long been a focus for modern drug discovery, the concept of using enzyme activators as drugs has found limited traction. Examples of enzyme activators as drugs are few and far between (Turberville et al. 2022). Enzyme activation as a therapeutic tool is underexploited. Identification of good enzyme activators will help utilize both activators and inhibitors in combination. Enzyme activators are classified as nonessential (also termed allosteric regulation)

N. S. Punekar, *ENZYMES: Catalysis, Kinetics and Mechanisms*, https://doi.org/10.1007/978-981-97-8179-9_36

and essential (or obligatory) activation. In nonessential activation, the enzyme can turn over the substrate in the absence of the activator, although at a lower rate than in its presence. The analysis of activation kinetics is very similar to inhibition analysis except that the effects are in the opposite direction! We may look at the relevant treatment given earlier on this topic (Sect. 21.4; Table 21.1). As with simple inhibitor mechanisms, the activator can bind either E (competitive), ES (uncompetitive), or both (noncompetitive). The allosteric activators of aspartate transcarbamylase belong to the noncompetitive category. Glucokinase is an example where the activator stabilizes the ES complex and activates the enzyme. Modulating glucokinase is an attractive therapeutic target for regulating type II diabetes. In another example, ligands have been identified to mimic the behavior of a protein-binding partner of phosphoinositide-dependent protein kinase 1. This binding leads to activation of the enzyme.

As a part of practical relevance of enzyme kinetics (Chap. 26; see Sect. 26.3), we alluded to enzyme-targeted inhibitors as drugs. Combining the chemical and kinetic tools, now we are in a position to understand how enzyme inhibitors may be rationally designed. This will form the subject for the rest of this chapter.

36.2   Enzyme Inhibitors as Drugs

The drug discovery process has evolved over time from serendipitous way to screening mode to rational design. The endeavor has built itself from accumulated earlier knowledge. The process is iterative and closely tied to structure–activity relationship study. Efforts to summarize the general requirements of good drug molecules have resulted in defining their desirable parameters viz., Lipinski rules (Leeson 2012) and quantitative estimate of drug-likeness. Once a lead compound/target enzyme has been identified, the process of screening takes over. Excellent strategies to design and execute high-throughput screening of molecular libraries are now available (Lloyd 2020). Such advances in instrumentation and assay designs support the measurement of kinetic parameters (association rate, dissociation rate, residence time, etc.) in real time. Greater reliance on inhibitor kinetic studies and structural biology methods have rationalized the new drug discovery. Of course, success of a molecule as a drug also depends on bioavailability, pharmacokinetics, toxicity, etc.

Briefly, the drug discovery process, for any potential enzyme inhibitor, involves the following steps:

Stage 1: Molecules (library of compounds—natural or synthetic) are screened and potential hits are evaluated for their action. The lead molecule identified is taken up for further characterization in terms of proposed binding mode, etc. The lead molecule is then subjected to optimization by evaluating many of its structural variants. A detailed structure–activity relationship analysis is performed to get an optimal structure. This molecule is again characterized for binding mode, biochemistry, and preliminary metabolic studies. Such a promising compound is taken up for synthesis and scaleup. Once a reasonable amount of this pure compound is made available, it is further taken up for detailed metabolism and preclinical (animal models!) pharmacokinetic properties.

Stages 2 and 3: These stages of drug discovery involve detailed preclinical studies involving in silico, in vitro, and in vivo experiments. This is where the drug molecule features like pharmacokinetics in terms of body clearance, availability and attaining sufficient plasma levels, first-pass effect, etc., are addressed. Modifications to the drug structure are explored to increase/decrease clearance rates, oral availability, etc. For example, enalaprilat is an excellent inhibitor of angiotensin-converting enzyme, but has poor oral absorption. However, its ethyl ester (enalapril) is effective and acts as a prodrug (McLeish and Kenyon 2003). A derivative of pitofenone (HL 752) works well by avoiding a liver esterase (the first pass effect) as a potent and long-acting antispasmodic agent (Bal-Tembe et al. 1997).

Stage 4: This step involves phase I, II, and III of clinical trials (in humans). Here again, the pharmacokinetics, dose, toxicity, and efficacy are evaluated in human volunteers.

Stage 5: In this final phase of drug development, the data is thoroughly reviewed and approved for use. The step also involves the post-market monitoring of drug effects.

No effective chemical entity as discovered is directly used as drug. It undergoes many chemical iterations before reaching the market. Typically, the entire process from concept to the market takes 12–15 years for a new drug.

As expected, the drug potency is often directly correlated with the kinetic parameters of the target enzyme (inhibitor k_{off} rates are better correlated than with its k_{on} rates, viz., slow tight binding inhibitors). The nature of enzyme inhibition also is a matter of choice. Competitive inhibitors are rarely useful because large concentrations are required for inhibition, and their inhibition is readily overcome by any buildup of substrate (drug resistance!). Molecules that show noncompetitive and uncompetitive inhibitors are better suited. However, designing drugs that act uncompetitively, particularly with single substrate enzymes, is a challenge. Inhibitors that target enzymes of bi-substrate reactions can be uncompetitive with respect to the bound substrate and block the subsequent chemistry of enzyme catalysis. Examples of such drugs include methotrexate (an inhibitor of dihydrofolate reductase), mycophenolic acid (inhibitor of inosine 5′-monophosphate dehydrogenase), and finasteride (inhibitor of steroid 5α-reductase). In these three cases the inhibitor binds target enzymes only after the redox cofactor has bound.

36.3 Isosters/Bioisosters

Unless an enzyme offers an additional site, distinct from the active site for binding, the molecular architecture of the inhibitor should resemble either the substrate or the product or something in between (an intermediate or the transition state). Although there are examples of inhibitors binding outside of the target enzyme active site, they are scarce (Wei et al. 2019). Substrate/product analogs are rarely useful as enzyme

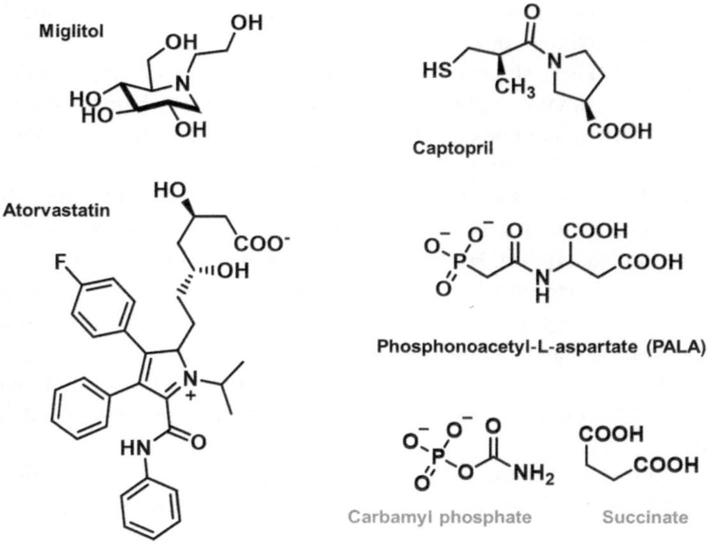

Fig. 36.1 Examples of structural analogs of substrates and products as enzyme inhibitors. Miglitol—α-glucosidase; Atorvastatin—3-hydroxy-3-methylglutaryl CoA reductase; Captopril—angiotensin-converting enzyme; Phosphonoacetyl-L-aspartate (PALA)—aspartate transcarbamylase (structures of carbamyl phosphate and succinate structures are shown for comparison)

inhibitory drugs but are often useful probes for determining enzyme specificity and mechanism (Fig. 19.3; Chap. 19). One could create isosteric structures by introducing unnatural functional groups and they may become more potent inhibitors (Meanwell 2023). A comprehensive sampling of carboxylic acid isosters and others may be found in this review by Meanwell. Typically, the use of an isostere converts the cleavable bond into a non-cleavable bond.

Creative ways of designing bioisosters includes incorporation of functional groups isosteric with carboxylic acids, sp^2 carbon, etc. In this regard, nitro and nitroso compounds, fluorine and fluorinated motifs, sulfoximine functionality, and boron derivatives have found utility in chemical design. For instance, iminosugars are sugar mimics where the ring oxygen of a typical monosaccharide is replaced by a nitrogen atom. The structural modifications can dictate the hybridization state of nitrogen from sp^3 (pyramidal) to sp^2 (planar), thereby affecting enzyme inhibition and biological activity. Miglitol (Glyset) (Fig. 36.1) is a potent inhibitor of the α-glucosidases in the digestive tract, is the first iminosugar-based drug for diabetes (Sanchez-Fernandez et al. 2020). Other examples are individually discussed below.

36.4   Multi-substrate Ligand Inhibitors

Multi-substrate analog inhibitors mimic the simultaneous binding of two or more substrates at the active site of the enzyme. These are clearly isosteric ligands except that they incorporate the structural features of more than one substrate of that enzyme. They compete with the corresponding substrates for the active site. An ideal bi-substrate analog inhibitor can bind up to 10^8 times more tightly than the product of the two substrate-binding constants. Typically, the K_I value for a bi-substrate analog inhibitor can be expected to approximate the product of the K_I values of the two substrate analogs. Linking the two groups effectively overcomes the unfavorable entropic barrier through reduced molecularity of the interaction. Because of the tightness of binding such multi-substrate ligand inhibitors may end up acting as slow, tight-binding inhibitors. Glycinamide ribonucleotide transformylase (GAR transformylase) catalyzes the transfer of a formyl group from N-formyltetrahydrofolate to glycinamide ribonucleotide. The bi-substrate inhibitor (β-thioGARdideazafolate) for this enzyme combines the features of both its substrates (the two substrate analogs linked by a stable thioether bridge) and had a K_I value of 250 pM. Two other examples of successful multi-substrate ligand inhibitors are atorvastatin (a 3-hydroxy-3-methylglutaryl CoA reductase inhibitor) and captopril (a competitive inhibitor of angiotensin-converting enzyme, and the first such inhibitor to be marketed) (Fig. 36.1).

Aspartate transcarbamoylase condenses carbamyl phosphate and L-aspartate to produce N-carbamyl-L-aspartate. N-Phosphonoacetyl-L-aspartate (PALA) combines two fragments, an analog of carbamyl phosphate and succinate (Fig. 36.1). The tight binding of PALA also suggests that it may be a potential transition-state analog. Thus, at times, it may be difficult to distinguish between the roles of a multi-substrate analog and a transition-state analog.

Clearly, there are two challenges of designing multi-substrate analog inhibitors— (a) the complexity of chemical synthesis and economy of its scaleup and (b) issues with the bioavailability of such elaborate structures (Broom 1989).

36.5   Special Functional Groups in Drug Design

36.5.1 Sulfoximine- and Phosphinate-Based Inhibitors

Both sulfoximine and phosphinate groups offer isosteric functionalities related to carboxylate and its activated derivatives. The sulfoximine moiety has a chiral, hydrophilic core that can accept H-bonds via the oxygen or nitrogen atoms, while N–H sulfoximines present an H-bond donor. The functionality was first discovered with methionine sulfoximine, whose N-phosphorylated derivative (transition-state mimic, formed in vivo) is a glutamate mimetic and potent mechanism-based inhibitor of glutamine synthetase. Subsequently, many similar sulfoximine moiety-bearing inhibitors have been developed (Hiratake 2005; Sirvent and Lecking 2017). The N-adenylated sulfoximine compound is a potent slow binding inhibitor asparagine

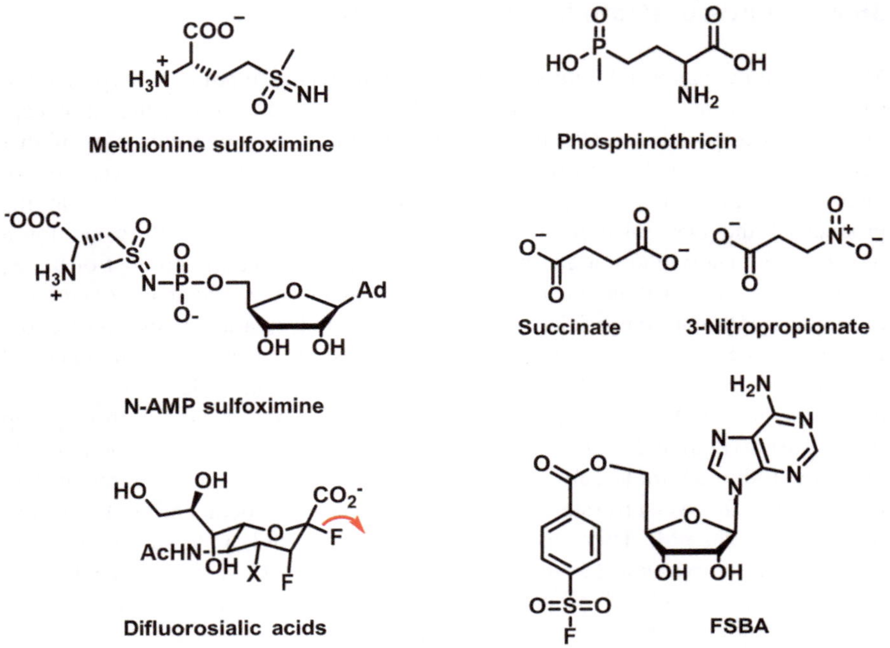

Fig. 36.2 Structures of inhibitors with sulfoximine, phosphinate, nitro, and halo functional groups. Methionine sulfoximine and Phosphinothricin—glutamine synthetase; N-AMP sulfoximine—asparagine synthetase A; 3-Nitropropionate—succinate dehydrogenase; Difluorosialic acids (X=OH, NH$_2$ etc.)—neuraminidase; 5′-p-fluorosulfonylbenzoyl adenosine (FSBA)—pyruvate kinase and other ADP/ATP requiring enzymes

synthetase A. It binds ($K_I = 67$ nM) the enzyme 25,000-times better than its substrates (Asp and ATP) (Fig. 36.2).

Another class of compounds that inhibit ATP-dependent ligases are phosphinate-based inhibitors. Both phosphinates (C–PO(OR)–C; derived from H$_2$PO$_3$) and phosphonates (C–PO(OR)$_2$; derived from H$_3$PO$_3$) contain C–P bond functionality. Phosphonamides are better mimics of peptide bond and are potential inhibitors of proteases/peptidases. Phosphinate anions are more basic than phosphonate anions and hence are better phosphorylated by ATP. Accordingly, phosphonate analogs are relatively weak inhibitors of ligases. In general, the inhibition potency is roughly proportional to the nucleophilicity of the inhibitor and the stability of its phosphorylated product. The sulfoximines (mentioned above), being more nucleophilic and less acidic nucleophiles, are therefore more potent inhibitors. The best example of a phosphinate-based inhibitor is phosphinothricin (Fig. 36.2)—an inhibitor of glutamine synthetase (Eisenberg et al. 2000). It is currently marketed as an effective herbicide. The phosphinyl oxygen of phosphinothricin is phosphorylated by ATP to form a phosphorylated phosphinate-bound tightly in the enzyme active site as a transition-state mimic.

The phosphorylation of the phosphinyl group (P–O$^-$) catalyzed by ADP-forming ligases is akin to the phosphorylation of the carboxy group. An ATP-dependent formation of a tightly bound phosphorylated species is generally observed with the phosphinate-based inhibitors in these enzymes viz., glutamine synthetase, D-Ala:D-Ala ligase, folylpoly-γ-glutamate synthetase, γ-glutamylcysteine synthetase, glutathionylspermidine synthetase, and many peptidoglycan biosynthetic enzymes (Mur ligases).

36.5.2 Nitro- and Nitroso-Group-Based Inhibitors

The nitro group closely resembles the carboxylate group. This similarity is further increased upon ionization of nitroalkanes to the nitronate state. The acidity of nitroalkanes is enhanced by electron-withdrawing substituents. The ionizable nitro compounds may serve as close analogs of carboxylic substrates for enzymes (Alston et al. 1983). This should permit the rational design of numerous potent and selective inhibitors of enzymes. Because it is more reactive than carboxy group, such inhibitors may often inactivate the enzyme. The nitro and nitroso compounds may serve as reversible inhibitors and in certain cases the ionized nitro compounds appear to function as "transition-state analogs."

Some nitro alkyl compounds like 3-nitropropanoic acid occur naturally. 3-Nitropropanoate is an isoelectronic analog of succinate (Fig. 36.2) and blocks cellular respiration by inhibiting succinate dehydrogenase. 3-Nitropropanoate establishes a chemical equilibrium with its conjugate base, the nitronate, by releasing a proton from the α carbon (Francis et al. 2013). This ionization is unusual in that it occurs with a pK_a value of 9.1 (shifted significantly down from a pK_a value of 25 for the deprotonation of CH bonds) and it occurs relatively slowly (requires minutes to hours to reach equilibrium). It is the nitronate form that irreversibly inactivates succinate dehydrogenase. A few other nitro analogs (as their nitronates) of carboxylic substrates and their target enzymes are listed in Table 36.1. The nitronate inhibitors of fumarase and aspartase possibly mimic their respective

Table 36.1 Nitro analogs as inhibitors of their target enzymes

Enzyme	Nitro compound (nitronate)	Remarks
Succinate dehydrogenase	3-Nitropropionate	Irreversible inactivation
Fumarase	3-Nitrolactate	Binds ~900 times tightly than malate
Aspartase	3-Nitroalanine	Binds ~1600 times tightly than aspartate
Aconitase	2-Hydroxy-3-nitropropane-1,2-dicarboxylic acid	Binds ~2700 times tightly than citrate
	1-Hydroxy-2-nitropropane-1,3-dicarboxylic acid	Binds ~72,000 times tightly than isocitrate; extremely tight-binding inhibitor ($K_I = 7 \times 10^{-10}$ M)

transition-state analogs, for nitronates are isoelectronic with α-carbanions of carboxylates. These enzymes probably abstract protons from the C-3 position of their substrates to transiently generate such α-carbanions. Similarly, aconitase is exquisitely sensitive to the nitronate analogs of citrate and isocitrate. In many such cases, the inhibitors are acting as extremely tight-binding inhibitors (Alston et al. 1983). Lastly, some of the nitro compounds, by virtue of being substrate isosters, can act as substrates for those enzymes.

36.5.3 Fluoro/Halo Group Containing Inhibitors

In terms of size, F atom is a good mimic of H atom and has only a small steric effect at the binding site. But due to its electronic properties, F enters into electrostatic and hydrogen-bond interactions. Depending on the context, it can thus be a versatile substitute and functions as a bioisoster of a C–H, C–OH, C=O, C–CN, or an S=O moiety (Gupta 2019; Wang et al. 2017). Fluorine substitution greatly increases the lipophilicity of the molecule. At present, the fluorinated drugs contribute about 20% of the drugs in the pharmacy. This is due to the special effects of fluorine chemistry. A prominent fluorinated motif extensively used in medicinal chemistry is the $-CF_2H$ moiety. This group has been deployed in a variety of structural backgrounds that take advantage of either its H-bond donor properties or its resistance to metabolic hydroxylation. A $-CF_3$ group can function as the carbonyl group of an amide and provide a metabolically nonbasic amine. Such a replacement can mimic tetrahedral transition-state analogs of ester and amide substrates. It can still have excellent hydrogen-bonding capability due to the strong inductive effect.

Atorvastatin (Fig. 36.1), the most prescribed drug given for the treatment of hypercholesterolemia, contains only one fluorine atom. Rosuvastatin, another member of this statin group of 3-hydroxy-3-methylglutaryl CoA reductase inhibitors, also contains a fluorine atom in its structure.

Suitably placed halo substituents can act as good leaving groups thereby covalently modifying/inactivating the enzyme. Besides the common chemical modification agents like iodoacetamide, bromoethylamine, diisopropylfluorophosphate (DIFP), phenylmethanesulfonyl fluoride (PMSF), etc., best example of an active site-directed irreversible inhibitor (affinity label) is N-4-toluenesulfonyl-L-phenylalanine chloromethylketone (TPCK) (Fig. 20.3; Chap. 20). Similarly, analogs of the various nucleosides and nucleotides are possible. 5'-p-Fluorosulfonylbenzoyl adenosine (5'-FSBA, Fig. 36.2) is an analog of ADP or ATP—is employed to label the NAD and ATP sites in enzymes. The difluorosialic acids (Fig. 36.2), as mechanism-based drugs against neuraminidase (sialidases), contain fluorine atoms that serve two important roles: (a) the fluorine at C-2 provides a good leaving group, and (b) the electronegativity of the fluorine at C-3 inductively destabilizes the oxocarbenium ion transition states for formation and hydrolysis of the covalent adduct, thereby slowing down hydrolysis.

Targeted covalent inhibitors are experiencing a growing resurgence in drug design (Feral et al. 2023). The discovery of new electrophilic groups that react selectively with specific amino acid residues is therefore highly desirable (De Cesco et al. 2017; Ray and Murkin 2019).

36.5.4   Organoborane Derivatives

Boronic acids can act as strong Lewis acids because of their open boron shell. They can be converted from a trigonal planar sp^2 geometry to a tetrahedral sp^3 geometry by substitution at boron. This interconversion mimics remarkably the formation of tetrahedral intermediates (of their carbon counterparts) of hydrolytic enzymes. The active site nucleophiles (viz., Ser, Thr) bind to the electron-deficient boron atom. Suitable boronic acid derivatives can thus act as potent inhibitors in the femtomolar range. Bortezomib, the first FDA-approved boronic acid-based drug (Fig. 36.3), inhibits the proteasome degradation pathway (Touchet et al. 2011).

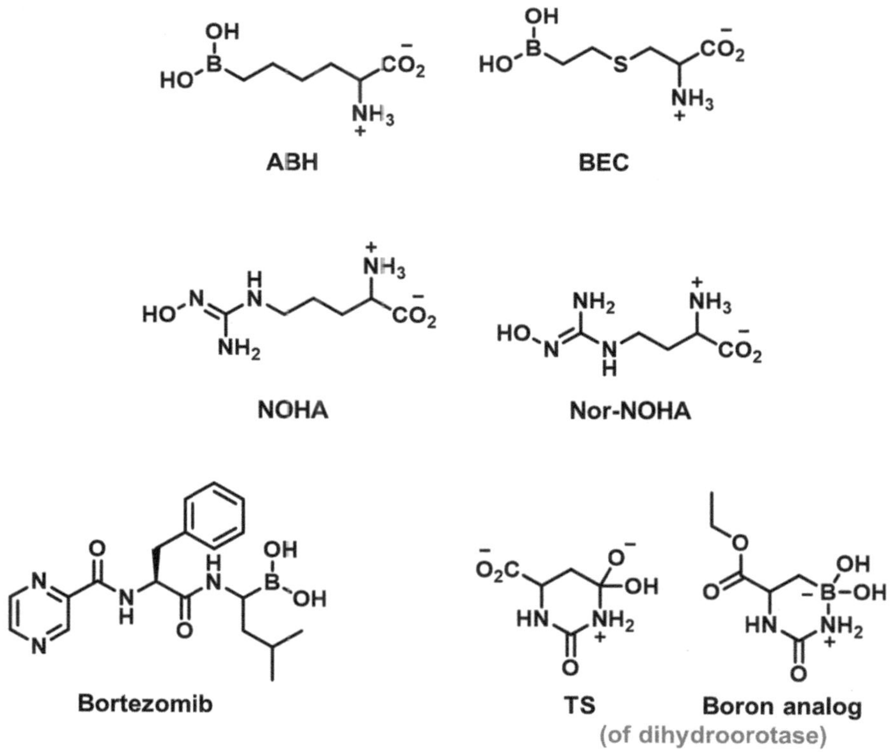

Fig. 36.3 Structures of boron-based inhibitors. Bortezomib—proteasome; 2-(*S*) amino-6-boronohexanoic acid (ABH) and *S*-(2-boronoethyl) L-cysteine (BEC)—arginase; boron-containing analog of carbamyl aspartate—dihydroorotase (unstable, tetrahedral-activated complex is shown for comparison). (NOHA and nor-NOHA are arginase inhibitors without boron, shown for comparison)

L-Arginine is a common substrate for arginase and nitric oxide synthase. Inhibition of arginase has implications to regulation of nitric oxide synthase and the biosynthesis of NO. The discovery of N^{ω}-hydroxy-L-arginine (NOHA), a stable intermediate during nitric oxide synthase catalysis, as an endogenous inhibitor of arginase led to synthesis of potential arginase inhibitors for therapy. Besides nor-NOHA (a lower homolog of NOHA) two boronic acids namely, 2-(S) amino-6-boronohexanoic acid (ABH) and S-(2-boronoethyl) L-cysteine (BEC) were designed as potent inhibitors of arginases (Fig. 36.3). The two boronic acids are competitive inhibitors of arginase but are slow-binding inhibitors at higher pH. The X-ray crystal structure of ABH-bound arginase reveals the electron-deficient boron atom attacked by the metal-bridging hydroxide ion to yield the tetrahedral boronate anion. This clearly mimics a chemical step of arginase catalysis.

A boron-containing analog of carbamyl aspartate rearranges to the stable, tetrahedral boronic acid derivative (Fig. 36.3; structure of its ethyl ester is shown) and mimics the tetrahedral-activated complex (TS of dihydroorotase; a highly charged, unstable sp^3 carbon species) of the dihydroorotase reaction. It has a tenfold greater affinity for the enzyme than that of the substrate carbamyl aspartate.

36.6 Suicide Substrates

This term was coined by Abeles and coworkers to emphasize active catalysis by enzymes of their own inactivation. A suicide substrate is built to exploit the high specificity of the enzyme, yet it contains features that generate reactive group/groups during catalysis to capture and inactivate the enzyme. This permits suicide substrates to be highly specific enzyme inactivators. The key features for such inhibition should include (a) occurrence of a catalytic step, (b) no release of the activated species before enzyme inactivation, and (c) an excellent partition ratio. While the first two (points a and b) are self-evident, the "partition ratio" needs an explanation. The partition ratio is the ratio of product release to enzyme inactivation and is a measure of the efficiency of the suicide substrate. It is an indicator of how many turnovers the enzyme goes through with the suicide substrate before getting inactivated. The most efficient suicide substrates will have partition ratio of zero i.e., the enzyme is killed in a single turnover (also see Sect. 20.3 in Chap. 20). Few examples of suicide substrates as drugs may be found in Table 20.2 and Table 26.4.

Rational design of suicide substrates is somewhat of a challenge because it draws heavily on the knowledge of the mechanism of catalysis by that enzyme—which is not always the case. Some of the successful drugs (Fig. 36.4) that covalently modify their target enzymes include—prostaglandin synthase inhibitor aspirin (acetylsalicylic acid); D-Ala–D-Ala carboxypeptidase inhibitor penicillin; UDP-N-acetylglucosamine-3-enolpyruvyltransferase inhibitor Fosfomycin; β-lactamase inhibitor clavulanic acid; GABA transaminase inhibitor vigabatrin; H^+, K^+ ATPase inhibitor omeprazole; and monoamine oxidase inhibitor selegiline. Many of them were released to the market without the knowledge that their mechanism of action

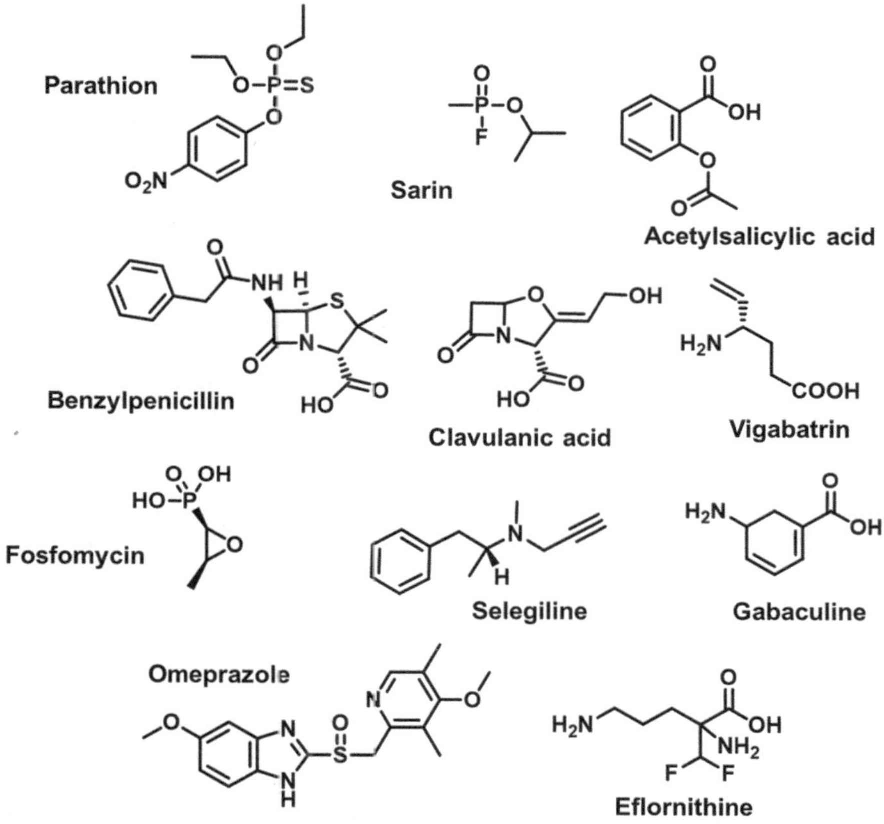

Fig. 36.4 Structures of suicide substrates. Parathion and Sarin—acetylcholinesterase; acetylsalicylic acid—prostaglandin synthase; Benzylpenicillin—D-Ala–D-Ala carboxypeptidase; Clavulanic acid—β-lactamase, Fosfomycin—UDP-N-acetylglucosamine-3-enolpyruvyltransferase; Selegiline—monoamine oxidase; Omeprazole—H⁺/K⁺ ATPase; Vigabatrin and Gabaculine—GABA transaminase; Eflornithine—ornithine decarboxylase

involved covalent target modification. These molecules were not designed a priori to act as covalent inhibitors.

Many of the pyridoxal phosphate (PLP)-dependent enzymes are well characterized and hence it was possible to successfully design suicide substrates for them. Mechanisms of GABA transaminase inhibition by gabaculine and vigabatrin (γ-vinyl-GABA) inhibition are well understood. The enzyme chemistry begins with PLP Schiff base formation. This is followed by removal of an α-proton by an active site base to form the reactive electrophilic intermediate. Subsequently, inactivation ensues when gabaculine gets aromatized during catalysis to trap the enzyme whereas vigabatrin forms a stable covalent enzyme adduct through Michael-type addition to its γ-vinyl group. Yet another currently used drug is eflornithine

(α-difluoromethylornithine, a $-CHF_2$ compound). It inactivates ornithine decarboxylase by generating electrophilic conjugated imine, after elimination of CO_2 and F^-.

Acetylcholinesterase is inhibited by parathion (insecticide) and sarin (nerve poison) (Fig. 36.4). They both react with the enzyme to form the active-site serine-phosphate esters. Hydrolysis of this covalent ester linkage is very slow, thereby making the inhibition effectively irreversible.

36.7   Transition-State Analogs

Pauling made the seminal remark that enzymes bind their transition states better than their substrates or products (Pauling 1946, 1948). Mimicking the structural and electronic features of the high-energy reaction intermediate or the transition state is a general and rational method for drug design. A good understanding of the reaction mechanism and of the transition state is thus a prerequisite. This is done by inspecting substrate/substrate analog and product/product analog structures in context and with detailed kinetic isotope effect study (Sect. 25.3 in Chap. 25). Transition-state analysis is further supported through computational approaches incorporating quantum mechanical and molecular mechanics (QM/MM) modeling/ simulation. We have emphasized the theory of how to assemble the probable transition-state structure earlier (in Chap. 25) but now elaborate how such analogs are built into potential drug molecules through examples. Some of them even turn out to be slow tight-binding inhibitors (Morrison 1982).

We have noted earlier on that C–P bond containing compounds can be good mimics of tetrahedral transition states. The N-methyl phosphonate derivative of D-glutamate (Fig. 36.5) resembles the tetrahedral intermediate that would be formed during N-acetyl-D-glutamate hydrolysis by a deacetylase. It is a potent competitive inhibitor with a K_I value of 460 pM. Tetrahedral and negatively charged phosphorus-containing peptide mimics are good inhibitors of metallo- and aspartic proteases. Transition-state inhibitors generally bind their enzyme targets with very high affinity (nanomolar to femtomolar, and theoretically even attomolar) and good specificity, and these have become predominant among active-site-directed enzyme inhibitor in clinical use. The inhibitor of HIV protease like saquinavir (Fig. 36.5) contains functionalities (hydroxyethylamine isoster moiety) that mimic the tetrahedral intermediate of the enzyme reaction. Likewise, pepstatin contains an unusual amino acid statin (see Fig. 4.9) which mimics the tetrahedral intermediate in pepsin catalysis. Captopril was the first rationally designed enzyme-targeted drug and is a transition-state inhibitor of angiotensin-converting enzyme. It contains thiolate and carboxylate groups (Fig. 36.1) that coordinate to the active site zinc and mimic the coordinated tetrahedral intermediate of the peptide substrate. The phosphonamidate peptide derivatives also mimic the tetrahedral transition state to inhibit metalloprotease like thermolysin.

Tetrahedral transition states are also found in dihydroorotase (see Fig. 36.3 above) and adenosine deaminase catalysis (see Fig. 4.9). Corresponding structural mimics are excellent inhibitors of these enzymes. The R isomer of

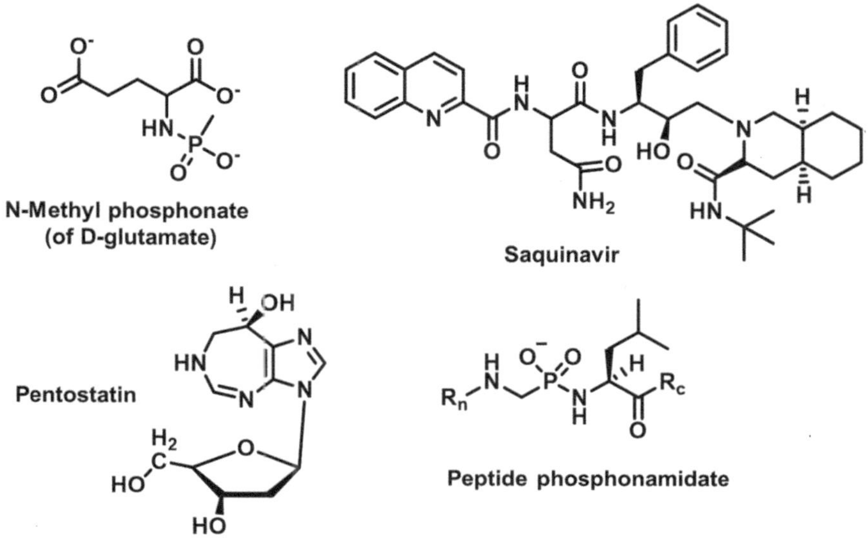

Fig. 36.5 Structures of some enzyme transition-state analogs. N-Methyl phosphonate derivative of D-glutamate-N-acetyl-D-glutamate deacetylase; Saquinavir—HIV protease; Pentostatin—adenosine deaminase; Phosphonamidate peptide derivative—thermolysin

2′-deoxycoformycin (pentostatin; Fig. 36.5) is a potent inhibitor of adenosine deaminase ($K_I = 2.5$ pM). The S isomer of the same tetrahedral transition state mimic is a much poorer inhibitor ($K_I = 33$ μM). This translates to a 10^7-fold difference in binding affinities and enzyme discrimination between the R and S isomers!

Excellent research on dissecting the transition states of enzymes to create potent TS analogs has come from Schramm and coworkers (Schramm 2015). The focus has been on 5′-methylthioadenosine/S-adenosylhomocysteine nucleosidases (Singh et al. 2005; Thomas et al. 2012) and phenylethanolamine N-methyltransferase (Mahmoodi et al. 2023). Femtomolar transition-state analog inhibitors were synthesized for the first enzyme. Knowledge of rate constants associated with such tight-binding inhibitors is closely related to drug design. Inhibitor residence times (off-rates) are better correlated with pharmacological efficacy rather than the equilibrium dissociation constants. Hence, experimental analyses of dissociation rates are a useful guide in optimizing their physiological relevance and action (Brown et al. 2023).

Transition-state analogs and substrate analogs share many structural similarities. Nevertheless, the enzyme active sites close around the transition-state structure to maximize affinity (Wolfenden 2003). This obviously involves precise conformational adjustments at the enzyme active site. For us, it should be possible to discriminate between an enzyme substrate analog versus the corresponding transition-state analog through kinetics. A change in the structure of a substrate may alter k_{cat}/K_M without altering the nonenzymatic rate of reaction. An analogous structural change in the transition-state mimic should bring about a similar change in its K_I. A linear relationship between the values of K_I for the transition-state analog and k_{cat}/K_M for the corresponding substrate is thus expected. Clearly, if K_I values for a series of isosters correlate strongly with K_M but not K_M/k_{cat}, then it indicates that these isosteric inhibitors are ground-state analogs.

36.8   Structure-Based Drug Design

Recent advances in structural biology and molecular modeling, combined with biochemical and mechanistic enzymology data, provide a potent mechanism for inhibitor discovery and optimization. Structure-based drug design programs mostly rely on available high-resolution protein crystal structures, particularly those bound to substrate/inhibitor. However, functionally active enzyme form(s) may be from one or more dynamic conformational states. While the need to account for protein and ligand flexibility is well appreciated, this drastically increases the computational time required. Several popular docking methods to bolster drug design and their discovery are now available (Simmons et al. 2010). Capturing most enzyme conformations, viz., through cryo-electron microscopy, may ease this situation (Chap. 6).

At one end of the spectrum, structures of substrates and known inhibitors are used to design novel molecules through complementarity of their interactions at the enzyme active site. The prospective leads are put through virtual screening protocols. Databases containing the structures of small molecules are used to dock them into a region of interest in silico and scored for best interactions within the target site. Such virtual high-throughput screening is assisted by different docking algorithms. Lastly, inhibitor scaffolds may be designed de novo. Functional groups (see above) and molecular fragments are optimally positioned in the active site of the target protein and then linked in silico to generate complete molecules. These molecules can then be scored and ranked for effectiveness. A typical procedure for de novo structural design of small molecules involves (a) define a protein structure, (b) identify binding site(s), (c) dock atoms, groups, or fragments to these sites, (d) suitably connect them to form a molecule, and (e) evaluate the complete skeleton for activity.

The power of structure-based drug discovery was first experienced through the success of captopril—an antihypertensive drug (Fig. 36.1). Its structure was rationally designed from a binding site model, based on the 3D information of carboxypeptidase A (a closely related zinc proteinase). The structural knowledge of the HIV protease enabled the successful design and development of saquinavir (Fig. 36.5), within a few years after its 3D structure was available (Kubinyi 1998). Another example of structure-based drug design is celecoxib—the cyclooxygenase 2 inhibitor.

36.9 Summing Up

Looking at the discovery of Molidustat, a drug candidate for the inhibition of HIF-prolyl hydroxylase, provides insights to the overall process of drug design and development (Beck et al. 2018). While screening (both real and virtual) of molecular libraries continues to be followed, rational drug design has picked up pace. In drug design, surprises are not uncommon! What are thought of as isosters or TS analogs may end up being slow tight-binding inhibitors. Many of the earlier drugs were released to the market without the knowledge that their mechanism of action. Enzyme inhibitor design is a hot research field as it encompasses frontier research on enzyme mechanisms as well as has direct relevance to pharmaceutical industry dealing with drug discovery. This chapter has covered a few examples (with an author bias!) of useful enzyme inhibitors from pharmaceutical and agriculture industry. Drug discovery is a subject of intense study and further details on many more cases may be had from the published literature.

Finally, ligand design (viz., enzyme inhibitor design) is not drug design! Many drug candidates may be highly active in vitro but inactive and in vivo. Pure enzyme active site-based optimization of structures neglects important biological properties like bioavailability and metabolic stability. Important parameters such as absorption, distribution, metabolism, and excretion need to be considered in the early phases of the drug discovery process.

References

Alston TA, Porter DJT, Bright HJ (1983) Enzyme inhibition by nitro and nitroso compounds. Acc Chem Res 16:418–424

Broom AD (1989) Rational design of enzyme inhibitors: multi-substrate analogue inhibitors. J Med Chem 32:2–7

De Cesco S et al (2017) Covalent inhibitors design and discovery. Eur J Med Chem 138:96–114

Eisenberg D et al (2000) Structure–function relationships of glutamine synthetases. Biochim Biophys Acta 1477:122–145

Francis K et al (2013) The biochemistry of the metabolic poison propionate 3-nitronate and its conjugate acid, 3-nitropropionate. IUBMB Life 65:759–768

Gupta SP (2019) Roles of fluorine in drug design and drug action. Lett Drug Des Discov 16:1089–1109

Hiratake J (2005) Enzyme inhibitors as chemical tools to study enzyme catalysis: rational design, synthesis, and applications. Chem Rec 5:209–228

Kubinyi H (1998) Structure-based design of enzyme inhibitors and receptor ligands. Curr Opin Drug Discov Dev 1:4–15

Leeson P (2012) Chemical beauty contest. Nature 481:455–456

Lloyd MD (2020) High-throughput screening for the discovery of enzyme inhibitors. J Med Chem 63:10742–10772

Meanwell NA (2023) Applications of bioisosteres in the design of biologically active compounds. J Agric Food Chem 71:18087–18122

Pauling L (1946) Molecular architecture and biological reactions. Chem Eng News 24:1375–1377

Pauling L (1948) Nature of forces between large molecules of biological interest. Nature 161:707–709

Ray S, Murkin AS (2019) New electrophiles and strategies for mechanism-based and targeted covalent inhibitor design. Biochemistry 58:5234–5244

Robertson JG (2005) Mechanistic basis of enzyme-targeted drugs. Biochemistry 44:5561–5571

Sanchez-Fernandez EM et al (2020) sp²-Iminosugars as chemical mimics for glycodrug design. Chapter 7. In: Small molecule drug discovery. Elsevier Inc, pp 197–222

Schramm VL (2015) Transition states and transition state analogue interactions with enzymes. Acc Chem Res 48:1032–1039

Simmons KJ, Chopra I, Fishwick CWG (2010) Structure-based discovery of antibacterial drugs. Nat Rev Microbiol 8:501–510

Touchet S et al (2011) Aminoboronic acids and esters: from synthetic challenges to the discovery of unique classes of enzyme inhibitors. Chem Soc Rev 40:3895–3914

Turberville A et al (2022) A perspective on the discovery of enzyme activators. SLAS Discov 27: 419–427

Wang B-C et al (2017) Application of fluorine in drug design during 2010–2015 years: a mini-review. Mini Rev Med Chem 17:683–692

Wei J et al (2019) An allosteric mechanism for potent inhibition of human ATP-citrate lyase. Nature 568:566–570

Further Reading

Bal-Tembe S et al (1997) HL 752: a potent and long-acting antispasmodic agent. Bioorg Med Chem 5:1381–1387

Beck H et al (2018) Discovery of Molidustat (BAY85-3934): a small-molecule oral HIF-prolyl hydroxylase (HIF-PH) inhibitor for the treatment of renal anemia. ChemMedChem 13:988–1003

Brown M, Tyler PC, Schramm VL (2023) Residence times for femtomolar and picomolar inhibitors of MTANs. Biochemistry 62:1776–1785

Copeland RA, Harpel MR, Tummino PJ (2007) Targeting enzyme inhibitors in drug discovery. Expert Opin Ther Targets 11:967–978

Feral A et al (2023) Covalent-reversible peptide-based protease inhibitors. Design, synthesis, and clinical success stories. Amino Acids 55:1775–1800

Mahmoodi N et al (2023) Cell-effective transition-state analogue of phenylethanolamine N-methyltransferase. Biochemistry 62:2257–2268

McLeish MJ, Kenyon GL (2003) Approaches to the rational design of enzyme inhibitors. In: Methods in drug discovery and discovering lead molecules, Burger's medicinal chemistry and drug discovery series. Wiley. https://doi.org/10.1002/0471266949.bmc015

Morrison JF (1982) The slow-binding and slow, tight-binding inhibition of enzyme-catalysed reactions. TIBS 7:102–105

Singh V et al (2005) Femtomolar transition state analogue inhibitors of 5-methylthioadenosine/S-adenosylhomocysteine nucleosidase from *Escherichia coli*. J Biol Chem 280:18265–18273

Sirvent JA, Lecking U (2017) Novel pieces for the emerging picture of sulfoximines in drug discovery: synthesis and evaluation of sulfoximine analogues of marketed drugs and advanced clinical candidates. ChemMedChem 12:487–501

Thomas K et al (2012) Femtomolar inhibitors bind to 5′-methylthioadenosine nucleosidases with favorable enthalpy and entropy. Biochemistry 51:7541–7550

Wolfenden R (2003) Thermodynamic and extra thermodynamic requirements of enzyme catalysis. Biophys Chem 105:559–572

Exploiting Enzymes: Technology and Applications

37

Much before enzymes were identified as discrete biochemical entities, they found favor through their useful properties. Early applications included use of enzyme preparations in meat tenderizing and starch hydrolysis. From the very beginning, commercial enzyme applications have largely belonged to a group of hydrolytic reactions. But a few oxidative enzymes were also exploited. While this trend holds even today, examples of designer enzymes and catalysts for more complex chemical processes are being developed. The first application of diastase (α-amylase) was by Jokichi Takamine. His 1894 patent (US Patent No. 525823) describes a process to make Taka-diastase from *Aspergillus oryzae*. This α-amylase was useful as a digestive aid, in eliminating starchy material from textiles and laundry. In a short but succinct paper, E. F. Leuchs (1931) described "the action of saliva on starch." The possible practical utility of such activity was clearly anticipated by him. The last line of his report reads "it will be possible to use saliva and gastric juice of killed animals very successfully in cases of defective digestion."

The quantity and quality of an enzyme are two critical parameters that define their application and extent of use. Industrial-scale processes require enzymes (often in crude form) in tons whereas precise clinical use mandates extreme purity and minimal or no contaminating factors. Accordingly, the enzyme production costs for different end objectives vary—they can be of high volume and low cost or low volume but of high cost. For instance, medically valuable products like streptokinase and asparaginase need to be very pure and are therefore expensive. Enzyme catalysts of practical import are sought by industries in many different ways. Significant among these are screening for useful activities from the naturally abundant diversity, modifying already available enzyme properties to suit our requirements, and genetically engineer desirable properties into these catalysts. We will briefly touch upon the applications of enzymes and industrial strategies with suitable examples in this chapter. Applications of enzymes and enzyme technology can occupy volumes and many authoritative books are available for the interested reader.

37.1   Natural Enzyme Diversity

The rich biodiversity on earth goes hand in hand with a naturally vast array of catalytic activities. A cleverly designed screen almost always leads to an enzyme with desired properties. Thermo-stable protease (from *Bacillus* strains) and DNA polymerase (from *Thermus aquaticus*) are two examples of enzymes chosen for high-temperature stability. The range of natural diversity is obvious from the number of enzymes that have found niche applications in the processing of carbohydrate polymers, proteins, and lipids.

37.1.1 Enzymes for Bioprocessing

Polysaccharides are the major biomolecules that comprise biomass on this planet. They serve two important functions—energy storage (such as starch) and structural rigidity (such as cellulose). It is therefore not surprising that enzyme technology took its roots through processes to hydrolyze these sugar polymers. Microbes (bacteria and fungi) constitute an abundant source of amylases and cellulases. Controlled hydrolysis of starch to sweeteners (and sugar substitutes) is a well-developed industry (Fig. 37.1). Various enzymes used in the starch saccharification process are α-amylases, β-amylases, glucoamylases, pullulanases, and glucose isomerase. Despite certain limitations, conversion of glucose to fructose through glucose isomerase is central to many sucrose substitutes—with distinct economic and manufacturing advantages.

Although there is an abundance of cellulose in nature, transforming cellulosic biomass into sugar has been a challenge. Concerted action of a bunch of enzymes (that constitute the "cellulase complex") is required for this (Payne et al. 2006). Significant advances in enzymatic processes to breakdown cellulose into fermentable sugars are being made (Beckham 2015). In the meanwhile, individual components (Table 37.1) of the cellulase complex have found application in textile and paper industry. More recently, the utility of auxiliary enzymes like lignin peroxidase, Mn-peroxidase, laccases, and lytic polysaccharide monooxygenases (LPMOs) are being explored to achieve more complete saccharification of biomass.

Proteases and lipases are next in order of significance in enzyme industry. Apart from the historical significance of papain and digestive enzymes (like trypsin and chymotrypsin), this class of enzymes have found wide ranging applications in foods, detergents, and tanning of leather. Bacteria and fungi are ideal sources for the large-scale production of proteases (Li et al. 2013). Most important alkaline protease producers are *Bacillus* strains and fungi belonging to genus *Aspergillus*. Subtilisin is the best-known bacterial protease additive of modern detergents. It has been extensively selected/modified for features like pH optimum and temperature stability. Chymosin (also known as rennin) is a milk-coagulating enzyme from calf stomach, which is used for generations in cheese making. An equivalent enzyme from a microbial source was sought and several *Mucor* strains were chosen to produce rennin substitutes.

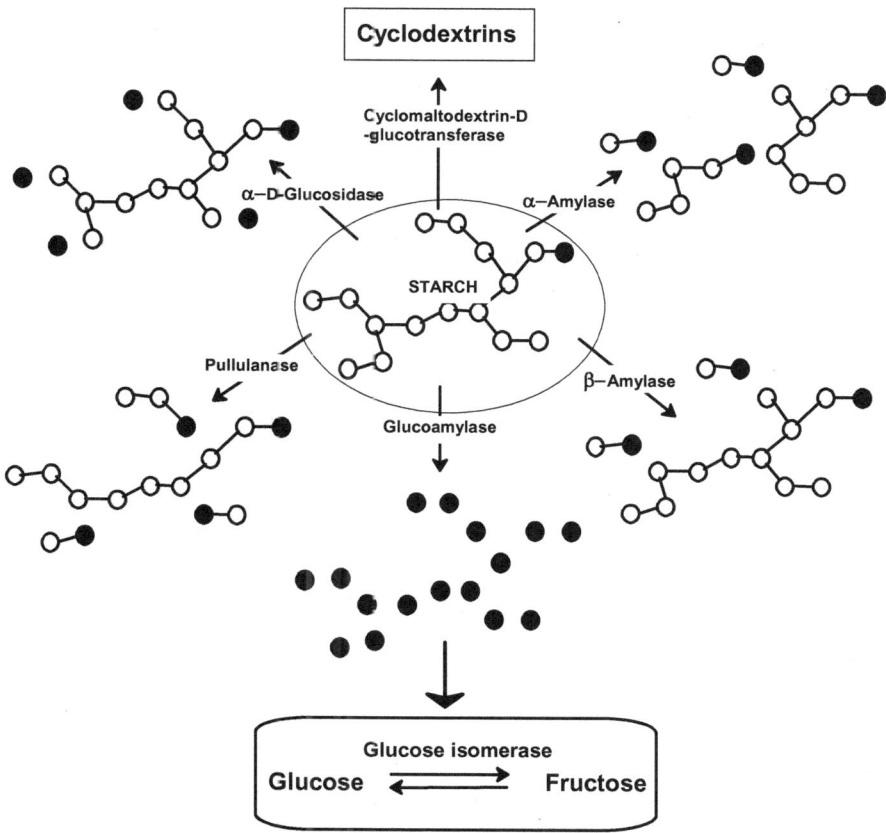

Fig. 37.1 Significant steps and enzymes employed in starch processing. Glucose residues of starch are schematically represented as circles. Filled circles indicate glucose residues whose C_1–OH has not entered into a glycosidic linkage (free reducing ends). Besides glucose isomerase, all the possible enzymatic modes of dismantling starch are shown. Some combinations of these enzymes are commercially available as industrial formulation

Many other enzymes including lipases and pectinases are also available on the industrial scale. A representative list of enzymes commonly used in industry is given in the table below (Table 37.2).

37.1.2 Enzymes in Pharma and Medical Applications

Pharmaceutical industry is another big beneficiary of applied enzymology. Enzymes isolated from natural sources as well as those cloned and expressed (through genetic engineering) are in use. Enzymes and their critical study serve multiple purposes in drug discovery and development.

Table 37.1 Component activities of cellulase complex and their applications

Cellulase component	Substrate specificity	Application
β-Glucosidase (cellobiase)	Cellobiose → glucose	Saccharification
Cellobiohydrolase I (CBH1)	Cellulose → cellobiose (exo—non reducing end)	Biomass conversion
Cellobiohydrolase II (CBH2)	Cellulose → cellobiose (both exo and endo)	Biomass conversion
Endoglucanase I (EG1)	Cellulose (endo)	Textile/fabric softening, biopolishing
Endoglucanase II (EG2)	Cellulose (endo)	Textile/fabric softening, biopolishing
Xylanase	Xylan	Paper pulp de-inking
All components	Cellulose and xylan	Feed/fodder, biomass conversion

Table 37.2 Large-scale use of enzymes in industry

Enzyme	Application
Acting on carbohydrates	
• Amylases	Starch processing
• Cellulase complex	Biomass conversion, textile industry
• Pectinases, esterases	Food industry, fruit juice, brewing
• Glucose isomerase, invertase	High fructose syrups, invert sugar
Acting on proteins	
• Papain, pepsin	Meat and leather processing, treating dough
• Rennin, chymosin	Cheese making
• Subtilisin	Detergents, leather and wool processing
Acting on lipids and esters	
• Lipases	Food and detergent industry, cocoa butter
Acting on antibiotics	
• Penicillin acylase	Produce 6-aminopenicillanic acid (6-APA)

- Active principles of many effective drugs are enzyme inhibitors (Robertson 2005). An enzyme, critically located in the intermediary metabolism, may provide an excellent target to screen for such inhibitors. A few successful examples of drugs have panned out from such enzyme screens.

The concept of screening for enzyme inhibitors was first adopted by Hamao Umezawa's group in Japan (Umezawa 1982). Since then, many enzyme inhibitors have been discovered (few are listed in Table 37.3) and are in use. For instance, preventing absorption of dietary fat (triglycerides) can be a possible strategy to control obesity. An appropriate lipase from the digestive juices could serve as a target for this screening (Fig. 37.2).

Often, the active chemical entity obtained from an enzyme screen may not find direct application. These lead compounds (inhibitors) are suitably altered/derivatized

Table 37.3 Examples of enzyme-targeted screens for active principles

Enzyme target	Screening outcome	End use
Pepsin	Pepstatin	Ulcers
Angiotensin-converting enzyme	Captopril	Hypertension
HMG CoA reductase	Lovastatin	Hypercholesteremia
α-Amylase	Acarbose	Diabetes
Triacylglycerol lipase	Orlistat (lipostatin)	Obesity
Acetylcholine esterase	Rivastigmine	Alzheimer's disease
β-Lactamase	Clavulanic acid	Combination therapy

Fig. 37.2 Flow chart outlining the design of a lipase inhibitor screen

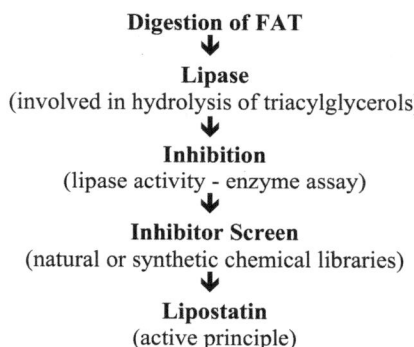

Digestion of FAT
⬇
Lipase
(involved in hydrolysis of triacylglycerols)
⬇
Inhibition
(lipase activity - enzyme assay)
⬇
Inhibitor Screen
(natural or synthetic chemical libraries)
⬇
Lipostatin
(active principle)

to achieve better bioavailability and reduce toxicity. An in-depth kinetic analysis of enzyme inhibition (concluded in Chaps. 26 and 36) is at the heart of modern drug discovery programs.

- Besides being targets for inhibitor screens, many enzymes are employed as catalysts for synthesis. The whole range of β-lactam antibiotics available today includes a large number of semisynthetic penicillins and cephalosporins. While penicillin G is produced by fermentation, 6-aminopenicillanic acid—an important precursor for semisynthetic penicillins—is derived from it (Fig. 37.3). Penicillin acylase is a valuable commodity in the large-scale production of 6-aminopenicillanic acid.

- Because of their catalytic potential coupled with specificity, many enzymes are used as exquisite analytical tools. Alkaline phosphatase and peroxidase are two reporter enzymes of extensive history in ELISA (enzyme-linked immunosorbent assay). Here, the desired specificity of interaction (through antibodies) is coupled to the signal amplification provided by enzyme catalysis. Enzymes as antibody-conjugates find routine use in detection of DNA/RNA/protein on blots. *Taq* DNA polymerase is extensively employed for DNA amplification through polymerase

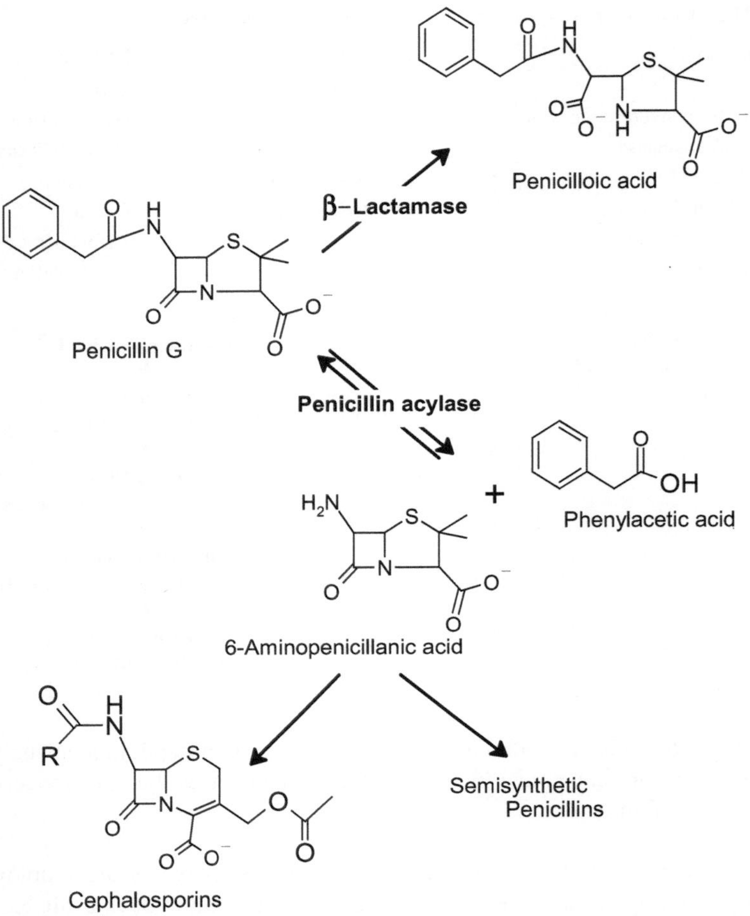

Fig. 37.3 Enzymes and steps relevant to penicillin (β-lactams) industry. Antibiotic resistance is often due to a β-lactamase; better antibiotics may be evolved by screening for novel structures that are not acted upon by the β-lactamase. Penicillin acylase is used to produce 6-aminopenicillanic acid (6-APA). 6-APA is an important precursor to make semisynthetic penicillins, both through enzymatic and chemical routes

chain reaction (PCR). A number of metabolites are analyzed (in a clinical setting) through assays involving enzymes. A list of more commonly used enzymes and the corresponding analytes are given in Table 37.4. Some of these enzymes are also employed as components of biosensors (see below).

• Enzymes have found medical applications in terms of diagnosis as well as therapy. A few enzymes find direct application as therapeutic agents in medicine.

Table 37.4 Examples of enzymes for metabolite analysis

Enzyme	Analyte detected/estimated
Catalase	Hydrogen peroxide
Glucose oxidase	Glucose
Hexokinase	Glucose
Alcohol dehydrogenase	Ethanol
Lactate dehydrogenase	Lactate/pyruvate
Luciferase	ATP
Urease	Urea
Cholesterol oxidase	Cholesterol

Table 37.5 Examples of enzymes as clinical markers

Enzyme	Used as marker for
Lactate dehydrogenase (H4 isoform)	Heart diseases
Glutamate-oxoglutarate transaminase (SGOT), glutamate-pyruvate transaminase (SGPT)	Liver function
Creatine kinase	Myocardial infarction, skeletal muscle damage
Lactase	Lactose intolerance
Hexosaminidase A	Tay-Sachs disease
Acid phosphatase	Prostate cancer
Phenylalanine hydroxylase	Phenylketonuria (PKU)

Best known examples from the market include diastase (α-amylase; digestive aid), asparaginase (leukemia, antitumor therapy), rhodanase (cyanide poisoning), and streptokinase (medication to dissolve blood clots). However, many enzymes are routinely monitored as clinical markers (Table 37.5). Enzyme profiles from serum, amniotic fluid, urine, etc., are monitored and their levels are often correlated with disease conditions and aid in the diagnosis of disease.

37.1.3 Enzymes and Issues of Safety

Industrial preparation and use of enzymes comes with its own safety and regulatory issues. Potential hazards from exposure to large quantities of a given enzyme include allergenicity, functional toxicity, chemical toxicity, and source-related contaminants. Large-scale manufacture of enzymes often ends with partially pure yet enriched preparations. Such material could contain potentially toxic chemicals (like mycotoxins) carried over from the source. Not all microorganisms are safe and their trace contamination in the final enzyme preparations requires attention. For enzymes to be used in food ingredients, they must be GRAS (*Generally Regarded As Safe*). Some enzymes like proteases are potentially dangerous—particularly upon exposure of sensitive tissues to concentrated preparations. Since enzymes are proteins, they can be potent allergens. Repeated exposure through inhalation or skin contact can trigger severe allergic responses. Enzyme preparations especially handled in the form of dust, dry powder, or aerosol are harmful and must be avoided. Many issues of safety regarding free enzyme preparations may be overcome by using them in the immobilized, granulated, or encapsulated form.

Many fungal and bacterial enzymes are routinely used in the food processing industry today. The safety aspects of these ingredients are of importance. The Enzyme Technical Association (ETA) periodically updates this list and maintains it on their web site, http://www.enzymetechnicalassoc.org/. Microbial enzymes are typically sold as preparations—with preservatives and stabilizers—that contain not only the desired enzyme activity but also other metabolites from the production strain. Thoroughly characterized nonpathogenic, nontoxigenic microbial strains are given the GRAS status. In turn, these organisms are the logical candidates for generating a safe strain lineage. Improved strains may be derived from them through traditional/classical or rDNA strategies. Enzyme safety evaluation mechanisms have been proposed to accommodate advances in applied enzymology (Pariza and Johnson 2001; Sewalt et al. 2016).

37.2 Modifying Enzymes to Suit Requirements

Despite the vast natural diversity of biological catalysts, significant technology has developed to alter the properties of available enzymes. This tinkering has involved facets of their immobilization, chemical modification, genetic engineering, or their use in nonaqueous solvents.

37.2.1 Immobilization for Better Use

Most natural enzymes isolated are in water-soluble state. They cannot be stored in this form for long, often due to instability. Their immobilization is one way to enhance their shelf life (Mateo et al. 2007). In addition, immobilized enzymes are easy to recover and amenable to repeated use. This is an important consideration when the cost of enzyme is very high. The characteristics of the matrix are very critical in determining the performance of the immobilized enzyme system. These supports may be inorganic or organic according to the nature of their chemical composition. The physical characteristics like mean particle size, swelling behavior, mechanical strength, etc., decide the technical conditions in which the system is used. Enzymes may be quarantined on the matrix either irreversibly (by covalent bonding, entrapment, microencapsulation, cross-linking, etc.) or reversibly (by adsorption, ionic binding, affinity binding, disulfide bonds, or chelate/metal binding). The cost associated with the process of immobilization determines whether it is economically viable to do so. Enzyme immobilization is an important innovation to achieve biocatalyst stability during the scale-up process. Good strategies for immobilization are still needed to facilitate optimal use of enzymes in terms of process development and economics of scaleup. The success of new-generation immobilized biocatalysts will be defined by critical inputs of protein

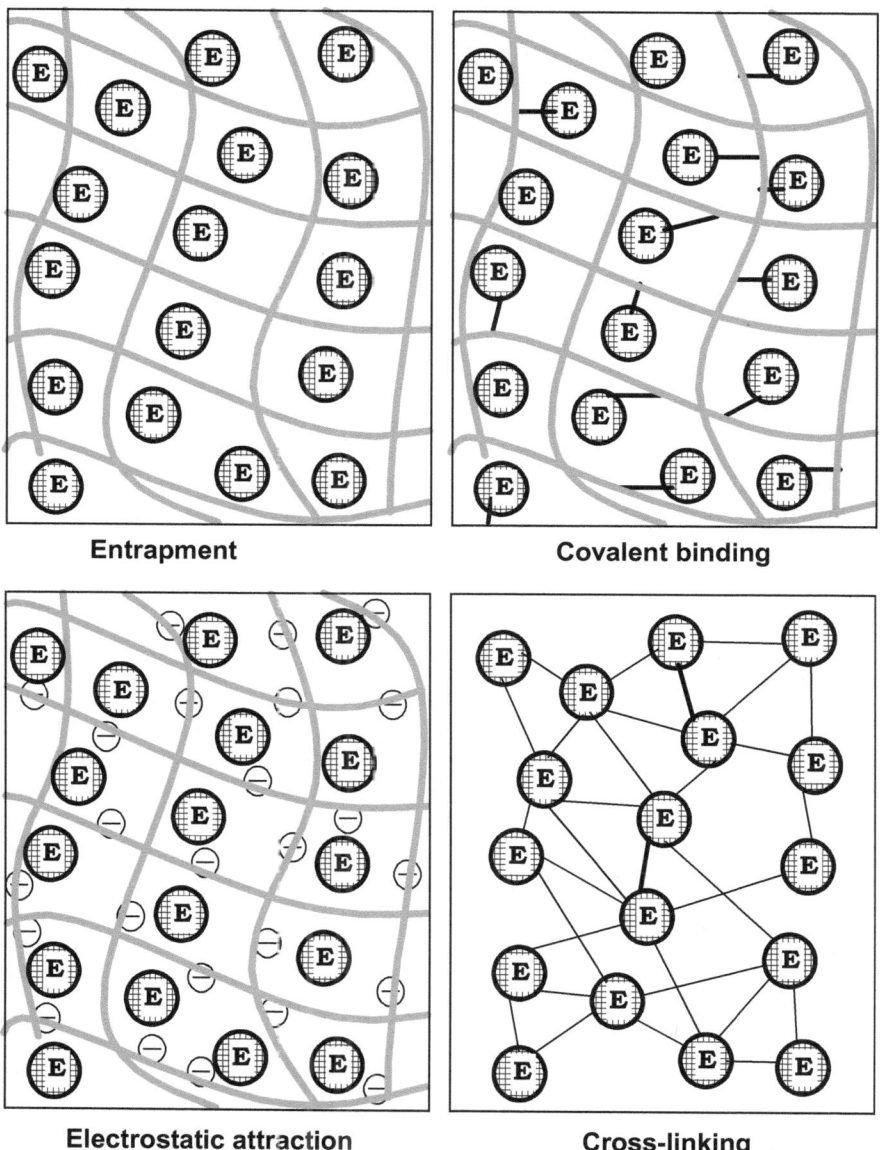

Fig. 37.4 Four different modes of enzyme immobilization

engineering, enzyme immobilization, process engineering, and life cycle analysis (Chapman et al. 2018).

Different means are adopted for enzyme immobilization in practice and only a few are represented in Fig. 37.4. The field of enzyme immobilization technology and

Table 37.6 Industrial uses of immobilized enzymes

Enzyme	Product
Aspartate ammonia lyase	L-Aspartic acid
Hydantoinase	L- and D-amino acids
Thermolysin	Aspartame
Lactase	Lactose-free milk and whey
Invertase	Invert sugar
Glucose isomerase	High-fructose syrup
Lipase	Cocoa butter substitute
Penicillin acylase	6-APA and penicillins
Nitrile hydratase	Acrylamide

its applications has grown vastly over the years. Many books and volumes (Methods in Enzymology series) are available for detailed reference (Brena and Batista-Viera 2006). This applied aspect of enzymes is briefly covered here, and the reader is encouraged to refer to the more specialized literature.

The choice of immobilization method depends on the type of enzyme and the nature of applications in question. Non-covalent confinement (like physical entrapment, micro-encapsulation, or electrostatic adsorption) methods at times may lead to enzyme leaching during operation. Covalent anchoring of enzymes on the other hand requires bifunctional cross-linking reagents and suitable functional groups on the enzyme surface. These functional groups must not be critical for enzyme activity, however. A great deal of sophisticated chemistry has been developed to activate inert organic/inorganic polymers for subsequent enzyme immobilization. Carrier-bound (covalently linked to polymers; Figure above) penicillin acylase is highly effective in the preparation of 6-APA and permits economic recycling of the catalyst. Penicillin acylase-immobilized cassettes are available that function at >99% conversion efficiency even after 1500 cycles of use. Glutaraldehyde is used to cross-link glucose isomerase—the cross-linked material can be reused many times in the commercial process for production of fructose syrups from glucose. Detergent protease components act as allergens and this problem was overcome through microencapsulation; the dustless protease preparations have reduced this risk. Some of the enzymes successfully immobilized for application are shown in Table 37.6.

It is expected that the substrate specificity and the catalytic potential of the enzyme should not be unfavorably affected upon immobilization. The changed microenvironment could affect the enzyme stability to heat, pH, and proteolytic enzymes. A reduction in its activity may also occur due to associated conformational changes in the enzyme. The pH optimum of an immobilized enzyme may change by as much as 2.0 pH units due to microenvironment effects. It is observed that an anionic carrier (matrix) pushes the pH optimum to a more alkaline value while a cationic carrier moves it to a more acidic value. These matrix (electrostatic field) effects are due to changes in the degree of ionization of amino acid residues on the enzyme; this effect is usually not observed in medium with high ionic strength—as salt ions counter the charges on the carrier.

The kinetic behavior of an immobilized enzyme can differ significantly from the free enzyme in solution. Both maximal velocity (V_{max}) and Michaelis constant (K_M—apparent affinity for the substrate; see Chap. 14 for a detailed treatment) of the enzyme may be affected. The apparent K_M significantly decreases when the carrier used is of opposite charge to that of the substrate—largely due to electrostatic interactions causing the substrate to be at a higher concentration near the carrier. In general, partition effects (arising from ionic, electrostatic, or hydrophobic interactions) cause changes in the local concentration of the ligands/molecules which in turn affect the kinetic constants. The apparent K_M of an immobilized enzyme is also affected by diffusion factors. Both the diffusion of the substrate to and the diffusion of the product away from the enzyme influence enzyme activity. Restricted diffusion of bulk substrate to the immobilized enzyme leads to significantly higher K_M value. Poor product diffusion, away from the enzyme, obviously results in an inhibited enzyme. If diffusion constrains the highest substrate concentration achievable, then the immobilized enzyme may display an apparent V_{max} lower than the free soluble enzyme.

The objective of using enzyme-catalyzed reactions in industry is to convert given amount of the substrate maximally into the product. The integrated form of the rate equation (Michaelis–Menten equation) is better suited to analyze such systems. However, the integrated form of the rate equation is also valid only if— (a) substrate concentrations far exceed that of the enzyme, (b) the overall reaction is irreversible, (c) enzyme is stable over the time period, (d) no product or substrate inhibition occurs, and (e) the system remains properly mixed (see Chap. 16 for a detailed treatment). Some of these factors can be tackled through choice of enzyme reactor configuration and process design. Continuous industrial processes could employ immobilized enzyme in stirred-tank, packed-bed, or fluidized-bed reactors or the enzyme immobilized on membranes, hollow fibers, or tubes. The flow of the reaction mixture may be exploited to input fresh substrate, remove the product formed or adjust pH, etc. Membrane cassettes of penicillin acylase have made the production of 6-APA very efficient—the reaction mixture pH can be maintained by dosing NaOH (pH stat) and continuously removing the product formed.

37.2.2 Enzyme Biosensors

Biosensors are used to determine/monitor the concentration of substances (often metabolites) of biological interest. Enzymes play a major role in such analytical devices—to convert a biological response into an electrical signal. Specificity and signal amplification (catalysis) of an immobilized enzyme are ideally suited for this purpose. Besides the fact that they may be reused, the immobilized enzymes (with an elevated K_M value) give proportional change in reaction rate over a substantial linear range of the substrate concentrations. Also, often these rates are independent of pH, temperature, ionic strength, and inhibitors—features advantageous in metabolite measurements in real analytical samples.

The critical component of the biosensor is the transducer which converts the outcome of an enzyme reaction into a measurable signal. Biosensors may have a transducer to exploit enzyme reactions with (1) heat generation (calorimetric biosensor—to measure H_2O_2 with catalase), (2) release or absorption of ions (potentiometric biosensor—glucose with glucose oxidase and urea with urease), (3) production of a current (amperometric biosensor—glucose with glucose oxidase, alcohol with alcohol oxidase, and cholesterol with cholesterol oxidase), and (4) absorption or emission of light (optical biosensor—peroxides with horseradish peroxidase and ATP with luciferase). Paper enzyme strips are also in use to measure/ detect substances through colorimetry. Self-powered biosensors that use a flavoenzyme or a $NAD(P)^+$-dependent enzyme as an electro-biocatalyst have been developed. The advantages of such self-powered biosensors are no external voltage is to be applied, biological fluids support implanted invasive sensing device, the operation of the biosensor device is specific, and a single concentration of the substrate is enough for calibration (Katz et al. 2001). Enzyme biosensors today occupy a substantial analytical market in health care, food industry, and environmental monitoring.

37.2.3 Function in Organic Solvents—Nonaqueous Enzymology

Living systems are intimately linked with their aqueous environment. Being biological catalysts, enzymes are exquisite products of long biological evolution. Enzymes are easily denatured by organic solvents. However, some enzymes can tolerate high concentrations of water-miscible organic solvents in their aqueous surroundings. Even in a water-immiscible nonpolar solvent, enzymes do need a monolayer of water molecules covering their exposed surfaces and the active site (Halling 2004). In this arrangement—with an essential water layer—they can continue to function as catalysts even in an organic solvent. The early work of Bourquelot and others (since 1913) showed that few enzymes could act in the presence of >80% of organic solvents such as ethanol or acetone. This example may very well be the forerunner of nonaqueous enzymology and use of enzymes in organic solvents for synthetic applications. The synthesis of a glucoside (by maltase, in 1898) highlighted very early the synthetic capabilities of an enzyme.

Enormous interest in biocatalysis in nonaqueous phase was triggered due to the merits of good enantioselectivity, reverse thermodynamic equilibrium, and no water-dependent side reactions. For most organic reactions, water is not an ideal solvent. But enzymes have evolved for the catalysis of reactions in water. It would be very useful to have enzymes perform catalysis in nonaqueous media. There are added advantages if this is made possible:

- Many substrates are more soluble in organic solvents than in water. Large initial concentrations can be achieved with hydrophobic substrates that are sparingly soluble in water. They continuously diffuse into the active site from the bulk solution (which is nonaqueous). Some products are labile in aqueous media and thus water-dependent side reactions can be minimized.

- Enzymes naturally find themselves in an aqueous environment with 55.5 M of water. Introducing them into nonaqueous environment means a drastically reduced water concentration! There are important consequences of lower water activity—particularly with reactions in which water is added or removed. For instance, peptide bond synthesis involves elimination of water—a thermodynamically uphill task in water. However, this step is favored in a nonaqueous environment. Consequently, a hydrolytic enzyme could be endowed with synthetic potential. Most hydrolytic reactions—generally irreversible in water—can be made reversible. Esterases, glycosidases, and proteases can be used for synthesis.

All water-soluble enzymes possess a small but significant amount of strongly bound water—thereby resulting in a two-phase system within the nonaqueous bulk medium. The *water activity*—a_w—is defined as the partial vapor pressure of water in a substance divided by the standard-state partial vapor pressure of water. It is indicative of the extent of water content around the enzyme molecules. Enzyme activity in nonaqueous media depends on the magnitude of a_w because it affects the extent of this bound water. Lower a_w may lead to a rigid enzyme with limited thermal motion and associated thermal stability. The miniscule water pool around the enzyme in an organic solvent (between 50–500 water molecules per enzyme molecule) retains the pH of the last aqueous solution from where it is derived. Therefore, it appears as if the enzyme remembers and functions in that pH (*pH memory*).

The enzyme is inactivated if the tightly bound water layer is stripped off or diluted by the organic solvent phase. The stability or inactivation of an enzyme is thus dictated by the polarity of the solvent used (Wang et al. 2016). A useful measure of this polarity is **log *P***—the logarithm of the partition coefficient of the organic solvent (X) between *n*-octanol and water. The partition coefficient *P* is a measure of hydrophobicity of organic solvent. The lower the *P* value, the more polar (hydrophilic) is the solvent.

$$\log P = \log\left(\frac{[X_{\text{Octanol}}]}{[X_{\text{Water}}]}\right).$$

As an empirical rule, enzymes are generally inactivated by solvents with log $P < 2$ but are little affected by more hydrophobic solvents with log $P > 4$.

Industrial focus on the use of enzymes in nonaqueous environments and reverse micelles systems is on the rise, with applications in foods, medicine, and industry. Being chiral catalysts, they are often valuable in resolving racemic mixtures of important drugs and intermediates in pharmaceutical industry. In practice, it is common to find applications of nonaqueous enzymology in transglycosylation, transesterification, and transpeptidation reactions (Klibanov 2001). An important example of the reversal of peptide bond hydrolysis is aspartame (α-L-aspartyl-Lphenylalanyl-O-methyl ester) synthesis. The protease thermolysin is used to condense L-aspartic acid with the methyl ester of L-phenylalanine and produce aspartame (sweetener and a sugar substitute). Similarly, suitable glycosidases are used to synthesize cyclodextrins. Inter-esterification reactions with lipases in nonaqueous solvents have found many applications. Consider an esterase catalyzing the hydrolytic reaction:

$$R\text{-}COOX + H\text{-}OH \rightarrow R\text{-}COOH + X\text{-}OH.$$

The same enzyme, in the presence of another alcohol (Y–OH), but in the absence of water (low a_w), can bring about efficient catalysis of an ester exchange shown below.

$$R\text{-}COOX + Y\text{-}OH \rightleftharpoons R\text{-}COOY + X\text{-}OH.$$

For example, isoamyl acetate (a banana fruit flavor) is produced from ethyl acetate and isoamyl alcohol. Yet another interesting, patented example of interesterification with *Mucor miehei* lipase is the preparation of cocoa butter substitute from palm oil (Harwood 1989).

37.3 Biocatalysis and Green Chemistry

Biotransformation and bioconversion through enzymes have found a firm place in the emerging biobased economy. Green chemistry has numerous applications as an alternative to chemical catalysis. The use of enzymes in organic synthesis, especially to make chiral compounds for pharmaceuticals, flavors, and fragrance, has substantially increased; biocatalysts are used on a large scale to make specialty and bulk chemicals. Immobilized penicillin acylase (Fig. 37.3 above) is an excellent example of enzyme use in producing an active pharmaceutical ingredient (API).

Development of enzyme-based synthetic processes in the industry has paralleled the progress made in industrial microbiology. Microbial enzymes have been at the forefront of applications. Screening and selection of such enzymes is driven by the proposed end use. Other concerns in their choice are substrate specificity, novelty of source, better features like stability towards heat, pH, and solvents. Biocatalyst development for specific industrial chemical reaction must satisfy a minimal set of criteria. The process should be sustainable and with good atom economy (minimal wasteful byproducts) while the starting substrate should be easily accessible. Second, the enzyme should be robust and readily recycled. The catalyst should be able to support high conversion/yield (>90% conversion) and volumetric productivity while achieving excellent enantioselectivity (>99% ee). These are stiff criteria, but experience shows that they are reachable and that is what makes the process competitive with the conventional chemical synthesis (Groger and Asano 2012).

A broad range of chemistries like hydrolytic reactions and redox reactions are effectively carried out with enzymes today (Winkler et al. 2021; Wu et al. 2021). Chemistry of C–C bond formation (and cleavage) is important in synthesizing drug molecules. The thioester condensation and thiolysis of 2-keto acids are central to the metabolism of fatty acids, polyketides, and in the non-ribosomal peptide synthesis; carbon chain elongation can be designed through suitable enzyme domains (Zhou et al. 2020). Aldolase and transaldolase reactions offer opportunities to build complex carbon structures (Voutilainen et al. 2020). Chiral amines and amino acids feature predominantly in APIs. Asymmetric reductive amination (Sharma et al. 2017) and transamination (Guo and Berglund 2017; Slabu et al. 2017; Kelly et al. 2018) are some of the most valuable reactions in the preparation of APIs. An excellent example is the asymmetric synthesis of sitagliptin, a chiral amine derived from the corresponding ketone (Savile et al. 2010).

The development of a unique catalyst typically begins with a search of available enzyme libraries. If the library does not contain the catalyst of interest, one may look for a closer match (in terms of reaction mechanism, specificity, etc.). We may have to engineer it for a specific substrate shape or even identify/evolve a new biocatalyst for a substrate for which no catalyst is available at all. Today, this activity is amply supported by bioinformatic tools in combination with mutational approach, rational design or directed evolution, and high-throughput screening tools. The development of a unique catalyst for sitagliptin synthesis provides a tour de force in purposeful development of a desired biocatalyst. Starting from an enzyme that had the desired catalytic machinery to perform expected chemistry, a transaminase with a marginal activity was created. This variant was then further reengineered via directed evolution for practical use in sitagliptin synthesis. In an earlier study, subtilisin E (a useful protease in laundry detergents) was evolved to be hundreds of times more active than its wild-type counterpart, in 60% aqueous dimethylformamide (Chen and Arnold 1993). For her seminal work on directed evolution, Frances H. Arnold was awarded the Nobel Prize in Chemistry in 2018. We will have more on directed evolution of enzymes later in this chapter. Remarkably, novel enzymes discovered in academic laboratories can be engineered within 5 years, for application and scaleup. Short development times and expanded enzymatic portfolio are contributing substantially towards green synthetic solutions. Many companies today are offering biocatalyst development services and enzymes tailored for specific applications. The technology of directed evolution has contributed significantly to the maturation of bespoken biocatalysts (Miller et al. 2022).

Enzymes, natural or evolved, have become indispensable in green chemistry because of their "benign by design" feature (Sheldon and Woodley 2018). They snugly fit into the major principles of green chemistry: atom efficiency, waste prevention instead of remediation; safer products by design through use of less hazardous materials and innocuous solvents and auxiliaries; shorter synthesis through catalytic rather than stoichiometric reagents; energy-efficient design with renewable raw materials; and degradable products leading to pollution prevention. Besides, enzymes provide exceptional solutions for the fabrication of functional materials, through direct enzymatic action on the surface or by their integration into the material itself (Richter et al. 2015).

37.4   Genetic Engineering and Enzymes

Large-scale production of enzymes is often a prerequisite for most applications. Obtaining them from animal and plant material—though of historical importance—has become progressively difficult for economic and ethical reasons. Therefore, the extant microbial biodiversity is routinely screened for enzymes with similar/desirable properties. Microbial rennin (a substitute for chymosin) produced from *Mucor* spp. is a case in point. Another exciting option is to produce the required enzyme through recombinant DNA technology, preferably in a microbial host.

The formidable tools of genetic engineering have allowed the expression and management of enzyme structures almost at will. Detailed recipes of recombinant DNA techniques are available in many texts and protocol books while we sketch a brief outline of these steps here. Systematic manipulation of the DNA sequence at the molecular level is the essence of genetic engineering. This means, we can cut and patch DNA fragments/gene(s) for any enzyme protein from diverse living organisms. These recombinant DNA molecules—capable of expressing a natural or a mutant enzyme protein—can then be moved into a suitable host for protein expression. When this genetic information is expressed (by transcription and translation) in the new host, it can produce an enzyme protein that is even foreign to it! A general strategy for genetic engineering is outlined in Fig. 37.5 below. There are essentially four stages to this powerful technique.

37.4.1 Isolation of the Gene/ORF/cDNA for the Enzyme Protein

Restriction endonucleases of different sequence specificities are employed to cut out the DNA fragment or the ORF (open reading frame) encoding the gene for your favorite enzyme (GFE), from a given source of genomic DNA. The required DNA fragment may be amplified by PCR (polymerase chain reaction) or the corresponding cDNA may be obtained through reverse transcriptase-PCR of the relevant mRNA. Alternatively, the required DNA may also be synthesized chemically—if the desired nucleotide or amino acid sequence is known.

37.4.2 Insertion of the Gene into an Expression Vector

The gene/ORF/cDNA/is integrated (linked) into another piece of DNA, the vector DNA, in order to promote its uptake and replication in a suitable host organism. In general, the DNA vectors are designed (engineered) to have a unique site for restriction endonuclease and carry a marker to facilitate selection of genetically modified host cells (transformants). The chimerical recombinant DNA is introduced in an appropriate host for expressing the enzyme protein. Typically, a bacterial plasmid vector is used to transform bacteria (Fig. 37.5). For animal and plant cells, their respective viral DNAs are often used as vectors.

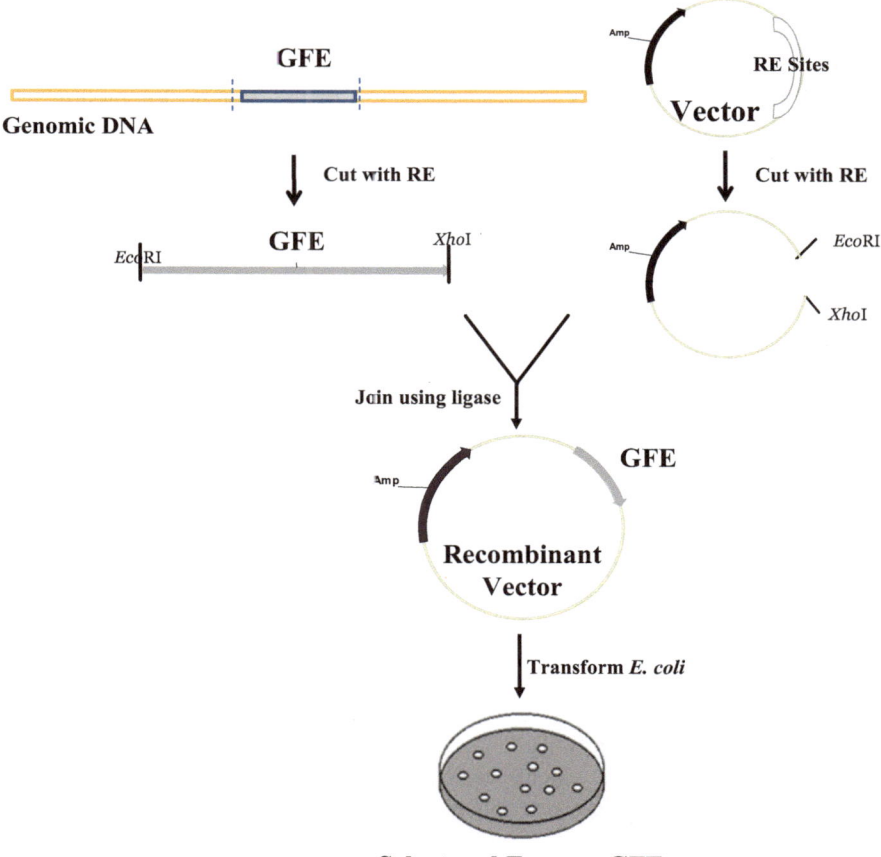

Fig. 37.5 A general genetic engineering strategy for enzyme expression. *GFE* gene for your favorite enzyme

37.4.3 Transformation of the Host Cell

The recombinant DNA vector is introduced into a host cell either directly (by the process of transformation) or by infecting it using a viral vector.

37.4.4 Detection of the Inserted Gene

The presence of the foreign DNA insert in the host is detected by molecular tools like Southern blotting (DNA–DNA hybridization) and PCR. The functional expression of the recombinant DNA may be directly monitored through enzyme activity and the presence of the protein.

At its simplest level, genetic engineering tools allow us to produce any enzyme protein in the common bacterium *E. coli*. One could choose from bacteria, yeast, plant, or mammalian cells as hosts for optimal expression and/or appropriate post-translational processing (like glycosylation, etc.) of the protein. Expression in a homologous host is generally successful. One of the principal reasons that an enzymologist manipulates DNA blueprint of an enzyme is to modulate the existing feature or to create new ones. Approaches like site-directed mutagenesis (SDM) and directed evolution of enzymes are routine nowadays (also see Chap. 40 Future of enzymology—An appraisal). In addition, genetic engineering has made optimal enzyme production possible in many ways. We will simply illustrate the field with a few examples.

1. Enzyme overproducing strains have been constructed by suitably overcoming regulation at the level of feedback inhibition, transcription, translation, or secretion of the enzyme protein. Expression of amylases is often under the control of carbon catabolite repression (glucose repression). Yet another bottleneck with respect to protease production is nitrogen metabolite regulation. Both bacterial and fungal strains, with mutations for deregulation, constitutive expression, and hypersecretion, find utility in enzyme production. Suitable genetically stable mutant strains combining many such features have found their place in enzyme industry.
   Individual component activities of the cellulase complex and their specific combinations find industrial applications (Table 37.1). Producer strains (like *Trichoderma reesei*) overexpressing individual activities as well as deleted in each one of them are available. Well-defined cocktail of cellulase components are suitable in textile industry and in biomass conversion.
2. Heterologous expression, even of mammalian or plant enzymes, in a convenient microbial host. The recombinant chymosin (bovine) was produced in a fungus by introducing a stable expression construct. The difficult challenges however include ability to obtain economically viable levels of secretion, stability of the expressed protein, and attaining proper post-translational modifications like glycosylation, if any.
3. Once the structural gene for the enzyme is cloned and expressed, it is feasible to generate mutant forms of this enzyme. Tinkering with enzyme properties like stability, pH optimum, specificity, and regulatory features is possible. Protein and enzyme engineering through site-directed mutagenesis has found direct applications in the enzyme industry. Subtilisin was engineered for better stability. Glucose isomerase was improved for its metal preference, substrate specificity, and pH optimum. DNA polymerase with high fidelity for PCR applications is another fruitful example.
   The impact of genetic engineering on the field of modern enzymology may be further gauged by examples presented below and in Chap. 40 (Future of enzymology—An appraisal).

37.5 Directed Enzyme Evolution

Evolution continues to be an excellent teacher of how to evolve/modify biological catalysts. Directed evolution as a tool accelerates the evolutionary process from millions of years to mere weeks! And natural enzymes are clearly not evolutionary dead ends. The field of directed evolution of enzymes has made rapid progress and provides a much larger scope and canvas than simple enzyme redesign (Bornscheuer et al. 2012). It has moved from being a tool for studying the relationship between sequence and function to being an extremely useful and efficient method for optimizing biocatalysts for industry. This Darwinian approach involves generation of random mutant library (generation of genetic diversity) followed by biological selection (of the fittest) for the desired activity (Moore and Arnold 1996; Arnold 2018). The greatest advantage of directed evolution is its independence from prior knowledge of enzyme structure. One need not know the nature of interactions between the enzyme and its substrate either.

The technology of directed evolution per say consists of (1) random gene library generation, (2) expression of these genes in a suitable host, and (3) screening/selection of libraries of variant enzymes for the bespoke property of interest (Fig. 37.6). It begins with the mutagenesis of the gene encoding the enzyme template of interest (GFE, Gene of your Favorite Enzyme). This important first step should

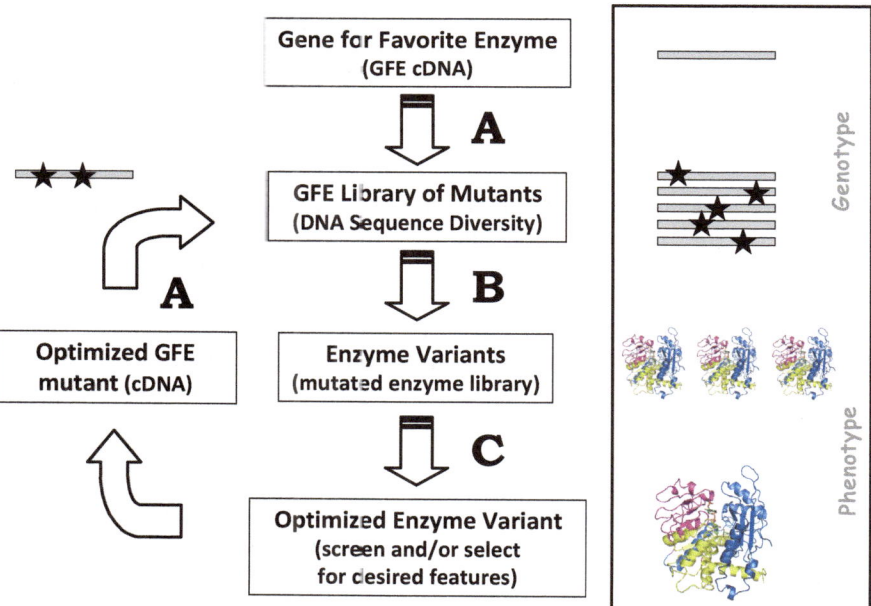

Fig. 37.6 Iterative steps in the directed evolution of an enzyme. Starting with the gene (cDNA) for the favorite enzyme (GFE), a random mutant library is generated (step A). These GFE variants are used to express respective enzyme variants (step B). The optimized variants are screened/selected (step C) and subjected to further cycles of directed evolution

Table 37.7 Methods to create GFE mutant libraries and generate diversity

Library diversification strategy	Examples and tools
Random mutagenesis	
Chemical and physical mutagenesis	EMS, MNNG, nitrous acid, UV
Error-prone PCR (epPCR)	Supplementing Mn^{2+} and/or unequal dNTP levels
Mutator strains	Mutagenesis plasmid (PACE), XL1-red strain of *E. coli*
Focused mutagenesis	
Site-directed saturation mutagenesis	NNK and NNS codons on mutagenic primers[a]
Computational strategies for high-quality library design	Rosetta design and computationally guided libraries, ISOR, consensus design, REAP, and SCHEMA
Recombination (gene shuffling)	
Homologous recombination	DNA shuffling, domain swapping, family shuffling, StEP, RACHITT, NExT, heritable recombination, ADO, and synthetic shuffling
Nonhomologous recombination	ITCHY, SHIPREC, NRR, SISDC, and overlap extension PCR (oePCR)

Adapted with permission from Packer and Liu, Nat Rev. Genet, 16:379–394. Copyright (2015) Springer Nature
ADO assembly of designed oligonucleotides, *EMS* ethyl methanesulfonate, *ISOR* incorporating synthetic oligonucleotides via gene reassembly, *ITCHY* incremental truncation for the creation of hybrid enzymes, *MNNG* N-methyl-N-nitrosoguanidine, *NExT* nucleotide exchange and excision technology, *NRR* nonhomologous random recombination, *PACE* phage-assisted continuous evolution, *RACHITT* random chimeragenesis on transient templates, *REAP* reconstructed evolutionary adaptive path, *SCHEMA* a computational algorithm, *SHIPREC* sequence homology-independent protein recombination, *SISDC* sequence-independent site-directed chimeragenesis, *StEP* staggered extension process, *UV* ultraviolet rays
[a]Where N can be any of the four nucleotides, K can be G or T, and S can be G or C

generate a representative yet exhaustive mutant library. The mutated gene library is then inserted into a suitable host (like *E. coli* or yeast) for expression and the transformants are plated on selective media. Respective protein variants are chosen from the single colonies growing on these plates. In most directed evolution studies, additional cycles of mutagenesis are necessary for obtaining an optimal catalyst. While many strategies are available (and continue to be developed) for generating exhaustive libraries, the success of a directed evolution experiment depends greatly on the method chosen for finding the best mutant enzyme. And, in directed evolution, "we simply get what we screen for"!

A range of gene mutagenesis tools are available for generating sequence diversity in a directed evolution experiment. A selection of them is listed in Table 37.7 (Packer and Liu 2015). Choice of a method is critical as each one of them has certain advantages as well as limitations. Features such as biased mutational spectrum, use of hazardous chemicals, uneven sampling of codon space, need for prior structural, biochemical or phylogenetic knowledge of each method become important considerations. Despite these diverse methods for gene diversification, it is impossible to cover the entire mutational space of a typical protein. For a polypeptide chain of "n" amino acids, 20^n combinations are possible as there are 20 different amino acids. Even nature has sampled a tiny fraction of these possible sequences over the huge evolutionary time scale. It is therefore best to begin with an existing enzyme sequence. The choice of the sequence diversification method will then depend on the nature of evolutionary trajectory required from the initial scaffold to the desired end point. For instance, directed evolution of enzyme robustness (for stability) may require simultaneous changes scattered across the sequence length.

There are two broad approaches to finding the best mutant enzyme, namely, screening and selection. Screening of the mutant library can be divided into two categories: (1) facilitated screening where mutants are distinguished based on distinct phenotypes, and (2) random screening in which we pick mutants blindly. Selection is always preferred over screening for its higher efficiency. However, selection requires a phenotypic functional link between the target gene and its encoding product that confers the selective advantage (such as better growth, etc.) (Payne et al. 2006).

Directed evolution has been successfully applied to achieve favorable changes in enzyme properties like stereo- and region-selectivity, expanding the substrate scope and/or activity, enzyme robustness, pH optimum, promiscuity as catalysts in organic synthesis (Jeschek et al. 2016; Reetz 2016). Select examples of evolved enzymes reported in the literature are listed in Table 37.8. Industrial-scale biocatalysis applications of optimized enzymes have focused primarily on hydrolases, a few ketoreductases (KREDs), transaminases, oxidative enzymes, aldolases, cofactor regeneration, and protein stability in organic solvents (Bornscheuer et al. 2012).

Table 37.8 Examples of directed enzyme evolution

Enzyme	Feature optimized
Monoamine oxidase	Deracemization of racemic amines
Ketoreductases (KREDs)	Chiral intermediates for pharmaceuticals
Laccase	Catalytic efficiency, neutral pH range
Glyphosate N-acetyltransferase	Catalytic efficiency for glyphosate resistance
Cephalosporin acylase	Catalytic efficiency towards adipyl-7-ADCA
ω-Aminotransferase	Specific activity and thermostability
Epoxide hydrolase	Enantioselectivity
Hydantoinase	Inverting enantioselectivity (L-met process)
β-Lactamase	Antibiotic resistance against cetotaxime
Xylanase, phytase and lipase	Thermostability
Endoglucanase	Thermal stability, alkaline pH range
Subtilisin	Stability in organic solvent
Halohydrin dehalogenase	Catalytic efficiency
Aldolases	Specificity, efficiency, thermostability
Nonribosomal peptide synthase–polyketide synthase (NRPS–PKS) hybrid	Produce broad-spectrum antibiotic (andrimid)

37.6   Summing Up

Being superbly crafted catalysts of nature, enzymes found their use very early in the game. The first patent for an enzyme application was in 1894, even before their chemical nature was known! Subsequently, bioprospecting for enzymes with unique properties has continued unabated. The entire industry to process starch has evolved by exquisite use of enzymes from across the three domains of life. In the clinical and pharmaceutical setting, enzymes have served as disease markers, analytical tools for metabolite measurements, biosensors, and targets for drug discovery. In an industrial process, the given amount of substrate has to be maximally converted to product. Immobilized enzymes have reduced the cost by permitting their reuse and reaction scaleup. Although evolved essential for catalysis in an aqueous environment, few enzymes can function in organic solvents. This has expanded the scope of their utility in terms of the types of reactions that can be carried out; many hydrolases may now be used to drive the reactions in reverse for synthetic purposes.

Having tasted their potential in industry, enzymes are being genetically engineered for desired features and also for efficient large-scale production. Directed evolution of enzymes has firmly established itself as a powerful technology and the future of enzymology has arrived. With a better understanding of how these catalysts work, the range of their applications in various industries has expanded. While it is beyond the scope for an elaborate coverage, the present chapter has attempted to provide a focused overview. For more detailed treatment, the reader may refer to the cited literature.

References

Arnold FH (2018) Enzymes by evolution: bringing new chemistry to life. Mol Front J 2:9–18

Beckham GT (2015) Fungal cellulases. Chem Rev 115:1308–1448

Bornscheuer UT et al (2012) Engineering the third wave of biocatalysis. Nature 485:185–194

Brena BM, Batista-Viera F (2006) Immobilization of enzymes. In: Guisan JM (ed) Immobilization of enzymes and cells. Humana Press, Totowa, pp 15–30

Chapman J, Ismail AE, Dinu CZ (2018) Industrial applications of enzymes: recent advances, techniques, and outlooks. Catalysts 8:238

Chen K, Arnold FH (1993) Tuning the activity of an enzyme for unusual environments: sequential random mutagenesis of subtilisin E for catalysis in dimethylformamide. Proc Natl Acad Sci USA 90:5618–5622

Groger H, Asano Y (2012) Introduction—principles and historical landmarks of enzyme catalysis in organic synthesis. In: Drauz K, Groger H, May O (eds) Enzyme catalysis in organic synthesis, 3rd edn. Wiley-VCH Verlag GmbH & Co. KGaA

Guo F, Berglund P (2017) Transaminase biocatalysis: optimization and application. Green Chem 19:333–360

Halling PJ (2004) What we can learn by studying enzymes in non-aqueous media? Philos Trans R Soc Lond B 359:1287–1297

Harwood J (1989) The versatility of lipases for industrial uses. Trends Biochem Sci 14:125–126

Jeschek M et al (2016) Directed evolution of artificial metalloenzymes for in vivo metathesis. Nature 537:661–665

Katz E et al (2001) Self-powered enzyme-based biosensors. J Am Chem Soc 123:10752–10753

Kelly SA et al (2018) Application of ω-transaminases in the pharmaceutical industry. Chem Rev 118:349–367

Klibanov AM (2001) Improving enzymes by using them in organic solvents. Nature 409:241–246

Li Q, Yi L, Marek P, Iverson BL (2013) Commercial proteases: present and future. FEBS Lett 587: 1155–1163

Mateo C et al (2007) Improvement of enzyme activity, stability and selectivity via immobilization techniques. Enzyme Microb Technol 40:1451–1463

Miller DC, Athavale SV, Arnold FH (2022) Combining chemistry and protein engineering for new-to-nature biocatalysis. Nat Synth 1:18–23

Moore JC, Arnold FH (1996) Directed evolution of a *para*-nitrobenzyl esterase for aqueous-organic solvents. Nat Biotechnol 14:458–467

Packer MS, Liu DR (2015) Methods for the directed evolution of proteins. Nat Rev Genet 16:379–394

Pariza MW, Johnson EA (2001) Evaluating the safety of microbial enzyme preparations used in food processing: update for a new century. Regul Toxicol Pharmacol 33:173–186

Payne CM et al (2006) Outlook for cellulase improvement: screening and selection strategies. Biotechnol Adv 24:452–481

Reetz M (2016) Directed evolution of selective enzymes: catalysts for organic chemistry and biotechnology. Wiley, Weinheim

Richter M et al (2015) Novel materials through nature's catalysts. Mater Today 18:459–467

Robertson JG (2005) Mechanistic basis of enzyme-targeted drugs. Biochemistry 44:5561–5571

Savile CK et al (2010) Biocatalytic asymmetric synthesis of chiral amines from ketones applied to sitagliptin manufacture. Science 329:305–309

Sewalt V et al (2016) The generally recognized as safe (GRAS) process for industrial microbial enzymes. Ind Biotechnol 12:295–302

Sharma M et al (2017) NAD(P)H-dependent dehydrogenases for the asymmetric reductive amination of ketones: structure, mechanism, evolution and application. Adv Synth Catal 359: 2011–2025

Sheldon RA, Woodley JM (2018) Role of biocatalysis in sustainable chemistry. Chem Rev 118: 801–838

Slabu I et al (2017) Discovery, engineering, and synthetic application of transaminase biocatalysts. ACS Catal 7:8263–8284

Umezawa H (1982) Low-molecular-weight enzyme inhibitors of microbial origin. Ann Rev Microbiol 36:75–99

Voutilainen S et al (2020) Substrate specificity of 2-deoxy-D-ribose 5-phosphate aldolase (DERA) assessed by different protein engineering and machine learning methods. Appl Microbiol Biotechnol 104:10515–10529

Wang S et al (2016) Enzyme stability and activity in non-aqueous reaction systems: a mini review. Catalysts 6:32

Winkler CK, Schrittwieser JH, Kroutil W (2021) Power of biocatalysis for organic synthesis. ACS Cent Sci 7:55–71

Wu S et al (2021) Biocatalysis: enzymatic synthesis for industrial applications. Angew Chem Int Ed 60:88–119

Zhou S et al (2020) Coenzyme a thioester-mediated carbon chain elongation as a paintbrush to draw colorful chemical compounds. Biotechnol Adv 43:107575

Further Reading

Kirk O, Borchert TV, Fuglsang CC (2002) Industrial enzyme applications. Curr Opin Biotechnol 13:345–351

van Beilen JB, Li Z (2002) Enzyme technology: an overview. Curr Opin Biotechnol 13:338–344

Part VI

Frontiers in Enzymology

नासतो वद्यिते भावो नाभावो वद्यिते सतः। उभयोरपि दृष्टोऽन्तस्त्वनयोस्तत्त्वदर्शभिः ᱯ
||2.16||

The unreal has no existence, and the real never ceases to be, the reality of both has been perceived by the seers of truth.

(Bhagavad Gita; 2.16)

Most organisms, however evolutionarily distant, contain a set of common metabolites. But their intracellular concentrations are unique to each individual species. This *metabolic identity of an organism* is the consequence of quantitative differences in relevant enzyme properties and their associated regulation. Features of metabolic regulation are unique to each organism—often within a closely related group of organisms. It is becoming increasingly clear that the concept of *unity in biochemistry* does not always extend to metabolic pathway control and enzyme regulation. Historically, metabolic regulation and control of enzyme activity have developed as closely related phenomena. What then is the justification to place ***regulation of enzyme activity*** here in "Frontiers in Enzymology"? Over the years, molecular developments in biology have outshone the progress made in physiological and system-level understanding of organisms. While the basic principles of enzyme/metabolic regulation may have been uncovered, novel modes of regulation continue to be discovered. Nature continues to surprise us with original ways of regulating enzyme activity. The novelty may be in the conceptual mechanism, or the regulatory ligands involved. For instance, fructose-2,6-bisphosphate as a regulator of glycolysis (at the phosphofructokinase step) was discovered much later (in early 1980s)—many decades after the complete description of glycolytic enzymes (Hers and Hue 1983). In this sense, the topic of regulation of enzyme activity will always be at the frontiers of enzymology.

The multitude of biochemical changes in metabolism are brought about by the battery of different enzymes. ***Regulation*** (also termed ***homeostasis***) is the ability to maintain metabolic constancy in the face of external perturbations. ***Control*** on the other hand is the ability to make changes to metabolism as and when necessary. Both these phenomena manifest through manipulation of enzyme activities. Without its regulation and control, the cell is essentially a bag of enzymes. However, *the cell is not just a bag of enzymes* (Mathews 1993). A remarkable degree of order is

N. S. Punekar, *ENZYMES: Catalysis, Kinetics and Mechanisms*,
https://doi.org/10.1007/978-981-97-8179-9_38

Fig. 38.1 Various factors that influence the overall rate of an enzyme-catalyzed reaction in cellular metabolism. Changes in the intensive or extensive properties listed may increase ($\uparrow$) or decrease ($\downarrow$) effective enzyme activity of that step

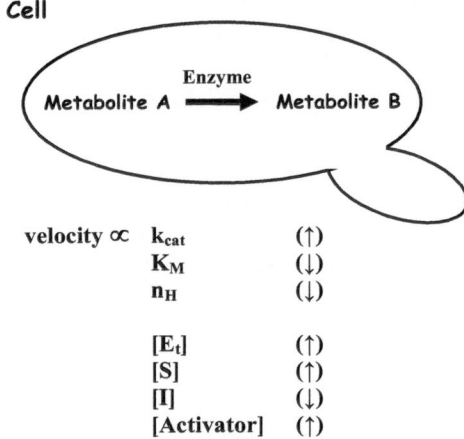

velocity $\propto$	k_{cat}	($\uparrow$)
	K_M	($\downarrow$)
	n_H	($\downarrow$)
	$[E_t]$	($\uparrow$)
	$[S]$	($\uparrow$)
	$[I]$	($\downarrow$)
	$[Activator]$	($\uparrow$)

maintained inside because of the stringent and very efficient regulation of enzyme activities. Regulation of enzyme activity is desirable and is accomplished because (a) the catalytic rates achieved, at times, may be too fast or too slow for the well-being of the cell and (b) more than one enzyme may share a metabolite as its substrate, thereby necessitating a logic for metabolic pathway flux distribution. In any case, it is not the metabolic fluxes that are being regulated, but the regulatory design is to regulate metabolite concentrations. The cellular economy works on supply and demand of various metabolites (Cornish-Bowden 2004).

At any given time, the reaction rate taking metabolite A to metabolite B depends on the in vivo activity of the corresponding enzyme. Effective intracellular enzyme activity is a function of several intensive (K_M, k_{cat}, and n_H) and extensive ([E_t], [S], [P], [I], and [activator]) properties (Fig. 38.1). The resultant rate is a consolidated response and an outcome of all these factors. Therefore, in principle, regulation of enzyme activity is possible by changing one or more of these properties.

Enzyme regulation can be achieved by either increasing/decreasing the number of enzyme molecules (through induction, repression, and turnover) or by modulating the activity of preexisting enzyme molecules (inhibition or activation). The former mechanisms respond relatively slowly to the changing external stimuli and provide for *long-term* control, while the latter respond rapidly to changing conditions and are *short-term* control mechanisms (Fig. 38.2). We will study representative examples for all these modes of enzyme regulation in this chapter. Looking at every variant of regulation would be arduous and would quickly grow into a textbook on metabolism. Instead of dwelling on the vast mechanistic permutations, case studies of well-established systems are highlighted. Furthermore, the emphasis will be on the examples of historical importance.

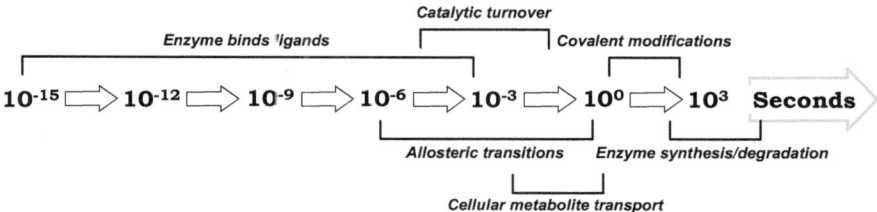

Fig. 38.2 Events relevant to regulation of enzyme activity. The timescales as indicated are approximate ranges

38.1 Control of Enzyme Concentration

We recall from Chap. 14 that maximal velocity (V_{max}) for an enzyme equals $k_{cat} \times [E_t]$. Of these, k_{cat} is an intensive property and is an intrinsic constant for a given enzyme. Enzyme concentration $[E_t]$ is an extensive property whose magnitude can be adjusted either by increasing or decreasing the number of enzyme molecules in the cell. This mode of regulation largely comes under genetic regulation but will be briefly mentioned here for the sake of completeness.

38.1.1 Induction and Repression

Enzymes involved in catabolic routes are "induced" in the presence of compounds that are destined to be degraded through these routes. The *lac* operon of *E. coli* and its induction by lactose is an excellent example. The enzymes of the biosynthetic pathway are "repressed" by the end-product of the pathway. In many bacteria, histidine is known to repress the expression of enzymes from *his* operon. The genetic control of pathways is brought about by inducer or repressor ("effector" in general) molecules which may be small molecular weight metabolites. There is no uniformity with regard to the effector identity for a given pathway in different organisms. This is where *unity in biochemistry* concept faces its biggest challenge. In the biosynthetic pathways, repression by the end-product and in the catabolic routes, induction by the initial substrate, affects all the enzymes of a given pathway. It is generally found that the majority of the metabolic routes are nonlinear, i.e., there is considerable branching (for biosynthetic routes) and convergence (for catabolic routes). Compared to linear metabolic sequences, branching and convergence produces an added dimension of complexity in regulation both at the biochemical as well as the genetic level. The control of enzyme synthesis occurs by sequential induction, and by multivalent repression, which are, in effect, variations of the same theme found in linear pathways.

Another prevalent genetic control mechanism, governing the number of enzyme molecules, is **catabolite repression**. The enzymes of catabolic routes may be repressed when glucose is abundant through *carbon catabolite repression*. This repression could be mediated through cAMP or some other effector. The *Nitrogen metabolite control* is an adaptive mechanism which imparts hierarchy to nitrogen sources—the ones most economically used generally being consumed first. *Nitrogen metabolite repression* is observed when ammonia is present as a nitrogen source. Ammonia acts by either inhibiting the uptake of other complex nitrogen sources by a phenomenon called 'inducer exclusion" or serves as a direct or indirect (such as via L-glutamine) repressor of the genes involved in the catabolism of complex nitrogen sources. Both carbon catabolite repression and nitrogen metabolite repression are broad modes of control acting globally across pathways.

38.1.2 Post-transcriptional Regulation of mRNA

The half-lives of the mRNA transcripts of key enzymes involved in metabolism are often determined by the cellular demands. The mRNA stability is known to increase during induction of a few structural genes in metabolic pathways. Stable, stored mRNAs of some microbial enzymes are translated in response to environmental cues. The mammalian ornithine decarboxylase antizyme is a protein regulator of the ornithine decarboxylase enzyme activity (see Sect. 0 below). This antizyme mRNA is significantly stable ($t_{1/2} = 12$ h). The synthesis of antizyme protein from its preformed mRNA is triggered by an increase in cellular polyamine levels—an example of translational control. The recent technological developments in RNAi and CRISPR-Cas are powerful genetic tools for regulation. RNAi (RNA interference) actually make use of mRNA stability and reduces gene expression at the mRNA level (knockdown). Whereas CRISPR-Cas system completely and permanently silences the gene at the DNA level (knockout).

38.1.3 Regulation by Protein Degradation

Cells continuously synthesize proteins from, and degrade them to, their component amino acids. This permits regulation of cellular metabolism by eliminating superfluous enzyme and other protein molecules. Remarkably, most rapidly turned over enzymes occupy important metabolic control points, whereas the relatively stable enzymes have nearly constant catalytic and allosteric properties so that cells can efficiently respond to environmental changes and metabolic needs. The half-lives of enzymes range from under an hour to more than 100 h. Enzyme protein turnover thus represents a regulatory mechanism belonging to longer time scales (Fig. 38.2). In general, the longer the life span of a cell (such as in eukaryotes) the more important is the process of enzyme turnover as a control mechanism.

The steady-state level of a given enzyme protein is a balanced outcome of its synthesis rate (which generally follows zero-order kinetics; $k_s \times [E]^0$ or simply k_s) and degradation rate (which normally obeys first-order kinetics; $k_d \times [E]$). Also, the degradation rate constant is related to half-life ($t_{1/2}$) by the following relation (see Chap. 8 for details):

$$k_{\mathrm{d}} = \frac{\ln 2}{t_{1/2}} = \frac{0.693}{t_{1/2}}.$$

The rate of change of enzyme level inside a cell is given by the following equation: $d[E]/dt = k_{\mathrm{s}} - k_{\mathrm{d}} \times [E]$. During the steady state, $d[E]/dt = 0$ and therefore we have the relation: $k_{\mathrm{s}} = k_{\mathrm{d}} \times [E]$.

The first step to establish the occurrence of enzyme turnover is to show that the protein level of that enzyme is changing. It is also important to show that the change in enzyme activity is not due to any other reason such as covalent modification or a conformational change. Enzyme turnover can be experimentally measured by using isotopically labeled amino acid precursors. The synthesis rate constant (k_{s}) may be evaluated by giving a single pulse of radiolabeled amino acid and subsequently following the incorporation (at short time intervals) of the label into the enzyme protein. For degradation rate constant (k_{d}) measurements, first the enzyme protein is labeled. Then the decay of specific radioactivity of this labeled protein is measured at fixed (often longer) time intervals. Among the liver proteins, ornithine decarboxylase is turned over much more rapidly ($t_{1/2} = 15$ min) than phosphofructokinase ($t_{1/2} = 168$ h). Many of the enzymes with short half-lives catalyze rate-limiting steps in metabolic pathways.

Intracellular proteins destined for degradation are either tagged (ubiquitinated) and taken to proteasomes or processed through autophagy. Proteasome is a large multi subunit protease found in all eukaryotes. This multi catalytic proteolytic complex degrades ubiquitin-tagged proteins in an ATP-dependent manner. A proteasome contains five different protease activities facing its lumen cavity—these are characterized as chymotrypsin-like, trypsin-like, post-glutamyl hydrolase, branched chain amino acid protease, and small neutral amino acid protease. Lactacystin is a selective proteasome inhibitor. Autophagy is the other route of protein turnover and involves lysosomes. Proteins/enzymes that are generally not ubiquitinated take this nonselective process of autophagy. Lysosomes contain more than 50 different hydrolytic enzymes, all with an acidic pH optimum. These include nine different cathepsins and nine more exoproteases. Most lysosomal proteases are inhibited by leupeptin.

Enzymes do have finite in vivo life spans. Besides being subjected to targeted proteolysis, it is not clear how long enzyme molecules remain functional in vivo. The cumulative deterioration (wear-out), sudden random failure, and other causes may affect enzyme's lifetime and force their repair or replacement (Bathe et al. 2021). The number of "catalytic cycles until replacement" is a potential metric for enzyme functional life span in vivo. It is the number of catalytic cycles that an enzyme mediates in vivo before failure or replacement (= metabolic flux rate/protein turnover rate). This number in different organisms ranges from 10^3 to 10^7. This yardstick may be used to assess what drives enzyme protein replacement and better metabolic design (Hanson et al. 2021).

38.2   Control of Enzyme Activity: Inhibition

Another level of regulation of metabolic pathways is to control the activity of a strategically placed enzyme by sensing metabolite concentrations. Compared to the slow, limited means of control of enzyme concentration, there are a plethora of mechanisms to control enzyme activity. This adds further diversity in regulation because one can modulate enzyme activity either positively (by activation) or negatively (by inhibition). Some such important control mechanisms are outlined below.

Inhibition of the activity of an enzyme by ligand binding is the quickest way of controlling its function. Being a non-covalent interaction, the binding equilibrium is concentration-driven and reversible. In this time scale (see Fig. 38.2), the enzyme concentration remains effectively constant. While examples of inhibitory ligands are plentiful, we do find instances of enzyme activation by ligands. The concept of *enzyme activation* is analogous to the more common *enzyme inhibition*. It is just that the effects are opposite! We will elaborate on the various modes of enzyme inhibition (and leave examples of enzyme activation mostly to the imagination of the reader).

Inhibitors could be structurally similar to either the substrate or the product of an enzyme. By virtue of this similarity, they may bind at the active site and exhibit their inhibitory effect. Such inhibitors are called *isosteric* inhibitors (see Chap. 19). Ligands that may or may not resemble the substrate (or the product) could bind to a site other than the enzyme active site. These inhibitors are termed *allosteric* inhibitors. It is worth noting that a substrate itself could be an allosteric activator (see "Cooperative/allosteric modulation" below).

Enzymes inhibited at higher substrate concentrations are rare. This phenomenon of *substrate inhibition* is known in enzyme kinetics (see Chap. 22). However, the regulatory significance of substrate inhibition is not well understood. Enzymes acting on metabolites that are toxic when accumulated often display substrate inhibition behavior. An aldehyde dehydrogenase interacting with its substrate aldehyde is an example. It appears that such enzymes have evolved to catalyze forward reactions at low substrate concentrations. The evolutionary pressure on catalysis in the reverse direction may have been insignificant on them. In any case, the toxic metabolic intermediate is quickly cleared by such enzymes.

Inhibition by the reaction product is the most common yet frequently overlooked mode of enzyme regulation. A product can access and bind at the enzyme active site. Product can inhibit the reaction by titrating out the active enzyme available for catalysis. It can also drive the reaction backwards by simple mass action. The extent to which a product inhibits an enzyme depends on its concentration and the binding constant. The lower the K_I, the greater the inhibition at any given product concentration. As a thumb rule, biosynthetic enzymes are much more sensitive to *product inhibition* than catabolic enzymes. This makes sense as catabolism should occur only when the corresponding substrate is in excess. Product inhibition ensures that a biosynthetic enzyme makes enough product (and is accumulated to satisfy the cellular needs) as is necessary, while wasteful metabolism is prevented.

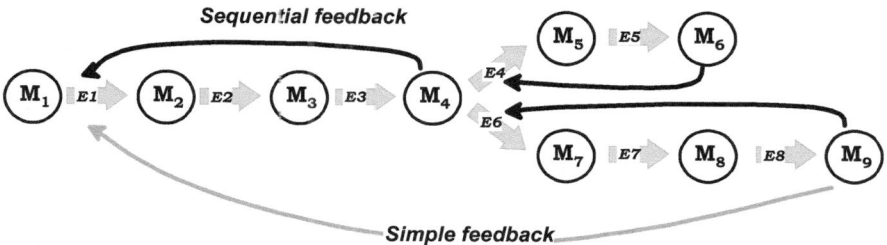

Fig. 38.3 Schematic of a branched anabolic pathway showing different modes of enzyme inhibition. Pathway metabolites are numbered M_1 through M_9 and enzymes are numbered $E1$ through $E8$. A simple feedback inhibition of $E1$ by end-product M_9 (gray arrow) is depicted. Sequential inhibition involving inhibitions of $E4$ by M_6 and $E6$ by M_9 with the resultant accumulation of M_4 and subsequent inhibition of $E1$ by M_4 (black arrows) is shown. Few other possible variations are described in the text

End-product inhibition is a common mode of metabolic regulation in biosynthetic pathways. Often a terminal metabolite, without any chemical analogy/reactivity to an earlier step, is a powerful inhibitor of its own synthesis. Since they have very little structural resemblance to the substrate(s) of the metabolic step they inhibit, these inhibitors invariably bind to an allosteric site of that enzyme (Pardee 2002; Pardee and Reddy 2003). In this mode of *feedback inhibition*, the end-product (or a near end-product) controls the metabolic flux by inhibiting the activity of one or more early enzymes of the pathway. Two examples are acetolactate synthase (in branched chain amino acid biosynthesis) feedback inhibited by L-valine and aspartate transcarbamylase (in pyrimidine biosynthesis) inhibited by CTP (Umbarger 1956; Monod et al. 1963). In branched pathways, the maximum inhibition is often attained only by the combined action of multiple end-products. This inhibition strategy circumvents the problem of completely shutting down a branched pathway by one end-product, thereby ensuring availability of other end-products to the organism. Figure 38.3 shows the schematic of a branched anabolic pathway to understand the various kinds of feedback inhibitions that may operate in biosynthetic pathways.

In branched pathways, often the end-product inhibits the first (or an early) enzyme of the respective branch. The branch point intermediate preceding the branch, in turn regulates the activity of the first enzyme common to all the end-products, thus maintaining a balance of the products formed (Fig. 38.3; M_6 inhibits $E4$, M_9 inhibits $E6$, and consequent higher levels of M_4 inhibits $E1$). Examples of such *sequential feedback inhibition* may be found in aromatic amino acid biosynthesis (e.g., pathway control by Trp, Phe, and Tyr in *B. subtilis*) and the biosynthesis of aspartate family of amino acids. Variations of this sequential feedback control are also possible. For instance, inhibition of the first (or important) common enzyme of each branch by the product at the branch point along with a simultaneous activation of the first enzyme after the branch point by the same or other intermediate of the pathway (Fig. 38.3; M_4 inhibits $E2$ and activates $E7$). Yet

38   Regulation of Enzyme Activity

another possible mode involves **compensatory activation and deinhibition,** where the first common enzyme is inhibited by one product and activated by the other product (Fig. 38.3; M_6 inhibits $E4$ and activates $E6$, and M_9 inhibits $E6$ and activates $E4$), thus maintaining a balance in the products formed. Examples of this kind may be found in biosynthesis of purine nucleotides (AMP and GMP) and of dNTPs.

Examples of an enzyme inhibited by more than one ligand are known—the so-called **multiple inhibition**. In the case of **concerted or multivalent inhibition,** the products of a branched pathway do not singly inhibit the first common enzyme. However, the presence of two or more products is essential for significant inhibition (Fig. 38.3; either M_6 or M_9 alone do not inhibit $E1$, but when both are present, $E1$ activity is markedly reduced). Threonine and lysine act in concert (but not individually) to inhibit *B. polymixa* aspartokinase. A **synergistic inhibition** is observed when mixtures of M_6 and M_9 at low concentrations bring about more inhibition (of $E1$) than the same total specific concentration of M_6 or M_9 alone. Interaction with AMP, histidine, and glutamine as inhibitors of *B. licheniformis* glutamine synthetase is one such case (where synergistic inhibition of the enzyme by Gln + His pair and AMP + His pair is reported). The concerted inhibition is thus an extreme case of synergism for inhibition between the inhibitors. By contrast, in **cumulative feedback inhibition**, there is no cooperation or antagonism between several inhibitors of an enzyme. Each end-product is a partial inhibitor and brings about the same percentage inhibition irrespective of whether other inhibitors are present or not. The *E. coli* glutamine synthetase provides an excellent case study of cumulative inhibition (also see later section) (Stadtman 2001). Suppose His alone at a given concentration yields 50% inhibition (enzyme retains 50% of the original activity) and Trp alone at a given concentration yields 30% inhibition (leaving 70% of the original activity). Then at same concentrations of "His + Trp," the enzyme retains 35% of the original activity (50% + [30% of 50%] or 30% + [50% of 70%] = 65% inhibition). If the nature of inhibition was additive, then the final inhibition reached would have been 80% (i.e., 50% + 30% = 80%).

If an enzyme is inhibited by more than one ligand, then understanding their mutual interaction is of interest. **Multiple inhibition analysis** (or the interaction of more than one inhibitor with the enzyme) is meaningfully done through Dixon analysis (Chap. 21). For example, when I and J competitively inhibit the enzyme, the double reciprocal form of the rate equation is

$$\frac{1}{v} = \frac{K_M}{V_{max}} \left(1 + \frac{[I]}{K_I} + \frac{[J]}{K_J} + \frac{[I][J]}{\alpha K_I K_J} \right) \frac{1}{[S]} + \frac{1}{V_{max}}.$$

Experimentally however, rates are measured at a fixed concentration of S while $[I]$ is varied. A $1/v \rightarrow [I]$ plot of this data is nothing but the Dixon plot. Now, in a similar setup, we can include different fixed concentrations of J and obtain a series of lines. The pattern of these lines is characteristic of the nature of interaction between the two inhibitors. The influence of one bound inhibitor on the binding of the other is estimated by the interaction constant, α. This constant is conceptually similar to the interaction term α, we used in describing noncompetitive inhibition in Chap. 21. An

α value of infinity indicates mutually exclusive binding of I and J (meaning only EI or EJ possible; EIJ does not form) and gives a parallel pattern in the Dixon analysis. A finite α value shows that both inhibitors can simultaneously bind the enzyme; values below unity are indicative of synergistic interaction between I and J.

38.3 Control of Enzyme Activity: Cooperativity and Allostery

In some enzymes, the ligand binding and/or catalytic activity follows a non-Michaelian saturation pattern. This feature allows the enzyme to function as a ligand concentration-dependent switch. At higher concentrations ligand binding may become progressively easier (positive cooperativity) or more difficult (negative cooperativity). Enzyme conformational changes accompany ligand-binding events in this mode of regulation.

38.3.1 Subunit Cooperativity and Switch Behavior

Most cooperative enzymes share a few features in common. These include:

1. Allosteric enzymes generally consist of multiple subunits (i.e., they are oligomeric).
2. The regulatory ligands (effectors) usually do not share any structural resemblance to the substrate(s) or product(s) of the enzyme reaction concerned.
3. Effectors may bind to an allosteric site distinct from the enzyme active site. It is thus possible to selectively destroy (by physicochemical or mutational methods) the allosteric site without affecting the catalytic site. Such a ***desensitized*** enzyme does not respond to allosteric effectors. For instance, upon limited heat treatment, *E. coli* aspartate transcarbamylase loses its ability to bind CTP.
4. Allosteric enzymes do not show Michaelian substrate saturation kinetics. Their $v \rightarrow [S]$ plots are sigmoidal rather than hyperbolic (see Fig. 14.4, Chap. 14). The sigmoid saturation curve indicates cooperative substrate binding—the binding of the first molecule facilitates the binding of subsequent molecules. The extent of cooperativity is measured by the value of h—the Hill coefficient (also denoted as n_H; see Chap. 14). An enzyme with $h = 1$ shows no cooperativity and is Michaelian. Negative cooperative enzymes have $h < 1$ whereas for those with positive cooperativity will have $1 < h < n$. If $h = n$ for an enzyme with n binding sites (each monomer with an active site), then such an enzyme will be extremely cooperative. We note that $h = 2.6$ for hemoglobin. The Hill coefficient for *E. coli* aspartate transcarbamylase is 2.0; it decreases to 1.4 in the presence of an allosteric activator (ATP) and increases to 2.3 in the presence of an allosteric inhibitor (CTP).

There have been several attempts to capture the phenomenon of cooperativity into a theoretical model. These include mathematical descriptors of allosteric behavior

(Hill equation and Adair equation, for example see box below) as well as physical models that incorporate enzyme structural information (see the two models by Monod, Wyman, and Changeux as well as by Koshland, Nemethy, and Filmer, briefly described later).

Oligomeric State, Subunit Cooperativity and Metabolic Switch Behavior

We define "Y" (fractional saturation) using the rearranged form of Hill equation (see Chap. 14) as

$Y = \frac{v}{V_{\max}} = \frac{K_{0.5}[S]}{1+K_{0.5}[S]}$ and on rearranging we get $[S] = \frac{Y}{K_{0.5}(1-Y)}$.

For dimeric ($n = 2$) enzymes this will be

$Y = \frac{v}{V_{\max}} = \frac{K_{0.5}[S]^2}{1+K_{0.5}[S]^2}$ and this gets rearranged to $[S]^2 = \frac{Y}{K_{0.5}(1-Y)}$.

In general, for an oligomeric enzyme consisting of "n" monomers, we arrive at a general form

$$[S]^n = \frac{Y}{K_{0.5}(1-Y)}.$$

Calculated $[S]_{0.9}/[S]_{0.1}$ ratios for oligomeric enzymes with various "n" values are shown below.

Oligomeric state (n monomers)	$[S]^1$ (n = 1)	$[S]^2$ (n = 2)	$[S]^3$ (n = 3)	$[S]^4$ (n = 4)
$[S]_{0.9}/[S]_{0.1}$ ratio	81	9	4.33	3

This analysis is based purely on the mathematical assumption of n subunits interacting with each other in an oligomer. Larger the "n" value, greater is the sensitivity to changing $[S]$; this is how typical *concentration-dependent switches* are expected to behave. However, as discussed above, only for extremely cooperative enzymes, n is equal to h.

Enzymes exhibiting cooperative/allosteric regulation are often multi-subunit proteins. But it is not necessary that oligomeric state is always associated with cooperativity. Besides the subunit cooperativity described above, there may be other reasons why oligomeric proteins are selected by evolution. Multiple subunit structure of an oligomeric enzyme may confer the following possible advantages:

(a) Multimeric nature may bestow structural stability to an otherwise unstable structural fold of a monomer. Lactate dehydrogenase is one such example.

(b) Different subunit types may be dedicated to bind different ligands (one to bind substrates—the active site, while others to bind regulatory ligands—the allosteric site). Examples include aspartate transcarbamylase and lactose synthase.

The earliest physical model to account for the behavior of allosteric proteins and enzymes was proposed by Monod et al. (1965). According to this model, in an oligomeric allosteric enzyme, the subunits occupy equivalent positions within the oligomer. Each monomer can exist in one of the two conformational states: either the R (for *Relaxed*—an active, high-affinity state with tighter binding to the ligand) or the T (for *Tense*—an inactive, low-affinity state with weak/no binding to the ligand) state. Further, the monomers are conformationally coupled to each other—when one subunit takes the R conformation, all others also change to R state such that the symmetry of the oligomer is maintained (Fig. 38.4). Hence, this model is known as the **symmetry model**. Allosteric ligands affect the $R \rightleftarrows T$ equilibrium and the subunits change their conformation in a concerted fashion. Therefore, it is also called the **concerted model**. Cooperative binding occurs when the ligand preferentially binds to the R state, thereby displacing the $R \rightleftarrows T$ equilibrium towards the R state. Sigmoidal oxygen binding to hemoglobin is a good example of this model. Nearly 100% of free hemoglobin occurs in T state while O_2 binds 70 times more tightly to the R state.

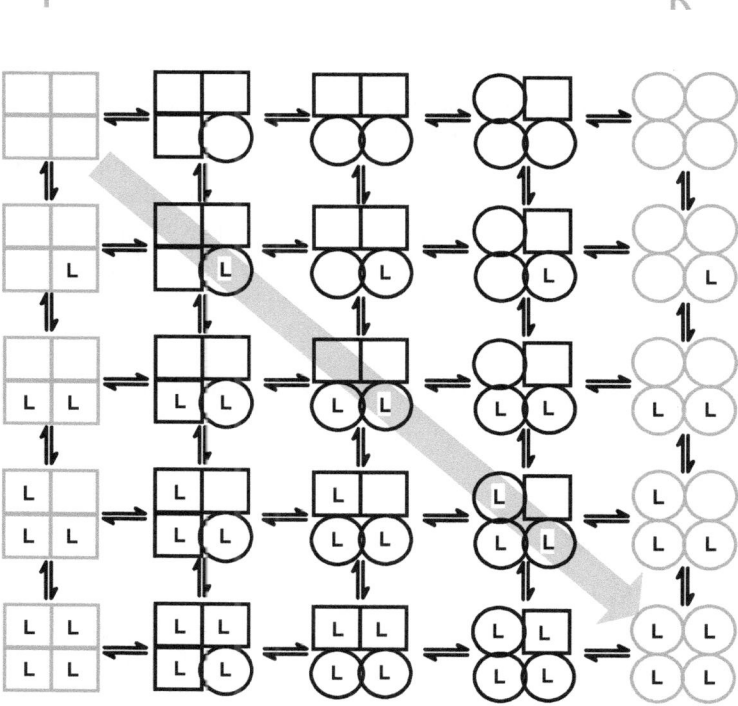

Fig. 38.4 Models of subunit cooperativity in a tetrameric enzyme. R represents a high affinity form (○) of the tetramer which is in equilibrium with T, the low affinity form (□) of the enzyme. The two vertical columns (in gray) show the species considered in the Monod, Wyman, and Changeux model. The species occurring along the diagonal (shown by the arrow) represent the forms considered by Koshland, Nemethy, and Filmer model. These two models are special cases of the more general Adair model (that includes all the enzyme species shown)

Koshland, Nemethy, and Filmer proposed another physical model to describe allosteric phenomena—the so-called *sequential model* (Koshland et al. 1966). This model is based on the concept of ligand binding by "induced fit" (Chap. 6). In the absence of the ligand, the oligomer exists in one conformational state (and not as equilibrium of R and T states). The subunits change their conformation sequentially as ligand molecules bind (Fig. 38.4). Conformational change in one subunit alters the interface of that subunit with its neighbors. This may result in more favorable (positive cooperativity) or less favorable (negative cooperativity) binding of the subsequent ligands. Unlike the symmetry model, this model can also account for and explain negative cooperativity.

More general models incorporating all possible conformational states for an allosteric enzyme have been proposed (Fig. 38.4), but with these the resulting kinetic treatment becomes extremely complex. The symmetry model (of Monod, Wyman, and Changeux) and the sequential model (of Koshland, Nemethy, and Filmer) have gained popularity over the years. The two models differ in the way ligand binding and conformational states are linked. Accordingly, they make specific predictions as to the allosteric behavior of an enzyme. The following experimental features are useful to distinguish between the two models:

(a) Observation of negative cooperativity in ligand binding points to a sequential model. For instance, rabbit muscle glyceraldehyde-3-phosphate dehydrogenase binds NAD^+ in a negative cooperative manner.

(b) Conformational changes accompany ligand binding according to both the models. However, they predict different patterns for these conformational changes. The sequential model predicts a one-to-one correspondence between the number of sites occupied by the ligand ($\overline{Y}$, the fraction of sites saturated) and the extent of conformational change ($\overline{R}$, the fraction of enzyme in the R state) observed. A plot of $\overline{R}$ against $\overline{Y}$ should therefore be linear (Fig. 38.5). However, this $\overline{R} \rightarrow \overline{Y}$ plot will be nonlinear in the case of symmetry model—at every stage of ligand binding there will be more than stoichiometric number of subunits in

Fig. 38.5 Plot of $\overline{R}$ against $\overline{Y}$ is different for symmetry and sequential models of cooperativity. The linearity of this plot supports the sequential model of Koshland, Nemethy, and Filmer while the nonlinear plot favors Monod, Wyman, and Changeux model. ($\overline{R}$ the fraction of enzyme in the R state, $\overline{Y}$ the fraction of sites saturated)

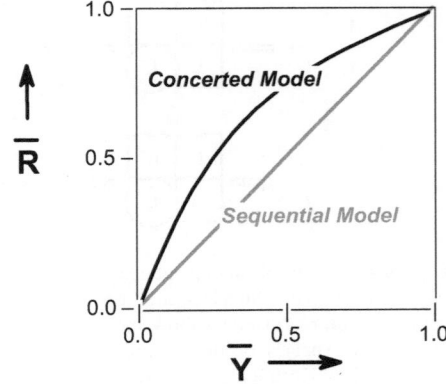

the R state. By this experimental criterion (the $\overline{R} \rightarrow \overline{Y}$ plot), the interaction of AMP with glycogen phosphorylase a conforms to the symmetry model while that of NAD^+ with glyceraldehyde-3-phosphate dehydrogenase follows the sequential model.

Structural information on allosteric enzymes and their various conformational states is necessary to understand the phenomenon of cooperativity. Ligand-induced conformational changes are the most valuable indicators. These changes could be assessed indirectly by spectroscopic probes or directly by X-ray crystallographic data. Best studied examples of distinct R and T states for allosteric proteins are hemoglobin (the honorary enzyme!), *E. coli* aspartate transcarbamylase, and phosphofructokinase.

The two original physical models account for cooperativity through ligand binding features. Direct ligand binding data (obtained by equilibrium dialysis, gel filtration, and/or ultracentrifugation) allows Scatchard analysis of the cooperative behavior. If the binding of a ligand to one subunit affects the affinity of another subunit for the same ligand, then such a cooperative interaction is termed **homotropic**. The sigmoid saturation of *E. coli* aspartate transcarbamylase by L-aspartate is an example. On the other hand, **heterotropic** allosteric effects are observed between substrates and effectors (other than that substrate). Allosteric inhibition by CTP observed on the L-aspartate saturation of aspartate transcarbamylase illustrates this point. The enzyme kinetic data (as opposed to direct ligand-binding data) often provides evidence of cooperativity. Since the enzyme activity is a manifestation of binding as well as catalysis one or both of these may account for the cooperative enzyme behavior. If an effector modifies the affinity of the enzyme for its substrate (i.e., it affects the $K_{0.5}$), then it is a **K system**. In this sense, CTP allosterically inhibits *E. coli* aspartate transcarbamylase by shifting the aspartate saturation curve to the right (the $K_{0.5}$ is increased without affecting the V_{max}). In a **V system**, the effector influences the V_{max} and not the $K_{0.5}$.

The structural biology today has offered strong evidence for multiple conformations in preexisting equilibrium for many enzyme proteins. This more dynamic view of enzyme structure needs to be addressed in describing allosteric phenomena. Accordingly, allosteric control may manifest by a *population shift* in the statistical ensembles of many states. This new outlook is discussed in a later section (Chap. 40).

38.3.2 Other Origins of Sigmoid Enzyme Kinetics

Enzymes displaying cooperative/allosteric regulation are often multi-subunit proteins. However, besides true subunit cooperativity described above, there may be other reasons why an enzyme displays sigmoid enzyme kinetics. Nature has evolved a varied set of mechanisms for generating sigmoidal effects for enzyme regulation.

(a) The enzyme activity may be a function of its state of aggregation (Lynch et al. 2017; Traut 1994). Enzyme monomers may be less active than the oligomeric aggregates. Further, this association–dissociation equilibrium may be influenced by substrate and/or regulatory ligands. Sigmoid kinetic behavior could arise from such association–dissociation. For instance, chicken liver acetyl CoA carboxylase monomers polymerize extensively to more active filamentous, polymeric aggregates; and this is helped by citrate (an activator) while palmitoyl CoA facilitates depolymerization (an inhibitor). Systems undergoing association–dissociation are amenable to experimental verification because protein concentration effects can be easily observed.

(b) Multiple kinetic paths exist for an enzyme exhibiting random mechanism. The reaction path followed at lower substrate concentrations may differ from the one favored at higher $[S]$. If the net rates of the two paths are different (and the one at higher $[S]$ is faster) a sigmoid kinetic curve ensues because of the substrate concentration-dependent reaction path switching!

(c) Although rare, there are examples of single subunit enzymes showing sigmoid kinetics. This kinetic behavior arises due to slow conformational change as a part of the enzyme catalytic cycle. Such systems with conformational memory are also called as *hysteretic* enzymes or *mnemonic* enzymes. For instance, glucokinase is monomeric but shows sigmoid kinetics with respect to its substrate glucose (Larion et al. 2012). The monomer undergoes a slow, glucose concentration-dependent conformational transition. An *h* value of 1.7 for glucose is reported.

(d) Complex non-Michaelian kinetics ensues when a mixture of isoenzyme forms with differing kinetic constants is analyzed. One isoform may saturate earlier than the other and the resultant $v \rightarrow [S]$ curve (overlap of the two!) could be sigmoid. We will look at isoenzyme regulation in some detail in a subsequent section.

38.3.3 How to Study Regulatory Enzymes

First and the foremost, one should ensure that the so-called regulatory kinetic behavior is not due to some artifacts of the assay. Many such caveats are discussed earlier (Chap. 11). Misleading sigmoid kinetics may be recorded when oxidation of enzyme or substrate, depletion of substrate due to complex formation, etc., occur during the enzyme assay. Before ascribing sophisticated regulatory mechanisms to unusual enzyme kinetic behavior such artifacts must be discounted.

In order to assess cooperative behavior, it is necessary to examine enzyme activity over a wide range of substrate concentration. The kinetic data may then be analyzed through suitable plots (like the Hill plot) to measure the degree of cooperativity. An idea about the enzyme architecture is required to ascertain the possibility of subunit interactions leading to cooperativity. Typical protein quaternary structure determination techniques such as study of molecular weight in the absence and presence of denaturing agents, subunit composition, and cross-linking studies are useful. The kinetic cooperativity may correspond to ligand-binding in an oligomeric protein. This can be tested through direct binding methods such as equilibrium dialysis, ultracentrifugation, etc. Once again, the ligand-binding cooperativity may be evaluated through Scatchard plot or Hill plot of the binding data (see Chap. 16). Conformational changes often accompany ligand binding. These can be scored directly by X-ray crystallography, cryo-EM, and NMR spectra. Indirect approaches with chromophores and fluorophores as reporters or accessibility of amino acid residues to chemical modification reagents are also useful. Finally, presence of allosteric sites (distinct from the active site) to bind regulatory ligands may be ascertained by desensitizing the enzyme through physicochemical and mutational tricks.

38.4 Isozymes and Regulation

Isozymes are multiple molecular forms of an enzyme catalyzing the same chemical reaction. They differ from each other in their primary sequence but often are of comparable size and are unique translational products of distinct genes. The covalent modification states (like the phosphorylated forms, etc.) of the same enzyme are not isoenzymes by this definition. Isozymes play critical roles in cellular and metabolic regulation. They may be found in the same cell but in different (a) metabolic states (such as NADP-glutamate dehydrogenase versus NAD-glutamate dehydrogenase), (b) organelles to integrate cellular metabolism (malate shuttle; mitochondrial and cytosolic malate dehydrogenase isoforms in heart muscle), (c) tissues to facilitate interorgan metabolism (lactate dehydrogenase in the skeletal muscle versus the liver), and (d) stages of development (e.g., laccases and trehalases during sporulation). Isoenzymes may be catalogued according to their distinguishing features and perceived metabolic significance (Table 38.1).

For the same chemical reaction, enzymes can be evolved that are more effective catalysts for one direction than the other. This is possible even though Haldane relationship (see Chap. 14) places certain constraints on the kinetic parameters of the enzyme. Recall that

Table 38.1 Isoenzymes grouped according to their metabolically significant features

Differing feature	Examples[a]
Michaelis constant	Hexokinase (m), aldolase (m)
Substrate and cofactor specificity	Glutamate dehydrogenase (f), isocitrate dehydrogenase (p)
Allosteric properties	Hexokinase (m), aspartate kinase (b)
Subcellular localization	Carbamyl phosphate synthetase (f), malate dehydrogenase (m)
Tissue/organ localization	Arginase (m), lactate dehydrogenase (m)
Catabolic or biosynthetic (inducibility)	Alcohol dehydrogenase (f), threonine deaminase (b)

[a]Isoenzymes observed in bacteria (b), fungi (f), mammals (m), and plants (p)

$$\frac{V_{\max f} \times K_{MP}}{V_{\max r} \times K_{MS}} = \frac{[P]_{eq}}{[S]_{eq}} = K_{eq},$$

and K_{eq} is an immutable thermodynamic parameter that the catalyst cannot tinker with. It is however perfectly possible to have more than one numerical solution to satisfy the above equation. For example, two isozymes may have the same $V_{\max f}$ values but different K_{MS}s. This is compensated by appropriate $V_{\max r}$ and K_{MP} values in the two cases, thereby resulting in an identical K_{eq} value. The isozyme with a lower $V_{\max f}/K_{MS}$ could either have a suitably lowered $V_{\max r}$, an elevated K_{MP}, or both. The forward reaction rate (the first-order rate with respect to $[S]$) is given by

$$v_f = \frac{V_{\max f}}{K_{MS}}[S] = \frac{k_{catf}[E]_t}{K_{MS}}[S].$$

On comparing the v_f for the two isozymes at a given $[S]$, the enzyme form with a lower K_{MS} performs better in this direction. Two general observations can now be made:

(a) As many enzymes never need to catalyze a reaction in the reverse direction in vivo, there is no evolutionary pressure to achieve catalytic perfection in that direction. If the active site is strictly complementary to the transition state, then the enzyme will be an optimized catalyst for both directions. Efficiency in one direction could however be preferentially improved by evolving an active site that binds either S or P better than it binds the transition state. Indeed, methionine adenosyltransferase is one such *one-way enzyme* (with its limiting-forward rate about 10^5 times greater than the reverse one).

(b) Since $V_{\max f} = k_{catf}[E_t]$, any unfavorable k_{catf} changes during catalyst design/evolution (arising out of thermodynamic constraints—such as Haldane relationship) can be compensated by the system. Despite having a lower k_{catf}, one can maintain the desired $V_{\max f}$ by increasing $[E_t]$. In reality, this implies that the cellular concentrations (abundance!) of various isozymes need not necessarily be maintained at the same level.

38.4.1 Isozyme Dedicated to a Pathway

Nature employs isozymes as a means to compartmentalize and regulate metabolism. These may be wired into metabolism to perform specific roles. Isozymes with dedicated function may be differently regulated at the genetic level. Bacteria elaborate two distinct isoforms of threonine deaminase: one for biosynthesis (with higher affinity for Thr; low K_M) and the other for catabolism (with lower affinity for Thr; high K_M). Similarly, the two carbamyl phosphate synthetases serve to feed the biosynthesis of arginine and pyrimidine, respectively. Multiple ω-amino acid transaminases are expressed in response to the availability of respective ω-amino acid inducers in the medium. They may be specific (such as GABA transaminase) or generic in their substrate specificity. There are two glutamate dehydrogenases in fungi to satisfy the cellular needs. They catalyze the same chemical reaction but in opposite directions! The NAD-glutamate dehydrogenase (catabolic enzyme) is induced upon nitrogen starvation and/or when glutamate is the sole nitrogen source available to the cell. The NADP-dependent enzyme is biosynthetic and is responsible for the synthesis of cellular glutamate. Interestingly, yeast displays yet another isoform (of NADP-glutamate dehydrogenase) during diauxic growth on ethanol—dedicated to make glutamate when the carbon source is switched to ethanol (and is not glucose).

38.4.2 Isozyme to Suit a Metabolic Demand

Isozymes may have evolved distinct kinetic virtues to suit the metabolic demands of an organism. The two mammalian lactate dehydrogenase isoforms are well suited for the requirement of converting lactate to pyruvate (H_4 form in the heart) or pyruvate to lactate (M_4 form in the skeletal muscle). In addition, pyruvate inhibits the muscle form of lactate dehydrogenase. On the whole, pyruvate is completely oxidized by the heart (glycolysis + Krebs cycle) while in the skeletal muscles, it is reduced to lactate and sent into the blood stream. The liver displays both isoforms and is able to reconvert lactate to pyruvate so that gluconeogenesis can occur. While the Cori cycle depends on their functionality, the proposed metabolic role of lactate dehydrogenase isozymes is far from conclusive.

Kinetic differences between isozymes may favor forward reaction in one tissue and the back reaction in another. Despite the immutability of the equilibrium constant, one can envision various absolute values for V_{maxf} and V_{maxr} as well as different V_{maxf}/V_{maxr} ratios. This is possible within the constraints of Haldane relationship (discussed above). Aldolase isozymes are good examples where such kinetic comparisons have been made. The liver isozyme (aldolase B) is clearly more effective in fructose-1,6-bisphosphate synthesis (and hence gluconeogenesis!). Table 38.2 summarizes the relevant kinetic data for the two aldolase isozymes.

Table 38.2 Kinetic features of rabbit aldolase isozymes

	Isozyme form	
Kinetic property	Aldolase A (muscle)	Aldolase B (liver)
V_{maxf} (fructose 1,6-bisphosphate cleavage)	$5300\ min^{-1}$	$250\ min^{-1}$
V_{maxr} (fructose 1,6-bisphosphate synthesis)	$10{,}000\ min^{-1}$	$2600\ min^{-1}$
K_M (fructose 1,6-bisphosphate)	$60\ \mu M$	$1\ \mu M$
K_M (dihydroxyacetone phosphate)	$2000\ \mu M$	$400\ \mu M$
K_M (glyceraldehyde 3-phosphate)	$1000\ \mu M$	$300\ \mu M$
(Fructose 1,6-bisphosphate)/(fructose 1-phosphate) activity ratio[a]	50	1

[a]The ratio favors the liver isoenzyme (aldolase B) for fructose metabolism via fructose 1-phosphate

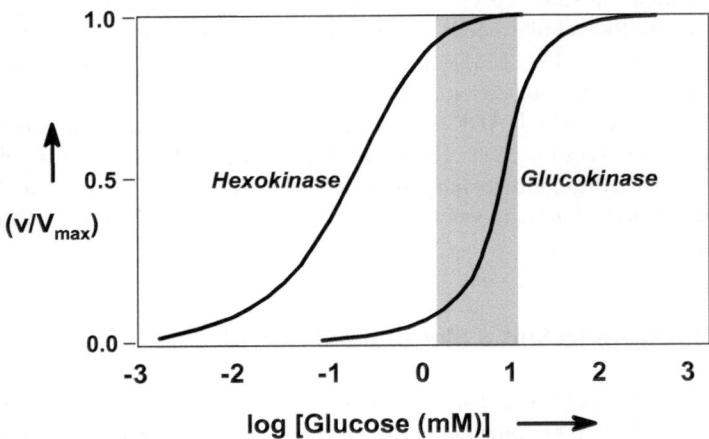

Fig. 38.6 Glucose saturation of hexokinase and glucokinase of liver. As the two isoenzymes show significantly different K_M for glucose (summary table above), a semilog plot of log [Glucose] with fractional velocity (v/V_{max}) on Y-axis, is shown (see Chap. 16)

Glucokinase occurs only in the liver while hexokinases are found both in the muscle and the liver. Further, hexokinase, but not glucokinase, is inhibited by the product glucose 6-phosphate. These features, in combination with their respective Michaelis constants, are in accordance with their role in glucose homeostasis in the body. The glucose concentration response of the two isoenzymes is shown in Fig. 38.6. It is obvious that liver (with its glucokinase) is able to respond and process high glucose presented to it in the blood.

Yet another example of isozymes tailored for catalysis in opposing directions is alcohol dehydrogenase (ADH) in yeast. Yeast growing in the absence of oxygen displays the constitutive ADH-I that is designed for aldehyde reduction. The ADH-II is induced upon aerobic growth and is well suited to oxidize alcohol to aldehyde for its further entry into Krebs cycle.

38.4.3 Isoenzymatic Regulation and Additive Inhibition

A single enzyme reaction leading to several end-products is a potential problem for regulation. Feedback inhibition by one of these products would not only affect its own formation but also interferes with that of the other products. Therefore, in some branched biosynthetic pathways, a number of discrete isoenzymes exist for the first committed step. Each one of them responds specifically and differently to inhibition by the various end-products. The aspartokinases from *E. coli* were the first examples described by Earl Stadtman's group at NIH. These isoforms are involved in the biosynthesis of L-aspartate family of amino acids (Lys, Thr, Met, and Ile). The multiplicity of aspartokinases is feedback inhibited by different end-products of the branched pathway. One isozyme is Thr-sensitive while the other is Lys-sensitive. Because of this, aspartokinase from *E. coli* extracts exhibits ***additive inhibition*** by Thr and Lys. Further, the Lys-sensitive isozyme is also under Lys repression. The Thr-sensitive isozyme is subjected to multivalent inhibition by Thr and Ile. A second interesting example of enzyme multiplicity is 3-deoxy-D-arabino-heptulosonate-7-phosphate (DAHP) synthase from *E. coli*. Formation of DAHP from erythrose-4-phosphate and PEP is the first step in the aromatic amino acid biosynthesis. Of the three isozymes of *E. coli* DAHP synthase, one is inhibited by Phe while the second by Tyr. The third minor isoform is inhibited by Trp. Although there are many sequence differences between the DAHP synthase (Phe) and the DAHP synthase (Tyr) of yeast, a single residue determines the sensitivity to feedback inhibition; the isozyme with Gly-226 is Tyr-sensitive whereas that with Ser-226 responds to Phe inhibition.

38.5 Covalent Modifications and Control

Many enzymes (and hence pathways) are controlled by mechanisms that involve posttranslational covalent modification of proteins (Huber and Hardin 2004; Walsh et al. 2005; Mesquita et al. 2024). Here, the actual enzyme protein concentration (i.e., $[E_t]$) remains unaffected but the activity of existing enzyme molecules is altered. We are interested in covalent modifications that lead to altered enzyme activity/specificity. The modification may result in a less active or more active form of the unmodified enzyme. Examples of covalent modification types amenable to regulation of enzyme activity are listed in Table 38.3. As we shall see shortly, such covalent modification may be reversible or irreversible.

38.5.1 Zymogen Activation by Limited Proteolysis is Irreversible

Proteolytic cleavage of polypeptides (and enzymes) is an essentially irreversible event. More often, we think of this process in terms of protein degradation. However, in some cases, limited and selective peptide bond cleavage can lead to activation of an inactive precursor—the so-called ***zymogen activation***. The well-known example of zymogen activation in the small intestine is a regulatory trick (safety mechanism) to guard against premature activation of pancreatic proteases. Enteropeptidase (earlier known as enterokinase) of the small intestine initiates the zymogen activation cascade (Fig. 38.7) by converting trypsinogen to trypsin. This is achieved by the cleavage of a single peptide bond of the zymogen. Once a small amount of trypsin is generated, the further conversion of trypsinogen to trypsin becomes autocatalytic (see inset, Fig. 38.7). A significant feature of zymogen activation is amplification of the initial signal. The irreversible cascades of zymogen activation also occur in blood clotting and activation of the complement system. In these two examples, each factor upon activation proteolytically cleaves the next zymogen in the sequence (for details see standard text books of biochemistry). These

Table 38.3 Well-characterized covalent modifications in enzymes

Type of modification	Target in protein	Examples
Irreversible (once only)	**Peptide bond**	
Limited (specific) proteolysis	(Asp–Lys–Ile–Val, Ala–Arg–Ile–Val, Ser–Arg–Ile–Val)	Pancreatic zymogens (trypsinogen, chymotrypsinogen, proelastase, procarboxypeptidases A and B)
	Arg–Ile, Arg–Gly	Blood clotting factors (factors XIIa, XIa, IXa, VIIa, Xa and prothrombin)
	Single site?	Complement system (C1r, C1s, C2)
Reversible (back and forth)	**Amino acid**	
Phosphorylation–dephosphorylation	Ser, Thr, Tyr, Lys	Glycogen synthase, glycogen phosphorylase, phenylalanine hydroxylase, triglyceride lipase, acetyl CoA carboxylase, cdc kinase, NAD-glutamate dehydrogenase
Nucleotidylation–denucleotidylation	Tyr, Ser	Glutamine synthetase and P_{II} protein (Gram negative bacteria)
Acetylation–deacetylation	Cys?	Citrate lyase (anaerobic bacteria)
Thiol-disulfide interchange	Cys	Plants (light activation)
ADP-ribosylation	Arg, Glu, Lys	EF-2 (diphtheria toxin), G protein (cholera toxin), glutamine synthetase (mammalian)
Methylation	Asp, Glu, Lys, His, Gln	Bacterial chemotaxis, histones

cascade systems are amenable to control and exhibit great signal amplification by using a catalyst to create more catalysts.

38.5.2 Enzyme Regulation by Reversible Covalent Modification

Many key enzymes of metabolism interconvert between two forms that differ in catalytic properties. While more than 100 different covalent modifications of amino acid residues are known, in terms of enzyme regulation, phosphorylation, nucleotidylation, and ADP-ribosylation are important and frequent. Regulation by thiol-disulfide interchange, well demonstrated in plants, may also be significant in other organisms. The best-documented example of adenylylation control is that of *E. coli* glutamine synthetase (see below). An extremely versatile mechanism of reversible covalent modification is via protein phosphorylation and dephosphorylation (Krebs and Beavo 1979). This is a predominant mode of control in eukaryotes. Reversible phosphorylation–dephosphorylation event occurs at the –OH of a Ser, Thr, or Tyr on the enzyme protein. The enzymes accomplishing phosphorylation and dephosphorylation are known as ***protein kinase***s and ***protein phosphatase***s, respectively. Kinase specificity is crucial to the fidelity of signaling pathways. This

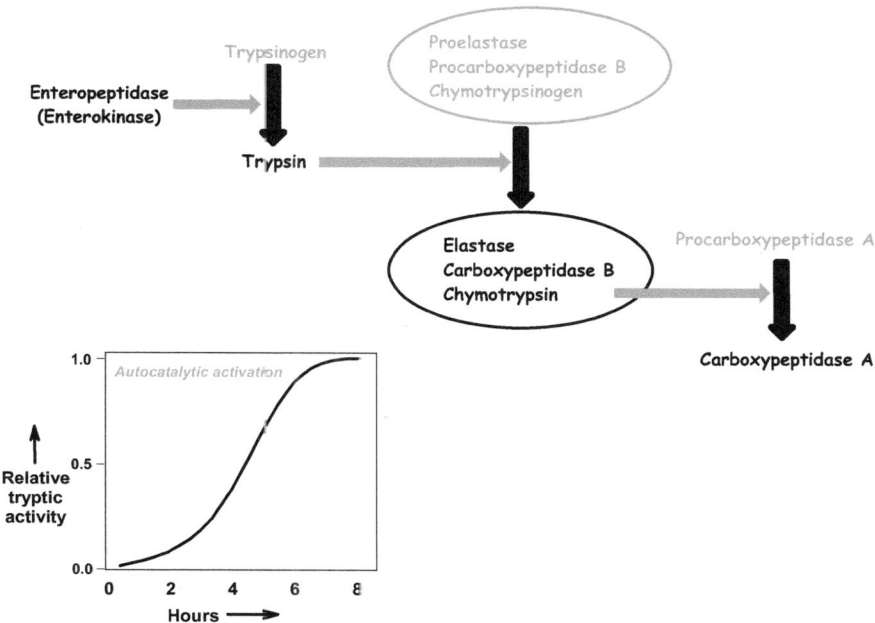

Fig. 38.7 Sequence of protease activation events associated with pancreatic zymogens. Activated enzyme forms are in black and zymogens are in gray. Black arrows indicate activation events catalyzed by the relevant enzyme (grey arrows). Autocatalytic activation of trypsinogen by trypsin is shown in the inset

specificity arises in part from the enzyme active site, but kinases are often tethered to their substrates and many phosphorylation reactions occur within macromolecular complexes. Tethering can enhance the rate by orders of magnitude which is also highly sensitive to the length of the intrinsically disordered linkers (Dyla and Kjaergaard 2020). The protein modification and its reversal are catalyzed by separate (converter) enzymes—a catalyst acts upon another—to create a *cascade* system. A well-studied example of an interconvertible enzyme is glycogen phosphorylase (Fig. 38.8). Glycogen phosphorylase along with its converter enzymes (the kinase and the phosphatase) defines a *monocyclic cascade*. In such a system, allosteric effectors can bind to either or both converter enzymes or directly to the interconvertible enzyme. One or more effectors can thus lead to an adjustment in the steady-state level of phosphorylation and the activity of glycogen phosphorylase. Monocyclic cascades can thus sense changes in the concentrations of many different metabolites and act as metabolic integration systems. When a converter enzyme (glycogen phosphorylase kinase) itself is subject to phosphorylation/dephosphorylation (a kinase kinase!), then a *bicyclic cascade* is defined. Glycogenolysis in the liver provided the first example of this kind (Fig. 38.8). Nucleotidylation control of *E. coli* glutamine synthetase (see below) provides yet another case of a bicyclic cascade (although a closed one!) (Stadtman and Chock 2014). Reader may refer to literature

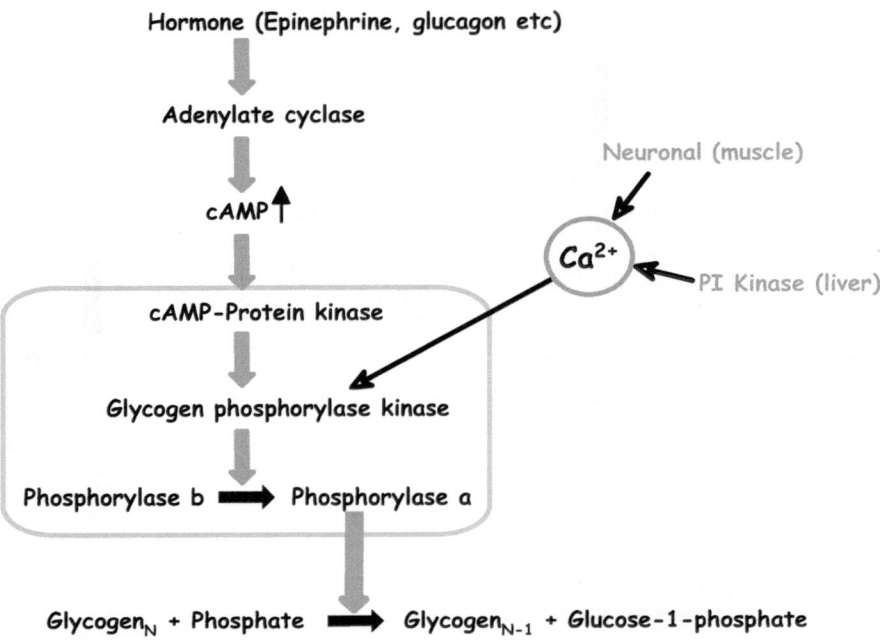

Fig. 38.8 Protein phosphorylation cascade of glycogen breakdown. Glycogenolysis in the liver and muscle is triggered by hormone action (also by neuronal stimulation in the muscle). Multiple protein phosphorylation steps allow exquisite control of glycogen phosphorylation (gray box). The phosphorylation cascade is further complicated by that (a) the phosphorylation state itself may also be determined by associated protein phosphatases and (b) at every level, both phosphorylation and dephosphorylation may be differently (reciprocally) affected by allosteric modulators like AMP and glucose-1-phosphate. This regulation also receives inputs from multiple second messengers like cAMP, Ca^{2+}, inositol-1,4,5-trisphosphate (IP_3), and diacylglycerol (DAG). The tissue-specific differences are not explicitly shown in this figure, but the reader is directed to excellent texts on this subject. Finally, a similar mechanism reciprocally controls glycogen synthase activity—phosphorylations leading to net decrease in its activity

on many other examples of excellent metabolic control by enzyme covalent modifications.

The phosphorylation of enzyme targets is a well-recognized mode of regulating enzyme activity (and hence metabolism) in eukaryotes. Typically, protein kinases recognize –Arg–Arg–X–Ser– or –Arg–Lys–X–Ser– sequences of the target enzyme and phosphorylate at the Ser–OH. Experimental analysis and interpretation of phosphorylation/dephosphorylation events is tricky because of the following: (a) A single protein kinase may be able to phosphorylate many different targets in vitro. Elucidating their in vivo relevance can be daunting. (b) Many target enzymes are subject to multiple phosphorylations at different sites. Not all of these phosphorylations may be directly involved in the control of activity. (c) The issues discussed for protein kinases (in the above two points) are equally relevant to the functioning of phospho-protein phosphatases. Nevertheless, some early discoveries like the phosphorylation/dephosphorylation cascade control of glycogen metabolism (Fig. 38.8) have been resolved to significant details.

Both positive and negative feedback loops are noted in enzymes that auto-catalyze covalent modifications. The poly ADP-ribosylating enzymes (PARP 1 and PARP 2) show auto-inhibition as a function of self PARylation and their time-course behavior displays an initial burst followed by constant reduction in activity. Auto-activation is often seen in kinases where the time-courses have an initial lag followed by continuous increase in slope.

38.5.3 Metabolic Significance of Covalent Modifications

The reversible covalent modification and cyclic cascade systems are of value in the regulation and integration of cellular metabolism. This mode of control has certain obvious advantages.

- The system develops a capacity for *signal amplification*. A small amount of signal (such as a hormone acting via cAMP; Fig. 38.9) can act on a much larger amount of target enzyme (via the converter enzyme(s)). For example, in the muscle tissue, the molar concentrations of cAMP protein kinase (0.2 μM), phosphorylase kinase (2.5 μM), and phosphorylase (80 μM) are widely different. Whereas, the intracellular cAMP levels do not exceed 2–3 μM. At these concentrations, cAMP obviously cannot activate phosphorylase directly! (Ferrell Jr 1996).
- They can act as *catalytic amplifiers* as a small amount of converter enzyme can act on a much larger amount of interconvertible enzyme target. For instance, the initial signal of enteropeptidase is quickly amplified by converting trypsinogen to trypsin.
- With multi-cyclic cascades, besides the capacity for signal and rate amplification, the *flexibility of response* of the system increases exponentially. Individual enzyme proteins, on an average, are not really big molecules. Their surface area is not unlimited and may accommodate a few additional sites for regulatory ligands besides the active site. Cyclic modification systems involve more enzymes and therefore can harbor more sites for effector binding. For example, *E. coli* glutamine synthetase responds directly to about 15 ligands, but indirectly (through the other players of the bicyclic cascade) to another 20 more regulators.

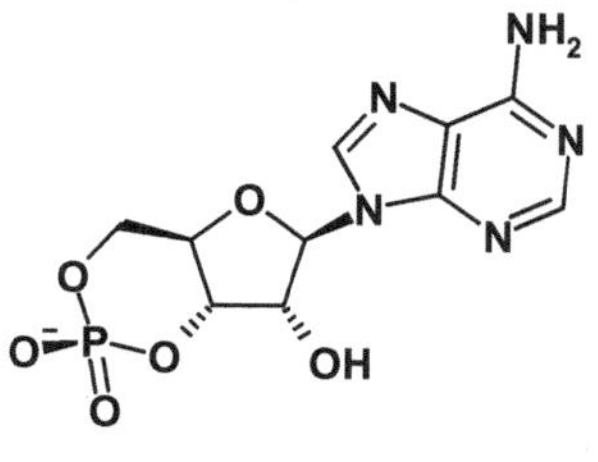

Fig. 38.9 Structure of cyclic AMP—the second messenger

Such systems have the capacity to respond to a range of signals thereby achieving *integration of metabolism* (Chock et al. 1980).

38.5.4 Role of Signal Molecules and Energy Charge

Protein phosphorylation is the most common (and prevalent) covalent modification observed in eukaryotic enzyme regulation. Protein kinases (that phosphorylate target enzymes) themselves respond to control signals. The major signal molecules include cyclic nucleotides (cAMP and cGMP), Ca^{2+}, Ca^{2+}/calmodulin, and diacylglycerol. Each one of these activates a unique class of protein kinases. The cAMP-dependent protein kinase is an important member of the hormonal activation pathway of glycogen phosphorylase (Fig. 38.8). Phosphorylation of phosphofructokinase-2 by cAMP-dependent protein kinase leads to its inhibition; this leads to decrease in fructose-2,6-bisphosphate levels and lowered phosphofructokinase and of glycolysis.

Similarly, diacylglycerol activates protein kinase C, which in turn catalyzes the phosphorylation of a range of proteins involved in cellular processes. The function of molecules like cAMP (Fig. 38.9) appears to be solely to act as signal in controlling enzyme activities. Sutherland called cAMP the *second messenger* for hormone action (Blumenthal 2012). Many hormones act on adenylate cyclase through G-proteins. Activation of adenylate cyclase leads to the synthesis and hence a rise in the intracellular levels of cAMP. Regulation by various second messengers and hormonal/neural control of cellular activities is beyond the scope of this book. The reader may refer to excellent and detailed account of this area in advanced texts on biochemistry.

It is well accepted that key glycolytic enzymes are responsive to adenylate compounds namely ATP, ADP, and AMP. Atkinson introduced the term *energy charge* to denote the relative concentrations of these three adenine nucleotides:

$$\textit{Energy Charge} = \frac{[ATP] + 1/2[ADP]}{[ATP] + [ADP] + [AMP]}.$$

According to this definition, the energy charge of a cell can take values from zero (when all the adenine nucleotide pool is present as AMP) to one (when all the adenine nucleotide pool is present as ATP). Typically, the energy charge of cells is maintained at or above 0.95. At least in the short term, the cellular concentration of adenine nucleotide pool (the sum [ATP] + [ADP] + [AMP]) remains constant while their relative concentrations are [ATP] >> [ADP], [AMP] (for instance, a 5 mM total of adenine nucleotides may be present in the ratio ATP:ADP:AMP as 4.48: 0.50:0.02). It appears that relative concentration change in the adenine nucleotides is a key determinant in sensing the state of metabolism, with AMP as the regulatory signal. This is because [AMP] is related to [ATP] through the equilibrium reaction catalyzed by adenylate kinase (also known as myokinase):

$$2\,\mathrm{ADP} \rightleftarrows \mathrm{ATP} + \mathrm{AMP}\,(K_{\mathrm{eq}} = 0.44)$$

Due to this equilibrium, small changes in [ATP] are quickly amplified many fold in terms of [AMP]. A 10% decrease in [ATP] would lead to more than fourfold increase in [AMP]. Again, it is the fractional change in [ATP] and [AMP] is critical (and not their absolute concentrations) for the regulation of target enzyme activity. AMP is an intracellular allosteric signal sensed by many enzymes. A specific protein kinase is activated by AMP in two ways: (a) AMP is an allosteric activator of this enzyme, and (b) AMP makes this protein more susceptible to phosphorylation by a kinase kinase.

Some of the regular cellular metabolites also act as critical signals, either through effects on metabolic pathways or via modulation of other regulatory proteins. These novel metabolite roles involve signaling processes outside metabolism such as nutrient sensing and storage, embryonic development, cell survival and differentiation, etc. (Baker and Rutter 2023).

38.6 Protein–Protein Interactions and Enzyme Control

Regulation of enzyme activity by reversible interactions with small molecular weight metabolites (as inhibitors or activators) is a common theme in metabolism. Many of them are described earlier in this chapter (Sect. 38.2). Enzyme–enzyme interactions may also play significant role in the regulation of metabolic reaction pathways (Srivastava and Bernhard 1986). What about protein binding to enzymes and regulating their activity? Although very few in number, there are indeed well-defined examples of this mode of enzyme control as well. A distinct regulatory protein may reversibly (or irreversibly) associate with an enzyme, thereby increasing or decreasing its overall activity (Table 38.4). In one sense these protein modulators could be viewed as *regulatory subunits* of the target enzyme. However, the concept is different from regulatory (R) subunits of allosteric enzymes, permanently associated with the catalytic (C) subunits of an enzyme (e.g., *E. coli* aspartate transcarbamylase with its $3R_2$–$2C_3$ architecture).

Protein modulators may regulate enzyme activity in different possible ways. This mode of regulation is often unique to an enzyme system or group of organisms. Examples include modifying the substrate specificity of an enzyme (lactalbumin), mediating the activation/inhibition of an enzyme upon receiving environmental cues (Ca^{2+}/calmodulin and α_s or α_i subunits of G-proteins), coupling a biosynthetic enzyme activity to end-product availability (epiarginasic control in *S. cerevisiae*), titrating out the active enzyme by presenting themselves as suicide substrates (serpins) (Silverman et al. 2001), or marking the enzyme for proteolysis (antizyme) (Small and Traut 1984). We have already noted the role of ubiquitination in targeting proteins for degradation. More recently, *Bacillus subtilis* has been reported to control unwanted glutamate breakdown through direct binding and inhibition of its glutamate dehydrogenase (GudB) by glutamate synthase (GltAB) (Jayaraman et al. 2022;

Table 38.4 Well-characterized protein modulators of enzyme activity

Regulator	Target enzyme	Comments
Lactalbumin	Lactose synthase (α subunit)	Lactalbumin is devoid of any catalytic activity; but it alters the specificity of the α subunit to synthesize lactose instead of N-acetyllactosamine
Calmodulin	Protein kinases	Many protein kinases (e.g., phosphorylase kinase) are activated by Ca^{2+}/calmodulin
G-proteins	Adenylate cyclase	Upon interacting with hormone receptor complex, α subunit exchanges GTP for GDP and detaches from the βγ dimer of the G-protein; adenylate cyclase is either stimulated by the $α_s$ or inhibited by $α_i$
Arginase	Ornithine transcarbamoylase	Yeast arginase and ornithine transcarbamoylase form a transient complex in the presence of ornithine/arginine. Ornithine carbamyltransferase activity is thereby inhibited (termed *epiarginasic control*)
Glutamate synthase	Glutamate dehydrogenase	Glutamate synthase forms a transient complex with glutamate dehydrogenase and inhibits its activity. Wasteful glutamate recycling is prevented
Serpins	Serine proteases	Serpins (*ser*ine *p*roteinase *in*hibitors) inhibit serine proteases by an irreversible suicide substrate mechanism. Well known examples include antithrombin (clotting), C1-inhibitor (complement activation) and antiplasmin (fibrinolysis)
Antizyme	Ornithine decarboxylase	Antizyme levels increase in response to elevated cellular polyamines. Upon antizyme binding, proteasome-mediated degradation of mammalian ornithine decarboxylase ensues

Hartmann 2022). This mode of regulation avoids wasteful recycling of 2-oxoglutarate at the interface of basic carbon and nitrogen metabolism.

38.7   Compartmental Regulation and Membrane Transport

We have seen early in this chapter that access to the substrate (as well as inhibitors and activators) is a major factor that determines the overall rate of an enzyme-catalyzed reaction in vivo. Controlling this access itself can be used as a regulatory feature. One way to go about this is to keep the substrate (or inhibitor or activator) and the enzyme in distinct compartments so that they do not see each other unless required. This separation is achieved in three principal ways.

Many enzymes are selectively expressed during specific phases of growth and/or development. However, they may not encounter corresponding substrates until much later. For instance, glutamate decarboxylase and trehalase are stored in fungal spores in latent forms. They get to see their substrates (glutamate and trehalose, respectively) only at the time when conditions are right for spore germination. Such *temporal compartmentalization* (separation of enzyme and its substrate in time) is also common in secondary metabolism. Fatty acid synthase and polyketide synthases share a few common substrates but rarely compete with each other. Polyketide synthases are generally expressed after the cessation of growth phase. Similarly, the sugar precursors are directed to aminoglycoside antibiotic synthesis as secondary metabolism is associated with post growth phase.

Essentially similar metabolites may serve as substrates for two distinct enzymes located in the same subcellular compartment. One way to dedicate their use by one or the enzyme is to differentially tag the metabolite and exploit enzyme specificity. The best example of this is selective use of NAD for catabolic purposes and NADP for biosynthetic reactions. For instance, despite the availability of NADH in the same compartment, the biosynthetic glutamate dehydrogenase uses NADPH. The two cofactors differ in a phosphate group but have essentially the same redox potential. Other instances of *chemical compartmentalization* include selective use of ATP or GTP as phosphate donor (and ADP or GDP as phosphate acceptor), UDP glucose for glycogen synthesis and glucose-1-phosphate from glycogen breakdown, CDP derivatives in phospholipid metabolism, etc. Fructose-1,6-bisphosphatase acts on the α anomer of fructose-1,6-bisphosphate whereas phosphofructokinase generates the β anomer of fructose-1,6-bisphosphate. Many steps of fatty acid biosynthesis and catabolism involve chemically similar intermediates. They are segregated in the same physical compartment of a prokaryotic cell by distinct chemical tags. All the intermediates of fatty acid biosynthesis are tethered to acyl carrier protein while the similar intermediates of catabolism are free as CoA derivatives; the stereochemistry of the β-hydroxy intermediates are also opposite. An interesting example of similar set of reactions but distinct chemical compartments is L-proline and L-ornithine biosynthesis in fungi. Glutamate is the precursor for proline biosynthesis, whereas N-acetylation of glutamate dedicates it for the biosynthesis of L-ornithine. The acetyl tag is removed from N-acetyl L-ornithine subsequent to its synthesis.

A prominent feature of eukaryotic cells is their reliance on achieving enzyme/metabolic regulation through *physical compartmentalization*. The presence of distinct membrane-enclosed organelles allows them to spatially separate otherwise competing reactions. N. crassa arginine metabolism is an excellent case in point. The L-ornithine biosynthesis occurs in mitochondria. Of the two carbamyl phosphate synthetase isozymes, the mitochondrial form is dedicated for ornithine/arginine biosynthesis. Excess cellular arginine is actively transported and sequestered into fungal vacuoles. The ornithine arising out of cytosolic arginase reaction cannot access entry into mitochondria—relevant membrane permease is sensitive to competitive inhibition by cytosolic arginine. Since participation of ornithine in the biosynthesis as well as catabolism of arginine is obligatory, wasteful cycling of ornithine is a distinct possibility. The overall compartmental organization of arginine metabolism in N. crassa ensures that futile cycling of ornithine does not occur. Besides regulation of enzyme activity, this is achieved by recruiting at least two specific transporters located on vacuolar and mitochondrial membranes. Yet another example of physical compartmentalization is that of nitrogen fixation in filamentous cyanobacteria. The oxygen generated by photosynthetic cells is toxic to nitrogenase. The nitrogenase and the nitrogen fixing apparatus are quarantined to specialized non-photosynthetic cells called heterocysts, which are fortified with defense enzymes like catalase, peroxidase, superoxide dismutase, etc.

38.7.1 Mediated Transport Versus Enzyme Kinetics

Cellular physical compartments most often are separated by phospholipid bilayers. Transport across these biological membranes is an important phenomenon. With the exception of a few molecules, most nutrients, metabolites, and ions are transported across such biological membranes through protein mediators. Various modes of transport and their characteristics are listed in Table 38.5. Rate of transport due to simple diffusion increases proportional to the existing concentration gradient—and no saturation is observed. On the other hand, like enzyme catalysis, mediated transport processes show saturation kinetics.

Mediated transport may be passive (facilitated diffusion) or active (input of energy to drive the transport against the prevailing concentration gradient). Both these forms of mediated transport are also amenable to general kinetic analysis used to analyze enzyme catalysis. Rate of transport may be saturated with the substance (denoted as A) transported—plots of initial rates of transport (transport flux, J_{tr}) versus substance concentration show a hyperbolic saturation (similar to the Michaelis–Menten kinetics for an enzyme; see Chap. 14).

$$J_{tr} = \frac{J_{tr\,max}\,[A]}{K_{tr} + [A]}.$$

Other important features of mediated transport that resemble enzyme kinetics are specificity towards the ligand (compound or ion) being transported, competitive inhibition, pH dependence of transport, and the ability to be modulated by inhibitory substances. For example, the glucose carrier of erythrocyte membranes facilitates the transport of glucose down the concentration gradient. This carrier (a) transports some other sugars like mannose and fructose, albeit less effectively and (b) is competitively inhibited by 2,4,6-trihydroxyacetophenone. The erythrocyte

Table 38.5 Different modes of transport across biological membranes

Nature of transport	Transporter mediated	$\Delta G°_{tr}$ ($= -RT \ln ([A]_{in}/[A]_{out})$)	Examples
Passive transport			
Simple diffusion	No	Negative (driven by concentration gradient)	Urea, water
Facilitated diffusion	Yes	Negative (driven by concentration gradient)	Glucose carrier of erythrocytes
Active transport			
	Yes	Positive (coupled to ATP hydrolysis, ion/pH gradient, etc.)	(Na^+K^+)-ATPase on plasma membrane, maltose transporter from *E. coli*

membrane also has a system for the facilitated transport of glycerol of which ethylene glycol is a competitive inhibitor.

While the $\Delta G°$ for a chemical reaction is related to equilibrium constant (K_{eq}), $\Delta G°$ for transport ($\Delta G°_{tr}$) is related to concentration ratio of the substance ($[A]_{in}/[A]_{out}$). Just as some enzymes catalyze endergonic reactions at the expense of ATP hydrolysis, in active transport a transporter pumps molecules against a concentration gradient when it is coupled to energy supply. This input of energy may be in the form of ATP hydrolysis or dissipation of a preexisting ion/pH gradient. In a nutshell, mediated transport may be analyzed similar to enzyme catalysis—the former measuring rates of transport from one compartment to the other (across the biological membrane) while the latter deals with reaction rates. It is no surprise that transporters have recently found their home in the EC 7 class! (see Chap. 7). One can visualize that the solute binds preferentially to a high-energy conformation of the transporter. Translocases are thus catalysts that facilitate the movement of ions or molecules across membranes or their separation within membranes. The three hallmarks of enzymes—rate acceleration, specificity, and regulation—come into play with translocases/transporters as well. They undergo most modes of regulation that enzymes are subjected to. Examples of various types of mediated transport may be found in the relevant literature dealing with biochemistry and bioenergetics.

38.7.2 Liquid–Liquid Phase Separation (LLPS)

Formation of biomolecular condensates via liquid–liquid phase separation is recently appreciated as a way of creating biochemically distinct sub-compartments. These membraneless regions when formed could partition distinct processes inside one cellular entity. The interior of such a biological condensate can resemble an organic rather than an aqueous phase and hence can partition certain components while excluding others. These "compartments without physical barriers," can achieve high local concentrations for molecular interactions, physically organize them, and make rapid chemical reactions feasible. The special material properties of the dense phase may have a variety of effects on substrate turnover and product release (Feng et al. 2019; Peeples and Rosen 2021; O'Flynn and Mittag 2021). Intrinsically disordered regions of the protein components are critical in the formation of biomolecular condensates which in turn are important for function. One example is the intrinsically disordered region from PP2C phosphatase whose phase separation feature functions in CO_2 sensing (Zhang et al. 2022).

The formation of the liquid-like condensate of ribulose-1,5-bisphosphate carboxylase/oxygenase (Rubisco) is mediated by dynamic interactions with the small subunit-like modules of the protein CcmM. These carboxysome condensates of cyanobacteria contain Rubisco (a complex of eight large and eight small subunits) and carbonic anhydrase. Similarly, within the eucaryotic algal pyrenoids (functional homologs of carboxysomes) Rubisco adopts a liquid-like state by interacting with an intrinsically disordered protein. The HCO_3^- can diffuse through the proteinaceous shell of the condensates but CO_2 cannot—due to which carbonic anhydrase generates high concentrations of CO_2 for fixation by Rubisco (Wang et al. 2019). Cells may thus use compartmentalization of enzymes through LLPS as a strategy to regulate metabolic pathways and increase their efficiency.

38.8 Glutamine Synthetase—An Anthology of Control Mechanisms

Glutamine is an essential metabolite and serves as a nitrogen donor in many crucial biosynthetic reactions. The enzyme that is responsible for glutamine synthesis is therefore central to cellular nitrogen metabolism. Glutamine synthetase is the sole de novo biosynthetic path to satisfy the glutamine needs of a cell. Along with glutamate synthase, glutamine synthetase offers an ATP-driven route to form glutamate from limiting concentrations of ammonia. In most organisms, therefore, glutamine synthetase occupies a pivotal role in the regulation of nitrogen metabolism. The enzyme is particularly well regulated in organisms that have to mobilize ammonia into organic nitrogen. According to their specific needs, the importance of glutamine synthetase and its modes of regulation vary (Table 38.6). In general, the enzyme levels are inversely related to the availability of favorable nitrogen source.

Glutamine synthetase from *E. coli* is an excellent example of enzyme regulation. It is regulated by the following mechanisms:

• The enzyme protein levels are subject to transcriptional control depending upon the availability of carbon and nitrogen compounds in the growth medium. Glutamine synthetase levels are induced under nitrogen-limiting conditions. Favorable nitrogen sources (like ammonia and glutamine) normally repress its expression.

Table 38.6 Glutamine synthetase from different organisms and its regulation

Organism	Oligomeric state	Regulation
Prokaryotes	Dodecamers	Genetic level (induction, repression, etc.); product and feedback inhibition; covalent modification cascade (such as in Gram negative bacteria); divalent metal ion-specific modulation
Yeasts and filamentous fungi	Tetramers and octamers	Genetic level (induction, repression, etc.); different monomeric subunits; aggregation–disaggregation equilibria; substrate (ammonia) availability; product and feedback inhibition
Plants	Dodecamers?	Product and feedback inhibition; substrate (ammonia) availability; covalent modification (symbiotic bacteria-associated enzyme)
Mammals	Tetramers and octamers	Availability of substrates (glutamate and ammonia); energy charge; divalent metal ion-specific modulation; inhibition by certain metabolites

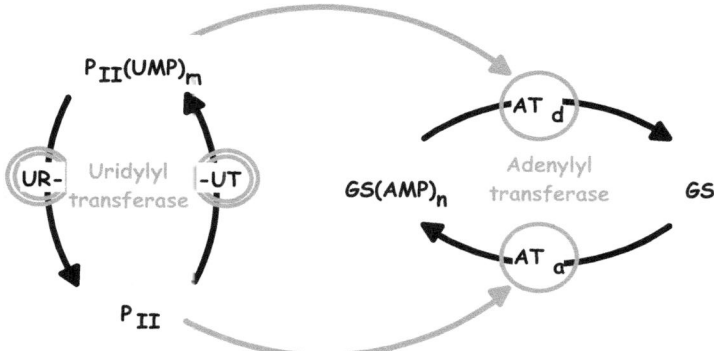

Fig. 38.10 The covalent regulation of *E. coli* glutamine synthetase. The uridylylation cycle and adenylylation cycle are linked through the regulatory protein P_{II}. Glutamine is an activator of UR activity while 2-oxoglutarate activates UT. The P_{II} protein activates AT_a while P_{II-UMP} activates AT_d. Ultimately, AT_a and AT_d are, respectively, responsible for the adenylylation and the deadenylylation of glutamine synthetase. AT_a transfers AMP from ATP to form one adenylyl-*O*-tyrosyl bond per each subunit of the dodecameric glutamine synthetase (where $n = 1$–12). Covalent modification by UT similarly involves the transfer of UMP from UTP to P_{II} (where $n = 1$–3)

- Regulation occurs through cumulative feedback inhibition by the multiple end-products of glutamine metabolism. Eight different metabolites are known to inhibit the enzyme with separate binding sites for each one.
- Divalent cations (Mg^{2+} and Mn^{2+} in particular) are known to bind and cause kinetically meaningful conformational changes. The native enzyme contains tightly bound Mn^{2+}.
- Through the work of Earl Stadtman's group, *E. coli* glutamine synthetase provides the best characterized example of reversible covalent modification. The biosynthetic ability of the enzyme is controlled by adenylylation and deadenylylation. The addition and removal of AMP moiety (on a surface exposed Tyr-OH group) modulates catalytic potential, susceptibility to feedback inhibition, and divalent cation specificity. The adenylylated glutamine synthetase ($GS_{(AMP)n}$ in Fig. 38.10) is much less active, has a lower pH optimum, requires Mn^{2+} for activity, and is more susceptible to feedback inhibition. The physiological significance of adenylylation–deadenylylation is apparent from the high levels of covalent modification and low biosynthetic activity on favorable/sufficient nitrogen availability; and the converse was found when the growth medium contained a limiting nitrogen source.

The covalent regulation of *E. coli* glutamine synthetase actually is a closed bicyclic cascade as shown in Fig. 38.10. A single adenylyltransferase catalyzes the adenylylation and deadenylylation reactions at separate noninteracting sites (AT_a and AT_d). This in turn is coupled to another nucleotidylation cycle in which P_{II} protein undergoes reversible uridylylation–deuridylylation. The uridylylation and deuridylylation of P_{II} protein is achieved by a bifunctional enzyme with separate catalytic centers (UT and UR, respectively). The unmodified form of P_{II} protein activates adenylylation (AT_a) of glutamine synthetase while the modified form ($P_{II\text{-}UMP}$) activates deadenylylation (AT_d). In this closed bicyclic cascade, all the covalent modification–demodification steps are dynamic processes and steady states get established rapidly. Glutamine synthetase is a dodecamer and each subunit can be covalently modified; therefore, average state of adenylylation can take values between 0 and 12, depending on the nutritional status of *E. coli*. Between UT, UR, P_{II}, $P_{II\text{-}UMP}$, AT_a and AT_d, the cellular levels of 2-oxoglutarate and glutamine are sensed. In fact, more than 40 different metabolites are known to affect the activities of one or more of the enzymes of this bicyclic cascade. Thus, *E. coli* glutamine synthetase is a very finely tuned enzyme for regulation of its activity and to sense the cellular nitrogen demands.

• Maintaining covalently modified glutamine synthetase for long periods is a poor investment for the cell. Such modified enzyme protein form ($GS_{(AMP)n}$ in Fig. 38.10) is oxidatively modified and leads to inactivation. This eventually renders the protein susceptible to proteolytic turnover.

38.9 Summing Up

Reining in a runaway chemical reaction is just as important as accelerating a sluggish one. While metabolism cannot do without enzyme catalysts, it can be catastrophic not to control reaction rates. Enzymes themselves provide the means for this modulation. Actual intracellular enzyme activity is determined by a combination of parameters like K_M, k_{cat}, n_H, $[E_t]$, $[S]$, $[P]$, $[I]$, and activators.

Steady-state cellular enzyme concentration is an outcome of gene expression at the level of induction, repression, mRNA stability, translation, and by protein degradation. These long-term modes of control often predominate in many prokaryotic microorganisms. On the other hand, preponderance of metabolic regulation via the control of enzyme activity is a common feature of eukaryotes. Control of preexisting enzyme activity is advantageous in terms of rapidity of the system response. This is typically achieved by ligand interaction phenomena such as enzyme inhibition, allostery, and cooperativity. Noncompetitive inhibition is important for the regulation of cell metabolism as the enzyme activity can be affected without a direct substrate analogy.

Sigmoid substrate saturation (the so-called homotropic allosteric interactions) can offer means to achieve significant changes in enzyme activity (and response) with just three to fourfold increase in $[S]$. With increasing n_H values, smaller fold increases in $[S]$ are required to give the same relative increase in enzyme activity. Since $[S]$ in vivo is often held constant (does not vary much), homotropic allostery alone is often unimportant. Its real value is in responding to relevant metabolite activator(s) through reversible shifts between hyperbolic and sigmoid substrate saturation kinetics.

Table 38.7 Citrate synthase from different organisms: function and its regulation

Organism	Function	Regulation
Mammalian cells (mitochondria; aerobic)	ATP production	Feedback inhibition by ATP
E. coli (mainly anaerobic)	Production of NADH and biosynthetic precursors	End-product inhibition by NADH
Germinating seeds (glyoxysomes)	Glyoxylate cycle; conversion of fatty acid to sucrose	Not regulated by either (as above)

Organisms solve similar metabolic problems in distinctly different ways. The same enzyme in two different organisms may be feedback inhibited by a different set of end-products. Aspartate transcarbamylase is inhibited by CTP in *E. coli* whereas UTP is the regulatory ligand in plants. Depending upon the organism, inhibition by the same group of end-products may be mechanistically distinct (cumulative, additive, concerted, or synergistic). The enzyme and corresponding substrate (or inhibitor or activator) may be sequestered in distinct compartments in different organisms—access itself may be a control feature. Interestingly, the same enzyme may perform different metabolic roles in different organisms; its regulation will accordingly be different in those organisms. Citrate synthase is a good case in point (Table 38.7).

Posttranslational covalent modification of enzymes, either reversible or irreversible, offers unique opportunities for metabolic regulation. Phosphorylation and proteolysis are more common. Besides the capacity for signal and rate amplification, covalent modification of preexisting enzymes allows a system-level mechanism to integrate a range of metabolic signals.

Nature is replete with examples of subtle variety in the mechanisms of control in different organisms—the "Unity of Biochemistry" concept does not always extend to metabolic pathway control. Most organisms, however evolutionarily distant, produce/utilize a set of common metabolites; but their concentrations are unique to each individual. This metabolic identity is the consequence of quantitative differences in relevant enzyme properties and their associated regulation. The patterns of enzyme regulation outlined in this chapter are common themes but are not exhaustive by any standard. It would be a surprise if we do not find novel variations of enzyme regulation in the future.

References

Baker SA, Rutter J (2023) Metabolites as signalling molecules. Nat Rev Mol Cell Biol 24:355–374

Bathe U et al (2021) The moderately (d)efficient enzyme: catalysis-related damage in vivo and its repair. Biochemistry 60:3555–3565

Blumenthal SA (2012) Earl Sutherland (1915–1974) and the discovery of cyclic AMP. Perspect Biol Med 55:236–249

Chock PB, Rhee SG, Stadtman ER (1980) Interconvertible enzyme cascades in cellular regulation. Annu Rev Biochem 49:813–843

Cornish-Bowden A (2004) The pursuit of perfection. Oxford University Press

Dyla M, Kjaergaard M (2020) Intrinsically disordered linkers control tethered kinases via effective concentration. Proc Natl Acad Sci USA 117:21413–21419

Feng Z et al (2019) Formation of biological condensates via phase separation: characteristics, analytical methods, and physiological implications. J Biol Chem 294:14823–14835

Ferrell JE Jr (1996) Tripping the switch fantastic: how a protein kinase cascade can convert graded inputs into switch-like outputs. Trends Biochem Sci 21:460–466

Hanson AD et al (2021) The number of catalytic cycles in an enzyme's lifetime and why it matters to metabolic engineering. Proc Natl Acad Sci USA 118:e2023348118

Hartmann MD (2022) A complex struggle for direction. Nat Chem Biol 18:119–120

Hers HG, Hue L (1983) Gluconeogenesis and related aspects of glycolysis. Annu Rev Biochem 52: 617–653

Huber SC, Hardin SC (2004) Numerous posttranslational modifications provide opportunities for the intricate regulation of metabolic enzymes at multiple levels. Curr Opin Plant Biol 7:318–322

Jayaraman V et al (2022) A counter-enzyme complex regulates glutamate metabolism in *Bacillus subtilis*. Nat Chem Biol 18:161–170

Koshland DE, Némethy G, Filmer D (1966) Comparison of experimental binding data and theoretical models in proteins containing subunits. Biochemistry 5:365–385

Krebs EG, Beavo JA (1979) Phosphorylation-dephosphorylation of enzymes. Annu Rev Biochem 48:923–959

Larion M et al (2012) Order–disorder transitions govern kinetic cooperativity and allostery of monomeric human glucokinase. PLoS Biol 10:e1001452

Lynch EM et al (2017) Human CTP synthase filament structure reveals the active enzyme conformation. Nat Struct Mol Biol 24:507–514

Mathews CK (1993) The cell—a bag of enzymes or a network of channels? J Bacteriol 175:6377–6381

Mesquita FS et al (2024) Mechanisms and functions of protein S-acylation. Nat Rev Mol Cell Biol 25:488–509

Monod J, Changeux J-P, Jacob F (1963) Allosteric proteins and cellular control systems. J Mol Biol 6:306–329

Monod J, Wyman J, Changeux J-P (1965) On the nature of allosteric transitions: a plausible model. J Mol Biol 12:88–118

O'Flynn BG, Mittag T (2021) A new phase for enzyme kinetics. Nat Chem Biol 17:627–631

Pardee AB (2002) Reflections: regulation, restriction, and reminiscences. J Biol Chem 277:26709–26716

Pardee AB, Reddy GP (2003) Beginnings of feedback inhibition, allostery, and multi-protein complexes. Gene 321:17–23

Peeples W, Rosen MK (2021) Mechanistic dissection of increased enzymatic rate in a phase-separated compartment. Nat Chem Biol 17:693–702

Silverman GA et al (2001) The serpins are an expanding superfamily of structurally similar but functionally diverse proteins. J Biol Chem 276:33293–33296

Small C, Traut TW (1984) Antizymes: inhibitor proteins or regulatory subunits? Trends Biochem Sci 9:49–50

Srivastava DK, Bernhard SA (1986) Enzyme–enzyme interactions and the regulation of metabolic reaction pathways. Curr Top Cell Regul 28:1–68

Stadtman ER (2001) The story of glutamine synthetase regulation. J Biol Chem 276:44357–44364

Stadtman ER, Chock PB (2014) Interconvertible enzyme cascades in metabolic regulation. Curr Top Cell Regul 13:57–97

Traut TW (1994) Dissociation of enzyme oligomers: a mechanism for allosteric regulation. Crit Rev Biochem Mol Biol 29:125–163

Umbarger HE (1956) Evidence for a negative-feedback mechanism in the biosynthesis of isoleucine. Science 123:848

Walsh CT, Garneau-Tsodikova S, Gatto GJ Jr (2005) Protein posttranslational modifications: the chemistry of proteome diversifications. Angew Chem Int Ed 44:2–33

Wang H et al (2019) Rubisco condensate formation by CcmM in β-carboxysome biogenesis. Nature 566:131–135

Zhang M, Zhu C, Lu Y (2022) The intrinsically disordered region from PP2C phosphatases functions as a conserved CO_2 sensor. Nat Cell Biol 24:1029–1037

Further Reading

Fell D (1996) Understanding the control of metabolism. In: Frontiers in metabolism. Portland Press, London

In Vitro Versus In Vivo: Concepts and Consequences

39

> (The biochemist's word) may not be the last in the description
> of life, but without his help the last word will never be said.
> *Sir Gowland Hopkins (1931)*

Biochemists enjoy the freedom to purify and study enzymes in isolation, saturate an enzyme with its substrate, trap/remove the products, as also provide optimal pH, ionic strength, etc. On the other hand, cell extracts are by their very nature "dirty enzymes"; intact cells and organisms are "dirtier" still. The cell by design is greatly constrained to provide a consensus medium to simultaneously support hundreds of diverse enzyme-catalyzed reactions. Only some of these enzymes may be operating under optimal conditions at any time. The context for an enzyme to function in vivo is very different from the well-defined conditions deliberately set up for its study in vitro. And classical biochemistry is founded on several assumptions valid in dilute aqueous solutions. These assumptions are often extended without question to the cellular milieu. But the cell interior is far away from being an ideal solution. The key features that differentiate the situation in vivo from that in vitro are cataloged below.

Organization and Compartmentalization The enzyme study in a test tube presupposes that the solution is homogeneous. This assumption is not valid for intact cells, tissues, and organisms. The cytoplasm may be better described as an aqueous gel than as a homogeneous solution. The presence of supramolecular organizations and membrane-bounded sub-compartments (in eukaryotic cells in particular) confers intrinsic inhomogeneities on the cell interior. Additionally, the organelles themselves are not randomly distributed due to cytoskeleton organization and intracellular transport. The presence of catalytically functional proteins alone is not adequate for their physiological function. They must be properly located. Also, subcellular enzyme relocalization can result in functional innovations in metabolism (Srere and Knull 1998; Schenck and Last 2020).

© The Author(s), under exclusive license to Springer Nature Singapore Pte Ltd. 2025

N. S. Punekar, *ENZYMES: Catalysis, Kinetics and Mechanisms*, https://doi.org/10.1007/978-981-97-8179-9_39

The Dilution Factor Due to the small, finite volume of cells and intracellular compartments, the actual number of a particular molecular species may be surprisingly low. Very few molecules per cell could translate into high, physiologically relevant molar concentrations. These concentrations may be far higher than those at which enzymes are usually assayed. Weak protein associations could well disappear at the far lower protein concentrations with which enzyme kineticists work in vitro.

Concentration of Enzymes, Substrates, and Other Ligands The total concentration of macromolecules inside cells is very high, with proteins being the most abundant species. Clearly the aqueous phase of the cytoplasm is crowded rather than dilute. Such crowded solutions are not amenable to the fundamental assumption of the physical chemistry of dilute solutions—interactions between solute molecules cannot be neglected. The concentration of a specific enzyme is generally much higher in cells than in conventional in vitro assays. This may be often much in excess of the K_D as determined in vitro. For instance, the K_D for interaction between calmodulin and myosin light chain kinase in vitro is around 1.0 nM. But the smooth muscle calmodulin concentration is in the order of 40 μM. Similarly, a substantial proportion of substrates may exist as complexes, making the availability of free substrate rate limiting.

Enzymes In Vivo are Components of an Open System Cell is not a bag of enzymes each working in isolation. The cellular metabolism is a complex web of enzyme pathways, many of which often compete for common substrates. To proceed efficiently under these conditions, channeling of substrates from one enzyme to another in a particular metabolic pathway may be necessary. Discrimination among competing interactions may be achieved by sequestering enzymes within an organelle or immobilizing them on a membrane.

We will now elaborate on a few of these key concepts and their consequences on enzyme action in vivo.

39.1   Why Michaelis–Menten Formalism Is Not Suitable In Vivo?

Much of the current paradigm to understand enzymes has been extrapolated from studies of dilute solutions containing a single enzyme and its cognate substrate—whose interaction is diffusion limited. Enzymology in the "test tube" is a careful observation and controlled study of an enzyme in isolation from the host of other interactions that otherwise make the understanding difficult. This reductionist approach has led to many valuable insights over the past century. However, accumulated knowledge on measurements of the physical properties of cells indicates that the interior of a cell departs from these ideal conditions in several important ways. Some of these are already listed above.

We recall that the Michaelis–Menten rate equation is derived for a well-defined set of initial conditions and making certain clear assumptions (see Chap. 14). Almost all of these may not be satisfied with respect to enzymes in vivo. To predict the kinetic behavior of an enzyme in vivo, it is sufficient to study it in the range of concentrations likely to occur inside the cell. However, one needs to study the effect of all the metabolites that potentially interact with that enzyme. Also, most enzymes operate in a regime far closer to fixed rates than fixed concentrations (as is usually studied in the test tube!). Both metabolic fluxes and metabolic concentrations are properties of the system and the enzyme in question does not operate in isolation but will just be a part of this system (Cornish-Bowden 1999).

39.1.1 Where Assumptions Fail

Enzyme being a catalyst, its assay concentration in vitro is usually held very much lower than that of the substrate. This permits us to approximate $[S_t] \approx [S]$, although $[S_t]$ actually equals "$[S] + [ES]$" while deriving the rate equation. Most often the substrate concentrations are held at least 1000 times higher than that of the enzyme (i.e. $[S_t] >> [E_t]$). The actual $[E_t]/[S_t]$ ratio becomes important in interpreting enzyme kinetic behavior under in vivo conditions. This ratio for various enzymes of glycolysis (in mammalian skeletal muscle) ranges from 0.016 (for phosphofructokinase) to 17.48 (glyceraldehyde 3-phosphate dehydrogenase) (Table 39.1). The assumption $[S_t] \approx [S]$ is valid only when this ratio is very low—such as in the case of phosphofructokinase. But when the ratio approaches/exceeds the value of 0.4, serious deviations from the Michaelis–Menten formalism occur and the corresponding rate equations are no longer valid.

Table 39.1 Concentration of individual glycolytic enzymes and intermediates

Enzyme	Active site (μM)	Metabolite (substrate)	Concentration (μM)
Phosphofructokinase	24.1	Fructose 6-phosphate	1500
Aldolase	809.3	Fructose 1,6-bisphosphate	80
Triosephosphate isomerase	223.8	Dihydroxyacetone phosphate	160
Glyceraldehyde 3-phosphate dehydrogenase	1398.6	Glyceraldehyde 3-phosphate	80
Phosphoglycerate kinase	133.6	1,3-Bisphospho-glycerate	50
Pyruvate kinase	172.9	Phosphoenolpyruvate	65
Lactate dehydrogenase	296.0	Pyruvate	380

Adapted from Srivastava & Bernhard, Curr Top Cell Regul, 28:1–68. Copyright (1986), with permission from Elsevier

39.1.2 Reaction Reversibility

Strictly the initial rate (velocity "v") has to be recorded in order to apply Michaelis–Menten formalism. This will be the unbiased rate when $[P] \approx 0$, whereas significant product concentrations are often observed in vivo. Products compete with substrates for the available enzyme; by reaction reversal they can also react to form substrate. Increasing $[P]$ can actually generate negative velocity, especially when $[S]$ is not high enough to force the reaction forward. Entire pathways, and most enzymes in such a pathway, are known to work in reverse under certain physiological states. For example, but for the two irreversible steps, most enzymes of glycolysis operate in reverse during gluconeogenesis.

39.1.3 Optimized In Vitro But Compromised In Vivo

An enzymologist's test tube is optimized for the best measurements of enzyme activity. The parameters like temperature, buffers, pH, ionic strength, cofactors, metal ion activators, etc., are rigorously controlled by the experimenter. A wide range of these conditions can be tested in vitro. However, a given physiological state of the cell provides a single consensus medium common for all its enzymes functioning in that compartment. Interestingly, new effectors were revealed by measuring enzyme activities under in vivo-like assay conditions (Garcia-Contreras et al. 2012). It is therefore rare that kinetic constants determined in vitro are accurate reflections on the enzymes in vivo. Rate laws determined in vitro with purified, dilute, homogeneous enzyme solutions may not reflect the enzyme–enzyme interactions that are important in vivo. Also, many small molecule regulators may be missed due to clean in vitro (test tube) assays. Optimal behavior for an isolated enzyme may mean nonoptimal behavior for the intact system, and vice versa. For instance, arginase functions inside a fungal cytoplasm at physiological (near neutral) pH and with in situ Mn[II] concentrations. This is very different from its alkaline pH optimum (of around pH 10.0) and the requirement for incubations with micromolar Mn[II] for optimal activity in the test tube. Regulatory interaction of yeast arginase with ornithine transcarbamylase (the so-called epiarginasic control) is also well documented (Table 38.4, Chap. 38).

In order to make the bottoms-up approach to *Systems Biology* meaningful, standardized in vivo-like conditions are being simulated to study enzyme function. One such standardized assay medium for all yeast cytosolic enzymes was recently described. Potential effects of macromolecular crowding (see Sect. 39.4 below) are still missed and not considered in this standardization effort (Table 39.2). Besides properly defined ionic strength, pH, and buffering, other components of the intracellular milieu should be included in enzyme assays to simulate in vivo conditions. Glutamate, glutathione, and phosphates are the main metabolite pool components that contribute to pH buffering in most cells. Effects of all these factors on k_{cat} as well as on K_M have to be accounted for. These measurements should also include data

Table 39.2 Standardized assay medium for yeast cytosolic enzymes

Component	Concentration
Potassium (K^+)	300 mM
L-Glutamate	245 mM
Phosphate	50 mM
Sodium (Na^+)	20 mM
Magnesium (Mg^{2+})	2 mM (free)
Sulfate	2.5–10.0 mM (depending on Mg^{2+} levels)
Calcium (Ca^{2+})	0.5 mM
Medium pH	6.8

Adapted with permission from van Eunen et al., FEBS Journal, 277:749–760. Copyright (2010) John Wiley & Sons Inc (van Eunen et al. 2010).

Table 39.3 Enzyme systems in vitro and in vivo: a comparison

Nature of	Enzyme in vitro	Enzyme in vivo
Variables	Few	Many
Interactions	Weak	Strong
Connectivity	Linear	Nonlinear
Processes	Additive	Associative
Aggregate behavior	Predictive	Emergent
System	Closed, simple	Open, complex

when [S] is well above K_M (effects on k_{cat}) as well as around/below K_M (effects on K_M).

A study of enzymes in vivo quickly moves into the realm of metabolism and complexity (Table 39.3). One could build an understanding of metabolism (i.e., enzymology in vivo) by studying individual enzymes one at a time and then upwardly integrating this in vitro knowledge (Smallbone et al. 2013). In practice, this is very complex—collecting such data for regulatory enzymes with multiple interactions becomes immensely difficult. This "bottoms up" approach is like describing the behavior of a gas by applying Newton's laws of motion to every individual molecule in the system. More often, enzymology in vivo takes a "top down" approach—akin to applying simple gas laws of thermodynamics at the macroscopic level. It is important to distinguish between kinetics as the study of molecular mechanisms and the kinetics of system dynamics. The very conditions that made Michaelis–Menten formalism produce the rate law for enzyme reactions in vitro tend to make it invalid for enzyme reactions in vivo. Power-law formalism as an alternative to capture in vivo behavior has been proposed (Savageau 1992). Interested readers may refer to this literature. Attempts to relate in vivo enzyme catalytic rates and their correspondence to in vitro k_{cat} measurements and extracting/predicting kinetic constants on a proteome-wide scale are being made with the aid of artificial intelligence tools (Davidi et al. 2016; Antolin and Cascante 2021).

39.2   Concentration of Enzymes, Substrates, and Their Equilibria

There is always a certain tension between the in vitro and in vivo approaches to enzymology. One aspect that biochemists have struggled with since the beginning is the extent to which in vitro enzyme data are relevant to in vivo metabolism. Cells and biological tissues are disrupted to extract and access enzymes prior to assay. Cell-free extracts are considerably diluted when compared to enzyme/metabolite concentrations in vivo (Albe et al. 1990; Bennett et al. 2009). Invariably, extrapolations from in vitro to in vivo involve assumptions and this often leads to poor conclusions. All the intracellular compartments are highly concentrated in terms of proteins and metabolites (Srere 1967, 1970). This also has a direct bearing on the physical features of the cytosol such as viscosity, diffusion rates, and excluded volume effects (discussed in detail a little later). We are just beginning to understand the influence of unique intracellular environment on enzyme activities.

A major experimental challenge in doing in vivo enzymology is the precise measurement and manipulation of ***absolute concentrations*** of enzymes and metabolites inside the cell. There are many issues that influence such numbers.

- Direct determination of metabolite concentrations is difficult for two reasons: (a) Concentrations may change rapidly during the time required for isolation, and (b) Low molecular weight substances diffuse out of organelles and redistribute during isolation. Based on the total ATP, ADP, creatine, and phosphocreatine concentrations measured in the cell, it was thought that creatine kinase reaction is far from equilibrium (towards phosphocreatine formation). Subsequently, the free concentrations of the phosphorylated compounds were directly measured in living cells using ^{31}P NMR. With this data, it was shown that the creatine kinase reaction is close to equilibrium in vivo.
- Cellular enzyme concentrations tend to be underestimated, mainly due to incomplete extraction and associated inactivation during isolation from crude extracts.
- Despite very accurate analytical methods for estimation, serious assumptions/ approximations are necessary to compute the final in vivo concentrations. For instance, the water content and the total soluble protein in different cell types (see Table 39.4) are averaged as if the cytoplasm is a homogeneous solution. Typically, in a gram of muscle tissue, all the cytoplasmic soluble components are found in about 0.75 mL of water.
- Because it is crowded with very high protein concentration, cytoplasm may be more of an aqueous gel and not a homogeneous solution. Accordingly, the cellular milieu lacks sufficient free water. For instance, 2-aminoacrylate generated by PLP-dependent eliminases can be spontaneously converted to pyruvate; it has a very short half-life of ~1.5-s in water. However, biochemical and genetic evidence shows that 2-aminoacrylate persists in vivo in the absence of an enamine/imine deaminase (Ernst and Downs 2018). Clearly, concentration of free water may be limiting in the cellular milieu.
- The metabolite concentrations are usually reported in the literature in units of μmol/g of dry or fresh weight. Such data can be converted to μM using suitable

Table 39.4 Dimensions and content of different cell types

Source	Cell dimensions (volume)	Water content (g/100 g wet cells)	Protein concentration (mg/mL)
E. coli	1×3 μm (2 μm^3)	70	235
Yeast	5 μm (66 μm^3)	65	280
Rat liver	10–20 μm (500–4000 μm^3)	69	313
Rat muscle		77	260
Human RBC	6–8 μm (90 μm^3)	65	300
Mitochondrion	1 μm (0.5 μm^3)	50	270–560

A volume of 3.2×10^{-15} L has been directly measured for *E. coli* cells
[1 μm$^3 = 10^{-15}$ L (1 femtoliter); 10^3 μm$^3 = 10^{-12}$ L (1 picoliter)]

conversion factors (see Table 39.4 above). Arriving at in vivo enzyme concentrations requires some more rough estimations: (1) turnover number (μmol substrate converted $\times$ min^{-1} $\times$ μmol enzyme^{-1}) is calculated from the specific activity (μmol substrate converted min^{-1} $\times$ mg protein^{-1}) of the most purified enzyme fraction available, the molecular weight of the holoenzyme and assuming that all protein represents active holoenzyme; (2) the V_{max} value (μmol substrate converted min^{-1} $\times$ liter cell volume^{-1}), assumed to represent in vivo activity, is calculated from the specific activity of a crude enzyme fraction using the conversion factors (see Table 39.4 above); (3) enzyme concentration is now obtained by dividing the V_{max} value (from 2) by the turnover number (from 1). The total enzyme site concentration can finally be represented by multiplying enzyme concentration by the number of subunits per holoenzyme, assuming one active site per subunit.

Despite inherent difficulties in measurements, there are some obvious take-home messages from the calculated in vivo concentrations of enzymes and metabolites (Srivastava and Bernhard 1986; Storey 2005). For instance, consider the computed concentrations for glycolytic enzymes and their substrates in the mammalian skeletal muscle (see Table 39.1 above). First, among the enzymes of the same pathway, some are much more abundant than the rest—aldolase and glyceraldehyde 3-phosphate dehydrogenase together constitute 40–50% of the total glycolytic enzymes. Molar concentrations of such enzymes in vivo are quite high. Notwithstanding this range, the sequence of enzymes sustains a common single flux through glycolysis. And further, the muscle glycolytic flux increases 100–1000-fold during "resting–working" transition without any change in the concentration of its intermediates. Second, the ratios of final calculated enzyme site concentration (expressed as μM) to corresponding free substrate (in μM) concentration(s) are also quite distinct (Free metabolite concentrations can be determined by solving simultaneous equations relating the free and bound concentrations of all metabolites; the corresponding Michaelis constants may be used to approximate corresponding equilibrium constants). The calculated ratio (of [enzyme site]/[free substrate]) ranges from very

Table 39.5 The K_{eq} for aqueous versus enzyme-bound reaction components

Enzyme reaction	K_{eq} when $[S] >> [E_t]$ (aqueous equilibrium) ($S \rightleftarrows P$)	K_{int} when $[E_t] \geq [S]$ (enzyme bound equilibrium) (ES $\rightleftarrows$ EP)
Hexokinase	2000	~1.0
Phosphoglucomutase	17	0.4
Triosephosphate isomerase	22	0.6
Pyruvate kinase	0.0003	1.0–2.0
Lactate dehydrogenase	10,000	1.0–2.0
Creatine kinase	0.1	~1.0

Adapted from Srivastava & Bernhard, Curr Top Cell Regul, 28:1–68. Copyright (1986), with permission from Elsevier

low for phosphofructokinase (0.004) to very high for glyceraldehyde 3-phosphate dehydrogenase (21). It is obvious that $[E_t] \geq [S]$ for at least a few enzymes in vivo (such as aldolase and glyceraldehyde 3-phosphate dehydrogenase) and consequently Michaelis–Menten formalism does not apply. More importantly, in such cases, a considerable portion of the substrate may be bound to the enzyme(s). However, calculated free substrate concentration (total minus bound substrate) may be employed in a general Michaelis–Menten analysis to predict the actual velocity of the reaction.

Besides very high total protein concentrations (Table 39.4), the cellular concentrations of some enzymes are also high (Table 39.1). An important consequence of high enzyme active site concentration (particularly when $[E_t] \geq [S]$) in vivo is its effect on the equilibrium position itself. In such cases, one has to consider the equilibrium between enzyme-bound S and P (i.e., $K_{int} = [EP]/[ES]$, the *internal equilibrium constant*) and not the aqueous equilibrium (i.e., $K_{eq} = [P]/[S]$). Table 39.5 lists some K_{eq} values, both for reaction within the enzyme active site (for all reaction components bound) and for the same reaction in aqueous solution. Note that the K_{int} for all reactions are closer to unity despite the wide range in K_{eq} values for the same reaction; this tendency is predicted on thermodynamic grounds in the evolution of "ideal" catalysts. Kinetically "perfect" enzymes (those that have evolved higher rates of turnover) bind substrates and products such that the bound complexes have nearly equal free energies (i.e., the K_{int} are closer to unity implies $\Delta G° = 0$). For example, overall equilibrium constants (K_{eq}) and the corresponding internal equilibrium constants (K_{int}) for several phosphotransferases may be found in Hassett et al. (1982).

We have seen earlier (refer to Chap. 14) that the K_{eq} (of an aqueous reaction chemical equilibrium) is related to corresponding enzyme kinetic constants through Haldane relationship. While this places limits on the overall kinetic behavior, certain features of the reaction catalyzed are still under the control of the enzyme and its properties. This is where the K_{eq} for a reaction within the enzyme active site (K_{int}) becomes very interesting! For example, consider the reaction catalyzed by methionine adenosyltransferase. The formation of S-adenosylmethionine is thought to be essentially irreversible with a forward rate 2×10^5 times faster than the reverse rate; this is despite the fact that the K_{eq} (for the unbound reactants) is near unity. Clearly, the internal equilibrium constant (i.e., $K_{int} = [EP]/[ES]$) should approximate 2×10^5 and that the enzyme has 2×10^5-fold higher affinity (approximated as inverse of K_M value) for P than for S.

39.3 Avogadro's Number Is a Very Big Number

One gram mole of a substance contains Avogadro's number (6.023×10^{23}) of molecules. These many molecules dissolved in 1 L of solvent make a molar (1.0 M) solution. However, cells and subcellular compartments (organelles) have finite, small volumes. We have already noted that typical volume of a bacterial cell is around 2 μm^3 (2×10^{-15} L) (Table 39.4); only about 70% of this is aqueous cytoplasmic volume and accessible to solutes. The presence of very few molecules can mean a significant concentration of that compound in such small volumes (Table 39.6). Assuming that free [H$^+$] contributes to intracellular pH in *E. coli*, just about 60 protons represent a pH of 7.0; addition of another 540 protons brings this intracellular pH down to 6.0. Confined to a limited cellular volume, a single molecule per bacterium (say *lac* operator DNA) implies a concentration of nearly 1.0 nM! At an experimental level, we are now able to manipulate volumes in attoliters (10^{-18} L) and reach detection limits down to zeptomoles (10^{-21} moles).

In general, for small, confined volumes, the number concentration of a particular molecular species (this includes enzymes!) may be more informative than its molar concentration (Halling 1989). The presence of a limited number of molecules within a cell/compartment has interesting implications. Some of these are numerically illustrated below.

Table 39.6 Molar versus number concentrations in confined cellular compartments

Compartment	Volume	Number of molecules	Concentration (approximate)
Vacuole (50 nm diameter)	6×10^{-20} L	1 H$^+$ ion	10^{-4} M (pH 4.0)
Vacuole (250 nm diameter)	7.5×10^{-18} L	1 H$^+$ ion 50 H$^+$ ions	10^{-6} M (pH 6.0) 10^{-5} M (pH 5.0)
E. coli	2×10^{-15} L	1 20 200	10^{-9} M (1 nM) 2×10^{-8} M (20 nM) 2×10^{-7} M (0.2 μM)
Mitochondrion	1×10^{-15} L	1	10^{-9} M
Yeast	7×10^{-14} L	15,000	10^{-7} M
Mammalian nucleus	5×10^{-13} L	300	10^{-9} M
Mammalian cell	2×10^{-12} L	1	10^{-12} M

As compartmental dimensions vary within a range, typical volumes are presented here. Concentrations are computed by assuming that the solute is uniformly distributed throughout the compartment. Concentration of over-expressed proteins in *E. coli* can be abnormally high and reach up to 5 mM

39.3.1 Number of Invertase Molecules Per Yeast Cell

It is the genius of JBS Haldane that the number of saccharase (popular as "invertase" now!) molecules per yeast cell was calculated with data available at that time and presented in the book "ENZYMES" in 1930. A similar calculation is presented here with recent data on invertase.

Invertase is a 60 kDa protein and constitutes 0.9% of total yeast protein.

Yeast cell with 5 μm diameter and density 1.1 weighs 7.2×10^{-11} g.

One gram of yeast consists of 0.65 g water. Considering a protein concentration of 280 mg/mL (Table 39.4), 1 g of yeast contains 182 mg protein. Hence,

1 g yeast contains $182 \times 0.9/100$ mg, i.e., 1.645 mg invertase

1 yeast cell (7.2×10^{-11} g) contains $1.645 \times 7.2 \times 10^{-11}$ mg, i.e., 1.2×10^{-14} g invertase

1.2×10^{-14} g invertase corresponds to $1.2 \times 10^{-14} \times 6.023 \times 10^{23}/60,000$ molecules of invertase = 120,000 molecules

Thus, one yeast cell contains 120,000 invertase molecules. This number is about ten time lower (12,000/cell) in uninduced yeast whereas in filamentous fungus *Aspergillus niger*, it may be only 1% of that number (1200/cell).

39.3.2 Small Number Concentrations Translate Into Discrete Molar Concentrations

Few enzymes may be found in very limited copies indeed. There is a single *oriC* locus in *E. coli* genome and just 10–20 copies of DNA polymerase III per cell to initiate replication. On the other hand, with its multiple origins of replication, a mammalian cell contains about 50,000 copies of this enzyme. A single molecule per bacterial cell typically corresponds to a molar concentration of 1.0 nM (Table 39.6); every additional molecule contributes to a discrete increment of 1.0 nM. In such situations, a bacterial cell may experience stepwise concentration increase—in 1.0 nM multiples (and none in between!) (Fig. 39.1). And changes in cellular volume then become important in accessing the full range of molar concentrations.

39.3.3 Dissociation Constants, Equilibrium Binding, and Stochasticity

The binding equilibrium between two molecular species depends on the equilibrium constant and the two concentrations. The binding of an enzyme to its substrate is no exception. However, the same binding phenomenon becomes stochastic when small "number concentrations" are involved. Consider a simplistic calculation of DNA polymerase III binding in vivo to *oriC* locus of *E. coli*. A single copy of *oriC* DNA (per chromosome) approximates 1.0 nM and ten copies of DNA polymerase III per cell would represent a concentration of 10.0 nM. If DNA polymerase III binds to

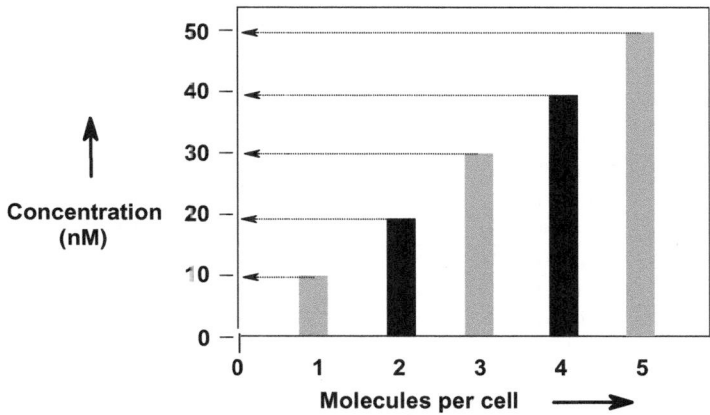

Fig. 39.1 Molar versus number concentrations in a confined cellular volume of *E. coli*

oriC DNA with a K_D of 1.0 μM, then we can evaluate the status of this binding equilibrium as follows:

$$oriC + \text{DNA pol III} \rightleftarrows oriC - \text{DNA pol III complex}$$

$$K_D = \frac{[oriC]\,[\text{DNA pol III}]}{[oriC - \text{DNA pol III complex}]}.$$

On rearranging,

$$\frac{[oriC]}{[oriC - \text{DNA pol III complex}]} = \frac{[oriC]_{\text{free}}}{[oriC]_{\text{bound}}} = \frac{K_D}{[\text{DNA pol III}]}.$$

Substituting the respective values,

$$\frac{[oriC]_{\text{free}}}{[oriC]_{\text{bound}}} = \frac{10^{-6}\text{M}}{10^{-8}\text{M}} = 100.$$

This implies 0.99 nM of the total of 1.0 nM *oriC* DNA is free (i.e., unbound). Since there is a single copy of *oriC* DNA, presenting the concentration of *oriC* DNA bound to DNA polymerase III as 0.01 nM (or 10^{-11} M) does not make any sense! We now enter the realm of probabilities, and the binding phenomenon is best considered as stochastic. A $[oriC]_{\text{free}}/[oriC]_{\text{bound}}$ ratio of 100 signifies that, at any given time, there is just 1% chance that DNA polymerase III is occupying the *oriC* locus. A few more representative calculations of this kind are informative (Table 39.7); lower the K_D (higher the affinity), greater is the probability of *oriC* DNA bound to the polymerase. At picomolar K_D value, the *oriC* DNA is fully occupied (nearly 100% probability) by DNA polymerase III. While these calculations are useful, actual DNA polymerase III binding to *oriC* locus in vivo is

Table 39.7 DNA polymerase III occupying the *oriC* locus in *E. coli*

K_D (*oriC*-DNA pol III complex)	Number of DNA pol III molecules per cell (concentration)	$\frac{[oriC]_{free}}{[oriC]_{bound}}$	*oriC* bound with DNA pol III (Probability %)
10^{-3} M (mM)	1 (10^{-9} M)	10^6	0.0009
	10 (10^{-8} M)	10^5	0.009
10^{-6} M (μM)	1 (10^{-9} M)	10^3	0.09
	10 (10^{-8} M)	10^2	0.99
10^{-9} M (nM)	**1 (10^{-9} M)**	**1.0**	**50.0**
	10 (10^{-8} M)	10^{-1}	90.9
10^{-12} M (pM)	1 (10^{-9} M)	10^{-3}	99.9
	10 (10^{-8} M)	10^{-4}	99.99

more complex and is supported by additional protein components. The *E. coli* DNA polymerase III (600 kDa protein) is an efficient (about 1000 nucleotides added per second) and highly processive enzyme. It remains bound to DNA template and does not fall off almost until the entire bacterial genome is replicated.

Another way of looking at the phenomenon of stochastic (probabilistic) binding is to compare the "on" and "off" rates. Note that the K_D may also be viewed as the ratio of the off and on rate constants (k_{off}/k_{on}; also see Sect. 20.4 in Chap. 20). Slower the rate of release ($k_{off} < k_{on}$) from the complex longer the protein resides on the DNA (since inverse of the first-order rate is time).

Finally, we note that stochastic (probabilistic) binding is encountered in many other cases such as when the tetrameric *lac* repressor (5–10 molecules per cell) binds (K_D, 10 pM) to the operator DNA of lactose operon. The study of single enzyme molecules, both in vitro and in vivo, is one research frontier in enzymology; we will briefly touch upon this topic in the next chapter.

39.4 Diffusion, Crowding, and Enzyme Efficiency

The physical event of substrate colliding with the enzyme molecule is diffusion controlled. As a catalyst, an enzyme can do very little to overcome this diffusion limit. The upper bound for catalytic rate acceleration is the prevailing diffusion rate and an enzyme is considered kinetically perfect when its k_{cat}/K_M approaches that diffusion limit (we have already noted this in Chap. 14) (Hasinoff 1984). The interior aqueous milieu of intact cells is a very crowded place. The diffusion of enzymes and their substrates are accordingly affected by the crowded and concentrated state of cytoplasm (Table 39.4) (Milo and Phillips 2016). Although the concentration of individual enzymes/proteins may not be very high, cells do contain a dense mixture of large and small molecules. This has a direct bearing on the viscosity of the cytoplasm. The ***macro-viscosity*** (or bulk viscosity as measured for example, by Brookfield viscometer) is affected by the polymeric solute molecules (cytoplasmic macromolecules and the synthetic ones like Ficoll or polyethylene glycol). This macroscopic flow property of the system does not necessarily correlate with effects on diffusion of small molecules. However, the small molecular weight solutes (like sucrose and glycerol) do influence diffusion rates of the system at the microscopic scale (the so-called *micro-viscosity*) (Verkman 2002). Techniques like fluorescent probe diffusion using photobleaching, correlation microscopy, and time-resolved anisotropy have provided a measure of micro-viscosity (fluid-phase viscosity) of cytoplasm. The viscosity of the cytoplasm has been estimated to be about 3–7 times that of water (Luby-Phelps 2000). From such data, the view of the cell interior has evolved from that of a viscous gel to that of a watery but crowded compartment. Cytoplasm has a low micro-viscosity and a high macro-viscosity. The mitochondrial matrix is even more crowded.

Three independent factors affecting solute diffusion in the cytoplasmic compartment are (a) specific binding to intracellular components, (b) slowed diffusion in fluid-phase cytoplasm (micro-viscosity), and (c) collision with intracellular components (macromolecular crowding). We have already noted in Table 39.1 that a significant proportion of substrates (metabolites) may be present in their enzyme-bound form. The fluid-phase viscosity of the cytoplasm (i.e., micro-viscosity) is not much greater than that of water. Therefore, the diffusion movement of small cytoplasmic solutes is similar to that in a dilute aqueous solution. However, macromolecular crowding leads to steric exclusion and is an important barrier to diffusion in the cytoplasm.

39.4.1 Macromolecular Crowding and Volume Exclusion

How much of the intracellular volume is available to macromolecules depends upon the numbers, sizes, and shapes of all the molecules present in that compartment. Cytoplasm is "crowded" rather than "concentrated" because no single macromolecular species occurs at high concentration (Ellis 2001). But, taken together, the macromolecules occupy a significant fraction (typically 20–30%) of the total volume. A simple illustration of this concept is shown in Fig. 39.2. The space occupied by macromolecules is physically unavailable to other molecules and the resultant volume exclusion has noteworthy consequences. Macromolecular crowding leads to (1) volume exclusion of reactants, (2) a reduction in the diffusion coefficient of macromolecules, and (3) a reduction in the degree of mixing of molecules with an increase in reactant segregation.

Molecular crowding may alter observed equilibrium constants and/or may profoundly affect the enzyme kinetic parameters. Some examples of the kinetic effects are listed below.

- Concentration of a molecule in the compact form is favored in the presence of a space-filling substance, and the effect increases exponentially with the concentration of inert molecule. If we crowd the solution, the system will change to minimize crowding—molecules will associate, thereby reducing the excluded volume. If the enzyme can associate to an oligomer (with the enzyme activity different from that of the monomer), then crowded solutions will favor the

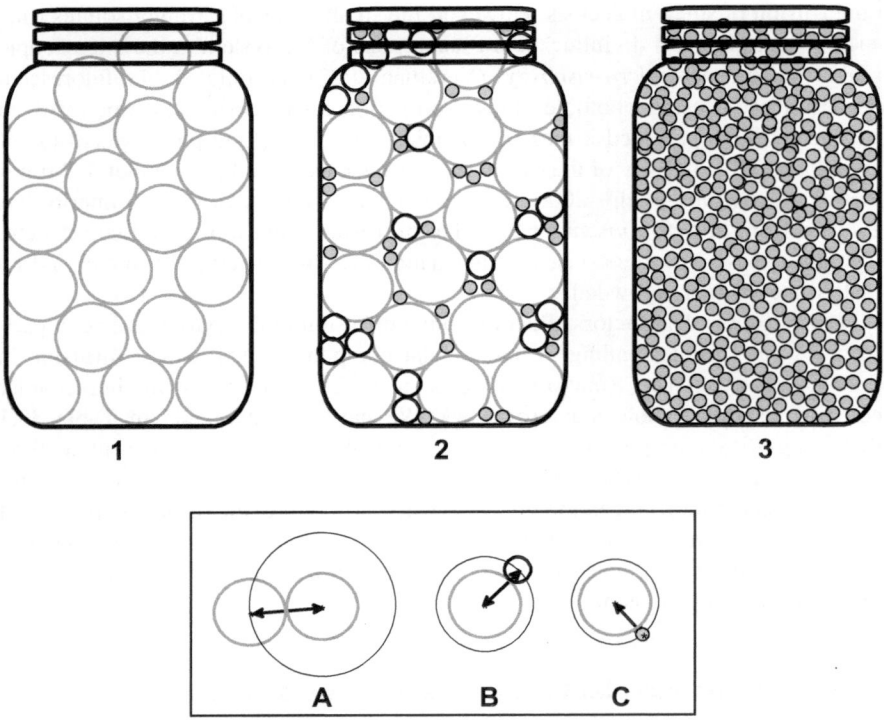

Fig. 39.2 Visualization of crowding effects. A glass jar can be filled by large (Jar 1), small (Jar 3), or medium-sized spheres. Jar 3 cannot further accommodate any of these spheres. However, Jar 1 can still accommodate medium-/small-sized spheres in between (Jar 2). This is because large spheres exclude more volume from around them. Similarly, a macromolecule in solution will exclude others from its neighborhood. The position of each molecule is specified completely by the position of its center. The closest any two molecules (assumed as spheres; box below) can approach is a distance equal to the sum of their radii. Around each molecule is a spherical volume from which the centers of all others are excluded. For instance, the volume around a macromolecule from which the center of another similar molecule is excluded is a sphere of radius twice (8 times the volume) that of a single molecule (case A in the box below)

oligomer (and alter the observed enzyme activity). The tetramer of glyceraldehyde 3-phosphate dehydrogenase is less active than the monomer—when crowded by other proteins (as is the case inside the cell) its activity is reduced which is consistent with excluded volume effects.

- Protein diffusion is slowed down by molecular crowding as other macromolecules become obstacles to be avoided (compare a person moving to the exit in a train station that is crowded versus empty). Weak protein associations (with $K_D > 10^{-4}$ M) that are otherwise functionally relevant, could well disappear in dilute enzyme assays with which enzyme kineticists work in vitro. Such weak protein–protein interactions may be specifically promoted by molecular crowding. Evolution seems to have conserved not only functional sites of protein molecules but also structural features that might determine the abilities of proteins to associate with one another.

- If the ES complex is more compact than E, then crowding will enhance the complex formation; this in turn lowers the K_M. For such enzymes, the in vivo K_M is unlikely to be simulated by kinetic measurements from routine test tube data. DNA replication in vivo requires the inclusion of high concentrations of polymer crowding agents like polyethylene glycol—this enhances the interaction between the DNA and the polymerase and other relevant proteins.
- During catalysis, the enzyme transition state may be expanded or contracted during catalysis. Therefore, in a crowded solution, the activation energy is raised or lowered, respectively; this in turn affects V_{max}. The measured V_{max} for pyruvate reduction by lactate dehydrogenase increases linearly with increasing concentrations of ovalbumin, serum albumin, or dextran—a result consistent with a decrease in the volume of the TS upon NADH binding.
- The beneficial effects of molecular crowding on ribozyme-catalyzed RNA assembly are known. Crowding stimulates ribozyme ligase activity at lower Mg^{2+} concentrations and stabilizes the ribozyme activity (Gupta et al. 2023).

After many failed attempts, Arthur Kornberg's group was successful in replicating the *oriC* plasmid in vitro, by including high concentrations of PEG in the incubation mixture (Kornberg 2000, 2003). As Kornberg put it "the PEG occupies most of the aqueous volume and excludes a small volume into which large molecules are crowded. This concentration is essential when several proteins are needed in the consecutive stages of a pathway." The effect of crowding on enzyme activity is reflected as one of his Ten Commandments—*thou shalt correct for extract dilution with molecular crowding*!

39.4.2 Enzyme Size Matters

The micro-viscosity of the medium, rather than the macro-viscosity, determines the velocity of a diffusion-limited enzyme reaction. The catalytic perfection of an enzyme can be benchmarked by comparing its k_{cat}/K_M against the prevailing diffusion rate constant (Chap. 14) (Knowles and Albery 1977; Burbaum et al. 1989). By this yardstick, most enzymes studied have already achieved "kinetic perfection" (see Table 14.2). Enzymes are some of the smallest phenotypic units on which evolutionary forces act. But then, is catalytic perfection the only feature of the enzyme selected by Nature? This quickly brings us to other questions like—Why are enzymes so big. Can we reduce the biosynthetic cost of an enzyme further? Are there tradeoffs while choosing between these features? It appears that other features like cost, stability, and regulation also figure significantly in the evolution of enzyme structure (Benner 1989). As Dobzhansky (1973) puts it—*Nothing in biology makes sense except in the light of evolution*. And enzymes are no exception.

Enzyme active sites most often occupy a small percentage of their total surface area. With the exception of those acting on polymeric substrates (like

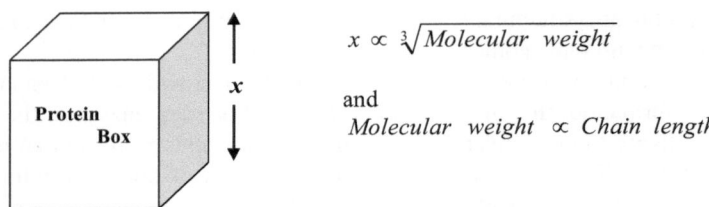

$$x \propto \sqrt[3]{Molecular\ \ weight}$$

and
$$Molecular\ \ weight \propto Chain\ \ length$$

Fig. 39.3 Relation between the three-dimensional size of an enzyme, its molecular weight, and the polypeptide chain length

Table 39.8 Catalyst size and relative biosynthetic cost

Form	Enzyme size (number of amino acid residues)	Notional catalytic efficiency [(k_{cat}/K_M)/diffusion rate]	Number of enzyme molecules required to achieve same relative flux	Relative biosynthetic cost
E1	10	0.001	10,000	1.000
E2	100	0.100	1000	1.000
E3	1000	1.000	100	1.000
E4	**500**	**0.800**	**125**	**0.625**

polysaccharides, polypeptides, DNA, or RNA), enzymes are relatively larger than their corresponding substrate(s). Is there no selection pressure to trim them to a smaller size while retaining catalytic function? In other words, are enzymes also evolved to optimize biosynthetic cost to the cell? Enzymes are three-dimensional protein boxes made of linear chain of amino acids linked to each other by peptide bonds. The dimensions of this protein box increase only with the cube root of its molecular weight (Fig. 39.3). A simple calculation will show that to achieve 5 times increase in the length "x" of the protein box, we need about 100 times increase in the molecular weight of that protein. Taken together with the relative sizes of the protein and the corresponding substrate that occupies the active site on its surface, it seems that enzymes are not really that big after all.

Almost 70% of the cellular energy is spent towards protein synthesis—larger enzymes are more expensive. Considering such high biosynthetic cost, it is not surprising that catalytic efficiency of an enzyme is not the only feature selected for by nature. Improvements in catalytic perfection beyond a point become counterproductive—an incremental increase in efficiency may require substantial increases in biosynthetic investment. This is conceptually illustrated through an imaginary example in Table 39.8.

Larger size of a polypeptide enzyme ensures the proper positioning of the active site residues required for most effective catalysis. Therefore, catalytic efficiency could be improved by accommodating better design through an increase in material input (increase in the number of amino acid residues per catalyst; *E*1–*E*3 in Table 39.8). Beyond a point, however, the gain in catalytic efficiency does not justify the large investment in the size increase. Natural selection will prefer such a cost-efficient catalyst. Yet an increase in the number of less-efficient enzyme molecules in principle could compensate for and support the required metabolic flux rate (see the imaginary example $E4$ whose k_{cat}/K_M is not diffusion-controlled; Table 39.8). For example, overexpression of less-efficient mutant enzyme forms is known to adequately complement the host defect. A case in point is L-methionine biosynthesis in *E. coli*—two different isoforms can support this pathway. The B_{12}-dependent enzyme is a larger polypeptide (130 kDa) than the corresponding non-B_{12}-dependent isoform (99 kDa). Although the B_{12}-dependent form is larger, this biosynthetic cost is offset by its 100 times superior k_{cat} value. Everything else being equal, in terms of cost, more than 75 non-B_{12}-dependent enzyme molecules are required to replace each B_{12}-dependent enzyme molecule for the same catalytic need (recall that $V_{max} = k_{cat} \times [E_t]$). The organism employs the larger isoform whenever it has access to B_{12} but otherwise resorts to many more molecules of the smaller isoform. Clearly, *an enzyme is also selected by nature for its biosynthetic cost efficiency*.

39.4.3 Then Why Are Enzymes Big?

It is obvious from the previous paragraphs that evolutionary pressures do act to optimize enzyme size (biosynthetic cost). Nonetheless, enzymes in general are perceived as large molecules compared to their substrates (Srere 1984). This is because we consider size in relation to linear dimensions rather than volume or mass. As noted earlier, however, the dimensions of a protein box increase only with the cube root of its molecular weight (Fig. 39.3). Polypeptides in the size range of 30–50 kDa make up more than 50% of the total cellular proteins; only 3–5% proteins are found above the 80 kDa range. No naturally occurring enzymes with polypeptide chains of less than 50 amino acid residue length are known. Some of the smallest enzymes known include the following: (1) 4-oxalocrotonate tautomerase consists of a 62 residue monomer but functions as a hexamer, (2) acylphosphatase consisting of a 98 residue monomer is one of the smallest enzymes known; it catalyzes the hydrolysis of acyl phosphates, and (3) the HIV protease functions as a homodimer; the active site lies between the identical subunits made up of 99 amino acids. Amongst synthetic organic chemists, proline is considered the smallest chiral catalyst, and it catalyzes asymmetric aldol reactions. The rate accelerations, however, are nowhere comparable to polypeptide-based enzymes. Lastly, if protein enzymes are considered big, RNA enzymes are even bigger. A short discussion on nonprotein catalysts may be found in the next chapter.

Hexokinase (~50 kDa protein) is about 70 times the combined molecular masses of glucose and ATP-Mg Moderately sized substrates like glucose make use of as many as 15 or more hydrogen bonds for proper binding to the enzyme. These considerations bring us to question the purpose of relatively larger size of an enzyme. The following points may be considered.

- Enzyme active sites most often occupy a fraction of their total surface area. A typical substrate (like glucose for hexokinase) covers 10–15% of the total enzyme surface area and occupies 2–3% of its total volume. But active site as a rigid entity is not acceptable. They should be flexible enough to bind and release substrate or product but rigid enough to best fit the transition state. The bulk of the enzyme that does not constitute the active site is needed to maintain the active site in geometry faithful to its transition-state structure. Enzymes have evolved for conformational flexibility, and this comes with a cost. Their large size ensures that the interaction of the substrate with the active site alters the global conformation of the enzyme in a meaningful way—the active site shifts from an initial substrate-specific geometry to a transition-state-specific geometry. In addition, in some cases, interactions (electrostatic!) between the enzyme and its substrate beyond the active site do contribute to rate accelerations.
- Some enzymes, in addition to acting as catalysts, also serve as sensors. To accommodate such regulatory features, the protein box is anticipated to display additional sites to bind the regulatory ligands. Larger proteins have enough surface area for multiple interactions through allosteric sites (such as glutamine synthetase; Chap. 38). Cooperative interaction between different sites requires that the binding information be transmitted across space through the conformational changes in the polypeptide. For example, hemoglobin (long considered an honorary enzyme!) has distinct sites and conformational states to bind oxygen, carbon dioxide, and 2,3-bisphosphoglycerate.
- Some enzymes harbor multiple functions on them and have multiple domains. Domain structure provides for combining catalytic and regulatory properties and protein–protein interactions. Protein–protein interactions are the rule rather than the exception. Through channeling (see below), metabolic advantages accrue to the cell besides the expected kinetic advantage.

We may conclude that the requirements for catalytic perfection, accommodation of regulatory site(s), and/or conformational flexibility are not necessarily congruent properties. One or more of these features may require that others are somewhat compromised. The ultimate design of an enzyme catalyst may be the result of tradeoffs between catalytic efficiency, protein stability, biosynthetic cost, and inclusion of regulatory features. Fortunately, the goal of a *Biochemist* and that of *Natural Selection* are not congruent. This leaves enough scope for enzyme engineering and redesign.

39.5   Consecutive Reactions and Metabolite Channeling

If one focuses on adjacent reaction steps inside a cell, the study of enzymes in vivo quickly becomes the study of a metabolic pathway. In coupled enzyme assays (for example see Fig. 11.2; Chap. 11), we can deliberately couple almost any two enzymes of our choice, provided they share a substrate–product pair. In the cellular context, enzymes of metabolism exist in pathways and they do not function in isolation. The product of the previous enzymatic step feeds into the next enzyme as its substrate.

39.5.1 Consecutive Steps in Metabolism

Enzymes catalyzing the consecutive steps of a metabolic sequence provide interesting insights into how metabolism is organized and how it responds. The driver for metabolism is the desire for reactions to reach equilibrium. Nature has exploited this principle to couple reactions of metabolic pathways; reactions are made spontaneous by adjusting the concentration of reactants and products. The direction of an equilibrium reaction is decided by suitably adjusting the mass action ratio (Γ). We have noted earlier (in Chap. 9) that continuous depletion of GA3P (by GA3P dehydrogenase) maintains the ΔG negative for triosephosphate isomerase (DHAP $\rightleftarrows$ GA3P) reaction and feeds DHAP into glycolysis.

So long as an enzyme is not saturated with the substrate, an increase in [S] could stimulate the rate of that reaction—this relationship is typically Michaelian. As a general rule, enzymes will operate with reactant concentrations in the region of their K_M or $S_{0.5}$. This has two implications: (a) the catalytic potential of the enzyme is better utilized and (b) the system tendency to revert to steady state (and stabilize [S]) is facilitated. Although individual enzymatic reactions are not at equilibrium in a cell, the metabolic pathways are believed to be at, or close to, steady state. Only when supply is balanced by demand, a steady-state concentration of intermediate is obtained. This means the concentrations of metabolic intermediates do not change appreciably while there is a flux (flow of $S \rightarrow P$) through the pathway.

Consider a simple two-enzyme system with two consecutive irreversible reactions:

$$A \xrightarrow{\text{E}_1} B \xrightarrow{\text{E}_2} C$$

Assuming that both E_1 and E_2 display Michaelian behavior, the velocity of the first step alone is given by

$$v_1 = \frac{V_{\max 1}\,[A]}{K_A + [A]}.$$

And of the second by

$$v_2 = \frac{V_{\max 2}\,[B]}{K_B + [B]}.$$

However, rate of change in [B] is given by

$$\frac{d[B]}{dt} = (v_1 - v_2) = \frac{V_{\max 1}[A]}{K_A + [A]} - \frac{V_{\max 2}[B]}{K_B + [B]}.$$

We can now consider three distinct cases. If $v_1 > v_2$, then B accumulates (increase in $[B]$). Second, if $v_2 > v_1$, then $[B]$ tends to zero; at the extreme, B may simply be transferred from the active site of E_1 to that of E_2 without any buildup of $[B]$. We will have more to say about this phenomenon of *channeling* in the next part. The third situation is where $v_1 = v_2$ and the so-called steady-state levels of $[B]$ are attained. This $[B]_{\text{steady state}}$ is given by

$$[B]_{\text{steady state}} = v_1 \left(\frac{K_B}{V_{\max 2} - v_1} \right).$$

The readers may look up the related treatment of a sequential two-step process (Fig. 9.4 and the dish washing analogy!) and the accumulation of intermediate presented in Chap. 9.

Coupled (natural or artificial) enzyme assays are often used in biochemical analysis. From the above equation it follows that E_2 should ideally have a smaller K_B and a larger $V_{\max 2}$. Enzymologists use much higher activity of the second enzyme (E_2 in the above case) to achieve very little or no lag time (the time before the system enters steady state).

39.5.2 Substrate Channeling

In the consecutive steps of a metabolic sequence (A → B → C, as above) the steady-state level of B is also determined by its *transit time*. This is the time required for product B of the first enzyme (E_1)-catalyzed reaction to diffuse to the active site of the next enzyme (E_2). And it depends on (a) distance between the two sequential enzymes (E_1 and E_2) of the pathway and (b) the exact diffusion coefficient of B in the medium between them. A range of possibilities exist: a metabolic intermediate like B may completely equilibrate with its pool in the surrounding medium, only a certain fraction of it may equilibrate or it may be directly transferred (channeled) to the next enzyme active site. The last possibility—a limiting case of direct transfer of B between active sites without any release into the bulk phase—with the shortest transit time is called *channeling*. Enzymes exhibiting multiple activities (see Table 13.3; Chap. 13) are obvious candidates to look for this phenomenon. Metabolite channeling in vivo may be achieved through a range of sequential active site interactions (Fig. 39.4) enumerated below.

- Direct channeling relies on the formation of protein tunnels that connect consecutive active sites, preventing metabolic intermediates from diffusing away. The first *molecular tunnel* within tryptophan synthase was discovered in 1988. Indole derived from the cleavage of indole-3-glycerol phosphate at one active site traverses 25 Å through a protein tunnel to the other site where it condenses with L-serine. The steady-state concentration of indole is extremely low as very

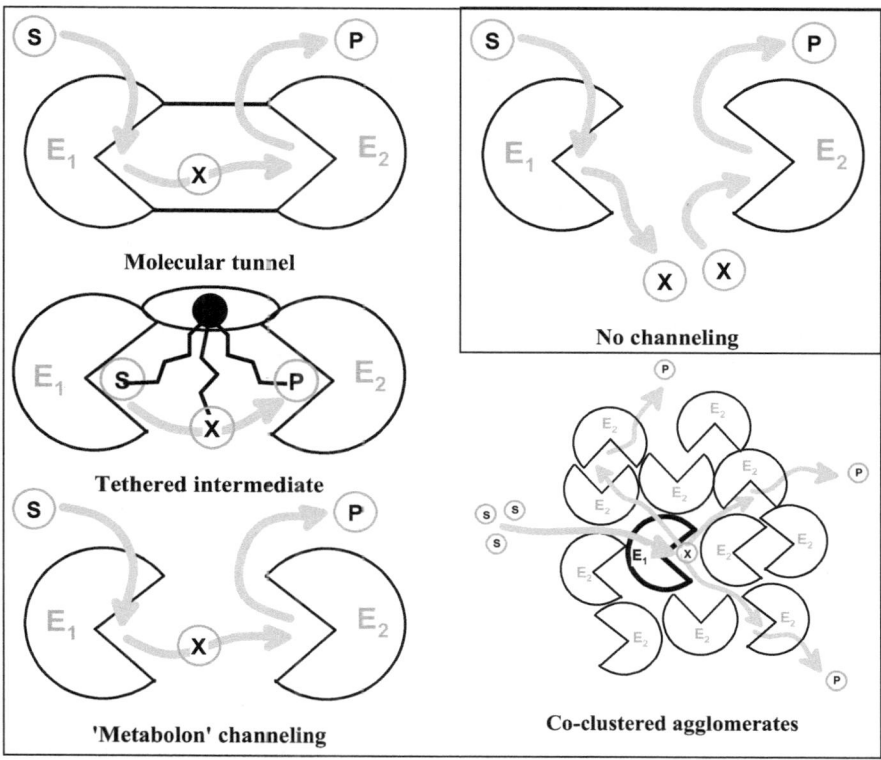

Fig. 39.4 Possible interactions involving consecutive active sites of sequential reactions. The product (X) from the first active site (E_1) may be taken to the next one (E_2) by a molecular tunnel or a covalent tether. Channeling of X may also occur from E_1 to E_2 by their proximity (forming a metabolon) or simple co-clustering (LLPS may offer one possibility; see Chap. 38). X equilibrates with the bulk metabolic pool only when there is no channeling

little leaves the enzyme into the surrounding medium. The translocation of ammonia, derived from the hydrolysis of glutamine, also occurs frequently through protein tunnels (Weeks et al. 2006). Channeling ammonia has been discovered within carbamoyl phosphate synthetase, asparagine synthetase, glutamate synthase, glutamine phosphoribosylpyrophosphate amidotransferase, imidazoleglycerol phosphate synthase, and glucosamine 6-phosphate synthase. Another example is the carbon monoxide channeling demonstrated in carbon monoxide dehydrogenase/acetyl-CoA synthase. It is interesting to note that in all these examples *the intermediate is not covalently bound to the enzyme but is simply shepherded to another active site*. The presence of molecular tunnels is becoming a recurring theme in structural enzymology (we will have more to say about them in the next chapter).

- Yet another strategy to ensure that metabolic intermediates are directed from one active site to the next is by tethering (Fischbach and Walsh 2006; Perham 2000).

Both multifunctional polypeptides and multienzyme complexes belong to this category. The ***tethered intermediates*** are held on a chemical leash for example through lipoate, biotin, pantothenate, etc. In this assembly-line strategy *swinging arms carry the reactant from one site to the other*. Examples include all α-keto acid dehydrogenase complexes, both type I and type II fatty acid synthases, polyketide synthases, and nonribosomal peptide synthases.

- Most high-affinity protein–protein complexes (typical $K_D < 10^{-6}$ M) are readily detectable. But when their interaction is weak or very weak (e.g., $K_D > 10^{-4}$ M), many conventional approaches fail to detect them. At high enzyme concentrations (and with crowded environment!) in vivo, ultra-weak interactions also become important, despite their transient nature and low stability. Enzyme–enzyme interactions can become more likely and the preference for direct metabolite transfer becomes accordingly far more favorable. Paul Srere coined the term ***metabolon*** to denote a complex of sequential enzymes, which may involve loosely or transiently associated proteins. Such a metabolon may have the ability to channel a metabolic pathway and it involves the preferential transfer of an intermediate from one enzyme to a physically adjacent enzyme, with restricted diffusion into the surrounding milieu. Carbamyl phosphate—dedicated for pyrimidine biosynthesis versus arginine biosynthesis—in *N. crassa* is an example. More recently, simple enzyme clustering was shown to accelerate the processing of intermediates through proximity channeling. Co-clustering multiple enzymes into compact agglomerates yields the same efficiency benefits as direct channeling (Castellana et al. 2014; Banani et al. 2017). How to test for definitive substrate channeling, using the aspartate aminotransferase/malate dehydrogenase system from *E. coli*, is reported (Geck and Kirsch 1999). However, simply fusing two enzymes together will not cause productive channeling (Sanyal et al. 2015).

Channeling is an example of first-level metabolic organization, indicating that cell is not a simple bag of enzymes (Mathews 1993). Regardless of how it is achieved, channeling does provide certain advantages to the cell. Namely, it could (1) serve to protect toxic, unstable, or scarce metabolites by maintaining them in the protein-bound state, (2) provide a metabolic advantage by maintaining concentration gradients, (3) protect the solvation capacity of cell water and reduce solute burden, (4) provide kinetic advantages in terms of rate accelerations beyond bulk diffusion rate limitation, (5) facilitate a quick response of the pathway to inhibitors and activators, and (6) provide a regulatory feature through the dynamic formation/destruction of the metabolon complex. Channeling in a limiting case (through molecular tunnels) leads to one-dimensional diffusion for reaction. This reduction of dimensionality increases the speed and economy of diffusion-controlled reactions (as expected intuitively!) (Hardt 1979; Berg and von Hippel 1985).

Channeling is an attractive concept that makes much in vivo metabolic sense. But to provide experimental proof of channeling is a challenge (Anderson 1999). Channeling is suspected if an endogenous intermediate produced in a pathway fails to mix (either partially or completely) with the same intermediate produced exogenously by an enzyme located elsewhere in the cell. Experimentally, channeling can be observed by providing a radiolabeled precursor to the pathway and monitoring the either of the product or of an intermediate; label mixing with pools of nonradioactive intermediates (and subsequent dilution) leads to diminished specific radioactivity (Wheeldon et al. 2016). Channeling implies facilitated transfer of channeled intermediates. Conversely, restricted access by exogenous intermediates to the pathway is expected. Confirmation of metabolite channeling demands multiple experimental approaches, with each approach failing to disprove it (rather than proving it!). Besides analysis of its advantageous kinetic features (like reduction in transient times, enhanced in vivo reaction flux rates, demonstration of direct transfer of intermediates, metabolite compartmentation in the absence of organelles etc.), several approaches could be used to establish the physical proximity of enzymes in a metabolon. These tools include—co-fractionation/co-localization of enzyme activities, use of bifunctional cross-linking reagents and immunoprecipitation and pull-down assays. Partial or complete reconstitution of a functional complex from purified protein components also offers reasonable support for metabolite channeling.

39.5.3 Metabolic Branch Points—Enzymes Competing for a Metabolite

When different substrates compete for the active site of the same enzyme, it can be used to glean useful kinetic insight to enzyme mechanism (see Chap. 22). Then again, two different enzymes may compete for the same substrate. For example, L-arginine is a substrate for both nitric oxide synthase and arginase in mammalian cells. There are effectively two kinds of enzyme–substrate competitions in vivo. A single metabolite may be substrate for more than one enzyme, or one enzyme may accept different metabolites as its substrate.

Many enzymes vying for the same substrate (metabolite) are found at metabolic branch points. Consider the oxaloacetate node. Intracellular oxaloacetate concentration is very low and much of it is enzyme bound. The mitochondrial oxaloacetate is available to citrate synthase, malate dehydrogenase, phosphoenolpyruvate carboxykinase, and aspartate aminotransferase. However, distinct cellular metabolic states decide the flow of oxaloacetate either into Krebs cycle (citrate synthase) or towards gluconeogenesis (phosphoenolpyruvate carboxykinase). A host of intensive and extensive properties influence the overall rate of an enzyme-catalyzed reaction in cellular metabolism (Fig. 38.1; Chap. 38). Of these, relative affinities for a common substrate and concentrations of competing enzymes (V_{max} values) significantly decide the fate of a branch point metabolite. The enzyme with a lower K_M will win this competition and operate closer to its full capacity. Such an enzyme binds more substrate because of its higher affinity; also, enzyme present at higher concentration hogs larger share of the common substrate. This in turn generates a steeper substrate gradient near its vicinity. In general, relative K_M values of the competing enzymes dictate the fate of that branch point metabolite. For instance, phosphofructokinase and glucose-1-phosphate uridyltransferase drain glucose-6-phosphate as their respective hexose phosphate substrates—fructose-6-phosphate and glucose-1-phosphate. But with a much lower K_M for its substrate, phosphofructokinase has a higher preference for hexose phosphate; significant glycogen synthesis begins only after phosphofructokinase is fully saturated. The following examples will illustrate this point further. Whereas many more equally interesting cases do exist, the four examples presented below carry a personal bias.

1. Fungal arginine metabolism: Filamentous fungi are capable of biosynthesizing arginine starting from glutamate and via ornithine. They can also utilize arginine supplied from outside. Most of the cellular arginine (and ornithine) is located in the vacuoles and the cytosolic arginine concentrations are much lower. Arginase, vacuolar arginine transporter, and arginyl-tRNA synthetase compete for this arginine pool (Fig. 39.5, panel **a**). At most cellular concentrations of arginine, the vacuolar transporter would be more saturated than arginase. This ensures that significant arginine catabolism ensues only after the vacuolar reserves are filled up. Low cytosolic concentration of arginine enables arginyl-tRNA synthetase to compete successfully with arginase; based on their relative K_M values, arginyl-tRNA synthetase will be operating at >90% maximum while arginase is hardly active. When the pool of arginine in the cytosol increases rapidly (due to externally supplied arginine) arginase comes into action.
2. Metabolic fate of 2-oxoglutarate: Cellular 2-oxoglutarate occupies a branch point that connects carbon and nitrogen metabolism. It is either oxidized to succinyl CoA (by 2-oxoglutarate dehydrogenase complex and taken through further steps of Krebs cycle) or gets reductively aminated to glutamate (by NADP-glutamate dehydrogenase and leads to biogenesis of glutamate family amino acids). Respective 2-oxoglutarate K_M values for the two enzymes dictate how it partitions between the two routes—cellular energy needs versus biomass (Fig. 39.5, panel **b**). The 2-oxoglutarate dehydrogenase complex is effectively fully saturated at high [2-oxoglutarate], and only then is significant glutamate synthesis expected. In another example, this split ratio is an important determinant in the glutamate fermentation by *C. glutamicum*.
3. Pyruvate branch point: Pyruvate largely originates from phosphoenolpyruvate in glycolysis. The flux away from pyruvate is defined by four enzyme activities in *A. niger*. Except malic enzyme, others (pyruvate carboxylase, alanine aminotransferase, and pyruvate dehydrogenase complex) have sub-millimolar affinity for pyruvate (Fig. 39.5, panel **c**). Scrutiny of their relative K_M values and concentrations of competing enzymes is a prerequisite to attempt diverting pyruvate flux to lactate. This fungus does not produce lactate. Introducing a lactate dehydrogenase with appropriate kinetic features could in principle facilitate lactate formation (Dave and Punekar 2015).
4. Competition for ATP: A "metabolic contest" to regulate the use of glucose and glycerol without nutrient sensing and signaling occurs in *Trypanosoma brucei* (Allmann et al. 2021). This competition depends on the fact that the two pathways operate in the same subcellular compartment and the respective enzymes (glycerol kinase and hexokinase) vie for the same substrate (ATP). Because glycerol

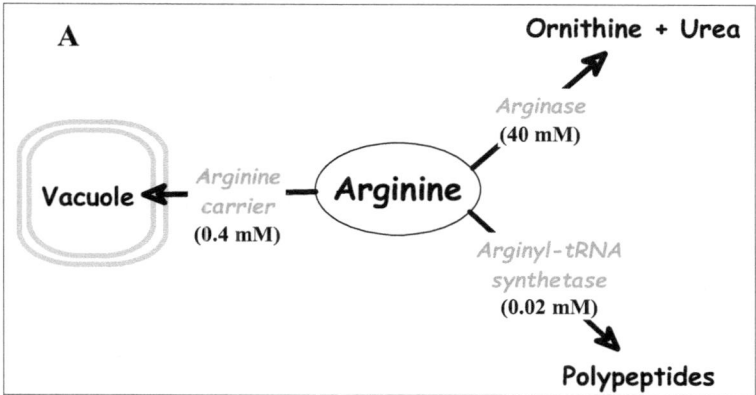

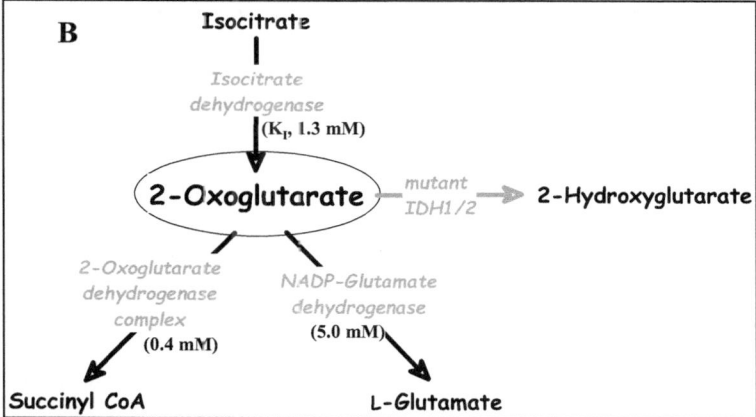

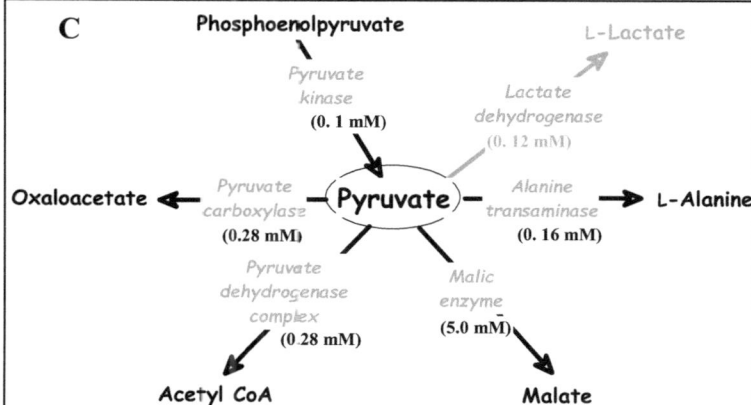

Fig. 39.5 Enzymes compete in vivo for a common substrate at the metabolic branch point. Competing reactions at L-arginine (**a**), 2-oxoglutarate (**b**), and pyruvate (**c**) nodes are illustrated. The K_M values shown for the respective enzymes (in gray) are from fungal literature (mostly from Aspergilli)

kinase is in large excess over hexokinase, the organism preferentially uses glycerol over glucose.

39.6   Underground Metabolism

Most enzymes recognize certain substrate analogs and are able to use them as alternative substrates. In some cases, these analogs are endogenous, natural metabolites. Under normal conditions, they are not accessible or acted upon by such enzymes. Reactions catalyzed by normal enzymes acting on substrate analogs which themselves are endogenous metabolites constitute *underground metabolism* (D'Ari and Casadesus 1998). Metabolic networks are inherently messy. Such underground metabolic activity (and enzyme promiscuity!) is sometimes useful and provides the driving force in the evolution of new metabolic pathways. Delbruck's "principle of limited sloppiness" may indeed be at play in Nature (Glasner et al. 2020). Harnessing underground metabolism in rewiring the pathways is an attractive possibility in biotechnology (Rosenberg and Commichau 2019). Some of the "abnormal" metabolites arising out of underground metabolism are reconverted to normal metabolites by repair enzymes. For instance, L-2-hydroxyglutarate dehydrogenase takes care of the mistakenly produced 2-hydroxyglutarate (see below). Such metabolite proofreading enzymes may be quite common, but many still remain unidentified (Van Schaftingen et al. 2013).

We will illustrate underground metabolism with a few examples. The biosynthetic pathways of arginine and proline involve analogous reactions on different substrates. *N. crassa* strains with deletions of the first two enzymes of proline synthesis (γ-glutamyl kinase and γ-glutamyl phosphate dehydrogenase) are viable because the deacetylase, which normally deacylates *N*-acetylornithine, promiscuously deacetylates *N*-acetylglutamate semialdehyde. Thus, the early part of arginine biosynthesis gets shortcircuited to generate proline using arginine biosynthetic enzymes.

Yet another example of underground metabolism is the reduction of 2-oxoglutarate to 2-hydroxyglutarate. 3-Phosphoglycerate dehydrogenase has an interesting dual function. The *S. cerevisiae* enzyme functions in the anabolic pathway of serine synthesis and may also reduce 2-oxoglutarate. The latter activity is manifest under physiological states when excess 2-oxoglutarate and reducing power coexist in vivo. Consistent with this idea, anaerobic growth of *E. coli* is also inhibited by external supply of 2-oxoglutarate, and this inhibition is reversed by serine addition. Accumulation of 2-hydroxyglutarate (considered an *oncometabolite*) can modulate the activities of 2-oxoglutarate-utilizing dioxygenases. Wild-type isocitrate dehydrogenase (IDH1 and IDH2 isoforms) catalyzes the NADP-dependent reversible conversion of isocitrate to 2-oxoglutarate. But cancer-associated gain-of-function mutations (at substrate-binding residues namely, Arg 132 in IDH1 and Arg 140 and Arg 172 in IDH2) enable mutant IDH1/2 to catalyze the NADPH-dependent reduction of

2-oxoglutarate to $R(-)$-2-hydroxyglutarate (D-2-hydroxyglutarate) (Fig. 39.5, panel **b**) (Dang et al. 2009).

Broader substrate specificity (enzyme promiscuity and moonlighting activities of some enzyme proteins; Chap. 13) of some enzymes manifests itself as underground metabolism. Promiscuous enzymes may be recruited by Nature to serve new functions—that manifest as novel catalysts important for fitness and/or survival (Copley et al. 2023). And it has been well argued that underground metabolism is a testing ground for evolution of metabolic pathways.

39.7 Summing Up

Enzyme study in vivo is much different from that in vitro due to subcellular organization and compartmentation. The aqueous phase of the cytoplasm is crowded and often has high enzyme concentrations. The Michaelis–Menten formalism is not suitable in vivo for many such enzymes (where $[S_t] >> [E_t]$), is not true). Further, the cell provides a single consensus medium common for all its enzymes functioning in that compartment whereas an enzymologist's test tube is optimized for the enzyme of his choice. Many interactions and regulatory features are simply not there in such a clean system. Rate laws determined in vitro with purified dilute homogeneous enzyme solutions may not reflect the enzyme–enzyme interactions that are important in vivo. Molar concentrations of many enzymes in vivo are quite high because of which a considerable portion of the substrate may be enzyme bound. For small, confined volumes (as we see in cellular compartments such as lysosomes, mitochondria, and peroxisomes) few molecules of enzyme or substrate may mean a significantly high molar concentration. Cytoplasm has a low micro-viscosity and a high macro-viscosity because of macromolecular crowding. Molecular crowding in turn may alter observed equilibrium constants and/or may profoundly affect the enzyme kinetic parameters. The ultimate design of an enzyme catalyst in vivo may therefore be the result of tradeoffs between catalytic efficiency, protein stability, biosynthetic cost and inclusion of regulatory features.

Erwin Chargaff held that—*But the cell is certainly more than a chemical slum*. What is this more? The study of enzymes in vivo quickly becomes the study of a metabolic pathway. The product of a previous enzymatic step feeds into the next enzyme as its substrate. This may occur with or without the intermediates freely equilibrating with the cellular metabolite pool. Metabolite channeling may occur through a range of sequential active site interactions. Many enzymes may compete for a single metabolite (at the branch point) or a single enzyme may accept more than one metabolite as its substrate. Enzymology in vivo thus merges into the complexity of cellular metabolism. A study in ***systems biology*** will not be complete without incorporating enzymes and their various properties.

References

Albe KR, Butler MH, Wright BE (1990) Cellular concentrations of enzymes and their substrates. J Theor Biol 143:163–195

Allmann S et al (2021) Glycerol suppresses glucose consumption in trypanosomes through metabolic contest. PLoS Biol 19:e3001359

Anderson KA (1999) Fundamental mechanisms of substrate channeling. In: Methods in enzymology, vol 308. Elsevier, pp 111–145

Antolin AA, Cascante M (2021) AI delivers Michaelis constants as fuel for genome-scale metabolic models. PLoS Biol 19:e3001415

Banani SF et al (2017) Biomolecular condensates: organizers of cellular biochemistry. Nat Rev Mol Cell Biol 18:285–298

Benner SA (1989) Enzyme kinetics and molecular evolution. Chem Rev 89:789–806

Bennett BD et al (2009) Absolute metabolite concentrations and implied enzyme active site occupancy in *Escherichia coli*. Nat Chem Biol 5:593–599

Berg OG, von Hippel PH (1985) Diffusion-controlled macromolecular interactions. Ann Rev Biophys Biophys Chem 14:131–160

Burbaum JJ et al (1989) Evolutionary optimization of the catalytic effectiveness of an enzyme. Biochemistry 28:9293–9305

Castellana M et al (2014) Enzyme clustering accelerates processing of intermediates through metabolic channeling. Nat Biotechnol 32:1011–1018

Copley SD, Newton MS, Widney KA (2023) How to recruit a promiscuous enzyme to serve a new function. Biochemistry 62:300–308

Cornish-Bowden A (1999) Enzyme kinetics from a metabolic perspective. Biochem Soc Trans 27: 281–284

Dang L et al (2009) Cancer-associated IDH1 mutations produce 2-hydroxyglutarate. Nature 462: 739–744

D'Ari R, Casadesus J (1998) Underground metabolism. BioEssays 20:181–186

Dave K, Punekar NS (2015) Expression of lactate dehydrogenase in *Aspergillus niger* for L-lactic acid production. PLoS One 10(12):e0145459

Davidi D et al (2016) Global characterization of in vivo enzyme catalytic rates and their correspondence to in vitro k_{cat} measurements. Proc Natl Acad Sci USA 113:3401–3406

Dobzhansky T (1973) Nothing in biology makes sense except in the light of evolution. Am Biol Teach 35:125–129

Ellis RJ (2001) Macromolecular crowding: obvious but underappreciated. Trends Biochem Sci 26: 597–604

Ernst DC, Downs DM (2018) Mmf1p couples amino acid metabolism to mitochondrial DNA maintenance in *Saccharomyces cerevisiae*. MBio 9:e00084–e00018

Fischbach MA, Walsh CT (2006) Assembly-line enzymology for polyketide and nonribosomal peptide antibiotics: logic, machinery, and mechanisms. Chem Rev 106:3468–3496

Garcia-Contreras R et al (2012) Why in vivo may not equal in vitro—new effectors revealed by measurement of enzymatic activities under the same in vivo-like assay conditions. FEBS J 279: 4145–4159

Geck MK, Kirsch JF (1999) A novel, definitive test for substrate channeling illustrated with the aspartate aminotransferase/malate dehydrogenase system. Biochemistry 38:8032–8037

Glasner ME, Truong DP, Morse BC (2020) How enzyme promiscuity and horizontal gene transfer contribute to metabolic innovation. FEBS J 287:1323–1342

Gupta SD, Zhang S, Szostak JW (2023) Molecular crowding facilitates ribozyme-catalyzed RNA assembly. ACS Cent Sci 9:1670–1678

Halling PJ (1989) Do the laws of chemistry apply to living cells? Trends Biochem Sci 14:313–353

Hardt SL (1979) Rates of diffusion controlled reactions in one, two and three dimensions. Biophys Chem 10:239–243

Hasinoff BB (1984) Kinetics of carbonic anhydrase catalysis in solvents of increased viscosity: a partially diffusion controlled reaction. Arch Biochem Biophys 233:676–681

Hassett A, Blättler W, Knowles JR (1982) Pyruvate kinase: is the mechanism of phospho transfer associative or dissociative? Biochemistry 21:6335–6340

Knowles JR, Albery WJ (1977) Perfection in enzyme catalysis: the energetics of triosephosphate isomerase. Acc Chem Res 10:105–111

Kornberg A (2000) Ten commandments: lessons from the enzymology of DNA replication. J Bacteriol 182:3613–3618

Kornberg A (2003) Ten commandments of enzymology, amended. Trends Biochem Sci 28:515–517

Luby-Phelps K (2000) Cryo-architecture and physical properties of cytoplasm: volume, viscosity, diffusion, intracellular surface area. Int Rev Cytol 192:189–221

Mathews CK (1993) The cell—a bag of enzymes or a network of channels? J Bacteriol 175:6377–6381

Milo R, Phillips R (2016) Cell biology by the numbers. Garland Science, Taylor & Francis Group LLC, New York

Perham RN (2000) Swinging arms and swinging domains in multifunctional enzymes: catalytic machines for multistep reactions. Annu Rev Biochem 69:961–1004

Rosenberg J, Commichau FM (2019) Harnessing underground metabolism for pathway development. Trends Biotechnol 37:29–37

Sanyal N et al (2015) First evidence for substrate channeling between proline catabolic enzymes. J Biol Chem 290:2225–2234

Savageau MA (1992) Critique of the enzymologist's test tube. In: Fundamentals of medical cell biology, vol 3A, chemistry of the living cell. Elsevier, Amsterdam

Schenck CA, Last RL (2020) Location, location! Cellular relocalization primes specialized metabolic diversification. FEBS J 287:1359–1368

Smallbone K et al (2013) A model of yeast glycolysis based on a consistent kinetic characterization of all its enzymes. FEBS Lett 587:2832–2841

Srere PA (1967) Enzyme concentrations in tissues. Science 158:936–937

Srere PA (1970) Enzyme concentrations in tissue II. Biochem Med 4:43–46

Srere PA (1984) Why are enzymes so big? Trends Biochem Sci 9:387–390

Srere PA, Knull HR (1998) Location—location—location. Trends Biochem Sci 23:319–320

Srivastava DK, Bernhard SA (1986) Enzyme–enzyme interactions and the regulation of metabolic reaction pathways. Curr Top Cell Regul 28:1–68

Storey KB (2005) Functional metabolism: regulation and adaptation. Wiley, Hoboken

van Eunen K et al (2010) Measuring enzyme activities under standardized in vivo-like conditions for systems biology. FEBS J 277:749–760

Van Schaftingen E et al (2013) Metabolite proofreading, a neglected aspect of intermediary metabolism. J Inherit Metab Dis 36:427–434

Verkman AS (2002) Solute and macromolecule diffusion in cellular aqueous compartments. Trends Biochem Sci 27:27–33

Weeks A, Lund L, Raushel FM (2006) Tunneling of intermediates in enzyme-catalyzed reactions. Curr Opin Chem Biol 10:465–472

Wheeldon I et al (2016) Substrate channelling as an approach to cascade reactions. Nat Chem 8:299–309

*If we wish to catch up with nature we shall need to use the
same methods as she does, and I can foresee a time in which
physiological chemistry will not only make greater use of
natural enzymes but will actually resort to creating
synthetic ones.*
Emil Fischer, 1902 Nobel Lecture

Enzymes are the only biomolecules that combine three hallmark features, namely—
catalysis, *specificity*, and *regulation*. No wonder that the study of enzymes will
continue to occupy the prime position in modern biology (and chemistry) (Editorial,
Closing in on catalysis 2009). This much is amply obvious from each and every
context and examples that we have come across in the preceding chapters. The study
of enzymes in isolation—most often in purified form—has occupied much of the
time in the past. Importance of enzyme function in vivo will form one of the key
frontiers with the present emphasis on systems biology as a pointer in this direction.
Enzymes in sequence, in combination with other enzymes and other cellular
components bring in interesting features often not manifested by an enzyme in
isolation—coupled reactions, regulatory networks, and distributed control of metabolism are some of them. While single enzyme studies have taken the back seat, the
knowledge gleaned from such research is important for a systems biology approach.
After all, *any biologist who follows their research interest to the finest level of detail
will become an enzymologist* (quote by Perry Frey).

Major metabolic pathways and the enzymes that function in them are well
documented. But many more novel reactions and corresponding enzyme catalysts
are being discovered on an almost regular basis. For instance, let us consider a few
recent representative cases:

1. Most bacteria and all archaea synthesize glutaminyl-tRNA indirectly. The glutamate charged on to tRNA^{Gln} is converted to glutamine—in a second step—by a

© The Author(s), under exclusive license to Springer Nature Singapore Pte
Ltd. 2025
N. S. Punekar, *ENZYMES: Catalysis, Kinetics and Mechanisms*,
https://doi.org/10.1007/978-981-97-8179-9_40

tRNA-dependent amidotransferase. The two enzymes form the "glutamine transamidosome" that also involves channeling of ammonia (Ito and Yokoyama 2010).

2. The dehydroalanine moieties of the lantibiotic nisin are formed from the Ser residues of the peptide by the action of lantibiotic dehydratase. This enzyme glutamylates Ser side chains using glutamyl-tRNA^{Glu} for activation; subsequent glutamate elimination results in the dehydroalanine formation (Ortega et al. 2015).

3. The biosynthesis of lincomycin A (a sulfur-containing lincosamide antibiotic) recruits two bacterial thiols (mycothiol and ergothioneine) where mycothiol acts as the sulfur donor after thiol exchange (Zhao et al. 2015).

4. Riboflavin (vitamin B2) is a well-known redox cofactor in a wide variety of flavoproteins. Addition of a fourth ring to its existing three-ring system generates a riboflavin derivative (Clarke and Allan 2015; Leys 2018). This previously unknown cofactor catalyzes new types of chemistry and is crucial for the decarboxylation of an intermediate in coenzyme Q biosynthesis.

5. A SAM-dependent enzyme-catalyzed pericyclic transformations leads to the formation of the natural product leporin. Such novel roles for SAM (S-adenosyl-L-methionine) are likely to be found in other examples of enzyme catalysis (Ohashi et al. 2017).

6. The Diels–Alder reaction is one of the most powerful chemical transformations. The concerted cycloaddition between a diene and an alkene to form a cyclohexene is used to build complex chemical structures. The first FAD-dependent Diels–Alderase is reported from mulberry plants (Gao et al. 2020).

7. A carbene-transfer chemistry was assembled in microbial biosynthetic pathways through engineered P450 mutants (Huang et al. 2023). Introducing unnatural carbene-transfer reactions expands the scope of cellular metabolism to produce diverse organic compounds.

8. The early phosphinothricin biosynthesis involves a highly unusual reaction for C–C bond cleavage with no precedent (Blodgett et al. 2007). The reduction of phosphono acetaldehyde may be a common step in other phosphonate biosynthetic pathways.

9. A unique enzymatic route to biologically rare alkyne functional group was discovered in *Streptomyces cattleya* (Marchand et al. 2019). Biosynthesis begins at lysine to produce a terminal alkyne containing amino acid.

These are just a few examples that represent the author's personal bias. But certainly, many more novel reactions and enzymes will continue to be discovered and reported.

We will now consider aspects of enzymology where rapid progress is being made. Understanding the rate enhancements of enzymes continues to be a fundamental challenge of mechanistic enzymology (Herschlag and Natarajan 2013). These and other topics are anticipated to attract much attention from enzymologists in the foreseeable future.

40.1 Transition State (TS) Analysis and Computational Enzymology

Conventional view of enzyme catalysis treats the transition state (TS) in thermodynamic terms. Equilibrium between reactants and the TS is assumed; tight binding to the enzyme active site would sequester the TS from solution and increase the reaction rate. Kinetic isotope effects can provide direct information on the enzymatic TS (Chap. 25). When the intrinsic kinetic isotope effects (KIEs) are available for an enzyme reaction, then the TS structure can be deduced in the usual physical organic chemistry sense. Interpretations of KIEs give detailed bond order and geometric features of the transition state for an enzyme. In turn, molecular electrostatic potential surfaces of these transition-state models guide chemical synthesis of transition-state analogs, yielding excellent, high affinity inhibitors (Schramm 2013; Frushicheva et al. 2014). Complexes of such inhibitors with their cognate enzymes provide structural models for computational analysis of enzymatic transition states. Taken together, KIEs and computational enzymology provide a conceptually complete picture of the TS in an enzymatic reaction.

There is a progressive appreciation that the enzymatic transition state is a state of maximum free energy. And this may not be simply captured as an equilibrium between the Michaelis complex, transition state, and products. Dynamic contributions of protein motion to enzyme-catalyzed reactions must be incorporated to provide a full understanding of catalysis. Besides applications of kinetic measurements, isotope effects, time-resolved spectroscopy, NMR spectroscopy, X-ray crystallography, and cryo-EM (electron microscopy), computational enzymology is making significant contributions in understanding the enzyme TS. Most experimental techniques have focused on conformational changes that occur in microsecond to millisecond timescales; these times are often correlated with catalysis. These timescales are much slower than the TS lifetimes that occur on a bond-vibrational timescale. Chemical bond forming/breaking steps of interest (in enzyme catalysis) have vibrational modes in the low femtosecond timescale (Fig. 10.3). They are valuable in understanding TS but otherwise are experimentally difficult to access. This is where developments in computational enzymology are making some headway. Bond vibrations during enzyme catalysis can be simulated with some accuracy in the femto-second to the nanosecond timescale. This provides computational dynamic access to the timescales of the TS lifetimes. However, commonly encountered enzymatic catalysis time scales (10^{-2} s) require enormous computational time; and this is far beyond continuous dynamic calculations possible at the present. This frontier in computational enzymology and of complex enzyme models was recognized with chemistry Nobel Prize in 2013 (to Martin Karplus, Michael Levitt, and Arieh Warshel). The emergence of the quantum mechanical/molecular mechanics (QM/MM) approach allows one to ask what the origins of the catalytic power of enzymes are (Warshel 2014). The enormous increase in computer power makes it virtually certain that computer simulations will increasingly contribute to modeling molecular enzyme catalysis (Garcia-Viloca et al. 2004). This approach also promises to capture the contributions of subtle protein molecular dynamics to enzyme rate accelerations.

40.2   Single Molecule Enzymology

The Michaelis–Menten equation (refer Chap. 14) is a highly satisfactory description of kinetic data involving a very large number of enzyme molecules in the assay. However, all enzyme molecules are not synchronized with each other in an ensemble-averaged kinetic measurement. The extraction of dynamic information from such an asynchronous assembly is complicated and difficult. A single enzyme molecule gives kinetic signals that reflect the dynamic states of that individual catalytic entity. This dynamic information is lost in the average signals from ensembles. Reactions involving single enzyme molecules can now be examined by the advances in fluorescence and related time-resolved spectroscopic techniques (Smiley and Hammes 2006). These techniques combined with computational approaches allow real-time access to study the dynamic behaviors of individual molecules.

Would the ensemble averaged enzyme kinetics also hold for a single enzyme molecule? This interesting question has attracted serious attention in recent times for the following reasons. Single-molecule behavior is a particularly powerful way of uncovering (a) mechanistic pathways and intermediates, (b) how enzyme conformational fluctuations affect catalytic activity, and (c) heterogeneities hidden in the ensemble average. The turnover events of a single enzyme molecule are intrinsically stochastic. We find the sequence of the time intervals between consecutive turnover events (the waiting times, denoted as τ), as a function of the turnover index number. The average waiting time (denoted by $\langle \tau \rangle$) of a single enzyme molecule plotted against $1/[S]$ mimics the linear Lineweaver–Burk plot (refer Chap. 16) recorded in an ensemble measurement. The rate equation for single enzyme molecule kinetics is shown below.

$$\frac{1}{\langle \tau \rangle} = \frac{\chi^2 [S]}{C_M + [S]} \left(\text{compare this with } v = \frac{k_{cat}[E_t][S]}{K_M + [S]} \right).$$

The reciprocal mean waiting time determined for a single enzyme molecule is thus related to enzyme catalyzed velocity in an ensemble measurement ($1/\langle \tau \rangle = v/[E_t]$). While the Michaelis–Menten relation continues to hold for the single enzyme molecule kinetics, interpretations of kinetic constants are different. Specifically, the k_{cat} (formally represented as χ^2 for single enzyme study) measured at saturating $[S]$ is a weighted harmonic mean of the different catalytic turnover rate constants represented in the single enzyme over time. Similarly, K_M [$= (k_{cat} + k_{-1})/k_1$ or C_M for a single molecule] also acquires an ensemble-averaged meaning (Min et al. 2005).

We notice that single-molecule and ensemble-averaged Michaelis–Menten kinetics can be reconciled (Walter 2006; English et al. 2006). An interesting insight from single molecule study is that the waiting times measured at high [S] levels follow an asymmetric probability distribution. The long-time span of the catalytic turnover rate constant (i.e., k_{cat}) indicates that the single enzyme molecule's catalytic velocity fluctuates over a broad range (from 10^{-3} to 10 s) of time scales. These catalytic fluctuations point to conformational isomers of an individual enzyme slowly interconverting over time. The conformational heterogeneity has been experimentally observed in the studies of cholesterol oxidase, staphylococcal nuclease, and a few other enzymes. Single enzyme molecule displays inevitable, stochastic fluctuations in its catalytic activity. The effects of such fluctuations would be less significant for a system comprising of a large number of enzyme molecules. However, many processes inside cells rely on the activity of a single enzyme molecule, such as in DNA replication, transcription, translation, and protein transport along the cytoskeleton. Stochastic fluctuations due to low copy number of enzymes have important physiological implications for cells/organelles. These are now being probed on a single-molecule basis in vivo.

40.2.1 Enzymes as Molecular Motors

We have noted earlier (Chap. 3) that enzymes surf the heat wave—and the heat generated by catalysis propels the enzyme molecules (Riedel et al. 2015). Enzymes as proteins are anisotropic structures and their active sites are usually buried. Catalytically relevant site-specific protein thermal networks may facilitate heat transfer to and from the reaction center to the bulk medium (Klinman 2023). Catalysis boosts the motion of enzymes, and the boosts are more frequent at high substrate concentrations. This reaction–motion coupling is a general principle of catalysis (Jee et al. 2018).

It now begs the question whether a single enzyme molecule can generate sufficient mechanical force through substrate turnover to cause its own movement. Also, whether the movement can become directional in the presence of a substrate concentration gradient. Both catalase and urease enzyme molecules diffuse/spread toward areas of higher substrate concentration, a form of chemotaxis at the molecular scale. Enzyme motion at the nano/microscale can be accomplished through catalysis-induced conversion of chemical energy to mechanical forces (Zhao et al. 2018). This has implications for the use of enzymes as chemo-mechanical transducers; one could power nano- and micromotors because of enzymes' great diversity and efficiency. For instance, surface-immobilized enzymes function as self-powered micropumps in the presence of their respective substrates, but with no ATP requirement.

40.3   Structure–Function Dissection of Enzyme Catalysis

As discussed earlier in this book (Chap. 6), both conformational flexibility and protein motion are very important for enzyme catalysis. A structure–function approach to understand enzyme action uses a combination of kinetic (including rapid, transient kinetics) and structural techniques. The ultimate goal of such a study is to develop an in-depth mechanistic understanding of enzyme function. The field has traditionally made use of protein chemical modifications, spectroscopic tools (like fluorescence, circular dichroism, etc.) and more recently molecular biology tools (like generation of site directed mutants, truncated or chimeric enzyme proteins). Whereas solution NMR has provided some information on enzyme molecular dynamics (although at timescales much slower than most events in enzyme catalysis), as of now this structural tool is limited to small-sized proteins. Mass spectrometry of larger polypeptides and oligomeric proteins is also coming of age. Much of direct structural information on enzymes has come from X-ray crystallography—mostly presenting to us a static picture of the enzyme. As aptly stated by the late Jeremy Knowles (biochemist and professor at Harvard University)—"*studying the photograph of a racehorse cannot tell you how fast it can run*"; and thus, there is a limit to what a snapshot protein structure can reveal. Although snapshots of enzyme bound to substrate, product, or transition state analogs are valuable, they do not capture structural dynamics of catalytic action. Nevertheless, the presence of molecular tunnels in enzymes like tryptophan synthase would not have been apparent without detailed structural inputs. Developments in cryo-EM are, however, increasingly contributing to capture most of the structural variety presented by an enzyme molecule.

Overwhelming biochemical, mechanistic and mutational data for any enzyme study must be supplemented by structural approaches. More recent developments in computational structure prediction (see below) provide complementary inputs. Structure-based drug discovery has also benefited from these approaches. One remarkable success in this venture is the development of HIV protease inhibitors (Wlodawer and Ericson 1993). These inhibitors were rationally designed from the knowledge of the structure and mode of action of aspartyl proteases. The discovery of Saquinavir, the first protease inhibitor, made use of the promising transition state mimic chemistry.

A fine balance between structural rigidity and conformational plasticity results in the unique catalytic power of enzymes. And structural enzymology also aims to address catalytic motions in detail (Ramanathan and Agarwal 2011). How enzymes achieve a catalytically competent state has become approachable only recently   · through experiments and computation.

40.3.1 Site Directed Mutagenesis and Crystal Structures

Application of molecular biology tools to probe enzyme function has matured over the last couple of decades. The reader may refer to standard textbooks of molecular biology and many protocol/recipe books on how to construct site directed mutants and to genetically engineer enzymes. The relevant cDNA and an expression system to obtain the mutant enzyme are all that are required (Fig. 37.5). The native and various mutant enzyme forms are then subjected to rigorous structural analysis through X-ray crystallography, circular dichroism spectra, and other tools. Excellent insights continue to be gathered on residues critical for catalysis, binding, and structural stability/flexibility for many enzymes.

Site-directed mutagenesis offers a powerful approach for the rational modification of an enzyme (Wagner and Benkovic 1990). It enables enzymologists to selectively replace active site residue (or any others) and ask some very interesting mechanistic questions. *Yet site-directed mutagenesis* does not fully account for enzymatic catalysis, because the effects of individual groups on catalysis are neither additive nor independent. At the resolution of amino acid residue level, one can check whether a given residue is relevant for binding and/or catalysis. This may not give unambiguous answers as—for many enzymes—substrate binding and catalytic residues often overlap; and they may not be clearly demarcated in the enzyme active site. In fact, a few examples show that specificity can reside beyond the amino acids directly contacting the substrate—requiring major structural changes with loop grafting, etc. Nature presents us with 20 naturally occurring amino acids in proteins; an opportunity exists to replace a given residue by any one of the other 19. But still the choice is limited because very few of these substitutions are more conservative than others. For instance, Val, Ile, and Leu are often interchangeably accepted; also, a Glu residue might replace Asp and vice versa. Alanine scanning mutagenesis is a mature tool and is often used to determine the contribution of a specific residue to the stability or function of a given protein. Alanine is used because its R group (methyl) is least disruptive and mimics the secondary structure preferences of many other amino acids. Such Ala replacements are usually done by site-directed mutagenesis or generated randomly by creating a PCR library.

Standard site-directed mutagenesis is largely limited to 20 natural, proteinogenic amino acid residues. Non-natural side chains often provide mechanistic insights. One approach has been to replace the relevant residue by Cys (through SDM) and then alkylate that Cys-SH by a suitable reagent. Interesting structural variations of imidazole side chains—with subtle pK_a changes—could then be tested (Earnhardt et al. 1999). In another example, the catalytic activity of glutamine synthetase R → C mutant was rescued back by chemically modifying the Cys-SH back to an arginine analog by covalent modification with 2-chloroacetamidine (Dhalla et al. 1994). More recently, sophisticated tools and technology are in place to directly incorporate non-protein amino acids into proteins in a position specific manner. Non-natural amino acids may be introduced into proteins by manipulating in vitro protein translation as well as through in vivo strategies by expanding the genetic code (Hendrickson et al. 2004). Pyrrolysine—the 22nd protein amino acid—was found at the active site of methyltransferases from methane-producing archaea. Like the 20 common amino acids, pyrrolysine is synthesized in the cytoplasm and incorporated at a specific position during the translation of the growing polypeptide chain (Atkins and Gesteland 2002; Ragsdale 2011). The pyrrolysine biosynthetic cassette could be used to incorporate other useful modified amino-acid residues into proteins.

Examples of SDM to probe enzyme function have rapidly grown. We will look at only a few case studies to illustrate the main issues and highlight some difficulties with this approach. The first significant effort to change the substrate specificity of an enzyme was with L-lactate dehydrogenase.

- Holbrook's group achieved a specific, highly active malate dehydrogenase by redesigning *Bacillus stearothermophilus* lactate dehydrogenase framework (Wilks et al. 1988; Clarke et al. 1989; Wagner and Benkovic 1990). This involved three amino acid replacements, namely, D197N, T246G, and Q102R.
- Human arginase II is highly specific and acts on arginine to produce ornithine and urea. A single amino acid replacement (the N149D variant) converts this enzyme into an agmatinase with almost no activity on arginine (Lopez et al. 2005).
- Glucose isomerase site-directed mutagenesis has resulted in enzyme forms with altered pH optima and altered divalent metal ion specificity.
- The rational modification of enzymes to *change* or extend their coenzyme *specificity* has also been possible (Moon et al. 2012). The Nicotinamide adenine dinucleotide phosphate (NADP)-dependent glutathione reductase was altered into a Nicotinamide adenine dinucleotide (NAD)-specific enzyme. Similar efforts in other dehydrogenases such as glutamate dehydrogenase are reported.
- There are many attempts to rationalize the enzyme stability as a useful parameter through site directed mutagenesis (Bryan 2000). These have been largely empirical because we do not yet fully understand what contributes to protein stability.
- Site-directed mutagenesis (SDM) provides enzymologists an opportunity to tinker with active site residues and ask some really interesting mechanistic questions. Almost every position in the subtilisin sequence has been subjected to SDM. The three critical active site catalytic residues S221, H64, and D32 of subtilisin were evaluated through this approach (see Chap. 35; Sect. 35.6). For the subtilisin S221A mutant (which still retains some activity), the reaction cannot proceed by the usual serine acyl-enzyme intermediate (ping-pong mechanism). While such a mechanistic change from an acyl-enzyme to a direct water attack may be possible, the converse is much more difficult because the essential nucleophile would be missing. It is very easy to lose enzyme function but much more difficult to gain new function through SDM.
- The H95 residue of triose phosphate isomerase is important for catalyzing the enolization of the substrates. The H95Q mutant is impaired in its ability to stabilize this reaction intermediate. There is an associated change in the pathways of proton transfer mediated by the mutant enzyme (Nickbarg et al. 1988).
- The cancer-associated isocitrate dehydrogenase mutant (R132H form of IDH1 isozyme) loses its native function but is able to catalyze the NADPH-dependent reduction of 2-oxoglutarate to $R(-)$-2-hydroxyglutarate (D-2-hydroxyglutarate) (Dang et al. 2009).

Site-specific mutagenesis approach has undoubtedly extended our knowledge of enzyme mechanism and function. The necessary first step in the kinetic analysis of a mutant enzyme is to show that any observed change in catalytic activity is due solely to the targeted alteration. For this, the enzymologist must be prepared to analyze both the structural and functional consequences of the mutation(s) made. To begin with, this should include a thorough evaluation of kinetic parameters like K_M, V_{max}, k_{cat}, and k_{cat}/K_M. Gross structural changes in a mutant enzyme can be discerned through techniques like circular dichroism and gel filtration chromatography. X-ray crystallography of the native and mutant enzyme forms (as well as their frozen structures bound to substrate, product, inhibitors, and transition-state analogs) offers valuable information. Of course, the pretty structures should conform to the hard data (Miller 2007). It is a distinct possibility that mutant enzymes might follow a different reaction pathway. This was highlighted above, with examples of triose phosphate isomerase and serine proteases. Therefore, a detailed mechanistic analysis is routinely needed for proper appreciation of the effects from site-directed mutations.

40.3.2 Changing Landscape of Enzyme Allostery

Historically, understanding the control of enzyme activity through feedback mechanisms led to the simultaneous discovery of enzyme cooperativity and allostery. Feedback inhibition and cooperativity over the years have got connected and appear to be two faces of the same coin. We have seen their implications to metabolic regulation in a previous chapter (Chap. 38). The non-Michaelian kinetic feature allows an enzyme to function as a concentration-dependent metabolic switch. Classical models invoke distinct enzyme conformations associated with allostery. These may involve multiple ligand binding sites (with associated conformational selection and induced fit) or polypeptide oligomerization. In most allosteric enzymes, the allosteric binding site lies far away from the active site. This implies that structural communication pathways (via the levers and pulleys of protein structure) must exist between these sites. The first mechanistic description for allosteric regulation was proposed more than 60 years ago; other possible mechanisms are being advanced regularly to describe these phenomena (Changeux 2013; Motlagh et al. 2014). Investigations at atomic detail (by high-resolution NMR of specifically labeled side chains) of glucokinase (a monomeric enzyme) for its kinetic cooperativity have provided recent insights. The enzyme molecule samples a range of conformational states in the absence of glucose. However, in the presence of glucose, the enzyme population shifts towards a narrow, well-structured ensemble of states in the presence of glucose (Larion et al. 2012).

Allostery is not always mediated by conformational changes that can be detected by standard techniques like X-ray and NMR relaxation measurements, determination of H/D exchange rates and isothermal titration calorimetry (ITC) experiments. These methods give a time-averaged snapshot of the 3D structure. However, recent advances in spectroscopy (probes to explore time-resolved dynamics of protein conformational changes) and the computational approaches (to study molecular dynamics simulations) indicate that multiple enzymatic conformational states exist even in kinetically simple Michaelis complexes. Also, the free enzyme itself is a

Table 40.1 Molecular mechanisms of allosteric regulation

Mechanism	Examples
Open/close active site	3-Phosphoglycerate dehydrogenase (the active site cleft closes upon binding of the end product, Ser)
Change active site conformation	3-Deoxy-D-arabinoheptulosonate-7-phosphate synthase [aromatic amino acid (the end product) binding leads to minor conformational modifications and prevents substrate binding]
Change active site electrostatic properties	Chorismate mutase [aromatic amino acid (the end product) binding brings a Glu residue into the active site; causes a major change in its electrostatics and repels the negatively charged substrate]
Affect protein–protein complex formation	ATP phosphoribosyltransferase [His (the end product) binding converts an active dimer to an inactive hexamer]
Change protein flexibility	Dihydrodipicolinate synthase (Lys (the allosteric ligand) binding affects distant sites via a change in the protein vibrational modes)
Population shifts in ensemble of conformers	– Glucokinase (its intrinsically disordered small domain samples a broad conformational ensemble; upon glucose binding, the population shifts toward a narrow, well-ordered ensemble) – Dimeric catabolite activator protein (binding of the first cAMP molecule lowers the affinity for the second cAMP molecule; first cAMP binding enhances motions within the protein, whereas the binding of second cAMP decreases these motions and its flexibility)

collage of protein conformational states. This protein disorder is clearly seen in the dynamic motion measured by distance- and time-resolved NMR studies. Present-day structural biology offers unequivocal evidence of multiple conformations in preexisting equilibrium for glucokinase, trypsin-like proteases, maltose-binding protein, etc. The new outlook on allostery incorporates this more dynamic view of the enzyme protein. Accordingly, allosteric control may manifest by a *population shift* in the statistical ensembles of many states, with some regions of low local stability and others of high stability. Ligand binding affects the relative free energies of these states; they in turn differ in their affinities for other ligands and/or their activity. This is different from the earlier concept of a few well-defined static conformational states (such as R and T states). Various mechanisms of allostery described to date are summarized in Table 40.1.

The new view of allostery encompasses a conformationally dynamic continuum of allosteric phenomena. With recent discoveries, we have thus moved toward increasing dynamics (disorder or fluctuations) starting from (a) rigid body movements, (b) side-chain dynamics, (c) backbone dynamics, (d) local unfolding, and to (e) intrinsically disordered structures.

Allostery, by definition, involves the propagation of signals between different sites in a protein structure. This may take place in the absence of detectable conformational changes and be exclusively mediated by transmitted changes in protein motions. The fact that the change in dynamics occurs in the absence of significant structural change suggests that dynamics alone may convey allosteric information. There is much interest to probe the existence of an entire channel/ network of amino acids through which allosteric signals are communicated. One approach that could map such paths (perturbations that travel across the structure) and implicate the interacting amino acid residues is through measurements of a double mutant cycle. The energetics of such residue interactions allows one to infer the degree of functional coupling between different sites of a protein. A complementary approach uses a sequence-based statistical analysis (Statistical Coupling Analysis, SCA) method for estimating the architecture of functional couplings in proteins (Reynolds et al. 2011). If two residues in a protein are functionally coupled, then they should have coevolved. This coevolution can be scored by statistically comparing homologous protein sequences. A combination of SCA and double mutant cycles along with functional, structural, and folding analyses can give insights into the existence of an entire wave, wire, channel, or network of amino acids through which allosteric signals are transmitted.

Transplantation of allosteric regulation has been possible in hemoglobin and glycerol kinase (Hardy and Wells 2004). Very few residues may be involved in the manifestation of allosteric regulation. The allosteric features of crocodile hemoglobin could be introduced into human hemoglobin by substitutions at 12 amino acid residues. Changing only 11 of 501 total residues (about 2%) converts an unregulated glycerol kinase to an allosterically regulated glycerol kinase. Small molecules can exert strong effects from unexpected locations and hence searching for new allosteric sites in enzymes is a challenge. It is hard enough to predict what ligand might bind in a binding pocket, but the presence of allosteric binding sites further complicates the matter. It may become possible to develop algorithms to distinguish allosteric from active sites with better databases in the future. It may also be feasible to predict allosteric regulation from protein structural data (Freire 2000).

The emerging radical view of enzyme function is that each catalytic step corresponds to an ensemble of thermodynamic and structural states (Wodak et al. 2019; McCullagh et al. 2024). Incidentally, allostery and catalysis no longer appear as distinct phenomena but as the manifestations of the same intrinsic protein dynamics. Important issues for further research include (a) mechanisms by which an allosteric effect is transmitted via amino acid networks, (b) how the distribution of protein conformations is altered, and (c) the timescales at which the redistribution of these conformations occur.

40.3.3 Predicting Enzyme Structure and Function

Computational tools have revolutionized the whole of biology. It is no wonder that substantial progress has taken place in enzymology, one of the more quantitative of the biological sciences. Incorporating new tools and technology to understand enzymatic catalysis is a recurring theme in enzymology. Computational enzymology has joined the earlier approaches like the application of kinetic measurements, kinetic isotope effects, crystallography, and distance- and time-resolved NMR. We have noted above that computational enzymology has made significant inroads into transition state analysis and molecular dynamics of enzyme action (Derat and Kamerlin 2022). Besides these, there is a rapid move toward enzyme structure and function prediction through computational approaches (Pearce and Zhang 2021; Akdel et al. 2022). This need has arisen as we (a) accumulate a large number of sequenced genomes; (b) come across orphan open reading frames, with no clues of their function; and (c) express sequences into proteins and even crystallize them without their actual functional demonstration (Cuesta-Seijo et al. 2011; Hai et al. 2015; Reyes et al. 2020).

Assigning valid functions to unknown (putative) proteins/enzymes identified in genome projects is a challenge (Kuznetsova et al. 2005). While experimental testing remains essential, computational approaches can help guide this experimental design. Bioinformatics approaches are being perfected to (a) identify informative sequence relationships using structure and genome context, (b) allow accurate high-throughput structure prediction through homology modeling, and (c) dock metabolites in silico to provide accurate and testable list of potential enzymes. Microbial metabolic pathways often are encoded by *"genome neighbourhoods"* (synteny and associated gene clusters and/or operons). Such positional information can provide important clues for enzyme function assignment. For instance, pathway docking is an efficient strategy for predicting in vitro enzymatic activities and allocating in vivo physiological functions. It has been possible to identify novel metabolites, enzyme activities, and biochemical pathways through this tactic.

In view of the exponential growth in genome sequence datasets (with significant proportion of sequences with unknown enzyme functions), an integrated strategy for functional assignment was recently proposed (Gerlt et al. 2011; Gerlt 2017). This enzyme function initiative (EFI; the multicentric program under National Institute of General Medical Sciences, USA) looks to predict the substrate specificities of unknown members of mechanistically diverse enzyme superfamilies—thereby predict their functions. The approach exploits conserved features within a given superfamily such as known chemistry, identity of active site functional groups, and composition of specificity-determining residues/motifs/structures. Initial targets chosen for this purpose include the amidohydrolase, enolase, glutathione transferase, haloalkanoic acid dehalogenase, and isoprenoid synthase enzyme superfamilies. Members of these enzyme superfamilies are functionally diverse (conserved partial reactions or chemical capability but with divergent overall function) which makes functional assignment difficult. Homology inferred from simple sequence comparisons alone cannot guide functional assignment in such cases. Therefore, an integrated approach involving following components is proposed:

- Perform bioinformatic analyses to cluster sequences into probable isofunctional groups, assign tentative functions for further investigation by structure determination, structural modeling/docking, and biochemical experimentation.
- Carry out homology modeling to expand the use of structural models; thereby guide functional assignment to proteins without experimentally determined structures.

- Employ computational docking methods to leverage structure to guide functional assignment by suggesting substrates and ligands for biochemical experimentation.

It has become increasingly obvious that enzyme function is defined not only through its chemical and kinetic competence, but also by its structural features. The change in the title and emphasis of the very well received book on enzymes by Alan Fersht (from "Enzyme Structure and Mechanism" for the first edition to "Structure and mechanism in protein science: A guide to enzyme catalysis and protein folding" in the later version) is a pointer in this direction. Structure is the necessary third leg— along with mechanism and function—of the secure stool to understand enzyme function.

40.4 Designing Novel Catalysts

Curiosity and the desire to imitate general principles of biological catalysis have led to many developments in the design and construction of artificial enzymes. Present approaches to create novel catalysts fall into three general categories—(a) de novo design and synthesis of catalysts from polypeptides and non-protein building blocks (such as macrocyclic compounds, poly-ethyleneimine, synthetic chemical/genetic polymers, and cyclodextrins), (b) modification/evolution of existing catalysts such as protein enzymes or ribozymes, by genetic or chemical methods, and (c) designer enzymes made to order. While antibodies and RNA as catalysts will be discussed subsequently, we will first focus on chemical models, enzyme mimics, and hybrid catalysts.

40.4.1 Chemical Models, Enzyme Mimics and Hybrid Catalysts

Enzyme models of increasing complexity have been designed and discovered. A small chiral molecule like proline may be considered the simplest enzyme. Recent reports on asymmetric catalysis by proline and its derivatives include activation of carbonyl compounds via nucleophilic enamine intermediates (MacMillan 2008). Several highly enantioselective important carbon–carbon bond-forming reactions (aldol additions and Mannich reaction) have been developed using this approach. The self-assembly of a single amino acid—phenylalanine—in the presence of Zn [II] zinc forms supramolecular structures that promote ester hydrolysis, and CO_2 hydration albeit much less efficiently than carbonic anhydrase (Makhlynets and Korendovych 2019). Many model systems anticipated their enzyme counterparts much ahead of time. Otto Warburg studied the oxidation of unsaturated fatty acids by combined action of iron and sulfhydryl (–SH) groups in 1925. Lipoxygenases— containing iron and –SH groups essential for their oxidative activity—were discovered much later. Aniline catalyzed rapid decarboxylation of acetoacetate via the "aniline-acetoacetate complex" is exemplified by JBS Haldane in his 1935 book on enzymes. More recently, chemical hydrogenase mimics with Co and Ni centers are reported for exploiting hydrogen as fuel. Models for enzymes performing free radical chemistry and other redox reactions are being sought to be incorporated in clean energy programs. The study of glutathione peroxidase anticipated the small molecular enzyme model Ebselen as well as an antibody enzyme—abzyme (see Table 40.2 in the next section for a full treatment). A small molecule mimicking protein disulfide isomerase function was designed. The compound $(\pm)$-*trans*-1,2-bis (2-mercaptoacetamido)cyclohexane had properties (i.e., $E^{\circ\prime}$ and pK_a) similar to protein disulfide isomerase, and it catalyzed protein folding both in vitro and in vivo (Woycechowsky et al. 1999). A dodecapeptide consisting of alternating L and D amino acids that binds a 4Fe−4S cluster was designed. It can withstand hundreds of redox catalysis cycles at room temperature (Kim et al. 2018).

As pointed out before, there are two reasons to learn about enzyme constructs and mimics. One objective of chemists and biologists is to elucidate the molecular basis of enzyme function. Second, using the available knowledge base (which is still far from complete), one could attempt to design and build novel catalysts—the so-called tailor-made enzymes. The chemical alteration of an existing enzyme by introducing additional functional groups is one route to rational enzyme design. The *semisynthetic enzyme* so generated can display very different catalytic activities from that of the natural enzyme. For example, papain (a thiol protease) was turned into an effective redox catalyst (an oxidoreductase) by appropriately tagging a flavin on a sulfhydryl group. In another example, new binding domains are selectively introduced to build/alter enzyme specificity. RNase A is a relatively nonselective enzyme hydrolyzing phosphodiester linkages of RNA. It was made specific to a definite RNA sequence by creating a chimeric RNase A—wherein a covalently attached single strand DNA confers specificity by annealing at the complementary RNA sequence (Fig. 40.1).

The oligonucleotide-tagged RNase A above is an effort in modifying the substrate specificity of an enzyme. However, an attempt was made to build a restriction endonuclease from the first principles. Several chemical agents intrinsically possess DNA cleavage activity. If these are incorporated into DNA/RNA binding proteins, a specific nuclease may be created. Indeed, one could design synthetic hybrid molecules with two components—one for specific DNA recognition and the other to cleave the DNA adjacent to it. The technique called *"affinity cleavage"* is based on the construction of such bifunctional molecules (Fig. 40.2). Many artificial metalloenzymes have emerged from similar approaches. Chemistry of EDTA-Fe [II] with oxygen is an excellent source of hydroxyl (HO·) radicals for DNA cleavage. This reaction can be turned into a catalytic cycle by adding ascorbate so that Fe[III] formed is reduced back to Fe[II]. Several chemical agents possess nuclease activity; rare earth metal ions are active in hydrolyzing phosphodiester linkages in DNA. For instance, DNA scission by Ce[IV] is hydrolytic and not oxidative. The fragments so generated can be re-ligated using a ligase. The *sequence recognizing moiety* brings the *molecular DNA scissors* close to the target phosphodiester bond. Binding to specific DNA sequences is provided either by a short oligonucleotide or by a specific DNA binding domain of the protein. Sequence recognition has also been achieved

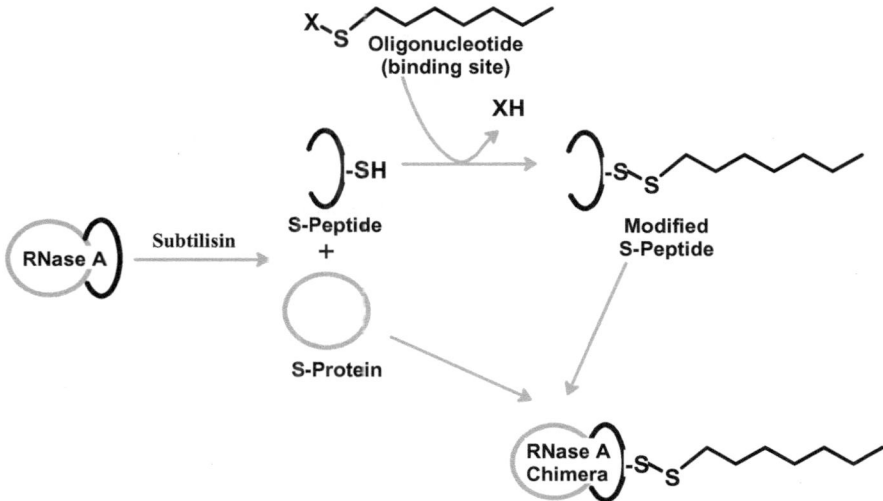

Fig. 40.1 Construction of a sequence specific RNase A chimera. The S-peptide was modified with an oligonucleotide sequence. The functional RNase A was reconstituted by combining the modified S-peptide with the S-protein (Zuckermann and Schultz 1988). [Adapted with permission from Zuckermann and Schultz, J Am Chem Soc, 110:6592–6594. Copyright (1988) American Chemical Society]

by pcPNA (a pseudo-complementary peptide nucleic acid). Artificial metalloenzymes are hybrid catalysts that result from the incorporation of an abiotic metal cofactor within a protein scaffold. They combine enzymatic with homogeneous catalysis to address the possible drawbacks associated with them (Davis and Ward 2019).

Most restriction enzymes used in molecular biology recognize a stretch of six bases and cut DNA at that site. Assuming random distribution of A, G, T, and Cs, there is a high probability (4^6, that is once every 4096 bp of DNA) of finding these restriction sites in DNA. To cut at a single unique position in human genome, we could use a 16mer (or longer) sequence recognizing moiety (4^{16}; for $>10^9$ bp of DNA in human genome). This site selectivity compares well with the molecular tools like (a) the *Achilles' heel cleavage* (only one of the many restriction sites in a genome is specifically protected from inactivation by a cognate methyltransferase; this in turn creates a unique restriction enzyme cleavage site) and (b) the genome editing by the CRISPR/Cas system. An artificial restriction enzyme that can cut only at one position in the human genome will require a 16mer (or longer) sequence recognizing moiety.

Novel catalysts may also be crafted on non-protein molecular framework—including synthetic macromolecules of non-biological origin (Bjerre et al. 2008; Wulff 2002). Poly(ethyleneimine) polymers possess intrinsic acid-base groups; along with pyridoxal phosphate or other suitable cofactors, some degree of catalysis was demonstrated. Nanomaterials with enzyme-like characteristics, termed

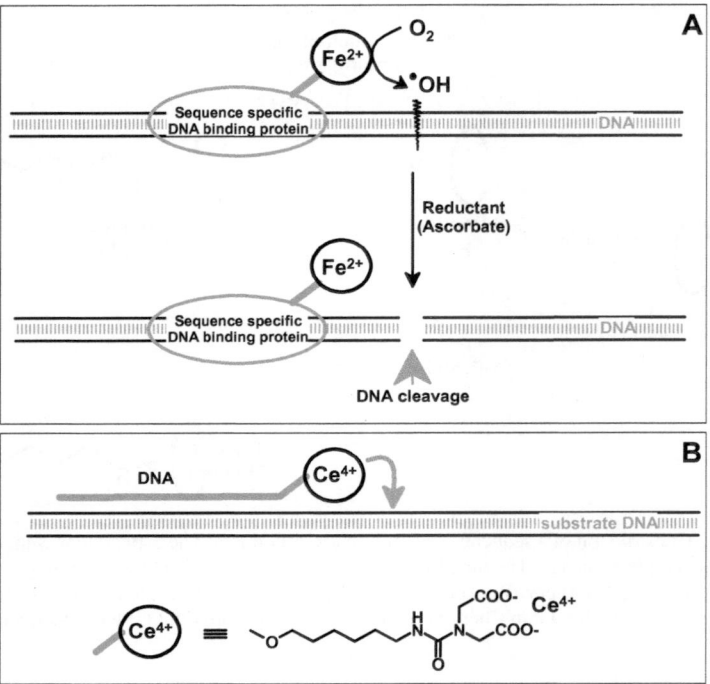

Fig. 40.2 Construction of artificial restriction enzymes. (**a**) Fenton chemistry of EDTA-Fe[II] with oxygen generates hydroxyl (HO·) radicals for local DNA cleavage. Specificity is provided by using a unique sequence recognizing moiety (e.g., *trp* or *lac* repressor or catabolite activator protein of *E. coli*). (**b**) A restriction enzyme construct of Ce^{4+} complex of iminodiacetate. This Ce^{4+} nuclease is juxtaposed to the specified DNA sequence for cleavage by an oligonucleotide sequence recognizing moiety

nanozymes, are also being explored for various applications (Wei and Wang 2013). Breslow's group has made extensive efforts to emulate enzyme catalysis using cyclodextrin scaffolds (Breslow 2005). While the cyclodextrin cavity offers a binding pocket, additional functional groups attached to the cyclodextrin ring offer new enzyme mimics (synzymes). The cavity size can be varied by choosing either β-cyclodextrin (seven glucose units in the ring) or γ-cyclodextrin (eight glucose units in the ring). Mimics of RNase A (β-cyclodextrin bis-imidazole) and chymotrypsin (for its esterase activity!) are demonstrated. A flexible capped cyclodextrin with the well-fitting substrate afforded a rate acceleration of 8×10^{7} fold in *p*-nitrophenyl ester hydrolysis (Fig. 40.3).

Finally, a word of caution on rate acceleration observed for enzyme mimics. Most protease/esterase models use *p*-nitrophenyl esters as substrates. The *p*-nitrophenyl esters are *not* protease/esterase substrates "in real life." With *p*-nitrophenyl esters as model substrates, partly the better leaving group effect contributes to the observed rate accelerations. Leaving groups whose pK_a's are above 9.0 lead to much lower

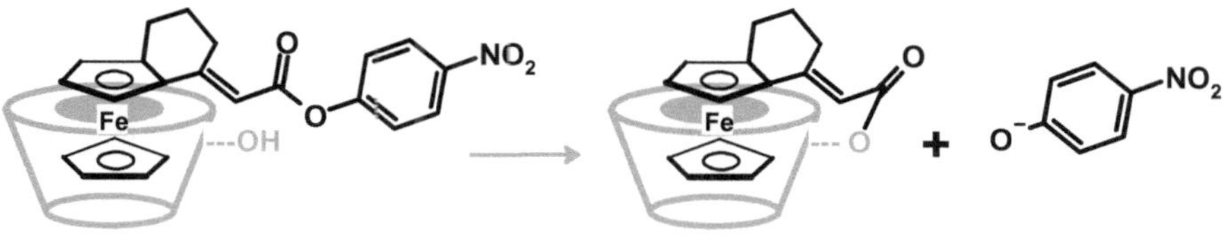

Fig. 40.3 A β-cyclodextrin mimic of chymotrypsin displaying catalysis of *p*-nitrophenyl ester hydrolysis. The cyclodextrin cavity binds the hydrophobic ferrocene core of the ester substrate

acceleration. One should account for such a *"p-nitrophenyl ester syndrome"* when evaluating the performance of synzymes (Menger and Ladika 1987).

40.4.2 Antibody Catalysts (Abzymes)

Linus Pauling recognized in 1940s that the ability of an enzyme to speed up a chemical reaction arises from the "complementarity" of its active site structure to the activated complex (i.e., the transition state). This has given rise to the productive field of catalytic antibodies (or *abzymes*). Antibody molecules represent a class of proteins with high affinity and exquisite selectivity; they could be raised against any small molecule (hapten) of our choice. Antibodies that can bind to transition state analogs of a substrate should therefore catalyze the conversion of substrate—through the transition state, to product. Abzymes production involves the following steps:

- Generating a stable transition state analogue using molecular design and chemical synthesis.
- Raising antibodies (monoclonal antibodies to be more precise) with the TS analog as the hapten.
- Isolating antibodies, which bind to the TS analog as potential catalysts for that reaction.

The seminal prediction of Pauling was verified by Lerner's research group in the 1980s. During an ester hydrolysis, the sp^2 hybridized carbonyl carbon is converted to a sp^3 hybridized center in the intermediate; and with the carbonyl oxygen resembles an oxyanion. The transition state presumably looks more like this unstable intermediate (sp^3, oxyanion). A phosphonate ester mimic, with a sp^3 hybridized phosphorous replacing the sp^2 hybridized carbonyl carbon (Fig. 40.4), was synthesized. While very resistant to hydrolysis, this phosphonate also has negatively charged oxygen similar to the intermediate during ester hydrolysis. The mouse antibodies (the monoclonal—6D4) against this phosphonate structure catalyzed the corresponding carboxylic acid ester hydrolysis.

More than 100 interesting examples of abzyme catalysis are known. They include many reactions that cannot be achieved by standard chemical methods (Benkovic 1992). Besides the ester hydrolysis mentioned above, these include pericyclic

Fig. 40.4 The phosphonate transition state analog used as hapten. The monoclonal antibodies (abzyme 6D4) against this hapten displayed esterase activity on the corresponding ester shown

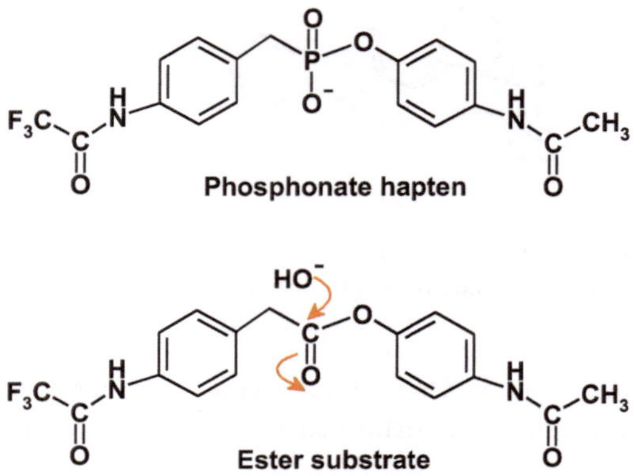

Phosphonate hapten

Ester substrate

Table 40.2 Antibody catalysts generated for different reaction types

Abzyme[a]	Reaction catalyzed
6D4	Carboxylic ester hydrolysis
48G7	p-Nitrophenyl ester hydrolysis
1F7	Claisen rearrangement of chorismate to prephenate
AZ-28	Oxy-Cope rearrangement
39-A11	Diels–Alder reaction
7G12	Ferrochelatase
33F12	Aldolase
34E4	E2 elimination of nitrobenzisoxazole
2F3 (scFv)	Glutathione peroxidase

[a]All the abzymes (except 2F3) are monoclonal antibodies (Hilvert 2000). The scFv (single chain Fragment variable) of abzyme 2F3 was activated by chemical modification of a reactive Ser by attaching Se to it (Ren et al. 2001)

processes, group transfer reactions, additions and eliminations, redox reactions, aldol condensations, and a few cofactor-dependent transformations. Some of these are listed in Table 40.2.

Virtually all experiments with abzymes employ monoclonal antibodies (Table 40.2). The monoclonals are a single homogeneous catalytic species (unlike the polyclonal sera) and their use greatly simplifies kinetic, mechanistic, and structural characterization of an abzyme. While expressing catalytic antibodies (or Fab fragments), folding of the two chains into native state is a challenge. One approach is to express and secrete them into the periplasmic space of *E. coli*, exploiting its diminished protease activity, ability to correctly fold and form disulfide bonds. The single chain Fragment variable (**scFv**) version of the antibody exhibits the same catalytic parameters as the parent monoclonal antibody. The scFv is a recombinant protein construct of a VL chain tethered to a VH chain with a polypeptide linker and is expressed efficiently by bacteria.

The fastest enzymes are diffusion limited catalysts. Even the less than *perfect* ones typically have apparent bimolecular rate constants (k_{cat}/K_M) between 10^6 and 10^8 M^{-1} s^{-1}. Catalytic antibodies have rate accelerations of many orders of magnitude below their enzyme rivals. By definition, abzymes are catalytic antibodies specific to (and raised against) the corresponding transition-state mimic. We note that stabilization of the transition state is *necessary but not sufficient* by itself for achieving good catalysis. Many other factors—like active site functional groups, conformational flexibility, shielding of the reaction intermediates, cofactor needs, etc.—substantially contribute to rate accelerations (refer Chap. 4). Transition states themselves have fleeting lifetimes and cannot be captured. Synthesis of effective analogs must therefore draw on our chemical intuition about the conformational, stereochemical, and electronic properties of the reaction under study (also see Chap. 36). Since no stable molecule can reproduce all characteristics of an actual TS, hapten design strategies have focused on incorporating the salient features of the TS. In the sum, limited catalytic ability of abzymes may be attributed to one or more of the following:

- Antibody scaffold is fixed, and this means limited structural space is explored for catalyst building. Real enzymes come in a variety of structural folds.
- The polypeptide sequence space exploited during antibody maturation is also limited.
- Catalytic antibodies may lack structural dynamics required for optimal catalysis.
- TS analogues are imperfect mimics of the actual transition state. The TS mimic designed may not capture the best options for discriminatory binding of the substrate versus TS.

Basic strategy to produce catalytic antibodies is indirect. Here the immune system is directed to evolve not for catalysis but towards binding tightly to an imperfect TS analog. To date, antibody enzymes display only modest catalytic activity and have not found significant practical utility (though a few have reached the market). Nonetheless they continue to be of considerable academic interest. Studying abzymes has yielded valuable insights into reaction mechanisms, catalysis, enzyme structure, and function. There is yet the promise of delivering tailored catalysts for difficult reactions for which natural enzymes do not exist. Such catalytic antibodies may be useful even if they do not attain enzyme-like efficiency.

40.4.3 RNA Catalysts (Ribozymes)

Substantial role for RNA as information molecule is well established in molecular biology. More recently, RNAs have assumed importance as components of (a) gene silencing through double stranded small interfering RNAs (*siRNA*) generated by Dicer, an RNase III like enzyme, and (b) genome editing through the CRISPR/Cas system where crRNA-guided interference is exploited. The catalytic role of RNA molecules (the so-called ribozymes) has expanded the realm of biological catalysis beyond proteins. RNA catalysts do satisfy the dual criteria of *catalysis* and *specificity* but are less impressive catalysts when compared to protein-based enzymes. They have had much more impact in understanding the origin of life problems and catalytic evolution. Like polypeptides, RNA molecules can fold into higher order structures that permit the formation of an active site. On the other hand, DNA is predominantly double helical, cannot fold into complicated shapes, and has limited repertoire of chemical groups (it lacks the 2′-hydroxyl group found in RNA) for catalysis. Nevertheless, DNA based catalysts (DNAzymes) were constructed (Baum and Silverman 2008). More recently, elaboration of different catalytic activities from synthetic genetic polymers (XNAs) was demonstrated (Taylor et al. 2015). XNAs (arabino nucleic acids, hexitol nucleic acids, or cyclohexene nucleic acids) fold into defined structures and bind ligands. XNAzymes were elaborated directly from random XNA oligomer pools, some of them exhibited *in trans* RNA endonuclease and ligase activities.

Single-stranded nucleic acid molecules are capable of folding into secondary and tertiary structures. Aptamers are short, single stranded nucleic acids which bind a variety of ligands with high affinity and specificity. DNA or RNA aptamers can be routinely isolated from synthetic combinatorial nucleic acid libraries by in vitro selection—known as "systematic evolution of ligands by exponential enrichment" (SELEX) (Tuerk and Gold 1990; Weigand and Suess 2009). Riboswitches are natural versions of aptamers discovered subsequently. The aptamer domains of most riboswitch classes are typically fewer than 100 nucleotides. Riboswitches with distinctive ligand recognition capabilities have been found in all domains of life; they occur with highest frequency within the 5′-UTRs of bacterial mRNAs and typically regulate genes involved in metabolism. For instance, glycine riboswitch consists of two different aptamer types that individually bind to a single molecule of glycine (Famulok 2004). Cooperative interaction between the two sites allows better sensing of this metabolite. Typically, aptamers and riboswitches are specific ligand binding RNAs with no catalytic function (Olenginski et al. 2024). The *glmS* ribozyme was originally considered a riboswitch on the basis of its binding to glucos-amine-6-phosphate. Since it acts by degrading the message, it is argued that the bacterial *glmS* RNA is a metabolite responsive ribozyme and not a riboswitch. Aptamers may fold or undergo a conformational change upon binding the cognate ligand. An aptamer may be fused to a ribozyme to generate *aptazymes*. The binding of its cognate ligand to the aptamer displaces it from the stand on which it is bound on the ribozyme, thereby providing for regulation of ribozyme function.

Catalytic RNA was discovered through RNA species capable of auto-cleavage reaction (Abelson 2017). Subsequently, other activities catalyzed by RNA have been reported (Doudna and Lorsch 2005; Wilson and Lilley 2009). While natural ribozymes catalyze mainly self-cleavage or ligation reactions, they can also acceler-ate other reaction types (Table 40.3) (Traut 2007). The RNA component of RNase P is responsible for catalytic processing of tRNA precursors. The self-splicing group I intron from *Tetrahymena thermophila* was engineered to perform as a multiple turnover RNA enzyme. The peptide bond formation is attributed to the RNA

Table 40.3 Kinetic constants for some catalytic RNAs

Catalyst	Substrate	k_{cat} (s^{-1})	K_M (M)	k_{cat}/K_M
Ribozyme	RNA (cleavage)	2.0×10^{-1}	2.0×10^{-8}	1.0×10^7
Self-splicing intron	RNA (splicing)	1.0×10^{-3}	1.1×10^{-9}	9.0×10^5
Ribosome	(Peptide bond formation)	5.0×10^0	5.0×10^{-3}	1.0×10^3
DNAzyme IV	Ornithine decarboxylase mRNA	1.0×10^{-3}	3.0×10^{-7}	3.0×10^3
Abzyme 34E4 (E2 elimination)	Nitrobenzisoxazole	6.6×10^{-1}	1.2×10^{-4}	5.5×10^3

Catalytic efficiencies of a DNAzyme and an abzyme are also listed for comparison

Table 40.4 Catalytic strategies of protein and RNA enzymes: a comparison

Catalytic strategy	Protein enzyme	RNA enzyme
Substrate orientation and approximation	Yes	Yes
General acid-base catalysis	Always	Deficient[a]
Metal ion catalysis	Many	Always
Organic cofactors	Many	None
Active site electrostatics (dielectric manipulated effectively)	Most proficient	Not proficient
Utilization of binding energy from interactions away from active site	Yes	Yes
Covalent catalysis	Yes	Yes

[a]RNA enzymes lack groups with pK_as around pH 7.0. The nucleolytic ribozyme GlmS provides an exception to the exclusive use of nucleobases in general acid-base catalysis by ribozymes; a molecule of glucosamine-6-phosphate specifically bound to the RNA structure serves as the general acid in GlmS

component of the ribosomes (Schmeing and Ramakrishnan 2009; Tirumalai et al. 2021).

There are similarities as well as differences between RNA and protein catalysts. Greater structural variety (of amino acids) allows better catalytic properties in protein enzymes than in RNA enzymes. Protein enzymes are superior catalysts and many of them work at the diffusion limit. Ribozymes, on the other hand, are rather slow with an apparent maximal rate constant of ~1 min^{-1} (Doudna and Lorsch 2005). However, ribozymes might be easier to produce than enzymes. Rigidity allows an enzyme to maximize specific (binding) interactions with the TS relative to the ground state and hence maximizes catalysis. Larger size ensures better positioning and rigidity within the active site—required for most effective catalysis (see Sect. 39.4.3 in Chap. 39). For the same catalytic function, RNA has to be much bigger than an enzyme protein. Protein enzymes are big and RNA enzymes are even bigger. RNA is clumsier than proteins in terms of functional groups, structural variety, and ability to fold. Hammerhead ribozyme and RNase A have similar sizes but RNase A (with best k_{cat}/K_M of 2.8×10^9 M^{-1} s^{-1}) achieves 10^5-fold higher maximal rates. A direct comparison of catalytic strategies (Table 40.4) available to protein enzymes and ribozymes is illustrative (Narlikar and Herschlag 1997).

40.5   Enzymes Made to Order

Until man duplicates a blade of grass, nature will laugh at his so-called scientific knowledge.—Thomas Edison

From the general theme of this book and earlier discussion in this chapter, it is obvious that much is known about how enzymes function as catalysts. One measure of how well we understand enzymes is to try and build similar catalysts from the first principles. In this sense, de novo enzyme design is an intellectual challenge, and the exercise serves two important objectives. It allows experimental validation of the principles of catalysis that we have learnt so far. Second, bespoken catalysts can be built for industrial applications, particularly for those reactions for which natural enzymes do not exist. The design of enzymes with new functions and properties has long been a goal of the protein engineer. However, in enzyme engineering (and de novo design), serendipity continues to outstrip design—a clear sign that our basic understanding of enzyme catalysis to date is far from complete.

Enzyme catalytic power results from a combination of multiple mechanistic strategies (Chap. 4). Therefore, valuable insight into the evolution of catalytic function can be gained through de novo design experiments. However, the methods of kinetic analysis discussed earlier in this book will be applicable to all catalysts (synthetic or natural) regardless of their chemical nature. We may note that foundations of enzyme kinetics were laid much before the chemical nature of enzymes as proteins was established by Sumner (using jack bean urease).

A major benefit of recombinant DNA technology is the ability to do protein, and hence, enzyme engineering. This includes the skill to precisely replace/delete/add one or more amino acids in a given enzyme. These designed yet specific mutations can be engineered on a desired gene (coding for the desired enzyme) with the help of synthetic oligonucleotide constructs as primers (see *Site directed mutagenesis and crystal structures* section, in this Chapter). It is very easy to lose an enzyme function through site directed mutations, but very difficult to gain a new function. Besides point mutations, functional elements of a protein scaffold could also be replaced/changed through available recombinant DNA tools (Fig. 37.5). The approach is powerful and is anticipated to deliver many tailored enzymes.

Engineering novel enzymes is a rapidly evolving field and the examples presented here are selective and only representative (also see Chap. 37). A foolproof and robust enzyme activity assay is at the heart of any enzyme engineering and design. While the methodological details are outside the scope of this book, various strategies to generate novel enzyme designs through recombinant DNA technology are outlined below.

40.5.1 Enzyme Redesign

One way to generate novel catalysts is to start from an existing enzyme scaffold and rationally alter its structure to cause a predicted change in function. Enzyme redesign

Table 40.5 Examples of de novo enzyme design

Designer enzyme	Comments on the design
Oxaloacetate decarboxylase (metal-free)	A rationally designed synthetic 14 amino acid residue cyclic peptide—oxaldie; decarboxylates oxaloacetate via an imine intermediate on its Lys NH_2. Catalytic efficiency comparable with abzymes (Johnsson et al. 1993)
Dihydrodipicolinate synthase	The N-Acetylneuraminate lyase scaffold was rationally redesigned to switch the activity towards dihydrodipicolinate synthase. The designed activity showed 19-fold increased specificity for the new substrate (Joerger et al. 2003)
β-Lactamase (from glyoxalase II)	Several loop grafting steps at the active site to achieve major switch in function. Evolved new activity on an existing glyoxylase II scaffold (Park et al. 2006)
Sesquiterpene synthase	A promiscuous sesquiterpene synthase scaffold was used to build seven novel terpene synthases, catalyzing the synthesis of different sesquiterpenes (Yoshikuni et al. 2006)
Retro-aldolase	Retro-aldol cleavage of a carbon–carbon bond. The mechanism involves enamine catalysis by lysine (via a Schiff base or imine intermediate) giving a catalytic proficiency of 10^4, which is far from natural enzymes (Jiang et al. 2008)
Diels–Alderase (inter-molecular)	Organic bimolecular reaction forming two carbon–carbon bonds and up to four new stereogenic centers in one step. No naturally occurring enzymes are known for this reaction. The designed enzyme is 20 times better than corresponding abzymes (Table 40.2) (Siegel et al. 2010)
Triose phosphate isomerase	The Rosetta enzyme design protocol demonstrated for the triose phosphate isomerase reaction as an example (Richter et al. 2011)
Kemp eliminase	A well-studied organic model system for proton transfer from carbon. In Kemp elimination, the deprotonation of substrate (5-nitrobenzisoxazole) leads to electronic rearrangements that break the C–H and N–O bonds while forming a C≡N triple bond. The designer enzyme accelerates the elementary chemical reaction (6×10^8-fold), nearly as efficient as natural enzyme like triose phosphate isomerase (Blomberg et al. 2013)

may be achieved through simultaneous incorporation and/or adjustment of protein functional elements—through deletion, insertion, loop grafting, and substitution of relevant active site loops, generating chimeras, etc. This could be followed by point mutations to fine-tune the enzyme activity. Using one such approach, a β-lactamase activity was introduced into the scaffold of glyoxalase II. The resulting enzyme completely lost its original activity but catalyzed the hydrolysis of cefotaxime (Table 40.5). Three other interesting examples include the following:

(a) The domain swap in bacterial glutamate dehydrogenases to change their pyridine nucleotide specificity (Sharkey and Engel 2009).

(b) Mixing and matching of different modules of polyketide synthases and non-ribosomal peptide synthases leading to product diversity and the production of hybrid or novel antibiotics (Penning and Jez 2001).

(c) The cancer-associated mutations from isocitrate dehydrogenases were extrapolated to homologous residues in the active sites of homoisocitrate dehydrogenases, for the catalytic conversion of 2-oxoadipate to (R)-2-hydroxyadipate, a critical step for adipic acid production (Reitman et al. 2012).

Enzyme redesign as an approach is expected to deliver more tailor-made enzymes in the near future.

Protein engineering to redesign a known enzyme has the potential to bring about (a) altered substrate/cofactor specificity and improve catalytic efficiency, (b) enhanced enantioselectivity, (c) a change in metal ion specificity, (d) a desired pH optimum of an enzyme, (e) increased enzyme stability, and (f) alteration of an existing site to catalyze a new chemical reaction. Both close and distant mutations appear similarly effective in improving enzymes in terms of thermostability and catalytic activity (Pinney et al. 2021). The mutations close to the active site are more effective than distant ones for changing enantioselectivity, substrate selectivity and alternate catalytic activity of an enzyme (Khersonsky et al. 2006). Besides the *ligand specificity first* view of enzyme redesign, it may be possible to take the more challenging *chemical mechanism first* approach. Accordingly, one may introduce catalytic residues into a ligand-binding site of a chosen protein to create an active site capable of catalyzing a chemical reaction. After all, the "acid test" of enzyme redesign is to engineer new catalytic activities and aim to change the reaction mechanism itself.

40.5.2 De Novo *Enzyme Design*

Designing an enzyme to catalyze the reaction of one's choice is a grand challenge (Nanda 2008). This strategy obviously requires detailed knowledge of the protein structure, structural basis of biological catalysis and computational tools for enzyme design. In principle the following steps may be envisaged:

- An appropriate catalytic mechanism is chosen for the target reaction.
- The transition state for this reaction is described.
- An idealized active site that positions the catalytic residues to maximize transition state stabilization is modeled.
- An appropriate protein scaffold is chosen from the available library.
- It is optimized in silico to best accommodate the reaction transition state and catalytic residues.
- The candidate "*theozyme*" polypeptides are actually created/produced.
- They are tested for catalytic activity.
- Best candidate designer enzyme may be fine-tuned by further sculpting around the transition state model.

Although the initial activities of de novo enzyme constructs are typically low, they can be significantly improved by directed evolution approaches (see Chap. 37). Computational enzyme design has emerged as a promising tool for generating made-to-order biocatalysts. The Rosetta de novo enzyme design protocol may be used to tailor enzyme catalysts for a variety of chemical reactions. There are constant efforts to improve the reliability of the Rosetta design cycle (Richter et al. 2011). Both the pre- and post-design structural analysis promises to play an increasingly important role here. The progress made in de novo enzyme (protein) design was recognized by conferring the 2024 Nobel prize to David Baker.

Genetic engineering is relatively easy and accessible. But to attempt rational protein engineering, a vastly improved understanding of protein structure-sequence relationship is required. Our appreciation of protein dynamics is still very limited, and this makes structure-function correlation very hard. Subtle changes in the active site geometry have tremendous, unexpected consequences for enzyme function. Hence, rational de novo design of a reasonably efficient enzyme continues to elude us. Analysis of enzymes designed so far suggests that our challenges lie in catalysis rather than binding. Few pioneering and brave attempts to build designer enzymes have been made. Relevant examples listed in Table 40.5 highlight the successes and limitations encountered in the de novo design of enzymes.

As can be seen from the representative cases listed above, most designer enzymes have not reached the expected catalytic performance (Bar-Even et al. 2011). This is because of our limited mastery of protein folding, structure, stability, dynamics, and catalysis. Indeed, subtle changes in the geometry of an active site suffice to generate tremendous unpredicted consequences for enzyme function. Poor catalytic performance of the de novo designs comes with its own caveats. Ensuring that the observed activity is really due to the designed enzyme, and not a contaminating activity from the expression host, is critical. Attempts to convert catalytically inert ribose-binding protein into an active triose phosphate isomerase ran into such difficulties and elicited the response—"It is a bush-league error not to purify your proteins well, especially in such work" (Hayden 2008).

De novo protein design allows us to explore the full sequence space (Lechner et al. 2018; Pan and Kortemme 2021). Computational methodology has progressed well for a wide range of structures to be designed from scratch with atomic-level accuracy (Huang et al. 2016). However, obtaining more active catalysts will require improved control over substrate binding and better pre-organization of the active site. Modifying existing protein scaffolds through rational redesign has been a more fruitful option so far—where the emphasis is on finding what works rather than predicting what works (Khersonsky et al. 2006; Copley 2021). The fine-tuning of engineered enzymes can only be fulfilled today by combinatorial approaches. The marriage of rational design and directed evolution (combinatorial redesign) seems to be the way to go at present.

Mirror enzymes have also become a part of de novo design efforts. Here the enzyme (protein) is built exclusively using D-amino acids—now they can only be made by chemistry. Peptides and small protein fragments are synthesized using solid phase peptide synthesis techniques and subsequently stitched together to form the full-length enzyme. The credit for the first total chemical synthesis of a D-enzyme goes to the enantiomer of HIV-1 protease. Expectedly, it was active and showed reciprocal chiral substrate specificity (deL Milton et al. 1992). The chemically synthesized D-Barnase also showed reciprocal chiral substrate specificity (Vinogradov et al. 2015). In an effort to replicate a mirror-image life, two key steps in the central dogma of molecular biology were demonstrated successfully. The L-DNA template-directed polymerization of DNA and its transcription into L-RNA by chemically synthesized D-amino acid polymerase are reported on an L-DNA template (Wang et al. 2016). In yet another step to copy mirror-image life, an active DNA ligase made of D-amino acids was synthesized (Weidmann et al. 2019). The D-protein ligase was active on D-DNA oligonucleotides (but not on L-DNA oligonucleotides) and required L-ATP for the ligation reaction. These demonstrations of D-enzyme activities are a small step in making mirror-image life forms (*D. coli - E. coli* bacteria based on D-amino acid building blocks!?).

40.6   Enzymes and Evolution

Nothing in biology makes sense except in the light of evolution—T. Dobzhansky (1973)

And enzymes are no exception! In the postgenomic era, the topic of enzyme evolution has attained prominence. This is because of its direct relevance to evolutionary events bridging the abiotic with the biotic origin of life issues. Second, it has to do with the recent successes of directed enzyme evolution in the laboratory/industry. While being key ingredients of biological evolution, enzymes themselves are continuously subjected to evolutionary pressures. Acquiring a new enzyme/pathway often contributes to favorable phenotypes and hence their selection. Useful features of an enzyme may be selected naturally over deep time or be forcefully chosen in a test tube. The latter—directed enzyme evolution—has already found many applications as alluded to in Chap. 37. Several thematic reviews have appeared on this topic in the recent past and continue to illuminate the field (Allewell 2012; Banerjee 2014; Tawfik 2020). As these provide a varied exposure to the field, we will limit the discussion and highlight a few aspects.

Enzyme promiscuity is where it all begins. Due to their catalytic promiscuity, enzymes are capable of catalyzing secondary (unrelated) reactions. The potential for catalytic promiscuity (see Chaps. 3 and 13) can be (a) an advantage in generating novel catalysts for industry and (b) a valuable playing field for evolution to work (Copley et al. 2021). Promiscuous enzymes could become part of another metabolic pathway or join hands with other enzymes to create novel metabolic routes (Jensen 1976). Such underground metabolic activity (and enzyme promiscuity) is sometimes useful and provides the driving force in the evolution of new metabolic pathways (see Chap. 39).

Catalysis and life are inseparable. There is much interest in how catalysis itself evolved. It is suggested that cofactors of modern enzymes may have served as small molecular catalysts during life's origin and early stages of evolution (Muchowska et al. 2020). These cofactors/coenzymes along with ancient ribozymes may have been transitioned to modern enzymes by replacement during the evolution of "translation." This hypothesis has garnered some support from prebiotic geochemistry, ribozyme biochemistry, and evolutionary biology. Accordingly, certain coenzymes and cofactors may bridge present-day biology with the past (Goldman and Kacar 2021; Piedrafita et al. 2021).

Finally, urzymology or the study of primitive (ancestral) enzymes. Urzymes are catalysts derived from invariant cores of protein superfamilies. The three-dimensional structural superposition is used to identify invariant cores. Such reconstructs represent legitimate experimental models for very early, ancestral enzymes (Carter Jr. 2014). They inform us on how the mutations set the stage for emergence of new activities in the cellular context. We could rerun evolution in vitro from robust ancestors, look at possible intermediate stages, and apply different, artificial selection pressures (Risso et al. 2018). Conversely, one can peek into the in vivo evolutionary constraints faced by that enzyme. Urzymes thus offer a robust platform to characterize even simpler ancestral protein catalysts of deep time as well as to test possible evolutionary paths for the future. The study is proving useful in exploring enzyme features like temperature stability (Thomson et al. 2022) and the evolution of old enzymes to perform new functions (Hendrickson 2018).

40.7 Summing Up

Our knowledge on and databases of genome sequences, metabolic pathways, protein sequences and their three-dimensional folds, enzyme active sites, and chemical reactions is expanding very fast. Sophisticated computational methods are expected to rationalize this vast amount of information and aid in predicting the changes required to alter one enzyme into another. Efficiently introducing a new enzymatic activity in a chosen protein scaffold may not be too far off in the future. In the meanwhile, rational de novo design of an enzyme continues to be a grand challenge (Editorial 2009).

This final section is demanding and difficult to cover—the subject matter of research is contemporary, and many reviews and new developments are reported on a very frequent basis. Almost every chapter, especially the last part (Part VI) of this book, becomes outdated in a short span of time. It is remarkable that for a subject so much undervalued and displaced away from the mainstream biology of today, very high-quality research gets added continuously to literature.

In the era of systems/synthetic biology, enzymology may not be fashionable, but it will continue to excite and motivate. As Pasteur famously stated, *there are no applied sciences but only applications of science*. This is so true with the study of enzymes. We rarely find an unemployed enzymologist. Surely, the future will be no different.

References

General

Blodgett JAV et al (2007) Unusual transformations in the biosynthesis of the antibiotic phosphinothricin tripeptide. Nat Chem Biol 3:480–485

Clarke CF, Allan CM (2015) Unexpected role for vitamin B2. Nature 522:427–428

Editorial (2009) Closing in on catalysis. Nat Chem Biol 5:515

Gao L et al (2020) FAD-dependent enzyme-catalysed intermolecular [4+2] cycloaddition in natural product biosynthesis. Nat Chem 12:620–628

Herschlag D, Natarajan A (2013) Fundamental challenges in mechanistic enzymology: progress toward understanding the rate enhancements of enzymes. Biochemistry 52:2050–2067

Huang J et al (2023) Complete integration of carbene-transfer chemistry into biosynthesis. Nature 617:403–408

Ito T, Yokoyama S (2010) Two enzymes bound to one transfer RNA assume alternative conformations for consecutive reactions. Nature 467:612–616

Leys D (2018) Flavin metamorphosis: cofactor transformation through prenylation. Curr Opin Chem Biol 47:117–125

Marchand JA et al (2019) Discovery of a pathway for terminal-alkyne amino acid biosynthesis. Nature 567:420–424

Ohashi M et al (2017) SAM-dependent enzyme-catalyzed pericyclic reactions in natural product biosynthesis. Nature 549:502–506

Ortega MA et al (2015) Structure and mechanism of the tRNA-dependent lantibiotic dehydratase NisB. Nature 517:509–512

Zhao Q et al (2015) Metabolic coupling of two small-molecule thiols programs the biosynthesis of lincomycin A. Nature 518:115–119

Transition State Analysis and Computational Enzymology

Frushicheva MP et al (2014) Computer aided enzyme design and catalytic concepts. Curr Opin Chem Biol 21:56–62

Garcia-Viloca M et al (2004) How enzymes work: analysis by modern rate theory and computer simulations. Science 303:186–195

Schramm VL (2013) Transition states, analogues, and drug development. ACS Chem Biol 8:71–81

Warshel A (2014) Multiscale modeling of biological functions: from enzymes to molecular machines (Nobel lecture). Angew Chem Int Ed 53:10020–10031

Single Molecule Enzymology

English BP et al (2006) Ever-fluctuating single enzyme molecules: Michaelis–Menten equation revisited. Nat Chem Biol 2:87–94

Jee A-Y et al (2018) Catalytic enzymes are active matter. Proc Natl Acad Sci USA 115:E10812–E10821

Klinman JP (2023) Dynamical activation of function in metalloenzymes. FEBS Lett 597:79–91

Min W et al (2005) Fluctuating enzymes: lessons from single-molecule studies. Acc Chem Res 38:923–931

Riedel C et al (2015) The heat released during catalytic turnover enhances the diffusion of an enzyme. Nature 517:227–230

Smiley RD, Hammes GG (2006) Single molecule studies of enzyme mechanisms. Chem Rev 106: 3080–3094

Walter NG (2006) Michaelis–Menten is dead, long live Michaelis–Menten! Nat Chem Biol 2:66–67

Zhao X et al (2018) Powering motion with enzymes. Acc Chem Res 51:2373–2381

Structure-Function Dissection of Enzyme Catalysis

Akdel M et al (2022) A structural biology community assessment of AlphaFold2 applications. Nat Struct Mol Biol 29:1056–1067

Atkins JF, Gesteland R (2002) The 22nd amino acid. Science 296:1409–1410

Bryan PN (2000) Protein engineering of subtilisin. Biochim Biophys Acta 1543:203–222

Changeux J-P (2013) 50 Years of allosteric interactions: the twists and turns of the models. Nat Rev Mol Cell Biol 14:819–829

Clarke AR, Atkinson T, Holbrook JJ (1989) From analysis to synthesis: new ligand binding sites on the lactate dehydrogenase framework. Part I. Trends Biochem Sci 14:101–105. Part II. Trends Biochem Sci 14:145–148

Cuesta-Seijo JA et al (2011) Structure of a dimeric fungal α-type carbonic anhydrase. FEBS Lett 585:1042–1048

Dang L et al (2009) Cancer-associated IDH1 mutations produce 2-hydroxyglutarate. Nature 462: 739–744

Derat E, Kamerlin SCL (2022) Computational advances in protein engineering and enzyme design. J Phys Chem B 126:2449–2451

Dhalla AM et al (1994) Regeneration of catalytic activity of glutamine synthetase mutants by chemical activation: exploration of the role of arginines 339 and 359 in activity. Protein Sci 3: 476–481

Earnhardt JN et al (1999) Introduction of histidine analogs leads to enhanced proton transfer in carbonic anhydrase V. Arch Biochem Biophys 361:264–270

Freire E (2000) Can allosteric regulation be predicted from structure? Proc Natl Acad Sci USA 97: 11680–11682

Gerlt JA (2017) Genomic enzymology: web tools for leveraging protein family sequence-function space and genome context to discover novel functions. Biochemistry 56:4293–4308

Gerlt JA et al (2011) The enzyme function initiative. Biochemistry 50:9950–9962

Hai Y et al (2015) Crystal structure of an arginase-like protein from *Trypanosoma brucei* that evolved without a binuclear manganese cluster. Biochemistry 54:458–471

Hardy JA, Wells JA (2004) Searching for new allosteric sites in enzymes. Curr Opin Struct Biol 14: 706–715

Hendrickson TL, de Crecy-Lagard V, Schimmel P (2004) Incorporation of nonnatural amino acids into proteins. Annu Rev Biochem 73:147–176

Kuznetsova E et al (2005) Enzyme genomics: application of general enzymatic screens to discover new enzymes. FEMS Microbiol Rev 29:263–279

Larion M et al (2012) Order–disorder transitions govern kinetic cooperativity and allostery of monomeric human glucokinase. PLoS Biol 10(12):e1001452

Lopez V et al (2005) Insights into the interaction of human arginase II with substrate and manganese ions by site-directed mutagenesis and kinetic studies: alteration of substrate specificity by replacement of Asn149 with Asp. FEBS J 272:4540–4548

McCullagh M et al (2024) What is allosteric regulation? Exploring the exceptions that prove the rule! J Biol Chem 300:105672

Miller C (2007) Pretty structures, but what about the data? Science 315:459

Moon H-J et al (2012) Molecular determinants of the cofactor specificity of ribitol dehydrogenase, a short-chain dehydrogenase/reductase. Appl Environ Microbiol 78:3079–3086

Motlagh HN et al (2014) The ensemble nature of allostery. Nature 508:331–339

Nickbarg EB et al (1988) Triosephosphate isomerase: removal of a putatively electrophilic histidine residue results in a subtle change in catalytic mechanism. Biochemistry 27:5948–5960

Pearce R, Zhang Y (2021) Toward the solution of the protein structure prediction problem. J Biol Chem 297:100870

Ragsdale SW (2011) How two amino acids become one. Nature 471:583–584

Ramanathan A, Agarwal PK (2011) Evolutionarily conserved linkage between enzyme fold, flexibility, and catalysis. PLoS Biol 9:e1001193

Reyes M-B et al (2020) Insights into the Mn^{2+} binding site in the agmatinase-like protein (ALP): a critical enzyme for the regulation of agmatine levels in mammals. Int J Mol Sci 21:4132

Reynolds KA, McLaughlin RN, Ranganathan R (2011) Hot spots for allosteric regulation on protein surfaces. Cell 147:1564–1575

Wagner CR, Benkovic SJ (1990) Site directed mutagenesis: a tool for enzyme mechanism dissection. Trends Biotechnol 8:263–270

Wilks HM et al (1988) A specific, highly active malate dehydrogenase by redesign of a lactate dehydrogenase framework. Science 242:1541–1544

Wlodawer A, Ericson JW (1993) Structure-based inhibitors of HIV-1 protease. Annu Rev Biochem 62:543–586

Wodak SJ et al (2019) Allostery in its many disguises: from theory to applications. Structure 27: 566–578

Designing Novel Catalysts

Abelson J (2017) The discovery of catalytic RNA. Nat Rev Mol Cell Biol 18(11):653–653

Baum DA, Silverman SK (2008) Deoxyribozymes: useful DNA catalysts in vitro and in vivo. Cell Mol Life Sci 65:2156–2174

Benkovic SJ (1992) Catalytic antibodies. Annu Rev Biochem 61:29–54

Bjerre J et al (2008) Artificial enzymes, "chemzymes": current state and perspectives. Appl Microbiol Biotechnol 81:1–11

Breslow R (2005) Artificial enzymes. Wiley-VCH, Weinheim

Davis HJ, Ward TR (2019) Artificial metalloenzymes: challenges and opportunities. ACS Cent Sci 5:1120–1136

Doudna JA, Lorsch JR (2005) Ribozyme catalysis: not different, just worse. Nat Struct Mol Biol 12: 395–402

Famulok M (2004) RNAs turn on in tandem. Science 306:233–234

Hilvert D (2000) Critical analysis of antibody catalysis. Annu Rev Biochem 69:751–793

Kim JD et al (2018) Minimal heterochiral de novo designed 4Fe–4S binding peptide capable of robust electron transfer. J Am Chem Soc 140:11210–11213

MacMillan DWC (2008) The advent and development of organocatalysis. Nature 455:304–308

Makhlynets OV, Korendovych IV (2019) A single amino acid enzyme. Nat Catal 2:949–950

Narlikar G, Herschlag D (1997) Mechanistic aspects of enzyme catalysis: lessons from comparison of RNA and protein enzymes. Annu Rev Biochem 66:19–59

Olenginski LT, Spradlin SF, Batey RT (2024) Flipping the script: understanding riboswitches from an alternative perspective. J Biol Chem 300:105730

Ren X et al (2001) Cloning and expression of a single-chain catalytic antibody that acts as a glutathione peroxidase mimic with high catalytic efficiency. Biochem J 359:369–374

Schmeing TM, Ramakrishnan V (2009) What recent ribosome structures have revealed about the mechanism of translation. Nature 461:1234–1242

Taylor AI et al (2015) Catalysts from synthetic genetic polymers. Nature 518:427–430

Tirumalai MR et al (2021) The peptidyl transferase center: a window to the past. Microbiol Mol Biol Rev 85:e00104–e00121

Traut TW (2007) Allosteric regulatory enzymes. Springer Science & Business Media, Boston

Tuerk C, Gold L (1990) Systemic evolution of ligands by exponential enrichment: RNA ligands to bacteriophage T4 DNA polymerase. Science 249:505–510

Wei H, Wang E (2013) Nanomaterials with enzyme-like characteristics (nanozymes): next-generation artificial enzymes. Chem Soc Rev 42:6060–6093

Weigand JE, Suess B (2009) Aptamers and riboswitches: perspectives in biotechnology. Appl Microbiol Biotechnol 85:229–236

Wilson TJ, Lilley DMJ (2009) The evolution of ribozyme chemistry. Science 323:1436–1438

Woycechowsky KJ et al (1999) A small-molecule catalyst of protein folding in vitro and in vivo. Chem Biol 6:871–879

Wulff G (2002) Enzyme-like catalysis by molecularly imprinted polymers. Chem Rev 102:1–28

Zuckermann RN, Schultz PG (1988) Hybrid sequence-selective ribonuclease S. J Am Chem Soc 110:6592–6594

Enzymes Made to Order

Bar-Even A et al (2011) The moderately efficient enzyme: evolutionary and physicochemical trends shaping enzyme parameters. Biochemistry 50:4402–4410

Blomberg R et al (2013) Precision is essential for efficient catalysis in an evolved Kemp eliminase. Nature 503:418–421

Copley SD (2021) Setting the stage for evolution of a new enzyme. Curr Opin Struct Biol 69:41–49

deL Milton RC, Milton SCF, Kent SBH (1992) Total chemical synthesis of a D-enzyme: the enantiomers of HIV-1 protease show demonstration of reciprocal chiral substrate specificity. Science 256:1445–1448

Hayden EC (2008) Designer debacle—news feature. Nature 453:275–278

Huang P-S, Boyken SE, Baker D (2016) The coming of age of de novo protein design. Nature 537: 320–327

Jiang L et al (2008) De novo computational design of retro-aldol enzymes. Science 319:1387–1391

Joerger AC, Mayer S, Fersht AR (2003) Mimicking natural evolution in vitro: an N-acetylneuraminate lyase mutant with an increased dihydrodipicolinate synthase activity. Proc Natl Acad Sci USA 100:5694–5699

Johnsson K et al (1993) Synthesis, structure and activity of artificial, rationally designed catalytic polypeptides. Nature 365:530–532

Khersonsky O, Roodveldt C, Tawfik DS (2006) Enzyme promiscuity: evolutionary and mechanistic aspects. Curr Opin Chem Biol 10:498–508

Lechner H, Ferruz N, Hocker B (2018) Strategies for designing non-natural enzymes and binders. Curr Opin Chem Biol 47:67–76

Menger FM, Ladika M (1987) Origin of rate accelerations in an enzyme model: the p-nitrophenyl ester syndrome. J Am Chem Soc 109:3145–3146

Nanda V (2008) Do-it-yourself enzymes. Nat Chem Biol 4:273–275

Pan X, Kortemme T (2021) Recent advances in de novo protein design: principles, methods, and applications. J Biol Chem 296:100558

Park H-S et al (2006) Design and evolution of new catalytic activity with an existing protein scaffold. Science 311:535–538

Penning TM, Jez JM (2001) Enzyme redesign. Chem Rev 101:3027–3046

Pinney MM et al (2021) Parallel molecular mechanisms for enzyme temperature adaptation. Science 371:eaay2784

Reitman ZJ et al (2012) Enzyme redesign guided by cancer-derived IDH1 mutations. Nat Chem Biol 8:887–889

Richter F et al (2011) De novo enzyme design using Rosetta3. PLoS One 6(5):e19230

Sharkey MA, Engel PC (2009) Modular coenzyme specificity; a domain swapped chimera of glutamate dehydrogenase. Proteins 77:268–278

Siegel JB et al (2010) Computational design of an enzyme catalyst for a stereoselective bimolecular Diels–Alder reaction. Science 329:309–313

Vinogradov AA, Evans ED, Pentelute BL (2015) Total synthesis and biochemical characterization of mirror image barnase. Chem Sci 6:2997–3002

Wang Z et al (2016) A synthetic molecular system capable of mirror-image genetic replication and transcription. Nat Chem 8:698–704

Weidmann J et al (2019) Copying life: synthesis of an enzymatically active mirror-image DNA-ligase made of D-amino acids. Cell Chem Biol 26:645–651

Yoshikuni Y, Ferrin TE, Keasling JD (2006) Designed divergent evolution of enzyme function. Nature 440:1078–1082

Enzymes and Evolution

Allewell N (2012) Thematic minireview series on enzyme evolution in the post-genomic era. J Biol Chem 287:1–2

Banerjee R (2014) Introduction to the thematic minireview series on enzyme evolution. J Biol Chem 289:30196–30197

Carter CW Jr (2014) Urzymology: experimental access to a key transition in the appearance of enzymes. J Biol Chem 289:30213–30220

Goldman AD, Kacar B (2021) Cofactors are remnants of life's origin and early evolution. J Mol Evol 89:127–133

Hendrickson TL (2018) Old enzymes versus new herbicides. J Biol Chem 293:7892–7893

Jensen RA (1976) Enzyme recruitment in evolution of new function. Annu Rev Microb 30:409–425

Muchowska KB, Varma SJ, Moran J (2020) Nonenzymatic metabolic reactions and life's origins. Chem Rev 120:7708–7744

Piedrafita G et al (2021) Cysteine and iron accelerate the formation of ribose-5-phosphate, providing insights into the evolutionary origins of the metabolic network structure. PLoS Biol 19:e3001468

Risso VA, Sanchez JM, Ozkan SB (2018) Biotechnological and protein-engineering implications of ancestral protein resurrection. Curr Opin Struct Biol 51:106–115

Tawfik DS (2020) Enzyme promiscuity and evolution in light of cellular metabolism. FEBS J 287: 1260–1261

Thomson RES et al (2022) Engineering functional thermostable proteins using ancestral sequence reconstruction. J Biol Chem 298:102435

Bibliography

Books:

General and Historical

Cornish-Bowden A (1997) New beer in an old bottle: Eduard Buchner and the growth of biochemical knowledge. Universitat de Valèrcia, Valencia. ISBN: 84-370-3328-4

Fruton JS (1999) Proteins, enzymes, genes: the interplay of chemistry and biology. Yale University Press. ISBN-13: 9780300076080

Haldane JBS (1965) Enzymes. MIT Press Classics, Cambridge. (originally publisher, Longmans, Green, 1930). ISBN: 0262580039, 9780262580038

Jencks WP (1987) Catalysis in chemistry and enzymology. Courier Corporation. ISBN: 0486654605, 9780486654607

Kornberg A (1991) For the love of enzymes: the odyssey of a biochemist. Harvard University Press, Cambridge. ISBN: 0674307763, 9780674307766

Lagerkvist U (2005) The enigma of ferment: 'from the philosopher's stone to the first biochemical Nobel prize'. World Scientific, Hackensack. ISBN: 9814338168, 9789814338165

Michal G, Schomburg D (eds) (2012) Biochemical pathways: an atlas of biochemistry and molecular biology, 2nd edn. John Wiley & Sons, Inc., Hoboken, NJ. ISBN: 978-0-470-14684-2

Tanford C, Reynolds J (2003) Nature's Robots: A history of proteins (Oxford Paperbacks). ISBN-10: 019860694X, ISBN-13: 978-0198606949

Enzyme Kinetics

Bisswanger H (2002) Enzyme kinetics: principles and methods (trans: Bubenheim L). Wiley-VCH Verlag GmbH, Weinheim

Cook P, Cleland WW (2007) Enzyme kinetics and mechanism. Garland Science, New York. ISBN: 1136844287, 9781136844287

Cornish-Bowden A (2012) Fundamentals of enzyme kinetics, 4th edn. Wiley Blackwell, Weinheim. ISBN: 978-3-527-33074-4

Engel PC (1981) Enzyme kinetics: the steady-state approach, 2nd edn. Springer-Science+Business Media, B.Y.. ISBN: 978-0-412-23970-0

Leskovac V (2003) Comprehensive enzyme kinetics. Kluwer Academic/Plenum Publishers, New York

© The Author(s), under exclusive license to Springer Nature Singapore Pte Ltd. 2025

N. S. Punekar, *ENZYMES: Catalysis, Kinetics and Mechanisms*, https://doi.org/10.1007/978-981-97-8179-9

Marangoni AG (2002) Enzyme kinetics: a modern approach. Wiley, Hoboken. ISBN: 978-0-471-15985-8

Palfey BA, Switzer RL (2022) Kinetics of enzyme catalysis. ACS In Focus ACS Publications. ISBN 978-0-8412-9939-9

Plowman KM (1971) Enzyme kinetics, McGraw-Hill series in advanced chemistry. McGraw-Hill, New York

Purich DL, Allison RD (1999) Handbook of biochemical kinetics: a guide to dynamic processes in the molecular life sciences. Academic. ISBN: 0080521932, 9780080521930

Segel IH (1993) Enzyme kinetics: behavior and analysis of rapid equilibrium and steady-state enzyme systems. Wiley, New York. ISBN: 978-0-471-30309-1

Taylor KB (2002) Enzyme kinetics and mechanisms. Springer, Dordrecht. ISBN: 978-1-4020-0728-6, 978-90-481-6065-5

Enzyme Chemical Mechanisms

Abeles RH, Frey PA, Jencks WP (1992) Biochemistry. Jones and Bartlett, Boston. ISBN: 0-86720-212-2

Breslow R (ed) (2005) Artificial enzymes. Wiley-VCH Verlag, Weinheim. ISBN: 978-3-527-31165-1

Bugg TDH (2012) Introduction to enzyme and coenzyme chemistry, 3rd edn. Wiley, Hoboken. Print ISBN: 9781119995951, Online ISBN: 9781118348970

Dugas H (1996) Bioorganic chemistry: a chemical approach to enzyme action, Springer advanced texts in chemistry, 3rd edn. Springer-Verlag, New York. eBook ISBN: 978-1-4612-2426-6, ISBN: 978-0-387-98910-5

Frey PA, Hegeman AD (2007) Enzymatic reaction mechanisms. Oxford University Press, New York. ISBN: 0195352742, 9780195352740

Kirby AJ, Hollfelder F (2009) From enzyme models to model enzymes. Royal Society of Chemistry, Cambridge. ISBN: 978-0-85404-175-6

Kyte J (1995) Mechanism in protein chemistry. Garland Science, New York. ISBN: 9780815317005– CAT# GS71

Silverman RB (2002) Organic chemistry of enzyme-catalyzed reactions, 2nd ed., revised edn. Academic, New York. ISBN: 0080513360, 9780080513362

Walsh C (1979) Enzymatic reaction mechanisms. WH Freeman, San Francisco. ISBN: 0716700700, 9780716700708

Woggon W-D (2023) Bioorganic and enzymatic catalysis. Wiley-VCH. ISBN: 978-3-527-67835-8

Practical Enzymology

Bisswanger H (2019) Practical enzymology, 3rd edn. Wiley-VCH Verlag GmbH & Co.KGaA

Burgess R, Deutscher MP (eds) (2009) Guide to protein purification: methods in enzymology, vol 463, 2nd edn. Academic, San Diego. ISBN: 9780080923178, ISBN: 9780123745361

Dawes EA (1980) Quantitative problems in biochemistry, 6th edn. Longman, London. ISBN: 0582444020, 9780582444027

Dawson RMC, Elliott DC, Elliott WH, Jones KM (1986) Data for biochemical research. Oxford Science Publications/Clarendon Press, Oxford. ISBN: 0198552998, 9780198552994

Deutscher MP (ed) (1990) Guide to protein purification: methods in enzymology, vol 182. Academic, San Diego. ISBN: 10 0-12-213585-7, ISBN: 13 978-0-12 213585-9

Eisenthal R, Danson MJ (2002) Enzyme assays: a practical approach, 2nd edn. Oxford University Press, Oxford

Engel PC (ed) (1996) Enzymology labfax (BD Hames and D Rickwood, Labfax series). BIOS Scientific, Oxford. ISBN: 0122388402, 9780122388408

Passonneau JV, Lowry OH (2008) Enzymatic anaysis: a practical guide. Springer Science+ Business Media LLC. eBook ISBN: 978-1-60327-407-4

Reymond J (ed) (2006) Enzyme assays: high-throughput screening, genetic selection and fingerprinting. Wiley, Weinheim. ISBN: 10 3-527-31095-9, ISBN: 13 978-3 527-31095-1

Segel IH (1976) Biochemical calculations: how to solve mathematical problems in general biochemistry, 2nd edn. Wiley, New York. ISBN: 978-0-471-77421-1

Enzymology Texts

Copeland RA (2004) Enzymes: a practical introduction to structure, mechanism, and data analysis, 2nd edn. Wiley, Hoboken. ISBN: 0471461857, 9780471461852

Fersht A (2000) Structure and mechanism in protein science: guide to enzyme catalysis and protein folding. WH Freeman & Co Ltd.. 3rd revised ed., ISBN 10: 0716732688, ISBN-13: 978-0716732686

Palmer T (1993) Principles of enzymology for technological applications, biotechnology by open learning, Biotol series. Butterworth-Heinemann, Oxford. ISBN: 0750606894, 9780750606899

Palmer T, Bonner PL (2007) Enzymes: biochemistry, biotechnology, clinical chemistry, 2nd edn. Woodhead Publishing. eBook ISBN: 9780857099921, ISBN: 9781904275275

Price NC, Stevens L (2000) Fundamentals of enzymology: the cell and molecular biology of catalytic proteins, 3rd edn. Oxford University Press, London. ISBN: 9780198502296

Suckling CJ (1990) Enzyme chemistry: impact and applications, 2nd edn. Chapman and Hall. ISBN-13: 978-94-010-7317-2 1990

Suzuki H (2015) How enzymes work: from structure to function. CRC Press, Boca Raton. ISBN: 9814463930, 9789814463935

Wharton CW, Eisenthal R (1981) Molecular enzymology. Springer Science+ Business Media, LLC. ISBN 978-1-4615-8534-3

Yon-Kahn J, Hervé G (2010) Molecular and cellular enzymology. Springer-Verlag, Berlin, Heidelberg. ISBN: 978-3-642-01227-3

Enzyme Regulation and Applications

Abramowicz DA (ed) (2013) Biocatalysis, Van Nostrand Reinhold electrical/computer science and engineering series. Springer. ISBN: 9401091242, 9789401091244

Barker RDJ (ed) (1993) Technological applications of biocatalysts, biotechnology by open learning, biotol series. Butterworth-Heinemann, Oxford

Chaplin MF, Bucke C (1990) Enzyme technology. Cambridge University Press, Cambridge. ISBN: 0521344298, 9780521344296

Cohen P (2013) Control of enzyme activity, outline studies in biology. Springer, Dordrecht. ISBN: 9400959818, 9789400959811

Currell BC, Dam-Mieras RCE (2014) Biosynthesis & integration of cell metabolism, biotechnology by open learning. Butterworth-Heinemann, Oxford. ISBN: 1483297322, 9781483297323

Fell D (1996) Understanding the control of metabolism, frontiers in metabolism. Portland Press, London. ISBN-10: 185578047X ISBN-13: 978-1855780477

Grunwald P (2009) Biocatalysis: biochemical fundamentals and applications. Imperial College Press, Hackensack. ISBN: 1860947441, 9781860947445

Heinrich R, Schuster S (2012) The regulation of cellular systems. Springer, New York. eBook ISBN 978-1-4613-1161-4

Ochs R (2018) Metabolic structure and regulation: a neoclassical approach. CRC Press Taylor & Francis Group, Boca Raton. ISBN-13: 978-1-4822-3608-8

Rider CC (2014) Isoenzymes, outline studies in biology. Springer My Copy UK. ISBN: 9401094489, 9789401094481

Svendsen A (ed) (2004) Enzyme functionality: design, engineering, and screening. Marcel Dekker, Inc., New York. ISBN: 0-8247-4709-7

Svendsen A (2016) Understanding enzymes function, design, engineering, and analysis. Taylor & Francis Group. ISBN-13: 978-981-4669-33-7

Traut TW (2007) Allosteric regulatory enzymes. Springer, Boston. ISBN: 0387728910, 9780387728919

Treven MD (1980) Immobilized enzymes: an introduction and applications in biotechnology. Wiley, New York. ISBN: 0835767116, 9780835767118

Vogel A, May O (eds) (2019) Industrial enzyme applications. Wiley-VCH Verlag GmbH & Co, KGaA, Weinheim, Germany. ISBN: 978-3-527-34385-0

Series:

Volumes Covering Advances in Enzymology

Anon (n.d.-a) Advances in enzymology and related areas of molecular biology, more than 200 volumes and continuing. John Wiley & Sons

Anon (2014) Current topics in cellular regulation; originally published in 1972 by Academic Press. Elsevier. (Revised editions)

Anon (2013) Enzyme handbook; volumes for each enzyme class and updates. Springer Science &Business Media

Anon (n.d.-b) Enzymes; series originally edited by PD Boyer. Academic Press

Anon (n.d.-c) Methods in enzymology; more than 500 volumes and continuing. Academic Press

Biochemistry Textbooks

For Background Material on Protein Structure, Metabolism and Gene Regulation

Berg JM, Stryer L, Tymoczko JL, Gatto GJ (2015) Biochemistry, 8th edn. WH Freeman, New York. ISBN-10: 1319153933, ISBN-13: 978-1319153939

Nelson DL, Cox M (2017) Lehninger principles of biochemistry, 7th edn. WH Freeman, New York. ISBN-10: 1319108245, ISBN-13: 978-1319108243

Voet D, Voet JG (2010) Biochemistry, 4th edn. Wiley, New York. ISBN-10: 0121822117, ISBN-13: 978-0470570951